PRINCIPLES OF SEQUENCE STRATIGRAPHY

SECOND EDITION

PRINCIPLES OF SEQUENCE STRATIGRAPHY

SECOND EDITION

OCTAVIAN CATUNEANU
Professor, University of Alberta, Edmonton, Alberta, Canada

ELSEVIER

Elsevier
Radarweg 29, PO Box 211, 1000 AE Amsterdam, Netherlands
The Boulevard, Langford Lane, Kidlington, Oxford OX5 1GB, United Kingdom
50 Hampshire Street, 5th Floor, Cambridge, MA 02139, United States

Notices
Knowledge and best practice in this field are constantly changing. As new research and experience broaden our understanding, changes in research methods, professional practices, or medical treatment may become necessary.

Practitioners and researchers must always rely on their own experience and knowledge in evaluating and using any information, methods, compounds, or experiments described herein. In using such information or methods they should be mindful of their own safety and the safety of others, including parties for whom they have a professional responsibility.

To the fullest extent of the law, neither the Publisher nor the authors, contributors, or editors, assume any liability for any injury and/or damage to persons or property as a matter of products liability, negligence or otherwise, or from any use or operation of any methods, products, instructions, or ideas contained in the material herein.

ISBN: 978-0-444-53353-1

For information on all Elsevier Science publications visit our website at
https://www.elsevier.com/books-and-journals

Publisher: Candice Janco
Acquisitions Editor: Peter Llewellyn
Editorial Project Manager: Judith Clarisse Punzalan
Production Project Manager: Paul Prasad Chandramohan
Cover Designer: Greg Harris

Typeset by TNQ Technologies

Contents

Preface to second edition

Sequence stratigraphy provides a process-based approach to stratigraphic analysis that enables insights into the patterns of sediment distribution during the evolution of sedimentary basins, as well as facies predictions in areas away from data control points. These applications appeal to a wide segment of the stratigraphic community, from academia to the industry. The methodology improved significantly since the 1970s, from a model-driven approach underlain by assumptions regarding the dominant role of eustasy on sequence development, with consequent assertions of global correlations, to a data-driven approach that promotes the use of local data and unbiased geological reasoning. The latter approach affords realistic constructions of local stratigraphic frameworks, which prove to be highly variable, not only from one sedimentary basin to another, but also between subbasins of the same sedimentary basin. The construction of basin-specific stratigraphic frameworks is a major breakthrough and departure from the early models.

The evolution of sequence stratigraphy involved advances in the understanding of the array of possible controls on sequence development, improvements in stratigraphic resolution, revisions to the definition and classification of sequences, and ultimately the emergence of a standard approach to methodology and nomenclature. Most significant to the conceptual developments and practical applications, the increase in stratigraphic resolution prompted a complete overhaul of the early principles of low-resolution seismic stratigraphy, and presented the opportunity to improve our insights into key areas such as the definition, the scales, and the classification of sequences. Old myths of seismic stratigraphy have been debunked in light of high-resolution data, making way for realistic principles that honor the stratigraphic variability generated by the interplay of local and global controls on accommodation and sedimentation. All in all, a new look at the principles of sequence stratigraphy, from the basic concepts to the field criteria that enable the application of the method, is timely.

A number of developments arose since the publication of the first edition of this book in 2006. Undoubtedly the most significant development was the endorsement of sequence stratigraphy by the International Commission on Stratigraphy in 2011, along with the publication of guidelines on methodology and nomenclature (Catuneanu et al., 2011). This concluded decades of work on the definition of a standard approach to sequence stratigraphy, following the failed attempts of two previous working groups led by A. Salvador (1995—2003) and A.F. Embry (2004—2007). The standard methodology relies on the observation of stratal stacking patterns in a manner that is independent of scale and the interpretation of the underlying controls. This ensures maximum flexibility and objectivity, and a consistent application of the method irrespective of geological setting and the types of data available. Once a sequence stratigraphic framework is constructed, the interpretation and testing of the underlying controls on sequence development may continue indefinitely.

A standard methodology does not imply a freeze on developments. While a standard approach to sequence stratigraphy is now in place, future developments are still needed to refine and diversify the criteria that afford the identification of systems tracts and bounding surfaces in the sedimentary record (e.g., the emerging trends of using geochemical and pale-ontological proxies in sequence stratigraphy, in conjunction with information derived from other independent datasets). Parallel progress in stratigraphic modeling will also continue in order to improve the accuracy of testing of sedimentary responses to the variety of possible controls on sequence development. However, methodology and modeling remain two distinct lines of stratigraphic research, with different goals and underlying data-driven vs. model-driven principles. Therefore, it is important to separate methodology from modeling in sequence stratigraphy. A standard methodology does not prevent future developments in the field of stratigraphic modeling.

Advances in computing power enhanced the ability to test the results of various combinations of controls on the stratigraphic architecture. While this can provide useful insights under realistic boundary conditions, uncalibrated simulations and overreaching conclusions can be misleading and even counterproductive. A dangerous extreme is to think that virtual reality (the outcome of numerical modeling) is more meaningful than the reality described by actual data. Software enthusiasts seem to assume so when 'demonstrating' stratigraphic scenarios that have little in common with the real world, due to the unrealistic selection of input parameters. Proposals that numerical simulations should become part of the sequence stratigraphic workflow are not only impractical, but a setback as they bring confusion between modeling and methodology. This undermines the progress made in the development of sequence stratigraphy as a data-driven method with an objective workflow that relies on observations rather than model-driven assumptions. While useful to test the possible controls on the stratigraphic architecture, modeling requires validation with real data and plays no role in the sequence stratigraphic methodology.

The separation of modeling and methodology is an important 'first amendment' in sequence stratigraphy. Mixing the methodology with the interpretation of controls on sequence development was a pitfall since the inception of sequence stratigraphy, which took decades of work to correct. The early models favored eustasy as the dominant control, which led to the assumption of global correlations in the 1970s and the 1980s, while the role of sediment supply on par with accommodation was only fully recognized since the 1990s (i.e., the 'dual control' of Schlager, 1993). Between these end members, the relative contributions of processes that control accommodation and sedimentation vary with the tectonic and depositional settings and are often difficult to quantify. Linking the methodology to any specific control on sequence development is ultimately misleading, as it is always an interplay of multiple controls that defines the stratigraphic architecture. For this reason, the methodology must remain neutral with respect to the interpretation of underlying controls, and any reference to a specific control (e.g., 'tectono-sequence stratigraphy') should be avoided.

The latest trend in numerical modeling is the shift from an overemphasis on accommodation to an overemphasis on sediment supply, which is equally deceptive. Beyond interpretations, the methodology remains grounded on field data and the model-independent observation of strata stacking patterns and stratigraphic relationships. The stratigraphic record is highly variable in terms of the architecture and composition of sequences, and it includes a mix of diagnostic and non-diagnostic elements with respect to the definition and identification of systems tracts and bounding surfaces. Variability rather than orderly patterns is the stratigraphic norm. This means that the methodology needs to follow an objective data-driven workflow that is independent of model assumptions, in which the construction of the sequence stratigraphic framework is guided by data rather than the model. An important part of this process is the distinction between the diagnostic stacking patterns that afford the identification of systems tracts and the nondiagnostic variability that can accompany the formation of any systems tract.

This second edition updates all key aspects of sequence stratigraphy, from the theoretical principles to the practical workflow. The book provides new insights and an in-depth analysis of all elements of the sequence stratigraphic framework, along with examples that illustrate the concepts and applications of sequence stratigraphy in all geological settings. Long-standing inconsistencies and 'grey areas' (e.g., the difference between depositional and geometrical trends in sequence stratigraphy; the scale of sequences and component systems tracts and depositional systems; the difference between high-frequency sequences and parasequences; the role of global vs. local, and allogenic vs. autogenic controls on sequence development; the role of various types of clinoform rollovers in the definition of the sequence stratigraphic framework) are addressed and clarified in a systematic manner. The result is a state-of-the-art presentation of the principles and guidelines that afford a consistent application of sequence stratigraphy across the entire range of geological settings, stratigraphic scales, and types of data available.

Acknowledgments

This work benefitted from the joint effort that led to the publication of formal guidelines on sequence stratigraphic methodology and nomenclature by the International Subcommission on Stratigraphic Classification (ISSC) of the International Commission on Stratigraphy (Catuneanu et al., 2011). I thank William E. Galloway, Christopher G.St.C. Kendall, Andrew D. Miall, Henry W. Posamentier, Andre Strasser, and Maurice E. Tucker for their contributions and support as members of the ISSC task group on sequence stratigraphy. Additional insights were provided by many experts over the years, including V. Abreu, J.P. Bhattacharya, M.D. Blum, I. Csato, R.W. Dalrymple, A.F. Embry, P.G. Eriksson, C.R. Fielding, W.L. Fisher, P. Gianolla, M.R. Gibling, K.A. Giles, W. Helland-Hansen, J.M. Holbrook, R. Jordan, D.A. Leckie, B. Macurda, O.J. Martinsen, J.E. Neal, D. Nummedal, L. Pomar, B.R. Pratt, J.F. Sarg, W. Schlager, B.C. Schreiber, K.W. Shanley, R.J. Steel, A.R. Sweet, C. Winker, and M. Zecchin. I am grateful to Steven Holland for contributing most of the section on Body Fossils in Chapter 2, and to James A. MacEachern, Luis A. Buatois, and Murray K. Gingras for their contributions to the section on Trace Fossils in Chapter 2.

These interactions helped clarify the confusions that plagued sequence stratigraphy for decades. Over the course of my career I worked with professionals embracing different models with respect to the selection of the sequence boundary and the delineation of systems tracts, and yet collaboration was never a problem. This is because all models have a common ground that transcends any arbitrary or nomenclatural differences. In practical terms, each sequence model originated from the observation of specific datasets; therefore, all "schools" have merits, and the choice between models is not a matter of right or wrong, but a function of the types of data available and the geological setting. This fact ultimately enabled the identification of the core principles that afford a unified approach to sequence stratigraphy. Perhaps the most pertinent advice I give people taking my courses is to forget about models, and let the data guide them to the actual architecture of the stratigraphic record, however "imperfect" that may be.

Octavian Catuneanu
University of Alberta
Edmonton, 2022

1

Introduction

1.1 Overview of sequence stratigraphy

1.1.1 Scope of sequence stratigraphy

Sequence stratigraphy is a type of stratigraphy that relies on stacking patterns for the definition, nomenclature, classification, and correlation of stratal units and bounding surfaces. The method examines the stratigraphic cyclicity and the related changes in sedimentation regimes that can be observed at the scales afforded by the resolution of the data available. Stratal stacking patterns are at the core of the sequence stratigraphic methodology, as they provide the criteria for the definition of all units and surfaces of sequence stratigraphy (Fig. 1.1); i.e., sequence stratigraphic units are bodies of sediment or sedimentary rocks defined by stratal stacking patterns and their bounding surfaces, and sequence stratigraphic surfaces are stratigraphic contacts which mark changes in stratal stacking pattern. The correlation of strata based on their stacking patterns sets sequence stratigraphy apart from other correlation methods that rely on similarities of units in terms of lithology (i.e., lithostratigraphy), fossil content (i.e., biostratigraphy), magnetic polarity (i.e., magnetostratigraphy), geochemical signatures (i.e., chemostratigraphy), or geological age (i.e., chronostratigraphy). Sequence stratigraphic units may or may not coincide with other types of stratigraphic units.

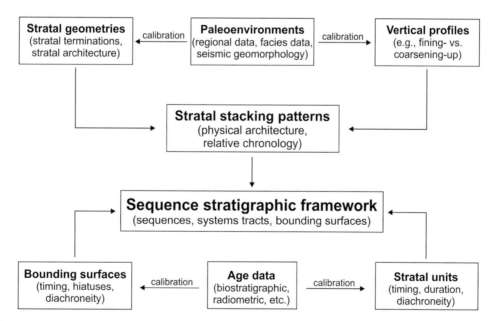

FIGURE 1.1 Construction of the sequence stratigraphic framework, based on the integration and mutual calibration of independent datasets. The meaning of stratal geometries and vertical profiles is best constrained within a paleo-depositional context. The reliability and the resolution of the constructed framework depend on the type(s) of data available. Where only seismic data are available, stratal stacking patterns are observed at scales above the seismic resolution, on the basis of seismic reflection terminations and architecture (i.e., seismic stratigraphy). Outcrop and well data afford the construction of higher resolution sequence stratigraphic frameworks at sub-seismic scales. Age data enhance the reliability of correlations, but the lack thereof (e.g., in most Precambrian and many Phanerozoic case studies) does not prevent the application of sequence stratigraphy. In the absence of age data, physical stratigraphic markers (e.g., volcanic ash beds, regional coal seams, or regional maximum flooding surfaces) can be used to aid the correlations. Stratal stacking patterns provide the basis for the definition of all units and surfaces of sequence stratigraphy.

The sequence stratigraphic methodology enables the construction of stratigraphic frameworks based on the observation of stratal stacking patterns at scales defined by the purpose of study or by the resolution of the data available. Interpretations in sequence stratigraphy have different degrees of relevance to the methodology. Intrinsic to the method is the rationalization of the sedimentary processes that generate the observed stacking patterns, by placing the data in the correct tectonic and depositional settings (steps 1 and 2 of the sequence stratigraphic workflow; details in Chapter 9). This affords predictions with respect to the sedimentological nature of seismic facies or stratal units in areas away from data control points such as outcrops or wells, within the paleogeographic context of linked depositional systems. The observations that afford the construction of sequence stratigraphic frameworks are independent of the interpretation of underlying controls on sequence development (e.g., the relative contributions of tectonism, sea-level changes, climate, and autogenic controls on stratigraphic cyclicity). The latter defines the scope of stratigraphic modeling (Fig. 1.2).

The correct understanding of sequence stratigraphy requires a clear distinction between the observational workflow of the methodology and the interpretations derived from stratigraphic modeling (Fig. 1.2). Case in point, the modeling and testing of the possible controls on sequence development can continue indefinitely after the construction of a sequence stratigraphic framework. Confusion between the two lines of research undermines the progress made in the development of sequence stratigraphy as an objective, data-driven methodology. Uncalibrated numerical models can generate unlimited results, which, in the absence of reality checks, remain an undifferentiated mix of realistic and unrealistic stratigraphic scenarios. The methodology restores the value of natural processes and facts, by outlining the field criteria that enable the identification of all elements of the sequence stratigraphic framework. The construction of a framework of sequences and component systems tracts explains the genetic relationships between same-age depositional systems, which afford insights into the patterns of sediment distribution across the basin.

Sequence stratigraphy provides the means to rationalize the stratigraphic relationships that develop at different scales within sedimentary basins placed in all tectonic settings, depositional settings, and climatic regimes. While the methodology is independent of geological setting, the sequence stratigraphic framework is variable in terms of timing, scales, and the systems-tract composition of sequences, reflecting the unique accommodation and sedimentation conditions of each sedimentary basin. For this reason, the methodology must be applied objectively, without any *a priori* assumptions, with the data rather than the model leading to the construction of the sequence stratigraphic framework. The sequence stratigraphic methodology integrates all available datasets that can be derived from surface (e.g., outcrops, modern environments) and subsurface

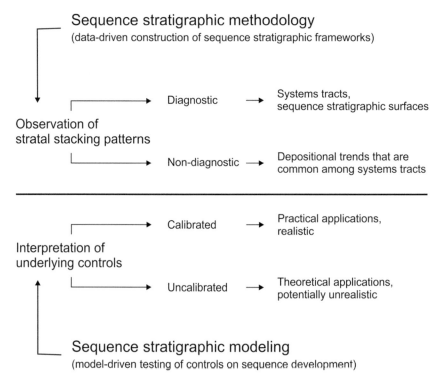

FIGURE 1.2 Methodology vs. modeling in sequence stratigraphy. The construction of a sequence stratigraphic framework is based on the observation of stratal stacking patterns, irrespective of the interpretation of the underlying controls. In contrast, modeling tests the possible controls on sequence development (e.g., the relative contributions of tectonism, sea/lake-level changes, climate, and sediment supply). The observed stacking patterns can be rationalized in terms of depositional processes by placing the data in the proper paleogeographic context (see workflow of sequence stratigraphy in Chapter 9). In downstream-controlled settings, diagnostic stacking patterns relate to shoreline trajectories, whereas non-diagnostic stacking patterns refer to depositional trends that can accompany the formation of any systems tract (e.g., the aggradation of fluvial topsets). In upstream-controlled settings, diagnostic stacking patterns relate to the dominant depositional elements (e.g., channels vs. floodplains in fluvial systems), irrespective of fluvial styles and the interpreted accommodation conditions at syndepositional time (details in Chapter 4). Modeling has no bearing on the methodology, and calibration with field data is required for realistic results (details in Chapter 9).

(e.g., borehole, seismic) data sources, and combines insights from all geosciences that contribute to basin analysis (e.g., sedimentary geology, geophysics, geomorphology, and all absolute and relative dating techniques; Fig. 1.1). The reliability of the sequence stratigraphic framework depends on the amount and quality of the data available. Sequence stratigraphic frameworks are typically "work in progress" as they are constantly improved and refined with the acquisition of more and higher resolution data.

Seismic stratigraphy, which is the precursor of modern sequence stratigraphy, set the early standards for the scales and applications of the methodology (Payton, 1977). Subsequently, the scope of sequence stratigraphy was expanded to include applications to all scales and datasets (from low-resolution seismic stratigraphy to high-resolution sequence stratigraphy at sub-seismic scales afforded by borehole and outcrop data; Amorosi et al., 2005, 2009, 2017; Catuneanu and Zecchin, 2013; Zecchin and Catuneanu, 2013, 2015, 2017; Magalhaes et al., 2015; Zecchin et al., 2017a,b; details in Chapters 2–7), depositional settings (from eolian to deep-water; Kocurek and Havholm, 1993; Catuneanu, 2020a; details in Chapter 8), tectonic settings (from "passive" to "active" basins; Bastia et al., 2010; Martins-Neto and Catuneanu, 2010; Maravelis et al., 2016, 2017; details in Chapter 8), and climatic conditions (from icehouse to greenhouse regimes; Bartek et al., 1991, 1997; Kidwell, 1997; Naish and Kamp, 1997; Saul et al., 1999; Fielding et al., 2000, 2001, 2006, 2008; Naish et al., 2001; Cantalamessa et al., 2005; 2007; Di Celma and Cantalamessa, 2007; Isbell et al., 2008; Csato and Catuneanu, 2012; Zecchin et al., 2015; details in Chapter 8), from Precambrian to Phanerozoic successions (Eriksson et al., 1998; 2004, 2005a,b, 2006, 2013; Catuneanu and Biddulph, 2001; Catuneanu et al., 2005; 2012; Sarkar et al., 2005; details in Chapter 9).

Beyond the fundamental research of the basin-fill architecture, sequence stratigraphy also provides a genetic framework to rationalize and predict the distribution of economic deposits that relate to sedimentary processes. The cyclicity, geographic extent, and the physical and temporal relationships of mineral placers, aquifers, coal beds, and petroleum systems guide the exploration and subsequent production development of natural resources. These natural-resource industries employ and benefit from the sequence stratigraphic methodology. The development and distribution of the various types of mineral placers, aquifers, petroleum reservoirs, and coal beds depend on the sequence stratigraphic surfaces and systems tracts with which they are associated (e.g., Hamilton and Tadros, 1994; Banerjee et al., 1996; Bohacs and Suter, 1997; Diessel et al., 2000; Catuneanu and Biddulph, 2001; Ketzer et al., 2003a,b; Fanti and Catuneanu, 2010). The emphasis on depositional processes also led to a shift in the focus of petroleum exploration from

structural traps to combined or purely stratigraphic traps (e.g., Bowen et al., 1993; Brown et al., 1995; Posamentier and Allen, 1999). An entire range of new types of petroleum plays thus emerged, and is now defined in light of the sequence stratigraphic concepts.

1.1.2 Sequence stratigraphy—a revolution in sedimentary geology

Sequence stratigraphy is the third of a series of major "revolutions" in sedimentary geology (Miall, 1995). Each revolution resulted in quantum paradigm shift that changed the way geoscientists understand sedimentary strata. The first breakthrough was marked by the development of the flow regime concept and the associated process/response facies models in the late 1950s and early 1960s (Harms and Fahnestock, 1965; Simons et al., 1965). This first revolution provided a unified theory to explain, from a hydrodynamic perspective, the genesis of sedimentary structures and their predictable associations within the context of depositional systems. Beginning in the 1960s, the incorporation of plate tectonics and geodynamic concepts into the analysis of sedimentary processes at regional scales, marked the second revolution in sedimentary geology. Ultimately, these first two revolutions led to the development of Basin Analysis in the late 1970s, which provided the scientific framework for the study of the origins and depositional histories of sedimentary basins. The conceptual breakthroughs in the fields of process sedimentology and basin analysis paved the way for the emergence of sequence stratigraphy as an interdisciplinary method and genetic approach to stratigraphic analysis.

As the most recent revolutionary paradigm in the field of sedimentary geology, sequence stratigraphy started in the late 1970s with the publication of AAPG Memoir 26 (Payton, 1977), even though its roots can be traced much further back in time as explained below. The concepts embodied by this discipline have resulted in a fundamental change in geological thinking and in particular, the methods of facies and stratigraphic analyses. Over the past few decades, this approach has been embraced by geoscientists as the preferred style of stratigraphic analysis, which has served to tie together observations from many disciplines. In fact, a key aspect of the sequence stratigraphic approach is to encourage the integration of datasets and research methods. Blending insights from a range of disciplines invariably leads to more robust interpretations and, consequently, scientific progress. Thus, the sequence stratigraphic approach has led to improved understanding of how stratigraphic units, facies tracts, and depositional elements relate to each other in time and space within sedimentary basins (Fig. 1.3). The applications of sequence stratigraphy range widely, from predictive exploration for natural resources

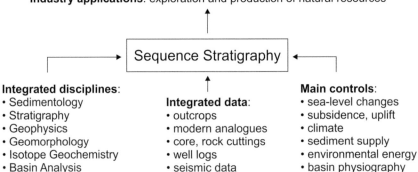

FIGURE 1.3 Sequence stratigraphy in the context of interdisciplinary research. The range of natural resources that can be rationalized in the context of sequence stratigraphy includes hydrocarbons, water, coal, and mineral deposits.

to improved understanding of Earth's geological record of local to global changes.

The conventional disciplines of process sedimentology and classical stratigraphy are particularly relevant to sequence stratigraphy (Fig. 1.4). The sedimentological component emphasizes on the processes that lead to the formation of facies and facies contacts within the confines of individual depositional systems. Some of these contacts represent event-significant stratigraphic surfaces that mark changes in stratal stacking patterns and associated sedimentation regimes, which are important for regional correlation. The study of stratigraphic contacts may not, however, be isolated from the facies analysis of the strata they separate, as the latter often provide the criteria for the identification of specific bounding surfaces. Owing to the genetic nature of the sequence stratigraphic approach, process sedimentology is an important prerequisite that cannot be separated from, and forms an integral part of sequence stratigraphy. At the smaller scales of depositional systems, sequence stratigraphy can be used to resolve and explain issues of facies cyclicity, facies associations and relationships, and reservoir compartmentalization, without necessarily applying this information for larger-scale correlations.

The importance of process sedimentology in sequence stratigraphy becomes evident in the workflow of identification of sequence stratigraphic surfaces in the rock record. Basic criteria for the identification of stratigraphic surfaces relate to the conformable vs. unconformable

nature of the contact, as well as the nature of the juxtaposed facies across the contact under analysis. Insights of process sedimentology are critical to understanding the origin of the various types of unconformity that may form in nonmarine, coastal, or marine environments, as well as the facies characteristics and variability across systems tracts. The stratigraphic component of sequence stratigraphy relates to its applicability to correlations, both within and beyond the confines of individual depositional systems, in spite of the lateral changes of facies that are common in any sedimentary basin. In addition to its sedimentological and stratigraphic affinities, sequence stratigraphy also brings a component of facies predictability which is particularly appealing to industry-oriented research (Fig. 1.4).

The success and popularity of sequence stratigraphy following the 1970s stems from its widespread applicability in both frontier and mature hydrocarbon basins, where lower and higher resolution predictions of facies development can be formulated, respectively. These predictive models have proven to be particularly effective in reducing lithology-prediction risks for hydrocarbon exploration and production, and have been subsequently employed for the exploration and production of other natural resources as well, including aquifers, coal beds, and mineral placers. In addition to its economic applications, sequence stratigraphy is also employed to resolve issues of fundamental research related to the evolution and stratigraphic architecture of sedimentary basins.

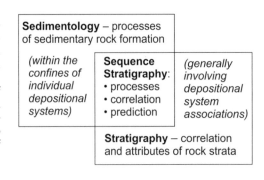

FIGURE 1.4 Sequence stratigraphy at the limit between process sedimentology and conventional stratigraphy (definitions modified from Bates and Jackson, 1987). Sequence stratigraphy makes use of the principles and methods of both process sedimentology and conventional stratigraphy, in addition to which it brings a new element of facies predictability.

1.1.3 Sequence stratigraphy—an integrated approach

The roots of sequence stratigraphy can be traced far back in the classic principles of sedimentary geology, which established the fundamental guidelines of sedimentological and stratigraphic analyses. These "first principles" set up the ground rules for the physics of flow and sediment motion, and the processes of sediment accumulation, bypass or erosion in relation to a shifting balance between sediment supply and the energy of the transporting agent (Fig. 1.5). These principles still represent the scientific backbone of sequence stratigraphy, which allows old and modern concepts to blend into a new way of looking at the sedimentary rock record. With this background, sequence stratigraphy emerged as an interdisciplinary approach that relies on the integration of multiple research methods and datasets (Fig. 1.3). At the same time, sequence stratigraphy also provides support for other lines of research such as basin analysis and source-to-sink modeling, which require a multidisciplinary approach.

Sequence stratigraphy has become an essential component of basin analysis. In the context of larger scale source-to-sink numerical models, sequence stratigraphy provides a reality check for the calibration of model results with field data. Beyond the data-based sequence stratigraphic analysis of a basin fill, source-to-sink studies integrate the analysis of sediment sources within numerical simulations of drainage systems and sediment delivery patterns from the provenance to the depocenters. This type of research extends the field of stratigraphic modeling to larger scales, and the reliability of its predictions depends on the calibration of model results with field data (Fig. 1.2). The development of source-to-sink modeling techniques does not change nor replace the need for sequence stratigraphic work. Sequence stratigraphy will continue to provide the means to rationalize the stratigraphic relationships within a basin fill, in a data-driven approach that is independent of model assumptions. The results of sequence stratigraphic analysis can be used to constrain realistic input parameters for the source-to-sink models.

The complexity and accuracy of geological models devised to resolve academic or economic issues improved over time in response to corresponding advances in concepts and technology. Classical geology

---

Principles of flow and sediment motion

All natural systems tend toward a state of equilibrium that reflects an optimum use of energy. This state of equilibrium assumes a balance between sediment removal and accumulation.

Fluid and sediment gravity flows tend to move from high to low elevations, following pathways that require the least amount of energy for fluid and sediment motion.

Flow velocity is proportional to the slope gradients. Flow discharge (subaerial or subaqueous) is equal to flow velocity times cross-sectional area.

The amount of sediment in the flow reflects the ability of source areas to produce sediment and the capacity of the flow to transport the sediment.

The transport capacity of the flow reflects the combination of flow discharge and velocity.

The mode of sediment transport (bedload, saltation, suspension) reflects the balance between grain size/weight and flow competence.

Principles of sedimentation

Walther's Law: within a relatively conformable succession of genetically related strata, vertical shifts of facies reflect corresponding lateral shifts of facies.

The direction of lateral facies shifts (progradation, retrogradation) reflects the balance between sedimentation rates and the rates of change in the space available for sediment to accumulate.

Processes of aggradation or erosion are linked to the shifting balance between energy flux and sediment supply: excess energy leads to erosion, excess sediment load leads to aggradation.

The bulk of clastic sediment is extrabasinal, derived from elevated source areas and delivered to sedimentary basins by river systems.

As environmental energy decreases, coarser grained sediments are deposited first.

FIGURE 1.5   First principles of sedimentary geology that are relevant to sequence stratigraphy (modified after Middleton, 1973; Posamentier and Allen, 1999).

remains the foundation of everything we know today, by providing the means to understanding the "first principles" of sedimentary geology (Fig. 1.5). This does not mean that sequence stratigraphy only presents old concepts in a new package, or that it developed as a stand-alone discipline in isolation from other methods. Due to its integrated approach (Fig. 1.3), sequence stratigraphy affords new insights into the genesis and architecture of sedimentary basin fills, which were not possible prior to the introduction of seismic stratigraphic concepts in the 1970s. The issue of facies predictability is a good example of a new insight that was made possible by the sequence stratigraphic approach, which is highly significant on both academic and economic grounds.

Technological advances in the fields of three-dimensional seismic data acquisition and processing resulted in the development of seismic geomorphology starting with the 1990s, in parallel with sequence stratigraphy. As defined by Posamentier (2000, 2004), seismic geomorphology deals with the imaging of paleogeographic elements, such as depositional systems and elements thereof, using three-dimensional seismic data. Seismic geomorphology can be performed as a stand-alone technique, but it can also be integrated with sequence stratigraphy for a three-dimensional control of the basin fill that combines 2D section-view insights from seismic stratigraphy (e.g., reflection geometries and terminations, stratigraphic discontinuities, seismic facies) with the 3D plan-view images of seismic geomorphology. This three-dimensional control on the stratigraphic architecture is important at any stage, from frontier exploration to production development, as it provides support and enhances the accuracy of facies predictions, including the interpretation of seismic facies in terms of sedimentary facies.

The resolution of the sequence stratigraphic work will continue to improve in parallel with technological advances in data acquisition and processing. Current efforts aim at reducing the error margin of stratigraphic models and interpretations, during both stages of exploration and production of natural resources, as well as the costs of exploration and production. As with the introduction of seismic geomorphology, technological advances will dictate the next cornerstone that can be achieved. For example, borehole imaging using electric logs (micro resistivity data, which simulate a virtual coring of boreholes) afford insights into process sedimentology (e.g., the visualization of sedimentary structures and paleoflow directions), thus eliminating the costs of mechanical coring. Such techniques and datasets will continue to be integrated into sequence stratigraphy in order to advance our knowledge and understanding of the evolution and architecture of sedimentary basin fills, from sedimentological to stratigraphic scales.

1.1.4 Sequence stratigraphy vs. other types of stratigraphy

Sequence stratigraphy is a type of stratigraphy that is uniquely focused on the identification and correlation of stratal stacking patterns in the sedimentary record (Fig. 1.6). This is fundamentally different from the correlation approaches that are employed by all other types of stratigraphy (Fig. 1.6). The distinction between the different types of stratigraphy is well defined, and yet confusions still arose between sequence stratigraphy and other types of stratigraphy, notably chronostratigraphy and allostratigraphy. Some of these confusions stem from early model assumptions (e.g., the assumption that sequence stratigraphic surfaces are time lines, hence the confusion between sequence stratigraphy and chronostratigraphy) or from faulty definitions of stratal units and bounding surfaces in the early days of sequence stratigraphy (e.g., the definition of parasequences and flooding surfaces, which employed allostratigraphic rather than sequence stratigraphic criteria). Although these confusions have been addressed and resolved, they still occasionally permeate the sequence stratigraphic practice and literature.

Figs. 1.7 and 1.8 present the classic definitions of sequence stratigraphy and of the main stratal units involved in sequence stratigraphic analysis. Subsequent to these developments, the concept of "sequence" continued to evolve and several types of stratigraphic sequence have been defined, depending on the selection of the "sequence boundary" and the definition and nomenclature of the component systems tracts (Figs. 1.9 and 1.10). For this reason, and in spite of having been widely embraced by the stratigraphic

Stratigraphy	Attribute of strata
Lithostratigraphy	lithological character of strata
Biostratigraphy	fossil content of strata
Magnetostratigraphy	magnetic polarity of strata
Chemostratigraphy	chemical properties of strata
Chronostratigraphy	absolute ages of strata
Cyclostratigraphy	cycles driven by orbital forcing
Allostratigraphy	lithological discontinuities
Seismic stratigraphy	seismic-reflection architecture
Sequence stratigraphy	stratal stacking patterns

Stratal stacking patterns are the basis for the definition and correlation of all units and surfaces of sequence stratigraphy

FIGURE 1.6 Types of stratigraphy, and the rock attributes that they use for correlation. Sequence stratigraphy is a type of stratigraphy which relies on stacking patterns for the definition, nomenclature, classification, and correlation of stratal units and bounding surfaces. Sequence stratigraphic studies highlight the stratigraphic cyclicity that develops in response to changes in relative sea level (accommodation) and base level (sedimentation).

Sequence stratigraphy (Posamentier et al., 1988; Van Wagoner, 1995): the study of rock relationships within a time-stratigraphic framework of repetitive, genetically related strata bounded by surfaces of erosion or nondeposition, or their correlative conformities.

Sequence stratigraphy (Galloway, 1989): the analysis of repetitive genetically related depositional units bounded in part by surfaces of nondeposition or erosion.

Sequence stratigraphy (Posamentier and Allen, 1999): the analysis of cyclic sedimentation patterns in stratigraphic successions, as they develop in response to variations in sediment supply and the space available for sediment to accumulate.

Sequence stratigraphy is a type of stratigraphy that is uniquely focused on the analysis of stratal stacking patterns and changes thereof in the stratigraphic record. Developments in sequence stratigraphy led to the definition of different types of stratigraphic sequences, as a function of the type of surface that is selected as the sequence boundary.

The more recent definitions of sequence stratigraphy de-emphasize the role of unconformities in the delineation of sequences, as stratigraphic cyclicity may also develop within conformable successions. However, all types of sequences consist of 'genetically related strata' that belong to the same stratigraphic cycle.

FIGURE 1.7   Definitions of sequence stratigraphy.

Depositional system (Fisher and McGowen, 1967): three-dimensional assemblage of lithofacies, genetically linked by active (modern) processes or inferred (ancient) processes and environments.

Depositional system (Galloway, 1989): three-dimensional assemblage of process-related facies that records a major paleo-geomorphic element.

Depositional systems are stratal units that represent the product of sedimentation in corresponding depositional environments (e.g., a delta system includes the body of sediment which accumulated within a deltaic environment).

Systems tract (Brown and Fisher, 1977): a linkage of contemporaneous depositional systems, forming the subdivision of a sequence.

Systems tracts are defined by specific stratal stacking patterns and bounding surfaces. Systems tracts and their bounding surfaces can be diachronous, reflecting changes in accommodation and sedimentation conditions across a basin.

Sequence (Mitchum, 1977): a relatively conformable succession of genetically related strata bounded by unconformities or their correlative conformities.

Sequence (Galloway, 1989): a succession of genetically related strata bounded by maximum flooding surfaces.

Sequence (Catuneanu, 2006): a stratigraphic cycle of change in stratal stacking patterns, defined by the recurrence of the same type of sequence stratigraphic surface in the sedimentary record.

Sequences may or may not be relatively conformable successions, but they always consist of genetically related strata that belong to the same stratigraphic cycle. Sequence boundaries may be conformable or unconformable, depending on the model of choice.

The concepts of sequence, systems tract, and depositional system are independent of scale, and may be observed at different hierarchical levels depending on the types and resolution of the data available. Well-log motifs are not part of the definition of sequence stratigraphic concepts, although general trends may be inferred from the stacking patterns of systems tracts. The magnitude of the log deflections typically varies with the hierarchical rank of the stratal units.

FIGURE 1.8   Building blocks of the sequence stratigraphic framework.

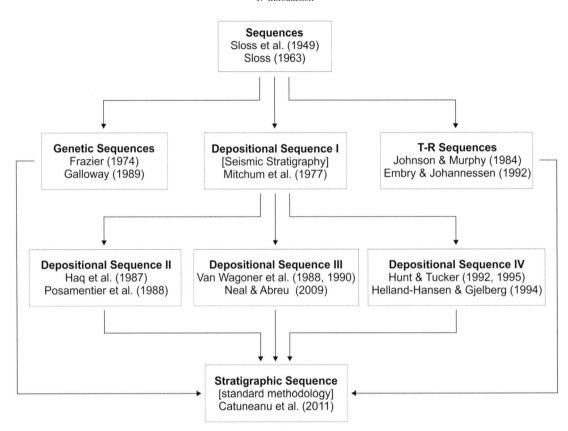

FIGURE 1.9   Evolution of sequence stratigraphy (modified from Catuneanu et al., 2011).

community, sequence stratigraphy remained the last type of stratigraphy to be endorsed with formal guidelines for methodology and nomenclature by the International Commission on Stratigraphy (Catuneanu et al., 2011). This formal endorsement was made possible by the identification of common-ground principles that enable a standard application of the method in a manner that is independent of model, as explained in this book.

Important to note is that the definition and application of sequence stratigraphic concepts are independent of scale. There is no reference to scale in the definition of concepts (Figs. 1.7 and 1.8), and the same terminology can be applied for depositional systems, systems tracts, sequences, and bounding surfaces that develop at different temporal and physical scales. Sequence stratigraphy thus applies to features as small as those produced in an experimental flume, formed in a matter of hours (e.g., Wood et al., 1993; Koss et al., 1994; Paola, 2000; Paola et al., 2001), as well as to those that are continent wide and formed over periods of millions of years. Nonetheless a distinction must be made between larger- and smaller-scale sequences, systems tracts, and stratigraphic surfaces. This is addressed through a hierarchy based on the use of modifiers such as first-order, second-order, third-order, etc., commonly in a relative rather than an absolute sense. Although this terminology is

often associated with specific time ranges (Vail et al., 1977a,b, 1991; Krapez, 1996), this has not always been common practice in the scientific literature (see discussions in Embry, 1995; Posamentier and Allen, 1999; Catuneanu et al., 2004, 2005). One reason for this is that we often do not know the scale (especially duration, but also lateral extent or thickness changes across a basin) of the stratal units we deal with within a given study area, so the use of specific names for specific scales may become quite subjective. Another advantage of using a consistent terminology regardless of scale is that jargon is kept to a minimum, which makes sequence stratigraphy more user-friendly and easier to understand across a broad spectrum of readership. These issues are tackled in more detail in Chapter 7.

Among the key concepts shown in Fig. 1.8, the term "depositional system" defines the largest stratal unit of sedimentology, and predates modern sequence stratigraphy. In contrast, "systems tract" and "depositional sequence" are specific sequence stratigraphic terms, introduced with the advent of seismic stratigraphy in the 1970s. A systems tract is a sum of laterally correlative depositional systems (hence, the use of plural: systems), which forms during a specific stage of a stratigraphic cycle (e.g., a "transgressive" systems tract forms during shoreline transgression). A sequence includes two or

Events and stages	Sequence model	Depositional Sequence I	Depositional Sequence II	Depositional Sequence III	Depositional Sequence IV	Genetic Sequence	T-R Sequence
HNR			HST	early HST	HST	HST	RST
end of T						——— MFS ———	
T	Sequence		TST	TST	TST	TST	TST
end of R							——— MRS ———
LNR			late LST (wedge)	LST	LST	late LST (wedge)	
end of RSL fall			- - - - - - - -		——— CC** ——— - - - - - - -		RST
FR			early LST (fan)	late HST	FSST	early LST (fan)	
onset of RSL fall		——— CC* ———	——— CC* ———	- - - - - - - -			
HNR			HST	early HST	HST	HST	

———— sequence boundary
———— systems tract boundary
············ within sequence surface
- - - - - within systems tract surface

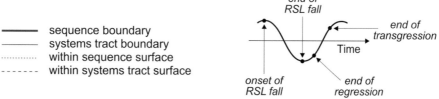

end of
RSL fall

end of
transgression

Time

onset of
RSL fall

end of
regression

FIGURE 1.10   Approaches to sequence stratigraphy: nomenclature of systems tracts and timing of sequence boundaries (from Catuneanu et al., 2011). While the concept of "systems tract" was introduced in the 1970s (Brown and Fisher, 1977), its usage in sequence stratigraphy only started in the 1980s (Posamentier and Vail, 1985). Abbreviations: CC*—correlative conformity in the sense of Posamentier et al. (1988), herein referred to as the "basal surface of forced regression"; CC**—correlative conformity in the sense of Van Wagoner et al. (1988), herein referred to as the "correlative conformity"; FR—forced regression; FSST—falling-stage systems tract; HNR—highstand normal regression; HST—highstand systems tract; LNR—lowstand normal regression; LST—lowstand systems tract; MFS—maximum flooding surface; MRS—maximum regressive surface; R—regression; RSL—relative sea level; RST—regressive systems tract; T—transgression; T–R—transgressive-regressive; TST—transgressive systems tract. See Fig. 1.9 for the proponents of the different models.

more systems tracts, depending on the number of stratal stacking patterns that develop during a stratigraphic cycle. The actual scale for sequence stratigraphic work is highly variable, ranging from depositional system scale (also highly variable) to the entire fill of a sedimentary basin, and beyond. When applied to the analysis of a depositional system (e.g., an ancient delta; Fig. 1.11), sequence stratigraphy is mainly used to resolve the details of facies relationships. Such studies are often performed to describe the degree of reservoir compartmentalization in the various stages of oil field exploration and production. When applied to the scale of depositional system associations, the issue of stratigraphic correlation becomes a primary objective, and provides the framework for the larger scale distribution of facies.

The principles outlined above provide a general idea about the range of potential outcomes and objectives of sequence stratigraphy as a function of scope and scale of analysis. There is a common misconception that sequence stratigraphy is always related to regional,

continental, or even global scales of observation (sub-basins, basins, and global cycles). This does not need to be the case, as sequence stratigraphy can be applied virtually to any scale. A good example of this is the study of the "East Coulee Delta" (Posamentier et al., 1992a), where an entire range of sequence stratigraphic elements (including systems tracts) have been documented at a centimeter to meter scale (Fig. 1.12). In recent years there have been numerous flume-based studies where sequences have been created under controlled laboratory conditions (e.g., Wood et al., 1993; Koss et al., 1994; Paola, 2000; Paola et al., 2001). Such studies have provided valuable insight as to variations on the general sequence model.

Almost any type of study of a sedimentary basin fill requires the construction of cross sections. The lines we draw on these two-dimensional representations are of two main types: (1) lines that build the chronostratigraphic or time framework of the studied interval, and (2) lines that illustrate lateral changes of facies or lithology. The chronostratigraphic framework is commonly

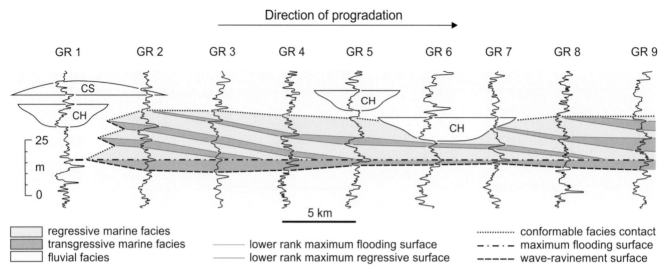

FIGURE 1.11 Stratigraphic architecture of a shallow-water system (Late Cretaceous, Alberta, Canada). Following the transgression of the seaway, the long-term regression is punctuated by higher frequency stages of progradation and retrogradation which delineate clinoforms separated by transgressive shales. This is an example of petroleum reservoir compartmentalization at a production development scale, in which each clinoform is a separate hydrodynamic unit. Abbreviations: GR—gamma-ray log; CH—fluvial channel fill; CS—fluvial crevasse splay.

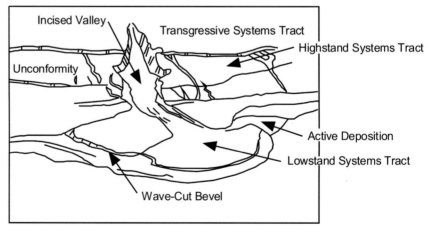

FIGURE 1.12 East Coulee Delta (from Posamentier et al., 1992a; image courtesy of Henry Posamentier), demonstrating the applicability of sequence stratigraphic concepts at virtually any scale. In this example, the highstand systems tract was incised during the fall in the water level (pond evaporation), followed by the progradation of the lower elevation lowstand delta. See Posamentier et al. (1992a) for a more detailed interpretation.

constructed by the correlation of surfaces of sequence stratigraphic significance, or true time markers such as bentonites or magnetic polarity boundaries. This is where some confusion can arise. Strictly speaking, sequence stratigraphic surfaces are not true time lines but in fact are to some degree time transgressive, or diachronous. However, because true time lines are not commonly observed, the geoscientist is relegated to using these surfaces as proxies for time lines, being pragmatic and accepting the notion that within the confines of most study areas they are at least close to being time lines and therefore, are fundamentally useful. The degree of diachroneity of sequence stratigraphic surfaces, as well as of other types of stratigraphic surfaces, is discussed in more detail in Chapters 6 and 7.

Sequence stratigraphic surfaces are not necessarily easier to observe than the more diachronous contacts that mark lateral and vertical changes of facies. Consequently the practitioner can be faced with the dilemma of where to begin a stratigraphic interpretation; in other words, what lines should go first on a cross section. The sequence stratigraphic approach yields a genetic interpretation of the basin fill, which clarifies by time increment how a basin has filled with sediment. To accomplish this, sequence stratigraphic surfaces are interpreted first, to produce a genetic framework within which other types of surface can be rationalized. Subsequently, the sections between sequence stratigraphic surfaces are interpreted by recognizing facies contacts. These two types of surfaces define sequence stratigraphy and lithostratigraphy, respectively (Fig. 1.13).

The inherent difference between lithostratigraphy and sequence stratigraphy is important to emphasize, as both analyze the same sedimentary succession but with the focus on different stratigraphic aspects or rock properties. Lithostratigraphy deals with the lithology of strata and with their organization into units based on lithological character (Hedberg, 1976). The boundaries between lithostratigraphic units are often highly diachronous facies contacts, in which case they develop within the sedimentary packages bounded by sequence stratigraphic surfaces (Fig. 1.13). The advantage of the sequence stratigraphic approach is that correlation can be carried out despite the lateral changes of facies that commonly occur across a basin, for which reason sequence stratigraphic surfaces are typically more extensive than the facies contacts. It is also important to note that facies analyses leading to the interpretation of paleoenvironments are much more critical for sequence stratigraphy than for lithostratigraphy, as illustrated in Figs. 1.14 and 1.15. These figures show that even along 1D vertical profiles, sequence stratigraphic units are often offset relative to the lithostratigraphic units due to their emphasis on different rock attributes. Understanding what constitutes a reasonable vertical and lateral relationship between facies within a time framework assists in correlating event-significant surfaces that mark changes in stratal stacking patterns through varying lithologies.

An example of a sequence stratigraphic—as contrasted with a lithostratigraphic—interpretation based on the same dataset is illustrated in Fig. 1.16. The

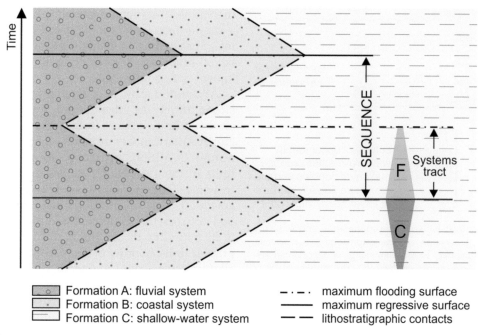

Formation A: fluvial system	– · – · maximum flooding surface
Formation B: coastal system	——— maximum regressive surface
Formation C: shallow-water system	– – lithostratigraphic contacts

FIGURE 1.13 Sequence stratigraphic vs. lithostratigraphic approaches to stratigraphic correlation and the definition of stratal units. Sequence stratigraphic surfaces mark changes in stratal stacking patterns (e.g., from progradation to retrogradation and *vice versa*), with or without changes in lithology. Lithostratigraphic surfaces mark changes in lithology, with or without changes in stacking pattern. As they are independent of lithology, sequence stratigraphic units and bounding surfaces are typically mappable over larger areas despite the lateral changes of facies, and therefore provide a superior method of stratigraphic correlation. Facies contacts are also important to map *after* the construction of the sequence stratigraphic framework, as they are best rationalized within the genetic context of systems tracts. Abbreviations: C—coarsening upward; F—fining upward.

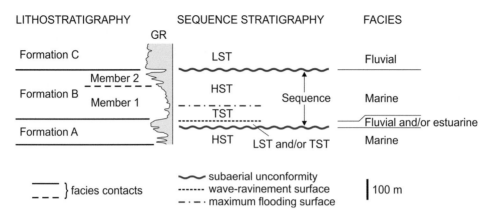

FIGURE 1.14   Lithostratigraphy and sequence stratigraphy of a facies succession (modified from Posamentier and Allen, 1999). Lithostratigraphy defines rock units on the basis of lithology, often irrespective of the depositional system. Sequence stratigraphy defines rock units based on stratal stacking patterns and changes thereof across their bounding surfaces. Lithostratigraphic and sequence stratigraphic surfaces may or may not coincide. The maximum flooding surface, which often provides the best datum for stratigraphic correlation, is commonly placed within undifferentiated lithostratigraphic units. Other sequence stratigraphic surfaces may also be missed within the lithostratigraphic framework if the facies below and above share a similar lithological character. Abbreviations: GR—gamma-ray log; LST—lowstand systems tract; TST—transgressive systems tract; HST—highstand systems tract.

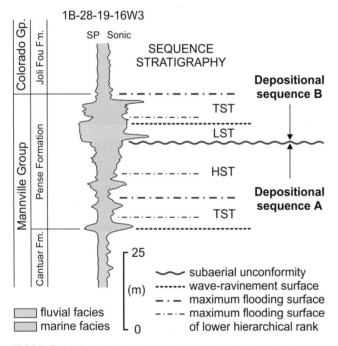

FIGURE 1.15   Contrast between lithostratigraphy and sequence stratigraphy in the delineation of stratigraphic units (Cretaceous, Western Canada Sedimentary Basin). Knowledge of depositional systems is only critical to sequence stratigraphy. The systems tracts of depositional sequences A and B include lower rank stratigraphic cycles. Abbreviations: SP—spontaneous potential; LST—lowstand systems tract; TST—transgressive systems tract; HST—highstand systems tract.

interpretation of sequence stratigraphic surfaces is based on two fundamental observations: the type of stratigraphic contact, conformable or unconformable; and the nature of facies (depositional systems) which are in contact across the stratigraphic surface. The reconstruction of paleodepositional environments is critical in sequence stratigraphy. In contrast, the lithostratigraphic cross section does not require knowledge of paleoenvironments, but only mapping of lithological contacts. Some of these contacts may coincide with sequence stratigraphic surfaces; others may only reflect lateral changes of facies. As a result, the lithostratigraphic units (e.g., formations A, B, and C in Fig. 1.16) provide only descriptive information of lithologic distribution, which in some instances could combine the products of sedimentation of various depositional environments. Thus a simple map of lithologic distribution may give little insight as to the general paleogeography, and as a result be of little use in predicting lithologies away from known data points.

Allostratigraphy is a stratigraphic discipline with strong affinity to lithostratigraphy, but with emphasis on bounding surfaces rather than stratal units. The North American Commission on Stratigraphic Nomenclature (NACSN) introduced formal allostratigraphic units in the 1983 North American Stratigraphic Code to name "discontinuity-bounded units." As currently amended, "an allostratigraphic unit is a mappable body of rock that is defined and identified on the basis of its bounding discontinuities" (Article 58). Allostratigraphic units, in order of decreasing rank, are allogroup, alloformation, and allomember—a terminology that originates and is modified from lithostratigraphy. The fundamental unit is the alloformation (NACSN, 1983, Art. 58). The bounding discontinuities which define the allostratigraphic approach are represented by any mappable lithological contact, with or without a stratigraphic hiatus associated with it. In this approach, all lithostratigraphic and sequence stratigraphic surfaces that are associated with a lithological contrast may be used for allostratigraphic studies (e.g., Bhattacharya and Walker, 1991; Plint, 2000). However, not all allostratigraphic surfaces are of sequence stratigraphic

1. Data: vertical profiles and depositional systems

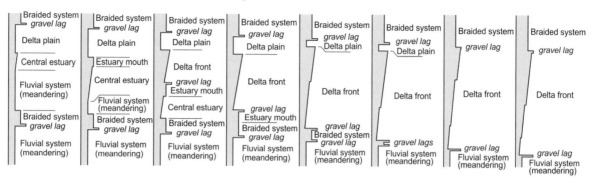

2. Sequence stratigraphic framework

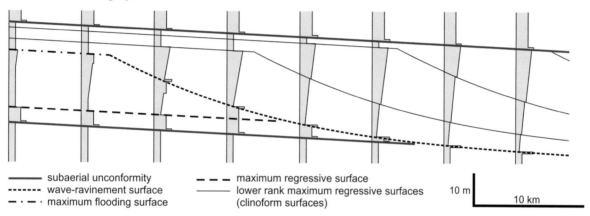

―――――― subaerial unconformity   ― ― ― maximum regressive surface

------------- wave-ravinement surface   ―――― lower rank maximum regressive surfaces

― · ― · ― maximum flooding surface   (clinoform surfaces)

10 m   10 km

3. Sequence stratigraphic framework, facies contacts, and depositional systems

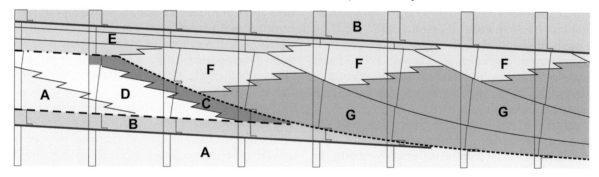

4. Lithostratigraphic framework

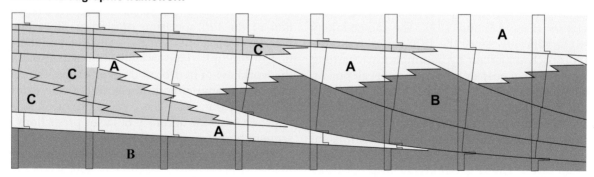

FIGURE 1.16   Sequence stratigraphic vs. lithostratigraphic frameworks. 1. The reconstruction of depositional systems and the observation of the scoured vs. conformable nature of stratigraphic contacts are important steps in the sequence stratigraphic workflow; 2. Sequence stratigraphic framework based on the observation of stratal stacking patterns; 3. Sequence stratigraphic framework with facies contacts and depositional systems: A—meandering system; B—braided system; C—estuary-mouth complex; D—central estuary; E—delta plain; F—upper delta front; G—lower delta front—prodelta; 4. Lithostratigraphic framework (e.g., formations): A—sandstone-dominated unit; B and C—mudstone-dominated units, with silty and sandy interbeds; units B and C are separated by unit A; additional lithostratigraphic units (e.g., members) may be defined as a function of variations in lithology and color.

significance (e.g., "flooding surfaces" that develop during transgression, with retrogradational stacking patterns below and above, are surfaces of allostratigraphy but not of sequence stratigraphy; Catuneanu, 2019a).

1.2 Development of sequence stratigraphy

Modern sequence stratigraphy started as "seismic stratigraphy" in the 1970s (Payton, 1977), whereby the method was applied specifically to seismic data. This evolved into the more generic "sequence stratigraphy" in the 1980s as the application of the method was extended to other types of data, such as those provided by wells and outcrops (e.g., Wilgus et al., 1988). Noteworthy, the development of sequence stratigraphy as a "new" type of stratigraphy in the 1970s and 1980s was preceded by the much earlier publication of several of its key concepts (e.g., Barrell, 1917; Sloss et al., 1949; Wheeler, 1964, Fig. 1.17). Some of these milestones include the concepts of base level (Powel, 1875; Gilbert, 1895; Barrell, 1917), unconformity-bounded sequence (Longwell, 1949; Sloss et al., 1949), correlative conformity (introduced as "continuity surface"; Wheeler, 1964), and depositional system (Fisher and McGowan, 1967). Building on this foundation, the refinements brought about by seismic stratigraphy include the definition of systems tracts (Brown and Fisher, 1977, Fig. 1.8), seismic-reflection terminations (Mitchum, 1977), a revision to the definition of a "sequence" (Mitchum, 1977, Fig. 1.8), and the usage of seismic data for a genetic interpretation of the stratigraphic architecture (Payton, 1977). From here, sequence stratigraphy bloomed and diversified in the 1980s and the 1990s, with several models being proposed (Figs. 1.9 and 1.10).

The evolution of sequence stratigraphy involved advances in the understanding of the array of possible controls on sequence development, improvements in stratigraphic resolution, revisions to the definition and classification of sequences, and ultimately the emergence of a standard approach to methodology and nomenclature (Figs. 1.9 and 1.17). The methodology improved significantly since the 1970s, from a model-driven approach underlain by assumptions regarding the dominant role of eustasy on sequence development, with consequent assertions of global correlations (Vail et al., 1977a,b; Haq et al., 1987), to a data-driven approach that promotes the use of local data without any assumptions regarding the underlying controls on sequence development (Miall, 1986, 1991, 1992; Catuneanu et al., 2011). The latter affords realistic constructions of local stratigraphic frameworks, which prove to be highly variable in terms of timing and scales of stratigraphic cycles, not only from one sedimentary basin to another but also among the sub-basins of the same

sedimentary basin (Catuneanu et al., 1999, 2000, 2002; Miall et al., 2008; Menegazzo et al., 2016). The construction of basin-specific stratigraphic frameworks is a major breakthrough and departure from the early models.

Concomitant with the improvements in methodology, refinements in stratigraphic modeling driven by advances in computing power enhanced the ability to test the response of sedimentary systems to any combinations of possible controls on sequence development. The insights enabled by numerical simulations are particularly useful when calibrated with field data (Euzen et al., 2004; Rabineau et al., 2005, 2006; Csato et al., 2013, 2015; Leroux et al., 2014; Catuneanu and Zecchin, 2016). In light of these advances, a clear distinction is required between methodology and modeling in sequence stratigraphy. The methodology is based on the observation of stratal stacking patterns in a manner that is independent of the interpretation of the underlying controls. Beyond the purpose of the methodological workflow, the modeling and testing of the possible controls on sequence development can continue indefinitely after the construction of a sequence stratigraphic framework. Stratigraphic modeling is an independent line of research that plays no role in, and it does not change the outcome of the methodological workflow (Catuneanu, 2020b, Fig. 1.2).

Most significant to its practical applications, the development of sequence stratigraphy was accompanied by a gradual increase in stratigraphic resolution, from the seismic scales of petroleum exploration (Payton, 1977) to the sub-seismic scales of petroleum production development (e.g., Van Wagoner et al., 1990; Homewood et al., 1992; Homewood and Eberli, 2000; Zecchin and Catuneanu, 2013, 2015; Magalhaes et al., 2015; Catuneanu, 2019b). Sequence stratigraphic frameworks can be constructed at different scales, depending on the scope of the study and the resolution of the data available. Seismic data are typically used to build larger scale, lower resolution frameworks, whereas borehole and outcrop data afford the construction of higher resolution frameworks at sub-seismic scales. The observation of the full spectrum of stratigraphic complexity that develops at intertwining scales relies on the integration of multiple datasets with different resolutions. In the workflow of hydrocarbon exploration, the construction of a stratigraphic framework typically starts with 2D seismic transects. The context provided by seismic data enhances the accuracy of facies predictions in subsequent higher resolution studies.

The development of seismic stratigraphy in the 1970s made use of established sedimentological concepts (e.g., the concept of "depositional system" as a three-dimensional assemblage of lithofacies linked by a common environment of deposition; Fisher and McGowan, 1967) to define new building blocks of the stratigraphic

1600s – 1900s: Basic principles

- Steno (1669): angular unconformities; principles of superposition and original horizontality
- Hutton (1788, 1795): unconformities in the context of the rock cycle; concept of uniformitarianism
- Jameson (1805): coins the term "unconformity"
- Smith (1819): the law of faunal succession; the principle of organic evolution
- Lyell (1833): the concept of sea-level change
- Gressly (1838): the concept of facies
- d'Orbigny (1842): the concept of biologic stage
- Oppel (1856): the concept of biologic zone (biozone)
- Walther (1894): the law of facies successions
- Grabau (1905): the concept of disconformity
- Grabau (1906): stratigraphic relationships of truncation and onlap
- Blackwelder (1909): the concept of nonconformity
- Willis (1910): the concept of paraconformity
- Udden (1912): the concept of cyclic sedimentation

1900s: Development of models

- Barrell (1917): concepts of base level, diastem, rhythms
- Milankovitch (1930): orbital forcing (astronomical cycles)
- Weller (1930), Wanless and Weller (1932), Wanless and Shepard (1936): concept of cyclothem
- Longwell (1949), Sloss et al. (1949): the concept of sequence as an unconformity-bounded unit
- Hedberg (1951): the concept of chronostratigraphic surface
- Rich (1951): the concept of clinoform
- Wheeler (1958, 1964): chronostratigraphic charts (Wheeler diagrams); continuity surfaces
- Frazier (1974): concept of depositional complex (precursor of the genetic stratigraphic sequence)
- Vail et al. (1977a,b): seismic stratigraphy; the concept of global cycles
- Mitchum (1977): correlative conformities as sequence boundaries; toplap, onlap, downlap, offlap
- Brown and Fisher (1977): the concept of systems tract
- Johnson and Murphy (1984): transgressive-regressive (T-R) sequence model
- Posamentier et al. (1988); Van Wagoner et al. (1988): depositional sequence models
- Galloway (1989): genetic-stratigraphic sequence model
- Hunt and Tucker (1992): four systems tracts in sequence stratigraphy

2000s – 2010s: Towards a unified methodology and nomenclature

- Catuneanu et al. (2009): first informal international effort towards a standard approach
- Catuneanu et al. (2011): first formal guidelines for a standard methodology and nomenclature (International Commission on Stratigraphy – subcommission on stratigraphic classification)

FIGURE 1.17 Milestones in the development of sequence stratigraphy. Modern sequence stratigraphy, which started as "seismic stratigraphy" in the 1970s (Vail, 1975; Payton, 1977), was dominated in the 1970s–1980s by the assumption that eustasy exerted the main control on sequence development; this led to the construction of reference global-cycle charts for stratigraphic cyclicity and correlations worldwide (Vail et al., 1977b; Haq et al., 1987, 1988; Posamentier et al., 1988). Subsequent developments in the 1990s changed the emphasis on tectonism as a major control on sequence development (e.g., Krapez, 1996: "The importance of eustasy in sequence stratigraphy should be de-emphasized"). The standard methodology that emerged in the 2000s–2010s is decoupled from the interpretation of the underlying controls, and it is based on the observation of stratal stacking patterns in the stratigraphic record.

framework (e.g., "systems tracts," as linkages of contemporaneous depositional systems; Brown and Fisher, 1977). Owing to the specific type of data used to develop the methodology (i.e., 2D seismic transects), seismic stratigraphy introduced by default a minimum scale for depositional systems, systems tracts, and sequences, which had to exceed the vertical seismic resolution (i.e., typically in a range of 10^1 m in the 1970s). As a result, the building blocks of the seismic stratigraphic framework are commonly observed at scales of 10^1–10^2 m. The perception that sequences and their component systems

tracts and depositional systems develop typically at scales of 10^1–10^2 m is an artifact of seismic resolution, but it dominated stratigraphic thinking for decades.

The reality of sequences, systems tracts, and depositional systems at sub-seismic scales has become evident with the advances in high-resolution sequence stratigraphy (e.g., Tesson et al., 1990, 2000; Lobo et al., 2004; Amorosi et al., 2005, 2009, 2017; Bassetti et al., 2008, Catuneanu and Zecchin, 2013; Nanson et al., 2013; Zecchin and Catuneanu, 2013, 2015, 2017; Csato et al., 2014; Nixon et al., 2014; Magalhaes et al., 2015; Zecchin et al., 2015, 2017a,b; Ainsworth et al., 2017, 2018; Pellegrini et al., 2017, 2018; Catuneanu, 2019a,b). Improvements in stratigraphic resolution demonstrated that unconformities may form over a wide range of scales, both below and above the seismic resolution, and therefore, unconformity-bounded units are not restricted to the scales of seismic stratigraphy (e.g., Miall, 2015; Strasser, 2016, 2018). It is now clear that most commonly, the building blocks of a seismic stratigraphic framework consist of higher frequency sequences that develop at sub-seismic scales; e.g., seismic-scale systems tracts $(10^1 - 10^2$ m) consist typically of sequences of $10^0 - 10^1$ m scales, which are different from and should not be confused with parasequences (Catuneanu, 2019a).

1.2.1 Unconformities

Sequence stratigraphy started to emerge as a method of stratigraphic analysis ever since the recognition of unconformities in the rock record, which allowed the subdivision of the sedimentary succession into units separated by breaks in deposition (Fig. 1.17). Early depictions of angular unconformities date as far back as the 17th century (Nicolaus Steno, 1669), but their geological meaning was only realized late in the 18th century (James Hutton, 1788, 1795). Hutton presented the concept of unconformity as part of the rock cycle, although he never actually used the term "unconformity"; instead, he used descriptive phrases in which the expression "conjunction of vertical and horizontal strata" was widely used (Tomkeieff, 1962). The term "unconformity" was coined by Robert Jameson in the early 19th century (Jameson, 1805), but it was only widely incorporated into the geological vocabulary following the work of Charles Lyell and Henry De la Beche (De la Beche, 1830; Lyell, 1830).

The term "unconformity" was synonymous with what we know today as "angular unconformity" until the beginning of the 20th century, when the term "disconformity" was proposed for a particular type of erosional unconformity where sedimentary strata above and below the contact are parallel to each other (Grabau, 1905). Other types of unconformities were recognized subsequently, including the contact between basement rocks and the overlying sedimentary rocks (i.e., known today as "nonconformity"; Blackwelder, 1909) and the non-depositional hiatus that assumes a gap in sedimentation without necessarily involving erosion (i.e., referred to today as "paraconformity"; Willis, 1910). All these types of unconformity assume a "substantial" break in the geologic record (Neuendorf et al., 2005), although the actual span of time of what is considered to be "substantial" (as opposed to "brief," as in a diastem) was never quantified.

The spectrum of hiatal stratigraphic contacts was completed with the definition of the "diastem" by Barrell (1917), essentially as a "minor" paraconformity. However, the distinction between a paraconformity and a diastem, both in terms of temporal significance and physical expression, remains elusive to the present day. Further discussion and recommendations on the usage of the diastem concept are presented in Chapter 7. The definition of the various types of stratigraphic contacts is summarized in Fig. 1.18.

1.2.2 Unconformity-bounded units

As early as the eighteenth century, Hutton (1788, 1795) recognized the alternation through time of processes of erosion and deposition, setting up the foundation for what is known today as the "rock cycle" (Fig. 1.17). Hutton's observations may be considered as the first account of stratigraphic cyclicity, whereby unconformities provide the basic subdivision of the rock record into repetitive successions. The link between unconformities and base-level changes was subsequently explained by Barrell (1917), who stated that "sedimentation controlled by base level will result in divisions of the stratigraphic series separated by breaks" (Fig. 1.19).

Following the recognition of unconformities in the rock record, stratigraphic hiatuses have become natural candidates for correlation and the subdivision of the stratigraphic succession into units characterized by relatively continuous sediment accumulation. The nomenclature of unconformity-bounded units included terms such as "rhythms" (Barrell, 1917; to describe sedimentary cycles of various scales), "cyclothems" (Weller, 1930; Wanless and Weller, 1932; Wanless and Shepard, 1936; to describe small-scale unconformity-bounded units of Carboniferous age in the mid-continental US), "sequences" (Longwell, 1949; Sloss et al., 1949; Sloss, 1963; to describe continental-scale unconformity-bounded units in North America), "depositional complexes" (Frasier, 1974; to describe units bounded by marine starvation surfaces), and "synthems" (Chang, 1975; to rename the concept of "sequence" of Sloss, 1963). The usage of the term "sequence" in a stratigraphic context was not without controversy (e.g., discussion in van Loon, 2000), but it eventually prevailed, and became widely used following the publication of the seismic stratigraphic methodology in the 1970s (Mitchum et al., 1977).

The unconformity-bounded sequence of Sloss (1963) provided the stratigraphic community with mappable units that could be used for regional correlation and the subdivision of the rock record into genetically-related packages of strata. The concept of "unconformity-bounded unit" was formalized in the International Stratigraphic Guide in 1994. The limitation of this

Unconformity = hiatus ± erosion

A break in the geological record, whatever its cause and magnitude, with or without accompanying erosion. Types of unconformities:

- **Disconformity** = hiatus + erosion

  An unconformity in which the bedding planes above and below the break are essentially parallel, ...and usually marked by a visible and uneven erosion surface of appreciable relief.

- **Paraconformity** = hiatus ± erosion

  An obscure or uncertain unconformity with no discernable erosion, in which the beds above and below the break are parallel to each other. 'Minor' paraconformities are also referred to as 'diastems'.

- **Angular unconformity** = hiatus, erosion, and tilt

  An unconformity between two groups of rocks whose bedding planes are not parallel or in which the older, underlying rocks dip at a different angle (usually steeper) than the younger, overlying strata.

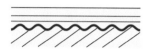

- **Nonconformity** = top of basement rocks

  An unconformity developed between sedimentary rocks and older basement rocks that had been exposed to erosion before the overlying sediments covered them.

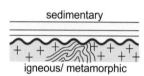

sedimentary

igneous/ metamorphic

Conformity = no hiatus

Undisturbed relationship between adjacent sedimentary strata that have been deposited in orderly sequence. True stratigraphic continuity in the succession of beds.

FIGURE 1.18 Types of stratigraphic contacts (modified after Bates and Jackson, 1987; Neuendorf et al., 2005; Catuneanu, 2006). The scale-independent definition of an unconformity as a stratigraphic hiatus of any duration eliminates the need for the concept of diastem. The usage of diastems may be restricted to sedimentology, or be eliminated altogether from sedimentary geology (see text for details).

type of unit becomes evident at smaller scales, as unconformities of lesser magnitudes tend to be restricted to the basin margins (Fig. 1.20). In such cases, the number of sequences decreases in a downdip direction, hampering stratigraphic correlation. This limitation required a conceptional innovation whereby sequence boundaries could be extended beyond the termination of unconformities. The breakthrough was the concept of "continuity" surface (Wheeler, 1964), which in physical terms referred to the paleodepositional surface (e.g., seafloor) at the time when the unconformity reached its maximum extent. The "continuity" surface of Wheeler (1964) was subsequently rebranded as the "correlative conformity" in the context of seismic stratigraphy (Mitchum, 1977). The advantage of the modern sequence, bounded by a composite surface that may include a conformable portion, lies in its greater areal extent, thus with improved potential for stratigraphic mapping and correlation (Fig. 1.21).

1.2.3 Concept of "sequence"

In parallel with the development of the "sequence" concept in a stratigraphic context, sedimentologists in the 1960s and 1970s also used the term "sequence" to define vertical successions of facies that are "organized in a coherent and predictable way" (Pettijohn, 1975), reflecting the natural evolution of a depositional environment. This practice was entrenched in landmark publications by Reading (1978) and Selley (1978a), whereby facies sequences were used to illustrate the Walter's Law in conformable sedimentary successions. Examples of facies sequences in a sedimentological sense include coarsening-upward successions of deltaic facies, or the repetition of channel-fill, lateral-accretion,

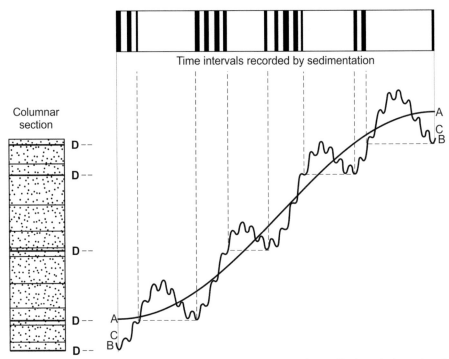

FIGURE 1.19   Concept of "base level" (redrafted from Barrell, 1917: p. 796). Fluctuations in base level over different time-scales explain the patterns of sediment preservation (black bars at the top) and the formation of unconformities of different magnitudes (white intervals between the black bars). In any depositional setting, the base level is a surface of equilibrium between sedimentation and erosion (e.g., a graded fluvial profile, or a graded seafloor profile). The base level is a descriptor of sedimentation; changes in base level, which control the rates of sedimentation at any location, depend on all factors that modify sediment supply and the energy of the sediment-transport agents, including accommodation, climate, source-area tectonism, and autogenic processes. Deposition takes place every time the base level rises, but preservation is only possible where the periods of deposition occur during a longer term rising trend.

Columnar section

Time intervals recorded by sedimentation

Sedimentary record made by harmonic oscillations in base level

A-A. Primary curve of rising base level.
B-B. Diastrophic oscillations, giving disconformities D-D.
C-C. Minor oscillations, exaggerated and simplified, due largely to climatic rhythms

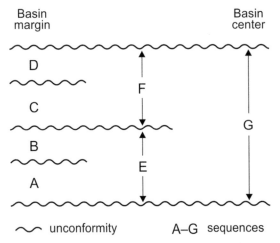

FIGURE 1.20   Unconformity-bounded sequences in the sense of Sloss et al. (1949) and Wheeler (1958, 1964) (modified after Wheeler, 1958). As more unconformities develop in proximal areas, the number of sequences decreases in a downdip direction, posing a problem for stratigraphic correlation.

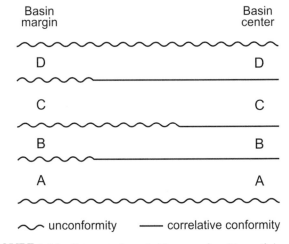

FIGURE 1.21   Sequences bounded by unconformities *or their correlative conformities* (modified after Wheeler, 1958; 1964, and Mitchum, 1977). Wheeler (1964) recognized "continuity" surfaces beyond the termination of unconformities, but did not consider them as part of the sequence boundary. The inclusion of correlative conformities in the definition of a sequence (Mitchum, 1977) improved stratigraphic correlation from the basin margin to the depocenter. A–D: sequences.

and overbank architectural elements in meandering river systems. In a stratigraphic context, such facies sequences are only part of the sedimentological makeup of systems tracts. The development of seismic and sequence stratigraphy in the late 1970s and 1980s revitalized the use of the term "sequence" in stratigraphy, which remained the dominant approach to date. It is therefore important to distinguish between the "sequence" of sequence stratigraphy and the "facies sequence" of sedimentology. The latter would be better termed "facies succession." Herein, "sequences" are used in a stratigraphic sense.

The concept of stratigraphic "sequence" evolved over time, in parallel to the gradual increase in the resolution of stratigraphic studies (Fig. 1.22). The early definition of a sequence in the 1940s referred to unconformity-bounded units observed at continental scales (Longwell, 1949; Sloss et al., 1949), by considering only interregional unconformities as sequence boundaries. As a result, the Phanerozoic sedimentary cover of North America was subdivided into only six sequences (Sloss, 1963). These continental-scale sequences included "unconformity-bounded masses of strata of greater than group or super-group rank" (Krumbein and Sloss, 1951), which restricted the applicability of the "sequence" concept only to regional-scale stratigraphic studies. Moreover, the large scale of these sequences, which reach ±1,000 m in thickness, renders them unusable for economic exploration as, for example, they can include multiple petroleum systems.

Subsequent work by Wheeler (1958, 1959, 1964) depicted stratigraphic cyclicity in a time domain (depth-to-time Wheeler transformation, or "Wheeler diagram" as it is known today; Qayyum et al., 2014, 2015), and recognized sequence-bounding unconformities of smaller magnitude and smaller areal extent than those considered by Sloss (1963). In doing so, Wheeler (1964) introduced the concept of "continuity" surface beyond the termination of an unconformity, which later became the "correlative conformity" in the 1970s (Mitchum, 1977). This was the first step toward increasing the resolution of stratigraphic studies, by decreasing the scale of a sequence. This trend materialized in the 1970s with the definition of sequences at seismic scales (Payton, 1977),

and continued subsequently with the definition of sequences at sub-seismic scales (i.e., high-resolution sequence stratigraphy; Catuneanu and Zecchin, 2013; Csato et al., 2014; Magalhaes et al., 2015). These developments made sequence stratigraphy relevant to petroleum exploration in the 1970s (i.e., in the context of seismic stratigraphy, with sequences observed at scales of 10^1-10^2 m), and subsequently to petroleum production development (i.e., in the context of high-resolution sequence stratigraphy, with sequences observed at sub-seismic scales of 10^0-10^1 m) (Fig. 1.22).

The evolution of the concept of sequence is reflected in the revisions to the definition, which, as stratigraphic resolution improved, recognized the increasingly important role of conformable surfaces at the sequence boundary (Fig. 1.22). Following this trend, the definition changed from "an unconformity bounded unit" (1940s) to "a unit bounded by unconformities or their correlative conformities" (1970s), and eventually to "a unit bounded by any recurring surface of sequence stratigraphy" (2010s) (Fig. 1.22). This trend highlights the fact that as the scale of observation decreases, the magnitude and the areal extent of unconformities decrease as well, and conformable surfaces become increasingly important to delineate sequences. The result is that at high-resolution level, and also depending on the type of sequence stratigraphic surface selected as the "sequence boundary" (Fig. 1.10), a sequence may no longer require unconformities at its boundaries.

Sequences are defined by their bounding surfaces. Different types of sequence have been proposed (i.e., "depositional," "genetic stratigraphic," and "transgressive-regressive"), depending on the selection of the sequence boundary (Fig. 1.9). The sequence of seismic stratigraphy was introduced as the

Decade	Definition of 'sequence'	Resolution
1940s	Rock-stratigraphic unit bounded by interregional unconformities [1]	10^2-10^3 m
1970s	A relatively conformable succession of genetically related strata bounded by unconformities or their correlative conformities [2]	10^1-10^2 m
2000s	A stratigraphic cycle defined by the recurrence of the same type of sequence stratigraphic surface in the rock record [3]	10^0-10^1 m

FIGURE 1.22 Evolution of the concept of "sequence," in response to the increase in stratigraphic resolution and the need to accommodate all sequence stratigraphic approaches. The recurrence of the same type of sequence stratigraphic surface in the rock record delineates cycles of change in stratal stacking patterns. Different types of stratigraphic sequences can be defined, depending on the specific type of sequence stratigraphic surface that is selected as the sequence boundary (Fig. 1.10). Sequences may or may not be "relatively conformable" successions, but they always consist of "genetically related" strata that belong to the same stratigraphic cycle. The identification of sequences and bounding surfaces is based on the observation of stratal stacking patterns, independently of the interpretation of underlying controls. References: (1) Longwell (1949), Sloss et al. (1949), Sloss (1963); (2) Mitchum (1977); (3) Catuneanu (2006), Catuneanu et al. (2009, 2011).

"depositional sequence." The unconformable portion of the depositional sequence boundary generally relates to processes of subaerial exposure, whereas the correlative conformity is typically a marine surface (i.e., a paleoseafloor) located downdip of the termination of the unconformity. Criteria for the identification of these unconformities and correlative conformities have been defined in relation to seismic-reflection terminations and architecture (e.g., offlap; Mitchum, 1977). The genetic stratigraphic and transgressive-regressive sequences use other types of bounding surfaces, which develop at the end of transgressions and regressions, respectively (Fig. 1.10). Regardless of the type of sequence, all sequences correspond to stratigraphic cycles in the sedimentary record (i.e., cycles that start and end with the same type of sequence stratigraphic surface; Fig. 1.22).

The revisions to the definition of a "sequence" reflect conceptual advances and the need for an inclusive approach that accommodates all sequence models (Figs. 1.9, 1.10; Catuneanu et al., 2009, 2010, 2011). The latest definition of a sequence in Fig, 1.22 is independent of model, as it accommodates all sequence stratigraphic approaches (Catuneanu, 2019a).

1.2.4 Sequence models

The development of sequence stratigraphy following the emergence of seismic stratigraphy in the 1970s took place informally for more than three decades, without guidance from any international stratigraphic commission. The result was the proliferation of an unnecessarily complex terminology (e.g., several synonymous terms for the same concept), as well as the misusage of terms (e.g., the same terms used with different meanings by different authors). The methodology also evolved in different directions, reflecting the divergent views of authors with respect to the definition of a "sequence" (e.g., dependent or independent of scale and the resolution of the data available); the selection of the "sequence boundary" (i.e., what surfaces should be elevated to the rank of "sequence boundary"); the dominant controls on sequence development (e.g., global eustasy vs. local tectonism); and the approach to the classification of stratigraphic cycles that develop at different scales (i.e., the definition of a hierarchy system for sequence stratigraphic units and bounding surfaces).

Some of the reasons for this variety of opinion include the type of basin from which models were derived (e.g., passive margins vs. forelands or rifts, each dominated by different patterns and mechanisms of subsidence), and the types of data used to construct the models (e.g., seismic vs. well logs or outcrops, each with different resolution and mapping potential). As the proponents of the various models drew their insights from different settings and datasets, their views also diverged

in key aspects of sequence stratigraphy, including the sequence stratigraphic surfaces that are most suitable as sequence boundaries (e.g., correlative conformities vs. maximum flooding surfaces or maximum regressive surfaces); the scale of sequences and systems tracts (i.e., typically larger for studies based on seismic data); and the development of systems tracts within sequences (e.g., with or without lowstand systems tracts, depending on the geological setting). Ultimately, each sequence model works best in the context in which it was defined, and no one model is applicable to the entire range of case studies; therefore, none can be adopted over the others as a standard model.

The unconformity-bounded sequence of Sloss (1963) was widely embraced by the stratigraphic community, and formalized in the 1994 International Stratigraphic Guide, because the concept of unconformity was noncontroversial. The modification of the original concept of "sequence," by introducing correlative conformities as part of the sequence boundary, triggered both progress and debates. An initial source of contention was the timing of these correlative conformities, which split the school of seismic stratigraphy (i.e., the "depositional sequence I" in Fig. 1.9) into two competing approaches in the 1980s (i.e., depositional sequences II vs. III in Figs. 1.9 and 1.10). Beyond this conceptual debate, a more practical aspect of the methodology related to the mappability of correlative conformities on various datasets. While mapping these surfaces on dip-oriented seismic lines is generally straight forward, based on the architecture of seismic reflections, the same cannot be said about their identification in outcrop or well data, where their physical signature is more cryptic. For this reason, in the absence of seismic data, sequence stratigraphic surfaces with stronger expression in outcrop and wells became natural candidates for the definition of sequences outside of the realm of seismic stratigraphy (i.e., "genetic stratigraphic" and "transgressive-regressive" sequences; Figs. 1.9 and 1.10).

The debate between the supporters of the two alternative depositional sequence models in the 1980s was exacerbated by conflicting choices with respect to the nomenclature of systems tracts. While both groups agreed that a sequence was to be subdivided into three systems tracts (i.e., lowstand, transgressive, and highstand), the different timing of their correlative conformities (i.e., the onset vs. the end of relative sea-level fall; Fig. 1.10) resulted in a different usage of the systems tract nomenclature. Specifically, as the depositional sequence boundary was supposed to be placed at the base of the lowstand systems tract, the depositional sequence II model assigned the deposits of relative sea-level fall to the lowstand systems tract, whereas the depositional sequence III model assigned the same deposits to the highstand systems tract (Fig. 1.10). This generated significant confusion, which could only be resolved by the separation of the falling-stage deposits as a distinct systems tract in the 1990s (i.e., the depositional sequence IV model in Figs. 1.9 and 1.10).

The introduction of the fourth systems tract marked a significant step forward in the development of sequence stratigraphy, on both conceptual and practical grounds. Conceptually, this resolved a significant nomenclatural controversy. Practically, the strata that accumulate during relative sea-level fall generate a unique stacking pattern that is different from any of the stacking patterns that develop during relative sea-level rise, which affects the distribution of sediment across a basin. Therefore, the falling-stage systems tract has unique characteristics that impact the development of petroleum systems and other economic deposits, which makes it important to separate from other systems tracts in the process of exploration for natural resources. The identification of the falling-stage systems tract is afforded by sedimentological and stratigraphic data, as detailed in Chapters 4 and 5, and provides the means to reconstruct the history of relative sea-level changes during the evolution of sedimentary basins, which is an important aspect of basin analysis.

The different sequence models in Figs. 1.9 and 1.10 represent alternative ways of describing the same stratigraphic data, only using different approaches for their packaging into sequences and systems tracts (i.e., the selection of the sequence boundary, and the nomenclature of systems tracts); therefore, common ground does exist in the form of the stacking patterns of the strata under analysis. Stratal stacking patterns provide the field criteria for the identification of all elements of the sequence stratigraphic framework. The identification of the stacking patterns and bounding surfaces that can be observed in a dataset is more important than the nomenclature of systems tracts or the selection of the sequence boundary. These principles are at the core of the model-independent approach to sequence stratigraphy.

1.2.5 Standardization of sequence stratigraphy

By the mid-1990s, the informal developments in sequence stratigraphy led to the proposal and parallel usage of several sequence models, each promoting a different approach to sequence delineation and to the nomenclature of systems tracts (Figs. 1.9 and 1.10). This triggered methodological and nomenclatural confusion, with a negative impact on the effectiveness of communication of ideas and results between practitioners embracing alternative approaches. At that point, it became evident that work was needed to streamline the methodology and nomenclature, to the benefit of all practitioners. The International Commission on Stratigraphy initiated this effort in 1995, by appointing task

groups on sequence stratigraphy in charge of finding the common ground among the different approaches. This proved to be a difficult undertaking, which took 16 years of work and three task groups to complete (Salvador et al., 1995—2003; Embry et al., 2004—2007; Catuneanu et al., 2008—2011). Following the unsuccessful attempts of the first two working groups, a solution eventually emerged, leading to the publication of formal guidelines on methodology and nomenclature (Catuneanu et al., 2011).

The common ground that enables a consistent, model-independent application of sequence stratigraphy is defined by the observation of stratal stacking patterns in the sedimentary record, at all scales permitted by the data available, in an objective manner that is independent of the interpretation of the underlying controls (Fig. 1.2). The outcome of this core workflow is a sequence stratigraphic framework that explains the sediment distribution across a basin, and therefore, it affords facies predictability. Beyond the construction of this framework, the nomenclature that is applied to units and bounding surfaces observed at different scales, as well as the selection of sequence boundaries, take a subordinate role and add no further value to the practical outcome of a sequence stratigraphic study. The interpretation of the underlying controls may follow the construction of a sequence stratigraphic framework, but it plays no role in the sequence stratigraphic workflow and methodology.

None of the alternative approaches to sequence delineation proved to provide the "best practice" in sequence stratigraphy, as the applicability of each approach is a function of geological setting and the types of data available. Decades of research reveal that the stratigraphic record is far more complex than any model can predict. Sequences may consist of variable combinations of systems tracts (e.g., Csato and Catuneanu, 2012, 2014, Fig. 1.23), which may or may not conform to model predictions. The geological setting controls the patterns of accommodation and sedimentation, which impact the development of systems tracts and bounding surfaces (e.g., Martins-Neto and Catuneanu, 2010). Consequently, stratigraphic frameworks may not include the entire array of systems tracts and sequence stratigraphic surfaces. Moreover, the mappability of the different kinds of "sequence boundary" may vary with the types of data available, which also affects the applicability of any particular model. As a result, the data often dictate what approach is best suited to a particular case study. The lesson learned is that the construction of sequence stratigraphic frameworks needs to be performed on a case-by-case basis, based on the observation of local data, without any *a priori* model in mind.

The resolution of sequence stratigraphic studies has also increased to unprecedented levels (e.g., Mawson

Systems tracts	Conditions of development
FSST – LST – TST – HST	• Accommodation cycles: negative – positive • Sedimentation within the range of accommodation
FSST – TST – HST	• Accommodation cycles: negative – positive • Sedimentation within the range of accommodation • Accommodation starts with high rates (no LST)
FSST – TST	• Accommodation cycles: negative – positive • Accommodation > sedimentation at all times (no NR)
FSST – LST	• Accommodation cycles: negative – positive • Sedimentation > accommodation (no transgression)
TST – HST	• Positive accommodation only (no falling stage) • Sedimentation within the range of accommodation

FIGURE 1.23   Combinations of systems tracts that can define sequences. There is no correlation between scale and the systems-tract composition of sequences. Both small- and large-scale sequences may consist of similar combinations of systems tracts, and sequences of similar scales may vary in terms of the component systems tracts. The scales and the systems-tract composition of sequences are basin specific, reflecting the local conditions of accommodation and sedimentation. Abbreviations: FSST—falling-stage systems tract; LST—lowstand systems tract; TST—transgressive systems tract; HST—highstand systems tract; NR—normal regression (lowstand and highstand systems tracts).

and Tucker, 2009; Csato et al., 2014; Magalhaes et al., 2015; Amorosi et al., 2017), requiring a complete rethinking of the scale of sequence stratigraphic units (i.e., the scale of sequences and of their component systems tracts and depositional systems) relative to the norms of seismic stratigraphy in the 1970s. Scale is a key issue in sequence stratigraphy, with implications for methodology, nomenclature, and practical applications. The scale of observation may be selected by the practitioner as a function of purpose of study (e.g., petroleum exploration vs. production development), or it may be constrained by local parameters such as the geological setting and the resolution of the data available. Scales are basin specific, reflecting the natural variability in local accommodation and sedimentation conditions. Therefore, the methodology and nomenclature must remain independent of scale and all local variables, for a consistent application of sequence stratigraphy across the entire range of geological settings, stratigraphic scales, and types of data available.

Despite the progress to date, different groups of practitioners are still compelled to adhere to traditional approaches, for various reasons. This is often the case with petroleum companies, whereby company policies enforce a common language and workflow for all employees, with the risk of becoming isolated and out of sink with the methodology and nomenclature used by other groups. Such policies are difficult to change, and a lag time can be expected between conceptual developments and the implementation of the new concepts. For example, the ExxonMobil methodology (Abreu et al., 2010) still follows the depositional sequence III model of the 1980s (Fig. 1.10), and in spite of a few cosmetic upgrades, it is outdated by several key developments (details in Chapters 4, 5, and 7). The reluctance to change perpetuates confusion with respect to a number of issues, including the scale of sequences and component systems tracts and depositional systems; the difference between high-frequency sequences and parasequences; and the classification of sequence stratigraphic units and bounding surfaces.

The standard approach to sequence stratigraphy is based on model-independent core principles which provide a simple and robust platform for the application of the method. The basin-specific nature of the sequence stratigraphic framework is a fundamental reality that validates only the simplest approach to methodology and nomenclature; i.e., the approach that ensures consistency under all circumstances and irrespective of any local variables, including the resolution of the data available. The field criteria that afford the identification of stratal units and bounding surfaces take precedent over any model assumptions, and lead to the construction of local sequence stratigraphic frameworks that may be unique in terms of the timing, scales, and the systems-tract composition of sequences. Therefore, the data rather than the model dictate the outcome of a sequence stratigraphic analysis.

CHAPTER

2

Data in sequence stratigraphy

The most successful application of sequence stratigraphy requires the integration of various datasets and methods of data analysis into a unified, interdisciplinary approach (Fig. 1.3). Each type of data (e.g., outcrop, borehole, seismic) presents unique advantages but also limitations (Figs. 2.1 and 2.2), and only their integration and mutual calibration can lead to the most reliable and complete sequence stratigraphic framework. For example, outcrops provide the opportunity for facies observations and sampling, but their correlation can be difficult without the subsurface imaging provided by seismic data. At the same time, the use of seismic data without calibration with well data can lead to erroneous interpretations of the seismic facies. Similarly, the lack of calibration of well logs with rock data (cuttings, core, or nearby outcrops), and their correlation without the support provided by seismic imaging, can also lead to erroneous interpretations, as log motifs are not necessarily diagnostic of any depositional setting (i.e., similar log motifs can be encountered in different depositional systems). The integration of all available datasets is therefore key to the most effective and reliable application of sequence stratigraphy.

This chapter presents a brief account of the main methods which, data permitting, need to be integrated into a comprehensive sequence stratigraphic analysis. This includes the facies analysis of sediments and sedimentary rocks, the analysis of well-log motifs, the analysis of seismic data, and the acquisition of absolute or relative age data (Fig. 1.1). Each of these methods forms the core of a specialized discipline, so this presentation only reiterates aspects that are particularly relevant to sequence stratigraphy.

2.1 Geological data

Access to geological data, whether from field work (outcrops, modern environments) or boreholes (core, rock cuttings), provides the opportunity to observe sedimentary facies and to sample the sediments and sedimentary rocks for various types of analyses. The acquisition of geological data also provides the means to calibrate geophysical data (well-log, seismic), in order to constrain the geological meaning of geophysical information such as the well-log motifs and the seismic facies and reflection geometries.

2.1.1 *Sedimentology*

2.1.1.1 Facies analysis

Facies analysis is a fundamental sedimentological method of characterizing bodies of sediment and sedimentary rocks with unique physical and biological

Key: √√√ good, √√ fair, √ poor	Rock data			Geophysical data		
	Outcrops		Core	Well logs	Seismic data	
	Large-scale	Small-scale			2D	3D
Tectonic setting	√√	√	√	√√	√√√	√√
Lithofacies	√√√	√√√	√√√	√√	√	√√
Depositional elements	√√√	√√	√√	√√	√	√√√
Depositional systems	√√√	√√	√√	√√	√√	√√√
Stratal terminations	√√√	√	√	√	√√√	√√√
Stacking patterns	√√√	√√	√√	√√	√√√	√√
Nature of contacts	√√√	√√√	√√√	√√	√√	√√

FIGURE 2.1 Utility of different datasets in sequence stratigraphy. Seismic data and large-scale outcrops provide continuous imaging of stratigraphic sections. In contrast, small-scale outcrops, core, and well logs provide information from discrete locations within the basin.

Principles of Sequence Stratigraphy, Second Edition
https://doi.org/10.1016/B978-0-444-53353-1.00006-7

Dataset	Main applications /contributions to sequence stratigraphic analysis
Seismic data	Continuous subsurface imaging; tectonic setting; structural styles; regional stratigraphic architecture; imaging of depositional elements; geomorphology
Well-log data	Vertical stacking patterns; grading trends; depositional systems; depositional elements; calibration of seismic data
Core data	Lithology; textures and sedimentary structures; nature of stratigraphic contacts; physical rock properties; paleocurrents in oriented core; calibration of well-log and seismic data
Outcrop data	Lithofacies and facies architecture; depositional elements; depositional systems; all other applications afforded by core data
Geochemical data	Depositional environment; depositional processes; diagenesis; absolute ages; paleoclimate
Paleontological data	Depositional environment; depositional processes; paleoecology; relative ages

FIGURE 2.2   Contributions of datasets to the sequence stratigraphic analysis. Integration of insights afforded by independent datasets leads to the most reliable results.

attributes relative to the adjacent deposits. This method is commonly applied to describe the sediments and sedimentary rocks observed in outcrops, core, and modern environments. Facies analysis is of paramount importance for a sequence stratigraphic study, as it provides critical clues for paleogeographic and paleoenvironmental reconstructions, as well as for the definition of sequence stratigraphic surfaces. As such, facies analysis is an integral part of sedimentology and sequence stratigraphy, as both disciplines assume a process-based approach to the study of sedimentary successions (Fig. 1.4). The ultimate goal of facies analysis is the reconstruction of depositional systems, which are the largest units of sedimentology, and the building blocks of systems tracts in stratigraphy.

A depositional system (Fig. 1.8) is the product of sedimentation in a specific depositional environment; hence, it consists of a three-dimensional (3D) assemblage of strata whose facies and geometry are related by processes that operate within a common environment of deposition. Important to note, the concept of depositional system is not tied to any specific temporal and physical scales. Depositional systems can be observed at different scales, depending on the purpose of study and the resolution of the data available. High-resolution studies of the Holocene identified depositional systems at scales of 10^2–10^3 yrs and 10^0–10^1 m (e.g., Amorosi et al., 2005, 2009, 2017; Pellegrini et al., 2017, 2018), which accumulated during the lifespan of their corresponding environments of deposition. At the opposite end of the spectrum, depositional systems

can also be observed at the scale of entire sedimentary basin fills, in low-resolution studies (see details on depositional systems in Chapter 5).

At each scale of observation, depositional systems provide the link between the scopes of sedimentology and stratigraphy (i.e., the end result in sedimentology, and the starting point in stratigraphy). The study of depositional systems is intimately related to the concepts of facies, facies associations, and facies models, which are defined in Fig. 2.3. Facies analysis is an essential method for the reconstruction of paleodepositional environments, as well as for understanding broader aspects that influence the evolution of a sedimentary basin, such as the subsidence history and the underlying climatic conditions. More specifically for sequence stratigraphy, the knowledge of depositional systems is essential for the correct identification of sequence stratigraphic surfaces, as explained in detail in Chapter 6 (e.g., what defines an unconformity as 'subaerial' is the presence of continental deposits on top). Therefore, studies of the depositional setting must precede the construction of the sequence stratigraphic framework (see the workflow of sequence stratigraphy in Chapter 9).

2.1.1.2 Rock constituents

The observation of sedimentary facies in outcrops or core is often enough to constrain the position of sequence-bounding unconformities, where such contacts juxtapose contrasting facies that are genetically unrelated (Fig. 2.4). The larger the stratigraphic hiatus associated with sequence boundaries, the greater the

Facies (Bates and Jackson, 1987): the aspect, appearance, and characteristics of a rock unit, usually reflecting the conditions of its origin; esp. as differentiating the unit from adjacent or associated units.

Facies (Walker, 1992): a particular combination of lithology, structural and textural attributes that defines features different from other rock bodies.

Facies are generated by sedimentary processes that operate in particular areas of the depositional environments. Hence, facies analysis is instrumental in the reconstruction of syn-depositional conditions of sedimentation.

Facies Association (Collinson, 1969): groups of facies genetically related to one another and which have some environmental significance.

Facies associations are key to the reconstruction of paleo-depositional environments and to the identification of sequence stratigraphic surfaces (details in Chapter 6).

Facies model (Walker, 1992): a general summary of a particular depositional system, involving many individual examples from recent sediments and ancient rocks.

Facies models assume a predictable evolution of depositional environments, leading to diagnostic vertical profiles and lateral changes of facies. However, the natural variability of allocyclic and autocyclic processes can lead to departures from the standard models.

FIGURE 2.3   Concepts of facies, facies associations, and facies models.

FIGURE 2.4 Subaerial unconformity (arrows) at the contact between the Burgersdorp Formation and the overlying Molteno Formation (Middle Triassic, Karoo Basin, South Africa). The succession is fluvial, with an abrupt increase in energy levels across the contact. The fluvial styles change from meandering (with lateral accretion) to braided (with amalgamated channels). The unconformity is associated with a c. 7 My hiatus (Catuneanu et al., 1998b), and separates fluvial sequences that are genetically unrelated.

chance of mapping these surfaces by simple facies observations. There are however cases, especially in proximal successions composed of coarse, braided fluvial deposits, where subaerial unconformities are 'cryptic' and difficult to distinguish from any other channel-scour surface (Miall, 1999). Such cryptic depositional sequence boundaries may occur within thick fluvial successions consisting of unvarying facies, and may well be associated with substantial breaks in sedimentation. In the absence of abrupt changes in facies and paleocurrent directions across these sequence boundaries, petrographic studies of cements and framework grains may provide the only solid evidence for the identification and mapping of sequence-bounding unconformities. The Late Cretaceous Lower Castlegate Sandstone of the Book Cliffs (Utah) provides an example where a subaerial unconformity was mapped within a continuous braided fluvial sandstone succession only by plotting the position of subtle changes in the detrital petrographic composition, interpreted to reflect corresponding changes in provenance in relation to tectonic events in the Sevier highlands (Miall, 1999).

Besides changes in provenance and the related composition of framework grains, subaerial unconformities may also be identified by the presence of secondary minerals that replace some of the original sandstone constituents via processes of weathering under subaerial conditions. For example, it has been documented that subaerial exposure, given the availability of sufficient amounts of K, Al, and Fe that may be derived from the weathering of clays and feldspars, may lead to the replacement of calcite cements by secondary glauconite (Khalifa, 1983; Wanas, 2003). Glauconite-bearing sandstones may therefore be used to recognize sequence-bounding unconformities, where the glauconite formed as a replacement mineral. Hence, a distinction needs to be made between the syndepositional glauconite of marine origin (framework grains in sandstones) and the secondary glauconite that forms under subaerial conditions (coatings, cements), which can be resolved with petrographic analysis.

The distribution patterns of early diagenetic clay minerals such as kaolinite, smectite, palygorskite, glaucony, and berthierine, as well as of mechanically infiltrated clays, may also indicate the position of sequence stratigraphic surfaces (Ketzer et al., 2003a,b; Khidir and Catuneanu, 2005; Figs. 2.5—2.7). As demonstrated by Ketzer et al. (2003a), 'changes in relative sea-level and in sediment supply/sedimentation rates, together with the climatic conditions prevalent during, and immediately after deposition of sediments control the type, abundance, and spatial distribution of clay minerals by influencing the pore-water chemistry and the duration over which the sediments are submitted to a certain set of geochemical conditions' (Figs. 2.5 and 2.6). The patterns of change in the distribution of early diagenetic clay minerals across subaerial unconformities may be preserved during deep-burial diagenesis, when late diagenetic minerals may replace the early diagenetic ones (e.g., the transformation of kaolinite into dickite with increased burial depth; Fig. 2.7).

2.1.1.3 Grading trends

Grading trends refer to the fining- or coarsening-upward profiles that can be observed in outcrops, core, or borehole litholog constructed from rock cuttings.

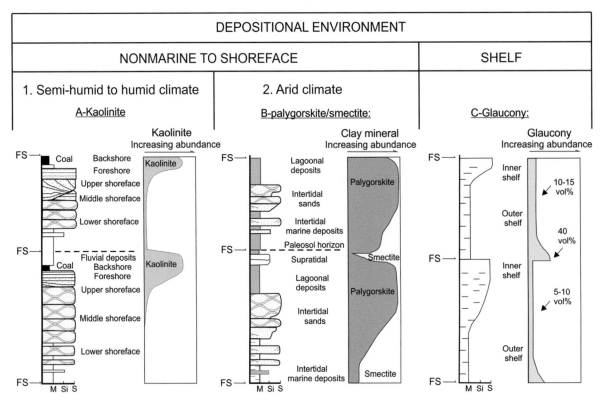

FIGURE 2.5 Distribution of early diagenetic clay minerals in fluvial to shallow-water deposits (modified from Ketzer et al., 2003a). A—kaolinite content increases toward the top of clinoforms where continental facies are exposed to extensive meteoric water flushing under semi-humid to humid climatic conditions; kaolinite content increases in the presence of unstable silicates and organic matter, as the degradation of the latter generates acidic fluids; B—palygorskite content increases toward the top of clinoforms capped by evaporitic deposits, under arid climatic conditions; C—in fully marine successions, autochthonous glauconite is most abundant at the base of clinoforms, and decreases gradually toward the top of clinoforms. Abbreviation: FS—flooding surface; M—mudstone; Si—siltstone; S—sandstone.

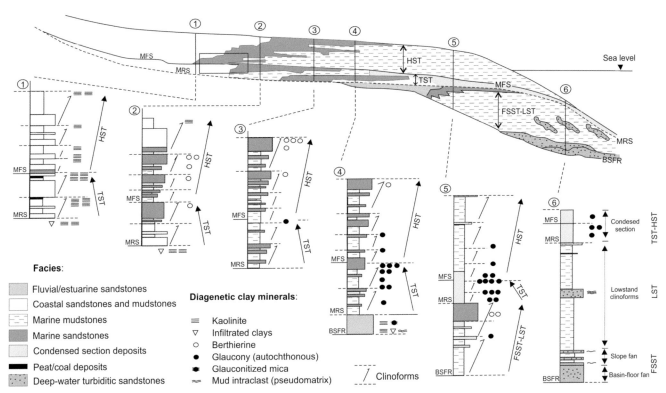

FIGURE 2.6 Distribution of diagenetic clay minerals in a sequence stratigraphic framework (modified from Ketzer et al., 2003a). Abbreviations: MFS—maximum flooding surface; MRS—maximum regressive surface; BSFR—basal surface of forced regression; HST—highstand systems tract; TST—transgressive systems tract; LST—lowstand systems tract; FSST—falling-stage systems tract.

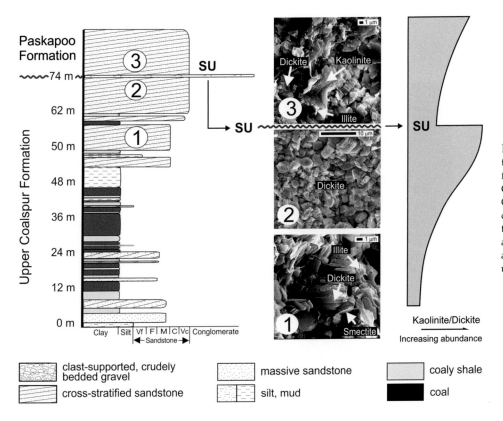

FIGURE 2.7 Pattern of change in the distribution of clay minerals in a fluvial succession (from Khidir and Catuneanu, 2005; Paleocene, Western Canada Basin). Kaolinite/dickite content increases gradually toward the top of a sequence, and decreases abruptly across the sequence boundary. Abbreviation: SU—subaerial unconformity.

Vertical profiles are an integral part of sequence stratigraphic analyses, and are used to identify progradational and retrogradational trends in shallow-water successions, or to delineate fluvial depositional sequences in continental settings. Fluvial sequences, for example, commonly display fining-upward trends that reflect aggradation in an energy-declining environment (e.g., Eberth and O'Connell, 1995; Hamblin, 1997; Catuneanu and Elango, 2001; Fig. 2.8). Sequence boundaries (subaerial unconformities) in such fluvial successions are placed at the base of the coarsest units, which often consist of amalgamated channels (Fig. 2.4). In tectonically active basins, the influx of coarser sediment that marks the onset of a new sedimentation cycle is typically linked to renewed uplift in the source area. In other settings,

similar fluvial sequences may be generated by climate cycles or even autocyclic shifts of alluvial channel belts (Miall, 2015; Catuneanu, 2019a).

A reliable interpretation of vertical profiles can only be made within a proper paleogeographic context. For example, the lithological contact (dashed line) in Fig. 2.9 was historically interpreted as an unconformity, due to the abrupt shift in grain size across it. If the entire section was fluvial, that interpretation would have been correct (e.g., braided channels on top of the floodplain fines of a lower energy river system). However, the underlying fine-grained facies are lacustrine; this changes the depositional context and the stratigraphic meaning of this contact, which is a diachronous surface at the limit between the fluvial topset and the lacustrine

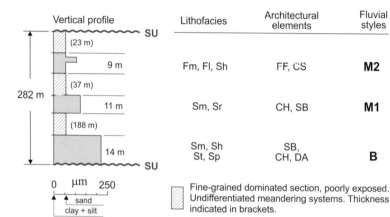

FIGURE 2.8 Unconformity-bounded depositional sequence in an upstream-controlled fluvial setting (Late Permian Balfour Formation, Karoo Basin, South Africa; from Catuneanu and Elango, 2001). Abbreviations: SU—subaerial unconformity; B—braided system; M1—sand-bed meandering system; M2—fine-grained meandering system; St—trough cross-bedded sandstone; Sp—planar cross-bedded sandstone; Sm—massive sandstone; Sh—sandstone with horizontal stratification; Sr—ripple cross-laminated sandstone; Fm—massive mudstone; Fl—horizontally laminated mudstone; CH—channel fill; DA—downstream-accretion macroform; SB—sandy bedforms; FF—floodplain fines; CS—crevasse splays.

FIGURE 2.9 Unconformity (solid line) at the contact between two fine-grained lacustrine systems (Maastrichtian, southwestern Saskatchewan, Western Canada Basin; from Sweet et al., 2003, 2005, and Catuneanu and Sweet, 2005). The unconformity originated as a surface of subaerial exposure, subsequently replaced by a wave-ravinement surface in a lacustrine setting. The limit between the upper lacustrine system and the overlying fluvial deposits is a conformable facies contact ('within-trend normal regressive surface') which separates the fluvial topset from the lacustrine foreset and bottomset of one prograding system. Notably, the unconformity is lithologically more subtle than the facies contact; this cautions against the *a priori* interpretation of lithological discontinuities as stratigraphic hiatuses, and demonstrates the value of age-dating techniques in the documentation of unconformities.

foreset and bottomset of one prograding system. Biostratigraphic and magnetostratigraphic data confirm the quasi-conformable nature of this lithostratigraphic contact. The true unconformity in this section is within the fine-grained succession, at the limit between the two lacustrine systems (solid line in Fig. 2.9), where a stratigraphic hiatus was demonstrated with palynological and magnetostratigraphic data (Sweet et al., 2003, 2005; Catuneanu and Sweet, 2005).

In spite of the potential limitations, the observation of grading trends remains a useful method of highlighting cyclicity in the stratigraphic record. As long as data are available (i.e., access to outcrops, core, or rock cuttings), plots reflecting vertical changes in grain size can be constructed by careful logging and textural analysis. The vertical profiles may display the bed-by-bed changes in grain size, or smoothed out curves that show the overall statistical changes in grain size (e.g., moving averages of overlapping intervals). The latter method is often preferred because it eliminates abnormal peaks that may only have local significance. The technique of constructing vertical profiles can also be adapted as a function of case study. The grain-size logs may be plotted using arithmetic scales, where fluctuations in grain size are significant, or on logarithmic scales where the succession is more monotonous. The latter technique works best in fine-grained successions, where logarithmic plots enhance the differences in grain size, but is less efficient in coarser deposits (Long, 2021).

The construction of grain-size logs is generally a viable method of identifying cycles at specific locations, but matching such trends across a basin, solely based on the observed grading trends, is not necessarily a reliable correlation technique. Changes in sedimentation patterns across a basin due to variations in subsidence and sediment supply make it difficult to know which cycles are of the same age when comparing vertical profiles from different sections. Ideally, age data (biostratigraphic, magnetostratigraphic, radiometric, marker beds) would provide the perfect solution to this problem. However, such age data are often missing, especially in the study of older successions, and in the absence of time control, other sedimentological observations have to be integrated with the petrographic data in order to constrain correlations (Fig. 1.11). In addition to the paleogeographic context, measurements of paleoflow directions from bedforms and related sedimentary structures help constrain changes in the dip direction within a basin, usually related to episodes of tectonic tilt. The documentation of such changes provides additional criteria for correlation and the identification of events in the evolution of a basin which often result in the formation of sequence-bounding unconformities.

2.1.1.4 *Paleoflow directions*

The major breaks in the stratigraphic record are potentially associated with stages of tectonic reorganization of sedimentary basins, and hence with changes in tilt direction across sequence boundaries. This is often the case in tectonically active basins, such as grabens, rifts, or forelands, where stratigraphic cyclicity is commonly controlled by cycles of subsidence and uplift. Other basin types, however, such as the 'passive' continental margins, are dominated by long-term thermal subsidence, and hence they may record little change in the tilt direction through time. In such cases, stratigraphic cyclicity may be mainly controlled by fluctuations in sea level, and paleocurrent measurements may be of little use to constrain the position of sequence boundaries.

In the case of tectonically active basins, where fluctuations in tectonic stress regimes match the frequency of cycles observed in the stratigraphic record (e.g., Cloetingh, 1988; Cloetingh et al., 1985, 1989; Peper et al., 1992), paleocurrent data may prove to provide the most compelling evidence for sequence delineation, paleogeographic reconstructions, and stratigraphic correlations, especially when dealing with lithologically monotonous successions that lack any high-resolution time control. An example is the overfilled portion of the Early Proterozoic Athabasca Basin in Canada, which consists of fluvial deposits that show little variation in grain size at any location. In this case, vertical profiles are equivocal, the age data to constrain correlations are missing, and the only reliable method to outline genetically related packages of strata is the measurement of paleoflow directions. Based on the reconstruction of fluvial drainage systems, the Athabasca basin fill has been subdivided into four second-order depositional sequences separated by subaerial unconformities across which significant shifts in the direction of tectonic tilt are recorded (Ramaekers and Catuneanu, 2004).

Overfilled foreland basins provide a classic example of a setting where fluvial sequences and bounding unconformities form in isolation from eustatic influences, with a timing controlled by orogenic cycles of thrusting (tectonic loading) and unloading (Catuneanu and Sweet, 1999; Catuneanu, 2019c). In such basins, fluvial aggradation occurs during stages of differential flexural subsidence, with higher rates toward the center of loading, whereas bounding surfaces form during stages of differential rebound. As the thrusting events are generally shorter in time relative to the intervening periods of orogenic quiescence, foredeep fluvial sequences preserve only a fraction of the geological time (Catuneanu et al., 1997a; Miall, 2015). Renewed thrusting in the orogenic belt marks the onset of a new sedimentation cycle in the basin. Due to

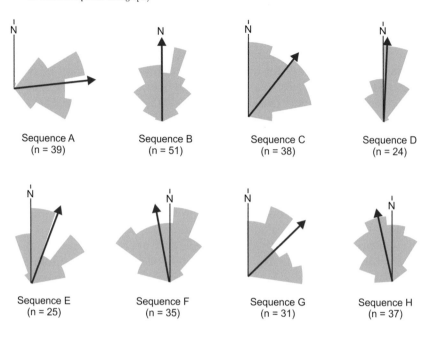

FIGURE 2.10 Paleoflow directions for the eight depositional sequences in an upstream-controlled fluvial succession (Late Permian Koonap-Middleton formations, Karoo Basin, South Africa; from Catuneanu and Bowker, 2001). The succession spans 5 My and measures 2630 m. The number 'n' of paleoflow measurements is indicated for each sequence. Sequence boundaries are marked by tectonic tilt (i.e., change in the dip direction) and abrupt shifts in fluvial styles from lower to higher energy systems.

the strike variability in orogenic loading, which is the norm rather than the exception, abrupt changes in tilt direction are usually recorded across sequence boundaries (Fig. 2.10). In the absence of other unequivocal criteria (e.g., as in the case of the Athabasca Basin discussed above), such changes in tectonic tilt may be used to delineate fluvial sequences with distinct drainage systems, and to map their bounding unconformities.

2.1.2 Pedology

2.1.2.1 Soils and paleosols

Pedology (soil science) deals with the study of soil morphology, genesis, and classification (Bates and Jackson, 1987). The formation of soils refers to the physical, biological, and chemical transformations that affect sediments and rocks exposed to subaerial conditions (Kraus, 1999). Paleosols (i.e., fossil soils) are buried or exhumed soil horizons that formed in the geological past on ancient landscapes. Pedological studies started with the analysis of modern soils and Quaternary paleosols, but have been vastly expanded to the pre-Quaternary record in the 1990s due to their multiple geological applications. Notably, some of these geological applications include (1) interpretations of ancient landscapes, from local to basin scales; (2) interpretations of ancient surface processes (sedimentation, nondeposition, erosion), including sedimentation rates and the controls thereof; (3) interpretations of paleoclimates, including estimations of mean annual precipitation rates and mean annual temperatures; and (4) stratigraphic correlations, and the cyclic change in soil characteristics in relation to base-level changes (Kraus, 1999). All these applications, and particularly the latter, have relevance to sequence stratigraphy.

The complexity of soils, and thus of paleosols, can only begin to be understood by looking at the diversity of environments in which they may form; the variety of surface processes to which they can be genetically related; and the practical difficulties to classify them. Paleosols have been described from an entire range of nonmarine settings, including alluvial (Leckie et al., 1989; Wright and Marriott, 1993; Shanley and McCabe, 1994; Aitken and Flint, 1996), palustrine (Wright and Platt, 1995; Tandon and Gibling, 1997), and eolian (Soreghan et al., 1997), but also from coastal settings such as deltas (Fastovsky and McSweeney, 1987; Arndorff, 1993) and coastal plains formed by the subaerial exposure of paleoseafloors due to stages of relative sea-level fall (Lander et al., 1991; Webb, 1994; Wright, 1994).

Irrespective of depositional setting, soils may form in relation with different surface processes, including sediment aggradation (as long as sedimentation rates do not outpace the rates of pedogenesis), sediment bypass (nondeposition), and sediment reworking (as long as the rates of scouring do not outpace the rates of pedogenesis). Soils formed during stages of sediment aggradation occur within conformable successions, whereas soils formed during stages of nondeposition or erosion are associated with stratigraphic hiatuses, marking diastems or unconformities in the stratigraphic record. These issues are particularly important for sequence stratigraphy, as it is essential to distinguish between paleosols with the significance of subaerial unconformities (i.e., depositional sequence boundaries) and paleosols that occur within sequences and systems tracts. Theoretical and field studies (e.g., Wright and Marriott, 1993; Tandon and Gibling, 1994, 1997) show that the paleosol types observed in the rock

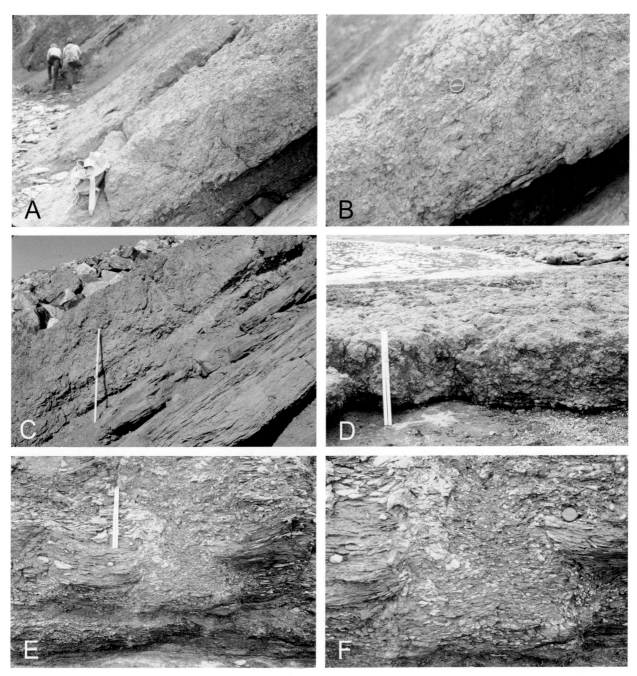

FIGURE 2.11   Calcareous paleosols formed during relative sea-level fall and subaerial exposure (photographs courtesy of Martin Gibling; Pennsylvanian Sydney Mines Formation, Sydney Basin, Nova Scotia; for more details, see Gibling and Bird, 1994; Gibling and Wightman, 1994; Tandon and Gibling, 1994, 1997). A—calcrete, marking a 'subaerial unconformity' (depositional sequence boundary) within coastal plain deposits. The carbonate soil implies a semi-arid climatic period, suggesting that lowstands in relative sea level were drier than the peat-forming periods of the overlying transgressive and highstand systems tracts; B—close up of calcrete in image A, showing well-developed vertic and nodular fabric; C—calcrete in image A, with strong nodular texture. Note the undisrupted nature of the siltstone below; D—calcrete exposed on wave-cut platform, with strong vertic fabric (scale: 50 cm); E—upright tree cast, partially replaced by carbonate beneath a calcrete layer. This occurrence suggests that carbonate-rich groundwaters caused local cementation through conduits below the main soil level; F—close up of carbonate-cemented tree in image E.

record depend in part on base-level changes, thus allowing one to assess their relative significance in a sequence stratigraphic context. For example, the Upper Carboniferous sequences in the Sydney Basin of Nova Scotia are bounded by mature calcareous paleosols (i.e., calcretes; Fig. 2.11) formed during times of increased aridity and lowered base level, whereas vertisols and hydromorphic paleosols occur within sequences, being formed in aggrading fluvial floodplains during times of increased humidity and rising base level (Fig. 2.12; Tandon and Gibling, 1997). The former have greater importance from a sequence stratigraphic standpoint.

FIGURE 2.12   Coastal plain successions showing calcrete horizons overlain by red calcic vertisols (images courtesy of Martin Gibling; Pennsylvanian Sydney Mines Formation, Sydney Basin, Nova Scotia). The top of the calcrete paleosols (arrows) mark subaerial unconformities formed during relative sea-level fall. The red vertisols (dryland clastic soils) reflect the resumption of sedimentation during transgression under conditions of high sediment supply, as transgression allowed sediment storage on the floodplain (Tandon and Gibling, 1997). A—calcrete paleosols pass upward into dryland clastic soils, marking the renewal of clastic supply to the coastal plain during relative sea-level rise; B—concave-up, slickensided joints (mukkara structure) in red vertisols. The sediments immediately below the calcrete horizons are calcite cemented.

The classification of soils and paleosols has been approached from different perspectives, and no universal scheme of pedologic systematics has been devised yet. The classification of modern soils relies on properties such as texture, structure, color, amount of organic matter, mineralogy, cation exchange capacity, and pH (Soil Survey Staff, 1975, 1998; Tabor et al., 2017; Fig. 2.13). The pitfalls of this approach, when applied to paleosols, are twofold: (1) the taxonomy of modern

soils does not emphasize the importance of hydromorphic soils (i.e., 'gleysols', common in aggrading fluvial floodplains, defined on the basis of soil saturation; Fig. 2.13); and (2) the classification is dependent on soil properties, some of which (e.g., cation exchange capacity or amount of organic matter) are not preserved in paleosols. For these reasons, Mack et al. (1993) devised a classification specifically for paleosols (Fig. 2.13), based on mineralogical and morphological properties that are preserved as a soil is transformed into a paleosol. Due to differences in the classification criteria, the two systems are not directly equivalent with respect to some groups of soils and paleosols (Fig. 2.13).

2.1.2.2 Paleosols in sequence stratigraphy

From a sequence stratigraphic perspective, paleosols may provide key evidence for reconstructing the syndepositional conditions (e.g., level of water table, rates of sedimentation, paleoclimate) during the accumulation of systems tracts, or about the temporal significance of stratigraphic hiatuses associated with sequence-bounding unconformities. The types of paleosols that may form in relation to the interplay between sedimentary surface processes (sedimentation, erosion) and pedogenesis are illustrated in Fig. 2.14. Stages of nondeposition or erosion, typically associated with sequence boundaries, result in the formation of mature paleosols along unconformity surfaces. Stages of sediment accumulation, typically associated with the deposition of sequences, result in the formation of less mature and generally aggrading paleosols of compound, composite, or cumulative nature, whose rates of aggradation match the rates of sedimentation (see Kraus, 1999, for a review of these paleosol types).

Soil systematics (Soil Survey Staff, 1975, 1998)	Paleosol systematics (Mack et al., 1993)
Entisol	
Inceptisol	Protosol
Vertisol	Vertisol
Histosol	Histosol
-	Gleysol
Andisol	-
Oxisol	Oxisol
Spodosol	Spodosol
Alfisol	
Ultisol	Argillisol
-	Calcisol
-	Gypisol
Aridisol	-
Mollisol	-
Gelisol	-

FIGURE 2.13   Comparison between the soil and paleosol classifications of the United States Soil Taxonomy (Soil Survey Staff, 1975, 1998) and Mack et al. (1993). Due to differences in the classification criteria, not all soil or paleosol types have equivalents in both systems (see text for details).

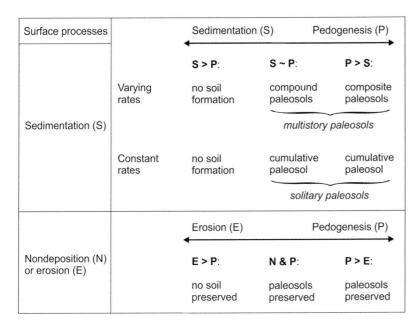

Surface processes		Sedimentation (S)		Pedogenesis (P)
		S > P:	**S ~ P:**	**P > S:**
	Varying rates	no soil formation	compound paleosols	composite paleosols
Sedimentation (S)			*multistory paleosols*	
	Constant rates	no soil formation	cumulative paleosol	cumulative paleosol
			solitary paleosols	
		Erosion (E)		Pedogenesis (P)
		E > P:	**N & P:**	**P > E:**
Nondeposition (N) or erosion (E)		no soil preserved	paleosols preserved	paleosols preserved

FIGURE 2.14 Interplay between pedogenesis and surface processes of sedimentation, sediment bypass (nondeposition) or erosion (modified from Morrison, 1978; Bown and Kraus, 1981; Marriott and Wright, 1993; Kraus, 1999). Pedogenesis concurrent with sedimentation generates soil horizons within conformable successions (e.g., the red vertisols in Fig. 2.12). Paleosols associated with nondeposition or erosion have the significance of subaerial unconformities (e.g., the top of calcrete horizons in Figs. 2.11 and 2.12).

Paleosols associated with sequence boundaries are generally strongly developed and well-drained, reflecting prolonged stages of sediment cut-off and lowering of the water table (Fig. 2.11), commonly during periods of base-level fall. In addition to the base level, climate also has a strong influence on the nature of sequence-bounding paleosols (e.g., a drier climate promotes evaporation and the formation of calcic paleosols). Base level and climate are sometimes linked, as climatic cycles driven by orbital forcing (e.g., eccentricity, obliquity, and precession cycles, with periodicities in a range of tens to hundreds of thousands of years; Fig. 2.15; Milankovitch, 1930, 1941; Imbrie and Imbrie, 1979; Imbrie, 1985; Schwarzacher, 1993) are a primary control on sea-level changes at the temporal scale of Milankovitch cycles. In such cases, stages of base-level fall may reflect times of increased climatic aridity (e.g., see Tandon and Gibling, 1997, for a case study). However, base-level changes may also be driven by tectonism, independently of climate changes, in which case base-level cycles, and the timing of sequence-bounding paleosols, may be offset relative to the climate cycles.

Irrespective of the primary control behind a falling base level, the cut-off of sediment supply is an important parameter that defines the conditions of formation of sequence-bounding paleosols. Stages of sediment cut-off during the depositional history of a basin may be related to either allogenic (e.g., sediment trapping within incised valleys as a result of tectonic uplift) or autogenic (e.g., lateral shifts of alluvial channel belts) processes. The stratigraphic hiatus associated with sequence-bounding paleosols varies greatly with the

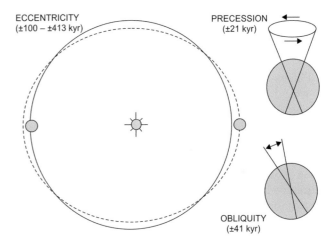

ECCENTRICITY
(±100 – ±413 kyr)

PRECESSION
(±21 kyr)

OBLIQUITY
(±41 kyr)

FIGURE 2.15 Main components of orbital forcing, showing the causes of Milankovitch-band (10^4–10^5 yrs) cyclicity (modified from Imbrie and Imbrie, 1979, Plint et al., 1992).

rank of the sequence and the underlying controls, and it is generally in a range of 10^4 yrs (for the higher-frequency Milankovitch cycles) to 10^5–10^7 yrs for the higher-rank sequences (Summerfield, 1991; Miall, 2000). Sequence-bounding unconformities are commonly regional in scale, as opposed to the more localized diastems related to intra-sequence processes (e.g., local channel avulsion), and depending on the paleolandscape, can be surfaces with highly irregular topographic relief along which the amount of missing time may vary considerably (Wheeler, 1958). Accordingly, the paleosols associated with sequence-bounding unconformities can show lateral changes that may be used to interpret lateral variations in topography and missing time (Kraus, 1999).

FIGURE 2.16 'Wet' and immature paleosol of gleysol type, formed in close association with a coal bed during a stage of relative sea-level rise. This example comes from the Upper Cretaceous Castlegate Formation in Utah, which consists of amalgamated braided fluvial channels (lowstand topset: positive but low rates of creation of accommodation). Such immature paleosols develop commonly within sequences, over timescales of ≤10^3 yrs (Fig. 2.17). The formation of wet and immature soils versus coal beds may reflect fluctuations in climatic conditions and fluvial discharge (exposure vs. flooding of fluvial overbank) rather than relative sea-level changes.

Paleosol type / Features	Sequence-bounding paleosols	Paleosols within sequences
maturity	strongly developed	weakly to well-developed
water table	low	higher
soil saturation	well-drained	wetter
timescales	commonly ≥10^4 yrs	commonly ≤10^3 yrs
controls	typically allogenic	autogenic (e.g., avulsion)
extent	regional	local
surface process	bypass or erosion	aggradation
significance	unconformity	conformity
architecture	solitary	commonly multistory

FIGURE 2.17 Comparison between sequence-bounding paleosols and the paleosols developed within sequences.

Paleosols that form within sequences may be weakly to well-developed, but are generally less mature than the sequence-bounding paleosols (Figs. 2.12 and 2.16). They form during stages of base-level rise, when surface processes are dominated by sediment aggradation. As the water table commonly follows the rise in base level, these paleosols tend to be 'wetter' relative to the sequence-bounding paleosols, to the extent of becoming hydromorphic (gleysol type) around maximum flooding surfaces which mark the timing of the highest water table in the nonmarine environment. Such 'wetter' and immature paleosols form over shorter periods of time, and are often seen in close association with coal beds (Fig. 2.16). Fig. 2.17 summarizes the main contrasts between the sequence-bounding paleosols and the paleosols that form within sequences. The latter type may show aggradational features, often with a multistory architecture due to unsteady sedimentation rates (Fig. 2.14), but may also be associated with hiatuses where autogenic processes such as channel avulsion lead to a cut-off of sediment supply in restricted overbank areas. As the periodicity of avulsion is commonly in a range of 10^3 yrs (Bridge and Leeder, 1979), the stratigraphic hiatuses of intra-sequence paleosols are generally at least one order of magnitude smaller than the time gaps associated with sequence-bounding paleosols (Fig. 2.17).

Fig. 2.18 illustrates a generalized model of paleosol development in relation to a base-level cycle. As a matter of principle, the higher the sedimentation rates, the weaker the development of the paleosol. Hence, the most mature paleosols are predictably associated with depositional sequence boundaries (zero or negative sedimentation rates), and the least developed paleosols are expected to form during transgressions, when aggradation rates and the water table are commonly highest. Due to the high water table in the nonmarine environments during transgression, hydromorphic paleosols are often associated with regional coal beds within the interval that includes the maximum flooding surface (Fig. 2.18; Tandon and Gibling, 1994).

It can be concluded that paleosols are highly relevant to sequence stratigraphy, as they complement the information acquired via other methods of data analysis. Pedologic studies are routinely performed on outcrops and core (Leckie et al., 1989; Lander et al., 1991; Platt and Keller, 1992; Caudill et al., 1997), and to a lesser extent on well logs (Ye, 1995), and may be applied to a wide range of stratigraphic ages, including strata as old as the Early Proterozoic (Gutzmer and Beukes, 1998).

2.1.3 Body fossils

2.1.3.1 General principles

Body fossils encompass a wide variety of fossil types, from microscopic invertebrate tests to large vertebrate bones. Key descriptors of body fossils that are relevant to sequence stratigraphy include: fossil concentrations (e.g., concentrations of benthic foraminifera, shell beds, or bone beds); biofacies (fossil assemblages and diversity); first and last appearances of species in the stratigraphic record; the mode of preservation (taphonomy); the % of in situ fragmentation of benthic foraminifera; the ratio between distal and proximal benthic foraminifera; and the ratio between planktonic and benthic foraminifera (Patzkowsky and Holland, 2012; Zecchin et al., 2021). Analysis of these parameters is best approached from an interdisciplinary perspective,

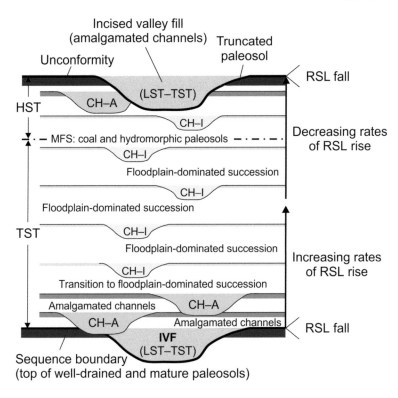

FIGURE 2.18 Paleosol development in relation to relative sea-level changes (modified after Wright and Marriott, 1993). In this model, the rates of fluvial aggradation (and the degree of channel amalgamation and paleosol maturity) are linked to the rates of relative sea-level rise. Low rates of sedimentation (early and late stages of relative sea-level rise) enable channel amalgamation and the formation of well-developed paleosols; high rates of sedimentation lead to weakly developed paleosols within floodplain-dominated successions. Abbreviations: RSL—relative sea level; LST—lowstand systems tract; TST—transgressive systems tract; HST—highstand systems tract; IVF—incised-valley fill; CH-A—amalgamated (multistory) channels; CH-I—isolated channels; MFS—maximum flooding surface.

whereby the paleontological data are integrated with sedimentological data and placed in a broader stratigraphic context.

Among the many factors that govern fossilization, the preservation of body fossils is strongly controlled by the environments in which organisms lived and by sedimentation rates, both of which are tied to sequence stratigraphic architecture (Kidwell, 1986, 1991a,b; Brett, 1995, 1998; Holland, 1995; Patzkowsky and Holland, 2012). The relationships between the fossil record and sequence stratigraphic architecture is the subject of *stratigraphic paleobiology* (Patzkowsky and Holland, 2012), which studies the occurrence of dense fossil concentrations, the patterns of fossil occurrences in stratigraphic columns, and the modes of fossil preservation. Although these principles have been most extensively explored for shelly marine fossils (particularly shell beds), they also apply to invertebrate tests, marine vertebrates, and continental fossils. For paleontologists, understanding the principles by which the fossil record is controlled by the sequence stratigraphic architecture is necessary for correctly interpreting the biological history that the fossil record preserves, such as patterns of morphological evolution (Webber and Hunda, 2007) and the tempo of mass extinctions and recoveries (Holland and Patzkowsky, 2015; Holland, 2020). For stratigraphers, the preservation of body fossils offers

an additional line of evidence for the construction of sequence stratigraphic frameworks.

2.1.3.2 *Fossil concentrations*

The development of fossil concentrations, commonly called 'shell beds' or 'bone beds', was the first aspect of the fossil record shown to be controlled by the sequence stratigraphic architecture (Kidwell, 1984, 1986, 1989, 1991a,b). In general, the abundance of fossils reflects the ratio of two rates: the rate of skeletal input (R-hardparts) and the rate of input of non-skeletal sediment (R-sediment; Kidwell, 1986).

If the rate of shell production is relatively constant, trends in sedimentation rate and the presence of an erosion or omission (i.e., nondeposition without erosion) surface imply four main types of shell beds (Fig. 2.19; Kidwell, 1986, 1991b). Where the rate of sedimentation decreases over time, the abundance of shells will increase, leading to type I or type II shell beds, depending on whether the shell bed is capped by an omission (type I) or erosion surface (type II). Similarly, where sedimentation rate increases over time, shelliness will decrease, creating a type III or type IV shell bed, depending on whether the shell bed is underlain by an omission (type III) or erosion surface (type IV). Changes in sedimentation rate are also reflected in the condition and age distribution of shells, with lower sedimentation

Bounding surface

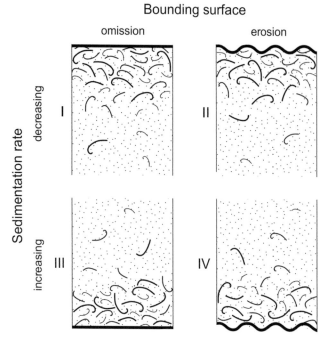

FIGURE 2.19 Four types (I–IV) of shell beds, derived from combinations of patterns of increasing or decreasing sedimentation rates, and the presence of a surface of erosion or omission (nondeposition) (diagram courtesy of Steven Holland; adapted from Kidwell, 1986).

rates fostering not only greater amounts of shell breakage, abrasion, bioerosion, and encrustation but also greater amounts of time-averaging, that is, the mixing of shells of different ages. Changes in sedimentation rate also affect the ecological composition of shell beds, with decreasing sedimentation rates favoring greater

proportions of shell-gravel species and fewer soft-bottom species. Approaches for the description, classification, and interpretation of shell beds have been provided by Kidwell et al. (1986), Kidwell (1991a), and Kidwell and Holland (1991; Fig. 2.20).

Because variations in sedimentation rates are tightly coupled with the sequence stratigraphic architecture, six common settings promote fossil concentrations: marginal settings indicated by toplap and coastal onlap; starved settings characterized by downlap/bottomset or backstepping/backlap; perched highs; and silled basins (Fig. 2.21; Kidwell, 1991a,b). Toplap and coastal onlap represent proximal areas that achieve low rates of sedimentation through sedimentary bypass: owing to low rates of accommodation, sediment bypasses these shallower water zones as it is delivered to more distal settings. Toplap shell beds may be present, for example, within the uppermost portion of highstand systems tracts (Banerjee and Kidwell, 1991) and are typically type I or II shell beds (Kidwell, 1991b). Onlap shell beds occur in transgressive positions, such as at transgressive wave-ravinement surfaces, and they are particularly common (Fig. 2.22A,C,D; Kidwell, 1984; Cattaneo and Steel, 2003; Zecchin et al., 2019). Onlap shell beds are commonly of type III or type IV (Kidwell, 1991b). Low sedimentation rates are achieved in downlap/bottomset and backstepping through starvation: sediment is deposited in more proximal areas, with little sediment being supplied to distal areas. Both contexts are associated with maximum flooding surfaces, where supply of sediment to distal areas is at a minimum (Fig. 2.22D; Banerjee and Kidwell, 1991). Backstepping

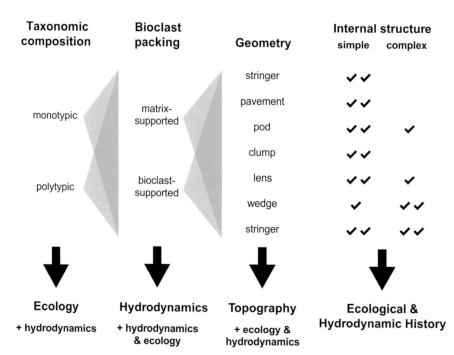

FIGURE 2.20 Guide to the description of shell beds (diagram courtesy of Steven Holland; adapted from Kidwell, 1986). Taxonomic composition, packing of bioclasts, geometry, and internal complexity are the principal clues to the underlying ecological, hydrodynamic, and topographic controls on the formation of a shell bed. Internal complexity in particular can help to distinguish relatively simple shell beds that represent single depositional events from the highly complex shell beds that reflect extended periods of formation at sedimentary hiatuses.

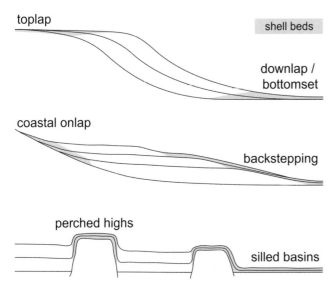

toplap

shell beds

downlap /
bottomset

coastal onlap

backstepping

perched highs

silled basins

FIGURE 2.21   Zones of stratal thinning or stratigraphic condensation, which are common sites of fossil concentrations (blue shading), such as shell beds and bone beds (diagram courtesy of Steven Holland; adapted from Kidwell, 1991a,b).

2.1.3.3 *Depth gradients in marine systems*

Both today and in the past, species are distributed along ecological gradients. In the marine realm, this primary ecological gradient is correlated with water depth (Fig. 2.23; Holland, 1995, 2000; Patzkowsky and Holland, 2012). Although this gradient is most apparent for benthic species, nektic (swimming) and pelagic (floating) species are also commonly stratified by water depth. Although water depth itself does not control the occurrence of most species, water depth is strongly correlated with many physical and chemical factors that are important for species, such as grain size, bed shear stress, light, oxygen, and food availability. As a result, the distribution of species can be described by water depth, even though water depth itself does not exert a direct control. In marine settings, studies of the modern as well as the ancient past have shown that at the scale of a sedimentary basin, water depth is the most important of the environmental gradients along which species are distributed.

As sediment accumulates during shoreline transgressions and regressions, these lateral gradients translate into vertical changes in the species composition of fossil assemblages (Fig. 2.23). For example, the abundance of a species will increase through a stratigraphic column as the facies that records its preferred environment is approached (see bold and dashed curves in Fig. 2.23, for example), and its abundance will decrease as facies record environments progressively more dissimilar to the preferred environment of a species. Although changes in fossil preservation among facies may alter these patterns somewhat, patterns of fossil occurrence at least for larger (>2 mm) body fossils largely reflect where those species lived because out-of-habitat transportation is uncommon (Kidwell, 2013). For example, Pleistocene molluscan assemblages show a strong correspondence with the modern onshore-offshore distribution of those species (Scarponi and Kowalewski, 2004). In some settings, however, downslope transport of microfossils can be substantial. The presence of depth-correlated ecological gradients, coupled with the sequence stratigraphic architecture, has important implications for the first and last occurrences of fossils, and for stratigraphic changes in the species composition of biofacies.

2.1.3.4 *First and last occurrences*

Because the abundance of species and therefore their probabilities of collection reflect sedimentary facies, the first occurrence of a species in a stratigraphic column generally does not indicate its time of origination or migration into the region, but instead reflects the local availability of suitable facies. Similarly, the last occurrence of a species seldom indicates the time of global

shell beds tend to be of type II, and downlap shell beds tend to be of type III (Kidwell, 1991b). Perched highs and silled basins represent cases in which sediment supply is reduced owing to topography, such as the inability to deliver sediment onto topographic highs that stand well above the surrounding seafloor, or isolated basins in which a topographic high or sill prevents the delivery of sediment (Fig. 2.22E).

Four of these (toplap, downlap, coastal onlap, and backstepping) have a strong connection to the sequence stratigraphic architecture and the criteria that afford the delineation of stratigraphic sequences on seismic lines. As a result, fossil concentrations can be important clues for the identification of disconformities in marine rocks, especially where facies variability is limited or where unconformities are otherwise subtle (Kidwell, 1984, 1989, 1997). Even where sequence stratigraphic units are more apparent, shell beds provide another way to identify sequence stratigraphic surfaces, particularly those associated with transgression (Beckvar and Kidwell, 1988; Banerjee and Kidwell, 1991; Kidwell, 1993; Cattaneo and Steel, 2003; Nawrot et al., 2018; Zecchin et al., 2019).

The potential for shell bed generation has increased during the Phanerozoic as the organisms that produce bioclasts have evolved toward greater abundance with more durable shells that have less organic matter (Kidwell and Brenchley, 1994). For example, shell beds from the Ordovician and Silurian are seldom greater than 0.5 thick, but Neogene shell beds commonly have thicknesses up to 4−8 m. Consequently, shell beds have the greatest utility in the sequence stratigraphic analysis of Mesozoic and especially Cenozoic successions.

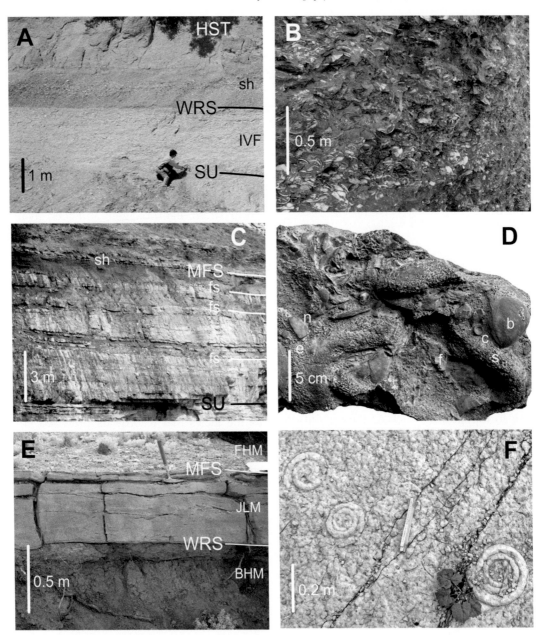

FIGURE 2.22   Examples of shell beds (images A–E courtesy of Steven Holland; image F courtesy of Guido Roghi). A—Onlap shell bed (sh) from the Miocene Choptank Formation of Virginia. The subaerial unconformity (SU) is overlain by tidal incised-valley fill deposits (IVF), truncated by a transgressive wave-ravinement surface (WRS), in turn capped by the Drumcliff shell bed and overlying muddy sands of the highstand systems tract (HST). B—Onlap shell bed from the Neogene Cobham Bay Member of the Eastover Formation, Virginia. Shell bed is several meters thick and consists of densely packed bivalves and gastropods, with rarer shark teeth and marine mammal bone. C—Composite backstepping–downlap shell bed (sh) from the Mississippian Pennington Formation of Tennessee. The shell bed straddles a maximum flooding surface (MFS) at the top of a retrogradational set of high-frequency carbonate sequences bounded by combined maximum regressive and maximum flooding surfaces of lower hierarchical rank (fs). The base of this set is a karstic subaerial unconformity (SU). The shell bed is pyritic and phosphatic, with a diverse fossil assemblage, traceable for >200 km along depositional strike. D—Bedding-plane view of shell bed in image C, showing sponges (s), bivalves (b), nautiloids (n), fenestrate bryozoans (f) and crinoids (c). Much of the shell material consists of highly abraded bioclasts the size of very coarse sand (e). E—Onlap shell bed from the Cretaceous Juana Lopez Member of the Carlile Shale (JLM), overlying offshore mudstone of the Blue Hill Member of the Carlile Shale (BHM) and underlying shallow-water limestone of the Fort Hays Member (FHM) of the Niobrara Formation, Colorado. This shell bed is bounded at the base by a transgressive wave-ravinement surface (WRS) that reworks an older unconformity, and at the top by a maximum flooding surface (MFS). The shell bed is complex, with numerous internal discontinuity surfaces, and consists of comminuted fragments of the bivalve *Inoceramus*, oysters, ammonites, shark teeth, and bone fragments. F—Shell bed from a Tethyan perched high in the Upper Cretaceous–Lower Eocene Scaglia Rossa, Italy, with nodular fabric and well-preserved ammonites.

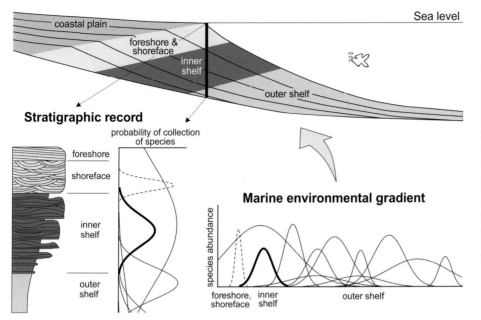

FIGURE 2.23 Marine environmental gradients in the occurrence of species, and their expression in the stratigraphic record (adapted from Patzkowsky and Holland, 2012). Because abundances of benthic species change with water depth, the probability of occurrence of those species changes systematically with stratigraphic position, reflecting the environments of deposition. Although species may have existed at a particular moment in time, their presence in a local stratigraphic column depends on the depositional environment at that location at that time.

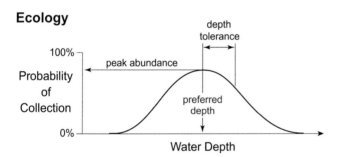

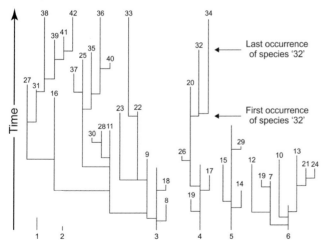

FIGURE 2.24 Core ecological elements that control the occurrence of marine fossils within stratigraphic columns (adapted from Holland, 1995). Species evolve through time via a random-branching process, and have first and last occurrences (lower diagram). Distributions in modern oceans define the ecology of extant species in terms of preferred water depth, depth tolerance, and peak abundance (upper diagram).

or regional extinction. Insights into the ecology and evolution of species shed light on the controls on their stratigraphic occurrence (Fig. 2.24) and reveal that four processes with relevance to sequence stratigraphy can create clusters of first and last occurrences in the stratigraphic record (Fig. 2.25; Holland, 1995, 2000; Patzkowsky and Holland, 2012).

First, abrupt changes in facies create sharp changes in the probability of collection of species, increasing it for some species and decreasing it for others. Across maximum regressive surfaces, particularly where they are merged with maximum flooding surfaces, the probability of occurrence of deeper water species abruptly increases stratigraphically and the probability of occurrence of shallower water species abruptly declines. As a result, these surfaces are commonly indicated by clusters of last occurrences of shallower water species and clusters of first occurrences of deeper water species. Transgressive wave- and tidal-ravinement surfaces have similar clusters. Surfaces of wave scouring that develop during forced regression (i.e., the regressive surface of marine erosion) display the opposite pattern, with clusters of last occurrences of deeper water species overlain by clusters of first occurrences of shallower water species. In all cases, greater contrast in the water depths represented by the facies at these surfaces causes clusters of first and last occurrences to be more pronounced. As a result, the maximum regressive/transgressive surfaces of high-frequency sequences within higher rank transgressive systems tracts, which are typically more prominent and have greater facies contrasts than those that form within other systems

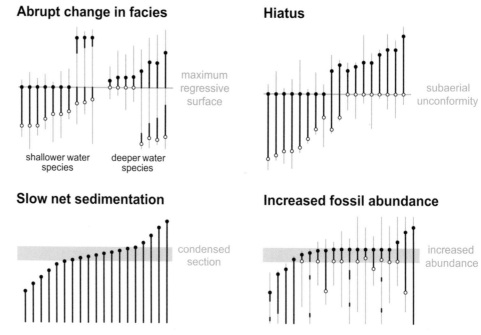

FIGURE 2.25 Four stratigraphic mechanisms of producing clusters of first and last occurrences (diagram courtesy of Steven Holland). Each image illustrates the occurrence of multiple species in a single stratigraphic column (black lines), as well as stratigraphic positions where a species may be present in the sedimentary basin, but not preserved locally (gray lines), generally owing to facies control or rarity of the species. For simplicity, the image showing slow net sedimentation depicts only last occurrences, but it similarly also generates clusters of first occurrences.

tracts, tend to display the most developed clusters of first and last occurrences.

Second, subaerial unconformities (or any hiatus) can generate clusters of first and last occurrences. Any species that goes extinct during the hiatus can have a last occurrence no higher than the unconformity, creating a cluster of last occurrences. Similarly, any species that originates or migrates into the basin during the hiatus can have a first occurrence no lower than the unconformity, potentially creating a pulse of first occurrences. The strength of these clusters directly reflects the duration of the hiatus, and a long hiatus may be reflected by complete species turnover at an unconformity.

Third, prolonged slow net rates of deposition (stratigraphic condensation) can also create pulses of first and last occurrences. Where last and first occurrences would normally be spread over a stratigraphically thick interval, prolonged slow net deposition can cause these first and last occurrences to be confined to a thin stratigraphic interval, or even a single surface. The duration of slow rates of deposition controls the development of these clusters of first and last occurrences.

Fourth, changes in the overall abundance of fossils can systematically change the probability of occurrence of many species. For example, shell beds locally raise the probability of occurrence of species, which can create pulses of first or last occurrences, especially of normally rare species (Nawrot et al., 2018). Intensified collection effort over a stratigraphic interval will also raise the probability of collection, artificially creating a pulse of first and last occurrences, again, especially of rare species.

These four processes often combine at particular horizons, increasing the intensity of clusters of first or last occurrences (Smith et al., 2001; Holland and Patzkowsky, 2015). For example, where a subaerial unconformity and transgressive wave-ravinement surface lie at the same position, the effects of the hiatus combine with the facies contrast caused by transgression, amplifying the number of first and last occurrences. If significant stratigraphic condensation occurs, it can also raise the numbers of first and last occurrences, as could the formation of a shell bed.

Clusters of first and last occurrences therefore show a strong relationship with the sequence stratigraphic architecture, a pattern noted by biostratigraphers (Fig. 2.26; Armentrout, 1987, 1991, 1996; Armentrout et al., 1991). Such sequence stratigraphic control is not limited to relatively abundant invertebrates, and it is also important for rarer vertebrates, which likewise tend to occur preferentially at significant sequence stratigraphic surfaces (Peters et al., 2019; McMullen et al., 2014).

Not only does the sequence stratigraphic architecture control the position of first and last occurrences, it controls the variation in the age of the first or last occurrences of a species across a sedimentary basin. This variation, known as diachrony (Spencer-Cervato et al., 1994) or range offset (Holland and Patzkowsky, 2002), is a principal source of biostratigraphic uncertainty. Many numerical methods have been developed to minimize this source of error, such as graphic correlation (Shaw, 1964), ranking and scaling (RASC; Agterberg

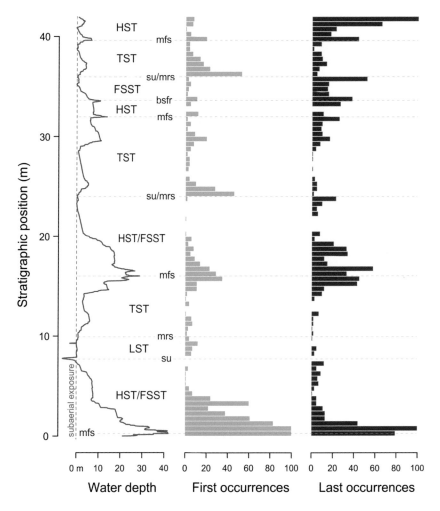

FIGURE 2.26 First and last occurrences of species within a stratigraphic column, under controlled experimental conditions (see Holland and Patzkowsky, 1999, 2002, 2015 for details). The probability of species extinction is constant through time and all clustering of first and last occurrences is solely the result of stratigraphic architecture: hiatuses, changes in sedimentation rate, and changes in facies coupled with the facies preferences of species. Note the tendency for clusters of first and last occurrences near the bounding surfaces of systems tracts. Abbreviations: su—subaerial unconformity; mrs—maximum regressive surface; mfs—maximum flooding surface; bsfr—basal surface of forced regression; LST—lowstand systems tract; TST—transgressive systems tract; HST—highstand systems tract; FSST—falling-stage systems tract.

and Gradstein, 1999), and constrained optimization (CONOP; Sadler et al., 2003). Intervals of persistent directional change in facies, as well as unconformities and intervals of stratigraphic condensation, increase the amount of diachrony or range offset, and therefore the precision of biostratigraphic correlations. This precision varies systematically and predictably with the sequence stratigraphic architecture (Holland and Patzkowsky, 2002), providing guidance on when to place more weight on biostratigraphic versus sequence stratigraphic correlations. For example, diachrony and range offset are typically highest in updip areas with numerous unconformities of long duration. They are also large at the prominent maximum regressive/ transgressive surfaces of high-frequency sequences within higher rank transgressive systems tracts.

2.1.3.5 Biofacies

Not only does the occurrence of species change systematically with the sequence stratigraphic architecture, so does the abundance of species (Fig. 2.27). Owing to the depth gradient of species distributions, depositional

environments and their facies are characterized by repeating associations of fossils, called biofacies (Ludvigsen et al., 1986). In some cases, biofacies can subdivide environments more finely than lithofacies, which allows for more detailed delineation of sedimentary environments (Brett, 1998; Holland et al., 2001; Scarponi and Kowalewski, 2004). This is particularly true in offshore settings, where relatively homogeneous mudstone facies mask biologically important environmental variation. Importantly, fossil associations largely reflect the environments in which these species lived, and most evidence now suggests that out-of-habitat transport is the exception, at least for benthic invertebrates (Kidwell and Flessa, 1996; Kidwell, 2013).

Biofacies provide an additional means for recognizing changes in depositional environments, particularly in cases where facies belts are broad, such as the offshore, or where facies may have few distinguishing features, such as widespread shallow-marine bioturbated sands in the Cenozoic. Whether as the sole or supporting criterion for recognizing facies changes, biofacies can aid in the identification of stacking patterns

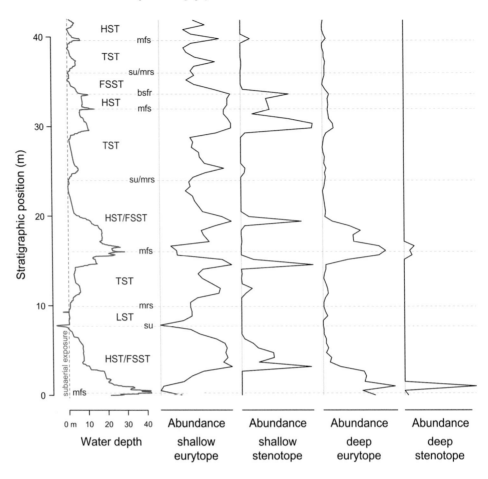

FIGURE 2.27 Abundance of four species that differ in terms of preferred habitat (shallow vs. deep) and tolerance for other environments (eurytopes: wide tolerances; stenotopes: narrow tolerances) (see Holland and Patzkowsky, 1999, 2002, 2015 for details). There is a strong correlation between the occurrence of a species and water depth, which will be reflected in sedimentary facies. Abrupt changes in species abundance occur at systems-tract boundaries. Abbreviations: su—subaerial unconformity; mrs—maximum regressive surface; mfs—maximum flooding surface; bsfr—basal surface of forced regression; LST—lowstand systems tract; TST—transgressive systems tract; HST—highstand systems tract; FSST—falling-stage systems tract.

and the sequence stratigraphic architecture. Moreover, some biofacies reflect responses to particular sedimentary conditions such as firmgrounds or hardgrounds, much like the substrate-specific *Glossifungites* and *Trypanites* Ichnofacies. Firmgrounds and hardgrounds can shift the taxonomic composition of biofacies away from infaunal species and toward epifaunal species, such as encrusting and cementing forms, free-lying species, and byssally attached bivalves, much as can be seen in shell beds (Kidwell, 1991b). Particular sequence stratigraphic surfaces, such as transgressive wave- and tidal-ravinement surfaces, maximum regressive surfaces, maximum flooding surfaces, as well as regressive surfaces of marine erosion may therefore bear a distinctive epifaunal-dominated biofacies (Brett, 1998).

2.1.3.6 Taphonomy

Because the sequence stratigraphic architecture controls not only the distribution of sedimentary facies within a basin but also rates of sedimentation, the taphonomic mode of fossil preservation is closely coupled to the architecture of the stratigraphic framework (Kidwell and Bosence, 1991; Brett, 1995; Kidwell and Holland, 2002). As such, two important controls on

taphonomy need to be distinguished. The first controls are those that reflect relatively normal or background sedimentary conditions, and the second are those that reflect unusual conditions of sedimentation such as those recorded at sequence stratigraphic surfaces.

Central to the taphonomy of background conditions is the concept of taphofacies, that sedimentary facies are characterized by a particular style of taphonomy, including the amounts of disarticulation, fragmentation, abrasion, bioerosion, dissolution, and mineralization (Brett and Baird, 1986). The first four of these reflect sedimentary dynamics such as bed shear stress, as well as how much time a fossil spends in the taphonomically active zone, that is, the upper portion of the sediment column in which a fossil is likely to be repeatedly exhumed and buried. Owing to the mineralogy of skeletons and how parts of a skeleton are attached to one another, major fossil groups experience disarticulation, fragmentation, abrasion, and bioerosion at different rates. Dissolution and mineralization primarily reflect chemical conditions during early diagenesis, especially pH and Eh, and these often vary with grain size and the abundance of organic matter. As a result, sedimentary facies commonly differ in the quality of fossil preservation, and stratigraphic changes in preservation often reflect facies change and sequence stratigraphic architecture.

Sedimentary dynamics can be substantially different during the formation of important sequence strati-graphic surfaces, including subaerial unconformities, maximum regressive and maximum flooding surfaces, and transgressive and regressive ravinement surfaces. This can markedly affect fossil preservation at these surfaces. For example, erosion at subaerial unconfor-mities can exhume fossils and rework them into younger deposits. Moreover, meteoric diagenesis at subaerial un-conformities can greatly accelerate rates of dissolution, completely destroying calcareous (especially aragonitic) fossils in uncompacted sediments but creating molds in compacted sediment and rock.

At maximum regressive and maximum flooding surfaces, the principal effect is that fossils are main-tained within the taphonomically active zone for greater amounts of time, potentially increasing the amounts of disarticulation, breakage, abrasion, and bioerosion of at least part of the fossil assemblage. Because organisms continue to produce bioclasts during the formation of these surfaces, a continuous supply of fresh bioclasts is provided, and the resulting fossil assemblage tends to include components in a wide variety of taphonomic states. Moreover, the presence of bioclasts on the seafloor provides the hard substrates necessary for other types of organisms, especially cementers, byssally attached bivalves, encrusters, and other epifaunal species. This can potentially increase the input of bioclasts in a process known as taphonomic feedback (Kidwell and Jablonski, 1983), as well as shift the ecological composition of the fossil association. As the duration of nondeposition during the formation of these surfaces increases, the amount of taphonomic damage will increase, and this is often especially apparent when comparing to the taphofacies that would typically be expected for lithofacies (Brett, 1995).

Surfaces that develop in relation to transgressive or regressive ravinement processes have similar tapho-nomic effects, but their formation also involves substan-tial erosion of the seafloor (Hunt and Tucker, 1992; Zecchin et al., 2019), causing greater reworking of fossils into younger strata. Owing to this erosional reworking and the prolonged nondeposition that accompanies the formation of ravinement surfaces, bioclasts at these surfaces are commonly completely disarticulated, highly fragmented, and strongly abraded. This creates shelly lags composed of the most chemically resistant and physically robust bioclasts (Brett, 1995; Zecchin et al., 2019). Reworking may also exhume bioclasts that have undergone early diagenesis, producing horizons of shells and internal molds that have pyritized and phosphatized (Brett, 1995).

2.1.3.7 *Continental settings*

While the study of the sequence stratigraphic controls on the fossil record started with the analysis of marine systems, the principles of stratigraphic paleobiology also apply to continental settings (Holland and Loughney, 2020). Most studies in continental settings have focused on issues of taphonomy, including the occurrence of fossil concentrations (dominantly bone beds) at discontinuity surfaces and the preservation of plants. It has become evident that the spatial and tempo-ral variability of accommodation and sedimentation in sedimentary basins exerts an important control on the stratigraphic architecture and the distribution of conti-nental fossils. Among the factors that link the occurrence of continental fossils to the stratigraphic framework, base-level fluctuations play a key role as they trigger changes in elevation and related variations in tempera-ture and precipitation, which affect the composition of animal and plant communities. Since base-level changes also control the deposition of sequences and the formation of subaerial unconformities in continental settings, the cyclic patterns in the preservation, concen-tration, and community composition can be linked to the elements of the sequence stratigraphic framework (Holland and Loughney, 2020).

Continental systems have discontinuity surfaces that have the potential to be sites of fossil concentrations, owing to an increase in the R-hardpart:R-sediment ratio (Rogers and Kidwell, 2000). The two most promising sites for continental fossil concentrations are scours of fluvial channels and subaerial unconformities. In a study of 55 vertebrate skeletal concentrations from the Campanian Two Medicine and Judith River formations of Montana, most continental fossil concentrations are associated with fluvial scours (including tidally influenced rivers) and transgressive wave ravinements that incised into fluvial deposits. In contrast, bone occur-rences are rare at subaerial unconformities, suggesting that destruction of bone counteracts its ongoing supply at the surface during the hiatus (Rogers and Kidwell, 2000). This study also found that the number of verte-brate occurrences was increased substantially above an expansion surface (Martinsen et al., 1999), at which the ratio of floodplain to channel facies abruptly increases. Intervals of increased floodplain aggradation are also likely to preserve abandoned channel and pond deposits that may contain rare species not otherwise found within the depositional system (Loughney et al., 2011). Collectively, these initial studies suggest that the principal control of continental stratigraphic architec-ture on the fossil record may lie in the changing relative proportions of facies in which fossils are preserved, such as channel versus floodplain, rather than at disconform-able surfaces, which are a comparatively important aspect of the marine fossil record.

Because plant fossil preservation typically requires that organic matter not be oxidized, the sequence stratigraphic architecture plays an important role in their fossil record (Gastaldo and Demko, 2011). For example, times of faster relative sea-level rise are typically accompanied by higher water table, enhancing plant preservation relative to times in which the relative sea level rises slowly or falls. As a result, plant preservation can be better in the transgressive systems tract than in the highstand, falling-stage, and lowstand systems tracts. Moreover, falling relative sea level can lower the water table such that organic-rich facies are newly exposed to flushing with oxidizing meteoric water, destroying organic matter in those stratigraphic intervals. Highstand systems tracts are commonly subjected to this secondary destruction. Because wood and palynomorphs are more resistant to oxidation than leaves and reproductive bodies, the types of plant fossils can also vary with the systems tract. Intervals that have been subjected to oxidizing conditions may be dominated by relatively resistant wood and palynomorphs, whereas those characterized by reducing porewaters (e.g., the transgressive systems tract) may also contain a rich record of delicate plant parts.

2.1.3.8 Discussion

Body fossils are relevant to sequence stratigraphy in several ways. First, *fossil concentrations*, such as shell beds and bone beds, develop typically at sites of onlap, downlap, toplap, and backstepping, providing important clues particularly in rocks that are otherwise fossil-poor or display little variation among lithofacies. Second, *biofacies* of distinct fossil assemblages may characterize lithofacies or even allow for more finely subdivided depositional environments. These can be used to interpret trends in water depth and stacking patterns, particularly where lithofacies display little variation. Substrate-specific biofacies can also provide evidence of significant stratigraphic surfaces. Third, *clusters of first and last occurrences* commonly develop at surfaces of abrupt facies change, hiatuses, intervals of slow net deposition, and intervals of increased fossil abundance, providing biostratigraphic evidence of key sequence stratigraphic surfaces such as subaerial unconformities, maximum regressive surfaces, and maximum flooding surfaces. Such clusters also identify levels of greater biostratigraphic uncertainty and diachrony, where sequence stratigraphic correlation may be more reliable than biostratigraphic correlation. Fourth, the *taphonomy* of fossils is strongly controlled by sedimentary dynamics. This generates distinctive modes of preservation within sedimentary environments, called taphofacies. It also creates unusual and contrasting styles of preservation at sequence stratigraphic surfaces.

Thus, taphonomy of fossils can provide another line of evidence in distinguishing sedimentary facies, from which stacking patterns may be inferred, as well as evidence of important discontinuity surfaces that define elements of the sequence stratigraphic framework.

In addition to shell beds and bone beds, *foraminiferal studies* can further constrain the position of key surfaces in the stratigraphic record, such as maximum flooding surfaces and surfaces of maximum water depth, beyond what is possible to achieve with sedimentological studies alone (Zecchin et al., 2021). Differences between strata1 stacking patterns and bathymetry, which are commonly overlooked, can be demonstrated by means of micropaleontology and help clarify the relationship between the physical stratigraphic architecture and the environmental conditions at syndepositional time (details in Chapter 7). Key parameters of foraminiferal studies include the abundance, diversity, and % fragmentation of benthic foraminifera; the ratio between distal and proximal benthic species; and the ratio between planktonic and benthic foraminifera (i.e., the plankton/benthos ratio) (Zecchin et al., 2021). Abundance and diversity are influenced both by shoreline shifts and paleoecology, and hence they are not always in tune with the physical stratigraphic framework. The % fragmentation, which reflects environmental energy conditions, and the distal/proximal ratio, which correlates with the distance from the shoreline, are indicators of shoreline shifts. The plankton/benthos ratio is more indicative of water mass variations than substrate conditions, and can be used as a proxy for bathymetric changes (Zecchin et al., 2021). These analyses help constrain the interval that contains lithologically cryptic surfaces within condensed sections, as well as pinpoint the stratigraphic level of deepest water, which is usually above the maximum flooding surface, within the highstand clinoforms (Figs. 2.28 and 2.29; details in Chapter 7).

2.1.4 Trace fossils

2.1.4.1 General principles

Ichnology is the study of traces made by organisms, including their description, classification, and interpretation (Pemberton et al., 2001). Such traces may be ancient ('trace fossils', or 'ichnofossils'—the object of study of paleoichnology) or modern (recent traces—the object of study of neoichnology), and generally reflect basic *behavior patterns* (e.g., resting, locomotion, dwelling, deposit feeding, or grazing—all of which can be combined with escape or equilibrium-adjustment structures; Ekdale et al., 1984; Frey et al., 1987; Pemberton et al., 2001). These behavioral patterns can be directly linked to a number of *(paleo)ecological controls* (e.g., substrate consistency, physical energy, sedimentation rates, nutrient availability,

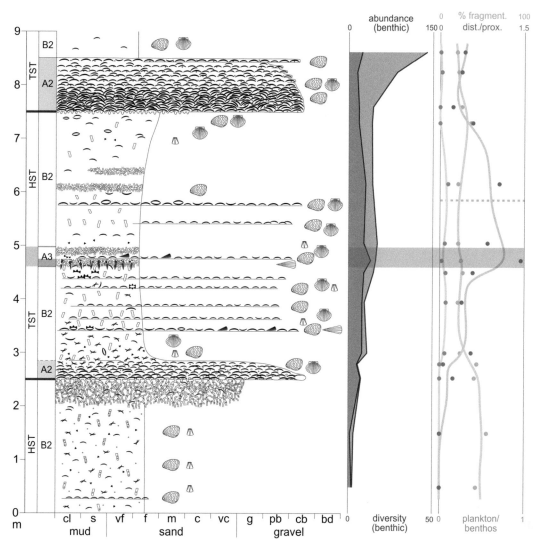

FIGURE 2.28 Placement of maximum flooding and maximum water-depth surfaces based on foraminiferal studies (Plio-Pleistocene, NW part of Crotone Basin, Italy; see Fig. 2.29 for the key to symbols and abbreviations; from Zecchin et al., 2021). The key parameters include the abundance, diversity, and % fragmentation of benthic foraminifera; the ratio between distal and proximal benthic species; and the ratio between planktonic and benthic foraminifera (see text for details). This section is in a proximal location relative to the section in Fig. 2.29. Note that the maximum water depth is recorded above the maximum flooding surface, within the highstand clinoforms.

salinity, oxygenation, water turbidity, or temperature), and implicitly to particular *depositional environments* (Seilacher, 1964, 1978; MacEachern et al., 2010; Buatois and Mángano, 2011; Knaust and Bromley, 2012). Ichnofossils have been documented in all depositional settings, from eolian (e.g., Bordy et al., 2004) through to deep water (e.g., Uchman and Wetzel, 2012), as well as from greenhouse carbonates (e.g., Knaust et al., 2012) to glacial environments (Netto et al., 2012).

Trace fossils include a wide range of biogenic structures wherein the results of organism activities are preserved in sediments or sedimentary rocks, but not the organisms themselves nor any body parts thereof. Ichnofossils also exclude molds of the body fossils that

may form after burial, but include imprints made by body parts of active organisms (Pemberton et al., 2001). Trace fossils are commonly found in successions that are otherwise unfossiliferous, and bring a line of evidence that can be used toward the reconstruction of paleoecological conditions and paleodepositional environments. As with any research method, data may be equivocal in some cases; e.g., when a group of biogenic structures appears to record combinations of behaviors that are at odds with one another and/or with the facies observations of the successions in which they are enclosed. For this reason, it is best to use ichnological data in conjunction with other clues provided by classical paleontology and sedimentology. Integration

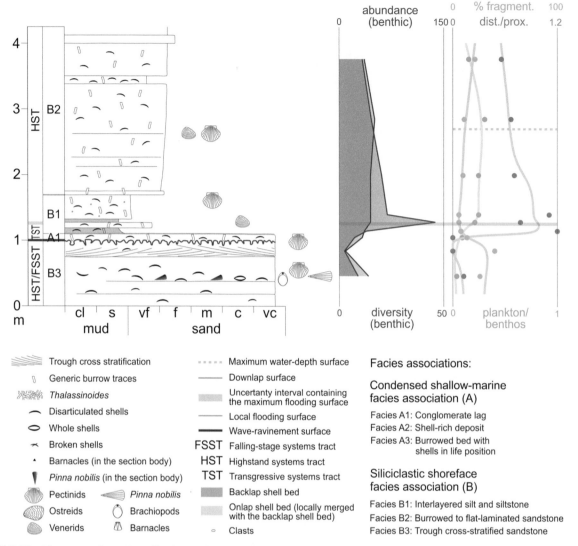

FIGURE 2.29 Placement of maximum flooding and maximum water-depth surfaces based on foraminiferal studies (Plio-Pleistocene, NE part of Crotone Basin, Italy; from Zecchin et al., 2021). The key parameters include the abundance, diversity, and % fragmentation of benthic foraminifera; the ratio between distal and proximal benthic species; and the ratio between planktonic and benthic foraminifera (see text for details). This section is in a distal location relative to the section in Fig. 2.28. Note that the maximum water depth is recorded above the maximum flooding surface, within the highstand clinoforms, and that the offset between the two surfaces increases in a basinward direction (compare Figs. 2.28 and 2.29).

and mutual calibration of complementary techniques allows one to better constrain paleoenvironmental interpretations. A list of basic principles of ichnology is provided in Fig. 2.30.

Ichnological datasets afford the definition of several key concepts which, in the order of increasing stratigraphic significance, include ichnofabrics, ichnocoenoses, and ichnofacies. *Ichnofabrics* record the details of animal-sediment responses at the bed scale, such as their abundance and disposition, which characterize the texture and internal structure of the deposit (Bromley and Ekdale, 1984). *Ichnocoenoses* correspond to trace-fossil suites, which can include multiple ichnofabrics spatially, and effectively record animal-

sediment responses at facies scales. *Ichnofacies* are summaries or syntheses of a paleocommunity's response to the depositional environment, regardless of specific trace-fossil genera, and reflect the facies model at depositional system scales. Lateral and vertical shifts in ichnofacies can be used to interpret changes across a sedimentary basin as well as through time in paleodepositional environments, based on the inferred shifts in paleoecological conditions (Knaust and Bromley, 2012).

Ichnofacies can be considered to recur throughout the Phanerozoic, regardless of the specific ichnogenera that comprise the suites. For example, even though ichnogenus *Cruziana* is restricted to the Paleozoic, the

Basic principles of Ichnology:

1. Trace fossils generally reflect the activity of *soft-bodied organisms*, which commonly lack hard (preservable) body parts. In many environments, such organisms represent the dominant component of the biomass.

2. Trace fossils may be classified into structures reflecting *bioturbation* (disruption of original stratification or sediment fabric: e.g., tracks, trails, burrows); *biostratification* (stratification created by organism activity: e.g., biogenic graded bedding, biogenic mats); *biodeposition* (production or concentration of sediments by organism activity: e.g., fecal pellets, products of bioerosion); or *bioerosion* (mechanical or biochemical excavation by an organism into a substrate: e.g., borings, gnawings, scrapings, bitings).

3. Trace fossils reflect *behavior patterns*, and so they have *long temporal ranges*. This hampers biostratigraphic dating, but facilitates paleoecological comparisons of rocks of different ages. Basic behavior patterns include resting, locomotion, dwelling and feeding, all of which can be combined with escape or equilibrium structures.

4. Trace fossils are sensitive to water energy (hence, they may be used to recognize and correlate event beds), substrate coherence, and other *ecological parameters* such as salinity, oxygen levels, sedimentation rates, luminosity, temperature, and the abundance and type of nutrients.

5. Behavior patterns depend on ecological conditions, which in turn relate to particular depositional environments. Hence, trace fossils tend to have a *narrow facies range,* and can be used for interpretations of *paleo-depositional environments.*

6. Trace fossils tend to be *enhanced by diagenesis*, as opposed to physical or chemical structures which are often obliterated by dissolution, staining or other diagenetic processes.

7. An individual trace fossil may be the product of *one organism* (easier to interpret), or the product of *two or more different organisms* (composite structures, more difficult to interpret).

8. An individual organism may generate *different structures* corresponding to different behavior in similar substrates, or to identical behavior in different substrates. At the same time, *identical structures* may be generated by different organisms with similar behavior.

FIGURE 2.30 Principles of ichnology (compiled from Seilacher, 1964, 1978; Ekdale et al., 1984; Frey et al., 1987; Pemberton et al., 2001).

Cruziana Ichnofacies is still being produced today. That is because the combination of animal-sediment responses that defines the ichnofacies continues to occur, even while the trace makers and the particular ichnogenera may change with time. This is also the reason why ichnofacies can be employed for interpreting depositional environments, regardless of the age of the rock or the particular basin in which it forms.

The concept of ichnofacies was developed originally based on the observation that many of the environmental factors that control the distribution of traces change progressively with increased water depth (Seilacher, 1964, 1967). It is important to realize, however, that the ecology of an environment reflects the interplay of multiple factors (Fig. 2.30), and therefore the types and number of organisms that inhabit a particular area (and implicitly the resultant trace fossil suites and ichnofabrics attributable to an ichnofacies) do not necessarily translate into specific water depths, distance from shore, or tectonic or physiographic setting (Frey et al., 1990; Pemberton and MacEachern, 1995). For example, the *Zoophycos*

Ichnofacies typically forms under deeper marine conditions, lying below the storm wave base, but may also be found in other low energy settings such as fully marine lagoons in coastal environments (Pemberton and MacEachern, 1995). Similarly, trace fossils like *Ophiomorpha*, traditionally regarded as diagnostic of shallow water, may also occur in sandy submarine fans because bathymetry is not the primary control, but rather the combination of physico-chemical conditions which only *tend* to change with water depth (e.g., Frey et al., 1990). This calls for caution when interpreting the paleobathymetry, or the syndepositional transgressions or regressions of a shoreline, based solely on ichnofacies data.

2.1.4.2 Ichnofacies classification

The classification of ichnofacies is fundamentally based on the food resource paradigm, which is tied to animal feeding strategies. To some extent, feeding strategies can be related to the substrate type and consistency, because delivery of food to the environment is largely determined by physical energy which is tied to the substrate type. Sandy particulate substrates require sufficient physical energy to mobilize sand, which in turn inhibits food settling to the substrate. Correspondingly, animals adapt to live in permanent burrows and strain food from the water. Muddy substrates that are cohesive represent low energy conditions with fine-grained sediment and food settling to the sea floor. Animals may opt to be mobile on and in the sediment searching for deposited food, or be sessile and probe the substrate for food. Hence, there is a strong linkage between feeding behaviors and associated ethology and substrate consistency. However, it is not the substrate *per se* that dictates the ichnofacies; it is rather the combination of animal behaviors that the environment permits that determines the ichnofacies. The more dynamic or variable the environment, the less the corresponding ichnofacies is controlled by substrate consistency (e.g., in deltaic settings or brackish-water environments).

Ichnofacies can be associated with various types of substrates, which in broad terms are either 'soft' (shifting or stable, but generally unconsolidated sediment) or 'resilient' (consolidated, at least in part). Both soft and resilient substrates can be found in all depositional environments, from continental to fully marine. Softground substrates generally indicate active sedimentation (low to high rates) on moist to fully subaqueous depositional surfaces, and hence are associated with conformable successions. Therefore, ichnofacies that develop on soft substrates (i.e., softground ichnofacies) record paleocommunities that are broadly contemporaneous with deposition. Examples include the *Scoyenia* and *Mermia* ichnofacies in continental settings (Fig. 2.31, and the *Psilonichnus*, *Rosselia*, *Phycosiphon*, *Teichichnus*, *Skolithos*, *Cruziana*, *Zoophycos*, and *Nereites* ichnofacies in marginal-marine and marine settings (Fig. 2.32). Softground substrates are variable in terms of sediment caliber, and certain types have received specific names (e.g., sandy continental substrates are commonly referred to as 'loosegrounds', which is the case with the *Scoyenia* Ichnofacies). Softground ichnofacies are effective tools for the reconstruction of paleodepositional environments, particularly where calibrated with independent biostratigraphic and sedimentological data.

Resilient substrates include 'woodgrounds' (*in situ* and laterally extensive carbonaceous substrates, such as peats or coal beds), 'firmgrounds' (semi-consolidated substrates, which are firm but unlithified), and 'hardgrounds' (consolidated, or fully lithified substrates) (Fig. 2.32). These substrates are particularly important for stratigraphic analyses, as being most commonly associated with hiatal surfaces in the rock record. The resilient substrates are either erosionally exhumed or the product of various processes during times of sediment starvation (pedogenesis, nondeposition, or incipient cementation). Woodgrounds form in continental to marginal-marine settings, but are

Ichnofacies	Environment		Substrate	Examples of traces
Celliforma	Subaerial	Carbonate-rich paleosols, palustrine environments, sub-humid to sub-arid climate	Paleosols	*Celliforma*, *Palmiraichnus*, *Pallichnus*, *Rebuffoichnus*, *Rosselichnus*, *Taenidium*
Coprinisphaera		Paleosols associated with herbaceous plants; from dry and cold to humid and warm		*Coprinisphaera*, *Celliforma*, *Eatonichnus*, *Chubutolithes*, *Ellipsoideichnus*, *Fontanai*
Termitichnus		Paleosols marked by termite nests, in forest ecosystems; warm and humid conditions		*Termitichnus*, *Masrichnus*, *Fleaglellius*, *Krausichnus*, *Vondrichnus*
Camborygma		Paleosols associated with marshes, swamps, wetlands; warm and humid conditions		*Camborygma*, *Loloichnus*, *Dagnichnus*, *Edaphichnium*, *Cellicalichnus*
Octopodichnus –Entradichnus		Eolian environments, arid deserts, low nutrients	Looseground	*Octopodichnus*, *Digitichnus*, *Arenicolites*, *Paleohelcura*, *Palaeophycus*, *Taenidium*
Scoyenia	Intermittent flooding	Subaerial, adjacent to surface water (river banks, lake margins, interdunes)	Softground/ Firmground	*Scoyenia*, *Acripes*, *Taenidium*, *Fuersichnus*, *Hexapodichnus*, *Cylindricum*, *Diplichnites*
Mermia	Freshwater	Fully aquatic: shallow to deep lakes, ponds, fjords	Softground	*Mermia*, *Gordia*, *Circulichnis*, *Cochlichnus*, *Helminthopsis*, *Archaeonassa*, *Treptichnus*

FIGURE 2.31   Classification of continental ichnofacies based on substrate type and consistency, as well as depositional environment (modified from Buatois and Mángano, 2011).

Ichnofacies	Environment		Substrate	Examples of traces
Psilonichnus	Marginal marine	Backshore ± foreshore	Softgrounds (marginal marine)	Psilonichnus, Macanopsis, Aulichnites, Lockeia, Planolites
Rosselia		Deltaic foreset		Rosselia, Thalassinoides, Planolites, Teichichnus, Schaubcylindrichnus
Phycosiphon		Deltaic bottomset		Phycosiphon, Spirophyton, Cosmorhaphe, Scolicia, Asterosoma, Rhizocorallium
Teichichnus		Brackish-water: estuaries, lagoons, restricted tidal flats		Teichichnus, Planolites, Cylindrichnus, Gyrolithes, Psilonichnus, Skolithos
Teredolites		Estuaries, deltas, bays, lagoons, incised valleys	Woodground	Teredolites, Thalassinoides
Trypanites	Marine	Foreshore - shoreface - shelf	Hardground	Trypanites, echinoid borings (unnamed), Caulostrepsis, Entobia
Glossifungites			Firmground	Gastrochaenolites, Skolithos, Diplocraterion, Arenicolites, Thalassinoides, Rhizocorall
Skolithos		Shoreface - foreshore	Softgrounds (marine)	Skolithos, Diplocraterion, Arenicolites, Ophiomorpha, Rosselia, Conichnus
Cruziana		Inner shelf ± lower shoreface and outer shelf		Phycodes, Rhizocorralium, Thalassinoides, Planolites, Asteriacites, Rosselia
Zoophycos		Outer shelf - slope		Zoophycos, Lorenzinia, Spirophyton
Nereites		Slope - basin floor		Nereites, Taphrhelminthopsis, Paleodictyon, Helminthoida, Cosmorhaphe, Spirorhaphe

FIGURE 2.32 Classification of marginal marine and marine ichnofacies based on substrate type and consistency, as well as depositional environment (modified from Knaust and Bromley, 2012).

colonized by the Teredolites Ichnofacies in marine or marginal-marine environments. Firmgrounds and hardgrounds may also form in a variety of depositional settings, from continental to fully marine, but the classification of ichnofacies that colonize them is based on the environment in which the tracemakers lived. The Glossifungites and Trypanites Ichnofacies reflect marine colonization of firmgrounds and hardgrounds, respectively, in contrast to the continental ichnofacies that relate to pedogenic processes and colonization under subaerial conditions (i.e., the paleosol-related Termitichnus and Coprinisphaera Ichnofacies).

Ichnofacies that develop on resilient substrates are often referred to as 'substrate-controlled' (Ekdale et al., 1984; Pemberton et al., 2001), as they tend to be associated with specific substrate types. However, a number of softground ichnofacies are also founded on substrate types,

in the sense that they are linked directly to specific unconsolidated substrates (e.g., the Scoyenia and Skolithos ichnofacies form typically on sandy substrates, whereas ichnofacies such as Mermia, Cruziana, Zoophycos, and Nereites are found on finer grained substrates in lower energy settings. Moreover, the 'substrate-controlled' ichnofacies are not as specific as initially thought; e.g., a firmground can form in almost all the range of settings as the softground ichnofacies, with the difference that the softgrounds have been split up into numerous ichnofacies. For example, the Glossifungites Ichnofacies can be subdivided into high-energy (Skolithos-like) suites, moderate energy (Cruziana-like) suites, and low-energy (Zoophycos-like) suites, and so the concept is broader than originally envisaged, as it can apply to a variety of settings from submarine canyon margins, through the shallow marine and into the backshore.

Most workers consider that the fundamental separation between the different ichnofacies types resides as to whether they are continental or marine. If the salinity barrier was not a factor, then the lacustrine *Mermia* Ichnofacies would have been called *Cruziana* or *Zoophycos*, and the looseground *Scoyenia* Ichnofacies would also have been called *Skolithos*. The ichnofacies in Figs. 2.31 and 2.32 are listed in order of increasing marine influences, from fully continental to marginal-marine and fully marine environments.

2.1.4.3 Continental ichnofacies

Continental environments afford the formation of both resilient and softground substrates (Fig. 2.31). Resilient substrates are generally represented by paleosols which, in terms of consistency, qualify as firmgrounds or hardgrounds. However, in contrast to the marine ichnofacies that are associated with firmgrounds and hardgrounds (i.e., *Glossifungites* and *Trypanites* ichnofacies, respectively), paleosols are colonized under subaerial conditions. Those mature and regional paleosols that are the product of prolonged sediment starvation and pedogenesis serve as depositional sequence boundaries in continental settings (i.e., subaerial unconformities; *Coprinisphaera* and *Celliforma* ichnofacies). Other paleosols may form under high-watertable conditions, and therefore may develop within sequences (Figs. 2.17 and 2.18; *Termitichnus* and *Camborygma* ichnofacies). Apart from the paleosol-related ichnofacies, the other continental ichnofacies are associated with softgrounds and require the presence of surface freshwater, at least to some degree. Irrespective of their placement, within sequences or at the sequence boundary, all ichnofacies afford paleoenvironmental reconstructions which are valuable to the sequence stratigraphic analysis.

The range of continental ichnofacies and associated environments extends from fully subaerial (e.g., low-watertable alluvial and coastal plains subject to pedogenesis; dry eolian environments) to fully subaqueous (i.e., lakes). In subaerial settings, it is the availability of moisture (position of the water table, soil moisture) and food that are the primary controlling factors for ichnofacies. The substrate consistency is also important, but only plays a secondary role. For example, eolian settings are typically loosegrounds like river margins or coastlines, but each setting is defined by its own discrete ichnofacies. At the opposite end of the spectrum, in lakes, moisture is no longer a variable factor, but substrate consistency appears to be fundamental, possibly related to energy and food delivery systems. From a sequence stratigraphic perspective, pedogenesis on sediment-starved landscapes under low watertable conditions may result in stratigraphic hiatuses which can be used to delineate depositional

sequences. However, all types of ichnofacies contribute to constrain the syndepositional conditions and the origin of systems tracts and bounding surfaces. A brief presentation of the continental ichnofacies follows below.

The *Scoyenia* Ichnofacies forms at the limit between subaerial and aquatic environments, being indicative of fluctuating water table (emergence—submergence cycles). Such environments may include ponds on fluvial floodplains that regularly experience desiccation; areas adjacent to lakes where the sediment is periodically exposed and inundated; and wet interdune areas in eolian systems (Frey et al., 1984; Buatois and Mángano, 1995, 1998, 2004; Pemberton et al., 2001; Fig. 2.33). Energy levels are low, and the colonized surfaces are often palimpsest, amalgamating exposure surfaces and flooding surfaces. Under these conditions, the *Scoyenia* Ichnofacies is associated with a moist to wet substrate consisting of silty to sandy sediment, which records subaerial animal communities that reside adjacent to available surface water (e.g., river banks, lake margins, or wet interdunes). The degree of substrate consolidation is variable, ranging from soft to firm (Buatois and Mángano, 2002). In this regard, the *Scoyenia* Ichnofacies may be subdivided into two distinct suites: one consisting of meniscate, backfilled structures without ornamentation formed in softgrounds; and other containing striated traces cross-cutting the former and developed in a firm substrate (Buatois et al., 1996; Savrda et al., 2000; Buatois and Mángano, 2002, 2004). The *Scoyenia* Ichnofacies may be intergradational with the *Psilonichnus* Ichnofacies, on the marginal-marine side of its environmental range, and with the *Mermia* Ichnofacies toward fully subaqueous lacustrine areas (Buatois and Mángano, 1995, 2004). Under arid conditions, the *Scoyenia* Ichnofacies may grade into the *Octopodichnus-Entradichnus* Ichnofacies. Representative ichnotaxa include *Acripes, Beaconites, Cochlichnus, Cruziana, Cylindricum, Diplichnites, Fuersichnus, Hexapodichnus, Merostomichnites, Palaeophycus, Permichnium, Planolites, Rusophycus, Scoyenia, Taenidium, Umfolozia,* and *Skolithos,* in addition to a wide diversity of tetrapod trackways.

The *Mermia* Ichnofacies is characteristic of fully subaqueous and low-energy freshwater environments such as perennial lakes, floodplain water bodies, and fjords affected by extreme glacial melting (Buatois and Mángano, 1995, 2009; Buatois et al., 2006; Scott et al., 2012; Fig. 2.34). In lakes, this Ichnofacies may extend from shallow to deep bathymetric zones, with the best development in areas well oxygenated and with abundant food supply (Buatois and Mángano, 1995, 2009). In fluvial systems, the *Mermia* Ichnofacies illustrates subaqueous conditions in floodplain ponds that do not experience desiccation. The associated sediment is silty to sandy, with a soft substrate consistency. The *Mermia*

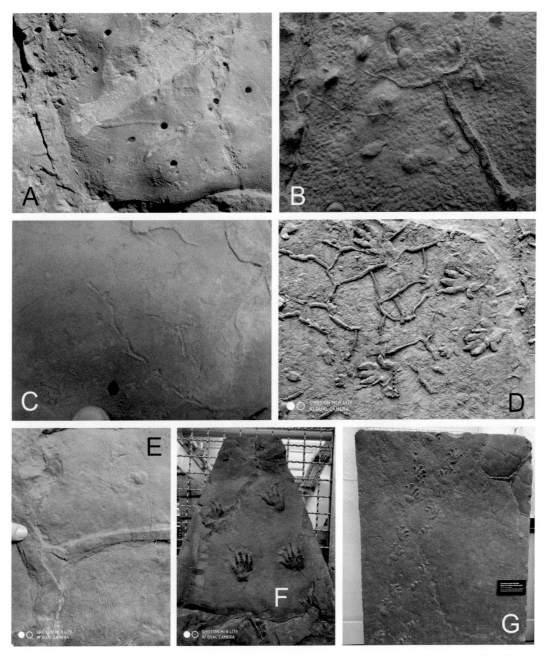

FIGURE 2.33   *Scoyenia* Ichnofacies (images courtesy of Luis Buatois and Gabriela Mangano). A—*Scoyenia gracilis* and desiccation cracks (floodplain deposits, Lower Jurassic, Kayenta Formation, Canyonland National Park, Utah, United States); B—*Tambia spiralis* (abandoned channel deposits, Lower Permian, Tambach Formation, Bromacker Quarry, Thuringian Forest, Germany); C—*Treptichnus bifurcus* (meandering point bar deposits, Lower Jurassic, upper Elliot Formation, Lower Moyeni tracksite, Mampoboleng, Lesotho); D—the tetrapod trackway *Chirotherium barthii* (overbank deposits, Middle Triassic, Solling Formation, replica of original *Chirotherium* type surface from Winzer Quarry, exhibited in the *Chirotherium* Monument of Hildburghausen, Thuringia, Germany); E—*Striatichnium bromackerense* (abandoned channel deposits, Lower Permian, Tambach Formation, Bromacker Quarry, Thuringian Forest, Germany); F—the tetrapod trackway *Dimetropus leisnerianus* (abandoned channel deposits, Lower Permian, Tambach Formation, Bromacker Quarry, Thuringian Forest, Germany); G—the tetrapod trackway *Ichniotherium sphaerodactylum* (abandoned channel deposits, Lower Permian, Tambach Formation, Bromacker Quarry, Thuringian Forest, Germany).

Ichnofacies may be intergradational with the *Scoyenia* Ichnofacies toward marginal lacustrine areas (Buatois and Mángano, 1995, 2004). Exhumed surfaces in lake-margin settings may show cross-cutting relationships revealing palimpsest surfaces with the *Mermia* suite overprinted by the *Scoyenia* suite (Scott et al., 2009). Vertical replacement of the *Mermia* Ichnofacies by the *Scoyenia* Ichnofacies is common in shallowing-upward lacustrine successions followed by exposure. Representative ichnotaxa are *Archaeonassa*, *Circulichnis*,

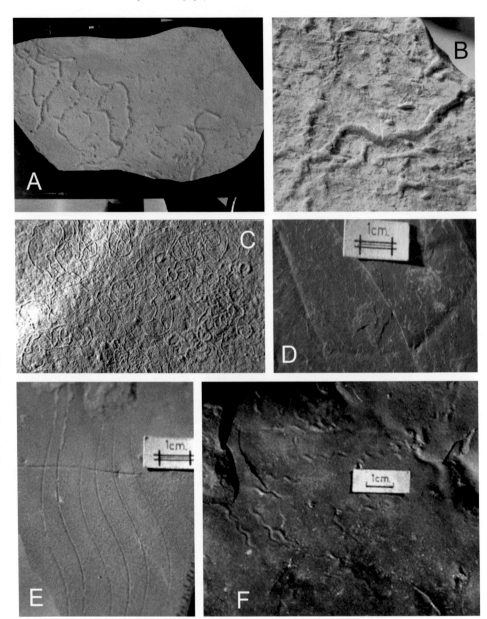

FIGURE 2.34 *Mermia* Ichnofacies (images courtesy of Luis Buatois and Gabriela Mangano). A—*Vagorichnus anyao* (lacustrine turbidites, Lower Jurassic, Anyao Formation, near Jiyuan city, Henan Province, China); B—*Helminthopsis hieroglyphica* (lacustrine turbidites, Lower Jurassic, Anyao Formation, near Jiyuan city, Henan Province, China); C—*Mermia carickensis* (lake margin deposits, Devonian, Old Red Sandstone, Dunure, Scotland); D—*Treptichnus pollardi* (underflow current deposits, Carboniferous, Agua Colorada Formation, Sierra de Narvaez, Argentina); E—the fish trail *Undichna insolentia* (underflow current deposits, Carboniferous, Agua Colorada Formation, Sierra de Narvaez, Argentina); F—*Cochlichnus anguineus* (pond deposits, Permian, La Golondrina Formation, Laguna Grande, Santa Cruz Province, Patagonia, Argentina).

Cochlichnus, Diplopodichnus, Gordia, Helminthoidichnites, Helminthopsis, Mermia, and *Treptichnus,* in addition to the fish trail *Undichna.*

The *Coprinisphaera* Ichnofacies is characteristic of paleosols associated with herbaceous plant communities, such as savannas, grasslands, prairies, and steppes (Genise et al., 2000). From a paleoenvironmental perspective, these types of soils may develop in many different depositional settings, including alluvial plains, overbank, and eolian environments (Fig. 2.35). This Ichnofacies may form under various climatic conditions ranging from dry and cold to humid and warm. Assessing the relative abundance of the different elements within the trace-fossil assemblage allows more

refined paleoclimate inferences. In general, traces produced by hymenopterans tend to be dominant under drier conditions, whereas termite nests become more common in more humid settings (Genise et al., 2000; Genise, 2017). The *Coprinisphaera* Ichnofacies may delineate subaerial unconformities (i.e., depositional sequence boundaries) in interfluves and related alluvial settings. Representative ichnotaxa are *Attaichnus, Celliforma, Chubutolithes, Coprinisphaera, Eatonichnus, Ellipsoideichnus, Fontanai, Monesichnus, Pallichnus, Palmiraichnus, Parowanichnus, Rosellichnus, Teisseirei,* and *Uruguay.*

The *Celliforma* Ichnofacies is archetypal of carbonate-rich paleosols (Genise et al., 2010; Genise, 2017). Although most examples correspond to palustrine

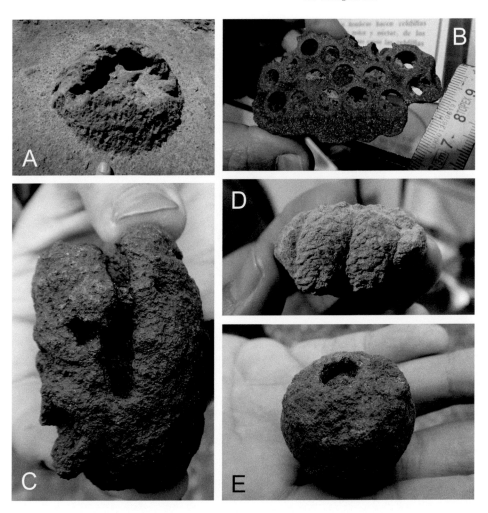

FIGURE 2.35 *Coprinisphaera* Ichnofacies (images courtesy of Luis Buatois and Gabriela Mangano). A—*Attaichnus kuenzelii* (nest of leaf-cutting ants, mollisols from a grass-dominated temperate setting; Miocene, Cerro Azul Formation, Salinas Grandes de Hidalgo, La Pampa Province, Argentina); B—*Uruguay auroranormae* (bee cells in laterites; Eocene, Asensio Formation, western Uruguay); C—*Monesichnus ameghinoi* (a cicada feeding burrow in laterites; Eocene, Asensio Formation, western Uruguay); D—*Chubutolithes gaimanensis* (flood-plain deposits, possibly a dung beetle pupation chamber; Eocene, Sarmiento Formation, Chubut Province, Patagonia, Argentina); E—*Coprinisphaera murguiai* (dung beetle nest in laterites; Eocene, Asensio Formation, western Uruguay).

environments, this ichnofacies is also present in calcretes (Genise et al., 2010). Within terrestrial (land-based) ecosystems, the *Celliforma* Ichnofacies indicates a drier climate than the *Coprinisphaera* Ichnofacies (Genise et al., 2010; Fig. 2.36). Associated vegetation includes scrubs and woodlands. In the case of palustrine environments, climatic conditions range from sub-humid to sub-arid (Alonso-Zarza, 2003). The *Celliforma* Ichnofacies may replace the *Scoyenia* Ichnofacies under continuous and progressive desiccation of the substrate. The *Celliforma* Ichnofacies may delineate depositional sequence boundaries in the subaerial portion of carbonate systems. Representative ichnotaxa are *Celliforma*, *Pallichnus*, *Palmiraichnus*, *Rebuffoichnus*, *Rosselichnus*, *Taenidium* and associated *Castrichnus* (Verde et al., 2007), and *Teisseirei*.

The *Termitichnus* Ichnofacies includes assemblages dominated by termite nests in paleosols formed in areas characterized by the establishment of closed forest ecosystems with abundant plant growth (Genise et al., 2000, 2010; Buatois and Mángano, 2011; Genise, 2017; Fig. 2.37). From a paleoenvironmental standpoint, this type of soil development is characteristic of abandoned channels and overbank areas (Genise and Bown, 1994). Climatically, this Ichnofacies indicates warm and humid conditions. Water table is typically high under these conditions (Wing et al., 1995), promoting sediment accumulation contemporaneous with pedogenesis. Therefore, the *Termitichnus* Ichnofacies may develop without a break in sedimentation, within depositional sequences. Representative ichnotaxa are *Fleaglellius*, *Krausichnus*, *Masrichnus*, *Termitichnus*, and *Vondrichnus*.

The *Camborygma* Ichnofacies is characteristic of paleosols formed in forests, scrubs, and open herbaceous communities, most notably in marshes, bogs, swamps, or wetlands (Genise et al., 2016; Fig. 2.38). From a paleoenvironmental perspective, these types of soils may develop in various depositional settings, including abandoned channels, floodplains, levees, crevasse splays, and ponds. This Ichnofacies indicates high and fluctuating water table, mostly under warm and humid climates (Genise et al., 2016; Genise, 2017). The *Camborygma* Ichnofacies may grade into the *Scoyenia* Ichnofacies under conditions of decreased humidity, when the water

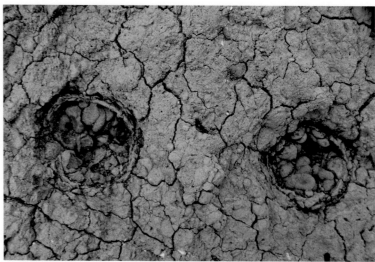

FIGURE 2.36 *Castrichnus incolumis* (an earthworm aestivation chamber, indicative of soils affected by drought periods; Pleistocene, Sopas Formation, northern Uruguay; Verde et al., 2007; image courtesy of Luis Buatois and Gabriela Mangano). This trace may be associated with the *Scoyenia*, *Celliforma*, and *Octopodichnus-Entradichnus* Ichnofacies.

FIGURE 2.37 *Termitichnus* Ichnofacies exemplified by Lower Jurassic termite nests (images courtesy of Emese Bordy; from Bordy et al., 2004, 2009). A—Freestanding vertical sandstone pillars are unevenly distributed and are more weathering resistant than the host sandstone of the Clarens Formation (note person for scale; Lower Jurassic, Tuli Basin, South Africa); B—heavily bioturbated pillar (hammer for scale, 28 cm long; Lower Jurassic, Tuli Basin, South Africa); C—c. 10-cm-thick and c. 1.2-m-long buttress tapers away from the northern side of a bioturbated pillar (hammer for scale, 28 cm long; Lower Jurassic, Tuli Basin, South Africa); D—side-view of a system of spheres that form both vertical columns and wall-like structures (hammer for scale, 33 cm long; Lower Jurassic, Toutle, Lesotho); E—side-view of a freestanding pillar with intricate network of randomly oriented, smooth-walled, cylindrical bioturbation structures (hammer for scale, 28 cm long; Lower Jurassic, Tuli Basin, South Africa); F—cross-sectional view of a bioturbated pillar that is partially enclosed in the host, massive sandstone (pen for scale, 7 cm long; Lower Jurassic, Tuli Basin, South Africa).

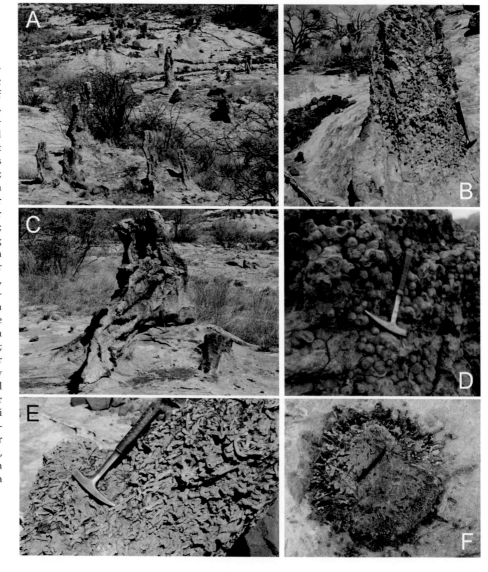

FIGURE 2.38 *Camborygma* Ichnofacies (images courtesy of Luis Buatois and Gabriela Mángano). A—The crayfish burrow *Camborygma litonomos* showing delicate scratch imprints (crevasse splay deposits, Oligocene, Lower Freshwater Molasse, Chli Sunnhalde, Switzerland); B—*Katbergia carltonichnus*, probably produced by freshwater decapod crustaceans (overbank deposits, Lower Triassic, Katberg Formation, New Lootsberg Pass, Karoo Basin, South Africa).

table is lowered. Representative ichnotaxa include *Camborygma, Cellicalichnus, Dagnichnus, Edaphichnium,* and *Loloichnus*.

The *Octopodichnus-Entradichnus* Ichnofacies typifies eolian environments, in particular sand dunes, dry interdunes, and sand sheets of wet eolian systems (Buatois and Mángano, 2011; Krapovickas et al., 2016). This Ichnofacies is typically present in arid deserts, but it may also occur in arid intervals of hyper-arid deserts. Reduced humidity and low nutrient availability are characteristic of these settings. Also, the *Octopodichnus-Entradichnus* Ichnofacies is indicative of mobile or temporary stabilized sandy substrates, subject to frequent erosion and deposition, and strong seasonality (Krapovickas et al., 2016). The *Octopodichnus-Entradichnus* Ichnofacies may grade into the *Scoyenia* Ichnofacies under conditions of increased humidity (Krapovickas et al., 2016; Buatois and Echevarría, 2019). Representative ichnotaxa are *Arenicolites, Digitichnus, Octopodichnus, Palaeophycus, Paleohelcura, Planolites, Skolithos,* and *Taenidium—Castrichnus* (Fig. 2.36).

2.1.4.4 Marginal-marine and marine ichnofacies

Marginal-marine settings include all coastal environments which experience both marine and continental influences, such as deltas, estuaries, bays, backbarrier lagoons, restricted tidal flats, and open-shoreline supratidal (backshore) and intertidal (foreshore) settings. The marine settings include all environments that are permanently submerged and subject to marine processes, starting with the shoreface (subtidal), and through to the shelf and the deep-water settings.

Most ichnofacies defined so far in marginal-marine to fully marine settings are related to softgrounds (i.e., the *Psilonichnus, Rosselia, Phycosiphon, Teichichnus, Skolithos, Cruziana, Zoophycos,* and *Nereites* Ichnofacies), and three more types are related to resilient substrates (i.e., the *Teredolites, Trypanites,* and *Glossifungites* Ichnofacies) (Fig. 2.32). Additional ichnocoenoses have been defined to record specific processes that may operate within marginal-marine and marine environments, such as *Arenicolites* which describes trace-fossil suites associated with the initial colonization of tempestites. These trace-fossil suites are useful to the identification of storm beds, which can be part of all depositional systems that accumulate above storm wave base.

Sedimentation rates on marine seafloors may vary greatly, as a function of sediment supply and energy conditions. Condensed sections assume very low sedimentation rates, and only some of them qualify as softground substrates, where the rates of sedimentation still outpace the rates of seafloor cementation (Bromley, 1975). Otherwise, seafloors in areas of stratigraphic condensation may be semilithified or even lithified (Loutit et al., 1988), in which case they become firmgrounds or hardgrounds, respectively. Therefore, softground substrates require a minimum rate of sediment accumulation, which must be higher than the rate of seafloor cementation, and so they are indicative of conformable successions. In contrast, resilient substrates are typically indicative of hiatal surfaces, commonly associated with specific sequence stratigraphic surfaces (see detailed discussion in Chapter 6).

FIGURE 2.39   *Psilonichnus* Ichnofacies (images courtesy of Stephen Hasiotis). A and B: *Psilonichnus* casts produced by ghost crabs (Delaware beach, upper foreshore, Atlantic Ocean, USA). C—*Psilonichnus* burrow cast from Fish Cay, protected beach, Crooked Island Platform, Bahamas. D and E: multiple *Psilonichnus* burrows in carbonate sandstone (Fish Cay Pleistocene beach berm—backshore, Bahamas).

The *Psilonichnus* Ichnofacies forms in marginal-marine softground substrates, and is typical of backshore (supra-tidal) environments (Frey and Pemberton, 1987; Fig. 2.39). Such settings are subject to intermittent marine flooding and hence, marked fluctuations in energy levels, being dominated by marine processes during spring tides and storm surges, and by eolian processes during neap tides and fairweather. As a result, the sediment composition of the substrate also varies greatly—from muds, silts, and immature sands to mature, well-sorted sands with a variety of physical and biogenic sedimentary struc-tures. Due to the occasional high-energy levels, the marginal-marine softgrounds are considered to be 'shift-ing substrates' (Pemberton and MacEachern, 1995), as clastic particles are commonly reworked by currents, waves, or wind. The *Psilonichnus* Ichnofacies may be intergradational with the *Scoyenia* Ichnofacies, on the continental side of its environmental range, and with the *Skolithos* Ichnofacies toward the foreshore (Pemberton et al., 2001).

The *Rosselia* Ichnofacies (MacEachern and Bann, 2020) records animal-sediment responses associated with the physico-chemical stresses imparted by fluvial sediment influx into shallow-water marine settings. The resulting facies are heterolithic, sand-dominated, and record marine delta front environments (Fig. 2.40). Sandstone beds are locally and periodically draped by layers of fluid mud. Deposition rates are generally high and beds may be erosionally amalgamated. Salinity in such settings may vary from normal marine to brackish, depending on the magnitude of river discharge. Storm events are generally accompanied by river floods, producing increased water turbidity, clay flocculation, and salinity reduction. These physico-chemical stresses imposed on tracemakers lead to a prevalence of vertical, inclined, and horizontal domiciles that are related to sessile surface detritus feeding, deposit feeding, and carnivory. Suspension-feeding behaviors are largely inhibited, due to elevated water turbidity. Most dwelling structures with spreiten show upward shifts during occupation,

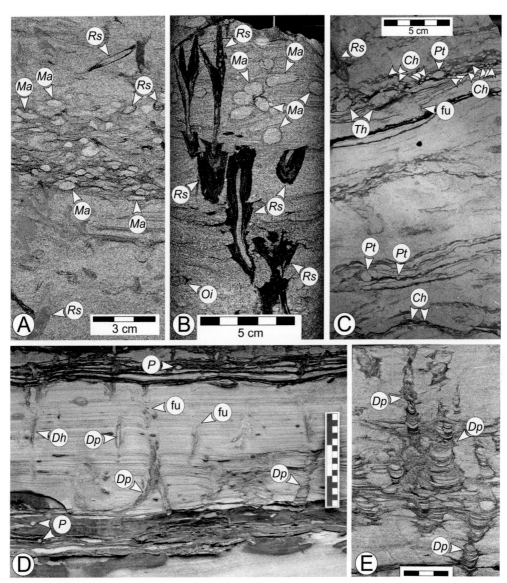

FIGURE 2.40 *Rosselia* Ichnofacies (images courtesy of James MacEachern). A—HCS sandstone beds of a wave-dominated delta front, with *Rosselia socialis* (*Rs*) and *Macaronichnus* isp. (*Ma*). Beds show BI 2-5. Upper Cretaceous Dunvegan Formation, Alberta, Canada. B—Stacked HCS sandstone beds of a wave-dominated delta front, with *Macaronichnus* isp. (*Ma*), *Ophiomorpha irregulaire* (*Oi*), and re-equilibrated (stacked) *Rosselia socialis* (*Rs*). Beds show BI 2-4. Lower Cretaceous Bluesky Formation, Alberta, Canada. C—Sandy heterolithic interval with thin mudstone drapes capping wavy parallel and oscillation ripple laminated sandstone of the delta front. The facies shows BI 1-4, with *Chondrites* (*Ch*), fugichnia (fu), *Palaeophycus tubularis* (*Pt*), *Rosselia socialis* (*Rs*), and *Thalassinoides* (*Th*). Jurassic Plover Formation, NW Shelf, Australia. D—Sandy heterolithic unit containing oscillation rippled and HCS sandstone beds intercalated with dark, carbonaceous mudstone beds of a mixed river- and wave-influenced distal delta front. The facies shows sporadically distributed burrowing (BI 1-2) with *Diplocraterion habichi* (*Dh*), *Diplocraterion parallelum* (*Dp*), fugichnia (fu), and *Planolites* (*P*). Permian Snapper Point Formation, southern Sydney Basin, Australia. E—Silty sandstone of a delta front, with *Diplocraterion parallelum* (*Dp*) showing both retrusive and protrusive spreiten, reflecting complex readjustment. Unit shows BI 3. Jurassic Tarbert Formation, Norwegian shelf.

reflecting the effect of rapid, episodic, and recurring sedimentation. Fugichnia and equilibrichnia are also common elements of the association, with navichnia confined mainly to thin intervening fluid-mud beds. Cryptic bioturbation is locally present, except where sandy event beds were draped by flocculated mud

during post-storm conditions. Owing to the prevalence of physico-chemical stresses in facies of the delta front, the *Rosselia* Ichnofacies may consist almost entirely of suites dominated by opportunistic (facies-crossing) elements. The presence of some specialized structures recording normal marine conditions is a diagnostic

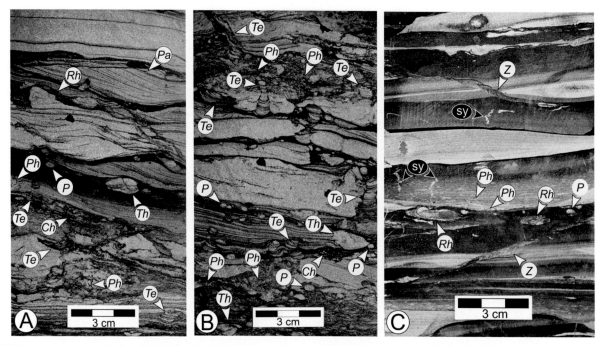

FIGURE 2.41 *Phycosiphon* Ichnofacies (images courtesy of James MacEachern; Jurassic Plover Formation, NW Shelf, Australia). A—Heterolithic succession in the prodelta, consisting of oscillation rippled and wavy parallel laminated sandstone beds draped by mud laminae. The facies shows BI 0-3 at the bed scale, and contains *Chondrites* (*Ch*), *Planolites* (*P*), *Palaeophycus heberti* (*Pa*), *Phycosiphon* (*Ph*), *Rhizocorallium* (*Rh*), and *Teichichnus* (*Te*). B—Sandy heterolithic interval of the proximal prodelta, with oscillation and current rippled sandstone draped with mudstone laminae. Unit shows BI 1-5 at the lamina-set and bed scales, and contains *Chondrites* (*Ch*), *Planolites* (*P*), *Phycosiphon* (*Ph*), *Teichichnus* (*Te*), and *Thalassinoides* (*Th*). C—Mudstone-dominated, lenticular-bedded composite bedset with syneresis cracks (sy) of the proximal prodelta. Unit displays BI 1-3, and contains *Planolites* (*P*), *Phycosiphon* (*Ph*), *Rhizocorallium* (*Rh*), and *Zoophycos* (*Z*).

indicator of the ichnofacies and differentiates it from estuarine (brackish-water) environments.

The *Phycosiphon* Ichnofacies (MacEachern and Bann, 2020) records animal responses to environmental conditions associated with river-sediment influx into low-energy, marine environments. The resulting facies are markedly heterolithic and mud-dominated, and record marine prodeltaic settings (Fig. 2.41). Associated trace fossil suites vary at the bed and bedset scales, recording spatial and temporal changes in physico-chemical conditions, consistent with river-sediment influx into low-energy marine environments. Trace fossil suites record generally low-energy conditions interrupted by periods of elevated energy, with rapid deposition of both mudstone and sandstone beds. The ichnofacies displays juxtaposition of suites recording normal marine salinities alternating with brackish-water or even largely freshwater conditions at the bed scale. Suites also record marked variations in substrate consistency. The resulting ichnological suites are dominated by mobile and sessile deposit-feeding structures and surface grazing structures. Fugichnia and equilibrichnia are common and indicate episodic deposition and heightened sedimentation rates, respectively. These spatial and

temporal variations in physico-chemical stress may lead to highly variable bioturbation intensities (BI 0–5) at the bed scale. Some tempestites contain cryptic bioturbation, while fluid mudstone drapes may contain navichnia (sediment-swimming structures). The recurring variations in environmental stress lead to trace-fossil associations expressed by juxtaposition of suites characterized by specialized open-marine ichnogenera with those dominated by strongly facies-crossing elements at the bed and bedset scales. Both trace-fossil suites should be present in order to be diagnostic of the ichnofacies.

The *Teichichnus* Ichnofacies (Pemberton et al., 2009; M.K. Gingras and J.A. MacEachern, pers. comm.) is the ichnological expression associated with brackish-water salinities (half marine salinity or less, but above zero) and sedimentary environments that suffer persistently fluctuating salinities. The sediments characteristics of this ichnological association are variably muddy, heterolithic, and sandy, with heterolithic facies the most strongly associated with the *Teichichnus* Ichnofacies (Fig. 2.42). The variable sedimentological expression is a reflection of the wide range of sedimentary environments that may be affected by salinity stresses, such as

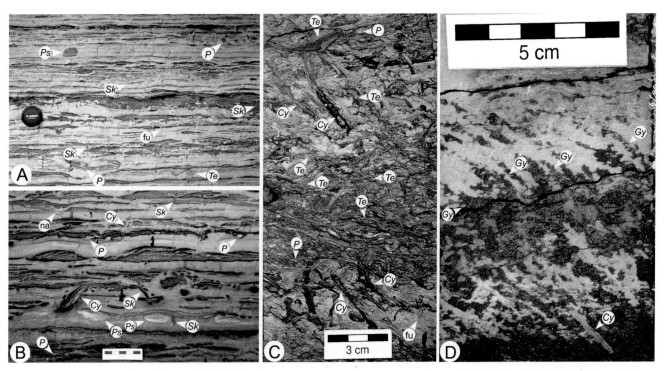

FIGURE 2.42 *Teichichnus* Ichnofacies (images courtesy of Murray Gingras). A and B: muddy heterolithic stratification with low burrowing intensities (BI 1-3), with lenticular to wavy bedded composite bedsets. Note that current ripples locally show bidirectional orientations. Pleistocene, Willapa Bay, Washington, USA. C and D: muddy heterolithic stratification with high bioturbation intensities (BI 4-5). The heterolithic character is preserved, despite the intense bioturbation, although no primary sedimentary structures have survived. Lower Cretaceous McMurray Formation, Western Interior Seaway, Alberta Canada. *Cylindrichnus* (*Cy*), fugichnia (fu), navichnia (na), *Gyrolithes* (*Gy*), *Planolites* (*P*), *Psilonichnus* (*Ps*), *Skolithos* (*Sk*), and *Teichichnus* (*Te*) are indicated.

wave- through tide-dominated estuaries, distributary channels of deltas, restricted embayments, and even some shallow seas. Nevertheless, the *Teichichnus* Ichnofacies is most prevalent in tidally influenced settings leading to a strong association with current-generated sedimentary structures, mud drapes as flaser through to wavy bedding, and common coalified plant debris. Fluid mud beds may be common. An interesting aspect of the tidal settings is that diurnal and semidiurnal changes in current velocity lead to rapid food settling, especially in the lower energy parts of the depositional environment. This means that although salinity stress restricts colonization by many types of marine invertebrates, those that are able to colonize brackish-water locales are supported by a plentiful and dependable food resource. As such the *Teichichnus* Ichnofacies has the following characteristics: trace fossils are smaller than marine counterparts (a reflection of optimization of ionic and osmotic regulation and high mortality rates), bioturbation intensities can be low to high even at the bed scale, and a range of feeding behaviors are successfully employed. Two main types of behaviors are observed: (1) spreite-bearing forms that are used to systematically exploit resource-rich strata; and (2) vertically oriented ichnofossils meant to permit access to food at

the sediment-water interface and capably keep up with the elevated sedimentation rates that are common to these settings.

The *Skolithos* Ichnofacies forms commonly in shoreface environments, where the energy level of waves and currents is relatively high, and the substrate consists of shifting particles of clean, well-sorted sand. In addition, this ichnofacies may also extend updip into the foreshore, where the intertidal area is highly dissipative and sheltered (see MacEachern et al. (2010, 2012) and Buatois and Mángano (2011), for summaries; Fig. 2.43). Storms exert a strong influence on the shoreface, and may condition much of its depositional record (Pemberton et al., 2012). Storm structures such as swaley cross-stratification are typically erosionally amalgamated, leading to a generally low degree of bioturbation. The *Skolithos* Ichnofacies is dominated by suspension-feeding behaviors over the deposit-feeding ethologies of the *Cruziana* Ichnofacies, although some traces of the latter may also be found in the shoreface as subordinate elements (Pemberton et al., 2012).

The *Cruziana* Ichnofacies is characteristic of the inner shelf, possibly extending into immediately adjacent environments (lower shoreface and outer shelf), where energy levels are moderate to low and the sediment on

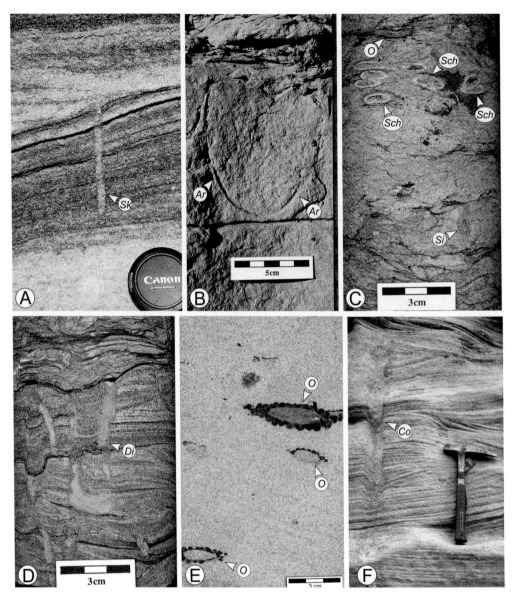

FIGURE 2.43   *Skolithos* Ichnofacies (images courtesy of James MacEachern). A—Upper shoreface trough cross-stratified sandstone showing BI 1 and isolated *Skolithos* (*Sk*). Upper Cretaceous Spring Canyon Formation, Book Cliffs, Utah, USA. B—Shallow-marine sandstone showing BI 1-2 and robust *Arenicolites* (*Ar*). Upper Cretaceous Sego Sandstone, Utah, USA. C—Thoroughly bioturbated (Bi 5) lower shoreface muddy sandstone with *Ophiomorpha* (*O*), *Siphonichnus* (*Si*), and robust *Schaubcylindrichnus coronus* (*Sch*). Lower Cretaceous Viking Formation, central Alberta, Canada. D—Low-angle wavy parallel laminated and oscillation ripple laminated shallow-marine sandstone with robust *Diplocraterion parallelum* (*Di*) and a BI of 2. Lower Cretaceous Paddy Member, central Alberta, Canada. E—Low-angle planar cross-stratified, lower shoreface sandstone containing *Ophiomorpha nodosa* (*O*) and displaying a BI of 2. Upper Cretaceous Sego Sandstone, Book Cliffs, Utah, USA. F—Well-sorted, trough cross-stratified upper shoreface sandstone with robust *Conichnus* (*Co*) and displaying a BI of 1. Upper Cretaceous Spring Canyon Formation, Book Cliffs, Utah, USA.

the seafloor is generally poorly sorted, consisting of admixtures of mud, silt, and sand (Fig. 2.44). This ichnofacies forms on stable, cohesive substrates, locally with sandy particular substrates, depending on energy levels and/or the deposition of tempestites (Pemberton and MacEachern, 1995). Within the environmental range of *Cruziana* Ichnofacies, the highest energy and proportion of sand are recorded above and near fairweather

wave base, on a shifting particulate substrate containing admixed mud, whereas both the energy conditions and sand contents decrease toward the storm wave base, where the substrate becomes more stable and cohesive.

The *Zoophycos* Ichnofacies is typically seen, according to the bathymetric schemes, as intermediate between *Cruziana* Ichnofacies and the deep-marine *Nereites* Ichnofacies, on stable seafloors that are below storm wave base

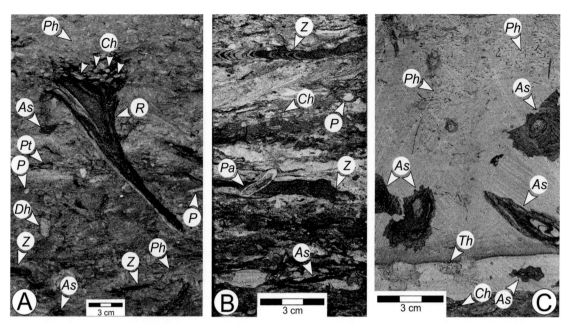

FIGURE 2.44 *Cruziana* Ichnofacies (images courtesy of James MacEachern). A—Thoroughly bioturbated (BI 5) sandy mudstone of the proximal inner shelf. The facies contains *Asterosoma* (As), *Chondrites* (Ch), *Diplocraterion habichi* (Dh), *Palaeophycus tubularis* (Pt), *Phycosiphon* (Ph), *Planolites* (P), *Rosselia* (R), and *Zoophycos* (Z). Permian Wasp Head Formation, southern Sydney Basin, Australia. B—Bioturbated (BI 4-5) sandy mudstone of the proximal inner shelf. The facies contains *Asterosoma* (As), *Chondrites* (Ch), *Palaeophycus heberti* (Pa), *Phycosiphon* (Ph), *Planolites* (P), and *Zoophycos* (Z). Upper Cretaceous Cardium Formation, Alberta, Canada. C—Bioturbated (BI 4-5) silty sandstone of the lower shoreface. The facies contains robust *Asterosoma* (As), with *Chondrites* (Ch), *Phycosiphon* (Ph), and *Thalassinoides* (Th) reburrowed with *Phycosiphon* (Ph). Upper Cretaceous Spring Canyon Formation, Book Cliffs, Utah, USA.

but free of sediment gravity flows (Wetzel and Uchman, 2012). Such environmental conditions typically occur on outer shelves and continental slopes, where the substrate is composed mainly of fine-grained sediments (Fig. 2.45). Muddy shelves are commonly poorly oxygenated, but this is not necessarily the case with tide-dominated shelves and slopes that experience contour currents or other sources of energy. Such higher energy settings may still host the *Cruziana* Ichnofacies or *Zoophycos* Ichnofacies (Hubbard et al., 2012). Beyond these common positions lying below storm wave base, the *Zoophycos* Ichnofacies can also form in other low-energy settings such as fully marine lagoons, where similar physico-chemical conditions are met (Pemberton and MacEachern, 1995).

The *Nereites* Ichnofacies is indicative of deep-sea slope to basin-floor settings, and it records the widest bathymetric range of all marine ichnofacies (i.e., 100+ to 5,000+ m). The most favorable depositional locale for this benthic community is in pelagites and hemi-pelagites, although the most common preservation of the patterned graphoglyptids casts is on the soles of thin-bedded turbidites (Fig. 2.46). The communities typical of *Nereites* Ichnofacies are characterized by slow deposition, limited food resources, and the need of tracemakers to set up farms, symbiotic associations with bacteria and organized surface mining. These

favorable, low-energy conditions may be disrupted by episodic turbidity flows which disturb the benthic community but enhance the preservation of their traces in the sedimentary record. Therefore, the formation and preservation of the *Nereites* Ichnofacies reflects a delicate balance between stable conditions and just enough turbidites to record those communities on bed soles (Pemberton and MacEachern, 1995).

The *Glossifungites* Ichnofacies (Figs. 2.47 and 2.48) develops on semi-cohesive (firm, but unlithified) substrates, best exemplified by dewatered muds. The process of dewatering typically takes place during burial, with subsequent erosional exhumation making the substrate available to tracemakers (MacEachern et al., 1992). Erosional exhumation of compacted mud may occur in a variety of settings, from fluvial (e.g., caused by channel avulsion or valley incision) to shallow-water (e.g., tidal channels or wave scouring) and deeper water (e.g., submarine channels eroding the seafloor) environments (Hayward, 1976; Fursich and Mayr, 1981; Pemberton and Frey, 1985). There are also cases where firmgrounds consist of semilithified condensed sections, where nondeposition allows for early cementation of seafloors (Bromley, 1975). Firmground conditions can also be generated by sub-aerial exposure. However, despite the wide range of

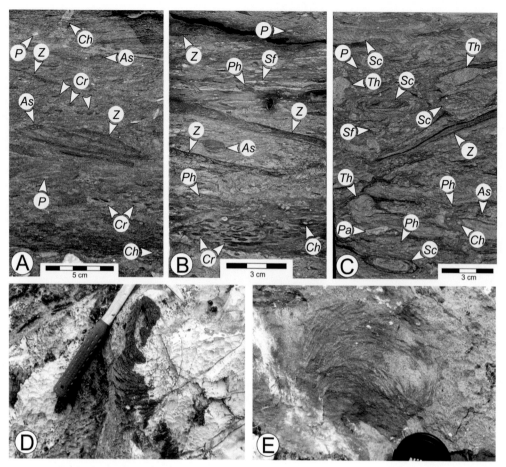

FIGURE 2.45 *Zoophycos* Ichnofacies (images A, B, and C courtesy of James MacEachern). A—Thoroughly bioturbated (BI 5-6) granule-bearing sandy to silty mudstone corresponding to a post-glacial muddy outer shelf environment. Trace fossils include *Asterosoma* (*As*), *Chondrites* (*Ch*), *Cosmorhaphe* (*Cr*), *Planolites* (*P*), and *Zoophycos* (*Z*). Permian Snapper Point Formation, southern Sydney Basin, Australia. B—Thoroughly bioturbated (BI 5), sandy mudstone, corresponding to a sandy slope margin setting. Trace fossils include *Asterosoma* (*As*), *Chondrites* (*Ch*), *Cosmorhaphe* (*Cr*), *Phycosiphon* (*Ph*), *Planolites* (*P*), *Schaubcylindrichnus freyi* (*Sf*), and *Zoophycos* (*Z*). Jurassic Nise Formation, offshore Norwegian shelf, Norway. C—Thoroughly bioturbated (BI 5) muddy sandstone, corresponding to a sandy inner or proximal outer shelf. Trace fossils include *Asterosoma* (*As*), *Chondrites* (*Ch*), *Cosmorhaphe* (*Cr*), *Palaeophycus heberti* (*Pa*), *Phycosiphon* (*Ph*), *Planolites* (*P*), *Schaubcylindrichnus freyi* (*Sf*), *Scolicia* (*Sc*), *Thalassinoides* (*Th*), and *Zoophycos* (*Z*). D and E—*Zoophycos* traces in plan view, concordant with the bedding planes, in a shelf setting (Mississippian, Etherington and Shunda formations, Jasper National Park, Alberta, Canada).

environments in which firmgrounds may form, the *Glossifungites* Ichnofacies reflects the colonization of firmgrounds in marine or marginal-marine settings (i.e., marine influence is required, in contrast to the continental ichnofacies related to pedogenic processes). From a sequence stratigraphic perspective, the *Glossifungites* Ichnofacies may demarcate scour surfaces cut by tidal currents in transgressive settings, waves in subtidal transgressive or forced regressive settings, estuarine incised valleys, submarine canyon margins, or maximum flooding surfaces associated with condensed sections or seafloor erosion (MacEachern et al., 1992, 2010, 2012; Pemberton and MacEachern, 1995).

The *Trypanites* Ichnofacies (Fig. 2.49) may form on a variety of fully lithified substrates, including rocky

coasts, reefs, fully cemented condensed sections (hardgrounds), or any type of exhumed bedrock (Bromley, 1975; Pemberton et al., 2001). Most often, hardground substrates are associated with significant stratigraphic hiatuses (±erosion), in which case they are important for the delineation of unconformities in the rock record, and implicitly for sequence stratigraphy. The generation or exposure of fully lithified substrates, such as the erosional exhumation of the bedrock, for example, may take place in any environment, from subaerial to subaqueous. The colonization of such substrates that lead to the development of the *Trypanites* Ichnofacies is, however, the product of marine conditions, and therefore the attributable trace fossil assemblages are typically associated with transgressive tidal- or wave-

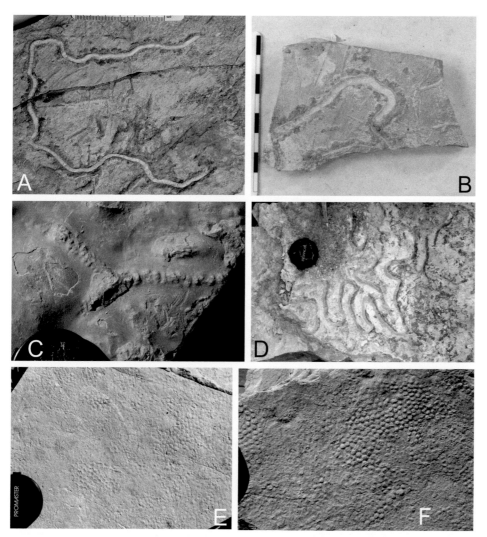

FIGURE 2.46 *Nereites* Ichnofacies (images courtesy of Stephen Hasiotis; Upper Eocene Benkovac Stone, Promina Basin, Dalmatia coast, Croatia). These traces are made by annelids, gastropods, and/or arthropods in an oligotrophic basinal setting. A, B, and C—*Nereites* traces, characterized by the central track with backfill when well preserved and the distinct lobes on the side of that track. D—*Cosmorhaphe* trace, defined by a flattened central tube with a thick, complete lining; the trace itself is winding with a recto-sinuous pattern. E and F—*Paleodictyon* traces, consisting of thin tunnels or ridges that usually form hexagonal or polygonal-shaped honeycomb-like networks.

ravinement surfaces, or with maximum flooding surfaces on the shelf. The environmental range of the *Trypanites* Ichnofacies is therefore relatively broad, similar to that of the *Glossifungites* Ichnofacies (Fig. 2.32).

The *Teredolites* Ichnofacies (Fig. 2.50) develops on woody substrates (woodgrounds), most commonly represented by driftwood pavements, peat or coal horizons (Bromley et al., 1984; Savrda, 1991; Pemberton et al., 2001). The woodgrounds themselves form in continental to marginal-marine settings, but the trace-makers that generate the *Teredolites* Ichnofacies (e.g., wood-boring bivalves and isopods) are marine or marginal marine. Woodgrounds may or may not require erosional exhumation prior to their colonization; they are resilient substrates, but differ from hardgrounds in terms of their organic nature. This characteristic makes them more flexible and readily biodegradable relative to the lithic substrates (Bromley et al., 1984). In the majority of cases, the *Teredolites* Ichnofacies is found in

marginal-marine settings (Fig. 2.32), where shoreline transgression brings marine tracemakers to the woodgrounds. In this context, and due to the resilient nature of woodground substrates, the *Teredolites* Ichnofacies may be preserved below transgressive tidal- or wave-ravinement surfaces. Where the *Teredolites* Ichnofacies occurs at the base of an incised-valley fill, it provides evidence that tidal- or wave-ravinement processes reworked the subaerial unconformity, which means that at least the lower part of the valley fill consists of transgressive deposits. Such analyses are important for the identification of sequence stratigraphic surfaces, and the clarification of the transgressive vs. regressive nature of incised-valley fills.

2.1.4.5 Discussion

It is important to note that many individual trace fossils are common amongst different ichnofacies. For example, *Planolites* may be part of both *Mermia*

FIGURE 2.47 *Glossifungites* Ichnofacies (images courtesy of Murray Gingras). A—*Glossifungites* Ichnofacies at the base of a tidal channel (arrow). The image shows *Thalassinoides* burrows descending into underlying intertidal deposits. The firmground is a transgressive tidal-ravinement surface, overlain by tidal channel and estuary channel point bar deposits with inclined heterolithic strata (Pleistocene, Willapa Bay, Washington, USA); B—*Glossifungites* Ichnofacies in a modern intertidal environment. The image shows burrows of *Upogebia pugettensis* (mud shrimp) descending into firm Pleistocene strata. The firmground is overlain by a thin veneer of unconsolidated (modern) mud, and has the significance of a transgressive tidal-ravinement surface (Goose Point at Willapa Bay, Washington, USA).

FIGURE 2.48 *Glossifungites* Ichnofacies (images courtesy of James MacEachern; Late Albian Viking Formation, Alberta, Canada). A—Tidal-ravinement surface (TRS) reworking a subaerial unconformity (SU), at the limit between underlying shelf mudstones and the overlying mouth complex of a wave-dominated estuary. The shelf mudstone hosts firmground *Rhizocorallium* (*Rh*) of the *Glossifungites* Ichnofacies. B—Wave-ravinement surface (WRS) reworking a subaerial unconformity (SU) in an interfluve location, at the top of distal inner shelf mudstones. The upper portion of the mudstone has been siderite cemented. The discontinuity is demarcated by firmground *Diplocraterion habichi* (*Dh*) and *Skolithos* (*Sk*) of the *Glossifungites* Ichnofacies, and it is overlain by a widespread conglomeratic transgressive lag.

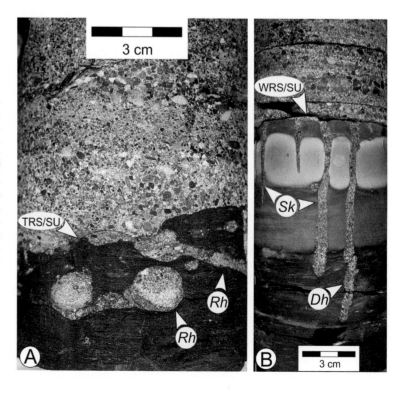

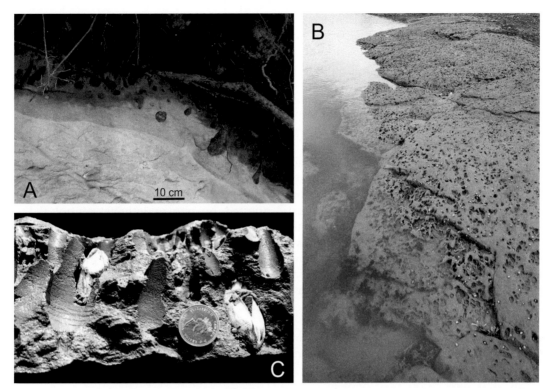

FIGURE 2.49 *Trypanites* Ichnofacies (images courtesy of Murray Gingras). A—Large *Gastrochaenolites* traces, which are borings made by pholad bivalves into the base of a Pleistocene tidal channel. The channel fill consists of organic-rich, unconsolidated sediment (dark color in the photograph). The underlying rock is a Miocene shoreface that belongs to the Empire Formation at Coos Bay, Oregon. The base of the channel corresponds to a transgressive tidal-ravinement surface. B—Modern intertidal environment. The traces shown are *Gastrochaenolites*. The hard-ground occurs as a scour cut into Triassic bedrock by tidal currents, and has the significance of a transgressive tidal-ravinement surface. Boring density may locally exceed 1250 borings per square meter. Location is near Economy, Nova Scotia (Bay of Fundy, Minas Basin). C—Modern intertidal environment (detail from B). The photograph shows the borings, the grooves cut by the bivalve (bioglyphs), and the tracemaker itself (*Zirfea pilsbyri*).

FIGURE 2.50 *Teredolites* Ichnofacies in a modern intertidal environment (Willapa Bay, Washington, USA; image courtesy of Murray Gingras). The borings are sand-filled, which provides their typical mode of preservation, and are made by the terenid bivalve *Bankia*. The woodground has the significance of a transgressive tidal-ravinement surface. The link between *Teredolites* and transgressive coastlines is generally valid for both *in situ* and allochthonous woodgrounds.

(freshwater) and *Cruziana* (marine water) assemblages; *Thalassinoides* may populate softground, firmground, or woodground substrates, etc (Fig. 2.32). Hence, the context and the association of individual traces, coupled with additional clues provided by body fossils, as well as physical textures and structures, need to be used in conjunction for the proper interpretation of both stratigraphic surfaces and paleodepositional environments.

In summary, the relevance of ichnology to sequence stratigraphy is twofold (MacEachern et al., 2012). Softground ichnofacies, which generally form in conformable successions, assist with the interpretation of paleodepositional environments and changes thereof with time. The vertical shifts in softground trace fossil suites are governed by the same Walther's Law that describes the principles of lateral and vertical facies variability in conformable successions of strata (Fig. 1.5), and therefore can be used to identify paleodepositional trends (progradation vs. retrogradation) in the rock record. The recognition of such trends, which in turn relate to the regressions and transgressions of paleoshorelines, is central to the sequence stratigraphic analysis.

Juxtaposition of trace fossil suites of nonadjacent ichnofacies (e.g., *Cruziana* Ichnofacies overlying *Scoyenia* Ichnofacies) may highlight the presence of a subtle discontinuity that has sequence stratigraphic significance.

Ichnofacies on resilient substrates (woodgrounds, firmgrounds, hardgrounds), which are genetically related to stratigraphic gaps and consist of omission suites (suites that record a break between deposition of a facies and its later, palimpsest colonization) assist with the identification of hiatal surfaces of different magnitudes in the rock record, and thus too have important applications for sequence stratigraphy. The actual meaning of the hiatal surface in sequence stratigraphic terms (e.g., a subaerial unconformity vs. a transgressive surface of erosion) can be evaluated further by looking at the nature and the direction of shift of the facies that are in contact across the omission surfaces. These aspects are discussed in more detail in Chapter 6, which deals with stratigraphic surfaces (see also Chapter 7 for a discussion of stratigraphic hiatuses that develop at different scales and hierarchical levels). As each individual method of data analysis may be equivocal to some extent, the integration of ichnology with conventional biostratigraphy and sedimentology provides the most reliable approach to facies analysis and sequence stratigraphy.

2.1.5 Geochemistry

The application of geochemical methods to constrain the sequence stratigraphic framework is generally underutilized, but is emerging as a powerful tool with great potential particularly in fine-grained successions where the conventional methods of facies analysis are less effective (e.g., Harris et al., 2013, 2018; Playter et al., 2018; LaGrange et al., 2020). Both organic and inorganic geochemistry have relevance to sequence stratigraphy and can be used to identify systems tracts and bounding surfaces in mud-dominated successions, and to reconstruct the history of relative sea-level changes during the evolution of a sedimentary basin (e.g., Slingerland et al., 1996; Harris et al., 2013, 2018; Dong et al., 2018).

2.1.5.1 Organic geochemistry

Early work on the variability of total organic carbon (TOC) in fine-grained successions established a general relationship between paleobathymetry and the amount of TOC in the sediment, leading to the conclusion that the higher TOC amounts correlate with stages of transgression and water deepening (Creaney and Passey, 1993). According to this model, the highest TOC is expected at maximum flooding surfaces, which mark

the end of transgressions, based on the idea that oxygen levels at the seafloor are controlled by water depth (i.e., as the water depth increases, so do the anoxia and the preservation potential of the organic matter that accumulates on the seafloor). Subsequent work demonstrated that in addition to syndepositional bathymetric conditions and oxygen levels, other factors such as nutrient supply to the water column and sediment supply to the seafloor can also influence significantly the amounts of TOC in the sedimentary record (e.g., Hay, 1995; Tyson, 2001; Arthur and Sageman, 2005; Katz, 2005). However, since all these factors can be affected by one underlying control (i.e., changes in relative sea level), the general conclusions of Creaney and Passey (1993) still stand, as they often match the observations in many case studies (e.g., Slingerland et al., 1996; Fleck et al., 2002; Arthur and Sageman, 2005; Bohacs et al., 2005; Turner et al., 2016; Dong et al., 2018; Harris et al., 2018).

In general terms, the accumulation and preservation of organic matter in organic-rich sediments require three main conditions (Dong et al., 2018):

1. Enhanced organic productivity (Pedersen and Calvert, 1990; Sageman et al., 2003; Tyson, 2001, 2005; Wei et al., 2012), which generally requires high nutrient supply to the water column (Hay, 1995). In turn, nutrient supply can be enhanced by several factors, including upwelling of bottom waters, river runoff, and nutrient cycling (Pedersen and Calvert, 1990; Arthur and Sageman, 2005; Algeo and Ingall, 2007), acting independently or in combination.
2. Enhanced preservation of organic matter (Demaison and Moore, 1980; Arthur and Sageman, 1994; Mort et al., 2007), which requires reducing conditions (i.e., oxygen deficiency) in the bottom waters.
3. Low sedimentation rate (Creaney and Passey, 1993; Tyson, 2005), which ensures a minimal dilution of the organic matter with detrital sediment. The effect of sediment supply on the accumulation and preservation of organic matter is, however, complex, as it can not only dilute the concentration of organic carbon (i.e., negative effect on the TOC amount) but also protect the organic matter on the seafloor from an oxygenated water column (i.e., positive effect on the TOC amount) (Tyson, 2001; Katz, 2005).

All three factors described above (i.e., productivity, preservation, and sediment supply) interact to constrain the TOC amounts found in organic-rich petroleum source rocks or unconventional petroleum reservoirs (Tyson and Pearson, 1991; Harris et al., 2004; Rimmer et al., 2004; Bohacs et al., 2005; Katz, 2005). Despite the complexity of the variables involved in the accumulation and preservation of organic matter, the stratigraphic analysis of the TOC in the sedimentary record is somewhat simplified by the fact that all three main factors have a common underlying control, namely the relative sea level. Case studies in different sedimentary basins ranging in age from Paleozoic to Holocene showed that relative sea-level changes can exert a direct control on bioproductivity, redox conditions, detrital sediment supply, and ultimately TOC concentrations (Fleck et al., 2002; Arthur and Sageman, 2005; Bohacs et al., 2005; Dong et al., 2018; Harris et al., 2018). For example, relative sea-level rise and transgression promote influxes of nutrient-rich upwelled water, which increase surface-water productivity; enhance water-column stratification, leading to the development of bottom-water anoxia; and result in the trapping of terrigenous sediment in coastal and nearshore depositional environments, which favors the development of organic-rich condensed sections offshore. Consequently, the TOC content is typically enriched in transgressive systems tracts, and is depleted in systems tracts that accumulate during regression (Dong et al., 2018; Harris et al., 2018).

Due to their common link to the relative sea level, the proxies for restriction of water masses (Mo/TOC), redox conditions (Mo/Al and S/Fe), and productivity of biogenic silica converge to the conclusion that TOC values are typically highest at or around maximum flooding surfaces (Harris et al., 2018). High-resolution geochemical data coupled with sedimentological data from core further document that the highest TOC values may in fact be recorded just above the highest rank maximum flooding surface within a studied section (Harris et al., 2018). This may be due to the fact that the horizon which corresponds to the deepest water at syndepositional time is often *above* the maximum flooding surface at any location (Catuneanu et al., 1998a; Zecchin et al., 2021; Figs. 2.28 and 2.29). At the other end of the spectrum, the lowest TOC values are most often recorded at times of lowstand in relative sea level (Fleck et al., 2002; Dong et al., 2018; Harris et al., 2018; Dominguez et al., 2020). Recent examples of TOC distribution within a sequence stratigraphic framework have been documented by Dominguez et al. (2020) and Reijenstein et al. (2020), who found the highest TOC values within transgressive systems tracts and the bottomsets of highstand systems tracts (Fig. 2.51).

2.1.5.2 Inorganic geochemistry

Besides TOC, inorganic geochemical proxies based on major, minor, and trace elements can also be used to build the sequence stratigraphic framework and to reconstruct changes in relative sea level (Harris et al., 2013, 2018; Dong et al., 2018; Playter et al., 2018; LaGrange et al., 2020). Basic observations that relate major element geochemistry to relative sea level include the aluminum enrichment in the deep-water setting at times of lowstand (i.e., due to the increase in clay delivery

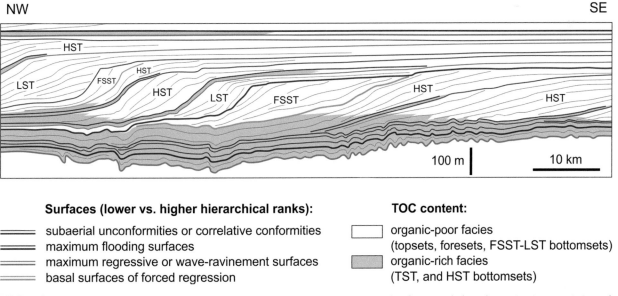

Surfaces (lower vs. higher hierarchical ranks):
- ═══ subaerial unconformities or correlative conformities
- ═══ maximum flooding surfaces
- ═══ maximum regressive or wave-ravinement surfaces
- ═══ basal surfaces of forced regression

TOC content:
- ☐ organic-poor facies (topsets, foresets, FSST-LST bottomsets)
- ▨ organic-rich facies (TST, and HST bottomsets)

FIGURE 2.51 Distribution of total organic carbon (TOC) within a sequence stratigraphic framework, based on seismic acoustic impedance data calibrated with well data (Quintuco–Vaca Muerta system, Tithonian–Lower Valanginian, Neuquén Basin, Argentina; modified after Dominguez et al., 2016, 2020, and Reijenstein et al., 2020). The highest TOC amounts are found within transgressive systems tracts and the bottomsets of highstand systems tracts (minimum dilution with detrital sediment). Sequence stratigraphic surfaces are shown at two hierarchical levels; thicker lines indicate surfaces of higher hierarchical rank. Abbreviations: FSST—falling-stage systems tract; LST—lowstand systems tract; TST—transgressive systems tract; HST—highstand systems tract.

during 'lowstand shedding'), and the increase in carbonate delivery to the deep water during 'highstand shedding' (Harris et al., 2018; Dong et al., 2018). These observations can be further constrained and calibrated with paired studies of detrital and clay proxies which, when used in conjunction, can demonstrate changes in relative sea level (Li et al., 2000; Playter et al., 2018). Within the context of each sedimentary basin, the proxies used for detrital input (e.g., Sc/Zn and Lu/Hf, higher during relative sea-level fall) and clay input (e.g., Zr/La, higher during relative sea-level rise and particularly during transgression) provide information about the basin-scale relative sea level rather than the global sea level (Playter et al., 2018). Documentation of trace element renewal (e.g., Mo) in partially restricted basins can also be related to the relative sea level, as communication between the restricted basin and the open ocean increases during stages of relative sea-level rise (Harris et al., 2013; Turner et al., 2016).

Challenges in identifying stratigraphic cyclicity in lithologically monotonous mudstone successions make the integration of chemostratigraphic data with conventional facies analyses increasingly common (LaGrange et al., 2020). Chemostratigraphy uses inorganic geochemical data, such as stable isotope signatures or elemental ratios, to define and correlate stratigraphic units (Pearce et al., 1999; Weissert et al., 2008). A recent review of the elements and elemental ratios that can be used as chemostratigraphic proxies for the sequence stratigraphic analysis of mudstone successions was provided by LaGrange et al. (2020). Among the elemental proxies that are most relevant to sequence stratigraphy, lows in terrigenous proxies (Ti/Al), minima in grain size proxies (Si/Al, Zr/Al, Zr/Nb, Zr/Rb, Y/Al, Y/Rb), and the increase in proxies for biogenic silica (Si/Al) are found to be closely associated with maximum flooding surfaces (Ratcliffe et al., 2012; Sano et al., 2013; Turner et al., 2016; Harris et al., 2018). At the opposite end of the scale, maximum regressive surfaces are characterized by peaks in terrigenous and grain size proxies, and minima in proxies for biogenic silica (Ratcliffe et al., 2012; Sano et al., 2013; Harris et al., 2018).

The correct interpretation of elemental proxies requires that elemental data are calibrated with mineralogical data, to ensure the proper usage of the former for the purposes of sequence stratigraphy (e.g., to discern between the depositional vs. diagenetic relevance of elemental data; LaGrange et al., 2020). So far, the most reliable application of geochemical proxies is for the identification of maximum flooding and maximum regressive surfaces, which delineate transgressive–regressive cycles (LaGrange et al., 2020). Other surfaces that form in deep-water settings (i.e., the basal surface of forced regression and the correlative conformity) are not as easily identifiable with geochemical data, but can be

identified based on different criteria (e.g., petrography of framework grains, or the geometry of seismic reflections; Catuneanu, 2020a; details in Chapter 6). Therefore, the integration of independent datasets always provides the best and most reliable approach to the identification of all elements of the sequence stratigraphic framework. For example, the position of a maximum flooding surface is best constrained where geochemical proxies are calibrated with biostratigraphic data (e.g., the highest abundance of microfossils and the highest ratio between distal and proximal benthic foraminifera within a condensed section; Gutierrez Paredes et al., 2017; Zecchin et al., 2021) and seismic stratigraphy (e.g., tracing downlap surfaces into the depocenters).

Much work is still needed to refine the geochemical criteria that afford the construction of a sequence stratigraphic framework, and integration with other methods such as sedimentary petrography, ichnology, and biostratigraphy can improve the reliability of interpretations. Furthermore, the geochemical proxies in the toolbox of sequence stratigraphy need to be adjusted to the *stratigraphic age* of the basin under analysis, as the chemistry of oceans and atmosphere, with impact on all geochemical parameters, changed over time (see discussion in Chapter 9). Additional variability is introduced by the *tectonic setting*, as water circulation and chemistry in restricted basins differ from the conditions in the open ocean, and by the *climatic regime* which modifies surfaces processes and associated detrital and clay proxies (Ruffell et al., 2002a,b).

2.1.6 Age dating

Age dating refers to the evaluation of the relative or absolute geological age of strata by means of biotic content (i.e., biostratigraphy) or physical properties such as magnetic polarity (i.e., magnetostratigraphy) or the relative abundance of parent/daughter radioactive isotopes (i.e., radiochronology). In addition, chronostratigraphic markers such as volcanic ash beds (Fig. 2.52) provide physical timelines that serve as reference horizons within the stratigraphic succession. The acquisition of age data is always desirable, and useful to constrain correlations at larger scales. However, the lack thereof does not prevent the application of sequence stratigraphy, which is based on the identification of stratal stacking patterns in the sedimentary record (Fig. 1.1).

Most significant to the accuracy of correlations, the resolution of the various dating techniques varies with the method, as well as with the age of strata under analysis. The resolution of age dating generally decreases with the increase in stratigraphic age, due to a number of factors including facies preservation, deformation,

FIGURE 2.52 Bentonite beds in a shallow-marine shelf succession (Late Cretaceous Bearpaw Formation; St. Mary River, Alberta, Western Canada Basin). Such bentonites extend for tens to hundreds of kilometers in the subsurface, providing chronostratigraphic horizons (timelines) which can be dated with radiometric methods.

diagenesis, metamorphism, and evolution of life forms. At the high-resolution end of the spectrum, radiocarbon dating of recent deposits has an error margin of only decades or centuries (Reimer et al., 2004), and infra-zonal stratigraphy in the Phanerozoic may allow subdivisions on millennial scales based on paleontological markers (Gladenkov, 2010). At the low-resolution end of the spectrum, the available age constraints for Precambrian successions may rely entirely on radiometric methods with error margins of millions or tens of millions of years (Dalrymple, 1994; Oberthur et al., 2002). However, even in the near-absence of chronological constraints, sequence stratigraphic frameworks can still be constructed based on a good knowledge of the paleoenvironments and facies relationships within the basin (Christie-Blick et al., 1988; Beukes and Cairncross, 1991; Krapez, 1993, 1996, 1997; Catuneanu and Eriksson, 1999, 2002; Eriksson and Catuneanu, 2004a; Fig. 1.1).

2.1.6.1 Biostratigraphy

Biostratigraphy is a type of stratigraphy that uses the fossil content of sedimentary strata for the subdivision of stratigraphic successions, correlations, and the assignment of relative ages. Early efforts to subdivide the stratigraphic record on the basis of fossils led to the introduction of the concepts of biological stage, as a unit defined by a major group of organisms (d'Orbigny, 1842), and biological zone (biozone), as an interval defined by a specific fossil taxon (Oppel, 1856). In that context, biozones served as subdivisions of stages, representing the first step toward refining the biostratigraphic resolution. Over the past century and a half, work focused on further increasing the resolution of biostratigraphy in order to improve the understanding of evolutionary lineages, and fine tune stratigraphic

schemes and correlations within and between sedimentary basins. In this process, the concept of 'zone' diversified in a deregulated manner, leading to inconsistencies in usage and interpretations. This confusion is also reflected in discrepancies between the definitions and recommendations promoted by the various national and international stratigraphic guides (Gladenkov, 2010).

Several types of biostratigraphic zones have been defined (Gladenkov, 2010): taxon-range zones (biozones, defined by the lifespans of specific taxa), concurrent-range zones (defined by the association of two taxa), lineage zones (segments of a specific phylogenetic line), interval zones (between the first occurrences of two partially overlapping taxa), abundance zones (defined by the maximum abundance of specific taxa), assemblage zones (defined by assemblages of three or more taxa), and ecozones (defined by the lifespan of specific ecological communities). While these types of zones have different meanings, they are often mixed in the stratigraphic schemes that define the biotic composition of various sedimentary basins. For example, the diatom zonal scheme of the upper Oligocene—Quaternary interval in the North Pacific includes interval zones, lineage zones, taxon-range zones, and concurrent-range zones (Barron and Gladenkov, 1995; Gladenkov, 2007). To some extent, these inconsistencies reflect attempts to provide the highest stratigraphic resolution, no matter what type of zone is most useful for a particular stage in the evolution of a sedimentary basin.

There are no standards for the duration of biostratigraphic zones, which in average range between 0.5—3 My (Gladenkov, 2010), depending on stratigraphic age and the types of fossils that define the zones. The types of zones used to develop stratigraphic schemes also depend on the depositional setting. Continental sections are typically subdivided using assemblage zones (Nichols and Sweet, 1993), and to a lesser extent lineage zones, ecozones, and abundance zones (Gladenkov, 2010). In contrast, marine sections are most commonly subdivided using interval zones, and to a lesser extent taxon-range zones, lineage zones, and concurrent-range zones (Barron and Gladenkov, 1995; Gladenkov, 2007). Therefore, different approaches are employed to develop stratigraphic schemes for continental and marine settings, and difficulties often arise in the correlation of these schemes. Correlations and timescales are further complicated by the fact that the boundaries of biostratigraphic zones are often diachronous, in response to climatic and ecological controls, and related population migrations. In some cases, the diachroneity of zone boundaries may exceed 0.5 My between tropical and boreal belts (Gladenkov, 2010), which, depending on the resolution of the stratigraphic study, may be unacceptable for chronostratigraphic correlations.

The increase in the resolution of stratigraphic schemes, which is most relevant to sequence stratigraphy, is at the forefront of biostratigraphic work. Biostratigraphic zones may correspond to 0.5 My (Cretaceous ammonite zonation in the Western Canada Basin; Obradovich, 1993), 1 My (upper Cretaceous nonmarine palynology in the Western Interior Basin of North America; Nichols and Sweet, 1993), 1.25 My (average diatoms zonation in the North Pacific during the Miocene—Quaternary interval; Barron and Gladenkov, 1995; Gladenkov, 2007), 1.3 My (average zonation of Miocene planktonic foraminifera; Bolli and Saunders, 1985), 1.5 My (average zonation of Miocene calcareous nannoplankton; Martini, 1971; Berggren et al., 1995), or 2 My (Permo-Triassic vertebrate fossils in the Karoo Basin; Rubidge, 1995). This precision is usually sufficient for the low-resolution seismic stratigraphy, but is insufficient for the high-resolution sequence stratigraphy at sub-seismic scales. Efforts to improve resolution include the subdivision of biostratigraphic zones into subzones by using paleontological markers ('bioevents') such as the first and last occurrences of taxa, distinctive occurrences, and evolutionary changes. For example, ostracod bioevents defined by first and last occurrences during the Burdigalian in the Kerala Basin of India established subzones with an average duration of 0.5 My (Bhandari, 2003). Exceptionally, subzones can record durations of only a few millennia (Gladenkov, 2010).

2.1.6.2 Magnetostratigraphy

Magnetostratigraphy is a type of stratigraphy in which stratigraphic divisions and correlations are made on the basis of magnetic properties of the rocks (e.g., magnetic polarity and susceptibility; Opdyke and Channell, 1996). What enables the delineation of magnetostratigraphic units is the fact that the Earth's magnetic field varied in its direction and intensity through time, making thus possible the definition of intervals with different magnetic properties. The dipole component of the Earth's magnetic field, which accounts for c. 90% of the observed field at the surface, has an irregular drift around the Earth's rotational axis, with an average cyclicity of 10,000 yrs (Ogg, 2012). The polarity of the dipole magnetic field can also reverse with irregular periodicity. The durations of polarity intervals vary from as little as 20—30 kyr to tens of millions of years, with an average frequency of geomagnetic reversals of 2 or 3 per My during the Cenozoic (Langereis et al., 2010; Ogg, 2012). Since 35 Ma, the mean duration of the polarity intervals was c. 300 kyr, with the last reversal recorded at 780 ka (Langereis et al., 2010). There were also long periods of time in the geological past during which the geomagnetic field did not reverse, such as the mid-Cretaceous 'normal quiet zone' that lasted for c. 40 My (124.5—84 Ma; Langereis et al., 2010).

Past magnetic fields are recorded by the orientation of iron-bearing minerals at the time of sedimentation, or at the time of cooling of volcanic rocks below the Curie temperature of their magnetic minerals. This process 'locks' a remanent magnetization within the rocks, potentially for billions of years as long as the rocks remain below the Curie points. Any overprinting by younger magnetic fields is typically weaker than the original magnetization, and can be removed by stepwise demagnetization before the primary paleomagnetic field is measured. By convention, the present-day polarity of the magnetic field is 'normal' (i.e., north-directed), whereas the south-directed lines of magnetic force define a 'reversed' polarity. The series of past reversals in the polarity of the Earth's magnetic field define the chronostratigraphic 'bar code' of alternating normal (black) and reverse (white) polarity chrons in the geological timescale, which was calibrated with radiometric methods and orbital tuning (Lourens et al., 2004). The unique pattern of this 'bar code' provides a reference geomagnetic polarity timescale (GPTS) relative to which the polarity reversal patterns obtained from field studies can be compared in order to infer the age of the rock successions.

Despite general trends that have been observed (e.g., the overall increase in the frequency of polarity reversals over the last 80 My, from approximately 1 reversal per My to 5 reversals per My in recent times; Langereis et al., 2010), the character of polarity reversals remains essentially random. It is this randomness that (1) defines the unique signature of the GPTS bar code, and (2) confers stratigraphic value to a measured series of polarity reversals in a conformable succession. This dating tool provided by magnetostratigraphy and the correlation to the GPTS applies to a variety of rock types, both sedimentary and volcanic, accumulated in a broad range of settings, from continental to lacustrine and marine (Langereis et al., 2010). However, matching the field data with the patterns of the reference scale can be equivocal, especially for sections that include unconformities. For this reason, calibration with independent age-dating techniques, such as biostratigraphy or radiochronology, is always recommended to ensure the accurate identification of specific polarity chrons. For example, the magnetostratigraphic study of continental sections in the Alberta Basin was done in conjunction with palynology; this proved to be an effective combination for integrated stratigraphic studies, which enabled a resolution of 0.4—0.5 My for the Maastrichtian—Paleocene interval (Lerbekmo et al., 1995; Lerbekmo and Sweet, 2000, 2008).

The resolution of magnetostratigraphy can be improved further by subdividing the polarity chrons into subchrons with durations as short as 20—30 kyr (Langereis et al., 2010). Smaller scale variations may also exist, referred to as 'tiny wiggles' (LaBrecque et al., 1977), 'cryptochrons' (Cande and Kent, 1992), or 'reversal excursions' (Merrill and McFadden, 1994) on timescales of 3—6 kyr (Langereis et al., 1997). However, these variations are associated with low paleointensities of the geomagnetic field, and are too short to be detected in most field studies; hence, they are not reliable for magnetostratigraphic correlations (Roberts and Winkelhofer, 2004). There is also a limit to the resolution of magnetostratigraphy, which is set by the period of time required to generate a polarity reversal; this quantifies the error margin of the methodology. Polarity boundaries are approximated as globally synchronous timelines in the GPTS, but in fact they take thousands of years to form, during which time the magnetic field is sign invariant (Langereis et al., 2010). Therefore, the highest resolution that can be achieved by means of magnetostratigraphy is practically in a range of 10^1 kyr. This is sufficient to support correlations in lower resolution studies (e.g., seismic stratigraphy), but too coarse for high-resolution studies that document stratigraphic sequences on millennial scales (e.g., Holocene studies, where radiocarbon dating provides a far better alternative; Amorosi et al., 2009, 2017).

2.1.6.3 Radiochronology

Radiometric dating is a technique used to determine the absolute age of materials such as rocks or carbon in which trace radioactive isotopes were incorporated at the time of formation. The basic condition for reliable dating is that neither the parent nuclide nor the daughter product can enter or leave the host material after its formation. Ensuring that this condition is met usually requires the dating of multiple samples from different locations of the rock unit, in order to demonstrate the consistency of results. Alternatively, wherever possible, it is recommended to apply different radiometric methods to the same sample, for mutual validation (e.g., the age of Archean rocks in Greenland was dated as 3.60 ± 0.05 Ga with uranium—lead, and as 3.56 ± 0.10 Ga with lead—lead, which is sufficiently consistent to validate the results; Dalrymple, 1994).

Radiochronology is fundamental to the absolute dating of the stratigraphic subdivisions of the geological timescale (e.g., Gradstein et al., 2012); however, these dates are subject to ongoing revisions and updates as the accuracy of the technique continues to improve. Several methods of radiometric dating are available, which differ in terms of the timescales (e.g., Precambrian vs. Quaternary) and the materials (e.g., 'whole-rock' samples, mineral samples, or carbon) to which they apply. Modern radiometric dating methods include uranium—lead, uranium—thorium, rubidium—strontium, potassium—argon, argon—argon (which supersedes potassium—argon in accuracy and ease of use), radiocarbon, fission track, and luminescence. Among the most utilized techniques are the radiocarbon dating, potassium—argon, and uranium—lead dating. Fission track and luminescence dating differ from the 'conventional' radiometric methods in the sense that they do not rely on parent/daughter isotope ratios. The fission track method measures the density of 'track' markings left by the spontaneous fission of uranium-238 isotopes on minerals or tektites, and can be applied to a wide range of geological ages. However, the tracks are 'healed' by temperatures over 200 C, which is a limitation of the method. Luminescence dating is based on the level of ionizing radiation absorbed by grains such as quartz and potassium feldspar in the rocks due to the background radiation emitted by radioactive elements. However, exposure to sunlight or heat de-ionizes the sample, resetting the clock to zero.

Accurate results of the 'conventional' radiometric methods generally require that the half-life of the parent isotope is long enough to ensure that significant amounts are still present at the time of the measurement, and that enough daughter material is produced to be measured accurately. For this reason, radiometric methods based on isotopes with different half-lives have different ranges of applicability, some best suited for relatively recent deposits (e.g., radiocarbon dating is effective for the last 60,000 years, due to the short half-life of carbon-14), others better suited for Precambrian deposits (e.g., rubidium—strontium, potassium—argon, uranium—lead). The precision of the dating methods also depends on the half-life of the parent isotope (i.e., the shorter the half-life, the higher the precision), from decades in the case of radiocarbon (Reimer et al., 2004) to millions or tens of millions of years in the case of uranium—lead (Dalrymple, 1994;

Oberthur et al., 2002; Manyeruke et al., 2004). However, the precision of the method is not simply inversely proportional to the half-life of the isotope, but it also depends on the match between half-life and the stratigraphic age of the rocks under analysis. For example, the uranium—lead method has an error margin of less than 0.1% for Proterozoic (Oberthur et al., 2002; Manyeruke et al., 2004), and 2%—5% for Mesozoic (Li et al., 2001), due to the long half-lives of ^{238}U (4.47 Ga) and ^{235}U (710 Ma), which exceed the duration of the Phanerozoic.

2.2 Well-log data

2.2.1 Introduction

Well logs represent geophysical recordings of various physical properties of the rocks encountered in boreholes, and can be used for geological interpretations. The most common log types that are routinely employed for facies analyses (lithology, porosity, fluid evaluation) and stratigraphic correlations are summarized in Fig. 2.53. Most of these log types may be considered 'conventional', as having been used for decades, but as technology improves, new types of logs have been developed. For example, the micro-resistivity logs combine the methods of conventional resistivity and dipmeter measurements to produce high-resolution

images that simulate the sedimentological details of actual mechanical cores. Such 'virtual' cores allow visualization of details at a millimeter scale, including sediment lamination, cross-stratification, bioturbation, etc., in three dimensions (Fig. 2.54). Therefore, the micro-resistivity logs afford insights into process sedimentology and ichnology even in the absence of mechanical core data. The continuous micro-resistivity imaging of boreholes eliminates the cost of mechanical coring and provides valuable information that can be used in conjunction with the motifs of conventional petrophysical logs and the lithological constraints provided by rock cuttings.

Well logs present both advantages and limitations relative to what outcrops can offer in terms of facies data. One major advantage of geophysical logs over outcrops is that they provide *continuous* information from relatively thick successions, often in a range of kilometers. This type of logging allows one to observe trends at various scales, from individual depositional elements within a depositional system to entire basin fills. For this reason, data provided by well logs may be considered more complete relative to the *discontinuous* information that may be extracted from the study of outcrops. Therefore, the subsurface investigations of facies relationships and stratigraphic correlations can usually be accomplished at scales much larger than the ones possible from the study of outcrops. On the other hand, nothing

Well log	Property measured	Units	Geological information
Spontaneous potential	Natural electric potential (relative to drilling mud)	Millivolts	Lithology, correlation, curve shape analysis, porosity
Conventional resistivity	Resistance to electric current flow (1D)	Ohm-metres	Identification of coal, bentonites, fluid types
Micro resistivity	Resistance to electric current flow (3D)	Ohm-metres and degrees	Borehole imaging, virtual core
Gamma ray	Natural radioactivity (e.g., related to K, Th, U)	API units	Lithology (including bentonites, coal), correlation, shape analysis
Sonic	Velocity of compressional sound waves	Microseconds/metre	Identification of porous zones, tightly cemented zones, coal
Neutron	Hydrogen concentration in pores (water, hydrocarbons)	Percent porosity	Porous zones, cross plots with sonic and density for lithology
Density	Bulk density (electron density) (includes pore fluid in measurement)	Kilograms per cubic metre (g/cm^3)	Lithologies such as evaporites and compact carbonates
Dipmeter	Orientation of dipping surfaces by resistivity changes	Degrees (azimuth and inclination)	Paleoflow directions (in oriented core), structural analyses
Caliper	Borehole diameter	Centimetres	Borehole state, reliability of logs

FIGURE 2.53 Types of well logs, properties they measure, and their use for geological analysis (modified from Cant, 1992).

FIGURE 2.54 Micro-resistivity logs combine resistivity with dipmeter data to produce 3D 'virtual cores'. Such detailed borehole imaging, with a vertical resolution of less than 8 mm, allows the observation of sedimentary structures in the absence of mechanical core (modified from data provided by Baker Atlas).

can replace the study of the actual sedimentary strata; hence, the wealth of details that can be obtained from outcrop facies analysis cannot be matched by well-log analysis, no matter how closely spaced the boreholes may be (Cant, 1992).

2.2.2 Geological uncertainties

Well logs describe physical rock properties which can be used to *infer*, rather than observe directly, geological information such as lithology, fluid content, or organic content. Spontaneous potential and gamma-ray logs are commonly used for the interpretation of siliciclastic successions in lithological terms, but one must remain aware of the potential pitfalls that may affect the interpretations. Changes in rock porosity and pore-water chemistry (fresh vs. sea water) may cause different responses on spontaneous potential logs, including deflections in opposite directions, even if the lithology is the same. Similarly, gamma-ray logs are often interpreted in grading terms (fining- vs. coarsening-upward), bathymetric terms (deepening- vs. shallowing-upward trends), or TOC content. In reality, gamma-ray logs simply indicate the degree of strata radioactivity which, even though it often correlates to the shaliness of the rocks and/or the amount of organic matter contained, still needs to be calibrated at least with the rock cuttings in order to validate the interpretations.

For example, sands derived from granitic sources may be radioactive, and so, without calibration with the actual rocks, may be confused with organic-rich shales.

Zones of high gamma-ray response may be found in a variety of depositional settings, from shelf and deep-marine to coastal plains, backshore marshes, and lacustrine environments. In fully subaqueous settings (marine or lacustrine), high gamma-ray responses may correlate to periods of restricted bottom-water circulation and/or with times of reduced sediment supply that enhance the TOC content. Such periods favor the formation of condensed sections, which commonly are associated with stages of shoreline transgression and maximum flooding surfaces (Galloway, 1989). However, due to the wide variety of environments which may result in the accumulation and preservation of organic matter and/or fine-grained sediment, the mere identification of high gamma-ray zones is not sufficient to unequivocally identify condensed sections (Posamentier and Allen, 1999). At the same time, condensed sections may also be marked by a variety of chemical and biochemical precipitates formed during times of sediment starvation (e.g., siderite, glauconite, carbonate hardgrounds, etc.), thus exhibiting a wide range of log motifs which may not necessarily fit the classic high peaks on gamma-ray logs (Posamentier and Allen, 1999).

The equivocal character of well logs, when it comes to geological interpretations, is also exemplified by the fact that different depositional systems may produce similar log motifs. Fig. 2.55 illustrates an example where comparable blocky log patterns formed in fluvial, estuarine, beach, shallow-marine, and deep-marine environments. Similarly, jagged log patterns are not diagnostic of any particular depositional system, and may be found all the way from fluvial to coastal (e.g., delta plains), shelf (particularly the inner shelf), and deep-water (slope and basin-floor) settings (Fig. 2.56). Such jagged log motifs simply indicate fluctuating energy conditions that lead to the deposition of alternating coarser and finer sediments (i.e., heterolithic facies), conditions which can be met in many depositional settings. Monotonous successions dominated by fine-grained sediments may also be common among different depositional systems, including deep-water 'overbanks' (areas of seafloor situated outside of channel-levee complexes or splay elements) and outer shelf settings (Fig. 2.57).

At the same time, one depositional system may produce different well-log signatures as a function of local variations in the sedimentation conditions (e.g., environmental energy, sediment supply, and landscape or seascape gradients). For example, sand-bed meandering fluvial systems consist typically of alternating fining upward channel fills, coarsening-upward crevasse splays, and mud-dominated floodplain

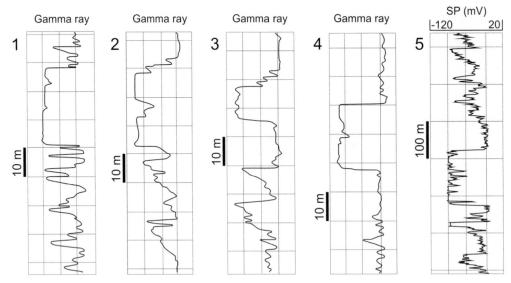

FIGURE 2.55   Well logs from five different siliciclastic depositional systems, each including a 'blocky' sandstone unit: 1—amalgamated fluvial channels; 2—estuary mouth complex; 3—sharp-based shoreface deposits; 4—deep-water channel filled with turbidites; and 5—beach deposits (modified from Posamentier and Allen, 1999, and Catuneanu et al., 2003). The reconstruction of paleodepositional environments requires integration of geological and geophysical data. Abbreviations: SP—spontaneous potential; mV—millivolts.

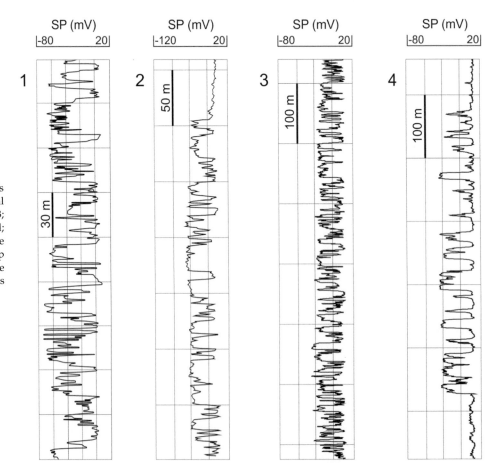

FIGURE 2.56   Jagged log motifs from different siliciclastic depositional systems (from Catuneanu et al., 2003; data courtesy of PEMEX): 1—fluvial; 2—delta plain; 3—inner shelf (above the storm wave base); and 4—deep water. Sand/mud ratio increases to the left. Abbreviations: SP—spontaneous potential; mV—millivolts.

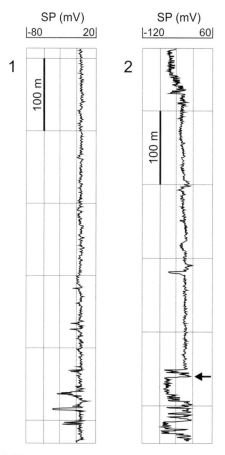

FIGURE 2.57 Fine-grained successions in different depositional settings (from Catuneanu et al., 2003; data courtesy of PEMEX): 1—deep-water basin floor; and 2—outer shelf (below the storm wave base, but above the shelf edge). The outer-shelf fines overlie shallower clinoforms (shoreface—inner shelf) and coastal deposits. The association of facies aids in the reconstruction of paleodepositional environments. The arrow indicates a flooding event on the shelf. Abbreviations: SP—spontaneous potential; mV—millivolts.

with time toward the location of the well may generate a coarsening-upward trend), and the actual depositional elements intercepted by the wells (e.g., channels, levees, splays, overbank) (Fig. 2.60).

Log patterns are therefore diverse, generally indicative of changing energy regimes through time but not necessarily diagnostic for any particular depositional system or architectural element. An entire range of log motifs has been described (e.g., Allen, 1975; Selley, 1978b; Anderson et al., 1982; Serra and Abbott, 1982; Snedden, 1984; Rider, 1990; Cant, 1992; Galloway and Hobday, 1996), but the most commonly recurring patterns include 'blocky' (also referred to as 'cylindrical'), 'jagged' (also referred to as 'irregular' or 'serrated'), 'fining-upward' (also referred to as 'bell-shaped') and 'coarsening-upward' (also known as 'funnel-shaped') (Cant, 1992; Posamentier and Allen, 1999). The blocky pattern generally implies a constant energy level (high in clastic systems and low in carbonate environments) and constant sediment supply and sedimentation rates. The jagged motif indicates alternating high and low energy levels, such as seasonal flooding in a fluvial system, spring tides and storm surges in a coastal setting, storms vs. fairweather in an inner shelf setting, or gravity flows vs. pelagic fallout in deep-water environments. Fining-upward trends can form in virtually any depositional environment, where there is a decline with time in energy levels. Conversely, coarsening-upward trends indicate an increase in energy levels through time.

It has been argued that the coarsening-upward trend is the least equivocal of all, especially for repeated sections 5–30 m thick (Posamentier and Allen, 1999). Typical examples of depositional elements that generate this log motif include prograding distributary mouth bars in deltaic settings and prograding shoreface deposits in open shoreline settings (Fig. 2.61). However, a similar log pattern may also characterize crevasse splays in fluvial settings (generally <5 m thick; Fig. 2.59), and even gravity-flow deposits in deep-water environments, such as at the distal edge of a seaward-building turbidite splay (Posamentier and Allen, 1999; Fig. 2.60). In the latter case though, the coarsening-upward trend is closely associated with a jagged log motif, which provides an additional clue for the identification of the deep-water setting (Fig. 2.60). Ultimately, well-log motifs need to be placed in the correct tectonic and depositional settings (e.g., regional 2D seismic lines can clarify whether the observed stratigraphic section accumulated above or below the shelf edge at syndepositional time, in the case of shelf-slope systems), and require calibration with independent datasets (e.g., information derived from core and rock cuttings) to ensure realistic interpretations of depositional systems and depositional elements.

deposits; braided fluvial systems are often composed of amalgamated channel fills, which appear 'blocky' on well logs; in contrast, other types of rivers, including fine-grained meandering or flashy ephemeral, can produce a more irregular, jagged type of motif on well logs (Figs. 2.58 and 2.59). Grading trends also depend on the scale of observation. Both fining- and coarsening-upward trends can develop at the smaller (±meter) scales of individual depositional elements, whereas fluvial successions observed at the larger stratigraphic scales of depositional sequences display typically a fining-upward (energy declining) profile. Slope and basin-floor deep-water systems may also generate a variety of log motifs, most commonly jagged or blocky, but also fining- and coarsening-upward, depending on the sediment transport mechanism (contourites vs. gravity flows; types of gravity flows), the lateral shifts of depositional elements (e.g., a channel-levee system shifting

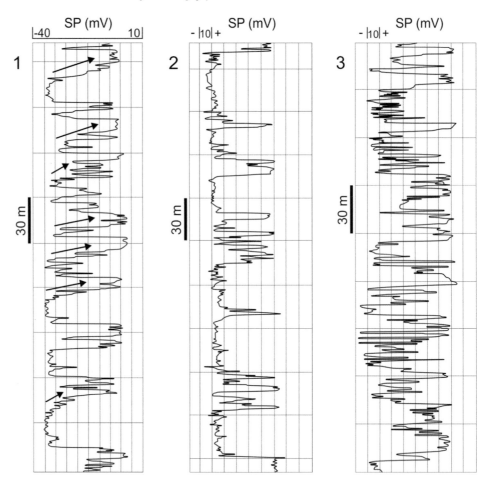

FIGURE 2.58 Log motifs in fluvial settings: 1—fining-upward trends; 2—blocky patterns; and 3—jagged pattern (modified from Posamentier and Allen, 1999). The sand/mud ratio increases to the left. Abbreviations: SP—spontaneous potential; mV—millivolts.

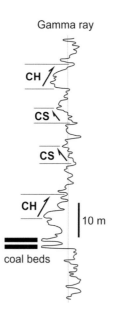

FIGURE 2.59 Log motifs in a low-energy fluvial system, including fining-upward (channel fills—CH) and coarsening-upward (crevasse splays—CS) trends (Maastrichtian, Horseshoe Canyon Formation, south-central Alberta, Western Canada Basin).

2.2.3 Well-log interpretations

The discussion in the previous section shows that the well-log interpretation of depositional systems, and implicitly of systems tracts and stratigraphic surfaces, is to a large extent speculative in the absence of actual rock data. Outcrop, core and well-cuttings data, along with the sedimentological, petrographic, biostratigraphic, ichnological, and geochemical analyses that they afford, provide the most unequivocal 'ground truth' information on depositional systems (Posamentier and Allen, 1999). It follows that geophysical data, including well-log and seismic, which provide only indirect information on the solid and fluid phases in the subsurface, must be calibrated with rock data in order to validate the accuracy of geological interpretations. Integration of all available datasets (e.g., outcrop, core, well cuttings, well-log, and seismic) is therefore the best approach to the correct identification of depositional systems, systems tracts, and bounding surfaces.

Well logs are generally widely available, especially in mature hydrocarbon exploration basins, and so they are

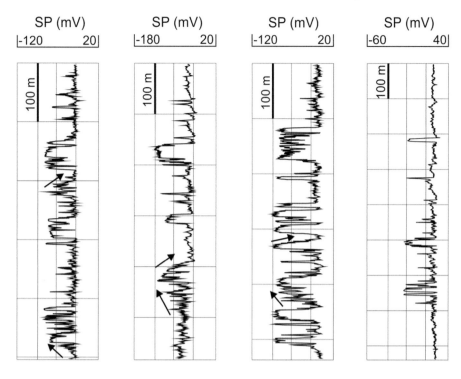

FIGURE 2.60 Log motifs in siliciclastic deep-water settings (the sand/mud ratio increases to the left) (from Catuneanu et al., 2003; data courtesy of PEMEX). Jagged and blocky patterns dominate, but fining- and coarsening-upward trends are also present. The sand-dominated units (grainflow deposits and turbidites) are interbedded with hemipelagic sediments. The identification of the types of gravity flows requires further analysis with 3D seismic data. Abbreviations: SP—spontaneous potential; mV—millivolts.

routinely used in stratigraphic studies. Seismic data are also available in most cases, as the seismic survey commonly precedes drilling. In the process of drilling, well cuttings are collected for information on lithology, porosity, fluid contents, and biostratigraphy (age and paleoecology). Mechanical coring is more expensive, so it is generally restricted to the levels of economic interest (e.g., petroleum reservoirs, unless the borehole is drilled for research or reference exploration purposes, in which case continuous mechanical coring may be performed). Nearby outcrops may be available in onshore basins, but are typically unavailable where drilling is conducted offshore. It can be concluded that, to a minimum, well logs can be analyzed in conjunction with seismic data and well cuttings, and to a lesser extent in combination with cores and outcrops. The calibration of petrophysical data with independent seismic and rock data is fundamental for a reliable geological interpretation of well logs. For example, two-dimensional (2D) seismic data provide critical insights into the tectonic setting (e.g., continental shelf vs. slope or basin floor) and the physiography of the basin. 3D seismic data provide additional information about the plan-view morphology of depositional systems or elements thereof. Such information, combined with any available rock data, helps to place the wells within the right tectonic and depositional context (see Chapter 9 for a discussion of the workflow of sequence stratigraphy).

The placement of a study area in the correct tectonic and depositional settings is crucial for the subsequent steps of well-log analysis. As noted by Posamentier and Allen (1999), 'correct identification of the depositional environment will guide which correlation style to use between wells. Thus, one style of correlation would be reasonable for prograding shoreface deposits, but a very different correlation style would be used for incised-valley-fill deposits, and still another style of correlation would be most reasonable for deep-water turbidites'. Well-log correlations are further aided by laterally extensive beds or groups of beds with a distinctive log response, which can be used as stratigraphic markers. Examples include bentonites and marine condensed sections, which (1) help to constrain correlations, and (2) have chronostratigraphic value (Fig. 2.52). Regional coal beds also serve as stratigraphic markers, although their chronostratigraphic significance needs to be assessed on a case-by-case basis. Coal beds that form at the time of maximum shoreline transgression represent the expression of maximum flooding surfaces within the continental realm and are essentially time lines (Hamilton and Tadros, 1994; Fanti and Catuneanu, 2010), whereas coals that originate in coastal settings during shoreline transgression or regression are commonly time transgressive (Banerjee et al., 1996; Bohacs and Suter, 1997).

An example of correlation style in a prograding shallow-water setting, with the base of the underlying condensed section taken as the datum, is presented in Fig. 2.62. This correlation accounts for the basic principles of stratal stacking in this setting, including the facts that clinoform surfaces slope seawards and downlap the underlying transgressive shales, and that

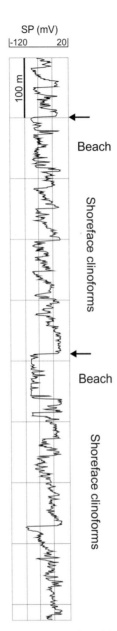

FIGURE 2.61 Coarsening-upward and blocky log motifs in a shallow-marine to coastal siliciclastic setting (from Catuneanu et al., 2003; data courtesy of PEMEX). Arrows indicate the most significant flooding events, but other episodes of abrupt water deepening are also recorded at the top of most clinoforms. Sand/mud ratio increases to the left. Abbreviations: SP—spontaneous potential; mV—millivolts.

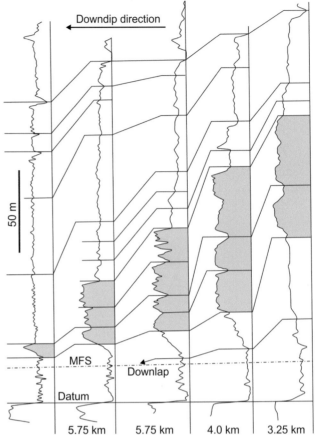

FIGURE 2.62 Gamma-ray cross-section showing correlation by pattern matching in a shallow-marine siliciclastic setting (Lower Cretaceous, Upper Mannville Group, British Columbia; redrafted from Cant, 1992, with permission from the Geological Association of Canada). The correlation lines mark clinoform surfaces which downlap onto the maximum flooding surface (MFS) in a downdip direction. Blue areas represent prograding shoreface sands.

clinoforms wedge out and become finer grained in a basinward direction. Without a good understanding of the depositional environment and the processes that operate within, a simple pattern matching may lead to errors in interpretation by forcing correlations across depositional time lines (clinoform surfaces in this example). Classic layer-cake correlations may still work where depositional energy and sediment supply are constant over large distances (e.g., in some distal basin-floor settings), but most environments tend to produce more complex stratigraphic architectures in

response to variations along dip and strike in energy levels and sedimentation patterns. In marginal-marine settings, for example, the selection of the best workflow and approach to correlation can be made within the paleogeographic context constrained with regional seismic data, including a knowledge of the depositional dip and strike directions: e.g., unraveling the architecture of a delta system must start with dip-oriented cross-sections, whereby the correlation is aided by the basinward plunge of clinoform surfaces (Fig. 2.63); once the component clinoforms are identified, the grid can be closed with strike-oriented cross-sections which may capture deltaic lobes wedging out in both directions (Fig. 2.64; e.g., Berg, 1982).

The analysis of well logs may serve several interrelated purposes including, with the increase in the scale of observation, the evaluation of solid and fluid phases in the subsurface, the interpretation of paleodepositional environments based on log motifs, and stratigraphic correlations based on pattern matching

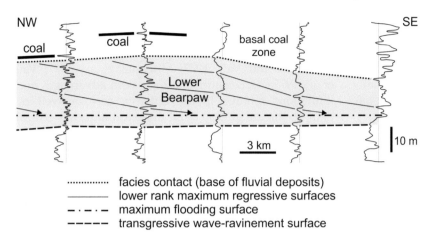

FIGURE 2.63 Dip-oriented gamma-ray cross-section through the shallow-marine facies of the Bearpaw Formation (blue area) in central Alberta (Upper Cretaceous, Western Canada Basin). The formation includes clinoforms prograding in a downdip direction and downlapping onto the maximum flooding surface. Each clinoform corresponds to a lower rank (higher frequency) transgressive–regressive cycle. The transgressive facies (fining-upward) are generally thinner than the regressive facies (coarsening-upward).

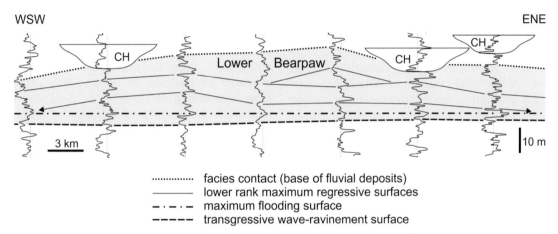

FIGURE 2.64 Strike-oriented gamma-ray cross-section through the shallow-marine facies of the Bearpaw Formation (blue area) in central Alberta (Upper Cretaceous, Western Canada Basin). Along strike, clinoforms may wedge out in either direction. For this reason, it is advised to start the stratigraphic analysis of shallow-water systems with dip lines, which afford a more precise correlation of clinoforms (Figs. 2.62, 2.63). Abbreviation: CH—fluvial channel.

and the recognition of marker beds. Different scales of observation may therefore be relevant to different objectives. Details at the smaller scales of individual depositional elements are commonly used for the petrophysical analysis of petroleum reservoirs (lithology, porosity, and fluid evaluation), regardless of the depositional origin of the stratigraphic unit. Such analyses are usually performed with cross plots that integrate information from two or more log types. This works particularly well where the succession is relatively homogeneous, consisting of only two or three rock types (e.g., mudstones, siltstones, and sandstones) (Miall, 2000). However, as the log motifs of individual depositional elements are generally nondiagnostic of the paleoenvironment, it is rather the larger-scale context within which these elements are observed that allows one to infer the original depositional setting (Posamentier and Allen, 1999). For example, blocky patterns can form in all environments, but they are commonly associated with fining-upward elements in fluvial settings, with coarsening-upward elements in coastal to

shallow-water settings, and with interbedded shales in deep-water settings (Fig. 2.61). But again, these conclusions need to be validated with seismic and rock data.

The validation of well-log cross-sections of correlation has a fundamental importance, and criteria for connecting the 'kicks' from one log to the next in ways that make most geological sense have been developed accordingly. For example, some basic 'rules' that apply to the correlation of shallow-marine successions include (Cant, 2004; Fig. 2.62): (1) prograding clinoforms always slope seaward; (2) shallow-marine regressive units tend to have lateral continuity along dip, and their number may only change along strike; (3) units tend to fine and thin seaward; (4) unit thicknesses do not vary randomly; where superimposed units show complementary thinning and thickening, the boundary between them is likely misplaced, unless tectonic inversion can be demonstrated; (5) strata may terminate landward by onlap, offlap, toplap, or truncation, and seaward by downlap (these types of stratal terminations are best seen on 2D seismic lines or in large-scale outcrops, and are reviewed in Chapter 4); and (6) where reasonable correlations cannot be made, the presence of an unconformity may be inferred—such contacts exert an important control on petroleum reservoirs, and may occur more frequently in the rock record than originally inferred at the low-resolution scales of seismic stratigraphy.

2.3 Seismic data

2.3.1 Introduction

Seismic data provide the means for the preliminary evaluation of a basin fill in the subsurface, usually prior to drilling, in terms of overall structure, stratigraphic architecture, and fluid content ('charge'). Seismic surveys are an integral part of hydrocarbon exploration, as they allow one to (1) assess the tectonic setting and the paleodepositional environments; (2) identify potential hydrocarbon traps (structural, stratigraphic, or combined); (3) evaluate potential reservoirs and seals; (4) evaluate source rocks and estimate petroleum charge in the basin; (5) evaluate the amount and the nature of fluids in individual reservoirs; (6) develop a strategy for exploration wells and subsequent production development; and (7) minimize the risk of petroleum exploration. The development of seismic exploration techniques prompted the emergence of seismic stratigraphy (the precursor of sequence stratigraphy—see Chapter 1 for

a discussion) in the 1970s (Vail, 1975; Payton, 1977), and led to the formulation of criteria for the interpretation of seismic data in sequence stratigraphic terms (Mitchum and Vail, 1977; Mitchum et al., 1977; Vail et al., 1977a,b). Seismic data have both advantages and limitations relative to the outcrop, core or well-log data, so the integration of all these techniques is critical for mutual calibration and the construction of reliable stratigraphic frameworks.

In the initial steps of a seismic survey, seismic data are collected along linear profiles, resulting in the acquisition of 2D (two-way travel time vs. horizontal distance) seismic lines. The information from a grid of 2D seismic lines can be interpolated to produce 3D seismic volumes (Brown, 1991; Fig. 2.65). Following the initial acquisition, the raw seismic information requires further processing (e.g., demultiplexing, gain recovery, static corrections, deconvolution, migration, etc.; Hart, 2000) before it is ready to be used for geological interpretations. Once available for analysis, the seismic lines provide continuous subsurface imaging over distances of tens of kilometers and depths in a range of kilometers. The continuous character of seismic data represents a major advantage of this method of stratigraphic analysis over well logs, core or outcrops, which only provide information from discrete locations within the basin. However, the seismic data are also subject to limitations, mainly in terms of vertical resolution (i.e., the thinnest package of strata that can be recognized on a seismic line

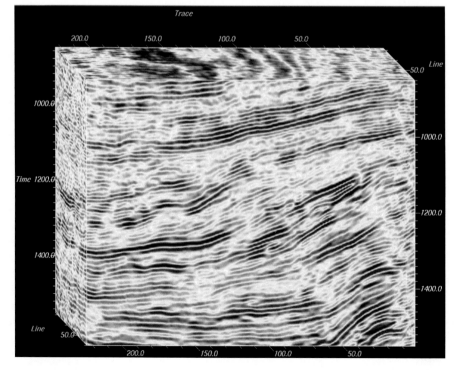

FIGURE 2.65 Three-dimensional seismic volume showing a prograding shelf-slope system in the Permian Basin, USA (from Hart, 2000; reprinted by permission of the Society for Sedimentary Geology). Seismic volumes can be sliced in different directions in order to observe depositional, stratigraphic, and structural features.

between two seismic reflections) and the nature of information (i.e., physical parameters as opposed to direct geology) that is represented on seismic lines.

2.3.2 Physical attributes of seismic data

The makeup of a seismic image reflects the interaction between the substrate geology and the seismic waves traveling through the rocks, modulated by the physical properties of the rocks. The seismic waves emitted by a source at the surface are characterized by specific physical attributes, including shape (wave form as depicted by a seismograph), polarity (direction of main deflection), frequency (number of complete oscillations per second), and amplitude (magnitude of deflection, proportional to the energy released by source). Except for frequency, which depends on the source of the seismic signal, all other attributes may change as the waves travel through the geological substrate. If the source emits a broad-band spectrum of frequencies, the dominant frequency received is lower than the dominant frequency of the source, as the high frequency end of the spectrum is filtered out with depth by the substrate. The physical properties which are most relevant to seismic data include the travel velocity of seismic waves, which is proportional to the density of the rocks, and the acoustic impedance (velocity multiplied by the rock's density) of the various layers. Contrasts in acoustic impedance between different units generate seismic reflections, which can signify changes in lithology, changes in fluid content or fluid pressure within the same lithosome, or contrasts in the degree of diagenetic overprint.

Often, however, seismic reflections do not correspond to discrete lithological or fluid contacts, but combine the response of several contacts, where the thickness of units is less than the vertical resolution of the seismic data. The vertical seismic resolution is primarily a function of the frequency of the emitted seismic signal. A high-frequency signal increases the resolution at the expense of the effective depth of investigation. A low-frequency signal can travel greater distances, thus increasing the depth of investigation, but at the expense of the seismic resolution. In practice, vertical resolution is generally calculated as a quarter of the wavelength of the seismic wave (Brown, 1991), so it also depends on travel velocity, which in turn is proportional to the rocks' densities. For example, the vertical resolution provided by a 30 Hz seismic wave traveling with a velocity of 2400 m/s is 20 m. This means that a sedimentary unit with a thickness of 20 m or less cannot be seen as a distinct package, as its top and base are tuned within a single reflection on the seismic line. Acquiring the optimum resolution for any specific

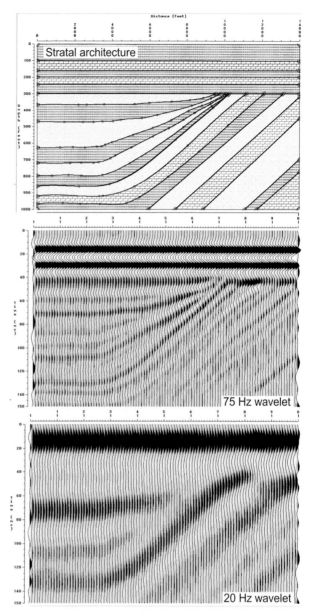

FIGURE 2.66 Seismic imaging of stratal architecture with 75 Hz and 20 Hz wavelets (from Hart, 2000, reprinted by permission of the Society for Sedimentary Geology). Lower resolution data generate seismic artifacts (e.g., the apparent onlap on the lower frequency image), as multiple geological horizons may tune into single seismic reflections.

case study requires therefore a careful balance between the frequency of the emitted signal and the desired depth of investigation (Fig. 2.66).

A seismic reflection that preserves the polarity of the original seismic signal (i.e., 'positive polarity') indicates an increase in acoustic impedance with depth across the geological interface, whereas a polarity reversal ('negative polarity') indicates a decrease in acoustic impedance with depth. The amplitude of the seismic reflection is usually proportional to the contrast in acoustic impedance across the geological contact.

Thus, high negative anomalies at the top of reservoir facies are encouraging hints for petroleum exploration, as they may be related to the presence of porosity and low density fluids (i.e., hydrocarbons) within the reservoir (e.g., shales sealing a charged unconventional reservoir). However, a similar response may mark the limit between compact sandstones (high acoustic impedance) and underlying shales (relatively lower acoustic impedance), or between hemipelagites with low water pressure and overpressured mudflow deposits below. Positive polarity reflections may also be equivocal, as they can be generated under various circumstances (e.g., shales overlying carbonates or compact sandstones, porous sandstones overlying shales, or the top of evaporites with high acoustic impedance). Therefore, studies of the depositional setting are needed to constrain the nature of the seismic reflections.

The origin of the seismic reflection (single contact vs. package of strata) adds another degree of uncertainty to the interpretation of polarity data in terms of rock and fluid phases. Where the vertical distance between seismic reflectors (geological contacts that mark a change in acoustic impedance) is greater than the vertical resolution (i.e., seismic reflections correspond to single geological contacts), the polarity of the reflections is more reliable in terms of geological interpretations. However, where seismic reflectors are closely spaced (i.e., separated by units that are thinner than the vertical seismic resolution), polarity interpretations become less reliable, as the seismic reflections combine the response of multiple reflectors into a composite waveform. Therefore, besides simple polarity and amplitude

studies, an entire range of additional techniques has been developed to assist with the fluid evaluation from seismic data, including the observation of bright spots (gas-driven high negative anomalies), flat spots (hydrocarbon/water contacts marked by horizontal high positive anomalies), and AVO (amplitude variance with offset) methods of data analysis that increase the chances of locating natural gas or light petroleum with a minimum of 5% gas.

The limitation imposed by the relatively coarse vertical resolution, commonly in a range of 10^1 m, has been a main hindrance to the use of seismic data in resolving details at the smaller (sub-seismic) scales of many individual petroleum reservoirs or depositional elements. For this reason, seismic data have been used traditionally to assess the larger scale structural and stratigraphic styles, but with limited applications to the smaller scales of high-resolution sequence stratigraphy. However, as technology evolved, the vertical resolution of seismic data improved from tens of meters to meters, particularly in the case of high-frequency datasets meant for shallower depths of investigation, enabling a detailed 3D seismic imaging of the subsurface. The quality of these images is typically better in the present-day offshore, which affords the acquisition of higher resolution datasets. In spite of this technological progress, seismic data still provide only indirect information on the solid and fluid phases in the subsurface, so calibration with borehole data is essential for determining the relationship between seismic facies and lithofacies, for velocity measurements, or for time—depth conversions (Fig. 2.67).

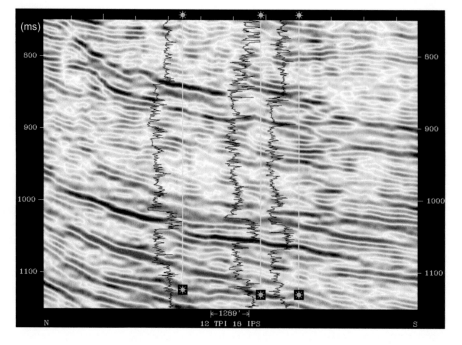

FIGURE 2.67 Seismic line with well-log overlay (gamma ray), showing slope progradation in a mixed carbonate—siliciclastic setting (Permian Basin, USA; from Hart, 2000, reprinted by permission of the Society for Sedimentary Geology). The high-amplitude seismic reflections correspond to lithological contrasts between clean carbonates (low gamma ray) and dolomitic sandstones and siltstones (higher gamma ray). The time/depth conversion affords the mutual calibration of well-log and seismic data.

2.3.3 Workflow of seismic data analysis

The common workflow in the analysis of seismic data includes an initial assessment of the larger scale structural and stratigraphic styles, followed by detailed studies in specific areas that contain features of potential economic interest. This is generally the approach taken in any stratigraphic study, tackling the 'big picture' first (e.g., tectonic setting, basin physiography) in order to have a context for the 'detail'. For this reason, the analysis of any basin starts with the acquisition of a grid of regional 2D seismic lines, the study of which guides the selection of smaller areas of special interest that warrant the acquisition of 3D seismic volumes for the detailed investigation of potential drilling sites. The following sections present the main steps in the workflow of seismic data analysis.

2.3.3.1 Reconnaissance studies

Regional 2D seismic lines afford the identification of key physiographic elements of a sedimentary basin, which can be used to narrow down the range of possible depositional environments that can be expected in specific areas. For example, in the context of a divergent continental margin, the shelf edge at each point in time is the limit between areas dominated by different environments of deposition and modes of sediment transport (Fig. 2.68). Updip from the shelf edge, on the continental shelf, depositional environments range from continental to shallow-marine, with the sediment being transported primarily by traction currents (e.g., rivers, longshore currents, tidal currents, storm surges). Downdip from the shelf edge, the slope, and the basin floor define deep-water environments in which gravity flows become the main processes of sediment transport and accumulation of conventional petroleum reservoirs. Therefore, mapping the location of the shelf edge at different timesteps in the evolution of the basin is fundamentally important before any detailed work at smaller scales is carried out.

Dip-oriented 2D seismic lines also afford the observation of clinoforms that develop at different scales, from the 10^1 m scales of deltaic systems on the shelf to the 10^2-10^3 m scales of slope systems beyond the shelf edge. This imaging helps constrain the morphology and gradients of paleoseafloors (clinoform surfaces), as well as estimate water depths at the time of deposition by using the scale of the clinoforms as a proxy for paleobathymetry (note caveats discussed in Chapter 4, regarding the subaerial vs. subaqueous nature of clinoform rollovers). Strike-oriented seismic lines afford the visualization of other types of processes, such as the cut-and-fill architecture of fluvial incised valleys on the shelf or of submarine canyons on the slope. Therefore, the observation of both dip- and strike-oriented transects is important in reconnaissance studies. Ultimately, preliminary regional studies lead to the identification of exploration opportunities on the shelf (e.g., fluvial incised valleys, siliciclastic coastal systems, shallow-water carbonates) and in the deep-water setting (e.g., submarine fans), providing the basis for the selection of specific areas that justify the acquisition of 3D seismic volumes.

FIGURE 2.68   Seismic line showing the long-term progradation of a divergent continental margin (from Catuneanu et al., 2003; image courtesy of PEMEX). The shelf-edge separates fluvial to shallow-marine systems on the continental shelf from deep-marine systems on the slope and basin floor. The slope clinoforms downlap the basin floor (yellow arrows), but due to the rise of a salt diapir (blue arrow) some downlap terminations may be confused with onlap (red arrows).

The reconnaissance analysis of a new 3D seismic volume (e.g., Fig. 2.65) starts with an initial scrolling through the data (side to side, front to back, top to bottom) in order to assess the overall structural and stratigraphic styles (Hart, 2000). In this stage, as well as in all subsequent stages of data analysis, the interpreter must be familiar with a broad range of depositional and structural patterns in order to determine what working hypotheses are geologically reasonable for the new dataset. Following the reconnaissance scrolling, the seismic volume is 'sliced' in the areas that show the highest potential, where structural or stratigraphic traps may be present. The occurrence of such traps is often marked by seismic 'anomalies' (e.g., Fig. 2.69), which can be further highlighted and studied by applying various techniques of data analysis. Slicing through the seismic volume is one of the most common techniques, and different slicing styles may be performed during the various phases of data handling (Fig. 2.70).

The easiest to obtain in the early stages of data analysis are the *planar slices* (horizontal time slices or inclined planar slices through the 3D seismic volume; Fig. 2.70), which can be acquired before seismic reflections are mapped within the volume. The disadvantage of planar slices, and of time slices in particular, is that they are usually time transgressive (i.e., features of different ages are shown on the same image), as it is unlikely that a paleodepositional surface (commonly associated with some relief, and potentially affected by subsequent tectonism or differential compaction) corresponds to a perfect geometrical plane inside the seismic volume. For this reason, planar slices are seldom true representations of paleolandscapes or paleoseafloors, unless the slice is obtained from recent sediments at shallow depths in relatively flat areas.

Once seismic reflections are mapped throughout a 3D volume, *horizon slices* can be generated by extracting various seismic attributes (e.g., amplitude, etc.) along the mapped reflections. By flattening a seismic horizon of interest, the stratigraphic sections immediately below and above the horizon can be studied further with additional parallel slices (Fig. 2.70). However, these parallel slices become increasingly time transgressive (i.e., lose usefulness) with the increase in the distance from the flattened reference horizon. Horizon slices are the closest images to the true paleogeography, and may reveal morphological details of past landscapes and seascapes that can provide key evidence for the interpretation of paleodepositional environments. This information also helps to constrain the meaning of well-log motifs in terms of depositional elements or larger scale depositional trends.

Still in the reconnaissance stage, the seismic anomalies highlighted by volume slicing can be studied further with additional techniques, such as voxel picking and opacity rendering, which can enhance geomorphological features. A voxel is a 'volume element', similar with the concept of pixel ('picture element') in remote sensing, but with a third dimension ('z') that corresponds to time or depth. The other two dimensions (measured along horizontal axes 'x' and 'y') of a voxel are defined by the bin size, which is the area represented by a single seismic trace. The vertical ('z') dimension is defined by the digital sampling rate of the seismic data, which is typically 2 or 4 milliseconds two-way travel time. Defined as such, each voxel is associated with a certain seismic amplitude value. The method of *voxel picking* involves auto-picking of connected voxels of similar seismic character, which can illuminate discrete depositional elements in three dimensions.

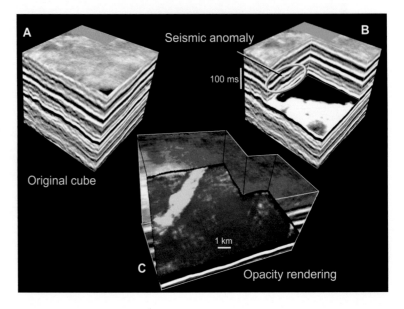

FIGURE 2.69 Reconnaissance study of a seismic volume (Western Canada Basin; images courtesy of Henry Posamentier). A—Original 3D volume, showing two section views and a plan view in the amplitude domain. B—Chair slice through the 3D volume, highlighting an amplitude anomaly. C—Opacity-rendered volume where only high-amplitude voxels are rendered opaque; all other voxels are rendered transparent. This allows for visualization of a linear amplitude anomaly, interpreted as a channel.

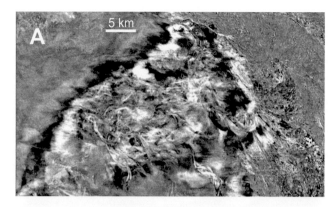
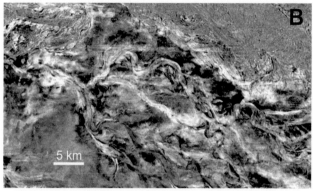
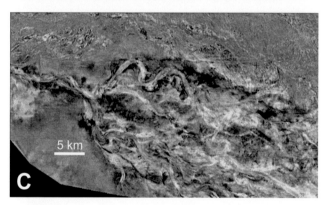

FIGURE 2.70 Reconnaissance study of a 3D seismic volume using different slicing techniques (images courtesy of Henry Posamentier). A—Time slice: amplitude extraction from a planar horizontal slice; shown here is the frontal splay of a deep-water turbidite system, eastern Gulf of Mexico. B—Dipping planar slice: amplitude extraction from a planar surface dipping at c. 2° to the east-southeast. C—Horizon slice: amplitude extraction from a horizon of interest mapped within the 3D volume. This type of slice yields the best imaging of depositional elements.

Similarly, *opacity rendering*, which makes opaque only those voxels that lie within a certain range of seismic values, can also bring out features of stratigraphic interest (Hart, 2000; Posamentier, 2004; Fig. 2.69).

2.3.3.2 Interval attribute maps

Once the stratigraphic objectives have been identified in the initial reconnaissance stages, the intervals bracketing sections of geological interest can be evaluated further by constructing interval attribute maps for

those seismic 'windows' (Fig. 2.71). Among the most common types of attributes used for this purpose, amplitude extraction maps, seismic facies maps, and seismic trace coherence maps are constructed in order to highlight different features of the depositional systems under analysis (Figs. 2.71–2.75). The optimal size of the window (e.g., 100 ms vs. 50 ms) varies with the case study, as it depends to some extent on the thickness of the depositional elements that are being investigated. The interpreter needs to experiment with different windows until the features of interest (e.g., channels, carbonate banks, etc.) are best visualized on the seismic image.

Amplitude extraction maps may display various amplitude attributes calculated over the selected interval (e.g., averages, positive polarity, negative polarity, cumulative amplitudes, amplitude peaks, square roots, etc.), and commonly reflect changes in the contrasts in acoustic impedance that may be interpreted in terms of lateral shifts of facies. Such maps often enable the interpreter to visualize geomorphological features that may be diagnostic of specific depositional systems, or even individual depositional elements within depositional systems (e.g., a fluvial channel fill in Fig. 2.72, or reef structures in Fig. 2.71).

Seismic facies maps also require the selection of an interval (e.g., 34 ms in Fig. 2.74), within which the shape of the seismic traces is analyzed by software algorithms and classified into a number of waveform classes. The color codes used to differentiate between the different waveform classes enable the construction of maps that can be interpreted in terms of sedimentary facies and depositional elements (Figs. 2.73 and 2.74). The underlying principle is that, as in the case of lateral changes in amplitudes across the study area, the change in seismic waveforms is influenced by lateral shifts of facies, and hence each trace shape may be associated with a specific lithology-fluid 'package'. Such a relationship needs to be calibrated with borehole data, although the morphology of depositional elements on the seismic facies maps is often evident enough to allow one to infer the lithofacies in the various areas of a depositional system. For example, classes 9 and 10 in Fig. 2.73 (encircled area) are thought to indicate the location of the best reservoir sands within the channel fill. Once waveforms are interpreted in terms of lithofacies, the visualization of depositional elements may be enhanced by highlighting only selected classes of trace shapes (Fig. 2.74).

The degree to which seismic traces correlate within a selected interval (window) may be further observed by constructing *coherence maps*, which provide additional means for the study of geomorphological features (Fig. 2.75). Coherence is a volume attribute that measures the degree of similarity of seismic traces—light colors are assigned to areas where seismic traces are

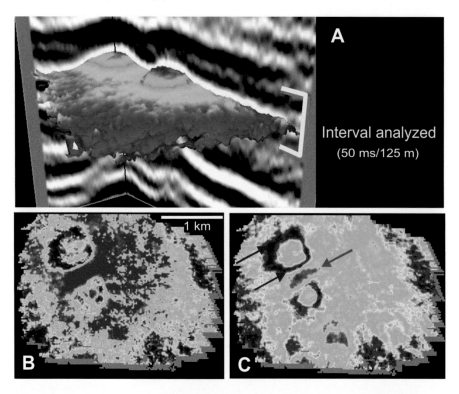

FIGURE 2.71 Devonian pinnacle reefs separated by a tidal channel (Western Canada Basin; images courtesy of Henry Posamentier). A—Section and three-dimensional view; colors on the map indicate time structure, with reds/greens representing highs and blue/purple representing lows; B and C—Interval attribute maps for the 50 ms window (B: maximum amplitude values; C: positive polarity total amplitude). The amplitude asymmetry around the reef structures may reflect the patterns of current circulation and the prevailing wind direction, with a landward and a leeward side. The amplitude anomaly between the reef structures indicates a different lithology, possibly calcarenites associated with enhanced tidal scouring.

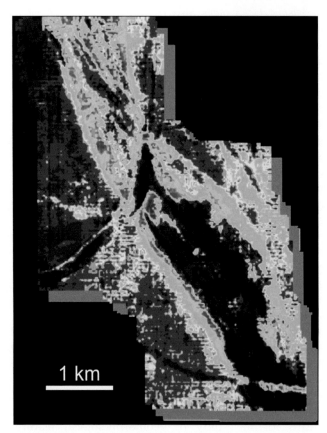

FIGURE 2.72 Interval attribute map (amplitude strength within a 40 ms window) of a fluvial channel with alternate bars (Cretaceous, Western Canada Basin; image courtesy of Henry Posamentier). Crosscutting relationships indicate that the two smaller channels in this image are younger than the main channel.

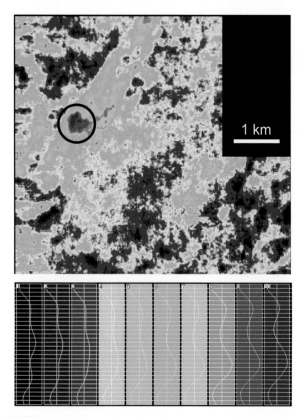

FIGURE 2.73 Seismic facies map based on a ten-fold classification of seismic traces (image courtesy of Henry Posamentier). This example shows a fluvial system in the Western Canada Basin. The black circle outlines a small structural high. Seismic class 10 is confined to this area, suggesting a possible accumulation of hydrocarbons within the channel at this location. The channel fill is dominated by classes 7—10, whereas the overbank is dominated by classes 1—6.

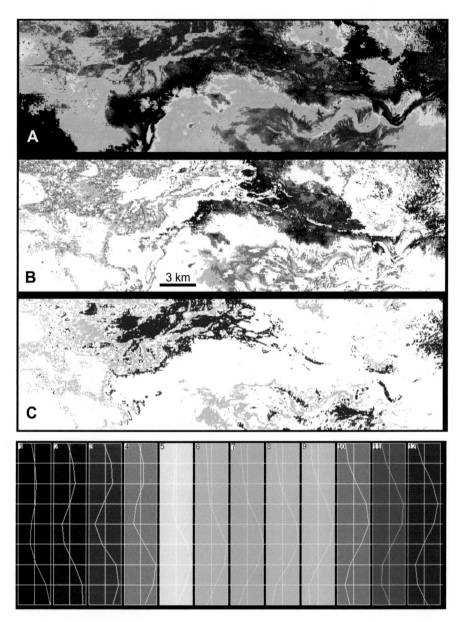

FIGURE 2.74 Seismic facies map of a deep-water system in a basin-floor setting (eastern Gulf of Mexico; images courtesy of Henry Posamentier). The map shows mudflow deposits (upper-left side of the image) overlying a turbidite system (lower-right side of the image). The analysis is based on a 34 ms interval, with twelve seismic classes defined. A—All classes are highlighted; B—only classes 2, 3, 4, and 9 are highlighted, revealing the sheet-like portion of the mass transport complex in the distal area; C—only classes 9 and 12 are highlighted, revealing the convolute part of the mass transport complex in the proximal area.

similar and correlate well, and dark colors indicate a lack of correlation. Coherence images highlight seismic edges, which may correspond to structural or depositional elements, and are particularly useful to delineate the position of faults. In the case of faults, seismic edges are associated with brittle deformation that affects the more rigid rock types (e.g., sandstones), in contrast to the more ductile lithologies (e.g., shales) which are able to accommodate stress without breaking.

Spectral decomposition maps can be generated from the frequency volume of a 3D seismic dataset, and record the response of the interaction between geobodies in the subsurface and seismic wavelets of different frequencies that are part of a broad-band seismic signal. These maps require the selection of an interval that is sufficiently large to include wavelengths of different

frequencies (e.g., 100 ms as a starting window), which can be subsequently modified in order to test the size of the optimal window for a specific dataset. The colors on the map correspond to specific frequencies (e.g., 20 Hz—blue; 40 Hz—green; 60 Hz—red; Fig. 2.76). As each frequency 'resonates' with depositional elements whose thicknesses match the wavelength of the seismic signal, the map highlights the elements that are in tune with the frequencies used to construct the image. The specific frequencies that are best suited for a dataset need to be tested and selected on a case-by-case basis, as they depend on the geobodies in the subsurface, as well as on the depth of investigation. The higher frequency colors dominate the shallower depths, where smaller geobodies can be detected. Spectral decomposition maps co-render images generated with different

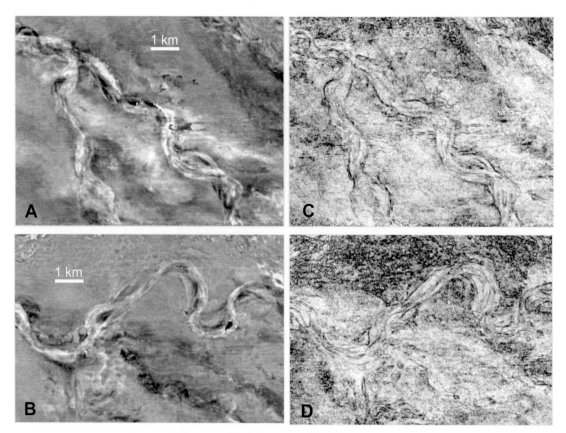

FIGURE 2.75   Interval attributes of a Plio-Pleistocene deep-water channel system in the eastern Gulf of Mexico (images courtesy of Henry Posamentier). A and B—amplitude extraction from two horizon slices; these images capture successive positions of the channel thalweg and shows episodes of channel avulsion. The shifting thalweg indicates meander loop migration to the right and concomitant flow in that direction. C and D—coherence slices of the same channel system shown in A and B. Coherence is a volume attribute that emphasizes the correlation of seismic traces: light colors are assigned where seismic traces correlate, and dark colors indicate a lack of correlation. Coherence highlights seismic edges (e.g., edges of depositional elements); in this example, it enhances the margins of the turbidite channels.

FIGURE 2.76 Spectral decomposition maps, co-rendering images generated with different frequencies: 20 Hz—blue; 40 Hz—green; 60 Hz—red (Mahanadi Basin, offshore India; 3D seismic timeslices between 3.5–3.8 sec below sea level; images courtesy of Reliance Industries, India). Spectral decomposition maps often yield the best imaging of depositional elements in the subsurface. These examples show the morphology of deep-water turbidite channels and splays in a basin-floor setting.

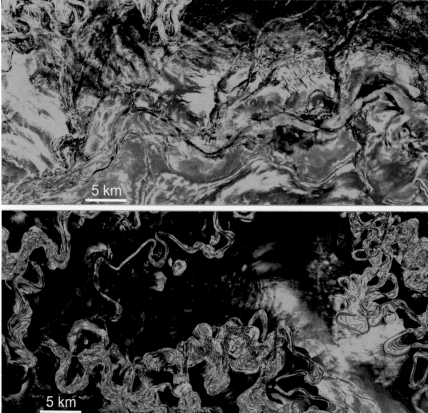

frequencies, and often yield the best imaging and the most detailed account of the features that define the depositional elements in the subsurface (Fig. 2.76).

2.3.3.3 Horizon attribute maps

Horizon attribute maps can be constructed by picking a specific stratigraphic horizon within the larger interval studied in the previous step, and extracting the desired seismic attributes along that horizon (e.g., amplitude, acoustic impedance, dip azimuth, dip magnitude, roughness, or curvature; Fig. 2.77). These maps may still have a narrow window bracketing the horizon itself, or can be constructed with zero thickness, and are horizon slices for the specific stratigraphic level of interest (Fig. 2.70C). Zero-thickness images are true 'slices', which show discrete values of the selected physical attributes across the study area (e.g., the maximum amplitude of the seismic reflection). In most cases, however, horizon attribute maps are constructed for narrow windows from which an average value of the physical attribute is calculated (e.g., the average amplitude for the selected window). Horizon attribute maps may enhance the visualization of geomorphological features and depositional elements across past landscapes or seascapes. If the mapping of seismic reflections is correct, these horizon slices should be very close to time lines, providing a snapshot of past depositional environments and paleogeography.

Time structure maps ('depth' in time of a horizon below a surface datum) may also be generated following the mapping of throughgoing stratigraphic horizons within a 3D volume, and add important information regarding the paleogeography, subsidence history, and the structural style of the studied area (Fig. 2.78). Interval or horizon attributes may be combined to enhance visualization effects, such as superimposing dip magnitude attributes on a time structure map (Fig. 2.79), or co-rendering coherence with amplitude data (Fig. 2.80). The importance of seismic attribute maps in stratigraphic analysis has

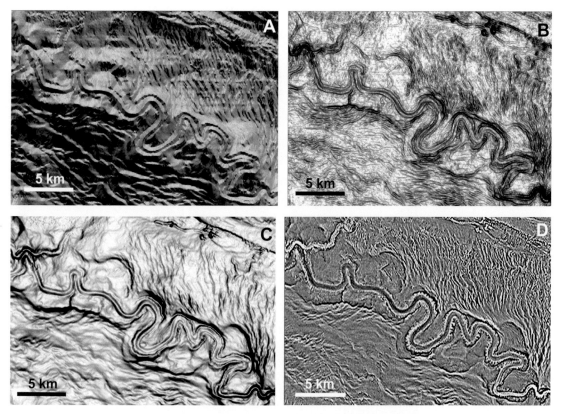

FIGURE 2.77 Horizon attributes of a mid—late Pleistocene deep-water channel in the northeastern Gulf of Mexico (images courtesy of Henry Posamentier). A—Dip azimuth map; this attribute shows the orientation of the image such that north-facing surfaces are assigned light colors, and south facing are assigned dark colors, creating a 3D perspective. Note the apparent knife-edge top of the raised channel, which is a seismic artifact. B—Surface roughness map; this attribute captures the roughness of a surface: rough areas are assigned dark colors, and smooth areas are assigned light colors. C—Dip magnitude map; this attribute shows changes in slope angles across the surface: steep angles are indicated in black, and gentle slopes are depicted in white. In this display, the raised channel no longer shows a knife-edged top, but a flat to rounded convex-up shape. D—Curvature map; this attribute highlights the curvature of the horizon, and outlines depositional elements by assigning dark colors to low-curvature (flat) areas and light colors to high-curvature edges of morphological elements. Detailed features that are less evident on the other attribute maps include slump scars on the inner levee flanks of the channel, and sediment waves on the overbank.

FIGURE 2.78   Time structure map of the channel shown in Fig. 2.77 (image courtesy of Henry Posamentier). The image captures the transition between the slope (more elevated: red color) and the basin floor (lower elevation: blue and purple), and the higher elevation of the channel (c. 65 m) relative to the adjacent overbank as a result of post-depositional differential compaction. The direction of flow was from left to right. The vertical scale is in ms below the sea level.

become increasingly evident with the improvements in seismic resolution, to the point that a dedicated discipline emerged as 'seismic geomorphology' (i.e., the plan-view imaging of depositional elements based on 3D seismic data; Posamentier, 2000, 2003, 2004; Posamentier and Kolla, 2003).

2.3.3.4 3D perspective visualization

3D perspective views add another degree of refinement to the information already available from the interval and horizon attribute maps. Such perspective views illustrate images extracted from 3D seismic data in x—y—z space. Interpreted horizons can be illuminated from different directions in order to highlight the relief and morphology of depositional elements. Figs. 2.81—2.83 illustrate examples of such 3D perspective images, which reconstruct paleolandscapes sculptured by fluvial systems (Fig. 2.81), paleoseascapes of carbonate platforms (Fig. 2.82), or basin floors in deep-water settings dominated by gravity flows (Fig. 2.83). Such seismic data provide tremendous support to the reconstruction of paleodepositional environments and to the calibration of borehole log motifs. The examination of geological features in three dimensions may be enhanced by changing the angle of view, or by changing the angle of incidence of the light source that illuminates a particular image (Fig. 2.84).

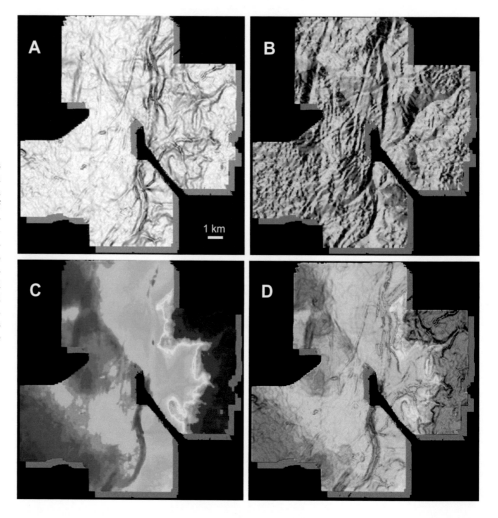

FIGURE 2.79   The base-Cretaceous unconformity in the Western Canada Basin, as depicted on four horizon attribute maps (images courtesy of Henry Posamentier): A—dip magnitude map; B—dip azimuth map; C—time structure map; and D—co-rendered time structure and dip azimuth map. This subaerial unconformity was generated by the flexural uplift of a forebulge within a foreland system (i.e., forebulge unconformity), and it is sculptured by fluvial systems (a main channel and tributaries are visible).

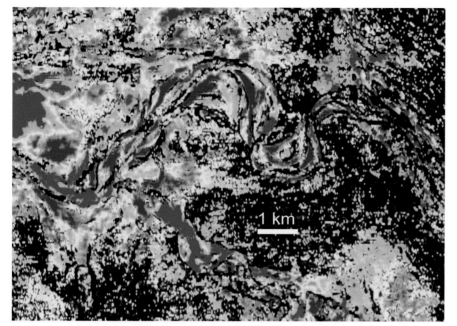

FIGURE 2.80 Co-rendered amplitude and coherence attributes of a Plio-Pleistocene deep-water leveed channel in the eastern Gulf of Mexico (image courtesy of Henry Posamentier). The image combines lithologic information inherent to the amplitude domain with edge effects that help delineating the channel. Multiple channel thalwegs are observed, with meander loop migration verging to the right, indicating flow from left to right across this area.

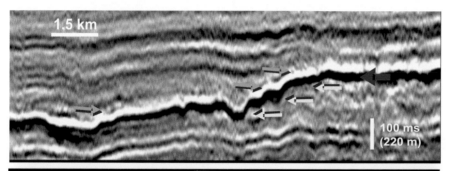

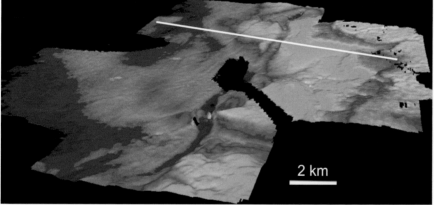

FIGURE 2.81 The base-Cretaceous unconformity in the Western Canada Basin, as depicted on a 2D seismic line and on a 3D perspective view map (images courtesy of Henry Posamentier). The unconformity (red arrow on the seismic line) separates Cretaceous strata above from Devonian deposits below, and is associated with significant erosion (yellow arrows indicate truncation) and change in tectonic and depositional settings. The unconformity is onlapped by the Cretaceous strata (blue arrows), and corresponds to a first-order sequence boundary that marks a change from a divergent continental margin to a foreland system. The top of the Devonian carbonates is incised by Cretaceous fluvial systems. The white line on the 3D perspective view map indicates the position of the 2D seismic line.

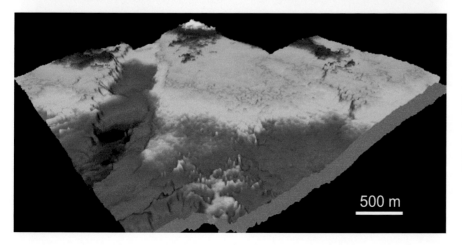

FIGURE 2.82 3D perspective view of pinnacle reefs separated by tidal channels (Devonian, Western Canada Basin; image courtesy of Henry Posamentier). The channels are filled with bioclastic calcarenites produced by tidal reworking on the carbonate platform.

FIGURE 2.83 3D perspective view of a Pleistocene deep-water channel in the basin-floor setting of the Gulf of Mexico (from Posamentier, 2003; image courtesy of Henry Posamentier). Differential compaction after burial modifies the top of the channel fills, providing clues about the lithology that can be expected within the channels. The avulsion channels are mud filled as indicated by the concave-up tops (more compaction within the channel relative to the adjacent areas), in contrast with the convex-up top of the sand-filled main channel (less compaction within the channel relative to the adjacent areas). The main channel is c. 625 m wide.

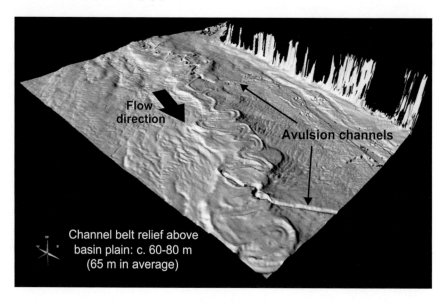

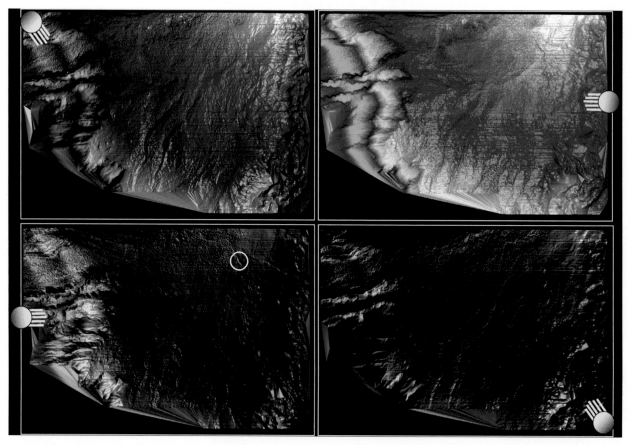

FIGURE 2.84 Illumination effects, such as changing the angle of incident light, may enhance the visualization of different features of the geological horizon of interest (images courtesy of Henry Posamentier). This example shows the modern deep-water seascape in the DeSoto Canyon area of the eastern Gulf of Mexico. For scale, the encircled channel is 300 m wide.

3

Controls on sequence development

Multiple allogenic and autogenic processes interplay to generate cyclicity in the stratigraphic record (Figs. 3.1 and 3.2). The rates and the relative importance of these processes may vary with stratigraphic age and geological setting, making it difficult to generalize their absolute or relative contributions to the development of sequences. With the exception of orbital forcing, the periodicity of processes in Fig. 3.1 can be highly variable and unpredictable. Even in the case of orbital cycles, whose periodicities are most regular, a one-to-one relationship with stratigraphic sequences is still difficult to demonstrate because the orbital signal may be distorted and overprinted by local or regional processes (Wilkinson et al., 2003; Strasser, 2018). However, the interplay of all controls on the stratigraphic architecture is quantified by two measurable variables, namely accommodation and sedimentation, which explain the formation of stratal stacking patterns irrespective of the dominant control(s) at syn-depositional time. This keeps the methodology focused on observational criteria, and independent of the interpretation of the underlying controls on sequence development (Fig. 3.3).

The emergence of seismic stratigraphy in the 1970s (Vail, 1975; Payton, 1977) was one of the most influential "revolutions" in stratigraphy. The concepts of seismic stratigraphy were introduced alongside a global cycle chart (Vail et al., 1977b), based on the assumption that eustasy is the main driving force behind sequence formation at all stratigraphic scales, albeit with different underlying causes at different scales (Fig. 3.4). Seismic stratigraphy and the global cycle chart were thus introduced as a seemingly inseparable package of new stratigraphic methodology. The global-eustasy model, as originally proposed, posed two challenges to the stratigraphic community: (1) that sequence stratigraphy, as linked to the global cycle chart, constitutes a superior standard of geological time to that assembled from conventional chronostratigraphic evidence; and (2) that stratigraphic processes are dominated by the effects of eustasy, to the exclusion of other allogenic mechanisms, including tectonism (Miall and Miall, 2001). Subsequent work showed that none of these implications were true.

At the opposite end of the spectrum, "tectonostratigraphy" (e.g., Winter, 1984) favored tectonism as the main driver of stratigraphic cyclicity. The weakness of either school of though is that an *a priori* interpretation of the underlying controls was automatically attached to the delineation of sequences, which undermined the empirical value of sequence stratigraphy and attracted considerable criticism (e.g., Miall, 1992). It became evident that sequence stratigraphy needed to be dissociated from any preconceived assumptions, including the global-eustasy model, and that an objective analysis needs to be based on empirical evidence observed in outcrop or the subsurface: "Each stratal unit is defined and identified only by physical relationships of the strata, including lateral continuity and geometry of the surfaces bounding the units, vertical stacking patterns, and lateral geometry of the strata within the units. Thickness, time for formation, and interpretation of regional or global origin are not used to define stratal units..., [which]... can be identified in well logs, cores, or outcrops and used to construct a stratigraphic framework regardless of their interpreted relationship to changes in eustasy" (Van Wagoner et al., 1990).

The debates over the controls on sequence development began as soon as unconformity-bounded units were defined, with both tectonics and sea-level change being recognized as major factors. Tectonism was generally inferred to generate stratigraphic cyclicity over longer timescales, as a background process for the higher frequency sea-level cycles. The idea of long-term tectonics vs. higher frequency eustatic oscillations as primary controls on sequence development persisted in the stratigraphic literature for several decades (e.g., Sloss, 1963; Vail et al., 1977a,b; Haq et al., 1987; Posamentier et al., 1988; Posamentier and Vail, 1988). It is now documented that tectonics and eustasy may generate stratigraphic cyclicity over similar timescales, both long- and short-term, and therefore their signatures

93

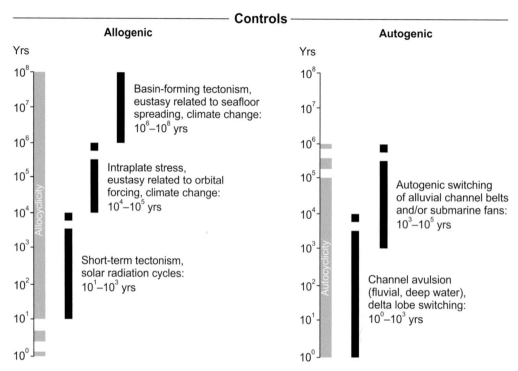

FIGURE 3.1   Controls on stratigraphic cyclicity (from Catuneanu, 2019a). Note the overlap between the timescales of allogenic and autogenic processes. Long-term climate changes on timescales of 10^{6-7} yrs relate to greenhouse—icehouse cycles; shorter term climate changes on timescales of 10^{4-5} yrs define glacial—interglacial cycles, which are most evident during icehouse regimes. Solar radiation cycles also induce climatic fluctuations associated with a complex interplay of ice-sheet dynamics, atmospheric circulation, and thermohaline circulation (Csato et al., 2014). Short-term tectonism includes stages of fault reactivation in fault-bounded sedimentary basins. In addition to tectonism and glacio-isostasy, the rates of subsidence are also affected by sediment loading and compaction during the entire evolution of sedimentary basins. The distinction between the allogenic and autogenic controls on stratigraphic cyclicity is often challenging and has no bearing on the sequence stratigraphic methodology. The observation of stratal stacking patterns, which the methodology is based upon, is decoupled from the interpretation of the underlying controls.

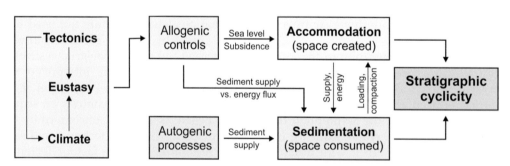

FIGURE 3.2   Development of stratigraphic cyclicity, in response to the interplay of allogenic and autogenic processes. All underlying controls on sequence development combine into the "dual control" of accommodation and sedimentation. *Arrows* indicate process—response relationships (e.g., eustatic changes are controlled both by climate and tectonics, etc.; see text for details).

cannot be distinguished based on the time duration of cycles alone (e.g., Cloetingh et al., 1985, 1989; Cloetingh, 1988; Peper et al., 1992; Catuneanu et al., 1997a,b, 1999, 2002; Isbell et al., 2008; Martins-Neto and Catuneanu, 2010; Miall, 2016, Fig. 3.1).

The discussion of the eustatic vs. tectonic controls on sequence development addresses only part of the natural complexity of the stratigraphic record. Other controls that contribute to the stratigraphic framework include

climate, with its influence on sea level, lake level, glacio-isostasy, energy flux (e.g., fluvial discharge, wind regime), and sediment supply; sediment loading and compaction, which add to the basin subsidence; and autogenic processes which modify sediment supply to particular locations within the sedimentary basin (e.g., fluvial avulsion, delta lobe switching, or autogenic switching of alluvial channel belts and submarine fans; Hajek et al., 2010; Hofmann et al., 2011; Csato and

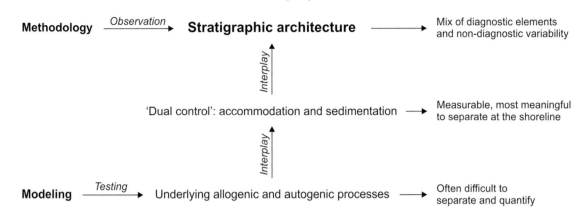

FIGURE 3.3 Development of the stratigraphic architecture, as a result of the interplay of all controls on sedimentation. The underlying allogenic and autogenic processes are often difficult to separate and quantify, but their interplay always boils down to two quantifiable variables (i.e., accommodation and sedimentation) which can be used to explain all aspects of the stratigraphic architecture. The methodology involves the observation of the stratigraphic architecture in order to identify the elements of the sequence stratigraphic framework. In contrast, modeling deals with the testing and interpretation of the underlying controls on sequence development. Modeling needs to be calibrated with field data for meaningful results.

Hierarchical order	Duration (My)	Cause
First order	200-400	Formation and breakup of supercontinents
Second order	10-100	Volume changes in mid-oceanic spreading centers
Third order	1-10	Regional plate kinematics
Fourth and fifth order	0.01-1	Orbital forcing

FIGURE 3.4 Tectonic and orbital controls on eustatic sea-level changes (modified from Vail et al., 1977b; and Miall, 2000). Local controls on accommodation and sedimentation influence the timing of sequences and offset the correlation of basin-fill stratigraphic frameworks with the global sea-level cycles.

Catuneanu, 2012; Catuneanu and Zecchin, 2013; Csato et al., 2014, Figs. 3.1 and 3.2). The allogenic controls on sequence development (i.e., eustasy, tectonism, and climate) operate across all stratigraphic scales, whereas the autogenic controls are typically restricted to the scales of high-resolution sequence stratigraphy (Fig. 3.1). Nevertheless, the formation of any sequence stratigraphic surface whose timing is controlled at least in part by sediment supply can be affected by autogenic processes, which therefore remain important at all hierarchical levels.

3.1 Allogenic processes

Allogenic processes are those external to a depositional system, such as eustasy, tectonism, and climate, which modify the total energy and sediment budget of the depositional environments hosted within a sedimentary basin (Einsele et al., 1991). The relative importance of these controls on sedimentation and stratigraphic architecture varies with the geological setting. Tectonism determines the type of sedimentary basin and the overall geometry of the basin fill, whereas the internal stratigraphic cyclicity and the sedimentological makeup of sequences depend on the interplay of all allogenic

controls. Tectonically active basins tend to be dominated by extrabasinal siliciclastic sediment with an internal architecture marked by tectonic events (Frostick and Steel, 1993a,b; Prosser, 1993; Martins-Neto and Catuneanu, 2010), whereas tectonically "passive" basins are more diverse in terms of sedimentary fill and more susceptible to the contributions of climate and sea-level changes to the stratigraphic architecture.

3.1.1 Eustasy

Eustatic fluctuations of global sea level are driven by plate-tectonic and climatic processes, over various time-scales (Fig. 3.4). Sea-level changes which result in shoreline transit cycles across continental shelves are typically recorded in response to orbital forcing at scales of 10^4-10^5 yrs (i.e., Milankovitch cycles; Fig. 2.15). At larger scales, global sea-level changes are associated with long-term Greenhouse–Icehouse cycles of climate change (10^6-10^7 yrs timescales), or with plate tectonic processes such as volume changes along mid-oceanic spreading centers (10^7-10^8 yrs timescales) and the formation and breakup of supercontinents (10^8 yrs time-scales; Fig. 3.4). At smaller scales, shorter term sea-level changes in a range of 10^1-10^3 yrs may be related to solar radiation cycles (Csato et al., 2014, Fig. 3.1).

The signature of the eustatic control on sedimentation may be recognized from (1) the tabular geometry of stratigraphic sequences, suggesting that accommodation was created in equal amounts across the basin; (2) the synchronicity of depositional and erosional stages across the basin, and beyond; and (3) the lack of source area rejuvenation, as it may be suggested by the absence of conglomerates along the proximal rim of the basin. The sea-level control on sedimentation has been documented in numerous case studies, with a degree of confidence that improves with decreasing stratigraphic age (e.g., Suter et al., 1987; Plint, 1991; Miller et al., 1991, 1996, 1998, 2003, 2004; Long, 1993; Locker et al., 1996; Stoll and Schrag, 1996; Kominz et al., 1998; Coniglio et al., 2000; Kominz and Pekar, 2001; Pekar et al., 2001; Posamentier, 2001; Olsson et al., 2002; Sweet et al., 2019).

Estimates of sea-level changes in the geological record have been obtained by means of backstripping, accounting for water-depth variations, sediment loading, compaction, basin subsidence, and foraminiferal $\delta^{18}O$ data. Studies of the "Icehouse world" of the past 42 Ma have demonstrated a relationship between depositional sequence boundaries and global $\delta^{18}O$ increases, linking the formation of many sequence boundaries to stages of glacio-eustatic sea-level fall (e.g., Miller et al., 1996, 1998). Even for the "Greenhouse world" of the Late Cretaceous—Early Cenozoic interval, backstripping studies on the New Jersey Coastal Plain, which was subject to minimal tectonic activity, indicate that sea-level fluctuations occurred with amplitudes of >25 m on timescales of <1 Ma (Miller et al., 2004). Such studies cast doubt on the assumption of a completely ice-free world during the Late Cretaceous, and have revamped the importance of sea-level changes on accommodation and sedimentation (e.g., Stoll and Schrag, 1996; Price, 1999; Miller et al., 2004).

The eustatic changes documented during periods of Greenhouse climatic regimes are inferred to be linked to the development of small, ephemeral ice sheets in Antarctica (Miller et al., 2003). Greenhouse sequences at scales of up to 100 m have been documented in the Late Cretaceous successions of the Western Interior Basin of North America, and have been linked to a glacio-eustatic control (Plint, 1991; Hampson et al., 2011) at the timescales of Milankovitch orbital parameters (Sethi and Leithold, 1994; Pattison, 2019). These sequences can include deposits accumulated during both stages of eustatic rise and fall (Pattison, 1995; Posamentier and Morris, 2000; Hampson et al., 2011; Catuneanu and Zecchin, 2020). In comparison with the glacio-eustatic sequences that developed on the same timescales during Icehouse climates, these Greenhouse sequences are generally thicker and less top-truncated. This is explained by the lower amplitude sea-level changes during the Mesozoic, which enhanced the

preservation of the Milankovitch-scale Greenhouse sequences (Catuneanu and Zecchin, 2013).

3.1.2 Tectonism

Subsidence is commonly attributed to tectonism, although additional processes such as crustal cooling (thermal subsidence), ice loading (isostasy), sediment loading and compaction, and salt dissolution or withdrawal (salt tectonics) may also bring important contributions to the total amount of subsidence in a sedimentary basin (Busby and Ingersoll, 1995; Waldron and Rygel, 2005; Miall, 2000; Allen and Allen, 2013; Gutierrez Paredes et al., 2018). Tectonism is the most important basin-forming mechanism; all sedimentary basins have a tectonic origin, which is why the classification of sedimentary basins is based on tectonic criteria (see Chapter 9 for details).

Tectonism is driven primarily by forces of internal Earth dynamics, which are expressed at the surface by plume or plate tectonic processes. There is increasing evidence that the tectonic regimes which controlled the formation and evolution of sedimentary basins in the more distant geological past were much more erratic in terms of origin and rates than formerly inferred solely from the study of the Phanerozoic record (e.g., Eriksson et al., 2004, 2005a,b). The more recent basin-forming processes are related to a rather stable plate tectonic regime, whereas the formation of Precambrian basins reflects a combination of competing mechanisms, including magmatic-thermal processes ("plume tectonics") and a more erratic plate tectonic regime (Eriksson and Catuneanu, 2004b). These insights indicate that time is largely irrelevant to the classification of stratigraphic cycles (see details in Chapter 7).

The signature of the tectonic control on sedimentation may be inferred from several traits, including (1) the wedge-shaped geometry of stratigraphic sequences, due to differential subsidence; (2) the accumulation of coarser-grained facies along the proximal rim of the basin in relation to the rejuvenation (uplift) of the source areas; (3) variations in the maximum burial depths across a basin, as determined from the study of late diagenetic minerals, fluid inclusions, vitrinite reflection, or apatite fission track; (4) changes in syndepositional topographic slope gradients, as inferred from the shift in fluvial styles through time; (5) changes in the direction of topographic tilt, as inferred from paleocurrent measurements; and (6) lateral variability in the systems-tract composition of sequences due to coeval subsidence and uplift within a sedimentary basin (Catuneanu et al., 2002; Gawthorpe et al., 2003). The role of tectonic mechanisms in the development of stratigraphic cycles and unconformities has been documented for sedimentary basins spanning all stratigraphic ages, from Precambrian to Phanerozoic. Early models assumed that tectonic processes operate mainly on long timescales, of $>10^6$ years (e.g., Vail et al., 1977b, 1984, 1991; Haq et al., 1987; Posamentier et al., 1988; Devlin et al., 1993), leaving eustasy as the likely cause of higher frequency cycles, at timescales of 10^6 years or less.

Advances in the understanding of tectonic processes led to the conclusion that tectonically-driven cyclicity may develop over a much wider range of timescales than originally inferred, both greater and less than 1 My (e.g., Cloetingh et al., 1985; Karner, 1986; Underhill, 1991; Peper and Cloetingh, 1992; Peper et al., 1992, 1995; Suppe et al., 1992; Karner et al., 1993; Eriksson et al., 1994; Gawthorpe et al., 1994, 1997; Peper, 1994; Yoshida et al., 1996, 1998; Catuneanu et al., 1997a, 2000; Catuneanu and Elango, 2001; Davies and Gibling, 2003). Even outcrop-scale cycles may be generated by tectonic processes, as in the case of the Pliocene Gilbert-type deltas of the Loreto Basin (Mexico), where the reactivation of normal faults was invoked to explain episodic delta deposition, resulting in stacked cycles (Dorsey et al., 1997). Local or regional tectonics can also shape the internal architecture of sequences generated by eustatic changes, such as in the case of growth folding (Gawthorpe et al., 1997; Ito et al., 1999; Castelltort et al., 2003; Zecchin et al., 2003), syn-sedimentary normal faulting (Gawthorpe et al., 1994; Howell and Flint, 1996; Zecchin et al., 2006; Zecchin, 2007), and in areas undergoing long-term uplift (Zecchin et al., 2010b, 2011).

It can be concluded that eustasy and tectonism may compete to generate stratigraphic cyclicity at all scales. The challenge in this situation is to evaluate their relative importance on a case-by-case basis. In this light, it has been noted that the amplitudes of sea-level changes reconstructed by means of backstripping (e.g., Miller et al., 1991, 1996, 1998, 2003, 2004; Locker et al., 1996; Stoll and Schrag, 1996; Kominz et al., 1998; Coniglio et al., 2000; Kominz and Pekar, 2001; Pekar et al., 2001) are in many cases lower than those interpreted from seismic data (e.g., Haq et al., 1987), questioning the accuracy of seismic data interpretations in terms of eustatic sea-level changes (Miall, 1986, 1992, 1994, 1997; Christie-Blick et al., 1990; Christie-Blick and Driscoll, 1995). Field observations also indicate that the amount of erosion associated with many sequence-bounding unconformities in tectonically active basins is often greater than the inferred amplitude of eustatic fluctuations, suggesting that the basinward shifts of facies associated with stages of relative sea-level fall are not necessarily related to changes in sea level (e.g., Christie-Blick et al., 1990; Christie-Blick and Driscoll, 1995). These insights reemphasize the importance of tectonism as a control on accommodation and sedimentation, which,

in tectonically active basins, may explain the observed cyclicity at virtually any timescale (Fig. 3.1).

3.1.3 Climate

Global climate changes are recorded over multiple timescales, in relation to a variety of external forces (orbital forcing, solar radiation) and internal processes of Earth dynamics (plate tectonics, volcanism). Regional climate changes may also be triggered by tectonic processes such as the formation of thrust-fold belts that may act as barriers for atmospheric circulation. Long-term climate changes are represented by Greenhouse—Icehouse cycles (10^6–10^7 yrs timescales), with shorter term fluctuations related to glacial—interglacial stages triggered by orbital forcing (10^4–10^5 yrs timescales). The expression of orbital cycles in the stratigraphic record is most evident during Icehouse regimes, when the magnitude of sea-level changes is greater than that recorded during Greenhouse conditions. Examples from the Permian marine record in eastern Australia indicate paleobathymetric changes of 70–80 m for sequences formed during glacial times, whereas similar units deposited in the absence of glacial influence show facies juxtapositions that indicate only 20–30 m shifts across sequence boundaries (Fielding et al., 2008). On even shorter timescales (10^1–10^3 yrs), solar radiation cycles induce climatic fluctuations associated with a complex interplay of ice-sheet dynamics, atmospheric circulation, and thermohaline circulation (Csato et al., 2014, Fig. 3.1).

Climate changes within the 10^4–10^5 years Milankovitch band are attributed to several separate components of orbital variation, including orbital eccentricity, obliquity, and precession (Fig. 2.15). Variations in orbital eccentricity, which refers to the shape (degree of stretching) of the Earth's orbit around the Sun, have major periods at around 100 and 413 kyr. Changes of up to 3° in the obliquity (tilt) of the ecliptic have a major period of 41 kyr. The precession of the equinoxes, which refers to the rotation (wobbling) of the Earth's axis as a spinning top, records an average period of about 21 kyr (Fig. 2.15; Imbrie and Imbrie, 1979; Imbrie, 1985; Schwarzacher, 1993). In addition to Milankovitch-band processes, other astronomical forces may affect the climate over shorter time intervals, from a solar band (tens to hundreds of years range; e.g., sun-spot cycles) to a high-frequency orbital band (e.g., nutation cycles of the motion of the axis of rotation of the Earth about its mean position, with a periodicity of about 18.6 years) and a calendar band (cyclicity related to seasonal rhythms, such as freeze-thaw, varves, or fluvial discharge cycles, and other sub-seasonal effects driven by the Earth—Moon system interaction) (e.g., Fischer and Bottjer, 1991; Miall, 1997).

Fluctuations in the syndepositional paleoclimate may be reconstructed by combining independent research methods such as (1) thin section petrography of the detrital framework constituents in sandstones, looking at the balance between stable and unstable grains; (2) the mineralogy of the early diagenetic constituents, assuming a short lag time between the deposition of the detrital grains and the precipitation of early diagenetic minerals; (3) the isotope geochemistry of early diagenetic cements; and (4) foraminiferal $\delta^{18}O$ data. Each of these techniques may be subject to errors in terms of paleoclimate interpretations, so their use in conjunction is recommended for more reliable conclusions (e.g., Khidir and Catuneanu, 2003). The role of climate as a major control on sedimentation has been emphasized in numerous case studies, including Blum (1994), Tandon and Gibling (1994, 1997), Miller et al. (1996, 1998), Blum and Price (1998), Heckel et al. (1998), Miller and Eriksson (1999), Ketzer et al. (2003a,b), and Gibling et al. (2005).

The effects of climate on the sedimentological makeup of stratigraphic sequences can be recognized in all depositional settings, from fluvial and eolian through to deep water. Climate affects the capacity of sediment-transport systems (e.g., increase in fluvial discharge during interglacial periods) but also sediment supply, by modifying the efficiency of weathering and erosion of extrabasinal source areas, the production of glaciogenic silt by glacial grinding, the precipitation of carbonates and other chemical or biochemical deposits, thus influencing the composition of sequences and the dominant depositional trends (Cecil, 1990; Paola et al., 1992; Leeder et al., 1998; Feldman et al., 2005). For example, Roveri and Taviani (2003) related the concentration of shells in the Mediterranean during the Plio-Pleistocene to climatic phases of reduced fluvial runoff and higher carbonate productivity and/or to hyperpycnal flows reworking and accumulating shell debris. Leeder et al. (1998) highlighted the influence of land vegetation on terrigenous supply to marine areas during the late Quaternary in both south-western USA and Mediterranean areas. Similarly, Massari et al. (2007) demonstrated the role of climate-modulated sediment supply in shaping Pleistocene sequences in the Crotone Basin (Italy).

Studies of eolian systems that are beyond marine influence demonstrated a strong relationship between the formation of loess-paleosol sequences and the glacial–interglacial cycles driven by orbital forcing on eccentricity timescales of 100 kyr (Fig. 3.5). The development of these sequences relates to variations in sediment supply between two extremes of high and low rates of eolian silt deposition (Muhs and Bettis, 2003). The production of eolian silt can be attributed to multiple processes, including glacial grinding, frost shattering, salt

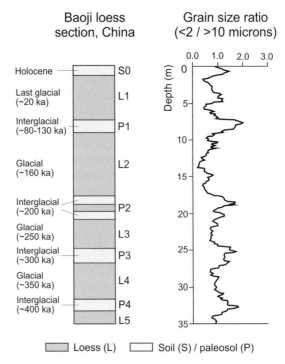

FIGURE 3.5 Relationship between loess-paleosol sequences and glacial–interglacial cycles on eccentricity timescales of c. 100 kyr (modified from Muhs and Bettis, 2003). Loess deposition occurred during glacial periods, whereas the reduced production of glaciogenic silt during interglacial periods enabled pedogenesis as the dominant process in inland areas.

weathering, fluvial and colluvial comminution, eolian abrasion, and ballistic impact (Muhs and Bettis, 2003). However, the close association between the timing of loess-paleosol sequences and the late Quaternary eccentricity cycles indicates that these sequences are largely climate-controlled, and that most of the eolian silt that contributed to the formation of loess is of glaciogenic origin, produced by glacial grinding during glaciations (Fig. 3.5; Muhs and Bettis, 2003). The loess-paleosol sequences qualify as "depositional" sequences in sequence stratigraphic terms, as paleosols are the expression of subaerial unconformities in continental settings subject to sediment starvation.

In terms of sequence architecture, differences between Icehouse and Greenhouse conditions are most evident in coastal to shallow-water settings, where sea-level changes have the most significant effect on the geometry of sequences. Icehouse sequences are typically thin (<50 m, and in many cases <10 m), often organized in stacks of several in succession, incomplete in terms of systems tracts, and severely top-truncated (Fig. 3.6). Examples of such glacial–interglacial cycles within the range of orbital forcing (10^4–10^5 yrs) come from late Paleozoic, Neogene, and Pleistocene successions formed under Icehouse conditions. The distinctive vertical stacking pattern was observed initially in the

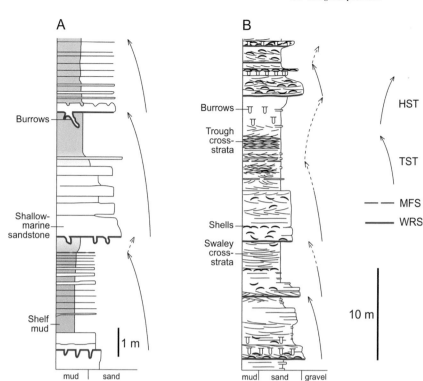

FIGURE 3.6 Stacked and top-truncated meter- to decameter-scale shallow-marine sequences dominated by transgressive deposits, generated by glacio-eustatic cycles during Icehouse conditions (modified from Catuneanu and Zecchin, 2013). A—The Miocene Calvert Cliffs succession, Maryland, USA (modified from Kidwell, 1997); B—The lower Pliocene Belvedere Formation, Crotone Basin, southern Italy (modified from Zecchin, 2005). Abbreviations: WRS—wave-ravinement surface; MFS—maximum flooding surface; TST—transgressive systems tract; HST—highstand systems tract.

Neogene continental margin successions around Antarctica (Bartek et al., 1991, 1997; Fielding et al., 2000; 2001; Naish et al., 2001), and it was also documented from Plio-Pleistocene sections in New Zealand (Naish and Kamp, 1997; Saul et al., 1999), the Miocene Chesapeake Group of eastern USA (Kidwell, 1997), the Miocene of western Chile and Ecuador (Di Celma and Cantalamessa, 2007; Cantalamessa et al., 2005, 2007), and the Lower Permian in the Sydney Basin of eastern Australia (Fielding et al., 2006).

The peculiar architecture of Icehouse sequences is attributed to the high-amplitude of the glacio-eustatic cycles, whereby erosional transgressions across shelves exposed during sea-level fall resulted in the removal of part or all of the highstand portion of sequences and the preservation of dominantly transgressive deposits (Fig. 3.6; Fielding et al., 2006). This trend is most evident along oceanic coasts, where wave erosion is stronger than that operating in interior basins, leading potentially to the removal of tens of meters of sediment during transgression (Demarest and Kraft, 1987; Leckie, 1994). Other conditions that lead to the development of thicker transgressive systems tracts include high sediment supply during slower transgressions, such as during the late Pliocene and early Pleistocene, when the glacio-eustatic cycles were dominated by the 40 kyr obliquity characterized by a relatively symmetrical shape of the sea-level curve that led to slower transgressions than those typifying the more asymmetrical late Quaternary

glacio-eustatic cycles with the same periodicity (Cantalamessa and Di Celma, 2004).

However, not all glacio-eustatic sequences are dominated by transgressive deposits. The late Quaternary high-frequency sequences related to the isotope substage cyclicity (10^4 yrs timescales) commonly show the dominance of regressive deposits, due to the rapid glacio-eustatic rises that inhibited sediment accumulation during transgressions (Fig. 3.7; Zecchin et al., 2010b). In this case, the thickness of transgressive deposits is reduced substantially, and sequences are dominated by highstand or falling-stage regressive deposits (Fig. 3.7). Lowstand deposits are typically absent in inner to middle shelf settings as they accumulated close to the shelf edge following the exposure of the shelf during the high-amplitude sea-level falls (Zecchin et al., 2011).

Irrespective of their dominant systems tracts, all glacio-eustatic sequences related to an Icehouse climatic regime share common features including reduced thicknesses, significant top truncation, and incomplete systems tract development (Posamentier et al., 1992b; McMurray and Gawthorpe, 2000; Fielding et al., 2006; 2008; Nalin et al., 2007; Lucchi, 2009; Zecchin et al., 2009a,b; 2010a,b, 2011). The reduced thickness of sequences is generally related to the limited time available for the deposition of transgressive and highstand deposits during sea-level rise, the marked stratigraphic foreshortening during sea-level fall, and the ravinement

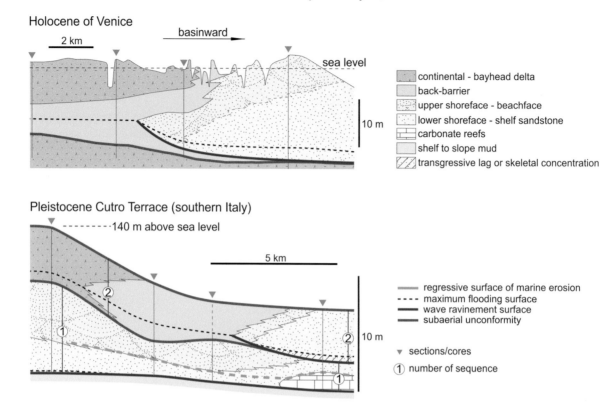

FIGURE 3.7 Glacio-eustatic sequences related to the marine isotope substage cyclicity during late Quaternary (modified from Zecchin et al., 2010b). The Holocene deposits of the Venice lagoon are interpreted by correlating cores, whereas the two superposed cycles of the middle Pleistocene Cutro Terrace are based on outcrop data. In both cases, sequences are dominated by regressive deposits (see text for details).

processes that occur during transgressions (Zecchin et al., 2010b). The systems tract composition of the high-frequency Icehouse sequences depends on the degree of asymmetry of the sea-level curve (i.e., the higher the asymmetry the less time for the accumulation of transgressive and highstand deposits), as well as the amounts of subaerial and transgressive wave erosion that affect the preservation of highstand deposits.

Additional controls on the composition of Icehouse sequences include the rates of local subsidence or uplift, the amounts of sediment supply, the basin physiography, and the environmental energy (e.g., wind, rivers, subaqueous currents) during the glacio-eustatic cycles (Catuneanu and Zecchin, 2013). Longer term cycles, such as those on 10^5 yrs eccentricity timescales, become less predictable in terms of internal architecture, as local controls unrelated to glacio-eustasy have more time to interfere with the climate-driven cycles. As such, the variable architecture exhibited by larger scale sequences is thought to reflect the increased influence of the tectonic setting (Zecchin et al., 2010b).

3.2 Autogenic processes

Autogenic processes are those internal to a depositional system, which lead to the redistribution of

sediment within a depositional environment without changes in the total energy and sediment budget of that environment (Einsele et al., 1991). Examples include the migration of depositional elements with various degrees of sedimentological complexity, from bedforms and macroforms to channels and channel belts, in response to changing equilibrium conditions within the depositional environment. Autogenic processes may be observed at different scales, from channel avulsion and delta-lobe switching (10^0–10^3 yrs timescales) to the relocation of alluvial channel belts and submarine fans (10^3–10^5 yrs timescales) (Fig. 3.1).

Beerbower (1964) was the first to differentiate between the internal and external controls on sedimentation. He termed the former "autocyclicity," although the term "autogenic" is now preferred since these processes are not necessarily cyclic in nature (Miall, 1996). A number of common patterns at sedimentological scales can be explained by such processes, including the fining-upward profiles of channel fills or point bars, and much of the sedimentological research in the 1960s and 1970s was focused on understanding the inner workings of autogenic processes (Miall, 1996). Subsequently, it has been realized that this internal reorganization of depositional systems is also important at stratigraphic scales in terms of the timing of sequence stratigraphic surfaces whose origins depend at least in part on sediment supply.

Autogenic changes in sediment dispersal patterns can generate cyclicity at the scales of high-resolution sequence stratigraphy (Catuneanu and Zecchin, 2013, Fig. 3.1), although the influence of autogenic processes on the timing of sequence stratigraphic surfaces can be observed at all stratigraphic scales (e.g., the timing of a first-order maximum flooding surface can also be modified by variations in sediment supply along the depositional strike; details in Chapter 7).

Autogenic processes can modify the direction of shoreline shift (progradation vs. retrogradation) in a manner that is independent of any allogenic control, in which case they have a direct impact on the formation of systems tracts and bounding surfaces. For example, delta lobe switching due to the avulsion of distributary channels on the delta plain can change significantly the amount of riverborne sediment transported by longshore currents to the shoreface environments on either side of the delta, influencing the direction of shoreline shift (e.g., Elliott, 1975; Pulham, 1989; Catuneanu and Zecchin, 2013, Fig. 3.8). Therefore, even though autogenic avulsion of distributary channels and delta lobe switching are processes intrinsic to the deltaic environment, their effects can influence depositional processes in all environments to which the sediment is transferred. Changes in the direction of shoreline shift in response to autogenic river diversions can affect areas of tens to hundreds of km along the coastline, and are difficult to differentiate from those that may be caused by variations in the rates of subsidence or sea-level rise (Amorosi et al., 2005; Stefani and Vincenzi, 2005, Figs. 3.9 and 3.10).

The link between contemporaneous depositional environments as interrelated components of a unitary sediment dispersal system indicates that autocyclicity is not necessarily "local cyclicity," as its effects may transcend the depositional-system limits, with impact on the stratigraphic architecture (e.g., Muto et al., 2007). Autogenic processes may affect areas as large as $10^1 - 10^2$ km along strike and dip, over timescales of $10^3 - 10^5$ yrs (e.g., Muto and Steel, 2002; Amorosi et al., 2005; Stefani and Vincenzi, 2005), which are comparable to, or even greater than the physical and temporal scales of some high-frequency sequences of allogenic origin. The larger the area affected by autogenic processes, the more difficult is to differentiate these processes from allocyclicity. The distinction between external and internal controls on sedimentation becomes increasingly difficult in the case of sparse exposures or cores, where the lateral development of stratigraphic units identified in vertical sections is poorly constrained.

3.3 Accommodation vs. sedimentation

The allogenic and autogenic variables in Fig. 3.1 combine into competing processes of creation of space

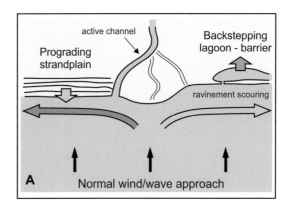

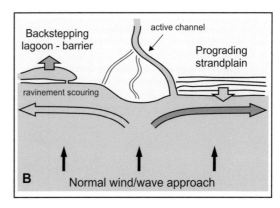

FIGURE 3.8   Shoreline shifts controlled by autocyclic river avulsion and delta lobe switching under conditions of relative sea-level rise in a shelf setting (from Catuneanu and Zecchin, 2013). Autocyclic shifts in the location of the active channel from time A to time B result in changes in the sediment supply delivered to the adjacent open shorelines by longshore currents. Consequent changes in the direction of shoreline shift from time A to time B generate concurrent maximum regressive and maximum flooding surfaces on the opposite sides of the delta.

for sediment to fill (accommodation) and consumption of space (sedimentation), which control the development of sequence stratigraphic frameworks in every sedimentary basin (e.g., Jervey, 1988; Schlager, 1993; Posamentier and Allen, 1999; Catuneanu, 2006; 2019a). This "dual control" on the stratigraphic architecture accounts for all underlying factors, whose relative contributions are otherwise difficult to separate and quantify (Fig. 3.3). The sequence stratigraphic methodology does not require the distinction between the various controls on stratigraphic cyclicity, but only the identification of stratal stacking patterns that result from the interplay of accommodation and sedimentation (Fig. 3.3). The correct understanding of the meaning of "accommodation" and "sedimentation" is critical for the proper application of sequence stratigraphy.

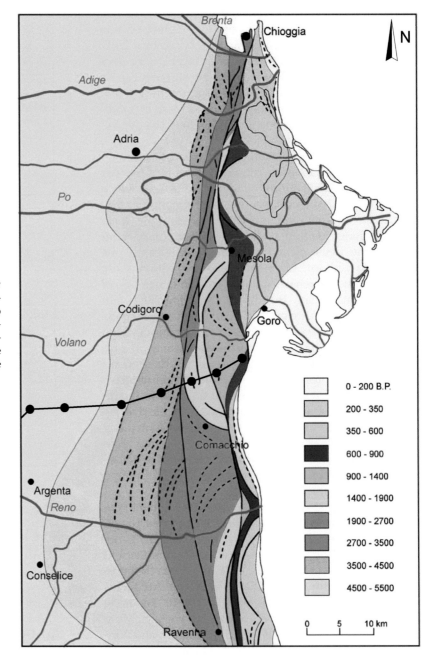

FIGURE 3.9 Distribution of high-frequency transgressive-regressive sequences developed during the long-term highstand progradation of the Po delta and adjacent shoreline (from Stefani and Vincenzi, 2005). Sediment from the Po delta was transported to the south by longshore currents. The transect on the map indicates the location of the cross section in Fig. 3.10.

3.3.1 Accommodation

Accommodation is the space made available for potential sediment accumulation, by subsidence, sea-level rise, or a combination of these two processes (Jervey, 1988; Posamentier et al., 1988; Van Wagoner, 1995; Neuendorf et al., 2005, Fig. 3.11). Accommodation is measured independently of sedimentation, allowing one to contrast these two controls on the stratigraphic architecture (i.e., the rates of creation vs. the rates of consumption of space). This is a fundamental tenet of conventional sequence stratigraphy, which explains the formation of stratal stacking patterns that define systems tracts in relation to the balance between accommodation and sedimentation *at the shoreline* (Catuneanu, 2006). The rates of accommodation and sedimentation are variable across a sedimentary basin, both along dip and strike directions. Only the rates along the shoreline are relevant to the timing of units and bounding surfaces that define the conventional sequence stratigraphic framework. The separation between accommodation and sedimentation becomes more difficult and less meaningful in areas remote

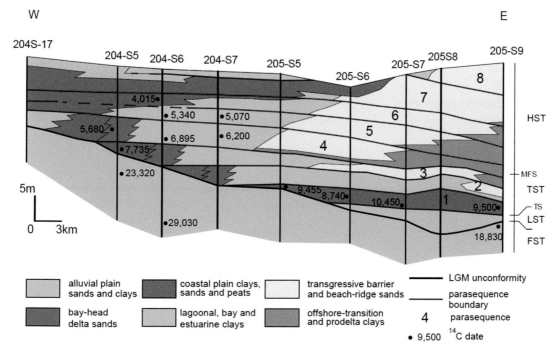

FIGURE 3.10 Stratigraphic cross section (location in Fig. 3.9) showing the Holocene architecture south of the present-day Po delta (from Amorosi et al., 2005). The timing of the TST boundaries (TS and MFS) is controlled by the interplay of allogenic (relative sea level) and autogenic (variations in sediment supply from the Po delta) factors. The higher frequency cyclicity within the TST and HST is largely controlled by the autocyclic shifts of the Po delta (Amorosi et al., 2005). The terminology applied to these high-frequency cycles ranges from "parasequences" to "small-scale cycles" and "high-frequency sequences." The latter term is preferred (details in Chapter 5). Abbreviations: FST—falling-stage systems tract; LST—lowstand systems tract; TST—transgressive systems tract; HST—highstand systems tract; TS—transgressive surface (= maximum regressive surface); MFS—maximum flooding surface.

from the shoreline, both in updip and downdip directions. The concept of accommodation also applies to lacustrine settings, by substituting "sea level" with "lake level."

The reference horizon that marks the top of available accommodation within a sedimentary basin is represented by the eustatic sea level, in basins connected to the global ocean, and by the elevation of the outflow ridge in interior basins disconnected from the global ocean (Fig. 3.12). Relative to this reference horizon, accommodation can be underfilled or overfilled (i.e., depositional surface below or above the reference horizon, respectively; Fig. 3.12). Sedimentary basins typically evolve from underfilled to overfilled, as accommodation is consumed by sedimentation. Available accommodation in underfilled basins may be subaqueous (below the sea/lake level) or subaerial (in interior basins where the sea/lake level is below the elevation of the outflow ridge; Fig. 3.12). However, the availability of subaerial accommodation takes a subordinate role in any sedimentary basin that includes subaqueous accommodation. Wherever present, the sea/lake level, due to its link to the shoreline, is the relevant reference for the amount of accommodation that is involved in the formation of conventional systems tracts and bounding surfaces. For this reason, *subaqueous* accommodation is

typically inferred when reference is made to "accommodation," in any sedimentary basin that includes a marine or lacustrine depocenter, and hence, a shoreline.

The role of sea/lake level as a control on accommodation is restricted to the submerged areas, whereas subsidence controls accommodation in all depositional settings across a sedimentary basin (Fig. 3.11). The contribution of eustasy to accommodation (<200 m) is typically at least one order of magnitude smaller than the contribution of subsidence (commonly reaching 10^3 m in depocenters). Sedimentary basins may be underfilled, where space created by basin-forming mechanisms is still available for sediments to fill (e.g., Gulf of Mexico, Dead Sea, Lake Malawi), or overfilled, where all accommodation generated by basin-forming mechanisms was consumed by sedimentation (e.g., Ganga Basin in front of the Himalayas). In the latter settings, aggradation may still continue in continental environments, driven by sediment supply that exceeds the transport capacity of wind and rivers. In the long term, however, the preservation potential of the sediment in excess of accommodation is poor, as denudation of the source areas proceeds following the cessation of tectonic activity.

In sedimentary basins connected to the global ocean (e.g., Gulf of Mexico; Fig. 3.12A), the sea level defines

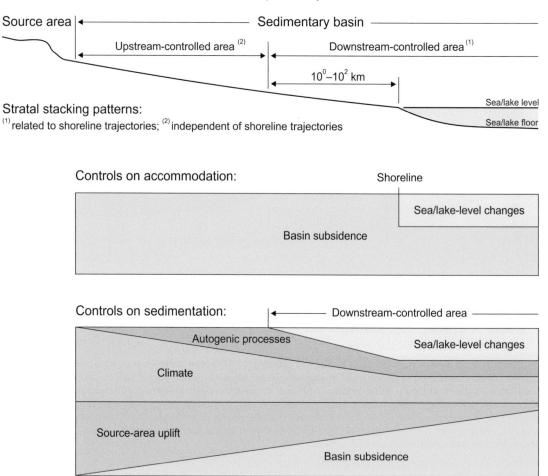

FIGURE 3.11   Controls on accommodation and sedimentation in downstream- and upstream-controlled settings. The downstream-controlled area includes continental, coastal, and marine (or lacustrine) systems which respond to changes in relative sea/lake level. The upstream-controlled area includes continental systems which are beyond the influence of relative sea-level changes. Accommodation is controlled by sea-level changes (downdip from the shoreline) and subsidence (all across the sedimentary basin). Subsidence relates to tectonism, with additional contributions from glacio-isostasy, sediment loading, and compaction. Sedimentation is controlled by all factors which modify sediment supply and the energy of the sediment-transport agents, including accommodation, climate, source-area uplift, and autogenic processes that affect the patterns of sediment distribution. Despite the dependence of sedimentation on accommodation, the two controls on the stratigraphic architecture can be measured independently of each other, which is particularly meaningful along the shoreline (see text for details). The amount of basin subsidence may increase in either updip or downdip directions, depending on the tectonic setting. Climate exerts a control on both accommodation and sedimentation, at multiple scales. The contribution of climate to accommodation is accounted for under subsidence (glacio-isostasy) and sea/lake-level changes. Climate also controls sedimentation by modifying sediment supply and the energy of the sediment-transport agents. Autogenic processes may also be observed at different scales, from channel avulsion and delta-lobe switching to the relocation of alluvial channel belts and submarine fans (Fig. 3.1).

the upper limit of available (i.e., unfilled) accommodation. In such settings, accommodation is underfilled in the marine environments, and overfilled in the continental environments above the sea level. At any given time, the shoreline marks the limit between the underfilled and the overfilled portions of the sedimentary basin. The location of the shoreline shifts through time in response to the changing balance between accommodation and sedimentation. River systems that are anchored to the sea respond to changes in the location of the shoreline, as well as to all other controls that modify fluvial energy and sediment load (Fig. 3.11).

In sedimentary basins isolated from the global ocean (e.g., Dead Sea, Lake Malawi; Fig. 3.12B,C), the local sea level (or lake level) substitutes the global (eustatic) sea level as the upper limit of subaqueous (marine or lacustrine) accommodation. The local sea/lake level may be either below (e.g., Dead Sea) or above (e.g., Lake Malawi) the eustatic sea level. Even though available accommodation in such inland basins is technically up to the elevation of the outflow ridge, which may be above the elevation of the local sea/lake level (Fig. 3.12B), the local sea/lake level remains the relevant reference horizon for any practical purposes,

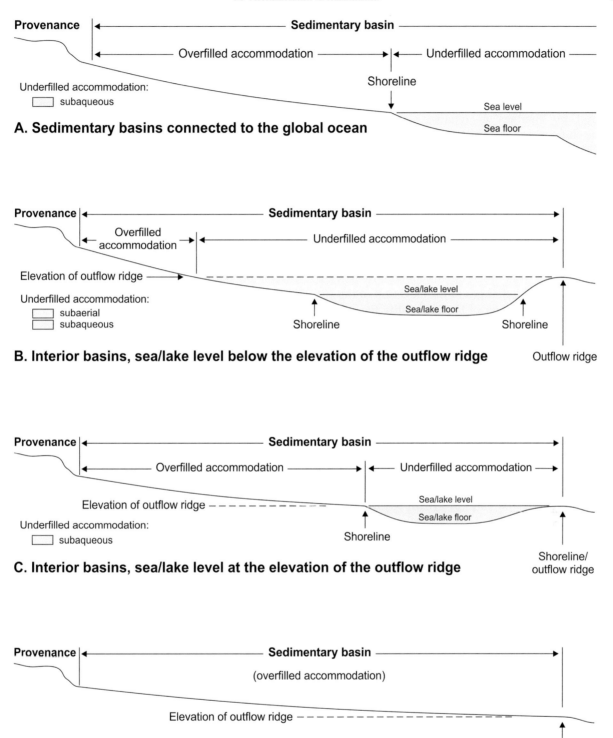

FIGURE 3.12 Underfilled vs. overfilled accommodation in sedimentary basins (from Catuneanu, 2017). Accommodation is measured up to the eustatic sea level in basins connected to the global ocean, and up to the elevation of the outflow ridge in interior basins that are isolated from the global ocean. In the case of basins connected to the global ocean (case A), as well as interior basins where the sea/lake level is at the elevation of the outflow ridge (case C), the shoreline marks the limit between underfilled and overfilled accommodation. In the case of interior basins where the sea/lake level is below the elevation of the outflow ridge (case B), accommodation can also be underfilled in continental settings. Sedimentary basins become entirely overfilled where all accommodation is consumed by sedimentation (case D). Sediment accumulation may continue in overfilled settings, driven by sediment supply that outpaces the energy of the sediment-transport agents. In a long term, the sediment in excess of accommodation has a lower preservation potential as provenance areas are subject to denudation.

since the stratigraphic architecture is tied to the shore-line and its shifts through time. Overfilled inland basins become entirely continental, with fluvial systems anchored to the outflow ridge (e.g., Ganga Basin; Fig. 3.12D).

In marine settings, changes in accommodation are measured by changes in relative sea level, which describe shifts in sea level relative to a subsurface reference horizon (the "datum" in Fig. 3.13; Jervey, 1988; Posamentier et al., 1988). The datum can be any mappable horizon in the subsurface (e.g., the basement top, or a stratigraphic marker within the basin fill), but the closer to the seafloor the better it incorporates the contributions of sediment loading and compaction to the total amount of subsidence. Changes in the elevation of the datum relative to the center of Earth quantify the amounts of subsidence or uplift in the basin, independently of sedimentation. Similarly, changes in the elevation of the sea level relative to the center of Earth quantify the magnitude of eustatic oscillations. The datum helps to define both accommodation (changes in the elevation of the sea level relative to the datum) and sedimentation (changes in the elevation of the seafloor relative to the datum; Fig. 3.13).

Relative sea-level changes account for the combined influence of eustasy and basin subsidence on accommodation. There are multiple combinations of subsidence/uplift and sea-level changes that can lead to rises or falls in relative sea level (Figs. 3.14 and 3.15). Accommodation is generated by relative sea-level rise, and it is consumed by sedimentation. At any point in time, and at any specific location, the space between the sea level and the seafloor (i.e., the water depth) defines the amount of accommodation that is still available for sediment to fill (unfilled accommodation: the balance between generation and consumption of space; Figs. 3.13 and 3.14). The rates of relative sea-level change are variable across a sedimentary basin, reflecting variations in the rates of subsidence. The interplay between accommodation and sedimentation *in coastal environments* controls the trajectory of the shoreline, which defines the stratal stacking patterns associated with "conventional" systems tracts (i.e., "lowstand," "transgressive," "highstand," and "falling-stage" systems tracts; details in Chapters 4 and 5).

Relative sea-level changes influence sedimentation on either side of the coastline, within the "downstream-controlled" portion of a sedimentary basin (Shanley and McCabe, 1994; Holbrook, 1996; Blum and Tornqvist, 2000; Holbrook et al., 2006, Fig. 3.11). The downstream-controlled area may extend for 10^0–10^2 km updip

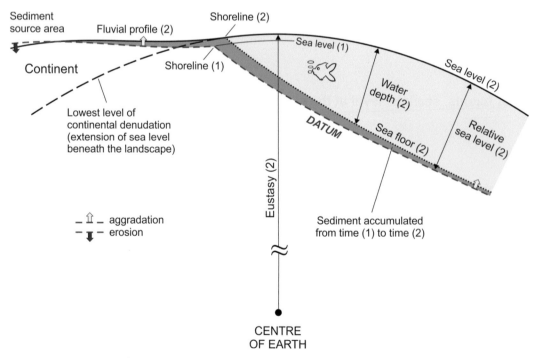

FIGURE 3.13 Concepts of relative sea level, sedimentation, and water depth (modified from Posamentier et al., 1988). The datum is a subsurface horizon (e.g., the depositional surface at time 1). Changes in the elevation of the datum relative to the center of Earth quantify the amounts of subsidence or uplift at any location. Changes in the elevation of the sea level relative to the datum define relative sea-level changes at any location. Water depth is the space still available for sediment accumulation below the sea level; i.e., the balance between the space created by relative sea-level rise and the space consumed by sedimentation (see text for details).

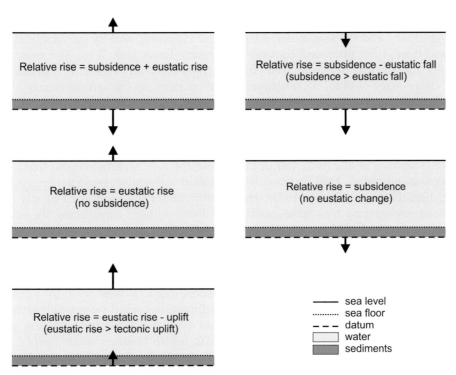

FIGURE 3.14 Combinations of subsidence/uplift and sea-level changes that lead to relative sea-level rise. The space created by relative sea-level rise can be consumed by sedimentation at higher or lower rates, leading to water shallowing or deepening, respectively.

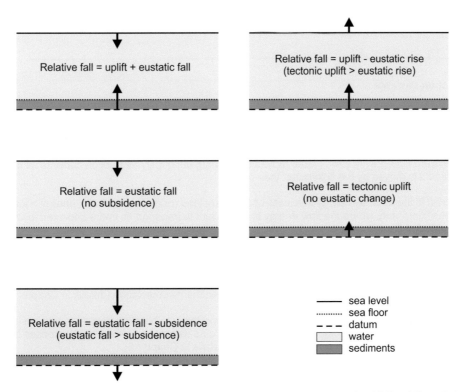

FIGURE 3.15 Combinations of subsidence/uplift and sea-level changes that lead to relative sea-level fall. A fall in relative sea level results in a loss of accommodation, which invariably leads to water shallowing.

from the shoreline into the continental setting, depending on topographic gradients and the size of the rivers (Blum, 1994; Blum and Price, 1998; Blum and Tornqvist, 2000), and potentially for 10^3 km downdip from the shoreline into the deep-water setting. Downstream-controlled settings play an important role in the development of sequence stratigraphy, as all conventional sequence stratigraphic models of the 1970s—1990s describe a stratigraphic architecture that is linked to the relative sea level (Mitchum et al., 1977; Vail et al., 1977a; Jervey, 1988; Posamentier et al., 1988; Posamentier and Vail, 1988; Sarg, 1988; Hunt and Tucker, 1992). The application of sequence stratigraphy was subsequently expanded to upstream-controlled settings where "unconventional" stratigraphic patterns form independently of relative sea-level changes (Shanley and McCabe, 1994; Blum and Price, 1998; Leckie and Boyd, 2003, Fig. 3.11).

3.3.2 Sedimentation

Sedimentation may be described in terms of influx of sediment to a depositional area (i.e., sediment supply), or in terms of rates of sedimentation. Sediment supply and sedimentation rates are two distinct parameters, whose relationship is mediated by the energy of the transport agents that distribute the sediment across the basin. In this context, "energy" refers to the flow capacity to transport sediment, driven by the motion of the transport medium in traction currents (wind, rivers, longshore currents, storm currents, tidal currents, contour currents), and by gravitational shear in sediment-gravity flows (Fig. 3.16). For example, high sediment supply does not necessarily translate into high sedimentation rates, as the sediment may bypass areas of high energy, and accumulate in areas of lower energy where the transport agent is unable to move its entire sediment load. Therefore, given the fact that the environmental energy flux can limit sediment accumulation, the descriptor of "sedimentation" that is relevant to the formation of stratal stacking patterns is the rate of sedimentation rather than sediment supply (Fig. 3.2).

Even without considering the role of environmental energy flux, volumetric calculations under theoretical conditions of constant relative sea-level rise and constant sediment supply indicate that stratal stacking patterns may still change from progradation to retrogradation, due to a decrease in the rates of sedimentation *if* sediment is spread over wider areas during progradation (i.e., the autoretreat process of Muto and Steel, 2002, which assumes water deepening at the toe of the clinoforms, leading to an increase with time in the areal extent of clinoform surfaces; Fig. 3.17). This further illustrates the point that the rate of sedimentation at the shoreline, rather than sediment supply, is the relevant variable in the formation of stacking patterns that define systems tracts. The difference between sediment supply and the rate of sedimentation is even more evident in the real world, where variations in the energy of the transport agents play a major role in the dispersal patterns

---

Traction currents

The *sediment* is dragged by the flow, irrespective of the mechanism that moves the transport medium (gravity, tides, or density variations of air or water). Examples: rivers, tidal currents, longshore currents, storm currents, wind, contour currents.

Rivers are fluid-gravity flows (water moved by gravity), but qualify as traction currents as the *sediment* is moved by the fluid rather than gravity.

Erosion is triggered by energy in excess of sediment load (underloaded flows: energy > load). Deposition occurs where the flow energy is insufficient to transport its entire sediment load (overloaded flows: load > energy).

Sediment-gravity flows

The *sediment* is moved by the action of gravity (gravitational shear). Examples: mudflows, grainflows, and turbidity flows, depending on the sediment composition (details in Chapter 5).

The erosive power of sediment-gravity flows is proportional to both the density (sediment load) and the velocity (energy) of the flow. Deposition occurs where the flow energy decreases below the level required to transport its sediment load (details in Chapter 5).

FIGURE 3.16   Types of flows, as a function of the control on sediment motion. Energy promotes erosion in both types of flows. Sediment load promotes aggradation in traction currents, but it contributes to erosion in sediment-gravity flows, particularly in the accelerating, higher energy parts of flows (details in Chapter 5).

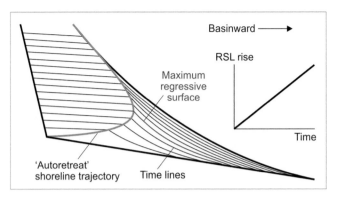

FIGURE 3.17 The "autoretreat" concept illustrated by an auto-genic change in the direction of shoreline shift (modified from Muto and Steel, 1997, 2002). Under conditions of constant relative sea-level (RSL) rise and sediment supply, progradation may be followed by shoreline retreat if water depth at the toe of the clinoforms increases during this process. This assumes that the bottomset aggrades slower than the topset of the prograding system, expanding the surface area of the clinoforms and reducing the rates of sedimentation as sediment is spread over wider areas with time. The model shows the difference between sediment supply and sedimentation, and that only the latter is relevant to the formation of stratal stacking patterns. The maximum regressive surface marks the change from progradational to retrogradational stacking patterns.

and the rates of sediment accumulation at any particular location within a basin.

In any depositional setting, the rates of sedimentation reflect the interplay of sediment supply and environmental energy flux. Both sediment supply and energy flux fluctuate over various timescales, resulting in highly variable sedimentation rates which tend to decrease with the increase in the scale of observation, as more hiatuses are incorporated within the measured sections (Miall, 2015). This is increasingly evident toward the basin margins, which are more susceptible to sediment bypass or erosion. At any scale of observation, changes in the balance between the energy of the sediment-transport agents (transport capacity) and their sediment load control the depositional trends of aggradation or degradation (Figs. 3.18 and 3.19). This principle holds true irrespective of the depositional setting and the nature of the transport agent (water, wind). A flow which has more energy than that required to transport its sediment load (i.e., underloaded flow) erodes the substrate, whereas a flow whose energy is below the level that is required to transport its entire sediment load (i.e., overloaded flow) results in aggradation.

The rates of sedimentation or erosion at any specific location are controlled by all factors which modify sediment supply and the energy of the sediment-transport agents, including accommodation (subsidence ± eustasy), climate, source-area uplift, and autogenic processes that affect the distribution of sediment within the basin (Figs. 3.2 and 3.11). These factors operate at the same time, with effects on depositional processes that

either enhance or cancel each other out. It is also possible that the same control may have opposite effects in different environments. For example, glaciations typically reduce sediment supply to river systems as provenance areas are covered by ice caps, but enhance sediment supply to eolian systems due to the production of glaciogenic silt by glacial grinding (Fig. 3.5). However, sediment supply is only important in relation to the environmental energy flux. While glaciations reduce sediment supply to the river systems, the discharge and transport capacity also decrease to the point that fluvial aggradation may occur. At the opposite end of the spectrum, interglacial periods increase the sediment load in rivers, but the even greater increase in discharge coupled with differential isostatic rebound (increase in slope gradients) may result in fluvial incision. These trends are often observed in upstream-controlled settings, where the timing of climate-controlled fluvial sequences is out of phase with the typical fluvial response to sea-level changes (Figs. 3.20 and 3.21).

Extrabasinal sediment supply (e.g., in the case of siliciclastic settings) is independent of accommodation, whereas intrabasinal sediment supply (e.g., in the case of carbonates and evaporites) depends in part on accommodation. Nonetheless, irrespective of depositional setting and the origin of the sediment, the rates of accommodation and sedimentation can be measured independently of each other (i.e., creation vs. consumption of space). The distinction between accommodation and sedimentation as separate controls on the stratigraphic architecture is most meaningful in downstream-controlled settings, where the definition of systems tracts is linked to shoreline trajectories (Fig. 3.11). In particular, this distinction is critical in coastal settings, where the interplay of accommodation and sedimentation controls the trajectory of the shoreline and the formation of "conventional" systems tracts. At any scale of observation, the balance between the rates of accommodation and sedimentation may change along a shoreline, resulting in the coeval deposition of different systems tracts along strike, and the formation of diachronous systems tract boundaries (e.g., Catuneanu et al., 1998a; Posamentier and Allen, 1999; Catuneanu, 2006; Csato and Catuneanu, 2014; Schultz et al., 2020; Zecchin and Catuneanu, 2020).

Along the shoreline, changes in accommodation (A) are measured by the relative shifts in the elevation of subaerial clinoform rollovers (i.e., shoreline upstepping vs. downstepping), while sedimentation (S) is quantified by the rates of sediment accumulation (i.e., changes in the elevation of the sea level and of the depositional surface, respectively, relative to a reference horizon; Fig. 3.13). At any location, the rates of accommodation and sedimentation are typically different, as

FIGURE 3.18 Surface processes that reflect the interplay of sediment supply and wind energy in eolian environments. Sediment supply exceeding the transport capacity of winds results in the accumulation of sand as sheets or dunes, depending on flow regimes. Winds stronger relative to their sediment load lead to erosion and the formation of deflation surfaces. A—sand dunes in the Namib Desert (Namibia), formed as a result of high sediment supply (sediment supply > wind energy; photo courtesy of Roger Swart); B—deflation surface on Mars, with lag deposits (wind energy > sediment supply; photo courtesy of NASA); C—deflation surface in the Namib Desert, Namibia (wind energy > sediment supply); D—detail from C, showing heavy minerals concentrated as lag deposits at the top of the basement rocks (Precambrian dolomites).

they depend on different controls (Fig. 3.11). For example, a rise in relative sea level in a coastal setting is quantified by the relative increase in the elevation of the shoreline, which depends on the rates of subsidence and sea-level change. At the same time, the rate of sedimentation may be higher or lower than the rate of creation of space, depending on the interplay between sediment supply and environmental energy flux (Fig. 3.2). This imbalance results in depositional trends of coastal progradation (S > A) or retrogradation (A > S).

The separation between accommodation and sedimentation is less meaningful away from the shoreline, in both updip and downdip directions, where sedimentation becomes the dominant control on the stratigraphic architecture irrespective of the local accommodation conditions. For example, processes of aggradation and degradation in the deep-water setting reflect adjustments of the seafloor profile in response to changes in sediment supply and energy flux, regardless of the amount of accommodation available. Within

downstream-controlled areas, the shoreline remains the reference for the formation of conventional stacking patterns and systems tracts, from continental systems through to deep-water systems (details in Chapters 4, 5, and 8). In upstream-controlled settings, cycles of aggradation and degradation generate depositional sequences with a timing constrained by the interplay of all controls on sedimentation, with or without a contribution from accommodation (Fig. 3.11). The relationship between sediment supply and energy flux is explained by the concept of base level, which controls processes of aggradation and degradation in all depositional settings, from sedimentological to stratigraphic scales.

3.4 Concept of "base level"

"Base level" is a surface of equilibrium which sedimentary processes strive to attain, at which neither erosion nor deposition takes place (Barrell, 1917). The

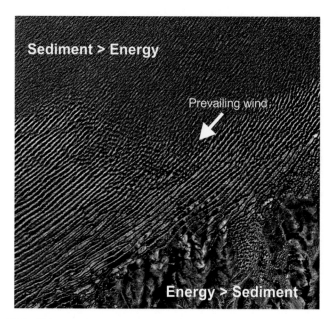

FIGURE 3.19 Satellite image of southern Arabian Peninsula showing a shift in the balance between sediment supply and wind energy across the area. Sediment supply exceeding the wind capacity to transport the sediment results in the accumulation of dunes, whereas the opposite leads to the exposure of the basement. The longitudinal dunes in this image are parallel to the prevailing wind, indicating an upper flow regime. This is higher energy than the lower flow regime that generated the transversal dunes in Fig. 3.18A.

ultimate base level for both continental and marine environments is the sea level, but *temporary* base levels exist before the ultimate base level is reached (Fig. 3.22). Barrell (1917) envisaged the base level fluctuating over a

wide range of timescales (his "harmonic oscillations in base level"), to explain the patterns of sediment preservation and the formation of unconformities of different magnitudes (Fig. 1.19). These "harmonic oscillations" refer to changes in the temporary base level (e.g., a seafloor, or a river bed) at various scales of observation, and therefore with various degrees of stratigraphic significance. As illustrated by Barrell (1917), sediment accumulation can only occur when the temporary base level rises, and preservation is only possible when the higher frequency oscillations take place during a longer term rising trend (i.e., deposition outpaces erosion in a long term; Fig. 1.19).

In any depositional setting, a temporary base level is a surface of equilibrium between sediment supply and the energy of the transport agents. Sediment supply promotes deposition, whereas the energy of the transport agents promotes erosion. The balance between deposition and erosion can be reached at different scales (Fig. 1.19), and changes thereof generate physical surfaces of different significance, from bedding planes to stratigraphic contacts. As such, the temporary base level provides the unifying concept that governs sedimentary processes of deposition and erosion at all scales, from sedimentology to stratigraphy (Fig. 3.23). Examples of temporary base levels include the seafloor, river bed, and interdune eolian surfaces. In a long term, all temporary base levels strive to attain the sea level, in response to continental denudation and subaqueous sedimentation which, ultimately, tends to consume all available (unfilled) accommodation. Attaining the ultimate base level in all settings will remain elusive for as long as

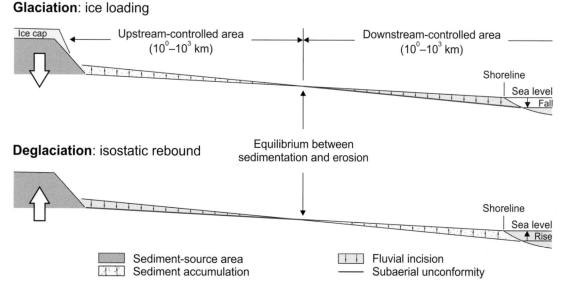

FIGURE 3.20 Fluvial responses to glacial–interglacial climate cycles. Glaciation leads to sea-level fall and common incision in downstream-controlled settings, but to aggradation in upstream-controlled settings in response to the decrease in discharge and topographic gradients. Deglaciation leads to the opposite trends. The scales and the relative development of upstream- and downstream-controlled areas are highly variable. In extreme cases, the upstream-controlled areas can extend to the shoreline (details in Chapter 4).

FIGURE 3.21 Aerial photograph of a modern incised valley (Red Deer River, Alberta; note farm houses for scale). Tributaries are also incised, which is a diagnostic feature of incised valleys. The incision was climate-driven, caused by the increase in fluvial discharge and topographic gradients associated with ice melting and isostatic rebound following the Late Pleistocene glaciation.

plate tectonics continue to operate, sustaining continental uplift and the formation of sedimentary basins.

Temporary base levels are fundamental to the stratigraphic architecture of sedimentary basins, at all scales of observation. Changes in the temporary base level may be observed from calendar timescales relevant to sedimentology (e.g., in response to seasonal changes in

sediment supply, or in response to fairweather-storm energy fluctuations) to geologic timescales relevant to stratigraphy (e.g., longer term changes in fluvial discharge and sediment supply in response to glacial—interglacial cycles of climate change). Consequently, temporary base levels may have different degrees of stratigraphic significance, from sedimentological contacts within stratigraphic units (e.g., bedding planes, local scours) to sequence stratigraphic surfaces that form at the end of depositional trends (e.g., a maximum flooding surface at the end of a backstepping trend, or a maximum regressive surface at the end of a progradational trend) (Fig. 3.23). Vertical changes in the temporary base level control the rates of sedimentation, which typically decrease with the increase in the scale of observation (Miall, 2015). These principles are independent of the origin and type of sediment (e.g., extrabasinal clastics or intrabasinal carbonates).

The concept of temporary base level (or simply "base level," as per the original usage of Barrell, 1917, Fig. 1.19) applies to both marine and continental environments, to include profiles of equilibrium of the depositional surface across an entire sedimentary basin (Fig. 3.22). At the limit between the marine and the continental portions of the base level, the shoreline represents the intersection between the temporary base level and the sea level, where the overall amounts of accommodation and sedimentation are in balance (i.e., the limit between the underfilled and the overfilled portions of a sedimentary basin; Figs. 3.11 and 3.22). The continental portion of the base level, commonly referred to as a "graded fluvial profile," is anchored to the shoreline, where changes in accommodation compete with all other

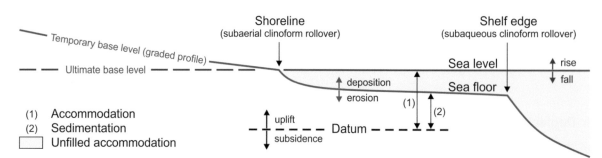

FIGURE 3.22 Accommodation vs. sedimentation in underfilled basins. Accommodation is space created by relative sea-level rise and consumed by sedimentation. Relative sea-level changes describe changes in accommodation, and account for the combined effects of sea-level change and subsidence. At any location, the amount of available (unfilled) accommodation (i.e., the water depth) reflects the balance between creation and consumption of space. Base level is the surface of equilibrium between sedimentation and erosion. The *ultimate* base level for subaqueous deposition and continental erosion is the sea level. Before the ultimate base level can be reached, *temporary* base levels are established as continental or marine graded profiles which the depositional surface strives to attain by means of sedimentation or erosion. The ultimate base level is linked to the concept of accommodation, whereas the temporary base level is a descriptor of sedimentation. Both subaerial and subaqueous clinoform rollovers may form within underfilled sedimentary basins. Only the trajectory of subaerial clinoform rollovers (i.e., the shoreline) matters to the definition of "conventional" systems tracts. The shoreline marks the limit between underfilled and overfilled accommodation, at the intersection between the ultimate and temporary base levels; i.e., the point of balance between the overall amounts of accommodation created and sediment accumulated within a sedimentary basin.

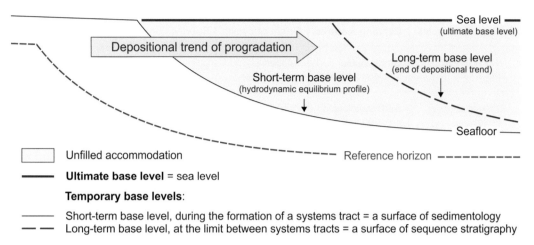

FIGURE 3.23   Base level—the unifying concept that explains processes of deposition and erosion from sedimentological to stratigraphic scales. Before the ultimate base level is reached, fluctuations in the temporary base level generate units and surfaces of sedimentology (i.e., short-term fluctuations during the formation of a systems tract), or units and surfaces of sequence stratigraphy (i.e., longer term fluctuations that involve changes in systems tract). Changes in the temporary base level relative to the reference horizon control the rates of sedimentation at both sedimentological and stratigraphic scales. The reference horizon is a datum relative to which changes in accommodation and sedimentation can be measured. Accommodation is space *made available* by relative sea-level rise, whereas water depth is space *still available* (i.e., unfilled accommodation: balance between creation and consumption of space). Changes in sea level relative to the reference horizon are referred to as "relative sea-level changes," to account for the effects of subsidence or uplift on accommodation. In contrast, changes in the temporary base level (or simply "base level," as per the original usage of Barrell, 1917) relative to the reference horizon are referred to as "base-level changes," as the depositional surface follows the subsidence or uplift of the reference horizon, and therefore sedimentation rates are measured independently of subsidence rates (i.e., the term "relative base-level changes" becomes redundant).

controls on sedimentation to determine shoreline trajectories and the depositional processes that accompany the shift of the shoreline (Fig. 3.11).

As a descriptor of sedimentary processes of aggradation and degradation, the base level accounts for all controls on sedimentation, including accommodation (subsidence ± eustasy), climate, source-area tectonism, and the autogenic redistribution of sediment (Fig. 3.11). In any depositional setting, these controls compete to determine the direction of shift of the temporary base level over various timescales, with a potential contribution from both allogenic and autogenic processes (Fig. 3.1). For this reason, the origin of sedimentary units bounded by hiatal surfaces, which are generated by base-level cycles, cannot be automatically interpreted in terms of allogenic vs. autogenic controls. The discrimination between the various allogenic and autogenic controls on sedimentary cyclicity, from sedimentological to stratigraphic scales, needs to be assessed on a case-by-case basis.

Short-term shifts in base level that occur during the formation of the lowest rank systems tracts generate units and surfaces of sedimentology at sub-stratigraphic scales (e.g., beds, bedsets, bedding planes; Fig. 3.23). These sedimentological cycles describe the internal architecture of the lowest rank sequence stratigraphic units. Short-term fluctuations in energy flux and sediment load may be related to both autocyclic and allocyclic processes, including the autogenic shift of geomorphic elements within depositional environments

(e.g., channel migration, bar and bedform migration), tidal cycles, fairweather–storm cycles, or seasonal changes in fluvial discharge and sediment supply (Figs. 3.24 and 3.25).

Longer term shifts in base level that involve changes in systems tract generate units and surfaces of sequence stratigraphy at the scales of the observed stratigraphic cycles. A long-term base level is the depositional surface at the end of a depositional trend observed at stratigraphic scales, and corresponds to a sequence stratigraphic surface (Fig. 3.23). For example, progradation occurs as the sedimentary system strives to reach a state of equilibrium at the maximum regressive surface, which marks the point of balance between progradation and retrogradation, and a change in coastal depositional system (e.g., from a delta to an estuary). Similarly, a subaerial unconformity starts forming when the depositional trend changes from aggradation to degradation in a continental setting. Long-term changes in base level are commonly attributed to allogenic controls, although autogenic processes are not excluded wherever sediment supply plays a role in the manifestation of a depositional trend (e.g., Catuneanu and Zecchin, 2013, Figs. 3.1 and 3.24).

Decoupling base-level changes (i.e., sedimentation) from relative sea-level changes (i.e., accommodation) provides the means to explain shoreline trajectories and the variety of depositional processes that may occur concurrently with the shifts of the shoreline. Due to the complex nature of sedimentary processes, which

Sedimentary cycles	Defining features	Origin of cycles	Subdivisions	Bounding surfaces
STRATIGRAPHY: sequences	Stratigraphic stacking patterns	Allogenic [1] or autogenic [2]	Systems tracts	Sequence stratigraphic
SEDIMENTOLOGY: bedsets	Sedimentological stacking patterns	Autogenic [3] or allogenic [4]	Beds, bedsets	Facies contacts

[1] Interplay of tectonism, climate, and eustasy; [2] Autogenic changes in sediment supply (channels, channel belts, deltas); [3] Migration of channels, macroforms, bedforms; [4] Tidal or fairweather-storm cycles, seasonal changes in energy/supply.

FIGURE 3.24   Classification of sedimentary cycles. Stratigraphic cycles (sequences) refer to cycles of change in *stratigraphic stacking pattern*, which involve changes in systems tract. Sedimentological cycles (bedsets) refer to cycles of change in *sedimentological stacking pattern*, within the lowest rank systems tracts. Both types of sedimentary cycles may display a nested architecture, and may form across wide ranges of overlapping scales (i.e., 10^0-10^1 m/10^2-10^5 yrs for high-frequency sequences, and $10^{-1}-10^0$ m/10^0-10^4 yrs for bedsets). However, the largest bedsets at any location are order(s) of magnitude smaller than the high-frequency sequence in which they are nested. The architecture of bedsets and sequences varies with the geological setting, reflecting the local conditions of accommodation and sedimentation. Criteria to discriminate between high-frequency sequences and bedsets in shallow-water settings have been discussed by Zecchin et al. (2017a,b). The distinction between stratigraphic and sedimentological cycles is more challenging in deep-water settings, whereby areas away from the paths of gravity-driven transport can accumulate pelagic sediment or contourites for periods of time that span multiple stratigraphic cycles. At such locations, the scale of sedimentological cycles defined by the recurrence of same-type depositional elements can be either smaller or larger than the scale of stratigraphic cycles (details in Chapter 8). The definition of stratigraphic and sedimentological cycles is independent of the interpretation of the underlying controls, as both types of sedimentary cycles can be produced by a combination of allogenic and autogenic processes (Catuneanu and Zecchin, 2013).

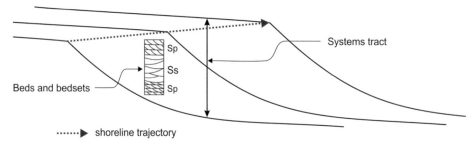

FIGURE 3.25   Bedsets in a sequence stratigraphic context. Bedsets are sedimentological cycles within the lowest rank systems tracts and component depositional systems. The formation of bedsets is attributed to variations in sediment supply or energy conditions, without changes in depositional environment and stratigraphic stacking pattern. In this example, bedsets form in relation to fairweather—storm cycles during a normal regression. Bedsets of different scales describe a nested architecture, commonly within ranges of $10^{-1}-10^0$ m and 10^0-10^4 yrs. The scale and the internal complexity of bedsets at any location depend on geological setting (i.e., local conditions of accommodation and sedimentation). Abbreviations: Sp—sandstone with planar cross-stratification; Ss—sandstone with swaley cross-stratification.

depend not only on accommodation but also on all other factors which modify sediment supply and environmental energy flux (Fig. 3.11), the timing of sediment aggradation or degradation (i.e., base-level rise or fall) at any location, including at the shoreline, may or may not correlate with the timing of positive and negative accommodation (i.e., relative sea-level rise and fall) (e.g., Summerfield, 1985; Pitman and Golovchenko, 1988; Butcher, 1990; Schumm, 1993; Posamentier and Allen, 1999; Blum and Tornqvist, 2000; Catuneanu, 2006; Catuneanu and Zecchin, 2016).

The rates and the direction of base-level changes (rise vs. fall) may also vary from one location to another, depending on the local conditions of sediment supply and energy flux; as a result, the base level at any given time describes a dynamic 3D surface that may be placed above (areas of sediment accumulation) or below (areas of erosion) the depositional surface (Barrell, 1917; Sloss, 1962; Cross, 1991; Catuneanu, 2006, 2017; Qayyum et al., 2017). The 3D morphology of the temporary base level is typically more complex in carbonates than it is in siliciclastic settings, due to the greater variations in the degree of syn-sedimentary lithification of carbonate seafloors (details in Chapter 8), but the same principles apply in all depositional environments. Moreover, at any scale of observation, the rates of sedimentation (i.e., the rates of base-level change) at any location are typically different from the rates of

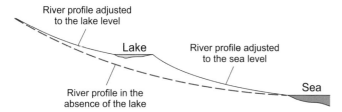

FIGURE 3.26 Marine and local base levels as illustrated by a river flowing into a lake and from the lake into the sea (modified after Press and Siever, 1986). In each river segment, the graded profile adjusts to the lowest level it can reach.

accommodation (i.e., the rates of relative sea-level change), resulting in constant changes in paleogeography (e.g., location of the shoreline) and paleobathymetric conditions.

The discussion in this section accounts for the presence of a sea within the basin under analysis. The same principles hold true for lacustrine environments, by substituting the "sea level" with the "lake level." In this case, the lake level becomes the "ultimate" base level for the lake floor and all fluvial systems draining into the lake (Fig. 3.26). Once the lake is filled with sediment, the next shoreline downstream becomes the new anchor for the fluvial graded profile (Fig. 3.26).

3.5 Allogenic vs. autogenic controls

The development of sequence stratigraphic units and bounding surfaces was traditionally attributed to allogenic controls (i.e., eustasy, tectonism, and climate). However, the role of autocyclicity (e.g., channel avulsion, delta lobe switching, migration of alluvial channel belts) in the formation of the sequence stratigraphic framework has become increasingly evident, particularly at (but not restricted to) smaller scales of observation (e.g., Muto and Steel, 2002; Catuneanu and Zecchin, 2013, Fig. 3.1). In essence, any time sediment supply plays a role in the formation of a sequence stratigraphic surface, whether marine or continental, there is potential for autogenic processes to interfere with, and partially control the timing of those surfaces. This is the case with all sequence stratigraphic surfaces, except those strictly controlled by relative sea-level fall (e.g., the basal surface of forced regression, or the regressive surface of marine erosion).

In downstream-controlled settings that are subject to relative sea-level rise, variations in sediment supply controlled by either allogenic or autogenic mechanisms may trigger changes in the direction of shoreline shift, as well as in coastal depositional system (e.g., from deltas to estuaries, or from prograding strandplains to backstepping lagoon-barrier island systems, and *vice*

versa), over different timescales (Fig. 3.8). Such variations play a critical role in the formation of systems tract boundaries such as maximum regressive and maximum flooding surfaces of different hierarchical ranks, even though the relative contribution of allogenic and autogenic controls may remain subject to interpretation. Even without variations in sediment supply, and under theoretical conditions of constant relative sea-level rise, processes of pure autogenic origin, such as "autoretreat," may still trigger changes in the direction of shoreline shift (e.g., Muto and Steel, 2002), and hence changes in stacking pattern and systems tract (Fig. 3.17).

In upstream-controlled settings, depositional sequences are generated by base-level cycles of aggradation and degradation which can also be observed at different scales (Fig. 1.19), with the smaller scale sequences being nested within larger scale systems tracts. At any scale, stratal stacking patterns and corresponding systems tracts are defined by the dominant architectural element (e.g., channel vs. overbank deposits in fluvial systems). The degree of channel amalgamation was traditionally interpreted in terms of accommodation conditions, with a "low-accommodation" systems tract dominated by amalgamated channels, and a "high-accommodation" systems tract dominated by floodplain deposits (Boyd et al., 2000; Zaitlin et al., 2000; 2002; Arnott et al., 2002; Wadsworth et al., 2002, 2003; Leckie and Boyd, 2003; Leckie et al., 2004). As accommodation in upstream-controlled settings is generated solely by basin subsidence (Fig. 3.11), the development of these "unconventional" systems tracts was linked directly to the rates of subsidence (i.e., low subsidence rates would promote channel amalgamation, whereas higher subsidence rates would be conducive to floodplain aggradation). However, the degree of channel amalgamation and the rates of floodplain aggradation depend not only on subsidence but on the interplay of several variables (Fig. 3.11), whose relative contributions are often difficult to quantify (see Chapter 4 for a full discussion).

It can be concluded that both allogenic and autogenic controls contribute to the development of the sequence stratigraphic framework. The discrimination between the two types of processes is challenging, as they operate and interplay across overlapping spatial and temporal scales, and may generate similar field expressions. Therefore, the definition and identification of stratal units and bounding surfaces must remain objective and based on the observation of stacking patterns, in a manner that is independent of the interpretation of the underlying controls (Catuneanu and Zecchin, 2013, 2016). The interpretation of the underlying controls on sequence development defines the objective of stratigraphic modeling, which is beyond the purpose of the sequence stratigraphic methodology (Fig. 1.2).

4

Stratal stacking patterns

Stratal stacking patterns describe the architecture of the sedimentary record, and provide the basis for the definition of all units and surfaces of sequence stratigraphy; therefore, they are key to the sequence stratigraphic methodology and nomenclature. Different types of stacking patterns can be defined in relation to the evolution of key geomorphic elements (e.g., the direction of shift of clinoform rollover points: forestepping, backstepping, upstepping, downstepping), in relation to depositional trends (progradation, retrogradation, aggradation, degradation), or based on the ratio between the depositional elements that comprise a depositional system (e.g., amalgamated channels vs. floodplain-dominated successions in fluvial systems). Not all types of stacking patterns are equally relevant to the definition of sequence stratigraphic units and bounding surfaces. For example, the trajectory of the subaerial clinoform rollovers (i.e., the shoreline) is more important to the definition of systems tracts and sequence stratigraphic surfaces than the trajectory of subaqueous clinoform rollovers (e.g., the shelf edge; discussion below). Also, the shoreline trajectory, rather than the coeval processes of aggradation or degradation, provides the key criteria for the definition of systems tracts and bounding sequence stratigraphic surfaces. Therefore, it is important to differentiate between the diagnostic and non-diagnostic stacking patterns that can be observed in the rock record (Catuneanu and Zecchin, 2016).

Stratal stacking patterns have different expressions in "downstream-controlled" settings (i.e., where the sequence stratigraphic framework relates to shoreline trajectories) vs. "upstream-controlled" settings (i.e., where the sequence stratigraphic framework is independent of shoreline trajectories) (Fig. 3.11). In downstream-controlled settings, the location and the trajectory of the shoreline depend on the interplay of relative sea-level changes (accommodation) and base-level changes (sedimentation) (Fig. 4.1). This interplay leads to depositional trends of coastal progradation or retrogradation,

accompanied by variable processes of sediment aggradation or degradation. In upstream-controlled settings, stratigraphic cyclicity relates to shifts in the temporary base level (combined influence of accommodation and all other controls on sedimentation; Figs. 3.11 and 3.22), which result in depositional trends of aggradation (accumulation of depositional sequences) and degradation (formation of subaerial unconformities). The internal architecture of upstream-controlled depositional sequences is described by "unconventional" stacking patterns defined by the dominant depositional elements (e.g., channels vs. floodplain in fluvial systems; Fig. 4.2).

The relative development of downstream- vs. upstream-controlled areas within a sedimentary basin depends on local circumstances. Overfilled basins are entirely upstream-controlled, with a stratigraphic architecture described by "unconventional" stacking patterns. Underfilled sedimentary basins, which still host a marine or a lacustrine environment, include both downstream-controlled and upstream-controlled settings (Fig. 3.11). The geographic location of the boundary between the two settings depends on multiple parameters (e.g., topographic gradients, the size of the rivers, tectonic setting, and the location of the provenance relative to the shoreline; Blum and Tornqvist, 2000), and is defined by the updip limit of the influence of relative sea-level changes on sedimentation (Fig. 3.11). It should be noted, however, that sedimentation, even in downstream-controlled settings, is the product of the interplay of multiple factors, including but not restricted to accommodation (Fig. 3.11). For this reason, the rates of sedimentation typically differ from the rates of accommodation in all depositional settings, whether downstream- or upstream-controlled. Where the provenance is located close to the shoreline (e.g., in a range of 10^0-10^1 km), the upstream controls may outweigh the effects of relative sea-level changes, leading to the extension of the upstream-controlled area all the way to the shoreline. In contrast, sedimentary basins with the provenance located far from the shoreline (e.g., range of

117

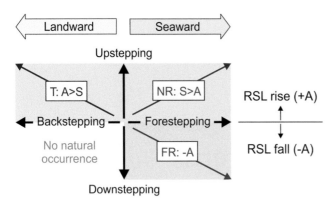

FIGURE 4.1 Shoreline trajectories, as defined by combinations of lateral (forestepping or backstepping) and vertical (upstepping or downstepping) shoreline shifts. All combinations are common in nature, except for transgression during RSL fall. The stratal stacking patterns that define systems tracts in downstream-controlled settings are linked to shoreline trajectories: normal regression (forestepping and upstepping), forced regression (forestepping and downstepping), and transgression (backstepping and upstepping). The amounts of shoreline upstepping and downstepping quantify the magnitudes of RSL rise and fall, whereas the extent of lateral shoreline shifts depends on the magnitude of RSL changes, sediment supply, and topographic gradients. Abbreviations: NR—normal regression; FR—forced regression; T—transgression; A—accommodation; S—sedimentation; RSL—relative sea level; +A—positive accommodation; −A—negative accommodation.

10^2–10^3 km) are prone to the development of wider downstream-controlled areas. In such cases, stratigraphic proxies that can be used to locate the updip limit of a downstream-controlled area include the head of incised valleys that form during relative sea-level fall, or the wedge-out limit of fluvial topsets that form during relative sea-level rise (Blum, 1994; Blum and Price, 1998; Blum and Tornqvist, 2000; Posamentier, 2001).

The same types of stratal stacking patterns can be observed at different scales, in relation to stratigraphic cycles of different magnitudes. Therefore, the same criteria can be applied consistently at all scales of observation (e.g., the retrogradation of a shoreline can be observed at different scales, describing transgressions of different magnitudes and hence transgressive systems tracts and maximum flooding surfaces of different hierarchical ranks). The internal architecture of the units defined by the observed stacking patterns becomes increasingly complex with the increase in the scale of observation (e.g., the internal makeup of a systems tract may vary from beds and bedsets to sequences of lower hierarchical ranks; Fig. 4.3). For example, a higher rank transgressive systems tract may include prograding systems, as well as maximum flooding surfaces, of lower hierarchical ranks (i.e., higher frequency transgressions and regressions during a long-term transgression, involving changes in coastal environment). This indicates that all types of sequence stratigraphic unit (i.e., sequences and their component systems tracts and depositional systems) can be observed at different scales (Catuneanu, 2017).

The variable internal makeup of systems tracts, from beds and bedsets to sequences of lower hierarchical ranks (Fig. 4.3), is evident in both downstream- and upstream-controlled settings. At each scale of observation, a cycle of change in stratal stacking patterns defines a stratigraphic sequence (Figs. 1.22, 3.24; Catuneanu and Zecchin, 2013). Within a sequence, each stratal stacking pattern defines a systems tract, and changes in the stacking pattern define systems-tract boundaries (i.e., sequence stratigraphic surfaces). These principles are at the core of the model-independent sequence stratigraphy, and keep the methodology and nomenclature independent of scale and data resolution, and hence, simple and objective. Given the variability and the basin-specific character of the stratigraphic record,

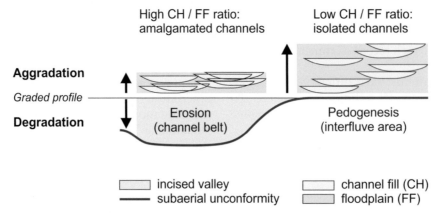

FIGURE 4.2 Depositional trends in upstream-controlled fluvial settings, where stacking patterns depend on (1) the rates of floodplain aggradation (proportional to the size of arrows in the diagram); (2) the ability of channels to shift laterally, which is a function of fluvial style; and (3) the frequency of avulsion. The processes and rates of aggradation and degradation depend on all factors which control sedimentation (i.e., sediment supply vs. energy flux), including accommodation, climate, source-area tectonism, and the autogenic controls on sediment distribution. The "Aggradation" part of the diagram illustrates the 7-step evolution of a channel under identical conditions of avulsion and lateral shift. In this case, the difference in stratigraphic architecture (i.e., amalgamated vs. isolated channels) is the result of differences in the rates of floodplain aggradation.

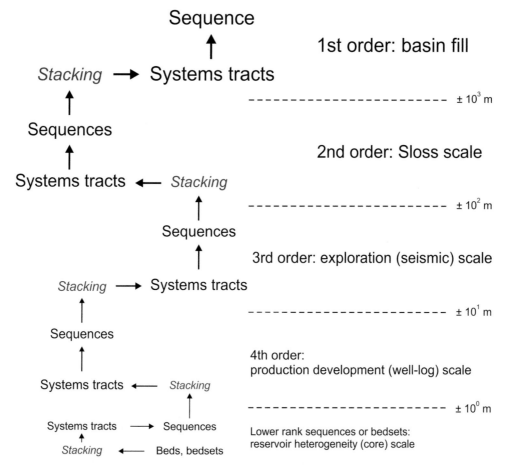

FIGURE 4.3   Stratigraphic scales, as defined by the purpose of study or the resolution of the data available. The stacking patterns of sedimentological units (beds, bedsets) define the lowest rank systems tracts, which are the smallest units of sequence stratigraphy. The same types of stratigraphic unit (sequences, systems tracts) can be observed at different scales; however, the stratigraphic architecture is not truly fractal, because sequences of different scales may have different underlying controls and internal composition of systems tracts. There are no standards for the temporal and physical scales of sequences and systems tracts, or for the lowest hierarchical rank that should be expected within a sedimentary basin. The scales in this diagram only represent statistical trends. Sequence stratigraphic frameworks are basin specific, and sequences of equal hierarchical ranks in different basins may differ in terms of timing, duration, thickness, geographic extent, and underlying controls.

any scale-dependent approach to the classification of stratal units and bounding surfaces is bound to be arbitrary and unrealistic. The observation of stratal stacking patterns, independent of scale and the interpretation of underlying controls, provides the flexible approach to a consistent application of sequence stratigraphy.

The criteria employed to identify stratal stacking patterns depend on the types of data available (e.g., outcrops, well data, seismic data), and generally rely on the observation of vertical profiles (e.g., log motifs) and/or stratal geometries (stratal terminations, stratal architecture) calibrated with facies data wherever possible (Fig. 1.1). The sequence stratigraphic framework is best constrained by the integration and mutual calibration of independent datasets (Fig. 1.1). Studies of the tectonic setting (type of sedimentary basin, subsidence mechanisms, dip and strike directions) and depositional setting (depositional environments and

paleogeography) must precede the interpretation of stratal units and bounding surfaces in the sequence stratigraphic workflow (details in Chapter 9). Vertical trends (e.g., fining- vs. coarsening-upward profiles) and stratal geometries (e.g., seismic-reflection terminations) need to be placed in a paleogeographic context for a reliable interpretation, as their meaning may change with the depositional setting. For example, fining-upward profiles and onlapping reflection terminations may develop in both continental and marine environments. Their stratigraphic significance (e.g., the topset of a regressive unit vs. the "healing-phase" wedge of a transgressive unit) can only be determined by placing them in the correct depositional setting.

4.1 Stratal terminations

Stratal terminations are defined by the geometric relationship between strata and the stratigraphic surface

FIGURE 4.4   Clinoform surfaces of a river-dominated delta front downlapping the finer grained bottomset (white arrows; Upper Cretaceous, Panther Tongue, Utah; note person for scale). The well-defined facies contact between the foreset and the bottomset of this prograding system (i.e., the "within-trend forced regressive surface," marked by red arrow; see Chapter 6 for definition) indicates rapid progradation, which is typical of forced regressions.

FIGURE 4.5   River-dominated delta showing prodelta fines overlain by delta front clinoforms and coal-bearing delta plain facies (Upper Cretaceous, Ferron Sandstone, Utah; the outcrop is c. 30 m high). Clinoform surfaces dip at an angle of 5—7 degree, and merge into the prodelta bottomset in a downdip direction (white arrows: apparent rather than real downlap). The gradual shift from prodelta to delta front indicates slow progradation, which is typical of normal regressions (compare with Fig. 4.4). The limit between foreset and topset (red arrow) is a diachronous facies contact (i.e., the "within-trend normal regressive surface"; definition in Chapter 6).

against which they terminate, and are best observed at larger scales, particularly on 2D seismic lines and in large-scale outcrops (Figs. 2.68, 4.4—4.6). The main types of stratal terminations are described by truncation, toplap, onlap, downlap, and offlap (Fig. 4.7). Except for truncation, which is a term stemming from classical geology (Grabau, 1906), all other stratal relationships have been introduced with the advent of seismic stratigraphy in the 1970s to define the architecture of seismic reflections (Mitchum and Vail, 1977; Mitchum et al., 1977). These terms have subsequently been incorporated into sequence stratigraphy in order to describe the stacking patterns of stratal units and to provide criteria for the recognition of the various surfaces and systems tracts

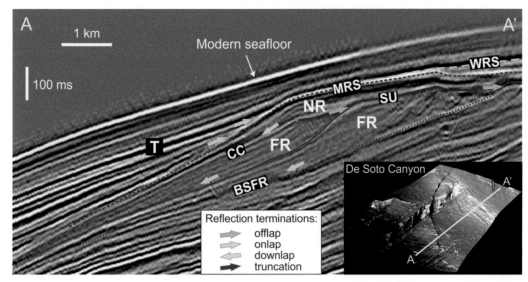

FIGURE 4.6   Stratal stacking patterns at seismic scale, in a shelf to slope setting (Pleistocene–Holocene, Gulf of Mexico; data courtesy of Henry Posamentier). The forced regression relates to the Late Pleistocene glaciation (sea-level fall and exposure of the continental shelf accompanied by fluvial erosion). The location of the shoreline at the shelf edge during the late stages of forced regression and lowstand normal regression enabled the progradation of the shelf-slope system, as seen on the 2D seismic line. The topset of the lowstand systems tract (NR) is a fluvial system at the seismic scale. However, this topset includes lower rank depositional systems at sub-seismic scales, from fluvial to shallow marine, related to higher frequency shoreline shelf-transit cycles. The clinoform surfaces within the seismic-scale systems tracts indicate episodic progradation during the higher frequency cycles. The 3D horizon is mapped at the base of the forced regressive deposits. Abbreviations: FR—forced regression (progradation with downstepping); NR—normal regression (progradation with upstepping); T—transgression (backstepping); BSFR—basal surface of forced regression; SU—subaerial unconformity; CC—correlative conformity; MRS—maximum regressive surface; WRS—wave-ravinement surface.

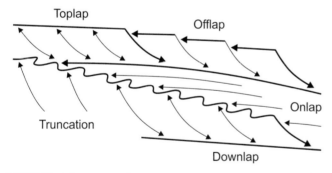

FIGURE 4.7   Types of stratal terminations: truncation, toplap, onlap, downlap, and offlap.

(e.g., Posamentier et al., 1988; Van Wagoner et al., 1988; Christie-Blick, 1991). The definitions of the key types of stratal terminations are provided in Fig. 4.8.

All types of stratal terminations in Figs. 4.7 and 4.8 can form in downstream-controlled settings in relation to the different types of shoreline shifts (Fig. 4.9), while truncation and fluvial onlap may also form in upstream-controlled settings. In some cases, the interpretation of stratal terminations in terms of syn-depositional shoreline trajectories is unequivocal; e.g., *coastal onlap* indicates transgression, and *offlap* is diagnostic of forced regression. In other cases, stratal terminations support alternative interpretations, as for example *downlap* may form in relation to either normal or forced regressions,

and additional criteria have to be used in order to arrive at unequivocal conclusions. In this example, the differentiation between normal and forced regressions that can generate *downlap* may be resolved by studying the trajectory of the subaerial clinoform rollovers in conjunction with sedimentological and stratigraphic criteria (Posamentier and Morris, 2000). *Onlap* terminations in continental and deep-water settings (i.e., fluvial and marine onlap, respectively) are also ambiguous in terms of their relationship to shoreline trajectories and relative sea-level changes (Fig. 4.9), and their precise origin can only be clarified by placing them in a broader stratigraphic context.

Processes of coastal aggradation during shoreline regression lead to the formation of topsets that may include delta plain (in prograding river-mouth settings), strandplain, and coastal plain (in open shoreline settings) deposits (Figs. 4.10–4.12). The topset is not a stratal termination, but a stratal unit consisting of nearly horizontal layers of sediments deposited on the top surface of a prograding coastline, which covers the edge of the seaward-lying foreset beds and is continuous with the landward alluvial plain (Bates and Jackson, 1987). Where topset thicknesses fall below the resolution of seismic data, a topset is expressed as *toplap* on 2D seismic lines (Fig. 4.13). In an ideal world, toplap may also form as a result of progradation during relative sea-level stillstand and perfect sediment bypass of the

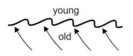
Truncation: termination of strata against an overlying erosional surface.

Truncation refers to the strata below an erosional surface, typically an angular unconformity.

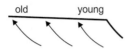
Toplap: termination of inclined strata against an overlying lower angle surface that shows no evidence of erosion; commonly, a foreset overlain by a surface of sediment bypass (no deposition, no erosion). Lapouts mark the updip depositional limits of the sedimentary units, and become progressively younger basinward.

Toplap may be a seismic artifact (a topset below the seismic resolution or truncation with indiscernible evidence of erosion), or it may mark a non-erosional unconformity (e.g., paleosol) at the top of prograding clinoforms.

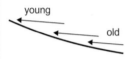
Onlap: termination of low-angle strata against an underlying steeper surface, in which each successively younger unit extends beyond the limit of the older unit on which it lies. A lapout with upstepping in the younging direction, which marks the updip termination of a sedimentary unit at its depositional limit.

Typically, the product of sedimentation during relative sea-level rise (either transgression or normal regression). Types of onlap:

• Marine onlap: deep-water sediment onlapping the continental slope, most commonly during transgression (e.g., 'slope aprons' of Galloway, 1989; 'healing-phase' deposits of Posamentier and Allen, 1993)
• Coastal onlap: coastal to shallow-water sediment onlapping the transgressive surface of erosion (tidal- or wave-ravinement surfaces) during transgression
• Fluvial onlap: fluvial sediment onlapping the basin margin as the area of fluvial sedimentation expands in a landward direction, most commonly during relative sea-level rise (normal regression and transgression)

Downlap: termination of inclined strata against an underlying lower angle surface. A baselap which marks the real or apparent downdip termination of a sedimentary unit at its depositional limit; a change from foreset to bottomset in a subaqueous environment.

Typically, the product of sediment progradation during relative sea-level rise or fall. A seismic artifact in the case of gradationally based prograding units with bottomsets that fall below the seismic resolution; a real termination in the case of sharp-based, forced regressive prograding shorefaces.

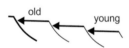
Offlap: termination of low-angle strata against an underlying steeper surface, in which each successively younger unit leaves exposed a portion of the older unit on which it lies. A lapout with downstepping in the younging direction, which marks the updip termination of a sedimentary unit at its depositional limit.

Typically, the product of sedimentation during relative sea-level fall, diagnostic of forced regression.

FIGURE 4.8   Stratal terminations: definitions and diagnostic features (modified after Mitchum, 1977).

coastal systems (i.e., neither deposition nor erosion in coastal environments). However, as neither of these two conditions is met on geological timescales (e.g., the relative sea level is always in motion due to the interplay of several independent controls which cannot be maintained in perfect balance), toplap remains an artifact of seismic resolution. Toplap may also develop during stages of relative sea-level fall (forced regressions) associated with minimum erosion, where the evidence of erosion is undetectable on seismic lines (Fig. 4.9).

A common error is to use the term *"truncation"* for all types of stratal terminations. In its correct usage,

truncation refers to the termination of strata *below* an unconformity (e.g., older strata cut by an incised valley, or by an angular unconformity). Toplap also assumes termination of strata *below* a seismic reflection, whereas all other types of stratal relationships (i.e., offlap, onlap, and downlap) refer to stratal terminations *above* an underlying surface or seismic reflection (Figs. 4.7 and 4.8). The type and origin of stratal terminations are best identified following regional studies that allow the placement of the study area in the correct tectonic and depositional settings. Without a proper geological context and an understanding of the regional

Stratal termination	Shoreline shift	Relative sea level
Truncation	FR, T	Fall, Rise
Toplap	NR, FR	Rise, Fall
Offlap	FR	Fall
Onlap, fluvial	NR, T ± FR	Rise, Fall
Onlap, coastal	T	Rise
Onlap, marine	T ± NR, FR	Rise, Fall
Downlap	NR, FR	Rise, Fall

FIGURE 4.9 Timing of stratal terminations in relation to shoreline shifts and relative sea-level changes. Abbreviations: R—regression; FR—forced regression; NR—normal regression; T—transgression.

1. Continental environments

• Colluvial and alluvial fans
• Fluvial environments
• Eolian environments

2. Coastal environments

• **River-mouth environments**
  - Deltas: regressive river mouths
  - Estuaries: transgressive river mouths

• **Open-shoreline environments**
  - Foreshore (intertidal)
  - Backshore (supratidal)

3. Marine/lacustrine environments

• **Shallow-water environments**
  - Shoreface (subtidal)
  - Shelf (inner and outer)

• **Deep-water environments**
  - Slope
  - Basin floor

FIGURE 4.10 Classification of depositional environments. Coastal environments are intermittently flooded, during tidal cycles or storms. Both types of coastal environments (river-mouth and open-shoreline) can be transgressive or regressive. The limits between the various types of coastal and shallow-water environments are defined in Fig. 4.11.

stratigraphic architecture, confusions can arise between different types of stratal terminations; e.g., between onlap and offlap in shelfal areas where prograding and downstepping clinoforms are detached, in which case the updip limit of each depositional unit is described by offlap rather than onlap (Fig. 4.14). Knowledge of the location of the shoreline at the time of deposition also allows differentiating fluvial from coastal onlap (Fig. 4.15), while the location of the shelf edge affords

the identification of marine onlap in the deep-water setting (Figs. 4.6 and 4.8).

The distinction between fluvial, coastal, and marine onlap is important for paleogeographic reconstructions, as it involves different types of facies that onlap the landscape or the seafloor at the time of sedimentation (Fig. 4.15). Truncation too may be caused by erosional processes in either continental or marine environments, and the clarification of the type of unconformity requires knowledge of the depositional system overlying the contact (details in Chapter 6). Post-depositional tectonic or halokinetic tilt may add another level of difficulty to the identification of stratal terminations, both in outcrop and on seismic lines. In particular, the recognition of onlap and downlap may be hampered by differential subsidence or uplift, which modifies the dip angle of strata and of the surfaces against which they terminate. For example, the motion of salt diapirs during the evolution of a basin may change the original inclination of strata, turning depositional downlap into apparent onlap, or vice versa (e.g., see the stratal terminations indicated by red arrows in Fig. 2.68, which resemble onlap geometries, but originated as depositional downlap related to the progradation of the continental slope in a divergent margin setting).

Where seismic data provide the only source of geological information, as it is often the case in "frontier" basins, most stratigraphic units thinner than several meters tune into single seismic reflections. For this reason, as noted by Posamentier and Allen (1999), "… because of limited seismic resolution, the location of stratal terminations, imaged on seismic data as reflection terminations, will, in general, not be located where the reflection terminations are observed. Coastal onlap as well as downlap terminations, in particular, can, in fact, be located a considerable distance landward and seaward, respectively, of where they appear on seismic data, because of stratal thinning." Another pitfall of the limited seismic resolution is that some stratal terminations depicted by reflection geometries are only artifacts of tuning of closely spaced bedding planes into single seismic reflections. This is often the case with toplap and downlap (Fig. 4.13), and also with onlap generated by strata that thin (but do not terminate) in an updip direction below the seismic resolution (Hart, 2000, Fig. 2.66).

A direct link between stratal terminations and relative sea-level changes can only be made in the case of coastal onlap and offlap (i.e., relative sea-level rise and fall, respectively), whereas all other types of stratal terminations may form during either relative sea-level rise or fall (Fig. 4.9). The correct understanding of the origin of stratal terminations can prove critical to unravelling the syn-depositional sedimentary processes, and hence to facies predictions. For example, clarifying the origin of downlap, which may be related to either

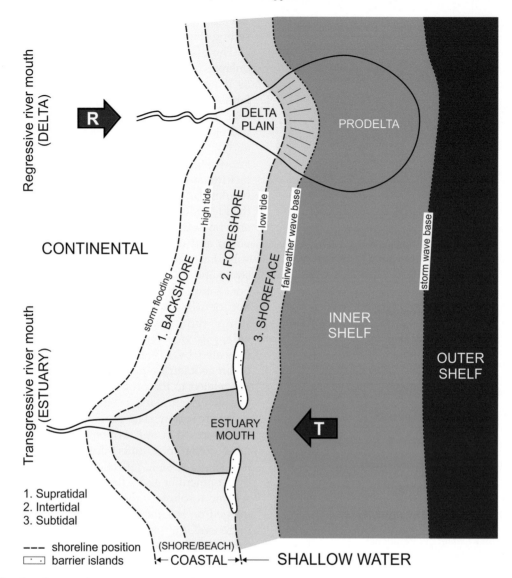

FIGURE 4.11   Classification of coastal to shallow-water environments. Between river mouths, the coastline is an open-shoreline system. The transgressive vs. regressive character of the shoreline may change along strike due to variations in subsidence and sedimentation rates. Abbreviations: R—regression; T—transgression.

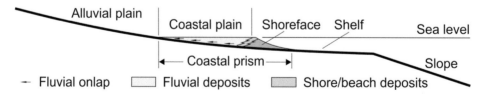

FIGURE 4.12   Development of coastal plains during normal regressions (modified after Posamentier et al., 1992b). "Coastal plain" is a geomorphological term that refers to a relatively flat area bordering a coastline and extending inland to the nearest elevated land (Bates and Jackson, 1987). Coastal plains may be tens to hundreds of kilometers wide, depending on sediment supply and the gradient of the onlapped landscape (e.g., the coastal plain of the Nueces River in Texas is c. 40 km wide: Blum and Tornqvist, 2000; the coastal plain of the Po River in Italy is c. 200 km wide: Hernandez-Molina, 1993; the coastal plain of the Mississippi River is at least 300—400 km wide: Blum and Tornqvist, 2000). Coastal prisms are typically part of lowstand and highstand systems tracts (i.e., normal regressions). The updip limit of a coastal prism was termed "bayline" by Posamentier et al. (1992b), and it shifts landward during normal regressions (fluvial onlap; Fig. 4.8). A lowstand coastal prism may be scoured by tidal- and/or wave-ravinement processes during subsequent transgression, whereas a highstand coastal prism is typically incised by rivers during subsequent relative sea-level fall. Both lowstand and highstand coastal prisms may be preserved in the rock record where the original thicknesses exceed the amounts of subsequent erosion.

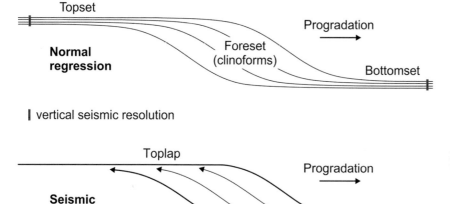

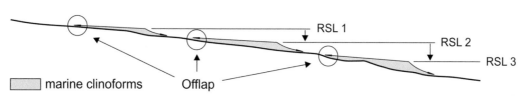

FIGURE 4.14 Detached forced-regressive clinoforms, prograding and downstepping in the younging direction. The updip termination of each clinoform can be confused with onlap, if observed out of context. In contrast to offlap, onlap is defined by stratal terminations upstepping in the younging direction (Figs. 4.8 and 4.15). Detachment is produced by steps of relative sea-level (RSL) fall with magnitudes that exceed the thickness of the previously deposited clinoforms. The discrete steps that generate offlap during forced regression are caused by variations in the rates of RSL fall, whether the changes in RSL are driven by uplift or sea-level fall.

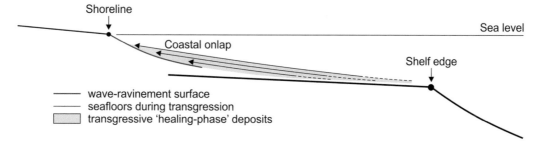

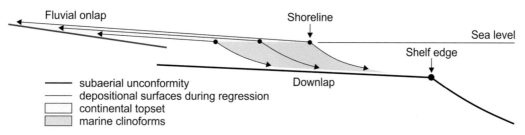

FIGURE 4.15 Coastal vs. fluvial onlap. The location of onlapping terminations relative to the coeval shoreline indicates the environment of deposition and the type of onlap. Marine onlap may also form downdip of the shelf edge, within the deep-water setting (Figs. 4.6 and 4.8).

normal or forced regressions, can make a significant difference to the prediction of facies that compose the clinoforms or accumulate downdip of the prograding clinoforms, as the transfer of riverborne sediment to the shallow- and deep-water environments is more efficient during forced regressions; hence, coarser sediment is generally expected within marine systems that accumulate during stages of relative sea-level fall. Downlapping clinoforms are also typically coarser grained than onlapping units, as the gradient of the seafloor reflects different energy levels and depositional processes (e.g., traction or gravity-driven currents in higher energy environments where downlap may form vs. sedimentation from suspension in lower energy environments, which generates onlap).

4.2 Stacking patterns in downstream-controlled settings

Downstream-controlled stacking patterns are described by the "conventional" sequence stratigraphy, whereby the sequence stratigraphic framework forms within the area of influence of relative sea/lake-level

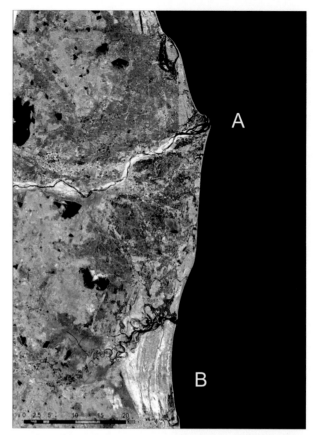

FIGURE 4.17 Examples of present-day shoreline trajectories: normal regression (progradation with upstepping; East coast of India). A—river-mouth setting; B—open-shoreline setting (prograding strandplain).

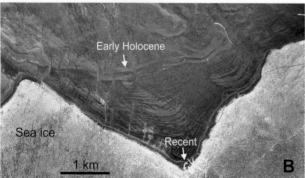

FIGURE 4.16 Examples of present-day shoreline trajectories: forced regression (progradation with downstepping; Canadian Arctic Archipelago). A—river-mouth setting; B—open-shoreline setting.

changes (Fig. 3.11). Within this context, the nomenclature of systems tracts makes reference to shoreline trajectories (e.g., "transgressive" systems tract) and/or to the relative sea level (e.g., "lowstand," "highstand," or "falling-stage" systems tracts). Even where the nomenclature is linked to the relative sea level, specific types of shoreline trajectory are still implied (e.g., a "falling-stage" systems tract implies a forced regression of the shoreline; a "highstand" systems tract implies a normal regression of the shoreline that follows a transgression of equal hierarchical rank; and a "lowstand" systems tract implies a normal regression of the shoreline that follows a forced regression of equal hierarchical rank). Therefore, the shoreline is a key element for the downstream-controlled sequence stratigraphic framework, which controls the formation and timing of all conventional systems tracts and bounding sequence stratigraphic surfaces (i.e., the "heart beat" of conventional sequence stratigraphy; Catuneanu, 2006). All three types of shoreline trajectory can be observed in the present-day environment (Figs. 4.16—4.18).

The architecture of the downstream-controlled stratigraphic framework is linked to the evolution of clinoform rollover points (i.e., forestepping, backstepping,

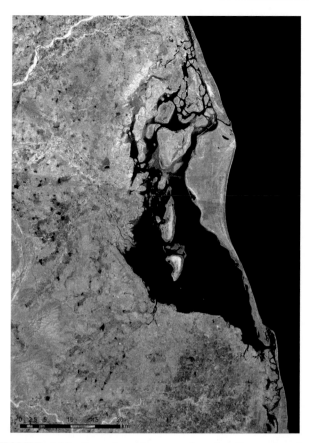

FIGURE 4.18 Examples of present-day shoreline trajectories: transgression (retrogradation and upstepping; lagoon—barrier island system, East coast of India).

upstepping, downstepping). Clinoform rollovers can be observed at different scales (e.g., from deltas to shelf-slope systems), and can form under variable conditions, from subaerial (i.e., shorelines) to subaqueous (e.g., subaqueous deltas, or shelf edges; Fig. 4.19). The stratal staking patterns that are relevant to sequence stratigraphy are those that form in relation to shoreline trajectories (i.e., the direction of shift of subaerial clinoform rollovers), which afford the delineation of systems tracts and bounding sequence stratigraphic surfaces (e.g., a "maximum flooding surface" forms in relation to the backstepping of the *shoreline*, and not of the shelf edge). Stratal stacking patterns that describe the architecture and evolution of other clinoform rollovers (e.g., the shelf edge, or subaqueous deltas on the shelf) may

form independently of shoreline trajectories (Fig. 4.20), and their role within a sequence stratigraphic framework needs to be rationalized in conjunction with the observation of shoreline trajectories.

The shelf edge is often erroneously referred to as the "offlap break." This is incorrect because "offlap" is a type of stratal termination that describes a specific geometrical relationship between prograding and downstepping clinoforms, which forms typically in relation to forced regression (Figs. 4.7 and 4.8). In contrast, the shelf edge is a geomorphic element that can shift in any direction, with or without the development of offlap. The shoreline may reach the shelf edge during late stages of relative sea-level fall, in which case the shelf-edge trajectory may generate offlap at the scale of the shelf-slope system (Fig. 4.6). However, the shelf edge is typically submerged at most other times, when its trajectory is variable and in- or out-of-phase with the direction of shoreline shift, depending on sediment supply and slope stability (Fig. 4.20). In a long term, the shelf edge most commonly undergoes progradation and aggradation (i.e., not offlap), leading to the growth of the shelf-slope system (Fig. 2.68).

4.2.1 Controls on stratal stacking patterns

The development of stratal stacking patterns that define the downstream-controlled sequence stratigraphic framework reflects the interplay of relative sea-level changes (accommodation) and base-level changes (sedimentation) *at the shoreline*. The rates of relative sea-level changes and base-level changes are variable across a sedimentary basin, both along dip and strike directions; only the rates of accommodation and sedimentation in coastal settings are relevant to the formation and timing of the downstream-controlled sequence stratigraphic framework (e.g., the formation of a transgressive systems tract, or the timing of a maximum flooding surface, are directly linked to the rates of accommodation and sedimentation at the shoreline). The shoreline is the only place within a sedimentary basin where accommodation and sedimentation need to be separated as distinct controls on the sequence stratigraphic framework. Anywhere else within the sedimentary basin, sedimentary processes (i.e., base-level changes) reflect the collective and undifferentiated

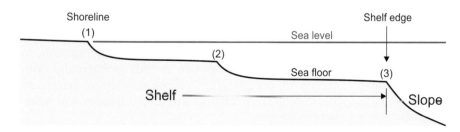

FIGURE 4.19 Clinoform rollovers: (1) subaerial (shoreline); (2) subaqueous, on the shelf (subaqueous "delta"); (3) subaqueous, at the shelf edge. The shoreline is the reference for the definition of downstream-controlled systems tracts and bounding surfaces.

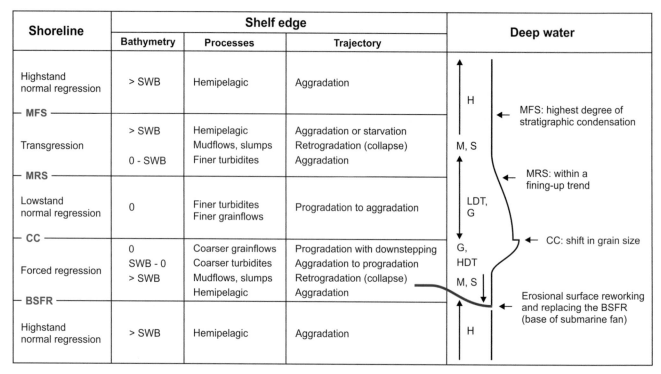

Shoreline	Shelf edge			Deep water
	Bathymetry	Processes	Trajectory	
Highstand normal regression	> SWB	Hemipelagic	Aggradation	
— MFS —				
Transgression	> SWB	Hemipelagic Mudflows, slumps	Aggradation or starvation Retrogradation (collapse)	
	0 - SWB	Finer turbidites	Aggradation	
— MRS —				
Lowstand normal regression	0	Finer turbidites Finer grainflows	Progradation to aggradation	
— CC —				
Forced regression	0 SWB - 0 > SWB	Coarser grainflows Coarser turbidites Mudflows, slumps Hemipelagic	Progradation with downstepping Aggradation to progradation Retrogradation (collapse) Aggradation	
— BSFR —				
Highstand normal regression	> SWB	Hemipelagic	Aggradation	

FIGURE 4.20   Shoreline vs. shelf-edge trajectories, and corresponding deep-water processes, in the case of siliciclastic settings that are subject to high-magnitude changes in relative sea level (i.e., relative sea level below the elevation of the shelf edge at the end of forced regression). The shelf edge may or may not backstep during forced regression and transgression, depending on its stability. The delivery of riverborne sediment to the deep-water setting is most effective when the bathymetry at the shelf edge is less than the SWB. Unfilled incised valleys across a submerged shelf may further enhance the transfer of sediment from river mouths to the shelf edge even during stages of highstand in relative sea level. In such cases, as well as in the case of atypically narrow shelves, riverborne sediment supply (both volume and grain size) to the deep-water setting may increase overall. This diagram illustrates common trends observed in the rock record. Any variability in the sedimentological makeup of systems tracts needs to be assessed in the context of each sedimentary basin. Abbreviations: BSFR—basal surface of forced regression; CC—correlative conformity; MRS—maximum regressive surface; MFS—maximum flooding surface; SWB—storm wave base; H—hemipelagic sediment; M—mudflows; S—slumps; HDT—high-density turbidity flows; G—grainflows; LDT—low-density turbidity flows.

contributions of accommodation and all other controls on sediment supply and the energy of the sediment-transport agents (Fig. 3.11).

The distinction between accommodation and sedimentation at the shoreline is key to understanding the formation of stratal stacking patterns in downstream-controlled settings (e.g., the manifestation of normal regressions vs. transgressions depends on the competing rates of accommodation and sedimentation at the shoreline; discussion below). At any point in time, the overall amounts of accommodation created and sediment accumulated within a sedimentary basin are in balance at the shoreline, where the sea level intersects the temporary base level (Figs. 3.11 and 3.22), and are out of balance updip and downdip from the shoreline (i.e., overfilled vs. underfilled accommodation, respectively). Through time, the shoreline is constantly shifting to new locations of equilibrium between accommodation and sedimentation, in response to changes in relative sea level and base level.

The depositional trends of coastal progradation or retrogradation are driven by the imbalance between the rates of accommodation and sedimentation at the shoreline. At the end of each depositional trend, the maximum regressive and maximum flooding surfaces mark points of equilibrium between the rates of creation and consumption of space (i.e., the depositional surface reaches a long-term base level, before the shoreline changes the direction of shift; Fig. 3.23). Along a shoreline, the timing of maximum regressive and maximum flooding surfaces is location-specific, and coeval progradation and retrogradation may be recorded in response to strike variations in the rates of accommodation and sedimentation (e.g., Wehr, 1993; Martinsen and Helland-Hansen, 1995; Catuneanu et al., 1998a; Catuneanu, 2006; Helland-Hansen and Hampson, 2009). On either side of the shoreline, fluctuations in base level in continental and marine environments control depositional trends of aggradation and degradation from sedimentological to stratigraphic scales. Coastal systems also experience aggradation or degradation, in response to base-level rise or fall, concurrently with trends of progradation or retrogradation driven by the interplay of relative sea-level changes and base-level changes.

Both the relative sea level and the base level are unstable at all scales of observation, from daily tides to

Forced regression

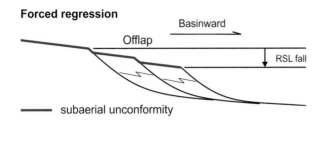

Stacking pattern: forestepping with downstepping

Origin: progradation driven by relative sea-level fall (negative accommodation). The coastline is forced to regress, irrespective of sediment supply.

Normal regression

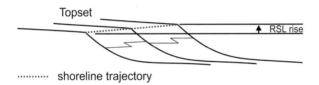

Stacking pattern: forestepping with upstepping

Origin: progradation driven by sediment supply. Sedimentation rates exceed the rates of relative sea-level rise (positive accommodation) at the coastline.

Transgression

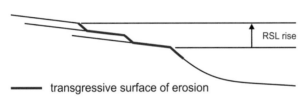

Stacking pattern: backstepping and upstepping

Origin: retrogradation and upstepping driven by relative sea-level rise. Accommodation outpaces the rates of sedimentation at the coastline.

FIGURE 4.21 Stratal stacking patterns in downstream-controlled settings: forced regression, normal regression, and transgression (from Catuneanu et al., 2011). These stacking patterns are defined by *geometrical trends* of the shoreline, and not by *depositional trends*. The diagram illustrates the most common depositional trends that accompany shoreline shifts (i.e., fluvial bypass or erosion during forced regression, and fluvial aggradation during transgression), but exceptions do occur (see text for details). The amounts of coastal upstepping during normal regression and transgression, and downstepping during forced regression, reflect the magnitude of relative sea-level changes at syn-depositional time.

processes that operate over geological timescales. Changes in relative sea level and base level trigger vertical and lateral shifts in the position of the shoreline, which combine to define specific types of shoreline trajectory: normal regression (forestepping and upstepping), forced regression (forestepping and downstepping), and transgression (backstepping and upstepping) (Figs. 4.1 and 4.21), which correspond to the ascending regressive, descending regressive, and transgressive shoreline trajectories of Løseth and Helland-Hansen (2001) and Helland-Hansen and Hampson (2009). Theoretical situations of pure aggradation (i.e., geographically stable shoreline: upstepping with no lateral shift) or pure progradation (i.e., regression during relative sea-level stillstand: forestepping with no vertical shift) have also been postulated (Løseth and Helland-Hansen, 2001; Helland-Hansen and Hampson, 2009), but are uncommon in nature as the location and the relative elevation of the shoreline depend on the interplay of multiple independent variables, and so are unlikely to be stable over any timescales, from "calendar" (relevant to sedimentology) to "geological" (relevant to stratigraphy). The stratal stacking patterns

described by normal regressions, forced regressions, and transgressions can be observed at different stratigraphic scales, from outcrop to seismic, in relation to shoreline shifts of different magnitudes (Figs. 4.6 and 4.22).

4.2.2 Geometrical vs. depositional trends

The architecture of the downstream-controlled stratigraphic framework may be described in terms of *geometrical trends* of the shoreline (i.e., shoreline trajectories: upstepping, downstepping, forestepping, and backstepping; Fig. 4.1), or in terms of *depositional trends* that accompany the shoreline shifts (i.e., aggradation, degradation, progradation, and retrogradation) (Fig. 4.23). The terms that describe the lateral components of the two types of trend can be used interchangeably (i.e., forestepping = progradation; backstepping = retrogradation), because they refer to the same regressions and transgressions of the shoreline, and are controlled by the same interplay of relative sea-level changes and base-level changes (Fig. 4.23). However, the vertical components of the geometrical and depositional trends

FIGURE 4.22   Stratal stacking patterns in downstream-controlled settings: outcrop examples. A—forced regression (Upper Cretaceous, Panther Tongue, Utah); B—normal regression (Upper Cretaceous, Ferron Delta, Utah); C—transgression (transgressive carbonates "T"; Upper Cretaceous, Northwest Basin, Argentina). Abbreviations: WRS/SU—wave-ravinement surface reworking and replacing a subaerial unconformity; FC—facies contact separating the foreset (below) from the topset (above) of a normal regressive succession (i.e., a within-trend normal regressive surface); MFS—maximum flooding surface.

may not necessarily correlate, due to differences in their causal mechanisms: the upstepping and downstepping of the shoreline are controlled by relative sea-level changes, whereas the depositional trends of aggradation and degradation are controlled by base-level changes (Fig. 4.23). This difference explains departures from the "norm," such as degradation during shoreline upstepping (e.g., transgressions accompanied by coastal and fluvial erosion: base-level fall during relative sea-level rise), or aggradation during shoreline

downstepping (e.g., forced regressions accompanied by coastal and fluvial aggradation: base-level rise during relative sea-level fall).

Numerous case studies indicate that the depositional trends in fluvial and coastal environments most often follow the geometrical trends of the shoreline (i.e., aggradation during shoreline upstepping, and degradation during shoreline downstepping), particularly where the provenance is far from the shoreline and/or the landscape gradients are low. However, as exceptions from this "norm" are also recorded, the construction and analysis of sequence stratigraphic frameworks must be performed on a case-by-case basis, without *a priori* assumptions. The potential offset between the geometrical trends of the shoreline and the depositional trends in coastal and fluvial environments indicates that the two types of trend have different degrees of relevance to the sequence stratigraphic methodology (Catuneanu and Zecchin, 2016). Diagnostic to the definition of systems tracts are the geometrical trends of the shoreline (i.e., shoreline trajectories; Figs. 4.1 and 4.21), whereas the depositional trends in coastal and fluvial environments are non-diagnostic (e.g., fluvial aggradation may accompany the formation of any systems tract; Fig. 1.2).

It can be concluded that the geometrical terms of "upstepping" and "downstepping," rather than the depositional terms of "aggradation" and "degradation," are the proper descriptors of the stratal stacking patterns that define systems tracts (Figs. 1.2 and 4.23). For example, forced regression (i.e., progradation during relative sea-level fall, leading to the deposition of the "falling-stage" systems tract) is more accurately described as "progradation with downstepping" than "progradation with degradation," because the downstepping of the shoreline during relative sea-level fall may be accompanied by either fluvial and coastal degradation or aggradation. The latter (depositional) processes are not diagnostic to the definition of the falling-stage systems tract. Similarly, the term "progradation with aggradation" is equivocal both in terms of definition of systems tracts and reconstruction of accommodation conditions at syn-depositional time, as it describes depositional trends that may develop during either normal regression (i.e., progradation during relative sea-level rise) or forced regression (i.e., progradation during relative sea-level fall; Fig. 4.24). The distinction between forced and normal regressions is important as they lead to different facies relationships, stratigraphic architecture, and patterns of sediment distribution across a basin (e.g., Posamentier and Morris, 2000; MacEachern et al., 2012; Catuneanu, 2017).

In terms of field expression, the geometrical trends of shoreline upstepping or downstepping are recorded by the relative changes in the elevation of subaerial clinoform rollovers (i.e., top of shoreface or delta front facies), which are controlled by relative sea-level changes. In contrast, the depositional trends of coastal aggradation or degradation refer to processes recorded within coastal environments (e.g., beach, delta plains, estuaries), which are controlled by changes in base level. This distinction is fundamental to the identification of systems tracts, which relies on the observation of diagnostic stratigraphic features. For example, diagnostic to transgression is the backstepping and upstepping of shallow-water facies (i.e., coastal onlap), irrespective of the depositional trends in fluvial and coastal environments. Similarly, diagnostic to forced regression is the progradation and downstepping of subtidal (shoreface or delta front) facies, irrespective of the depositional trends in fluvial and coastal environments. Criteria to differentiate between forced regressions, normal regressions, and transgressions involve the observation of facies relationships in high-resolution studies and of seismic-reflection geometries in low-resolution studies (e.g., Posamentier and Morris, 2000; Løseth and Helland-Hansen, 2001; Zecchin, 2007; Helland-Hansen and Hampson, 2009; Catuneanu et al., 2011; Catuneanu and Zecchin, 2016).

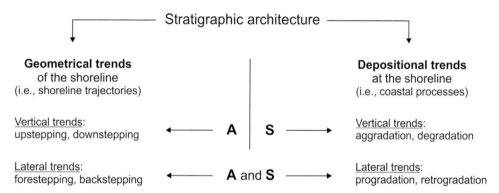

FIGURE 4.23 Controls on the stratigraphic architecture: accommodation (A) and sedimentation (S). The architecture of the downstream-controlled stratigraphic framework may be described in terms of *geometrical trends* of the shoreline (i.e., shoreline trajectories) or *depositional trends* at the shoreline (i.e., coastal processes that accompany the shoreline shifts). The lateral components of the two types of trends are equivalent to each other (i.e., forestepping = progradation; backstepping = retrogradation), as they are both controlled by the interplay of relative sea-level changes (A) and base-level changes (S). However, the vertical components of the two types of trends may be offset, as they are controlled by different processes (A vs. S). Relative sea-level changes control the upstepping (relative sea-level rise) and the downstepping (relative sea-level fall) of the shoreline, whereas base-level changes control the depositional trends of aggradation or degradation. While the temporary base level (graded profile) in fluvial and coastal environments often follows the change in relative sea level (i.e., degradation during shoreline downstepping, and aggradation during shoreline upstepping), exceptions may occur. Diagnostic to the definition of systems tracts are the geometrical trends that describe shoreline trajectories, irrespective of the depositional trends of aggradation or degradation that accompany the shoreline shifts (see text for details).

Geometrical trends of the shoreline	Stratal stacking patterns	Depositional trends in fluvial/coastal systems
Progradation with downstepping	Forced regression with fluvial/coastal degradation [1]	Progradation with degradation
	Forced regression with fluvial/coastal aggradation [2]	Progradation with aggradation
Progradation with upstepping	Normal regression	

FIGURE 4.24 Architecture of forced and normal regressions. Stratal stacking patterns may be described in terms of geometrical trends of the shoreline (i.e., shoreline trajectories) or depositional trends in fluvial to coastal systems. The geometrical trends (i.e., "downstepping" or "upstepping") provide the diagnostic criteria that define systems tracts and accommodation conditions at syn-depositional time. The recognition of forced vs. normal regressions may be based on facies data in high-resolution studies (e.g., to demonstrate the sharp- vs. gradationally based nature of shoreface deposits), or on the observation of the stratigraphic architecture in lower resolution studies (e.g., the downstepping vs. upstepping of subaerial clinoform rollovers). Notes: [1] most common in the stratigraphic record; [2] less common in the stratigraphic record

4.2.3 Diagnostic vs. non-diagnostic stacking patterns

The architecture of the downstream-controlled sequence stratigraphic framework includes elements that are diagnostic to the definition of systems tracts, as well as elements that are not diagnostic of (or unique to) any particular systems tract (Fig. 1.2; Catuneanu and Zecchin, 2016). For example, the combination of shoreline progradation and downstepping is diagnostic of forced regression, which defines the falling-stage systems tract (Fig. 4.1); at the same time, forced regression may be accompanied by variable processes of fluvial aggradation or degradation, which are not diagnostic to the definition of the falling-stage systems tract (Posamentier and Allen, 1999; Catuneanu, 2006; Catuneanu and Zecchin, 2016, Fig. 4.25). Therefore, it is important to recognize the relative significance of the various elements that build the stratigraphic record, and to differentiate between the stratal stacking patterns that are diagnostic vs. non-diagnostic to the definition of systems tracts (Figs. 1.2 and 4.23).

The diagnostic stacking patterns refer to elements of the stratigraphic architecture that are unique to particular systems tracts, therefore providing the criteria to the identification of systems tracts. In downstream-controlled settings, the diagnostic stacking patterns relate to shoreline trajectories (i.e., the trajectory of subaerial clinoform rollovers; Figs. 4.1 and 4.21): progradation and downstepping (i.e., forced regression), irrespective of coeval fluvial processes of aggradation, bypass, or erosion (Fig. 4.26); progradation and upstepping (i.e., normal regression), typically accompanied by fluvial aggradation (Fig. 4.27); and backstepping and upstepping (i.e., transgression), irrespective of coeval fluvial processes of aggradation, bypass, or erosion (Fig. 4.28). Notably, fluvial processes of aggradation or degradation are not specific to any particular systems

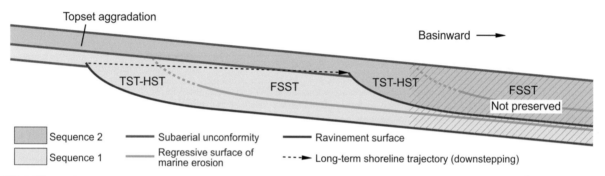

FIGURE 4.25   Architecture of a falling-stage systems tract (Pleistocene Cutro Terrace, southern Italy; horizontal scale: 10^1 km; vertical scale: 10^0 m; modified after Zecchin et al., 2011). The long-term shoreline trajectory is forced regressive (progradation with downstepping), with a gradient lower than that of the lower rank subaerial unconformities. This trajectory allowed topset aggradation (stacking of higher frequency sequences) during the long-term forced regression. The falling-stage systems tract consists of two lower rank sequences, each comprised of transgressive (TST), highstand (HST), and falling-stage (FSST) systems tracts.

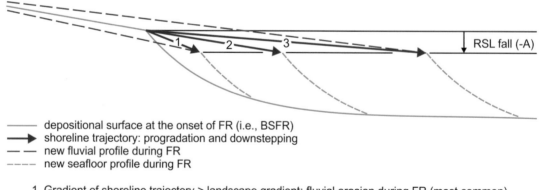

FIGURE 4.26   Depositional trends during forced regression (from Catuneanu and Zecchin, 2016). Diagnostic to forced regression is the combination of progradation and downstepping of the shoreline. Non-diagnostic depositional trends include fluvial processes of erosion (case 1), bypass (case 2), or aggradation (case 3), which depend on the gradient of the shoreline trajectory relative to the landscape gradient. Abbreviations: FR—forced regression; BSFR—basal surface of forced regression; RSL—relative sea level; −A—negative accommodation.

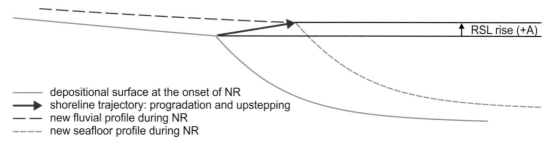

——— depositional surface at the onset of NR
——▶ shoreline trajectory: progradation and upstepping
— — new fluvial profile during NR
----- new seafloor profile during NR

FIGURE 4.27 Depositional trends during normal regression (from Catuneanu and Zecchin, 2016). Diagnostic to normal regression is the combination of progradation and upstepping of the shoreline. This shoreline trajectory leads to a lowering of the fluvial gradient with time, which results in the aggradation of a fluvial topset. Abbreviations: NR—normal regression; RSL—relative sea level; +A—positive accommodation.

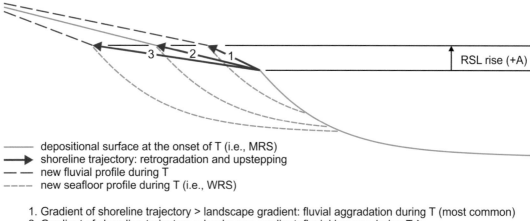

——— depositional surface at the onset of T (i.e., MRS)
——▶ shoreline trajectory: retrogradation and upstepping
— — new fluvial profile during T
----- new seafloor profile during T (i.e., WRS)

1. Gradient of shoreline trajectory > landscape gradient: fluvial aggradation during T (most common)
2. Gradient of shoreline trajectory = landscape gradient: fluvial bypass during T } SU forms during T
3. Gradient of shoreline trajectory < landscape gradient: fluvial erosion during T }

FIGURE 4.28 Depositional trends during transgression (from Catuneanu and Zecchin, 2016). Diagnostic to transgression is the combination of retrogradation and upstepping of the shoreline. Non-diagnostic depositional trends include fluvial processes of aggradation (case 1), bypass (case 2), or erosion (case 3), which depend on the gradient of the shoreline trajectory relative to the landscape gradient. Abbreviations: T—transgression; MRS—maximum regressive surface; WRS—wave-ravinement surface; SU—subaerial unconformity; RSL—relative sea level; +A—positive accommodation.

tract, and therefore they are not diagnostic to the definition of systems tracts (e.g., any systems tract may include fluvial deposits). It is also important to note that the subaerial unconformity may form during either forced regression or transgression, whereas normal regression promotes invariably fluvial aggradation.

The non-diagnostic stacking patterns are linked to depositional trends that can be associated with any systems tract, therefore not providing criteria to the identification of any particular systems tract. Non-diagnostic elements form in relation to changes in base level (i.e., aggradation or degradation), which may be in or out of phase with the upstepping and downstepping of the shoreline (e.g., Figs. 4.26 and 4.28). The degree of variability of the non-diagnostic trends of aggradation and degradation within the sequence stratigraphic framework depends in part on the depositional setting. This variability is higher in fluvial to coastal settings, as well as in deep-water settings, where processes of aggradation and degradation are less reliable for the

interpretation of "conventional" (i.e., shoreline-related) systems tracts, and it is lower in shallow-water settings, where depositional trends are more closely related to shoreline trajectories, and hence more reliable for the interpretation of "conventional" systems tracts. Therefore, the identification of systems tracts is not equally straight forward in all depositional systems, although process-response relationships do exist between the various depositional environments of a sedimentary basin (Fig. 4.20).

In fluvial and coastal settings, the depositional trends of aggradation or degradation can accompany both forced regressions and transgressions, depending on the gradient of the fluvial profile relative to the gradient of the shoreline trajectory (Figs. 4.26 and 4.28). Despite this non-diagnostic variability, processes of aggradation or degradation in fluvial and coastal environments have their own relevance to the sequence stratigraphic framework in terms of timing of formation of subaerial unconformities (i.e., during forced regressions or transgressions; Figs. 4.26 and 4.28), as well as in terms of sediment supply to the shoreline (i.e., higher during times of fluvial erosion or bypass, and lower during times of fluvial aggradation; Fig. 1.2).

Shallow-water systems respond more predictably to shifts in the position of subaerial clinoform rollovers, and therefore are more reliable for the reconstruction of shoreline trajectories and the identification of systems tracts. For example, transgression is typically accompanied by coastal onlap and the development of fining-upward trends within the shallow-water setting. Therefore, the observation of a retrogradational stacking

pattern in a shallow-water system is diagnostic for the transgressive systems tract, irrespective of the processes of fluvial and coastal aggradation or erosion that may occur at the same time (Fig. 4.29). Forced regression also results in the development of a unique architecture in shallow-water systems, which includes "sharp-based" shoreface deposits that are diagnostic for the falling-stage systems tract, irrespective of the coeval processes of fluvial and coastal aggradation or erosion (Fig. 4.30). Normal regression is also described by a unique stacking pattern in shallow-water systems, represented by gradationally based shoreface deposits

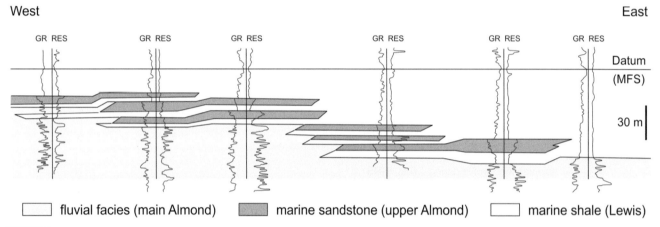

FIGURE 4.29 Long-term transgression punctuated by higher frequency transgressions and regressions (Campanian, Western Interior Seaway, Wyoming; from Catuneanu et al., 2011; modified after Weimer, 1966; and Martinsen and Christensen, 1992). The cross section is c. 65 km long. Abbreviations: GR—gamma ray; RES—resistivity; MFS—maximum flooding surface.

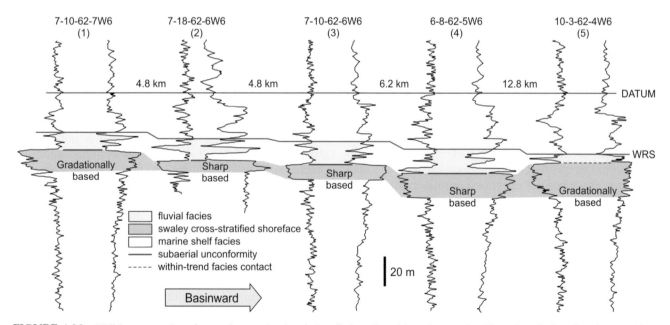

FIGURE 4.30 Well-log expression of normal regressive (gradationally based) and forced regressive (sharp-based) shoreface deposits (from Catuneanu et al., 2010; modified after Plint, 1988; Turonian, Cardium Formation, Western Interior Basin, Alberta). The normal regressive shoreface belongs to the highstand (well 1) and lowstand (well 5) systems tracts. The forced regressive shoreface belongs to the falling-stage systems tracts (wells 2–4). The well logs are gamma-ray (left) and resistivity (right). Abbreviation: WRS—wave-ravinement surface.

with prograding and upstepping clinoforms. Therefore, the observation of coarsening-upward and upstepping shoreface deposits with a gradational base is diagnostic of normal regressive systems tracts (i.e., lowstand and highstand systems tracts, Fig. 4.30; Catuneanu, 2002, 2006; Csato and Catuneanu, 2012, 2014).

Deep-water systems are less predictable in terms of types and stacking patterns of depositional elements, due to the basin-specific nature of sediment supply, mass-transport processes, and the sediment-dispersal pathways. However, beyond this non-diagnostic variability, diagnostic trends which afford the identification of systems tracts and bounding surfaces still emerge (Catuneanu, 2020a). Among the factors that control sedimentation and stratigraphic patterns in the deep-water setting, relative sea-level changes, the production of extrabasinal and intrabasinal sediment, and the physiography of the basin play major roles. Accommodation and sedimentation on the shelf, which depend on all these factors, define the "dual control" on shoreline trajectories on the shelf and sediment supply to the shelf edge. The sequence stratigraphic framework in the deep-water setting is defined by the relationship between the shoreline transit cycles on the shelf and the corresponding cycles of change in gravity flows beyond the shelf edge (Posamentier and Kolla, 2003; De Gasperi and Catuneanu, 2014, Fig. 4.20). Diagnostic to the definition of systems tracts and bounding surfaces are the changes in gravity flows during the shoreline shelf-transit cycles, rather than the actual types of flows (see Chapter 8 for details). The non-diagnostic variability depends on all basin-specific factors that modify sediment supply to the shelf edge during the shoreline transit cycles, including the physiography of the shelf (i.e., width, seafloor gradients, the presence or absence of unfilled incised valleys across the shelf); the magnitude and rates of relative sea-level changes at the shoreline; the rates of sedimentation at the shoreline; and the extrabasinal vs. intrabasinal origin of the sediment (Catuneanu, 2020a).

In any depositional setting, the diagnostic and non-diagnostic elements have different degrees of relevance to the sequence stratigraphic workflow (Fig. 1.2); the diagnostic trends must be observed first, and once the stratigraphic framework of systems tracts is constructed, the non-diagnostic elements can be placed and rationalized within that context (Catuneanu and Zecchin, 2016). The majority of sequence stratigraphic surfaces are associated with specific types of shoreline trajectory (e.g., forced regression: basal surface of forced regression, regressive surface of marine erosion, and correlative conformity; transgression: maximum regressive surface, transgressive surface of erosion, and maximum flooding surface). The notable exception is the subaerial unconformity, which can form during either forced regression (Fig. 4.26) or transgression (Fig. 4.28).

4.2.4 Shoreline trajectories

The interplay of relative sea-level changes and sedimentation controls the bathymetric trends, as well as the transgressive and regressive shifts of the shoreline (Fig. 3.13). As defined by Bates and Jackson (1987), the terms "shoreline" and "coastline" are often used synonymously, especially when referring to processes that occur over geological timescales. In the calendar band of time, however, there is a tendency to regard "coastline" as a limit fixed in position for a relatively long time, and "shoreline" as a limit constantly moving across the intertidal area (i.e., the intersection of a plane of water with the beach, which migrates with changes of the tide or of the water level) (Bates and Jackson, 1987). In the context of this book, reference is made to processes that operate over geological timescales, and therefore the terms "shoreline" and "coastline" are used interchangeably.

"Transgression" defines the landward migration of a shoreline, which is accompanied by a corresponding shift of facies, as well as water deepening *in the vicinity of the shoreline*. Transgressions result in retrogradational stacking patterns; i.e., distal facies overlying proximal facies in relative continuity of sedimentation (i.e., without breaking the paleogeographic trend at the selected scale of observation; Fig. 4.31). Within fluvial successions, transgression is commonly marked by the upward increase in the palynological marine index, the frequency of crevasse splaying along with the rise in the water table, and the occurrence of coal beds and tidal influences, e.g., sigmoidal cross-bedding, tidal (heterolithic wavy, flaser, and lenticular) bedding, oyster beds, and brackish to marine trace fossils (Shanley et al., 1992; Miall, 1997; Posamentier and Allen, 1999; Catuneanu, 2006). Retrogradation is the diagnostic stacking pattern for transgressions, defined as "the backward (landward) movement or retreat of a shoreline or of a coastline by wave erosion; it produces a steepening of the beach profile at the breaker line" (Bates and Jackson, 1987).

"Regression" defines the seaward migration of a shoreline, which is accompanied by a corresponding shift of facies, as well as water shallowing *in the vicinity of the shoreline*. Regressions result in progradational stacking patterns, e.g., proximal facies overlying distal facies in relative continuity of sedimentation (i.e., without breaking the paleogeographic trend at the selected scale of observation; Fig. 4.31). Progradation is the diagnostic stacking pattern for regressions, and is defined as "the building forward or outward toward the sea of a shoreline or coastline (as of a beach, delta, or fan) by nearshore deposition of river-borne sediments or by continuous accumulation of beach material thrown up by waves or moved by longshore drifting" (Bates and Jackson, 1987).

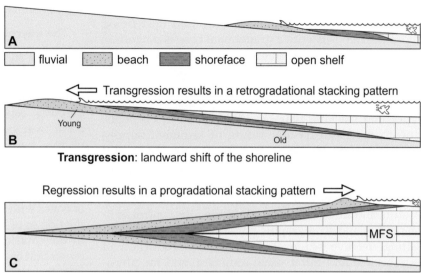

FIGURE 4.31 (A and B) Transgressions and (C) regressions, and the resulting retrogradational and progradational shifts of facies. The limit between transgressive and overlying regressive deposits is the maximum flooding surface (MFS).

The relationship between shoreline shifts (transgressions and regressions) and the bathymetric trends (water deepening and shallowing, respectively) is only safely valid for the shallow areas adjacent to the shoreline (i.e., shorefaces and laterally equivalent delta fronts). In offshore areas, the deepening and shallowing of the water may be out of phase with the coeval shoreline shifts, due to the variations in subsidence and sedimentation rates along dip directions (Catuneanu et al., 1998b). The Mahakam delta in Indonesia (Verdier et al., 1980) provides an example where the progradation (regression) of the shoreline is accompanied by a deepening of the water offshore, due to a combination of lower sedimentation rates and higher subsidence rates. This is often the case with tectonically active basins where depocenters experience water deepening irrespective of the direction of shoreline shift, due to sustained high rates of subsidence.

The occurrence of transgressions and regressions is explained by the interplay of relative sea-level changes and sedimentation at the shoreline (Fig. 4.32). The top sine curve in Fig. 4.32 idealizes the cyclic changes in relative sea level, with equal periods of rise and fall. However, asymmetrical shapes are more common in nature; e.g., glacio-eustatic cycles are dominated by stages of sea-level fall, as it takes more time to build than to melt ice caps; similarly, tectonic cycles in foredeep basins are dominated by stages of uplift (i.e., relative sea-level fall), as the episodes of thrusting and orogenic loading are typically shorter than the intervening stages of orogenic unloading and rebound (Catuneanu, 2004b, 2019c). Nevertheless, the principles illustrated in the diagram remain the same regardless the shape of the relative sea-level curve.

The lower part of the diagram in Fig. 4.32 indicates the rates of relative sea-level change and sedimentation. The rates of relative sea-level change (i.e., the first derivative of the top sine curve) are zero at the points of lowstand and highstand in relative sea level (i.e., the change from fall to rise and from rise to fall requires motion to cease), and are highest at the inflection points of the falling and rising legs. During relative sea-level fall, the shoreline is forced to regress irrespective of the rates of sedimentation. During relative sea-level rise, accommodation is created and consumed at the same time, leading to either normal regression (consumption outpaces the creation of space) or transgression (space is created faster than it is consumed). Sedimentation (i.e., consumption of space) tends to dominate the early and late stages of relative sea-level rise, when the rates of rise are low (i.e., increasing from zero and decreasing to zero, respectively), while the rates of relative sea-level rise (i.e., creation of space) are highest around the inflexion point where transgression is most likely to occur (Fig. 4.32).

There are no standards for the periodicity of the curves in Fig. 4.32; the same types of shoreline trajectories can be observed at all stratigraphic scales, depending on the resolution of the stratigraphic study. It is also important to distinguish between relative sea-level changes and transgressive-regressive cycles. It is a common confusion to equate regression with relative sea-level fall and transgression with relative sea-level rise, by neglecting the effect of sedimentation. In reality, the timing of the lowstands and highstands in relative sea-level is offset relative to the timing of maximum regressive and maximum flooding surfaces with the duration of normal regressions (Fig. 4.32). Within a cycle of

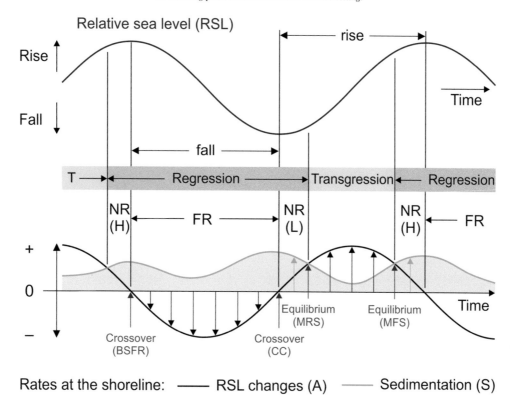

FIGURE 4.32 Concepts of transgression (T), normal regression (NR), and forced regression (FR), defined by the interplay of accommodation (A) and sedimentation (S) *at the shoreline*. The top curve shows changes in relative sea level, whereas the lower curves indicate the rates of relative sea-level change and sedimentation. Transgression occurs where positive accommodation outpaces sedimentation; normal regression occurs where sedimentation outpaces positive accommodation; forced regression is driven by negative accommodation. A normal regression that follows a forced regression is a lowstand (L) normal regression. A normal regression that follows a transgression is a highstand (H) normal regression. These principles are independent of the shapes of the relative sea-level and sedimentation curves. Additional abbreviations: BSFR—basal surface of forced regression; CC—correlative conformity; MRS—maximum regressive surface; MFS—maximum flooding surface.

relative sea-level change, the cumulative durations of forced and normal regressions account for more than half of the cycle, whereas transgression is a special case that may only occupy a portion of the rising leg. This implies that transgressions are shorter in time than the regressive portions of stratigraphic cycles, which is true in the case of symmetrical relative sea-level cycles and in the case of asymmetrical cycles domi nated by stages of relative sea-level fall (e.g., glacial-interglacial cycles, or tectonic cycles in foredeep basins).

However, the shorter lived character of transgressions compared to regressions of equal hierarchical rank cannot be generalized. Extensional basins dominated by long-term subsidence may lead to asymmetrical cycles dominated by stages of relative sea-level rise, in which case transgressions may potentially last longer than the regressive stages of the stratigraphic cycles. The duration of any of the stages illustrated in Fig. 4.32 is basin specific, reflecting the local conditions of accommodation and sedimentation. Where sedimentation rates are higher than the rates of thermal subsidence, as recorded in many divergent continental margins, normal regressions may dominate the

stratigraphic succession in spite of the prolonged stages of relative sea-level rise (Fig. 2.68). At the same time, fault-bounded extensional basins (e.g., rifts) which experience high rates of mechanical subsidence may favor transgressions at the expense of normal regressions in spite of the high rates of extrabasinal sediment supply. Therefore, no assumptions should be made at the outset of a stratigraphic study, and data need to be observed impartially for an objective, data-driven approach to stratigraphic analysis.

The ideal cycle in Fig. 4.32 depicts the most complete stratigraphic scenario, whereby a sequence develops during four distinct stages of shoreline shifts: one forced regression, two normal regressions (lowstand and highstand), and one transgression. Each of these stages results in the formation of a systems tract (i.e., falling-stage, lowstand, transgressive, and highstand). A sequence can include a maximum number of four systems tracts, but sequences with three or two systems tracts can also be encountered, depending on local circumstances that may prevent the formation or preservation of specific systems tracts. For example, the lowstand normal regression may be suppressed (and hence, the

lowstand systems tract may be missing) where accommodation is created rapidly at the onset of relative sea-
level rise (e.g., in fault-bounded basins such as rifts);
the transgression may be suppressed where the rates
of sedimentation outpace the rates of relative sea-level
rise at all times; and forced regressions (and implicitly
lowstand normal regressions) may not occur where

cyclicity develops during continuous relative sea-level
rise (Fig. 4.33).

It can be noted that the three types of shoreline trajectories (Fig. 4.1) lead to the formation of four systems
tracts, as there are two normal regressions within a relative sea-level cycle (Fig. 4.32). The distinction between
the lowstand and highstand normal regressions, and

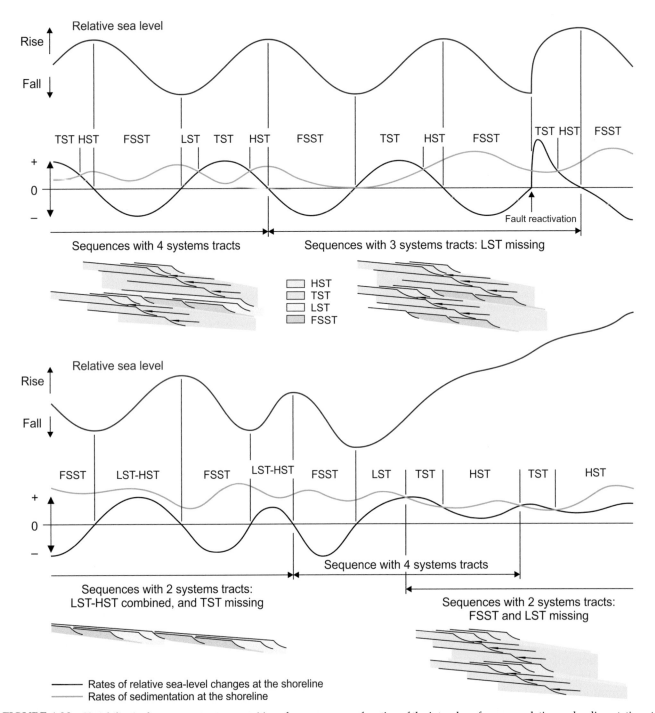

FIGURE 4.33 Variability in the systems-tract composition of sequences, as a function of the interplay of accommodation and sedimentation *at
the shoreline*. Abbreviations: FSST—falling-stage systems tract (forced regression); LST—lowstand systems tract (lowstand normal regression);
TST—transgressive systems tract (transgression); HST—highstand systems tract (highstand normal regression).

their corresponding lowstand and highstand systems tracts, is fundamentally based on the type of shoreline trajectory that precedes the normal regression (i.e., forced regression vs. transgression, respectively; Fig. 4.32); e.g., reference to a "lowstand" is only justified if there is a fall in relative sea level. The stratal stacking patterns associated with the three types of shoreline trajectories are presented below.

4.2.4.1 Forced regression

Forced regression is driven by relative sea-level fall (i.e., the shoreline is forced to regress by the falling relative sea level; Fig. 4.32), and it is defined by the progradation and downstepping of the shoreline (Figs. 4.1, 4.6, 4.16, 4.21, and 4.22). The architecture of the forced regressive stacking pattern is described by a vertical component (i.e., the amount of downstepping of subaerial clinoform rollovers) and a horizontal component (i.e., the dimension of the prograding lobes in a downdip direction) (Fig. 4.34). The vertical component is controlled by accommodation (i.e., the amount of relative sea-level fall), whereas the horizontal component is controlled by sediment supply and the amount of time available for

the deposition of each lobe. Forced regression results in the deposition of the falling-stage systems tract.

Criteria to recognize forced regression include downstepping subaerial clinoform rollovers and associated subtidal facies (Fig. 4.34); sharp-based, compressed shoreface deposits (Fig. 4.30); stratigraphic foreshortening in settings where the shoreline trajectory is steeper than the seafloor at the onset of forced regression (Figs. 4.35 and 4.36); and coarsening and steepening of clinoforms in a basinward direction due to the increase in topographic gradients and fluvial energy during relative sea-level fall (Fig. 4.36) (Hunt and Gawthorpe, 2000; Plint and Nummedal, 2000; Posamentier and Morris, 2000). Stratigraphic foreshortening may affect both shelf (Fig. 4.35) and slope (Fig. 4.36) clinoforms, and it is common in basins with low-gradient seafloors (e.g., continental shelves, forelands, etc.; Posamentier and Morris, 2000; Zecchin and Tosi, 2014; Zecchin et al., 2017c, Fig. 4.36). However, stratigraphic expansion may also occur in high-gradient ramp settings (e.g., rifts; Lickorish and Butler, 1996; Gawthorpe et al., 1997, 2000; Castelltort et al., 2003; Zecchin et al., 2003; Zecchin, 2007) or in areas affected by growth faults (e.g., Edwards, 1995; Zecchin et al., 2006) (Figs. 4.37 and 4.38).

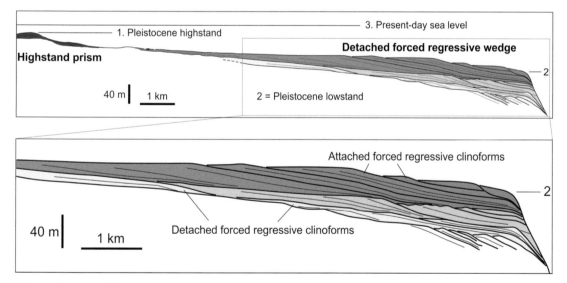

FIGURE 4.34   Architecture of forced regressive shallow-water deposits, based on a dip-oriented seismic line (Rhone shelf, offshore France; modified after Posamentier et al., 1992b). The three unconformity-bounded sequences correspond to climate-driven shoreline shelf-transit cycles of 10^4–10^5 yrs (i.e., glacial-interglacial cycles), and consist mainly of prograding and downstepping clinoforms. Only the falling-stage systems tracts are visible above the seismic resolution within the "forced regressive wedge." Each set of prograding and downstepping clinoforms is topped by a surface of subaerial exposure followed by the interglacial re-flooding of the shelf.

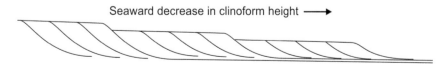

FIGURE 4.35   Decreasing clinoform height during relative sea-level fall in a shallow-water shelf setting (modified after Posamentier and Morris, 2000). This stratigraphic foreshortening occurs where the shoreline trajectory is steeper than the seafloor at the onset of forced regression.

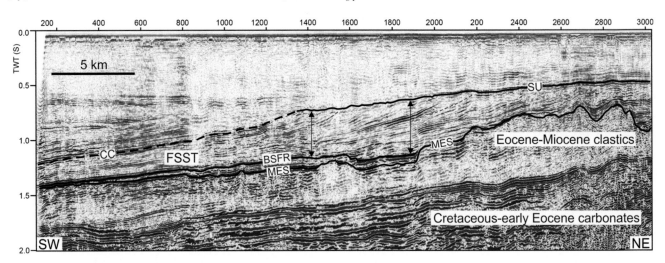

FIGURE 4.36 Dip-oriented seismic line in the northern Adriatic Sea, showing the architecture of a Pliocene forced regressive unit (FSST; modified after Zecchin et al., 2017c). The forced regressive clinoforms display stratigraphic foreshortening, as the trajectory of the forced regressive shoreline is steeper than the seafloor at the onset of forced regression, as well as coarsening and steepening in a basinward direction. Abbreviations: FSST—falling-stage systems tract; SU—subaerial unconformity (mid-Pliocene); CC—correlative conformity; BSFR—basal surface of forced regression; MES—Messinian unconformity.

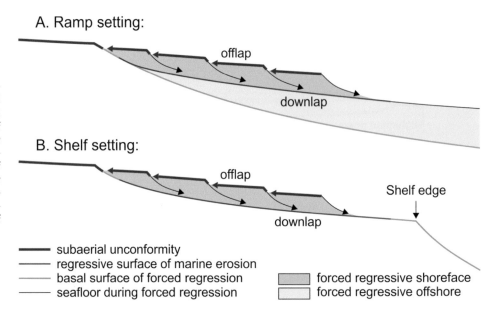

FIGURE 4.37 Degree of preservation of the basal surface of forced regression during relative sea-level fall, in ramp settings (gradient of shoreline trajectory < seafloor gradient at the onset of forced regression) vs. shelf settings (gradient of shoreline trajectory ≥ seafloor gradient at the onset of forced regression). Steeper seafloor gradients increase the preservation potential of offshore falling-stage sediments.

Forced regression may be accompanied by variable fluvial and coastal processes of erosion, bypass, or aggradation (Figs. 4.24 and 4.26). Fluvial and coastal erosion during forced regression is most common (case 1 in Fig. 4.26), leading to the formation of a subaerial unconformity at the top of downstepping and offlapping shallow-water deposits (Figs. 4.21 and 4.39A). This occurs where the trajectory of the shoreline is steeper than the fluvial profile at the onset of forced regression, leading to an increase in fluvial gradients and energy with time, and hence, erosion (Fig. 4.26). This is typically the case in sedimentary basins where the provenance, and the influence of upstream controls, are far from the shoreline, allowing the formation of a low-gradient fluvial topset during the preceding normal regression.

In such cases, the sedimentary basin includes a well-established, and potentially wide, downstream-controlled setting (Fig. 3.11). The heads of incised valleys that form during relative sea-level fall can be used as proxies to locate the updip limit of the downstream-controlled area (Blum, 1994; Blum and Price, 1998; Blum and Tornqvist, 2000; Posamentier, 2001). Forced regressions accompanied by the formation of subaerial unconformities assume nondeposition (erosion or bypass) in continental and coastal settings, and sediment accumulation in marine (or lacustrine) environments (Fig. 4.39A). In this case, riverborne sediment supply to the shoreline is highest during relative sea-level fall.

Less commonly, the fluvial profile at the onset of forced regression may be steeper than the shoreline

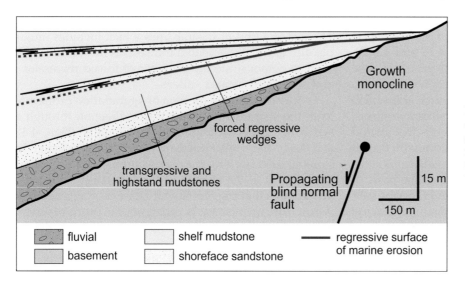

FIGURE 4.38 Forced-regressive shorefaces diverging in a basinward direction, due to syn-depositional tectonism (Miocene, Suez rift, Egypt; modified from Zecchin and Catuneanu, 2020; and Gawthorpe et al., 1997).

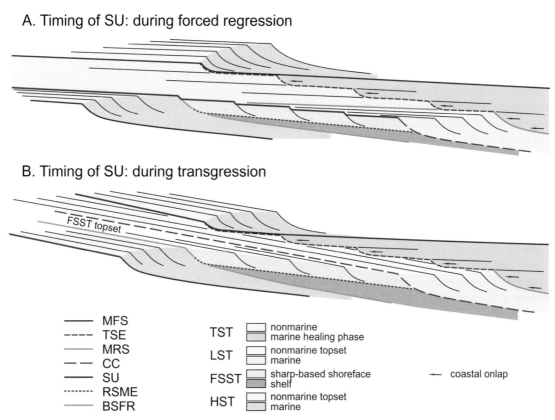

FIGURE 4.39 Architecture of systems tracts and bounding surfaces in siliciclastic shelf settings. A—Forced regressive and transgressive shoreline trajectories are steeper than the fluvial profile, leading to the formation of the subaerial unconformity during forced regression. This scenario is the most common in nature, and involves the dominance of downstream controls on the fluvial portion of systems tracts. In this case, the depositional trends of aggradation and degradation in fluvial to coastal environments follow the geometrical trends of shoreline upstepping and downstepping. B—Fluvial profiles are steeper than the forced regressive and transgressive shoreline trajectories, leading to the formation of the subaerial unconformity during transgression. This scenario is less common, and involves the dominance of upstream controls all the way to the shoreline. In this case, the entire stage of regression records progradation and aggradation in fluvial to coastal environments. However, forced and normal regressions can still be separated based on the downstepping vs. upstepping architecture of subaerial clinoform rollovers and associated shallow-water facies, among other sedimentological and stratigraphic criteria (see text for details). The strike variability in subsidence and sedimentation rates along the shoreline may result in the concurrent development of different systems tracts within same sedimentary basin. Abbreviations: BSFR—basal surface of forced regression; RSME—regressive surface of marine erosion; SU—subaerial unconformity; CC—correlative conformity; MRS—maximum regressive surface; TSE—transgressive surface of erosion; MFS—maximum flooding surface; HST—highstand systems tract; FSST—falling-stage systems tract; LST—lowstand systems tract; TST—transgressive systems tract. The TSE in this diagram is a wave-ravinement surface, onlapped by fully marine transgressive "healing-phase" deposits. In carbonate settings, the subaerial unconformity forms only during forced regression.

trajectory during forced regression, leading to a decrease in fluvial gradients and energy through time, and hence, aggradation (case 3 in Fig. 4.26; Figs. 4.25 and 4.39B). This is possible in settings where the provenance is close to the shoreline, generating steep topographic gradients. In such cases, upstream controls (e.g., source-area uplift) may outweigh the influence of downstream controls, resulting in the deposition of fluvial topsets that extend to the shoreline. Atypical forced regressions with fluvial topsets can still be separated from normal regressions, which typically include fluvial topsets, on the basis of the downstepping vs. the upstepping trajectory of sub-aerial clinoform rollovers and associated subtidal facies; the sharp-based character of the forced regressive clinoforms; the compressed nature of forced regressive shorefaces, in contrast to the expanded normal regressive shorefaces; and the stratigraphic foreshortening of forced regressive clinoforms, where the gradient of the forced regressive shoreline trajectory is steeper than the shelf gradient (Posamentier and Morris, 2000; Catuneanu and Zecchin, 2016, Figs. 4.36 and 4.40).

Forced regressions with subaerial unconformities that obliterate the evidence of offlap may be difficult to separate from normal regressions which do not preserve their fluvial topsets, based on stratal geometries alone (e.g., as observed on seismic lines). In such cases, sedimentological observations that document the downstepping vs. upstepping of shoreface facies in a downdip direction, along with the identification of sharp-based vs. gradationally based shoreface profiles, can prove critical to a reliable interpretation (Fig. 4.41). The evidence of offlap can be erased by processes of fluvial erosion within channel belts, or by strong wind degradation in the interfluve areas. The former are restricted to incised valleys, and the latter are rare in most settings. Consequently, the preservation potential of offlapping stacking patterns is generally high, and offlap is observed commonly on dip-oriented seismic transects. The use of a grid of seismic lines which includes multiple dip- and strike-oriented transects is most often sufficient to demonstrate the presence of forced regressions and associated incised valleys.

Processes of fluvial incision or aggradation propagate from the shoreline in an upstream direction through a series of landward-migrating knickpoints (Figs. 4.42 and 4.43). Each knickpoint marks an abrupt change in slope gradients along the fluvial profile at a particular time, and it is the change in fluvial energy across the knickpoints that triggers fluvial incision or aggradation. A downstream increase in valley slope is prone to fluvial incision (case A in Fig. 4.42; Fig. 4.43), whereas a downstream decrease in valley slope promotes fluvial aggradation (case C in Fig. 4.42) (Pitman and Golovchenko, 1988; Butcher, 1990; Posamentier and Allen, 1999). The fluvial response to changes in slope gradients is in fact more complex than depicted in Fig. 4.42, as rivers may, to some extent, adjust their flow parameters (e.g., the degree of channel sinuosity) in order to adapt to gradient changes without aggradation or incision (Schumm, 1993). However, the principles illustrated in Fig. 4.42 stand as general trends. In either case of fluvial incision of aggradation, the landward migration of knickpoints continues until the fluvial profile becomes graded. The distance of knickpoint migration is variable, depending on the circumstances of each sedimentary basin, generally within a range of 10^0-10^2 km.

In shallow-water settings, seafloor profiles are temporary base levels that reflect the balance between sediment supply and environmental energy (Figs. 3.22 and 3.23; see details in Chapter 3). Excess sediment supply leads to aggradation, whereas excess energy generates erosion. The lowering of the relative sea level and concomitantly of the fairweather wave base during forced regression leads to an increase in energy at the

Atypical forced regression

Typical normal regression

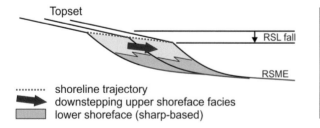
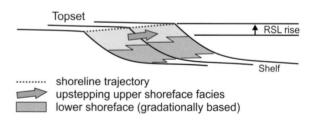

FIGURE 4.40   Criteria to separate atypical forced regressions with fluvial topsets from normal regressions that preserve their fluvial topsets. Forced regressions are characterized by downstepping and sharp-based shoreface facies, and true downlap of shoreface clinoforms against the RSME. Normal regressions are characterized by upstepping and gradationally based shoreface facies, and apparent downlap (i.e., seismic artifact) of shoreface clinoforms against the shelf bottomset facies. The fall in relative sea level during forced regression also results in the deposition of "compressed" shorefaces, with thicknesses limited by the depth of the fairweather wave base. In contrast, the rise in relative sea level during normal regression affords the development of "expanded" shorefaces, with thicknesses that can exceed the depth of the fairweather wave base. Abbreviations: RSL—relative sea level; RSME—regressive surface of marine erosion.

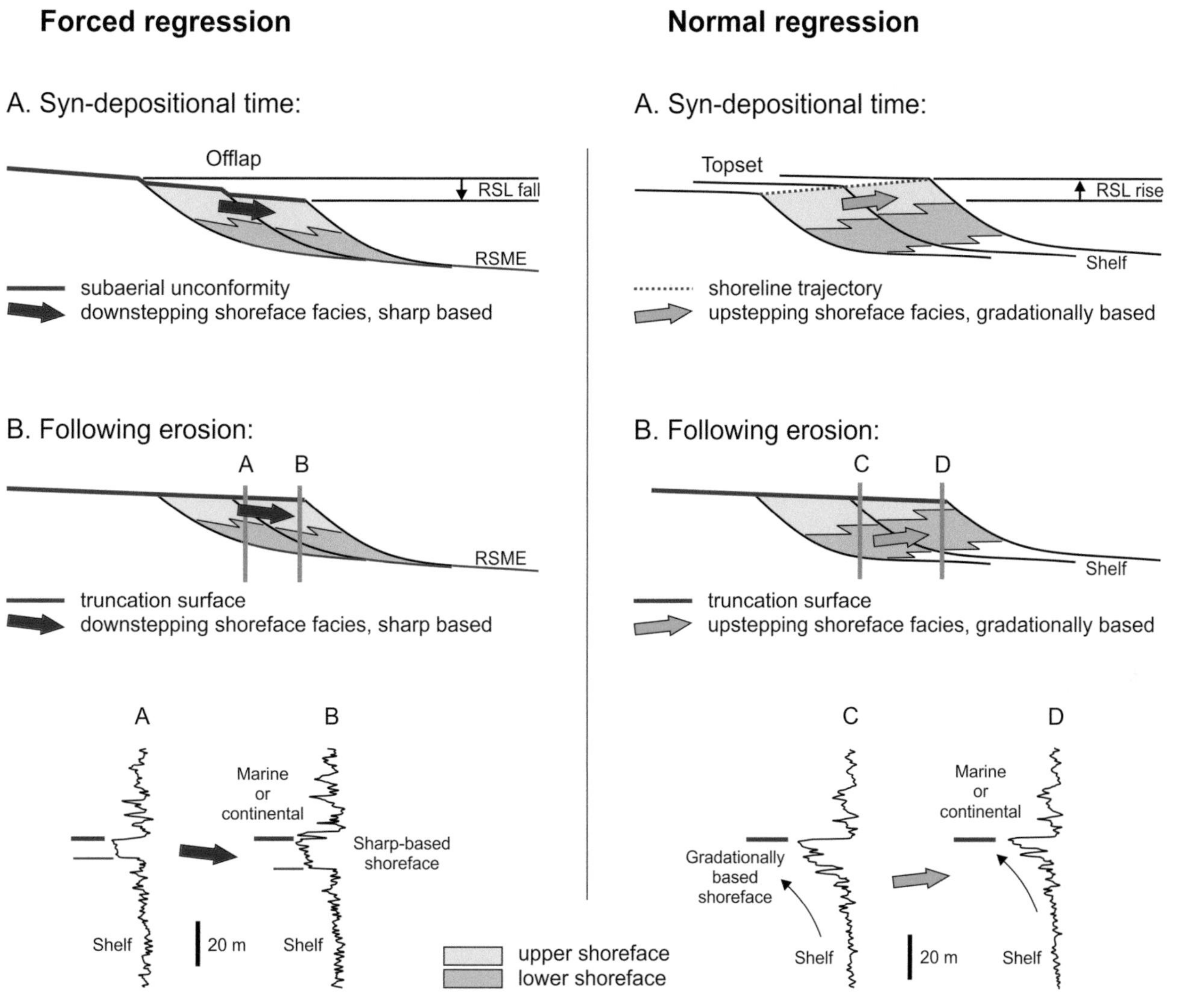

Forced regression

A. Syn-depositional time:

subaerial unconformity
downstepping shoreface facies, sharp based

B. Following erosion:

truncation surface
downstepping shoreface facies, sharp based

Normal regression

A. Syn-depositional time:

shoreline trajectory
upstepping shoreface facies, gradationally based

B. Following erosion:

truncation surface
upstepping shoreface facies, gradationally based

FIGURE 4.41   Criteria to separate forced from normal regressions, where erosion removes the evidence of offlap and fluvial topsets. Forced regressions are characterized by downstepping and sharp-based shoreface facies, and true downlap of shoreface clinoforms against the RSME. Normal regressions are characterized by upstepping and gradationally based shoreface facies, and apparent downlap (i.e., seismic artifact) of shoreface clinoforms against the shelf bottomset facies. Additional field criteria that can be used to identify forced vs. normal regressions have been discussed by Plint (1988), Plint and Nummedal (2000), Posamentier and Morris (2000), Tesson et al. (2000), Zecchin and Tosi (2014), Catuneanu and Zecchin (2016), Catuneanu (2019a), and Sweet et al. (2019). Abbreviations: RSL—relative sea level; RSME—regressive surface of marine erosion; A, B, C, D—wells, illustrated with gamma-ray logs.

seafloor in areas that were previously below the fair-weather wave base. This increase in energy is conducive to the scouring of the seafloor and the development of sharp-based shoreface clinoforms which, where present, are diagnostic of forced regression (Fig. 4.44A). However, the processes in the lower shoreface during relative sea-level fall depend also on sediment supply, which may allow or inhibit the seafloor scouring. Energy-dominated systems (i.e., energy > supply) tend to carve low-gradient seafloors, irrespective of the grain size and the natural angle of repose of the sediment, as the grains are continuously reworked and redistributed by the transport agents (waves, tides). Such settings are most susceptible to erosion

during relative sea-level fall (Figs. 4.44A and 4.45). In contrast, sediment-dominated systems (i.e., supply > energy) generate steeper seafloors, which reflect the angle of repose of the sediment, and are prone to aggradation despite the lowering of the wave base (Figs. 4.44B and 4.46).

In energy-dominated settings, such as wave- or tide-dominated open shorelines and deltas, maintaining the seafloor profile that is in equilibrium with the wave or tide energy during relative sea-level fall requires coeval deposition and erosion in the upper and lower parts of the subtidal area, respectively (Bruun, 1962; Plint, 1988; Dominguez and Wanless, 1991; Posamentier and Chamberlain, 1993, Fig. 4.44). As the shoreline

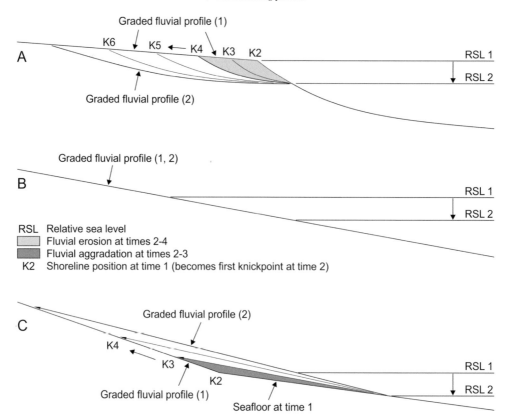

FIGURE 4.42 Fluvial responses to relative sea-level fall, as a function of the gradient contrasts between the fluvial profile and the exposed seafloor (modified after Summerfield, 1985; Pitman and Golovchenko, 1988; Butcher, 1990; Schumm, 1993; Posamentier and Allen, 1999; Blum and Tornqvist, 2000). A—fluvial incision (seafloor steeper than the fluvial profile); B—fluvial bypass (seafloor and fluvial profile dipping at the same angle); C—fluvial aggradation (fluvial profile steeper than the seafloor). Knickpoints (K) mark abrupt changes in fluvial gradients. A downstream increase in slope gradient (and corresponding fluvial energy) leads to fluvial erosion (case A). A downstream decrease in slope gradient (and corresponding fluvial energy) is prone to fluvial aggradation (case C). Knickpoints migrate in an upstream direction, resulting in the landward expansion of the subaerial unconformity (case A, more common in nature) or in the onlap and backfill of the landscape to the level of the new graded profile (case C).

FIGURE 4.43 Upstream-migrating fluvial knickpoint (arrow) along an actively incising "valley." Note the downstepping of the "coastal plain" as a result of water-level fall, and the morphology of the developing subaerial unconformity.

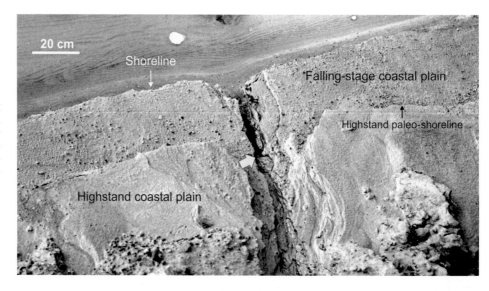

prograds, the upper subtidal clinoforms downlap the scour surface generated in the lower subtidal zone (Fig. 4.44). These "sharp-based" clinoforms, whether shorefaces in front of open shorelines or delta fronts in river-mouth settings, are diagnostic of forced regression, and commonly include well-sorted deposits with good lateral extent along strike, as they follow the paleoshorelines. The gradient of the clinoform surfaces is

Forced regressions:

A. Energy-dominated coastline (waves, tides)

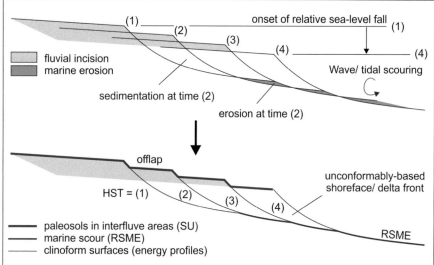

B. Supply-dominated coastline (riverborne sediment > wave/ tide energy)

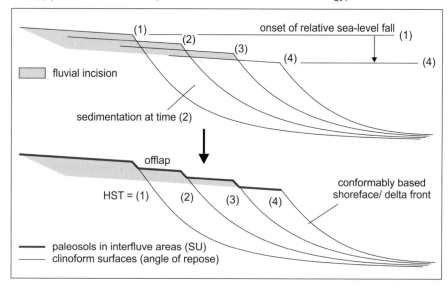

FIGURE 4.44 Sedimentary processes in forced regressive settings. Forced regressions are driven by relative sea-level fall, irrespective of sediment supply. Wave- and tide-dominated settings are prone to lower seafloor gradients due to the prevalent sediment reworking (energy > sediment supply). In such settings, maintaining the hydrodynamic equilibrium profile of the seafloor during relative sea-level fall requires erosion, which generates the RSME. Supply-dominated settings such as river-dominated deltas can sustain steeper seafloor gradients which reflect the angle of repose of the sediment. In this case, the RSME does not form (sediment supply > energy), and the seafloor aggradation concurrent with the fall in relative sea level results in the foreshortening of forced regressive clinoforms. In both cases, the forced regressive clinoforms are sharp-based (Figs. 4.4, 4.30, 4.45, 4.46, and 4.47). Abbreviations: HST—highstand systems tract; SU—subaerial unconformity; RSME—regressive surface of marine erosion.

typically lower than the angle of repose of the sediment, reflecting the energy profile of the seafloor, and therefore it cannot be used as a proxy for grain size where observed on seismic lines prior to drilling. The thickness of the sharp-based clinoforms cannot exceed the depth of the fairweather wave base; i.e., the space available for the accumulation of clinoforms is limited by the sea level at the top and by the locus of wave scouring at the base (Fig. 4.44).

In supply-dominated settings, such as rived-dominated deltas and shorefaces where the amount of riverborne sediment exceeds the capacity of waves and tides to rework it, the gradient of the clinoform

surfaces reflects the angle of repose of the sediment. The surplus of sediment relative to the available environmental energy prevents the scouring of the seafloor during relative sea-level fall, promoting aggradation instead. As a result, the clinoforms are conformably based (Figs. 4.44 and 4.46), as well as foreshortening due to the bottomset aggradation at the same time with the fall in relative sea level. Despite the lack of erosion of the bottomset, the clinoforms in supply-dominated settings are still "sharp-based" due to the high rates of progradation during relative sea-level fall, which promote a rapid shift from the bottomset (prodelta or shelf) to the overlying foreset (delta front

FIGURE 4.45 Sharp-based forced regressive shoreface (A) in a wave-dominated setting (Upper Cretaceous, Blackhawk Formation; Western Interior Basin, Utah). The shoreface sandstones overlie inner shelf deposits (B: inter-bedded sandstones and mudstones, 2 m thick in this image). The regressive surface of marine erosion (RSME) is an unconformity formed by wave scouring in response to the lowering of the fair-weather wave base during relative sea-level fall. The forced regressive shoreface is unconformably overlain by a lowstand topset of amalgamated fluvial channels (C: Castlegate Forma-tion). The intervening subaerial uncon-formity (SU), which formed as a result of exposure during relative sea-level fall, is covered by debris at this location. Both unconformities mark sharp increases in grain size, due to the abrupt shifts to more proximal deposits on top.

or shoreface) facies (Figs. 4.4, 4.46, and 4.47). In addi-tion to this, the forced regression can also be identified based on the pattern of progradation and downstep-ping of the subaerial clinoform rollovers and associated subtidal facies (Fig. 4.40).

The architecture of shallow-water forced regressive clinoforms is primarily a function of rates of relative sea-level fall, sediment supply, and gradient of the sea-floor (Ainsworth and Pattison, 1994; Posamentier and Morris, 2000). The interplay of these variables controls key features of the clinoforms, which may be attached or detached, stepped-topped or smooth-topped, and spread over short or long distances in the dip direction (Fig. 4.48). Detached clinoforms form compartmental-ized petroleum reservoirs or aquifers (i.e., independent flow units), with a potentially large extent along the depositional strike. Stepped-topped clinoforms generate offlap, and is common in the rock record (Figs. 4.6 and 4.34) as well as in modern environments such as in areas that are currently subject to post-glacial isostatic rebound at rates that exceed the present-day rate of sea-level rise (Fig. 4.16). The extent of forced regression along dip depends on the geological setting (e.g., shelf vs. ramp) and the magnitude of relative sea-level fall. The lower the gradient of the seafloor and the greater the magnitude of relative sea-level fall, the larger the area exposed during forced regression.

4.2.4.2 Normal regression

Normal regressions occur during relative sea-level rise, when sedimentation rates outpace the rates of rela-tive sea-level rise at the shoreline (Fig. 4.32), and are defined by the progradation and upstepping of the sub-aerial clinoform rollovers (Figs. 4.1, 4.17, 4.21, 4.22, and 4.49). This shoreline trajectory results in a decrease through time in the gradient of the fluvial profile, and hence in a decline in fluvial energy and the consequent topset aggradation (Figs. 4.6 and 4.49). During normal regressions, the accommodation created by relative sea-level rise is consumed entirely by sedimentation, aggradation is accompanied by sediment bypass (i.e., the surplus of sediment for which no accommodation is available), and a progradation of facies occurs (Fig. 4.49). This process results in the formation of conformable successions, which consist typically of coarsening-upward shallow-water deposits topped by coastal to fluvial facies (Fig. 4.50). Normal regressive successions may develop in both river-mouth (deltaic) and open-shoreline settings. In the former case, the ver-tical profile records a shift from prodelta, to delta front and delta plain facies (Fig. 4.50A), whereas in the latter setting the change is from shelf to shoreface and over-lying beach and coastal plain facies (Fig. 4.50B).

Normal regressions which preserve their fluvial top-sets can be separated from atypical forced regressions with fluvial topsets on the basis of facies relationships (e.g., gradationally based vs. sharp-based shorefaces), and the trajectory of subaerial clinoform rollovers (i.e., upstepping vs. downstepping). These contrasting trajec-tories also lead to the development of expanded (i.e., with thicknesses that can exceed the depth of the fair-weather wave base, in the case of normal regressions) vs. compressed (i.e., with thicknesses limited by the depth of the fairweather wave base, in the case of forced

FIGURE 4.46   Forced regressive, river-dominated deltaic succession (Upper Cretaceous, Panther Tongue; Western Interior Basin, Utah). A—Rapid shift of facies from prodelta to the overlying delta front deposits. The delta front clinoforms are sharp but conformably based as no scouring occurred during relative sea-level fall (see supply-dominated settings in Fig. 4.44); B—relatively steep delta front clinoforms (dip angle of 5—15 degree), reflecting the angle of repose of the sediment (sediment supply > wave energy). The delta front clinoforms are truncated by a wave-ravinement surface (arrow), overlain by a transgressive lag and transgressive shales. No delta plain topset accumulated during the progradation and downstepping of this delta system.

regressions) shoreface successions (Figs. 4.30 and 4.40). Aggradation takes place in all depositional environments during normal regression, leading to the accumulation of topsets, foresets, and bottomsets. For this reason, the toplap and the downlap seismic-reflection terminations that are observed in relation to normal regressions are artifacts, generated by topsets and bottomsets that fall below the seismic resolution (Fig. 4.13). This is in contrast with the forced regressive clinoforms that may be observed on high-resolution seismic lines in wave- or tide-dominated shallow-water systems, which are associated with true downlap against the regressive surface of marine erosion (Figs. 4.34 and 4.40).

The dip angle of clinoform surfaces depends on the dominant controls on sedimentation in the subtidal area, as well as on sediment supply. In the case of energy-dominated coastlines (wave- or tide-dominated settings), the angle of the seafloor equilibrium profile is typically very low, less than 1 degree (e.g., the mean seafloor gradient in a wave-dominated shoreface is c. 0.3 degree). This angle is steeper in the case of supply-dominated coastlines, such as river-dominated deltas and adjacent shorefaces, where seafloor gradients reflect the angle of repose of the sediment (e.g., up to 35 degree in the case of Gilbert-type deltas). In either case, the creation of accommodation during relative sea-level rise promotes aggradation on both sides of the shoreline, and hence no fluvial- or wave/tidal-cut unconformities are associated with this type of shoreline trajectory (Fig. 4.49). As a result, normal regressive shoreface or delta front deposits are "gradationally based" and commonly expanded (Figs. 4.30 and 4.50), in contrast

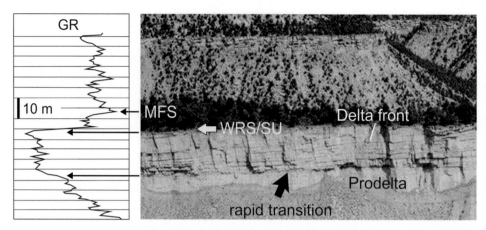

FIGURE 4.47   Outcrop and well-log expression of a forced regressive, river-dominated deltaic succession (images courtesy of Henry Posamentier; Upper Cretaceous, Panther Tongue; Western Interior Basin, Utah). The deltaic succession is conformable, albeit with a rapid transition from the prodelta to the delta front. This rapid shift is due to the typically high rates of progradation during relative sea-level fall (see also Figs. 4.4 and 4.46), and may be approximated with a facies contact (i.e., the "within-trend forced regressive surface"; details in Chapter 6). However, no single surface can be picked as a unique contact between the prodelta and delta front facies. In such supply-dominated settings, the delta front (c. 20 m thick in this example) may be thicker than the depth of the fairweather wave base, as the toe of the delta front clinoforms may reach below the fairweather wave base. Abbreviations: GR—gamma ray log; WRS/SUR—transgressive wave-ravinement surface reworking and replacing a subaerial unconformity; MFSR—maximum flooding surface.

FIGURE 4.48   Stratal architecture of shallow-water forced regressive clinoforms, as a function of sediment supply, rates of relative sea-level fall, and gradient of the seafloor. The interplay of these variables may generate different architectural styles, with clinoforms attached or detached, stepped-topped or smooth-topped, and spread over short or long distances (see Posamentier and Morris, 2000; for a detailed discussion).

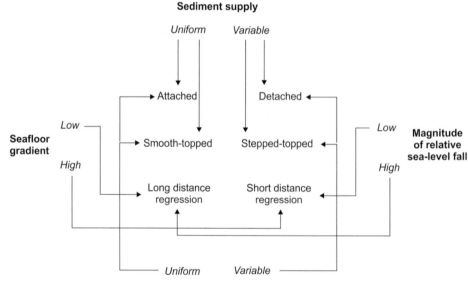

FIGURE 4.49   Sedimentary processes in normal regressive settings. Normal regressions are driven by sediment supply, where the rates of sedimentation outpace the rates of relative sea-level rise at the shoreline. These conditions are commonly met during early and late stages of relative sea-level rise, when the rates of accommodation are typically lower (Fig. 4.32). Progradation rates are generally low. Normal regressions are prone to aggradation in all depositional environments, due to the upstepping of the equilibrium fluvial and seafloor profiles following the progradation and upstepping of the shoreline. The normal regressive clinoforms are typically gradationally based (see examples in Figs. 4.5, 4.30, and 4.50).

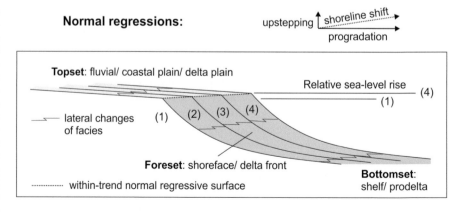

FIGURE 4.50 Normal regressive deposits in river-mouth (A—Upper Cretaceous, Ferron Delta; Western Interior Basin, Utah) and open-shoreline (B—uppermost Cretaceous, Bearpaw and Horseshoe Canyon formations, Western Interior Basin, Alberta) settings. The successions are conformable, with prominent facies contacts (arrows) at the limit between the foresets and the topsets of the prograding systems (i.e., the "within-trend normal regressive surface"; see Chapter 6 for definition).

with the compressed and "sharp-based" forced regressive clinoforms (Figs. 4.4, 4.30, and 4.45—4.47).

Normal regressive topsets include continental, supratidal, and intertidal deposits (Fig. 4.49), which wedge out in an updip direction (Fig. 4.39). The thickness of topset successions in coastal settings depends on the rates of coastal aggradation and the duration of normal regression, and it is a proxy for the magnitude of relative sea-level rise during the normal regression. Topsets can be identified on the basis of facies analysis in outcrops or boreholes, but their observation on seismic lines depends on their thickness relative to the vertical seismic resolution. Topsets that fall below the seismic resolution are expressed as toplap reflection terminations on dip-oriented seismic lines (Figs. 4.7, 4.8, and 4.13). In this case, other criteria must be used to separate normal regressions from forced regressive clinoforms with smooth tops, such as the expanded vs. foreshortened

nature of clinoforms, respectively (Fig. 4.36). The surface that separates the topset from the underlying subtidal foreset is typically a prominent within-trend facies contact (i.e., a surface of litho- rather than sequence stratigraphy; see Chapter 6 for details), which becomes younger basinward with the rate of shoreline regression (Figs. 4.49 and 4.50). The preservation potential of normal regressive topsets is variable, as they may be truncated by subaerial unconformities or transgressive ravinement surfaces, or may be overlain conformably by fluvial or estuarine deposits belonging to the overlying systems tracts (Figs. 4.6 and 4.39).

Normal regressions are classified into "lowstand" and "highstand," depending on the preceding stratal stacking pattern of equal hierarchical rank: a normal regression that follows a forced regression is designated as "lowstand," whereas a normal regression that follows a transgression is designated as "highstand" (e.g.,

Catuneanu, 2006; Csato and Catuneanu, 2012, 2014). A lowstand normal regression is typically characterized by a concave-up shoreline trajectory, which reflects a shift from a dominantly progradational to a dominantly upstepping trend, as a consequence of accelerating relative sea-level rise. In contrast, a highstand normal regression tends to display a convex-up shoreline trajectory, which is the result of decelerating relative sea-level rise and the consequent shift from a dominantly upstepping to a dominantly progradational trend (Catuneanu, 2006, Fig. 4.51). The ideal concave-up or convex-up shoreline trajectories are not always evident, so the most reliable protocol to differentiate lowstand from highstand normal regressions rests with the observation of stratigraphic relationships (i.e., the preceding stacking pattern of equal hierarchical rank). Furthermore, the geographic location of lowstand and highstand coastlines within the sedimentary basin (i.e., closer to the shelf edge vs. closer to the basin margin, respectively) provides additional clues to separate the two types of normal regression.

Other differences between lowstand and highstand normal regressions can be found in the architecture of their topsets. The rates of topset aggradation follow the rates of shoreline upstepping, and therefore they typically increase with time during lowstand normal regressions, and decrease with time during highstand normal regressions. These trends are reflected in the thickness of the higher frequency sequences that compose the topset units, and are particularly evident in carbonate systems where topsets consist of peritidal cycles generated by orbital forcing (Fig. 4.52). Contrasts between lowstand and highstand topsets are also evident in fluvial systems, due to differences in gradients and energy levels between the lowstand and the highstand rivers. The patterns of change in fluvial energy within a sequence depend on the timing of the subaerial unconformity, which may form during forced regressions or transgressions (Leckie, 1994; Posamentier and Allen, 1999; Catuneanu, 2006; Catuneanu and Zecchin, 2016, Figs. 4.26, 4.28, and 4.39). In both cases, the gradients of the fluvial profile increase during the formation of the subaerial unconformity, and decrease during the rest of the stratigraphic cycle when fluvial aggradation occurs.

Where the subaerial unconformity forms during forced regression (i.e., in settings where the shoreline trajectory is steeper than the fluvial profile; Fig. 4.26), river systems aggrade during the deposition of lowstand, transgressive, and highstand systems tracts (Fig. 4.39A). In this case, the lowstand topset includes the highest energy fluvial systems of the stratigraphic cycle, and the fluvial sequence records a decline in energy levels from lowstand to highstand. This scenario, which is most common in the stratigraphic record, is typically recorded where the provenance is far from the shoreline (e.g., 10^2-10^3 km), allowing for the development of low-gradient fluvial profiles within the downstream-

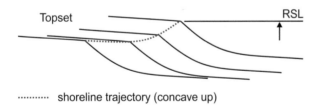

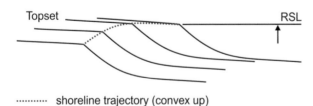

FIGURE 4.51 Architecture of lowstand vs. highstand normal regressions (from Catuneanu et al., 2009). In both cases, progradation is driven by sediment supply during relative sea-level rise (sedimentation > accommodation at the shoreline). Ideally, a lowstand normal regression records a concave-up shoreline trajectory (progradation to aggradation), whereas a highstand normal regression records a convex-up shoreline trajectory (aggradation to progradation). However, these ideal trends may or may not be observed in nature. The most reliable way to identify the type of normal regression is to observe the preceding stacking pattern: a normal regression that follows a forced regression is "lowstand"; a normal regression that follows a transgression is "highstand." Abbreviation: RSL—relative sea level.

FIGURE 4.52 Peritidal cycles within the topset of a highstand systems tract (Triassic, The Dolomites, Italy). The peritidal cycles correspond to high-frequency sequences bounded by exposure surfaces. Note the upward decrease in the thickness of peritidal sequences, reflecting the long-term decrease in the rates of accommodation. The highstand topset is overlain by the lowstand topset which displays a more continental character. Abbreviations: HST—highstand systems tract; LST—lowstand systems tract; SU—subaerial unconformity.

controlled settings. The actual fluvial styles (e.g., braided vs. meandering) are variable, depending on all factors that control the type of river in any particular location. If both unconfined (e.g., braided) and confined (e.g., meandering) fluvial styles are present within a depositional sequence, it is typical to find the higher energy braided systems within the lowstand topset, and the lower energy meandering systems within the highstand topset (e.g., Shanley et al., 1992; Kerr et al., 1999; Fanti and Catuneanu, 2009, 2010).

Where the subaerial unconformity forms during transgression (i.e., in settings where the fluvial profile is steeper than the shoreline trajectory; Fig. 4.28), the aggradation of river systems occurs during the deposition of highstand, falling-stage, and lowstand systems tracts (Fig. 4.39B). In this case, the highstand topset includes the highest energy fluvial systems of the stratigraphic cycle, and the fluvial sequence records a decline in energy levels from highstand to lowstand. This scenario is promoted by a proximal location of the provenance relative to the shoreline (e.g., 10^0-10^1 km), which leads to the development of steep landscape gradients and the potential dominance of upstream controls all the way to the shoreline (Catuneanu and Zecchin, 2016).

Irrespective of the timing of fluvial incision (i.e., during forced regression or transgression), the coarsest sediment is typically found above the subaerial unconformity, and the fluvial sequence displays a fining-upward profile that reflects the decrease in stream

energy and competence with time. This trend also explains the increased likelihood of occurrence of channel amalgamation at the base of depositional sequences, in relation to the steeper gradients which promote higher energy and potentially unconfined river systems. As such, the fluvial portion of depositional sequences in downstream-controlled settings often starts with amalgamated channels and coarser sediment at the base (i.e., "box" sands), grading upward into floodplain-dominated successions and finer-grained channel fills (Fig. 4.53). Similar trends are also found within fully fluvial depositional sequences in upstream-controlled settings, whether the sedimentary basins are overfilled or underfilled (e.g., Boyd et al., 2000; Catuneanu and Bowker, 2001; Catuneanu and Elango, 2001; Leckie et al., 2004).

Both lowstand and highstand normal regressive topsets wedge out toward the basin margin, irrespective of the dominant (downstream vs. upstream) controls on fluvial processes. Therefore, the downstream- vs. upstream-controlled nature of the continental setting is difficult to infer from stratal geometries alone. In sedimentary basins where the provenance is located far from the shoreline, and the landscape gradients are low, the lowstand and highstand topsets are typically confined to the downstream-controlled settings (Blum, 1994; Blum and Price, 1998; Blum and Tornqvist, 2000). In such cases, the updip limit of fluvial topsets can be used as a proxy for the boundary between downstream- and upstream-controlled areas. This interpretation can be corroborated with the observation of the fluvial response to forced regressions and transgressions, which is more closely tied to downstream vs. upstream controls (e.g., the formation of incised valleys during forced regression indicates downstream controls, and the updip extent of the incised valleys is a proxy for the limit between downstream- and upstream-controlled areas; in contrast, the formation of incised valleys during transgression can be linked to upstream controls; Leckie, 1994; Posamentier, 2001; Catuneanu and Zecchin, 2016).

Normal regressions which do not preserve their fluvial topsets, or with topsets and bottomsets below the seismic resolution, may be difficult to separate from forced regressions that lost the evidence of offlap, based on seismic reflection geometries alone (Fig. 4.41). In such cases, additional core and/or well-log data are necessary to document the downstepping vs. upstepping of shoreface facies in a downdip direction, the sharp-based vs. gradationally based nature of the shoreface clinoforms, and the compressed vs. expanded development of shoreface deposits (Fig. 4.41). The main features which afford the distinction between forced and normal regressive clinoforms are summarized in Fig. 4.54.

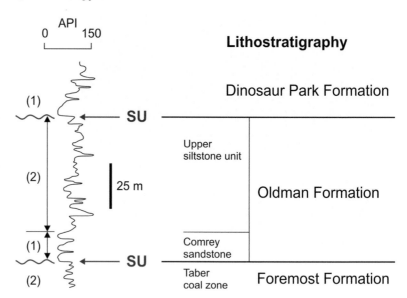

FIGURE 4.53 Unconformity-bounded fluvial sequence in a downstream-controlled setting (Upper Cretaceous, Belly River Group; Western Canada Basin, Alberta). (1) Amalgamated braided channels; (2) floodplain-dominated meandering system; Abbreviations: SU—subaerial unconformity; API—gamma-ray scale in American Petroleum Institute units.

4.2.4.3 Transgression

Transgression is driven by relative sea-level rise that outpaces the rates of sedimentation in coastal environments, and it is defined by the backstepping and upstepping of the shoreline (Figs. 4.1, 4.18, 4.21, and 4.22). Transgression may be accompanied by variable (nondiagnostic) fluvial processes of aggradation, bypass, or erosion, depending on the gradient of the shoreline trajectory relative to the landscape gradient (Fig. 4.28). Fluvial aggradation during transgression (case 1 in Fig. 4.28; Figs. 4.39A and 4.55A) occurs where the trajectory of the shoreline is steeper than the landscape profile, leading to a decrease in fluvial gradients and energy with time. This scenario is most common in nature, and it favors the preservation of estuarine deposits at the

river mouth. This is typically the case in sedimentary basins where the provenance, and the influence of upstream controls, are far from the shoreline (e.g., range of 10^2–10^3 km), allowing the development of low-gradient profiles in the downstream reaches of the fluvial system. In such cases, the fluvial portion of the transgressive systems tract is dominantly, if not entirely, downstream-controlled, and the landward limit of the transgressive systems tract can be used as a proxy for the updip extent of the downstream-controlled area (Shanley et al., 1992; Blum, 1994; Blum and Price, 1998; Kerr et al., 1999; Blum and Tornqvist, 2000).

Under particular circumstances, it is also possible that the shoreline trajectory matches the landscape gradient (case 2 in Fig. 4.28: fluvial bypass, with no change in

Clinoforms \ Features	Shoreline trajectory	Foreset facies trends	Foreset morphology	Basal contact	Development
Forced regressive, energy-dominated	Downstepping	Downstepping basinward	Sharp based	Unconformable (RSME)	Compressed
Forced regressive, supply dominated				Conformable (WTFRS)	Expanded to compressed
Normal regressive, energy and supply dominated settings	Upstepping	Upstepping basinward	Gradationally based	Conformable and gradual	Expanded

FIGURE 4.54 Characteristics of forced and normal regressive clinoforms in energy- and supply-dominated settings. Energy-dominated settings include wave- and tide-dominated deltas and shorefaces (energy > sediment supply). Supply-dominated settings include river-dominated deltas and shorefaces (sediment supply > wave/tide energy). The sharp-based character of the forced regressive clinoforms is due to the lowering of the wave base, as well as to the typically high rates of progradation during relative sea-level fall. Abbreviations: RSME—regressive surface of marine erosion; WTFRS—within-trend forced regressive surface.

Transgressive shorelines:

A. Fluvial aggradation

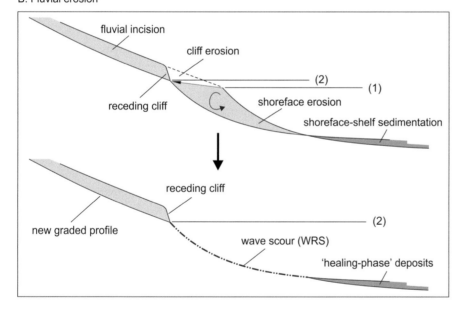

B. Fluvial erosion

FIGURE 4.55 Sedimentary processes in transgressive settings. Transgressions are driven by relative sea-level rise, where the rates of accommodation outpace the rates of sedimentation at the shoreline. Fluvial aggradation during transgression occurs where the shoreline trajectory is steeper than the landscape gradient; in this case, the fluvial system is downstream-controlled (i.e., fluvial aggradation driven by the rise in relative sea level). In contrast, fluvial erosion occurs where the landscape gradient is steeper than the shoreline trajectory; in this case, the fluvial system may be entirely upstream-controlled (e.g., source area uplift). In both cases, the hydrodynamic equilibrium profile of the seafloor during transgression is maintained by a combination of wave erosion and seafloor aggradation. As the coastline retreats, the areas of wave erosion and sea-floor aggradation backstep as well, resulting in the onlap of the wave-ravinement surface (WRS) by the shallow-water "healing-phase" deposits (coastal onlap).

the graded fluvial profile during transgression), or that the shoreline trajectory has a lower gradient than the fluvial profile (case 3 in Fig. 4.28; Fig. 4.55B). In the latter scenario, the slope of the channels steepens during transgression, leading to an increase in fluvial energy, and hence, degradation (Fig. 4.39B). This is typically the case in settings where the provenance is closer to the shoreline (e.g., range of 10^0–10^1 km), generating steep fluvial profiles. In this case, the upstream controls on fluvial processes may outweigh the downstream controls all the way to the shoreline, leading to the formation of subaerial unconformities during transgression

(Fig. 4.39B). The shoreline of transgressions accompanied by fluvial erosion or bypass becomes the updip limit of the transgressive systems tract and of the downstream-controlled area. Fluvial incision during transgression is typically accompanied by the formation of backstepping coastal cliffs in the interfluve areas (e.g., Fig. 4.56; Leckie, 1994; Catuneanu and Zecchin, 2016). This stratigraphic scenario has also been documented in Calabria, southern Italy, where the regional uplift of the source area (i.e., the Sila massif, located several tens of km inland from the shoreline) led to coastal cliff erosion and the formation of a subaerial unconformity

FIGURE 4.56 Fluvial incision and the development of coastal cliffs during transgression (Canterbury Plain, New Zealand). A—river-mouth setting; B—open-shoreline setting. These processes are promoted by a landscape gradient steeper than the shoreline trajectory, due to the proximity of the provenance (Southern Alps of New Zealand) to the shoreline. In this case, the upstream controls (i.e., source area uplift) may dominate fluvial processes all the way to the shoreline (case 2 in Fig. 4.55). The gravel on the beach (B) is in part riverborne, transported from river mouths (A) by strong longshore currents in this high-energy wave-dominated setting, and in part produced by the erosion of the receding cliff which consists of Quaternary diamictites.

during transgression, along with the exposure and deactivation of the Crotone Basin (Westaway, 1993; Zecchin et al., 2016).

The main processes that take place in the transition zone between continental and marine environments

during transgression are summarized in Fig. 4.55. These processes involve both sediment reworking and aggradation, depending on the balance between environmental energy and sediment supply at each location along the dip-oriented transect. Maintaining the equilibrium profile of the shoreface during transgression typically requires wave scouring which results in coastal erosion and the formation of a wave-ravinement surface that expands in a landward direction for as long as the shoreline transgresses. The combination of wave scouring in the shoreface and deposition on the shelf preserves the seafloor profile that is in equilibrium with the wave energy during transgression (Fig. 4.55, Bruun, 1962; Dominguez and Wanless, 1991). The amount of erosion in the shoreface is limited by the depth of the fairweather wave base, generally within a range of a few meters to 15—20 m. Exceptionally, in areas of high wind and wave energy, the erosion can remove up to 40 m of underlying sediment (e.g., along the Canterbury Plains of New Zealand; Leckie, 1994). The wave-ravinement surface is commonly draped by lag deposits and/or shell beds with average thicknesses of 10^{-2}—10^{-1} m (Fig. 4.57).

Irrespective of the variable processes of fluvial aggradation or erosion that may occur during transgression (Fig. 4.55), diagnostic to the transgressive systems tract is the development of onlapping "healing-phase" wedges in the marine (or lacustrine) environment (Fig. 4.6). Sedimentation from suspension during transgression "heals" the bathymetric profile of the seafloor which, following the deepening of the water, has a gradient that is steeper than the angle of repose of the finer grained transgressive sediments (Posamentier and Allen, 1993; Willis and Wittenberg, 2000). The healing-phase deposits re-establish equilibrium conditions at the seafloor by filling in the lows in both shallow-water settings (deposition associated with coastal onlap) and deep-water settings (deposition associated with marine onlap; Fig. 4.8). The observation of

FIGURE 4.57 Patterns of sediment distribution during shoreline transgression (modified from Posamentier and Allen, 1993; and Willis and Wittenberg, 2000). During fairweather, sediment from the shoreface contributes to the formation of coastal systems (e.g., beaches or barrier islands), while the coarser fraction mantles the wave-ravinement surface as a transgressive lag. The preservation of transgressive coastal systems depends on the balance between coastal aggradation and the subsequent wave erosion. The bathymetric profile of the youngest underlying clinoform is "healed" by the deposition of suspension sediment and tempestites.

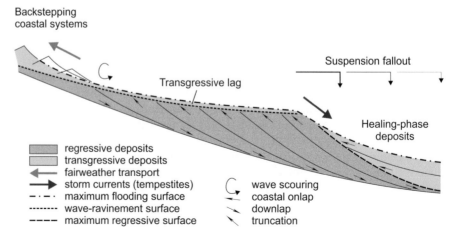

onlapping and fining-upward successions in shallow-water systems is particularly reliable for the identification of the transgressive systems tract.

The patterns of sediment distribution following the wave scouring in the shoreface during transgression are illustrated in Fig. 4.57. During fairweather, the bed-load and saltation load sediment that can be moved by waves in the shoreface is transported landward and contributes to the formation of backstepping coastal systems (backstepping beaches, estuary—mouth complexes, barrier islands; Swift and Thorne, 1991), whereas the suspended load, which is unable to accumulate in the high-energy foreshore and shoreface environments, drifts as hypopycnal plumes offshore, where it eventually flocculates and settles down from suspension as healing-phase deposits. During storms, coastal systems are subject to erosion and contribute coarser sediment to the healing-phase wedges; such tempestites are interbedded with the finer grained sediment accumulated during fairweather and generate heterogeneities, which materialize in seismic reflections where the healing-phase wedges are thicker than the vertical seismic resolution (Fig. 4.6). However, the healing-phase deposits are overall finer grained than the regressive clinoforms or the transgressive coastal systems, as being dominated by suspension sediment that onlaps the youngest prograding clinoform (Figs. 4.6 and 4.57).

The preservation of transgressive coastal systems depends on the balance between the competing processes of coastal aggradation and subsequent wave scouring as the shoreface shifts on top of the former coastline during transgression. Coastal erosion is inevitable, whether fluvial systems aggrade or incise during transgression (Fig. 4.55), but the preservation of at least part of a backstepping coastal system is still possible where the amount of coastal aggradation exceeds the amount of subsequent wave erosion. This is feasible where the trajectory of the transgressive shoreline is steeper than the landscape gradient (case 1 in Fig. 4.28; Figs. 4.39A and 4.55A). Coastlines dominated by aggradation lead to the preservation of estuarine or backstepping beach facies in the rock record. Coastlines dominated by erosion are commonly associated with processes of cliff formation and fluvial incision, leading to the development of subaerial unconformities during transgression (case 3 in Fig. 4.28; Figs. 4.39B and 4.55B). In this case, the stratigraphic hiatus of the subaerial unconformity is age-equivalent with the transgressive healing-phase wedge.

A modern example of a strongly erosional transgressive coastline is provided by the Canterbury Plains in the Southern Island of New Zealand (Leckie, 1994). The steep landscape gradients generated by the uplift of the Southern Alps (sediment provenance), coupled with the high wind and wave energy in the shoreface, led to the formation of incised valleys and coastal cliffs

during transgression. As a result, estuaries are incised into the coastal plain, and the open shorelines are marked by receding cliffs (Fig. 4.56). The high wave energy in this setting is enhanced by oceanic swell originating as far as 2000 km offshore. The wave-cut cliffs, which may be up to 25 m high, recede at a rate of c. 1 m per year. Coastal erosion lowers the gradient of the shoreline trajectory which, coupled with the steep landscape profile, result in fluvial incision at a rate of 1.5—4.2 mm per year in the vicinity of the coastline. The amount of incision gradually decreases inland from the coast, until it becomes minimal 8—15 km upstream (Leckie, 1994).

4.2.5 Stratigraphic scales in downstream-controlled settings

Downstream-controlled stacking patterns can be observed at different stratigraphic scales, in relation to shoreline trajectories of different hierarchical ranks. At any stratigraphic scale, the shoreline trajectory represents a trend which connects the maximum regressive shorelines of immediately lower hierarchical rank (e.g., a third-order shoreline trajectory connects the maximum regressive shorelines of the fourth-order cycles; Fig. 4.58). Therefore, with the exception of the highest frequency shoreline shifts recorded at sedimentological scales (e.g., during tidal cycles), shoreline trajectories are trends defined by composite rather than single physical surfaces, observed at different stratigraphic scales (Figs. 4.58 and 4.59). For example, the shoreline trajectory of a deltaic system can be mapped at the limit between delta plain (topset) and delta front (foreset), which connects the maximum regressive shorelines of lower hierarchical rank (Fig. 4.58). At the smallest stratigraphic scale, these maximum regressive shorelines are represented by the shoreline positions at low tide (i.e., the lowstand shorelines of the tidal cycles: the limit between intertidal and subtidal environments, which is the subaerial clinoform rollover of the lowest hierarchical rank; Fig. 4.58). At the opposite end of the spectrum, the trajectory of a shelf edge represents a first-order shoreline trajectory, which connects the lowstand shorelines of the second-order cycles (i.e., the subaerial clinoform rollovers of the highest hierarchical rank; Fig. 4.60; Catuneanu, 2019a,b).

The transit area of the shoreline during higher frequency transgressions and regressions is located updip from the shoreline trajectory of higher hierarchical rank (Figs. 4.58, 4.60, and 4.61). The updip extent of this transit area is a function of topographic gradients, sediment supply, and the magnitude of relative sea-level changes (e.g., at the smallest scale of observation, the updip extent of the transit area within a deltaic

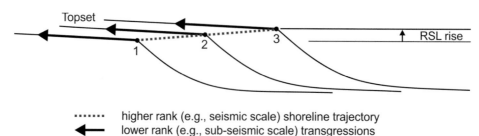

Topset

↑ RSL rise

······· higher rank (e.g., seismic scale) shoreline trajectory
◀─── lower rank (e.g., sub-seismic scale) transgressions
1, 2, 3 maximum regressive shorelines of lower hierarchical rank

FIGURE 4.58 Scale-independent stacking patterns: from deltas to shelf-slope systems. At any scale, the shoreline trajectory represents the trajectory of maximum regressive shorelines of immediately lower hierarchical rank. The lower rank shorelines transgress across the higher rank topset, affecting the preservation of continental deposits. The prevalent depositional system within the topset (continental vs. marine) depends on the balance between continental aggradation and the subsequent ravinement erosion and marine aggradation during the higher frequency transgressions. The shoreline transit time across the topset varies from diurnal (in the case of intertidal delta plains) to 10^4–10^5 yrs (in the case of continental shelves). In this context, a delta is a small-scale analogue of a siliciclastic shelf-slope system, whereby the progradation of clinoforms is enhanced during stages of lowstand when the clinoform rollovers become subaerial and riverborne sediment is delivered directly to the clinoform surface. At the opposite end of the spectrum, the trajectory of a shelf edge represents a first-order shoreline trajectory that separates a first-order topset (i.e., shelf setting) from a first-order foreset (i.e., slope setting). Abbreviation: *RSL*—relative sea level.

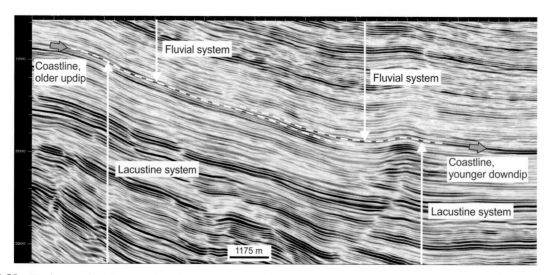

FIGURE 4.59 Diachronous limit between a lacustrine system (Pozo D-129 Formation) and a fluvial system (Castillo Formation), associated with the progradation of the shoreline during normal regression (Lower Cretaceous, Golfo San Jorge Basin, Argentina; image courtesy of YPF Argentina). The shingling of the seismic reflections is caused by the higher frequency transgressions that occurred during the long-term progradation of the shoreline.

environment—i.e., the intertidal area of the delta plain—depends primarily on the gradient of the delta plain and the tidal range). The transit time of the shoreline across the higher rank topsets varies with the scale of observation, from diurnal (tidal cycles, in the case of intertidal delta plains) to Milankovitch scales (10^4–10^5 yrs, in the case of continental shelves; Burgess and Hovius, 1998; Porebski and Steel, 2006). High-frequency sequences within higher rank topsets have been documented at timescales of 10^2–10^5 yrs, and thickness scales of 10^0–10^1 m (e.g., Pellegrini et al., 2017, 2018: 10^2–10^3 yrs, and 10^0–10^1 m; Nanson et al., 2013: 10^3 yrs, and 10^0–10^1 m; Ainsworth et al., 2017: 10^4–10^5 yrs, and 10^0–10^1 m).

At any scale of observation, the areal extent of sequences is typically greater than the transit area of the shoreline, as sequences extend both updip and downdip into the fully continental and fully marine settings (Fig. 4.60). The areal extent of sequences is highly variable, as a function of basin physiography and the controls on sequence development, and is not necessarily proportional to the thickness or the duration of cycles. In the case of sedimentary basins with large and shallow interior seaways, and with cyclicity controlled by relative sea-level changes, relatively thin sequences (± 10 m scale) can develop over distances of 10^2 km (e.g., Catuneanu et al., 1997a, 1999). In the case of narrower, fault-bounded sedimentary basins, and/or in cases where cyclicity is autogenic in origin, sequences may be much more restricted in terms of areal extent (e.g., Martins-Neto and Catuneanu, 2010; Catuneanu and Zecchin, 2013).

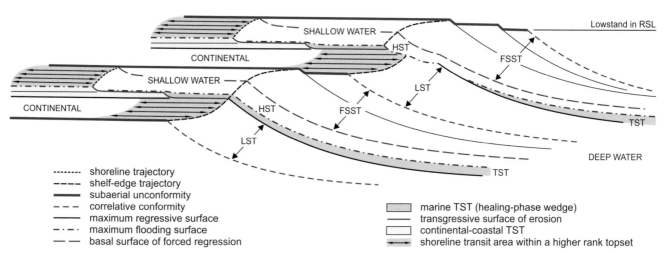

FIGURE 4.60 Shoreline vs. shelf-edge trajectories in siliciclastic shelf-slope settings (not to scale; slope height in a range of 10^2–10^3 m). Shoreline trajectories can be observed at different scales. At each scale of observation, the shoreline moves across a transit area located updip of the shoreline trajectory of immediately higher hierarchical rank. At the largest stratigraphic scale, the trajectory of the shelf edge is the first-order shoreline trajectory which marks the downdip limit of the second-order shoreline transit area within the first-order topset. In this example, the systems tracts shown are of second order, and the shoreline shifts within the second-order topsets are of third-order. At each hierarchical level, the width of the shoreline transit area depends on the gradient of the shelf, sediment supply, and the magnitude of RSL changes. Abbreviations: LST—lowstand systems tract; TST—transgressive systems tract; HST—highstand systems tract; FSST—falling-stage systems tract; RSL—relative sea level.

Shoreline trajectories of any hierarchical rank approximate the limit between continental and marine systems observed at that particular scale (Figs. 4.21 and 4.51). For this reason, the depositional systems located updip from the shoreline trajectory are generally labeled as "continental." However, relevant to the prediction of facies, processes within the transit area of the shoreline that operate at lower hierarchical levels (Figs. 4.58 and 4.60) may promote the deposition and preservation of any types of facies, from continental to shallow marine. For example, fourth-order transgressions during a third-order normal regression affect the preservation of continental deposits within the third-order topset. The actual contribution of continental systems to the topset depends on the balance between the amounts of continental aggradation and the amounts of subsequent ravinement erosion and marine aggradation during the higher frequency transgressions (Fig. 4.58). A topset may remain dominantly nonmarine where continental aggradation outpaces the effects of transgressive processes of ravinement erosion and healing-phase aggradation, or it may become dominantly marine where higher frequency transgressions rework the previously deposited nonmarine sediments. In the latter case, a third-order topset would consist of stacked fourth-order marine sequences bounded by fourth-order transgressive ravinement surfaces.

It can be concluded that stratal stacking patterns are independent of scale. The same types of stacking patterns can be observed at different scales, in relation to stratigraphic cycles of different magnitudes. Strati

graphic sequences of lower hierarchical ranks are nested within higher rank systems tracts, as illustrated by outcrop-scale sequences (10^0–10^1 m) that commonly build seismic-scale (10^1–10^2 m) systems tracts (Fig. 4.3). The scale of observation is set by the resolution of the data available, and/or by the purpose of the study (e.g., basin analysis: 10^2–10^3 m; petroleum exploration: 10^1–10^2 m; petroleum production development: 10^0–10^1 m; cyclostratigraphy of astronomical and solar radiation cycles: potentially reaching sub-meter scales). The architecture of sequences becomes increasingly complex with the increase in the scale of observation (Fig. 4.3). However, there are no standards for the scale and internal makeup of sequences of any hierarchical rank, due to the variability in accommodation and sedimentation conditions between different tectonic and depositional settings.

With the unprecedented increase in stratigraphic resolution, which is now approaching the scales of sedimentology, new criteria are needed to separate between stratigraphic and sedimentological stratal units. In downstream-controlled settings, the distinction between sequences and bedsets is based on the scale of the sedimentary cycle relative to the scale of the lowest rank systems tracts and component depositional systems (Figs. 3.24, 3.25, and 4.62). The stacking patterns relevant to sequence stratigraphy are observed at scales that afford the development of coastal depositional systems (i.e., building blocks of systems tracts: deltas vs. estuaries, or prograding strandplains vs. backstepping lagoon-barrier island systems), which commonly require minimum 10^2–10^3 yrs (e.g., Amorosi et al.,

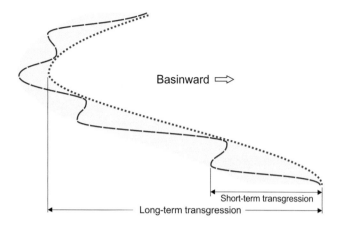

A. Paleogeography: long-term changes > short-term changes

```
·············  shoreline trajectory of higher hierarchical rank
– – – –       shoreline trajectory of lower hierarchical rank
[   ]         shoreline transit area for the lower rank cycles
```

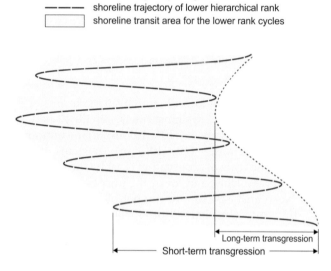

B. Paleogeography: short-term changes > long-term changes

FIGURE 4.61 Shoreline trajectories observed at different stratigraphic scales. Changes in paleogeography may be greater in the long term (A) or in the short term (B), depending on the scale of observation, topographic gradients, and the rates of accommodation and sedimentation. A shoreline transit area is located updip of the shoreline trajectory of higher hierarchical rank, and includes continental, coastal, and marine/lacustrine systems at the scale of the lower rank cycles. The same transit area is part of one depositional system at the scale of the higher rank trend. The available stratigraphic resolution determines the lowest rank depositional systems that can be observed (e.g., one seismic-scale fluvial topset vs. lower rank fluvial, coastal, and marine depositional systems at sub-seismic scales; Fig. 4.6).

2009, 2017; Nixon et al., 2014; Pellegrini et al., 2017, Fig. 4.62). Therefore, it is important to separate stratal stacking patterns observed at sedimentological scales (i.e., *sedimentological stacking patterns*: a stratal architecture that describes the internal makeup of a depositional system, without changes in systems tract), from stratal stacking patterns observed at stratigraphic scales (i.e., *stratigraphic stacking patterns*: a stratal architecture that

involves changes in systems tracts and component depositional systems). Stratigraphic stacking patterns have been documented starting with centennial scales (Amorosi et al., 2009; Pellegrini et al., 2017, Fig. 4.62). At the smallest stratigraphic scales (commonly within the range of $10^2–10^3$ yrs and $10^0–10^1$ m), systems tracts and component depositional systems consist of sedimentological cycles (i.e., bedsets); at any larger scales (i.e., higher hierarchical ranks), systems tracts and component depositional systems consist of lower rank sequences (Fig. 4.3).

4.3 Stacking patterns in upstream-controlled settings

Stratal stacking patterns in upstream-controlled settings are "unconventional" in the sense that the sequence stratigraphic framework forms outside of the area of influence of relative sea/lake-level changes (Fig. 3.11). In this context, the nomenclature of sequence stratigraphic units and bounding surfaces no longer makes reference to shoreline trajectories or relative sea/lake-level changes. The description of unconventional stacking patterns is a more recent development in sequence stratigraphy, firmly established starting with the 1990s (e.g., Kocurek and Havholm, 1993; Shanley and McCabe, 1994, 1998; Galloway and Hobday, 1996; Boyd et al., 2000; Zaitlin et al., 2002; Leckie et al., 2004; Holbrook et al., 2006). Stratal stacking patterns in upstream-controlled settings can be defined in both fluvial and eolian systems. This section focuses on fluvial systems, which are more widely documented and included in sequence stratigraphic studies. See Chapter 8 for a discussion of stacking patterns in eolian settings.

4.3.1 Controls on stratal stacking patterns

The definition of stratal stacking patterns in upstream-controlled settings is based on the dominant depositional element in the stratigraphic succession. In fluvial systems, the degree of channel amalgamation (i.e., amalgamated channels vs. floodplain-dominated successions) is key to the definition of upstream-controlled stacking patterns (Fig. 4.2; Shanley and McCabe, 1994; Boyd et al., 2000). The ratio between channel and floodplain elements is determined by the rates of floodplain aggradation, the degree of channel confinement, and the frequency of avulsion (Bristow and Best, 1993, Fig. 4.63). These processes reflect the interplay of all controls on fluvial sedimentation (i.e., accommodation, climate, source-area tectonism, and autogenic processes that affect sediment distribution; Fig. 3.11).

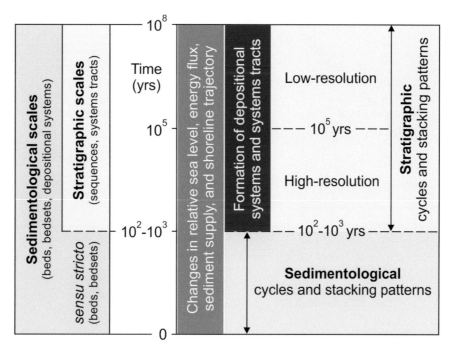

FIGURE 4.62 Timescales of sedimentological vs. stratigraphic cycles. Sedimentological cycles (beds, bedsets) are the building blocks of the lowest rank systems tracts and component depositional systems. In contrast, stratigraphic cycles (sequences) involve changes in systems tracts and component depositional systems. Both types of sedimentary cycles involve changes in stratal stacking pattern. *Sedimentological* stacking patterns describe the internal architecture of a depositional system, without changes in systems tract (e.g., cycles of change in the degree of amalgamation of storm beds). *Stratigraphic* stacking patterns refer to stratal architectures that define systems tracts. The minimum timescales required to form a depositional system are commonly in a range of 10^2–10^3 yrs. Changes in relative sea level, energy flux, sediment supply, and shoreline trajectory occur at all scales, and may accompany the formation of both sedimentological and stratigraphic cycles. At stratigraphic scales, shoreline trajectories are defined by the trajectory of subaerial clinoform rollovers, whose changes in the direction of shift result in changes in coastal environment (e.g., deltas vs. estuaries). At sedimentological scales, shoreline trajectories refer to shoreline shifts that occur without changes in coastal environment (e.g., in relation to tidal or fairweather-storm cycles). Criteria to differentiate between sedimentological and stratigraphic cycles have been summarized by Zecchin et al. (2017a,b).

Noteworthy, while important, accommodation is not the sole control on the formation of stratal stacking patterns, nor even the sole control on the rates of vertical aggradation (Bristow and Best, 1993; Holbrook et al., 2006; Miall, 2015). Early studies on the definition of stacking patterns in fluvial systems assumed a direct link between accommodation and the degree of channel amalgamation (i.e., amalgamated channels were interpreted as the product of "low accommodation" conditions, whereas floodplain-dominated successions were interpreted as the product of "high-accommodation" conditions; Boyd et al., 2000; Zaitlin et al., 2002; Leckie et al., 2004). As a result, the terminology of systems tracts made exclusive reference to accommodation conditions (i.e., low- vs. high-accommodation systems tracts; e.g., Boyd et al., 2000; Leckie and Boyd, 2003). This nomenclatural approach oversimplifies the process–response relationships in fluvial systems, as accommodation is only one of the several controls on sedimentation (Fig. 3.11; Miall, 2015; Catuneanu, 2017).

Part of the stratigraphic complexity of fluvial systems relates to the fact that fluvial styles and the associated depositional elements may change along the course of

the same river (Fig. 4.64). However, the degree of channel amalgamation may not necessarily change along dip, despite the changes in fluvial style, as being in part controlled by the rates of vertical shift of the graded fluvial profile (i.e., rates of floodplain and/or channel belt aggradation; Figs. 3.11, 4.2, and 4.63). If these rates are constant along the graded profile, fluvial stacking patterns and corresponding systems tracts may still correlate at regional scales despite the possible changes in fluvial styles. If the rates of shift of the graded profile are variable across the sedimentary basin, and/or coupled with marked changes in the degree of channel confinement and the frequency of avulsion, different types of stacking patterns may form at the same time between different areas of the basin, leading to a diachronous development of fluvial systems tracts. The diachroneity of systems tracts and bounding sequence stratigraphic surfaces remains a safe norm in both downstream- and upstream-controlled settings (e.g., Catuneanu, 2003, 2006).

The rates of floodplain and/or channel belt aggradation, which control to some extent the degree of channel amalgamation, may be linked to accommodation,

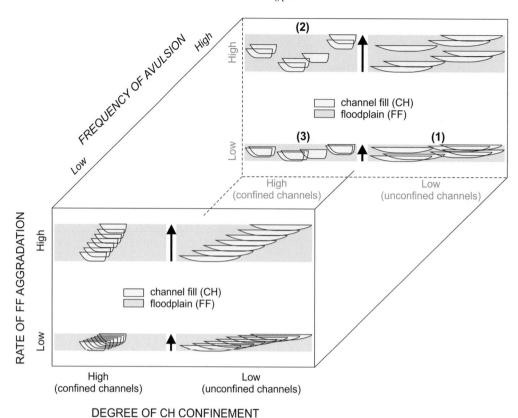

FIGURE 4.63 Fluvial architecture under variable conditions of floodplain aggradation, channel confinement, and avulsion frequency, as illustrated by the seven-step evolution of a channel (modified from Bristow and Best, 1993). The rates of fluvial aggradation depend on all factors which control sedimentation, including accommodation, climate, source-area tectonism, and autocyclic changes in sediment distribution. The occurrence of isolated channels is promoted by rapid aggradation coupled with frequent avulsion. The degree of channel amalgamation is proportional to the rate of lateral channel migration (higher in unconfined systems) and the frequency of avulsion (enhanced by sediment supply), and inversely proportional to the rate of aggradation. (1), (2), and (3) indicate the most common succession of stacking patterns, from the base to the top of a fluvial depositional sequence. All these common stacking patterns assume avulsion, but are different in terms of rates of aggradation and/or lateral channel migration.

FIGURE 4.64 Fluvial depositional sequences in an upstream-controlled setting (Late Permian, Balfour Formation; Karoo Basin, South Africa; modified after Catuneanu and Elango, 2001). The six sequences accumulated during the overfilled phase in the evolution of the Karoo Basin, over a timespan of 4 My. Each sequence displays a fining-upward profile, due to the change in fluvial styles from higher to lower energy systems. However, the succession records an overall coarsening-upward trend in response to the progradation of the orogenic front.

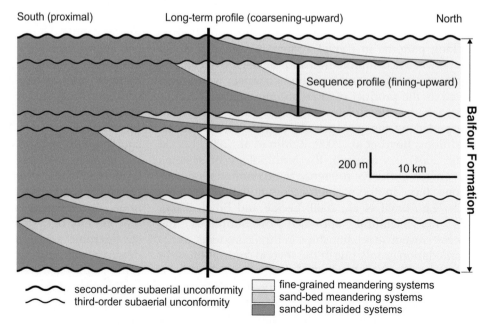

source-area uplift, climate, or autogenic processes (Fig. 3.11). The decoupling of sedimentation from accommodation is particularly evident in overfilled basins, where the rates of sedimentation are higher than the rates of subsidence. In such settings, cycles of aggradation and degradation may have various origins (e.g., tectonic, isostatic, climatic, autogenic), and result in the formation of unconformity-bounded sequences in fluvial or eolian systems under conditions of overfilled accommodation (Fig. 4.2; Catuneanu, 2006; Holbrook et al., 2006). Therefore, the observation of stratigraphic trends needs to be separated from the interpretation of underlying controls. The rates of sedimentation reflect the rates of change in the temporary base level (i.e., graded profile; Figs. 3.22 and 3.23), irrespective of the nature of the underlying control(s).

4.3.2 Depositional trends in upstream-controlled settings

The stratigraphic architecture of fluvial systems is the result of the interplay between the rates of vertical aggradation, the ability of channels to shift laterally, and the frequency of avulsion (Fig. 4.63). The degree of channel amalgamation is inversely proportional to the rate of vertical aggradation, and proportional to the ability of channels to migrate laterally (i.e., higher in unconfined rivers) and to the frequency of avulsion (Fig. 4.63).

The rates of vertical aggradation are constrained by the rates of shift of the graded profile, which depend not only on accommodation but also on all factors that modify the balance between sediment supply and energy flux at any particular location (i.e., accommodation, source-area uplift, climate, and autogenic processes; Fig. 3.11). Factors which increase the flow energy within river systems include the steepening of the topographic gradients (e.g., as a result of differential tectonic uplift or isostatic rebound) and the increase in discharge (e.g., as a result of ice melting during interglacial stages, or due to shifts to more humid climatic conditions). For example, the increase in flow energy during deglaciation, due to a combination of ice melting and differential isostatic rebound, explains the common formation of Holocene incised valleys in upstream-controlled settings, in spite of the global rise in sea level (Fig. 3.21). Similarly, the wind energy, coupled with the influence of climate on water table and sediment supply, are major factors which control the depositional trends of aggradation or degradation in eolian systems (Figs. 3.18 and 3.19).

The ability of channels to shift laterally relates to the degree of channel confinement (i.e., leveed channels vs. unconfined channels), which is a function of fluvial style. Topographic gradients play a major role in controlling fluvial styles, along with the influence of climate

and other factors (e.g., vegetation may consolidate banks). Low-gradient confined channels, such as within anastomosed or meandering rivers, are relatively stable and tend to preserve their floodplain deposits. For this reason, low rates of lateral channel migration are prone to the preservation of isolated channels, particularly under conditions of high rates of floodplain aggradation. Higher energy streams flowing on steeper gradients, such as within braided systems, display an unconfined character and are prone to channel amalgamation, particularly under conditions of low rates of vertical aggradation. It is evident that topographic gradients exert a significant control on the rates of lateral channel migration, and tend to be proportional to the degree of channel amalgamation. Most commonly, topographic gradients are steepest at the onset of fluvial aggradation, following the formation of a subaerial unconformity, and gradually decrease during the accumulation of a depositional sequence. For this reason, it is common to observe higher energy systems (e.g., braided river deposits), with coarser sediment and a higher degree of channel amalgamation, at the base of depositional sequences, and lower energy systems (e.g., meandering or anastomosed river deposits), with finer channel fills, toward the top of depositional sequences (Fig. 4.65). This trend is true for both fully fluvial sequences in upstream-controlled settings and for the fluvial portion of downstream-controlled sequences (e.g., Kerr et al., 1999; Catuneanu and Elango, 2001; Bordy and Catuneanu, 2001, 2002b; Fanti and Catuneanu, 2009; 2010).

Avulsion is different from lateral channel migration, and marks an abrupt change in the course of a stream as an old channel is abandoned and a new channel forms. Avulsion can be recorded at different scales, from the anabranch avulsion of a channel within a channel belt to the avulsion of an entire channel belt. Both types of avulsion can affect any type of river, irrespective of the degree of sinuosity or the degree of channel confinement that define the fluvial style (Bristow and Best, 1993, Figs. 4.66, 4.67). The frequency of anabranch avulsion is proportional to sediment supply (i.e., plugging of channels), whereas the frequency of channel-belt avulsion reflects the frequency of major floods or tectonic events (Bridge and Leeder, 1979). Typical avulsion frequency is documented at timescales of 500–1000 yrs (Allen, 1978, 1979; Leeder, 1978; Bridge and Leeder, 1979; Bridge and Mackey, 1993a,b; Mackey and Bridge, 1995; Stouthamer and Berendsen, 2007). The avulsion frequency, as well as the magnitude of avulsion (e.g., channel vs. channel belt), depends on climatic and tectonic controls which modify sediment supply, fluvial discharge, and the dip direction within the basin. Given the importance of external controls on avulsion, the average frequency may be similar for all types of fluvial styles. Avulsion is typically the norm in the

FIGURE 4.65 Fluvial stacking patterns in an upstream-controlled setting (Triassic, Karoo Basin, South Africa). A—Channel-dominated succession: high-amalgamation stacking pattern (Molteno Formation: higher energy, coarser grained braided channels). B—Floodplain-dominated succession: low-amalgamation stacking pattern (Burgersdorp Formation: lower energy, finer grained meandering river system). The arrows mark the position of the subaerial unconformity that separates the two stacking patterns. Note the abrupt change in fluvial energy across the subaerial unconformity (depositional sequence boundary).

FIGURE 4.66 Fluvial braidplain showing evidence of channel abandonment and anabranch avulsion within the channel belt (Melville Island, Canadian Archipelago; image courtesy of Chantel Nixon). This type of low-sinuosity, unconfined (high rate of lateral channel migration), multiple-channel system is prone to the development of a fluvial stacking pattern dominated by amalgamated channels. This stacking pattern is most common in the lower (higher energy) part of depositional sequences, whether downstream- or upstream-controlled.

evolution of a fluvial system, and hence the common fluvial stacking patterns are recorded on the high avulsion frequency side of the diagram in Fig. 4.63.

The depositional processes that control fluvial stacking patterns (i.e., floodplain aggradation, lateral channel migration, and avulsion frequency) may vary independently of each other, and therefore, multiple combinations are possible (Fig. 4.63). However, only few stacking patterns are common in the rock record, reflecting the following trends: (i) the rates of vertical aggradation first increase (following a stage of fluvial degradation), and then decrease (leading to a stage of fluvial degradation), during the accumulation of a depositional sequence (Wright and Marriott, 1993, Fig. 2.18); (ii) topographic gradients typically decrease during the accumulation of a depositional sequence, resulting in a decrease with time in fluvial energy, grain size, and the rates of lateral channel migration; and, (iii) avulsion is typically the norm throughout the accumulation of a fluvial sequence. The combination of these prevalent

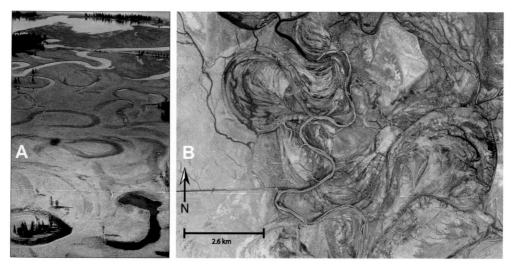

FIGURE 4.67 Meandering fluvial systems showing evidence of channel abandonment and anabranch avulsion within the channel belt (A—Pelican Creek, Yellowstone Park, USA; image courtesy of Jim Peaco; B—Colorado River, Baja California, Mexico; image courtesy of Rares Bistran). This type of high-sinuosity, confined (low rate of lateral channel migration), single-channel system is prone to the development of a fluvial stacking pattern dominated by floodplain deposits with isolated channels (A). Frequent channel avulsion may also result in local channel amalgamation (B), particularly where the rates of floodplain aggradation are low. This stacking pattern is most common in the upper (lower energy) part of depositional sequences, whether downstream- or upstream-controlled.

trends generates a succession of stacking patterns that is commonly observed within fluvial depositional sequences (Figs. 4.63 and 4.68):

(1) Low rates of floodplain aggradation, high rates of lateral channel migration, and high frequency of avulsion: this combination of trends results in the highest degree of channel amalgamation, which is typically found at the base of fluvial depositional sequences. This stacking pattern includes the *highest energy fluvial systems* (i.e., coarsest channel fills) of a depositional sequence, and is similar to the lowstand topset of a downstream-controlled

sequence, in cases where the subaerial unconformity forms during forced regression (Fig. 4.66).

(2) High rates of floodplain aggradation, low rates of lateral channel migration, and high frequency of avulsion: this combination of trends results in the lowest degree of channel amalgamation, which is typically found in the middle part of fluvial depositional sequences. This stacking pattern is similar to the fluvial portion of a transgressive systems tract in a downstream-controlled setting (Wright and Marriott, 1993), where the subaerial unconformity forms during forced regression (Fig. 4.67A).

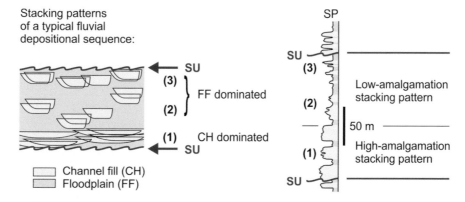

FIGURE 4.68 Stratigraphic architecture of a fluvial sequence in an upstream-controlled setting (Upper Cretaceous, Golfo San Jorge Basin; data courtesy of YPF Argentina). This example illustrates a common succession of fluvial systems: (1) high-energy unconfined channels, accumulated under conditions of low rates of floodplain aggradation; (2) low-energy confined channels and floodplains, accumulated under conditions of high rates of floodplain aggradation; and (3) low-energy confined channels and floodplains, accumulated under conditions of low rates of floodplain aggradation. Section (3) includes the lowest energy river systems with the finest channel fills of the fluvial sequence, and has the lowest preservation potential due to the development of the subaerial unconformity at the top. Deviations from this common architecture are possible, with end members represented entirely by either high-energy unconfined systems or low-energy confined systems. In the latter case, the degree of channel amalgamation may be low even at the base of the fluvial sequence. Abbreviations: SU—subaerial unconformity; SP—spontaneous potential.

(3) Low rates of floodplain aggradation, lowest rates of lateral channel migration, and high frequency of avulsion: this combination of trends may lead to a variable (low to high) degree of channel amalgamation at the top of fluvial depositional sequences, depending on the patterns of channel avulsion and the balance between the formation of new channels and the preservation of floodplain deposits. This stacking pattern includes the *lowest energy fluvial systems* (i.e., finest channel fills) of a depositional sequence, and is similar to the highstand topset of a downstream-controlled sequence, in cases where the subaerial unconformity forms during forced regression (Fig. 4.67B). While the theoretical models generally predict a relatively high degree of channel amalgamation for this uppermost segment of depositional sequences (Shanley and McCabe, 1993; Wright and Marriott, 1993; Emery and Myers, 1996), field examples show that the dominance of floodplain deposits is in fact common (Fig. 4.53). This indicates that conditions conducive to a low degree of channel amalgamation toward the end of a fluvial depositional cycle are often met, despite the low rates of floodplain aggradation, as afforded by a high degree of channel confinement coupled with a pattern of avulsion that maintains a low channel-to-overbank ratio.

The succession of fluvial stacking patterns in Fig. 4.68 depicts a typical fluvial depositional sequence with a common occurrence in the rock record. Section (1) is a channel-dominated unit (i.e., with a high degree of channel amalgamation; e.g., Fig. 4.69A), whereas sections (2) and (3) form an undifferentiated floodplain-dominated unit, despite differences in their rates of aggradation (e.g., Fig. 4.69B). For this reason, fluvial sequences in

upstream-controlled settings are commonly subdivided into only two systems tracts, even though the upper one may form under variable conditions of accommodation and/or sedimentation. This further illustrates the point that the nomenclature of fluvial stacking patterns and associated systems tracts needs to emphasize observable features (e.g., the degree of channel amalgamation) rather than interpreted controls (e.g., low vs. high accommodation conditions at syn-depositional time).

Section (1) in Fig. 4.68 is commonly referred to as a "low-accommodation" systems tract, by assuming that channel amalgamation is the consequence of low accommodation conditions, while sections (2) and (3) are combined into a "high-accommodation" systems tract. This nomenclature of unconventional systems tracts may be misleading, as fluvial stacking patterns are controlled not only by accommodation but also by all factors that modify fluvial sedimentation (i.e., accommodation, climate, source-area tectonism, and autogenic processes). While inconsistent to include section (3) in a "high-accommodation" systems tract, since the accommodation model predicts low accommodation conditions toward the top of a fluvial sequence, this is common practice as the field expression of sections (2) and (3) is similar. A more appropriate nomenclature for the unconventional systems tracts is high- vs. low-*amalgamation* (Catuneanu, 2017), which describes the actual nature of the stacking patterns.

Departures from the model in Fig. 4.68 may also be observed in the stratigraphic record, from sequences consisting entirely of high-energy unconfined systems to sequences consisting entirely of low-energy confined systems (e.g., Catuneanu and Elango, 2001). The thickness of fluvial depositional sequences may vary widely, typically from tens to hundreds of meters. Where

FIGURE 4.69   Fluvial stacking patterns and corresponding systems tracts in upstream-controlled settings: examples from the overfilled section of the Karoo Basin, South Africa. A—channel-dominated succession: high-amalgamation stacking pattern/systems tract (Early Triassic, Katberg Formation); B—floodplain-dominated succession: low-amalgamation stacking pattern/systems tract (Early-Middle Triassic, Burgersdorp Formation).

changes in the degree of channel amalgamation (i.e., stratigraphic cycles) are observed at different scales, the smaller scale sequences are nested within larger scale systems tracts, describing stratigraphic frameworks of different hierarchical orders (Fig. 4.70). A stack of high-frequency sequences dominated by amalgamated channels defines a higher rank "high-amalgamation" systems tract, whereas a stack of high-frequency sequences dominated by floodplain deposits defines a higher rank "low-amalgamation" systems tract (Fig. 4.70). At each scale of observation, depositional sequences display fining-upward trends in response to the decline in fluvial energy with time (Fig. 4.70).

In the case of nested cycles, the pattern of change in the thickness of the smaller scale sequences may reflect changes in the rates of aggradation of the larger scale sequence, assuming that the smaller scale cycles are of approximately equal duration. This may be difficult to demonstrate in fluvial settings, as compared, for example, with peritidal settings (Fig. 4.52). Nonetheless, as the rates of fluvial aggradation tend to initially increase and then decrease during the accumulation of a depositional sequence (Fig. 4.68), the thickness of the smaller scale sequences that are nested within the larger scale sequence may follow a similar trend (Fig. 4.70).

4.3.3 Types of stacking patterns

Stratal stacking patterns in upstream-controlled settings are defined by the dominant architectural elements in the stratigraphic succession. In fluvial systems, the degree of channel amalgamation is the defining factor that describes the stacking patterns and associated systems tracts in fully fluvial sequences (Figs. 4.2 and 4.70; Shanley and McCabe, 1994; Boyd et al., 2000).

4.3.3.1 High-amalgamation (channel-dominated) stacking pattern

Factors conducive to channel amalgamation include (i) low rates of floodplain aggradation; (ii) high sediment supply, fluctuating discharge, and episodic changes in dip direction (all conducive to avulsion); and (iii) steep topographic gradients, which promote unconfined rivers, with a high ability to shift laterally. Considering the average rates of avulsion and lateral channel migration, vertical channel amalgamation typically require timescales of minimum 10^2-10^3 yrs to develop. Moreover, the rate of floodplain aggradation needs to be low for vertical channel amalgamation to occur, such that channels have time to shift and return to the original position before any significant floodplain deposits can accumulate and be preserved on top. The conditions

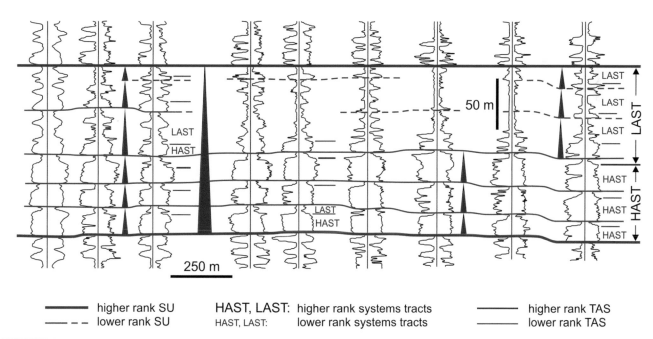

FIGURE 4.70 Nested architecture of fluvial sequences in an overfilled basin (Upper Cretaceous, Golfo San Jorge Basin; data courtesy of YPF Argentina). Well logs: spontaneous potential (left) and resistivity (right). Changes in the degree of channel amalgamation can be observed at different scales, defining stratigraphic cycles of different hierarchical ranks. At each scale of observation, depositional sequences display fining-upward trends (i.e., a decline in fluvial energy with time), and can be subdivided into high- and low-amalgamation systems tracts. A higher rank HAST consists of lower rank sequences dominated by amalgamated channels. A higher rank LAST consists of lower rank sequences dominated by floodplain deposits. Hierarchical ranks reflect the relative scales of sequences within the nested architecture of stratigraphic cycles. Abbreviations: SU—subaerial unconformity; TAS—top-amalgamation surface; HAST—high-amalgamation systems tract; LAST—low-amalgamation systems tract.

required by vertical channel amalgamation are also conducive to lateral channel amalgamation, assuming that avulsion does occur within the fluvial system.

A high-amalgamation stacking pattern is found commonly in the lower part of a depositional sequence, which typically includes the highest energy fluvial system of the stratigraphic cycle (Fig. 4.65). While the rates of aggradation and avulsion may be similar between the lower and the upper portions of a fluvial depositional sequence (i.e., sections (1) and (3) in Figs. 4.63 and 4.68), the degrees of channel amalgamation may still be different due to contrasts in the energy levels (highest vs. lowest, respectively) of the corresponding rivers (e.g., compare stacking patterns (1) and (3) in Figs. 4.63 and 4.68). Indeed, the rate of lateral channel migration increases with the energy of the river, which explains the difference in the degree of channel amalgamation between sections (1) and (3) in Figs. 4.63 and 4.68.

In summary, the development of a high degree of channel amalgamation in fluvial systems (i.e., high channel-to-overbank ratio) is promoted by (i) low rates of floodplain aggradation; (ii) high-energy, unconfined rivers; and (iii) a high frequency of channel avulsion (Bristow and Best, 1993, Figs. 4.63 and 4.68). This fluvial architecture of amalgamated channels was previously referred to as a "low-accommodation" stacking pattern. The "high-amalgamation" nomenclature is recommended as a more accurate descriptor of this stacking pattern, devoid of (a potentially erroneous) interpretation. High-amalgamation stacking patterns can be observed at different scales, in relation to stratigraphic cycles of different magnitudes (Fig. 4.70).

4.3.3.2 Low-amalgamation (floodplain-dominated) stacking pattern

Factors favorable to the development of a floodplain-dominated succession include (i) high rates of floodplain aggradation; (ii) low frequency of avulsion; and (iii) low topographic gradients, which promote low-energy confined rivers with a low ability to shift laterally (Bristow and Best, 1993). Such conditions, particularly (i) and (iii), are most typical of the middle part of depositional sequences (i.e., section (2) in Figs. 4.63 and 4.68). A low degree of channel amalgamation may still develop under conditions of low rates of floodplain aggradation and high frequency of avulsion, where fluvial systems include leveed channels with limited ability to shift laterally. This typifies the upper part of fluvial depositional sequences, which relates to the lowest energy rivers of the stratigraphic cycle (i.e., section (3) in Figs. 4.63 and 4.68).

The low-amalgamation stacking pattern typically dominates the middle and upper parts of a depositional sequence, which commonly combine into one systems tract despite the variability in the rates of floodplain

aggradation (i.e., sections (2) and (3) in Figs. 4.63 and 4.68). This stacking pattern was previously referred to as "high-accommodation," by assuming that the dominance of floodplain deposits is the product of high accommodation conditions. In reality, accommodation is only one of the several controls on fluvial sedimentation; therefore, the "low-amalgamation" nomenclature is recommended as a more accurate descriptor of this stacking pattern, devoid of (a potentially erroneous) interpretation.

Low-amalgamation stacking patterns can be observed at different scales, in relation to stratigraphic cycles of different magnitudes (Fig. 4.70). The ratio between the high- and low-amalgamation stacking patterns within the lower rank sequences defines the higher rank systems tracts. For example, a set of fourth-order sequences dominated by a high degree of channel amalgamation defines a third-order "high-amalgamation" systems tract, whereas a set of fourth-order sequences dominated by floodplain deposits defines a third-order "low-amalgamation" systems tract (Fig. 4.70). Notably, sequences of both scales start with a high-amalgamation followed by a low-amalgamation stacking pattern, which reflects the decline in fluvial energy with time at both scales of observation (Fig. 4.70).

4.3.4 Stratigraphic scales in upstream-controlled settings

The physical and temporal scales of the "unconventional" stacking patterns and related systems tracts in upstream-controlled settings depend on the rates of sedimentation of the depositional elements that define the stratigraphic stacking patterns, and the period of time over which the dominance of the diagnostic depositional elements can be maintained. In fluvial settings, the deposition of channel and overbank elements involves sedimentation rates of $10^{-1}-10^{2}$ m/ka and minimum timescales of 10^{2} yrs (Bridge and Leeder, 1979; Miall, 2015). This sets the lower limit of the temporal range required to form a fluvial stacking pattern and associated systems tract. How long this process can be sustained for (e.g., up to 10^{5} yrs, 10^{6} yrs, or even longer) depends on all variables that control sedimentation patterns in fluvial systems (i.e., the rates of floodplain aggradation, the frequency of avulsion, and the rates of lateral channel migration; Bristow and Best, 1993, Fig. 4.63), including accommodation (i.e., subsidence rates in upstream-controlled settings), climate, source-area tectonism, and autogenic processes (Catuneanu, 2017, 2019a,b; Fig. 3.11).

A comparison between the sedimentation rates of fluvial systems (i.e., $10^{-1}-10^{2}$ m/ka) and the average

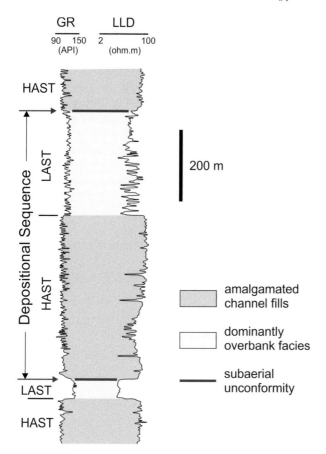

FIGURE 4.71 Stratal stacking patterns of a fluvial succession in an overfilled basin, describing cyclicity at timescales of 10^5 yrs (Miocene, Assam Basin, India; data courtesy of the Oil and Natural Gas Corporation, India). Sedimentation rates were in a range of 10^0-10^1 m/kyr. Abbreviations: HAST—high-amalgamation systems tract; LAST—low-amalgamation systems tract.

subsidence rates that operate at different timescales suggests that fluvial stacking patterns are most likely to develop within a 10^2-10^5 yrs time frame, which corresponds to the scale of high-frequency sequences (Miall, 2015). Within this time frame, the average subsidence rates in most tectonic settings match the typical rates of sedimentation of fluvial architectural elements, therefore providing suitable conditions for the development of unconventional systems tracts. In contrast, average subsidence rates on larger timescales (i.e., 10^6 yrs and longer) are orders of magnitude lower than the typical rates of sedimentation of the elements of the fluvial systems tracts (Miall, 2015). Nevertheless, fluvial sequences and component systems tracts are documented over a wide range of timescales, from short term (10^5 yrs and under; Fig. 4.71) to long term (10^6 yrs and over; Fig. 4.72). This indicates that controls other than subsidence need to be accounted for in order to explain sedimentation rates higher than the rates of subsidence on timescales of 10^6 yrs and longer. Indeed, subsidence is not the only control on sedimentation,

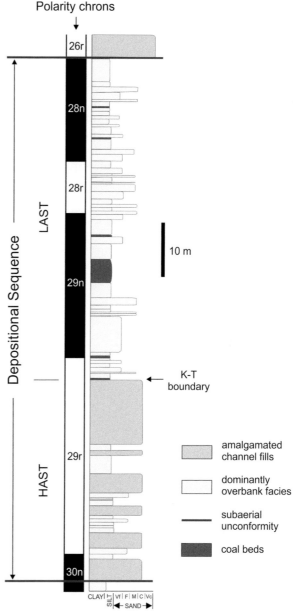

FIGURE 4.72 Stratal stacking patterns of a fluvial succession in an overfilled basin, describing cyclicity at timescales of 10^6 yrs (c. 3 My: 66-63 Ma, Alberta foredeep; Catuneanu and Sweet, 1999; Khidir and Catuneanu, 2003). This composite profile portrays the distal portion of a depositional sequence (referred to as Scollard, Coalspur or Willow Creek, depending on location), whose thickness increases to >1,000 m in the depocenter. Sedimentation rates were in a range of $10^{-1}-10^0$ m/kyr. The rates of accumulation of sequences and systems tracts in overfilled basins may outpace the rates of subsidence at syndepositional time (see text for details). Abbreviations: HAST—high-amalgamation systems tract; LAST—low-amalgamation systems tract.

and other factors may play a significant role in building the stratigraphic architecture (Fig. 3.11). The shift from underfilled to overfilled stages in the evolution of a sedimentary basin indicates that upstream controls such as climate and/or source-area uplift can sustain long-term aggradation at rates higher than the rates of subsidence. In such cases, fluvial systems tracts can develop over timescales of 10^6 yrs or longer, independently of the average subsidence rates recorded during that time interval (Fig. 4.72).

Cyclicity can be observed at different scales, with the higher rank systems tracts consisting of depositional sequences of lower hierarchical rank (Fig. 4.70). Sequences of different scales show consistent features, including a change from high- to low-amalgamation stacking patterns along with fining-upward trends, indicating that the decline in energy that characterizes a fluvial sequence is independent of scale and subsidence rates. At each hierarchical level, stages of increase in fluvial energy result in the formation of sequence-bounding unconformities, and stages of decrease in fluvial energy result in the deposition of sequences. These "energy cycles" (i.e., cycles of change in the temporary base level) that define depositional sequences in upstream-controlled settings may correspond to tectonic cycles, climate cycles, or cycles of autogenic migration of alluvial channel belts (Catuneanu and Bowker, 2001; Catuneanu and Elango, 2001; Hajek et al., 2010; Hofmann et al., 2011; Miall, 2015).

CHAPTER

5

Stratal units

The sequence stratigraphic framework consists of sequences and component systems tracts and depositional systems, which may form at all stratigraphic scales. At each scale of observation (hierarchical level), depositional systems are the building blocks of systems tracts, and systems tracts are the building blocks of sequences. Depositional systems are units of sedimentology (Fisher and McGowan, 1967), whereas systems tracts and sequences are units of sequence stratigraphy (Brown and Fisher, 1977; Mitchum, 1977). Parasequences (Van Wagoner et al., 1988, 1990) were also used as building blocks of seismic-scale systems tracts, but have become redundant with the advent of high-resolution sequence stratigraphy as high-frequency sequences that develop at parasequence scales provide a better and more reliable alternative for stratigraphic correlation (details in Chapter 5 below).

Sequence stratigraphic units (i.e., sequences and systems tracts) are defined by strata stacking patterns and bounding surfaces, and not by their inferred controls, sedimentological makeup, age, time span, or physical scales. The composition and scales of any type of sequence stratigraphic unit vary with the syndepositional conditions of accommodation and sedimentation, which are a function of tectonic and depositional settings. Stratigraphic sequences consist of genetically related strata that belong to the same cycle of change in stratal stacking pattern. Genetically related successions may be observed at different scales, in relation to stratigraphic cycles of different magnitudes. There are no temporal or physical standards for the scale of any type of sequence stratigraphic unit. At the smallest stratigraphic scales, systems tracts and component depositional systems consist of beds and bedsets (i.e., sedimentological cycles; Figs. 3.24 and 4.62). At any larger scales, systems tracts and depositional systems consist of higher frequency (lower rank) sequences (i.e., stratigraphic cycles; Fig. 4.3). The nested architecture of the sequence stratigraphic framework can be observed in both downstream-controlled (Fig. 5.1) and upstream-controlled (Fig. 4.70) settings.

The scale-independent nature of sequences and component systems tracts and depositional systems has become evident with the increase in the resolution of stratigraphic studies (e.g., Amorosi et al., 2005, 2009, 2017; Bassetti et al., 2008; Mawson and Tucker, 2009; Nanson et al., 2013; Nixon et al., 2014; Magalhaes et al., 2015; Ainsworth et al., 2017; Pellegrini et al., 2017; 2018; Zecchin et al., 2017a,b; Moran, 2020, Figs. 4.70 and 5.1). Seismic stratigraphy imposed by default a minimum scale to the concepts of sequence, systems tract, and depositional system, which had to exceed the vertical seismic resolution (i.e., 10^1 m in the 1970s). However, the same types of units can also be observed at sub-seismic scales in higher resolution studies (i.e., high-resolution sequence stratigraphy). Despite the scale-independent and nested nature of the sequence stratigraphic framework, the stratigraphic architecture is not truly "fractal," because sequences of different scales may have different underlying controls and internal makeup (e.g., different combinations and/or relative development of component systems tracts).

5.1 Depositional systems

The definition and scale of depositional systems are key to the classification and nomenclature of sequence stratigraphic units, and to the sequence stratigraphic methodology. As the largest units of sedimentology and the building blocks of systems tracts in stratigraphy, depositional systems hold a special place in the architecture of the sedimentary record, at the limit between the scopes of sedimentology and stratigraphy at any stratigraphic scale of observation.

5.1.1 Definition

A depositional system is a stratal unit that preserves the record of sedimentation of a depositional

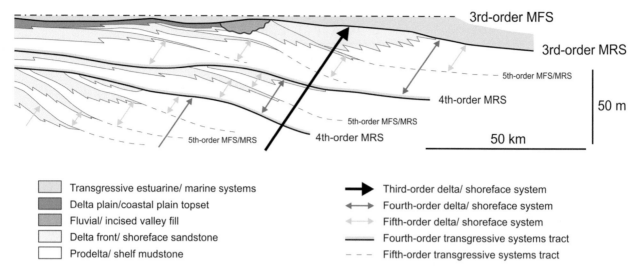

FIGURE 5.1   Stratigraphic architecture of a prograding system in a downstream-controlled setting (Cenomanian, Dunvegan Formation, Western Canada Basin; from Catuneanu, 2017, modified after Bhattacharya, 1993). Sequences, systems tracts, and depositional systems can be observed at different scales (hierarchical levels). The third-order "delta" is part of the Colorado second-order sequence, and includes several different depositional systems at the fourth- and fifth-order scales. In this example, the third-order delta approaches 10^6 yrs in duration, and the internal fourth- and fifth-order sequences developed over timescales of 10^5 and 10^4 yrs, respectively. Changes in relative sea level occurred at all scales, as demonstrated by the vertical stacking of sequences. Abbreviations: MFS—maximum flooding surface; MRS—maximum regressive surface, potentially reworked in part by the transgressive surface of erosion.

environment. The original definitions designate a depositional system as a "three-dimensional assemblage of lithofacies, genetically linked by processes and environments" (Fisher and McGowan, 1967), or as a "three-dimensional assemblage of process-related facies that record major paleogeomorphic elements" (Galloway, 1989). These definitions make no reference to temporal or physical scales. Depositional systems can be observed across a wide range of scales, from 10^0 m (Fig. 5.2) to 10^3 m (Fig. 5.3), depending on the purpose of study and the resolution of the data available (e.g., outcrops vs. seismic data; Catuneanu, 2017, 2019a,b, 2020c). At each scale of observation (i.e., hierarchical level), depositional systems have paleogeographic significance and correspond to specific or dominant environments of deposition.

Depositional systems form when the defining environments and related geomorphic elements (e.g., delta plain, delta front, and prodelta in a deltaic system) are established as dominant fairways for sediment distribution. The formation of depositional systems typically requires minimum timescales of 10^2 yrs, which afford the accumulation of the component architectural elements (e.g., Bridge and Leeder, 1979; Miall, 2015), and it may be sustained for $\geq 10^6$ yrs as long as the defining sediment fairways are maintained as dominant (but not necessarily exclusive) dispersal systems (Fig. 4.62). For example, a "delta" observed at a seismic scale is defined by a dominant trend of sediment progradation, even though, on shorter timescales, this trend is interrupted by stages of transgression (e.g., flooding of delta plain and the formation of estuaries) of lower hierarchical ranks (Fig. 5.1). Therefore, depositional systems can be

observed at different scales, depending on the resolution of the stratigraphic study, which enables the definition of systems tracts, sequences, and bounding surfaces at different hierarchical levels (Figs. 4.3, 4.70, and 5.1).

5.1.2 Scale of depositional systems

Seismic stratigraphy in the 1970s introduced by default a minimum scale to the concepts of sequence and systems tract (and, implicitly, depositional system), imposed by the vertical resolution of 2D seismic data. Consequently, depositional systems were recognized at larger scales, with minimum thicknesses of tens of meters. The "classic" sequence scale afforded by the low-resolution 2D seismic data is commonly referred to as of "third order" (e.g., Duval et al., 1998), and is relevant to petroleum exploration (Payton, 1977, Fig. 5.4). Subsequent refinements in stratigraphic resolution, afforded by the integration of seismic data with higher resolution datasets (e.g., well-log, core, outcrop), led to the identification of sequences, systems tracts, and depositional systems at smaller (sub-seismic) scales, often designated as of fourth-order or lower hierarchical ranks (Fig. 5.4). Therefore, a higher rank depositional system (e.g., a seismic-scale delta) may include sequences, systems tracts, and depositional systems of lower hierarchical ranks (e.g., marine, estuarine, and deltaic systems observed at sub-seismic scales; Fig. 5.1).

The Dunvegan deltaic complex in northwest Alberta illustrates the development of depositional systems at different scales (Fig. 5.1). The Dunvegan complex is a

FIGURE 5.2   High-resolution sequence stratigraphic framework of the lower Tombador Formation (Mesoproterozoic, Brazil; Magalhaes et al., 2015). These high-frequency sequences consist of fluvial and/or estuarine channels and bars (facies A) overlain by horizontally stratified shallow-marine deposits (facies B). The subaerial unconformities (red arrows) and transgressive ravinement surfaces (blue arrows) mark changes in depositional system, and represent the best fluid migration pathways. Note the meter-scale development of depositional systems at this hierarchical level. This is typical of the sub-seismic scales of high-resolution sequence stratigraphy.

seismic-exploration-scale delta, which matches the classic third-order hierarchical rank (Figs. 5.1 and 5.4). This delta system approaches 10^6 yrs in duration, and is subdivided into several 10^5 yrs scale lower rank sequences (Bhattacharya, 1988, 1991; Bhattacharya and Walker, 1991; Miall, 2010), designated as of fourth order in Fig. 5.1. In turn, the fourth-order stages of progradation record fifth-order changes in the direction of shoreline shift and associated depositional systems (e.g., deltas vs. estuaries), over 10^4 yrs timescales. Therefore, the long-term progradation of the third-order delta was interrupted by transgressions of different magnitudes accompanied by changes in depositional environment, leading to a nested architecture of stratigraphic cycles (Fig. 5.1). Sequences, as well as their component systems tracts and depositional systems, can be delineated at each scale of observation. In this example, the

scale of sequences decreases in terms of thickness, duration, and areal extent with the decrease in the hierarchical rank. Changes in relative sea level occurred at all scales, as evidenced by the vertical stacking of sequences.

Only the lowest rank depositional systems consist solely of process-related facies accumulated in specific environments. At larger scales (higher hierarchical ranks), depositional systems reflect dominant depositional trends rather than exclusive sedimentation in specific depositional environments. In the context of seismic stratigraphy, the "genetically linked" processes that operate within the confines of seismic-scale depositional systems may be interrupted by changes in depositional environment that occur at smaller scales, below the seismic resolution (Fig. 5.1). Therefore, with the exception of the smallest stratigraphic scales, whereby

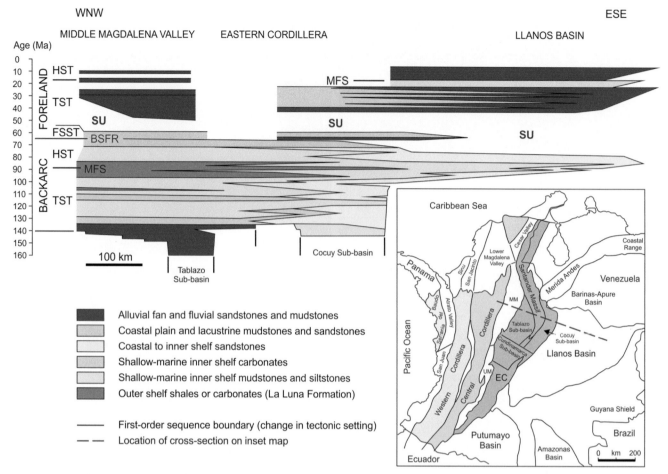

FIGURE 5.3   First-order depositional sequences, systems tracts, and depositional systems of the pre- and syn-Andean tectonic stages in Colombia (from Catuneanu, 2019b; modified after Sarmiento Rojas, 2001). Changes in the type of sedimentary basin (i.e., tectonic setting) mark the position of first-order sequence boundaries. These first-order sequences reach 10^3 m in the depocenters. The internal unconformities of the backarc and foreland sequences are negligible at the first-order scale of observation (i.e., they do not break the continuity in the paleogeographic evolution observed at the first-order scale); hence, first-order sequences are "relatively conformable" at the basin scale. The change from backarc to foreland took place under subaqueous conditions in the depocenters, where it is marked by a conformable BSFR (i.e., the onset of forebulge uplift). Abbreviations: EC—Eastern Cordillera; UM—Upper Magdalena Valley; MM—Middle Magdalena Valley; TST—transgressive systems tract; HST—highstand systems tract; FSST—falling-stage systems tract; MFS—maximum flooding surface; BSFR—basal surface of forced regression; SU—subaerial unconformity.

depositional systems consist only of sedimentological cycles (e.g., the fifth-order scale in Fig. 5.1), all higher rank depositional systems consist of lower rank (higher frequency) stratigraphic cycles that involve changes in systems tract and depositional system (e.g., the fourth- and third-order scales in Fig. 5.1). The scale at which changes in stratigraphic stacking pattern can be demonstrated is a function of data resolution, which defines the stratigraphic resolution that can be achieved in a case study (i.e., the lowest rank depositional systems and systems tracts that can be identified at any location).

Within the transit area of the shoreline, where changes in depositional environment are most frequent, the lowest rank depositional systems, which consist only of sedimentological units, can be referred to as depositional systems *sensu stricto*; in contrast, depositional systems of higher hierarchical ranks, which include lower rank stratigraphic cycles and associated changes in

depositional environment, can be referred to as depositional systems *sensu lato*. This distinction becomes less meaningful outside of the shoreline transit area, where stratigraphic cyclicity develops without changes in depositional environment. At any hierarchical level, the scale of depositional systems depends on the geological setting (i.e., local conditions of accommodation and sedimentation) and the lifespan of the specific (i.e., in the case of depositional systems *sensu stricto*) or dominant (i.e., in the case of depositional systems *sensu lato*) environments of deposition. The identification of depositional systems *sensu stricto* vs. *sensu lato* is a matter of data resolution. High-resolution studies indicate that depositional systems *sensu stricto* develop commonly below the scales of seismic stratigraphy (Figs. 5.1, 5.2, and 5.5).

Coastal settings are most susceptible to changes in depositional environment (e.g., Figs. 5.1, 5.2, 5.5, and

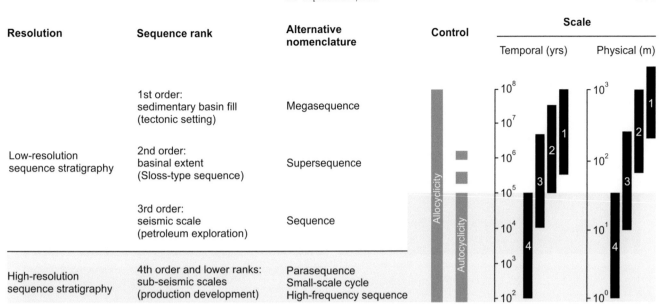

FIGURE 5.4   Classification of stratigraphic sequences. Temporal and physical scales compiled from Vail et al. (1977b, 1991), Williams (1988), Van Wagoner et al. (1990), Carter et al. (1991), Einsele et al. (1991), Reid and Dorobek (1993), Duval et al. (1998), Lehrmann and Goldhammer (1999), Schlager (2004, 2010), Miall (2010). "Megasequence" and "supersequence" nomenclature from Krapez (1996). Fourth-order and lower rank sequences are also termed "cyclothems" (Wanless and Weller, 1932), "cycles" (Heckel, 1986), or "simple sequences" (Vail et al., 1991; Schlager, 2010); third-order sequences are also termed "mesothems" (Ramsbottom, 1979), "megacyclothems" (Heckel, 1986), or "standard sequences" (Vail et al., 1991; Schlager, 2010); second-order sequences are also termed "composite sequences" (Abreu et al., 2010). A scale-independent nomenclature remains the most objective approach to terminology (see text for details).

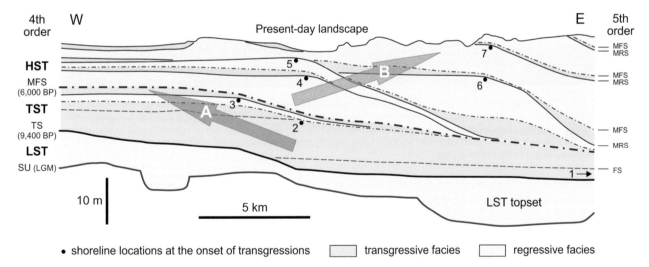

FIGURE 5.5   Stratigraphic architecture of the Holocene succession across the Po coastal plain, Italy (from Catuneanu, 2019a; modified after Amorosi et al., 2005; 2017). The fourth-order systems tracts record internal cyclicity at the fifth-order scale. Depositional systems can be observed at both scales. Fourth-order depositional systems (A—estuary; B—delta) are observed at 10^3 yrs timescales; fifth-order depositional systems were stable over 10^2–10^3 yrs timescales. Changes in coastal environment occurred rapidly, over ≤ 10^2 yrs, resulting in the formation of sequence stratigraphic surfaces. Abbreviations: BP—years before present; SU—subaerial unconformity; LGM—last glacial maximum; TS—transgressive surface; MFS—maximum flooding surface; FS—flooding surface; MRS—maximum regressive surface; LST—lowstand systems tract; TST—transgressive systems tract; HST—highstand systems tract.

5.6). The lifespan of depositional systems increases commonly in fully continental and fully marine settings, where stratigraphic cyclicity and associated changes in systems tracts may develop without changes in depositional environment (e.g., Figs. 4.70 and 5.7). In downstream-controlled settings, the lowest rank coastal systems determine the scale of the lowest rank systems tracts, while the highest frequency changes in coastal environments control the scale of the lowest rank stratigraphic sequences (Figs. 5.1, 5.2, and 5.5). In upstream-controlled settings, the scale of the lowest rank sequences and systems tracts is determined by the highest frequency changes in the dominant depositional element (Fig. 4.70; Catuneanu, 2017, 2019a).

FIGURE 5.6 Recent submergence along a modern coastline, marking a change from progradation to retrogradation and the formation of a maximum regressive surface (Eglinton Island, Canadian Archipelago; images courtesy of Chantel Nixon). In river-mouth settings, deltas are replaced by estuaries; in open-coastline settings, strandplains are replaced by backstepping beaches or lagoon-barrier island systems. The change in depositional system across the systems tract boundary takes place rapidly, over timescales of 10^1-10^2 yrs, resulting in the formation of sequence stratigraphic surfaces in the sedimentary record.

Therefore, the timescale of systems tracts at any location is quantified by the lifespan of depositional systems, within the transit area of the shoreline, and it is smaller than the lifespan of depositional systems in fully continental and fully marine settings.

The high-resolution studies of the Holocene show that changes between transgressive and regressive coastal environments can occur frequently, in relation to shoreline transit cycles of 10^2-10^3 yrs (Amorosi et al., 2009, 2017; Nixon et al., 2014; Pellegrini et al., 2017); i.e., one or two orders of magnitude below the 10^4 yrs scale of the fifth-order cycles in Fig. 5.1. The research of the Arno coastal plain in Italy (Amorosi et al., 2009) demonstrates that four stages of transgression separated by three stages of regression occurred between c. 12,690 yr BP and c. 7,820 yr BP, generating a c. 35 m thick succession. In this example, the transgressive–regressive cycles record average scales of 10 m and 1,400 yrs, with an average duration of

each stage of transgression or regression of 700 yrs. Therefore, coastal environments were stable on centennial scales, and full cycles of change in shoreline trajectory (i.e., high-frequency sequences) developed on millennial scales. Similar millennial-scale sequences which record changes in systems tracts and depositional systems have been discussed by Csato et al. (2014).

On the other side of the Italian Peninsula, the analysis of the Adriatic coastal plain reveals that the Holocene stratigraphic record includes two fourth-order systems tracts associated with (1) transgression (9,400 yr BP to 6,000 yr BP), followed by (2) highstand normal regression (6,000 yr BP to present day; Amorosi et al., 2005, 2017) (Fig. 5.5). The transgressive systems tract records a thickness of c. 10 m, and includes fluvial, estuarine, and marine depositional systems. The highstand systems tract reaches a thickness of c. 20 m, and includes fluvial, deltaic, and marine depositional systems. The fourth-order systems tracts record internal cyclicity at

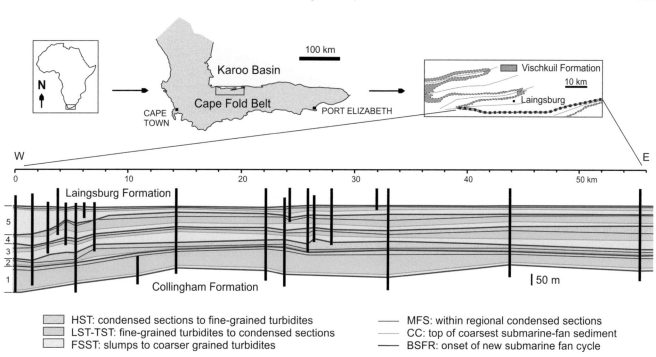

FIGURE 5.7 Sequence stratigraphy of the Permian Vischkuil Formation in a deep-water setting (Karoo Basin, South Africa; modified after van der Merwe et al., 2010). The Vischkuil Formation consists of five depositional sequences (1—5), which correspond to five submarine fan complexes. Stratigraphic cyclicity is controlled by glacio-eustatic fluctuations during an icehouse climatic regime, likely on timescales of 10^5 yrs. Note the lateral continuity of the stratigraphic architecture across the low-relief basin floor. Vertical bars represent measured outcrop sections. Abbreviations: FSST—falling-stage systems tract; LST—lowstand systems tract; TST—transgressive systems tract; HST—highstand systems tract; BSFR—basal surface of forced regression; CC—correlative conformity; MFS—maximum flooding surface.

scales of 10^0 m and $\pm 1,000$ yrs, associated with fifth-order transgressions and regressions across the Adriatic coastal plain (Fig. 5.5). Changes in depositional environment occurred as the fourth-order transgression was interrupted by fifth-order stages of progradation (e.g., deltas replacing estuaries on a short term), and as the fourth-order highstand progradation was interrupted by fifth-order stages of marine flooding and estuary development (Fig. 5.5).

The studies of the Arno and Adriatic coastal plains in Italy document high-frequency changes in the direction of shoreline shift, and hence in the type of transgressive vs. regressive coastal environments, in basins connected to the global ocean. Enclosed basins such as lakes or interior seas are prone to even higher frequency changes between transgressive and regressive shorelines, as their lake/sea levels are more susceptible to change in response to small variations in climatic conditions or local tectonism. For example, studies of the evolution of the Dead Sea during the Holocene reveal sea-level fluctuations of 10^1 m on timescales of 700—800 yrs, leading to the development of centennial-scale depositional sequences which correspond to full cycles of change in accommodation. In this case, individual transgressions and regressions occurred over 300—400 yrs timescales during the full shoreline transit cycles of 700—800 yrs (Bookman et al., 2004; 2006; Moran, 2020).

Changes in depositional environment observed at different scales justify the definition of depositional systems at different hierarchical levels, both below and above the resolution of seismic data (Figs. 5.2 and 5.3). The highest frequency changes in depositional system are recorded in coastal settings, which hold the key to the downstream-controlled sequence stratigraphic framework in terms of the timing of systems tracts and bounding surfaces. The lowest rank depositional systems consist of beds and bedsets (sedimentological cycles), which are commonly confined to the area of development of the hosting depositional system. The internal complexity of depositional systems increases with the scale of observation, from sedimentological cycles to stratigraphic cycles (sequences and component systems tracts) of lower hierarchical ranks (Fig. 5.1). At each scale of observation, depositional systems have paleoenvironmental significance and include process-related facies that reflect the dominant sediment dispersal patterns.

At sub-seismic scales, coastal environments are typically "stable" (e.g., a delta is maintained as a prograding river mouth before being replaced by an estuary) over 10^2—10^3 yrs timescales, and able to change rapidly across systems tract boundaries during periods of $\leq$ 10^2 yrs (e.g., Csato et al., 2014; Nixon et al., 2014; Amorosi et al., 2009, 2017; Pellegrini et al., 2017, Figs. 5.5 and 5.6). The period of stability represents the time during which a specific relationship between accommodation and sedimentation is maintained (e.g., the backstepping of a coastal system continues for as long as accommodation outpaces sedimentation at the shoreline). The change in coastal environment (e.g., from a backstepping to a prograding system, which marks the formation of a maximum flooding surface) corresponds to a brief time of balance between accommodation and sedimentation at the shoreline (Fig. 4.32). Such equilibrium cannot be maintained for any significant period of time, as both accommodation and sedimentation vary in response to the interplay of multiple independent controls. This explains the rapid changes in coastal environments that accompany the shifts from one type of stratigraphic stacking pattern to another, which result in the formation of sequence stratigraphic surfaces. Therefore, the development of systems tracts and component depositional systems typically takes orders of magnitude longer than the time required to change the stacking pattern across a systems tract boundary, even at the scale of the lowest rank systems tracts.

At seismic scales (commonly of third-order and higher ranks), the topset of a normal regressive unit may be defined as a fluvial system (Fig. 4.6). However, at higher resolution sub-seismic scales (e.g., in wells or outcrops: fourth-order and lower ranks), the seismic-scale fluvial system may include lower rank marine systems associated with higher frequency transgressions (e.g., the Drumheller Marine Tongue within the larger scale Horseshoe Canyon fluvial topset in the Alberta foredeep; Catuneanu et al., 2000). The observation of depositional systems at different hierarchical levels defines the overlap between the scales of sedimentology and stratigraphy. At each hierarchical level, depositional systems are the largest units of sedimentology, and the building blocks of systems tracts in stratigraphy. However, sedimentological cycles (beds and bedsets; Figs. 3.24 and 4.62) are always nested within the lowest rank systems tracts and component depositional systems, whose scale defines the highest resolution that can be achieved with a stratigraphic study at a specific

location, and the limit between sedimentology and stratigraphy *sensu stricto*.

5.2 Systems tracts

Sequences and component systems tracts develop in all tectonic and depositional settings, from underfilled to overfilled basins (Fig. 3.12). Underfilled basins include both downstream- and upstream-controlled settings (Fig. 3.11), whereas overfilled basins are entirely upstream-controlled. Systems tracts in downstream- and upstream-controlled settings have different origins and defining features, related to or independent of shoreline trajectories, respectively (Fig. 3.11), but serve the same purpose as subdivisions of stratigraphic sequences. This section reviews the characteristics of systems tracts that develop in downstream- and upstream-controlled settings, from definition to scales and identification criteria.

5.2.1 Definition

The concept of systems tract was introduced to define a linkage of contemporaneous depositional systems, forming the subdivision of a sequence (Brown and Fisher, 1977). This definition was coined for downstream-controlled settings, where continental, coastal, and marine environments coexist and respond to changes in shoreline trajectory (Fig. 3.11). "Conventional" systems tracts in downstream-controlled settings are defined by the trajectory of subaerial clinoform rollovers (i.e., shoreline trajectories), which can be observed at different stratigraphic scales. The concept of systems tract was subsequently extended to fully continental upstream-controlled settings, where "unconventional" systems tracts may form within a depositional system. In this case, the definition of systems tracts is based on the dominant architectural elements within the system (e.g., channel vs. floodplain deposits in a fluvial system; Figs. 4.70 and 5.8; see Chapter 4 for details). While the unconventional systems tracts are no longer "tracts" of

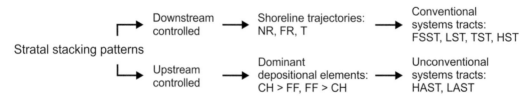

FIGURE 5.8 Classification of systems tracts. Stratal stacking patterns provide the basis for the definition of all units and surfaces of sequence stratigraphy. Downstream-controlled systems tracts are defined by stacking patterns associated with specific shoreline trajectories (i.e., NR, FR, T). Upstream-controlled systems tracts are defined by the dominant depositional elements. The definition of systems tracts is independent of temporal and physical scales. Sequences of all scales include component systems tracts, whose hierarchical level matches the one of the sequence to which they belong (e.g., Figs. 4.70, 5.1, and 5.3). Abbreviations: NR—normal regression; FR—forced regression; T—transgression; CH—channels; FF—floodplain fines; FSST—falling-stage systems tract; LST—lowstand systems tract; TST—transgressive systems tract; HST—highstand systems tract; HAST—high-amalgamation systems tract; LAST—low-amalgamation systems tract.

different depositional systems, they still form the subdivisions of sequences, and the "systems tract" term was retained for methodological and nomenclatural consistency.

The usage of the systems tract concept in both downstream- and upstream-controlled settings requires a more inclusive definition, to highlight the fundamental attribute of this type of stratal unit in all settings. In a most general sense, a systems tract is a stratal unit defined by a specific stacking pattern and bounding surfaces, forming the subdivision of a sequence. The definition of systems tracts is independent of physical and temporal scales; systems tracts can be observed at different scales, depending on the scope of the study and/or the resolution of the data available (Fig. 4.3). At each stratigraphic scale, systems tracts have genetic significance and consist of strata deposited within an interconnected sediment dispersal system (Galloway, 2004). Sediment dispersal systems are relative stable during the deposition of a systems tract, reflecting "a specific sedimentary response to the interaction between sediment flux, physiography, environmental energy, and changes in accommodation" (Posamentier and Allen, 1999). While sediment dispersal systems do evolve all the time, the most significant reorganizations occur across systems tract boundaries which mark changes in the balance between accommodation and sedimentation in downstream-controlled settings (Fig. 4.32), or in the sedimentation regimes in upstream-controlled settings (Figs. 4.63 and 4.68).

5.2.2 Scale of systems tracts

Sequences of all scales consist of systems tracts that match the hierarchical rank of the host sequence (Figs. 4.3 and 4.70). The usage of the "systems tract" concept at different hierarchical levels is consistent with the observation of depositional systems at different scales (Figs. 5.1, 5.2, and 5.3). The internal makeup of a systems tract varies greatly with the scale of observation, from a succession of beds and bedsets (i.e., sedimentological cycles within the lowest rank depositional systems; Fig. 5.2) to a set of higher frequency sequences of lower hierarchical ranks (Figs. 4.70, 5.1, and 5.5). The scale of the lowest rank systems tracts at any particular location defines the highest resolution that can be achieved with a sequence stratigraphic study.

Systems tracts of all scales may include hiatal surfaces of equal and/or lower hierarchical ranks (e.g., a third-order transgressive systems tract may include a third-order wave-ravinement surface, as well as other types of unconformities of lower hierarchical ranks). Even so, at each stratigraphic scale, the three-dimensional facies assemblages of systems tracts remain linked by the dominant depositional trends associated with the defining stacking pattern. For example, a transgressive systems tract is defined by a retrogradational stacking pattern, even though the transgression may be interrupted by higher frequency regressions of lower hierarchical ranks (Figs. 4.29 and 5.5). The change in stacking pattern across the systems tract boundary may be accompanied by a change in depositional system (e.g., a change from an estuary to a delta across a maximum flooding surface), or it may occur within a depositional system (e.g., a change from retrogradation to progradation within a shelf system across a maximum flooding surface). The highest frequency changes in depositional system associated with systems-tract boundaries are recorded within the transit area of the shoreline (Figs. 5.1, 5.2, and 5.5).

The development of stratigraphic stacking patterns (i.e., stratal stacking patterns observed at stratigraphic scales; Figs. 3.24 and 4.62) that define systems tracts in both downstream- and upstream-controlled settings starts typically from timescales of 10^2–10^3 yrs, which afford the formation of the lowest rank depositional systems (e.g., Bridge and Leeder, 1979; Amorosi et al., 2005, 2009, 2017; Nanson et al., 2013; Nixon et al., 2014; Miall, 2015; Bruno et al., 2017; Pellegrini et al., 2017; 2018; Catuneanu, 2019a,b, Moran, 2020, Figs. 4.62 and 5.5), and may continue for 10^7 yrs at the first-order basin scales (e.g., De Gasperi and Catuneanu, 2014, Fig. 5.3). In terms of physical scales, systems tracts can be observed from 10^0 m (Fig. 5.2) to 10^3 m (Fig. 5.3), depending on the resolution of the stratigraphic study. A full discussion of the range of scales and processes involved in the formation of "conventional" (downstream-controlled) and "unconventional" (upstream-controlled) stratigraphic stacking patterns and associated systems tracts is provided in Chapter 4.

5.2.3 Systems tracts in downstream-controlled settings

In downstream-controlled settings, "conventional" systems tracts form in relation to specific types of shoreline trajectories (i.e., forced regressive, normal regressive, and transgressive; Figs. 4.1, 5.8, and 5.9). Systems tracts may be observed at different scales, depending on the resolution of the data available, from seismic scales (higher hierarchical ranks; Fig. 4.6) to the higher resolution scales afforded by well-log, core, and outcrop data (lower hierarchical ranks; Figs. 5.2, 5.5, and 5.10). At any scale of observation (i.e., hierarchical level), sequences may consist of variable combinations of systems tracts, depending on the local conditions of accommodation and sedimentation (Csato and Catuneanu, 2012, 2014, Figs. 1.23 and 4.33). Systems-tract successions are basin or even sub-basin specific, and may or may not conform to the prediction of idealized models. For this reason, the construction of sequence

Systems tracts	Geometrical trends of the shoreline (i.e., shoreline trajectories)	Depositional trends at the shoreline (i.e., coastal processes)
HST	Forestepping and upstepping (NR following T)	Aggradation to progradation (convex-up shoreline trajectory)
TST	Backstepping and upstepping (T)	Retrogradation with aggradation * (SU forms during FR)
		Retrogradation with degradation ** (SU forms during T)
LST	Forestepping and upstepping (NR following FR)	Progradation to aggradation (concave-up shoreline trajectory)
FSST	Forestepping and downstepping (FR)	Progradation with degradation * (SU forms during FR)
		Progradation with aggradation ** (SU forms during T)

FIGURE 5.9   Definition of systems tracts in downstream-controlled settings. The defining criteria are provided by the geometrical trends of the shoreline (i.e., shoreline trajectories), which may or may not be followed by the depositional trends in fluvial to coastal systems. The depositional trends of coastal progradation and retrogradation coincide with the geometrical trends of forestepping and backstepping, respectively; therefore, these terms can be used interchangeably. In contrast, the depositional trends of aggradation and degradation (controlled by changes in base level) may or may not follow the geometrical trends of upstepping and downstepping (controlled by changes in relative sea level). Irrespective of the timing of the subaerial unconformity (during forced regression or transgression), field criteria are available to separate forced from normal regressions, as well as lowstand from highstand normal regressions (see text for details). Occurrence: * common; ** uncommon. Abbreviations: FR—forced regression; NR—normal regression; T—transgression; FSST—falling-stage systems tract; LST—lowstand systems tract; TST—transgressive systems tract; HST—highstand systems tract; SU—subaerial unconformity.

FIGURE 5.10   Sub-seismic scale systems tracts in a shallow-water lacustrine setting (Lower Cretaceous, Araripe Basin, Brazil). In this example, the falling-stage systems tract consists of evaporites (gypsum), whereas the lowstand, transgressive, and highstand systems tracts reflect the progradation (coarsening upward: red triangle) or the retrogradation (fining upward: blue triangle) of a mixed siliciclastic-carbonate shallow-water system. Climate exerted a dominant control on accommodation and sedimentation, with arid periods leading to evaporation (lake-level fall) and low terrigenous sediment influx, and humid periods leading to precipitation (lake-level rise) and higher terrigenous influx into the lake. Abbreviations: CC—correlative conformity; MRS—maximum regressive surface; MFS—maximum flooding surface; FSST—falling-stage systems tract; LST—lowstand systems tract; TST—transgressive systems tract; HST—highstand systems tract.

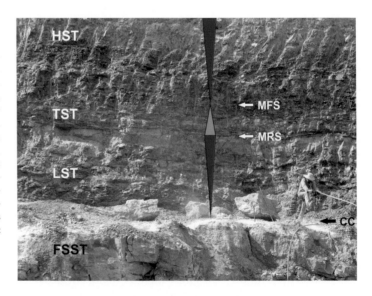

stratigraphic frameworks of systems tracts and bounding surfaces needs to be based on local data rather than model assumptions.

Sequence stratigraphic frameworks in downstream-controlled settings are potentially most complete in the sense that they may include the entire array of sequence stratigraphic surfaces and depositional systems (Fig. 4.39). Forced regressive, normal regressive, and transgressive deposits need to be separated as distinct systems tracts, as this is the key to the predictive aspect of sequence stratigraphy. Each type of stacking pattern involves a unique sediment dispersal pattern and distribution of economic deposits, and hence it presents different exploration opportunities. Merging different types of stacking patterns into a systems tract, as it is the case with the "lowstand systems tract" of depositional sequence II, the "highstand systems tract" of depositional sequence III, and the "regressive systems tract" of the transgressive-regressive sequence (Fig. 1.10) reduces the resolution of stratigraphic studies, and therefore is counterproductive to the efforts of exploration for natural resources.

Following the introduction of the "systems tract" concept in the 1970s (Brown and Fisher, 1977), the definition and nomenclature of different types of systems tracts were not without controversy (Figs. 1.9 and 1.10). Early sequence models developed for "passive" (divergent) continental margins in the 1980s accounted for four systems tracts, namely the "lowstand," "transgressive," "highstand," and "shelf-margin" (Vail, 1987; Posamentier and Vail, 1988; and Posamentier et al., 1988). These systems tracts were defined initially relative to a curve of eustatic changes (Posamentier and Vail, 1988; Posamentier et al., 1988), which was subsequently replaced with a curve of relative sea-level changes (Hunt and Tucker, 1992, 1995; Posamentier and James, 1993). The lowstand and the shelf-margin systems tracts were both related to the stage of fall—early rise of the sea-level cycle, so they were used interchangeably (Vail, 1987; Posamentier and Vail, 1988; Vail et al., 1991). A sequence composed of lowstand, transgressive,

and highstand systems tracts was defined as a "type 1" sequence, whereas a combination of shelf-margin, transgressive, and highstand systems tracts was designated as "type 2" sequence (Posamentier and Vail, 1988).

The distinction between the types 1 and 2 sequences, and implicitly between the lowstand and the shelf-margin systems tracts, was based on the development of the unconformable portions of sequence boundaries. According to Vail et al. (1984), a type 1 sequence boundary forms during a stage of rapid eustatic sea-level fall, when the rate of fall is greater than the rate of subsidence *at the shelf edge*. This scenario assumes complete exposure of the shelf by the end of sea-level fall, and the development of the subaerial unconformity all the way to the shelf edge. In contrast, a type 2 sequence boundary forms during a stage of slower sea-level fall, when the rate of fall is less than the rate of subsidence *at the shelf edge* (Vail et al., 1984). In this scenario, the subaerial unconformity is restricted to the proximal side of the shelf, and the fall in relative sea level at the shoreline is coeval with relative sea-level rise at the shelf edge. Therefore, a type 1 sequence boundary would include a "major" subaerial unconformity associated with significant erosion across the entire continental shelf, whereas a type 2 sequence boundary would only include a "minor" subaerial unconformity associated with minimal erosion and smaller extent (Fig. 5.11).

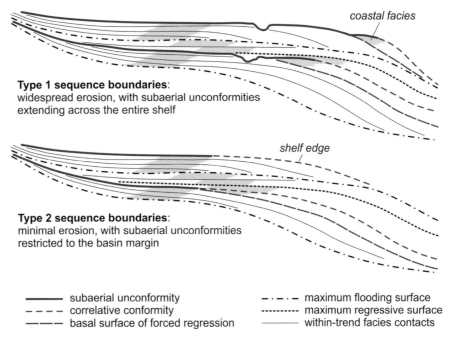

FIGURE 5.11 Definition of "type 1" and "type 2" sequence boundaries (modified from Vail et al., 1984, and Galloway, 1989). Both types 1 and 2 sequence boundaries consist of subaerial unconformities and correlative conformities. A type 1 sequence boundary includes a subaerial unconformity associated with widespread erosion across the entire continental shelf. A type 2 sequence boundary includes a subaerial unconformity restricted to the basin margin (minimal erosion and limited areal extent). The formation of subaerial unconformities requires relative sea-level fall *at the shoreline* (eustatic fall > subsidence). However, *at the shelf edge*, a type 1 sequence boundary assumes relative sea-level fall (eustatic fall > subsidence), while a type 2 sequence boundary assumes relative sea-level rise (subsidence > eustatic fall). Changes in the relative sea level between the shoreline and the shelf edge are enabled by differential subsidence. The two candidates for the conformable portion of the depositional sequence boundary include the "correlative conformity" *sensu* Hunt and Tucker (1992) and the "basal surfaces of forced regression" (i.e., the correlative conformity of Posamentier and Allen, 1999). Types 1 and 2 sequence boundaries are no longer in use (see text for details).

The types 1 and 2 sequences and sequence boundaries fell out of fashion eventually, due to practical limitations that prevented the universal application of the concept beyond the context of divergent continental margins. Most importantly, sedimentary basins placed in other tectonic settings may be missing a shelf-slope system, and implicitly a shelf edge, in which case the main criterion of classification of sequences into types 1 vs. 2 is lost. Another pitfall is that the classification of unconformities into two types is an oversimplification, as continental shelves can host unconformities of different magnitudes across a wide range of scales. After more than a decade of confusion and controversy, Posamentier and Allen (1999) advocated elimination of types 1 and 2 in favor of a single type of depositional sequence and sequence boundary. With the exclusion of the "type 2" unconformity from sequence stratigraphy, the "shelf-margin" systems tract, which was part of the "type 2" sequence, was also abandoned. As a result, the depositional sequence model was reduced to a tripartite scheme that included lowstand, transgressive, and highstand systems tracts as the subdivisions of a sequence (Posamentier and Allen, 1999).

The subdivision of sequences into three systems tracts compelled practitioners to combine forced regressive deposits with either lowstand or highstand normal regressive deposits, since a relative sea-level cycle may include *four* changes in stratal stacking patterns (Fig. 4.32). Posamentier and Vail (1988) assigned the forced regressive deposits to the lowstand systems tract, whereas Van Wagoner et al. (1988) assigned the same deposits to the highstand systems tract (Fig. 1.10). None of these choices was ideal, as different types of stacking patterns should not be merged into one systems tract (discussion above). Moreover, as the depositional sequence boundary is placed by convention at the base of the lowstand systems tract, the two options implied a different timing for the sequence boundary, at the onset vs. the end of forced regression (depositional sequence models II vs. III in Fig. 1.10). From a nomenclatural standpoint, neither approach is satisfactory because a stage of relative sea-level fall starts from the highstand in relative sea level and ends at the lowstand position. Therefore, neither the "lowstand" nor the "highstand" terms apply to the entire suite of forced regressive deposits.

This nomenclatural inconsistency was resolved by Hunt and Tucker (1992), who introduced the "forced regressive wedge" as a new systems tract that consists only of forced regressive deposits. The forced regressive wedge systems tract is also known as the "falling-stage" systems tract (Ainsworth, 1992, 1994; Plint and Nummedal, 2000), with the latter (and simpler) term being eventually endorsed by the International Commission on Stratigraphy (Catuneanu et al., 2011). The advantage of this approach is two-fold: first, forced regressive deposits are now separated from normal regressive deposits, thus increasing the resolution of stratigraphic studies; and second, the highstand and lowstand systems tracts are now restricted to the normal regressions associated with the late and early stages of relative sea-level rise, which develop close to the actual highstand and lowstand positions of the relative sea level, respectively (Fig. 4.32). In this approach, the timing of the sequence boundary is at the end of relative sea-level fall; i.e., the end of forced regression and the base of the lowstand systems tract (depositional sequence IV in Fig. 1.10). The schematic development of the four systems tracts is illustrated in Figs. 5.12 and 5.13.

The subdivision of a stratigraphic sequence into four systems tracts, as proposed by Hunt and Tucker (1992), satisfies all theoretical and practical requirements for an optimum application of sequence stratigraphy. Less commonly, all forced and normal regressive deposits within a sequence may be lumped together in a "regressive" systems tract, where the data available are insufficient for a higher resolution analysis (Fig. 1.10). The selection of the sequence boundary is less important than the identification of systems tracts, and it depends on the mappability of the sequence stratigraphic surfaces that are present within a study area. Systems tracts hold the key to the predictive aspect of sequence stratigraphy, as they afford insights into the distribution of sediment and natural resources within the basin. Field criteria are available for the identification of all stratal stacking patterns that define systems tracts, and involve the observation of facies and stratigraphic relationships (Fig. 5.14). The following sections describe the four systems tracts employed in the modern sequence stratigraphic analysis of downstream-controlled settings.

5.2.3.1 Falling-stage systems tract

The falling-stage systems tract corresponds to the "lowstand fan" of Posamentier et al. (1988) (Fig. 1.10), and it was separated as a distinct systems tract in the early 1990s, as a result of independent work by Ainsworth (1992, 1994), Hunt and Tucker (1992), and Nummedal (1992). The actual systems tract terminology varied from "falling-stage" (Ainsworth, 1992, 1994) to "forced regressive wedge" (Hunt and Tucker, 1992) and "falling sea-level" (Nummedal, 1992), with the simplest nomenclature of Ainsworth (1992, 1994) becoming most widespread and eventually endorsed by the International Commission on Stratigraphy (Catuneanu et al., 2011).

The falling-stage systems tract accumulates during relative sea-level fall, and it is defined by a forced regressive stacking pattern (i.e., progradation and downstepping of the subaerial clinoform rollovers; Figs. 4.1, 4.21, 4.32, 5.9, and 5.15; see discussion in Chapter 4). Where the subaerial unconformity forms during forced regression, the falling-stage systems tract is bounded

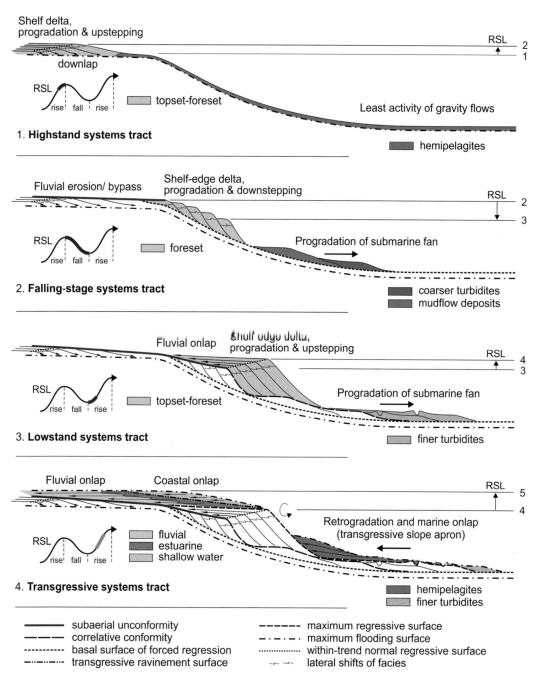

FIGURE 5.12 Regional trends of sedimentation in siliciclastic passive-margin settings. Departures from these trends may be recorded as a function of local accommodation and sedimentation conditions (e.g., the timing of the subaerial unconformity, or the shoreline location and the types of gravity flows during different stages of the relative sea-level cycle; details in Chapters 4 and 8). Clinoform surfaces are concave-up, whereas transgressive "healing phase" strata associated with coastal and marine onlap tend to be convex-up. Abbreviation: RSL—relative sea level.

at the base by a marine basal surface of forced regression (i.e., the paleoseafloor at the onset of forced regression), and at the top by the subaerial unconformity and its correlative conformity (Fig. 4.39A). In this case, with the exception of floodplain terraces and lateral accretion macroforms that may accumulate and be preserved during forced regression in the fluvial environment, the falling-stage systems tract consists solely of marine deposits.

Less commonly, forced regression may be accompanied by fluvial aggradation (Figs. 4.26 and 4.39B). In this case, the falling-stage systems tract is bounded at the base by a basal surface of forced regression with both continental and marine portions, and at the top by a conformity that marks the change in stacking pattern to the overlying lowstand systems tract (Fig. 4.39B). Forced regressions with fluvial topsets can be separated from normal regressions on the basis of

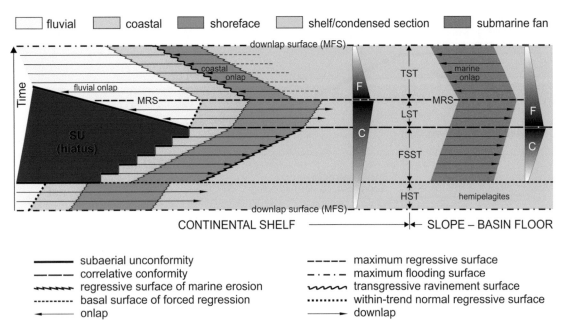

FIGURE 5.13 Time-domain (Wheeler) diagram illustrating the sedimentation patterns in Fig. 5.12. The subaerial unconformity expands basinward during relative sea-level fall, and it is onlapped by fluvial deposits during relative sea-level rise. The switch in grading trends, from coarsening- to fining-upward, is temporally offset between shallow- and deep-water systems. Even though the progradation of submarine fans continues throughout the regressive stage, the onset of relative sea-level rise marks a drop in sediment supply to the shelf edge as riverborne sediment is retained in the lowstand topset. Abbreviations: SU—subaerial unconformity; MRS—maximum regressive surface; MFS—maximum flooding surface; HST—highstand systems tract; FSST—falling-stage systems tract; LST—lowstand systems tract; TST—transgressive systems tract; F—fining-upward; C—coarsening-upward.

several field criteria, including the downstepping vs. upstepping of shoreface facies in a downdip direction, the sharp-based vs. gradationally based nature of shoreface profiles, and the compressed vs. expanded development of shoreface successions, respectively (Fig. 4.40). In areas where the overlying lowstand systems tract is missing, the conformity at the top of the falling-stage systems tract is replaced by surfaces associated with subsequent transgression (i.e., maximum regressive surface, transgressive surface of erosion, or subaerial unconformity). Where two or more sequence stratigraphic surfaces are superimposed, due to nondeposition or erosion, the name of the younger surface, which leaves the last imprint on the preserved contact, is typically used (Catuneanu, 2006).

The formation of subaerial unconformities during relative sea-level fall may involve a combination of processes, including fluvial incision, fluvial bypass, pedogenesis, and eolian deflation. Fluvial incision caused by relative sea-level fall occurs typically where the relative sea level is lowered below topographic breaks (e.g., depositional or fault scarps, the shelf edge, etc.; Figs. 4.42A and 5.16), thus exposing segments of former seascapes that are steeper than the fluvial graded profile (Schumm, 1993; Ethridge et al., 2001; Posamentier, 2001). In such cases, fluvial incision starts from the forced regressive shoreline and gradually propagates landward by the upstream migration of fluvial

knickpoints (Figs. 4.42A, 4.43, and 5.16). It is estimated that the migration rates of fluvial knickpoints are generally high, in a range of tens of meters per year, thus generating nearly isochronous unconformities at geological timescales (Posamentier, 2001). The extent of upstream migration of fluvial knickpoints may exceed 200 km, as documented on the Java Sea shelf that was exposed during the Late Pleistocene glaciation (Posamentier, 2001).

Highstand prisms of fluvial to shallow-water strata, abandoned on the exposed shelves behind forced regressive shorelines, provide classic examples of depositional scarps that are prone to fluvial incision during relative sea-level fall (Figs. 5.16A, 5.17, and 5.18). The resulting incised valleys characteristically have incised tributaries, which can be used as a diagnostic feature to separate incised vs. unincised rivers in subsurface studies (Fig. 5.18). Under the scenario in Fig. 5.16A, fluvial incision is restricted to the highstand prism whose forefront slope is steeper than the fluvial graded profile (Fig. 5.17). Similar processes of fluvial erosion on continental shelves may also be triggered by structural elements such as fault scarps (Fig. 5.19). In these examples, river incision is initiated by the increase in slope gradients and associated fluvial energy downdip of the topographic scarp, where the landscape is steeper than the fluvial graded profile. Where the exposed seascapes preserve the gradient of the fluvial graded profile, rivers

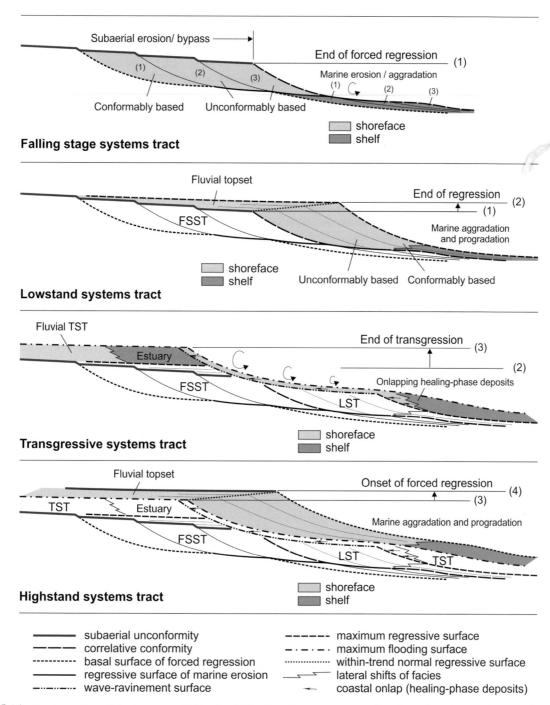

Falling stage systems tract

Lowstand systems tract

Transgressive systems tract

Highstand systems tract

—— subaerial unconformity	– – – maximum regressive surface
– – – correlative conformity	–·–·– maximum flooding surface
········ basal surface of forced regression	············ within-trend normal regressive surface
—— regressive surface of marine erosion	⌐⌐ lateral shifts of facies
–··–··– wave-ravinement surface	← coastal onlap (healing-phase deposits)

FIGURE 5.14 Stratigraphic architecture in a siliciclastic shelf setting. In wave-dominated settings, forced regressive shorefaces are unconformably based and normal regressive shorefaces are conformably based, with the exception of the earliest clinoforms of the falling-stage and lowstand systems tracts. Irrespective of their conformable or unconformable base, forced and normal regressive shorefaces can be differentiated on the basis of their downstepping vs. upstepping facies trends (details in Chapter 4).

may only bypass the shelf, without erosion or aggradation (Figs. 4.42B, 5.16A, and 5.19). This is commonly the case where the forced regressive shoreline remains inboard of the shelf edge during relative sea-level fall (Figs. 5.16A and 5.19).

Where fluvial systems incise following the onset of relative sea-level fall, the morphology of the low-

energy highstand rivers may be inherited by the younger and higher energy systems that become trapped within incised valleys. As highstand rivers are most commonly the lowest energy fluvial systems of a stratigraphic sequence (see discussion in Chapter 4), they are most often of meandering type, characterized by high-sinuosity channels. Erosion along such channels

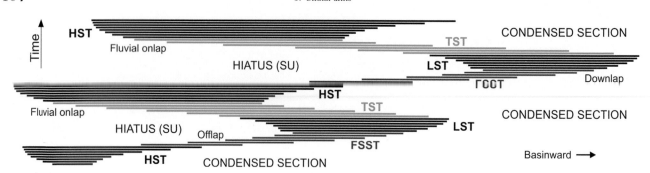

FIGURE 5.15 Stratal stacking patterns in downstream-controlled shelf settings, illustrated by the common architecture of seismic reflections in a time domain. The degree of preservation of the sedimentary record increases in a downdip direction, from continental to marine systems. The lowstands and highstands in relative sea level can be reconstructed in a model-independent manner, on the basis of stratigraphic relationships: the LST is a normal regression that follows a forced regression; the HST is a normal regression that follows a transgression. Abbreviations: FSST—falling-stage systems tract; LST—lowstand systems tract; TST—transgressive systems tract; HST—highstand systems tract; SU—subaerial unconformity. The condensed sections are marine.

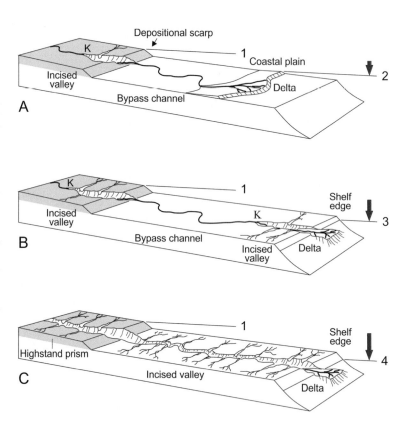

FIGURE 5.16 Incised and unincised (bypass) fluvial systems during relative sea-level fall (modified after Posamentier, 2001). In contrast to bypass or aggrading systems, incised valleys have characteristic V-shaped cross-sectional profiles and incised tributaries. Fluvial incision due to relative sea-level fall occurs where the subaerially exposed seascape is steeper than the graded fluvial profile. Fluvial incision propagates landward via the upstream migration of fluvial knickpoints (K). A—Early stage of relative sea-level fall, when the forced regressive shoreline is still inboard of the shelf edge. The highstand prism is subject to fluvial incision, but the rest of the exposed shelf may be bypassed by unincised rivers. B—As the shoreline falls below the elevation of the shelf edge, fluvial incision starts affecting the shelf. C—Late stage of relative sea-level fall, when the entire fluvial system is incised. Time 1 shows the sea level at the onset of relative fall (end of highstand); time 4 shows the sea level at the end of relative fall.

leads to the formation of incised meander belts that preserve the morphology of highstand rivers, even after the increase in slope gradients during subsequent relative sea-level fall (Fig. 5.20). For example, the Orange River in southern Africa, which started 100 Ma as an unincised meandering system, evolved into an incised meander belt in response to continental-scale tectonic uplift. The modern river still preserves the original meander pattern to a large extent, although a braided style emerged in areas of meander cut-off, valley widening, and erosion of valley walls (Fig. 5.21). The change from higher to lower sinuosity channels across a

subaerial unconformity occurs naturally in the case of unincised systems, as rivers that are not confined by an erosional landscape can adjust their morphology more readily to new energy regimes.

Unincised fluvial systems during the falling stage are more common in the rock record than originally inferred by the standard sequence stratigraphic models. This is especially the case in basins with gently sloping ramp margins, such as intracratonic basins and overfilled forelands, or in continental shelf settings where the forced regressive shoreline does not fall below the elevation of the shelf edge (Posamentier, 2001, Fig. 5.22). The

FIGURE 5.17 Oblique aerial photograph of a Pleistocene highstand coastline stranded behind the forced regressive shoreline of the Great Salt Lake, Utah (image courtesy of Henry Posamentier). The arrow points to localized fluvial incision that affects the highstand prism. The depth of incision decreases in a downstream direction, as the landscape gradient becomes in balance with the graded fluvial profile beyond the toe of the exposed highstand clinoform. The Holocene lake-level fall is due to evaporation.

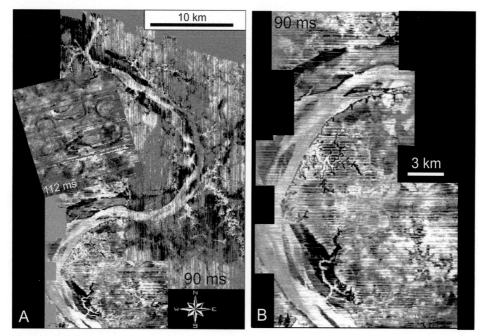

FIGURE 5.18   Time slices from the amplitude domain of a 3D seismic volume, illustrating a Late Pleistocene incised valley located c. 72 m (90 ms) beneath the modern seafloor, offshore Java (Indonesia) (modified from Posamentier, 2001; images courtesy of Henry Posamentier). Fluvial incision was caused by a relative sea-level fall in excess of 110 m, which led to the subaerial exposure of the entire continental shelf and upper slope. The main trunk of the valley, shown in image A, is c. 90 km long, and widens in a downdip direction. The valley system is characterized by short and incised tributaries with a dendritic pattern in plan view. The inset map of image A shows a time slice at a deeper level (c. 90 m/112 ms subsea) which captures the morphology of older, unincised highstand rivers. The unincised (highstand) and incised (falling-stage) systems have similar meander belt morphologies, except that the latter are associated with incised tributaries. The morphological similarity is due to the fact that the falling-stage valleys inherited the shape of the highstand rivers, which provided the starting point for incision at the onset of relative sea-level fall. Image B details the lower portion of image A.

under-representation of falling-stage bypass fluvial systems in the sequence stratigraphic literature may be due to a number of factors, including inadequate well spacing and a lack of high-resolution 3D seismic imaging. It also appears that many published examples of incised valleys may not, in fact, be incised systems (Posamentier, 2001). A look at the continental shelf of the Java Sea during the last 0.5 My reveals that extensive

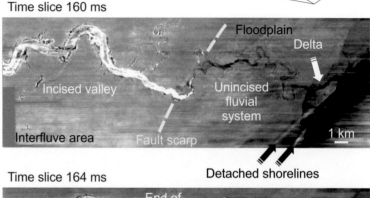

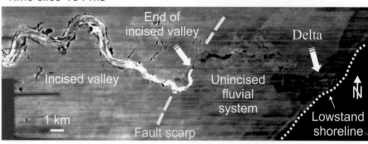

FIGURE 5.19 Falling-stage fluvial system, incised and unincised across a structural scarp (Late Pleistocene continental shelf, offshore Java, Indonesia; 3D seismic amplitude data, courtesy of Henry Posamentier). Fluvial incision occurred updip from the steeper fault plane. Since the shoreline at the end of forced regression remained inboard of the shelf edge, the downstream portion of the fluvial system is unincised. The fault scarp has the same effect on fluvial processes as the depositional scarp of a highstand prism (Fig. 5.16).

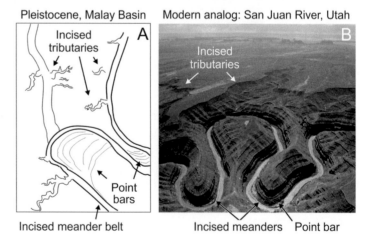

FIGURE 5.20 Incised meander belts formed during stages of relative sea-level fall. A—Subsurface example from a 3D seismic timeslice c. 77 m subsea in the Malay Basin, offshore Malaysia (Pleistocene sea-level fall; from Miall, 2002). B—Modern example of an incised meander belt, formed as a result of uplift (from Press et al., 2004). These incised meander belts inherit and preserve the morphology of pre-existing highstand rivers which become trapped within the valleys following the onset of falling-stage incision. Subsequent increases in slope gradients can no longer change the valley morphology.

fluvial valley incision, across the shelf, could only have occurred during three, rather short time intervals when the entire shelf and upper slope were exposed subaerially (Posamentier, 2001, Fig. 5.23). Consequently, incised valleys such as the one captured in Fig. 5.18 are exceptions, and the majority of Late Pleistocene fluvial systems across the Java shelf were unincised (Fig. 5.23).

The separation between incised and unincised fluvial systems is important for the evaluation of all depositional systems of the falling-stage systems tract. For example, stages of incomplete exposure of the shelf, as inferred from the presence of unincised fluvial systems, are prone to the accumulation of mud-rich sediments in the deep-water environment, as opposed to stages of full exposure of the shelf which likely result in the formation of deep-water conventional reservoirs (Fig. 4.20). The definition of criteria for the correct identification of incised vs. unincised fluvial systems is therefore highly important (Fig. 5.24). A detailed imagery of subsurface fluvial systems, as afforded by 3D seismic, well-log and core data, is necessary to document their stratigraphic makeup and architecture, aspect ratio, the characteristics of tributary systems, and the position of the fluvial deposits under analysis within the overall stratigraphic context. From the latter perspective, incised-valley fills form stratigraphic "anomalies," being genetically unrelated to the adjacent facies, whereas unincised channel fills integrate within the paleogeography of the juxtaposed and underlying depositional systems (Figs. 5.25 and 5.26).

Most commonly, the falling-stage systems tract is dominated by marine deposits, as fluvial systems incise or bypass the continental area behind the forced regressive shoreline (Figs. 5.27 and 5.28). However, non-diagnostic aggradation of fluvial and coastal systems during forced regression is also possible under special circumstances, where the influence of upstream controls extends to the shoreline (Figs. 4.26, 4.39B, 4.40, and 4.42C; see Chapter 4 for details). As pointed out by Plint and Nummedal (2000), the pattern of stratal offlap that is diagnostic of forced regression (Figs. 4.6 and 4.34) may also be obliterated by subsequent subaerial or transgressive ravinement erosion (Fig. 4.41). In such cases, the identification of falling-stage systems tracts may be based on the observation of sharp-based shoreface sandbodies (Fig. 4.45); compressed and downstepping shoreface facies (Fig. 4.41); detached prograding clinoforms (Figs. 5.27, 5.28, 5.29, and 5.30); long-distance progradation (Fig. 5.28); and stratigraphic foreshortening of clinoforms that coarsen in a basinward direction (Fig. 4.36) (Posamentier and Morris, 2000).

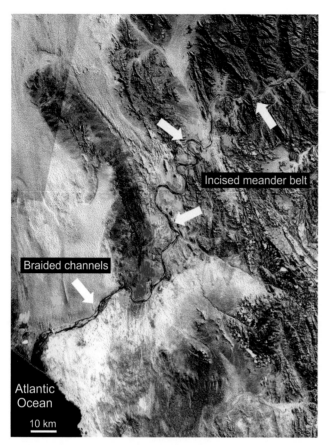

FIGURE 5.21 Satellite photograph of the Orange River along the Namibia—South Africa border (image courtesy of John Ward). The incised valley retains the meander pattern inherited from the earlier unincised river which flowed on a relatively flat and low-gradient landscape 100 Ma. The subsequent steepening of the slope gradient and the associated increase in fluvial energy did not change the valley morphology, but it is expressed in the braided character of the river outside of the areas of valley confinement.

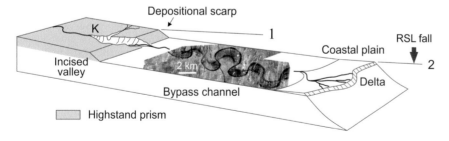

FIGURE 5.22 Unincised (bypass) fluvial system on a continental shelf that is not fully exposed by the fall in relative sea level (offshore Java, Indonesia; modified after Posamentier, 2001; seismic image courtesy of Henry Posamentier). The seismic amplitude horizon slice shows an unincised high-sinuosity channel across the Miocene shelf of the Java Sea, down-dip of the highstand prism. Abbreviations: RSL—relative sea level; K—fluvial knickpoint.

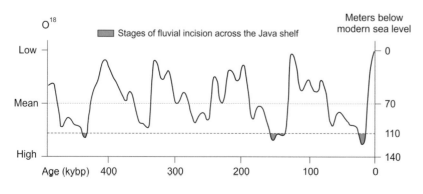

FIGURE 5.23 Late Pleistocene—Holocene sea-level curve based on oxygen isotope data (from Bard et al., 1990). The Java shelf edge is 110 m below the modern sea level. A fall of less than 110 m below the present level results in the formation of unincised fluvial systems (Figs. 5.16A and 5.22), whereas a fall in excess of 110 m results in the formation of incised valleys across the shelf (Fig. 5.16C; Posamentier, 2001). The last 0.5 My in the evolution of the Java shelf were dominated by unincised fluvial systems.

FIGURE 5.24 Criteria to differentiate between incised-valley fills and unincised fluvial systems. The presence of incised tributaries, best observed on 3D seismic horizon slices, is diagnostic of incised valleys (Figs. 3.21, 5.18, and 5.20). Incised-valley fills form stratigraphic "anomalies" which disrupt the continuity of stratigraphic markers and the facies associations predicted by Walther's Law (Fig. 5.25).

Criteria \ Systems	Incised-valley fills	Unincised fluvial systems
Stratigraphic architecture	Complex, involving fluvial, estuarine, and marine systems	Simple, commonly including only fluvial deposits
Width-to-thickness ratio	Low, commonly less than 200:1	High, potentially close to 1000:1
Tributaries	Incised	Unincised
Well-log response	Anomalous, showing a lack of correlation with adjacent units	Good correlation with adjacent units
Well-log markers	Commonly truncated by valley incision	Preserved in a relatively conformable succession
Petroleum potential	High	Average

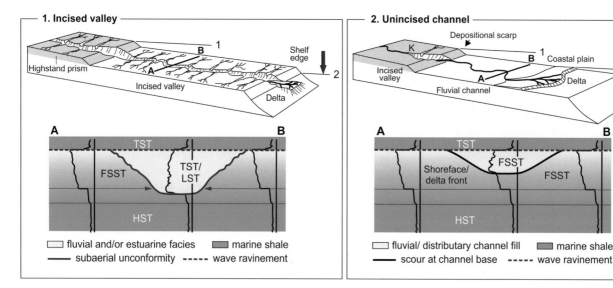

FIGURE 5.25 Synthetic gamma-ray logs illustrating the stratigraphic context of incised-valley vs. unincised-channel fills (not to scale; modified after Posamentier and Allen, 1999). Incised-valley fills are younger than the laterally juxtaposed deposits into which the valley is cut. Unincised channel fills are genetically related to the adjacent facies; in this example, the channel is part of a fluvial or delta plain environment which progrades over delta front facies. Abbreviations: RSL—relative sea level; K—upstream-migrating fluvial knickpoint; HST—highstand systems tract; FSST—falling-stage systems tract; LST—lowstand systems tract; TST—transgressive systems tract.

The shoreface clinoforms of the falling-stage systems tract are distinctively different from the normal regressive shoreface of the lowstand and highstand systems tracts (Fig. 4.54). The defining features, which can be observed consistently irrespective of any nondiagnostic variability, are the downstepping trajectory of the shoreline and shoreface facies, and the sharp-based morphology of the foreset deposits (Figs. 4.4, 4.30, 4.45, 4.46, 4.47, and 4.54). In energy-dominated settings, in which the majority of shoreface deposits accumulate, the forced regressive foresets are also unconformably based (Figs. 4.44A and 4.45), except for the earliest clinoform which overlies the proximal portion of the basal surface of forced regression, which is commonly preserved

(Figs. 5.14, 5.31, and 5.32). Progradation during relative sea-level fall commonly results in the development of compressed (i.e., with a thickness less than the depth of the fairweather wave base) shoreface successions, as opposed to the expanded (i.e., with a thickness greater than the depth of the fairweather wave base) normal regressive shoreface successions (Fig. 5.31). Forced regressive foresets in supply-dominated settings may also be expanded in the early stages of relative sea-level fall, when the toe of the clinoforms can reach below the wave base (Figs. 4.44B and 4.47), although concurrent bottomset aggradation and relative sea-level fall eventually leads to water shallowing and foreshortening (clinoform compression; Fig. 4.44B).

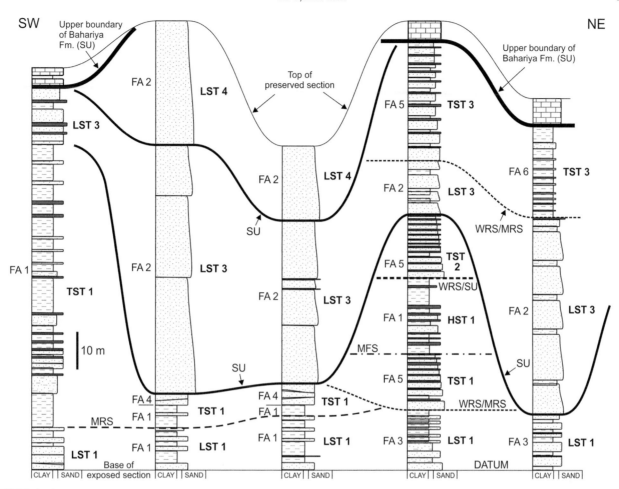

FIGURE 5.26 Sequence stratigraphic framework of the Lower Cenomanian Bahariya Formation in the Bahariya Oasis, Western Desert of Egypt (from Catuneanu et al., 2006; cross-section length: c. 100 km). The erosional relief generated by incised valleys explains the abrupt facies shifts that occur laterally, as well as the absence of some systems tracts in areas affected by fluvial erosion. The lowstand fluvial topsets contribute to the peneplanation of the incised-valley topography. Abbreviations: LST—lowstand systems tract; TST—transgressive systems tract; HST—highstand systems tract; SU—subaerial unconformity; MRS—maximum regressive surface; MFS—maximum flooding surface; WRS/MRS—wave-ravinement surface that replaces the maximum regressive surface; WRS/SU—wave-ravinement surface that replaces the subaerial unconformity. Facies associations (FA): FA 1—low-energy fluvial systems; FA 2—high-energy fluvial systems; FA 3—aggrading and prograding delta plain; FA 4—beach deposits; FA 5—shoreface facies; FA 6—glauconitic shelf shales.

The depositional elements which are most typical of forced regression, which include offlapping shoreface clinoforms, shelf deposits, shelf-edge deltas, and slope and basin-floor fans (Figs. 5.12, 5.27, 5.28, and 5.33), are not necessary found in every falling-stage systems tract. The elements that may accumulate during a forced regression depend on the tectonic and depositional settings, the magnitude of relative sea-level fall, the location of the shoreline relative to the shelf edge, and the bathymetry at the shelf edge. As the shoreline approaches the shelf edge during forced regression, the transfer efficiency of riverborne sediment to deep water improves and results in predicable changes in the dominant types of mass-transport processes (Figs. 4.20, 5.27, and 5.28). The preservation of forced regressive deposits depends on the magnitude

of subsequent erosion associated with processes of subaerial exposure, transgressive wave/tide ravinement, and deep-water mass transport. The shallow-water facies that accumulate during forced regression are particularly susceptible to subsequent erosion (fluvial incision, eolian deflation, transgressive ravinement) in high-energy settings, and where the relative sea level falls below the elevation of the shelf edge (Fig. 4.6).

In the majority of siliciclastic settings, the bulk of the falling-stage systems tract accumulates in the deep-water environment, and the bulk of the submarine fan complex belongs to the falling-stage systems tract; i.e., the process of "lowstand shedding," whereby most riverborne sediment bypasses the shelf and is shed to the slope during relative sea level fall (Posamentier and Allen, 1999; Posamentier and Kola, 2003; Sweet et al.,

Early falling-stage systems tract: shoreline progradation & downstepping

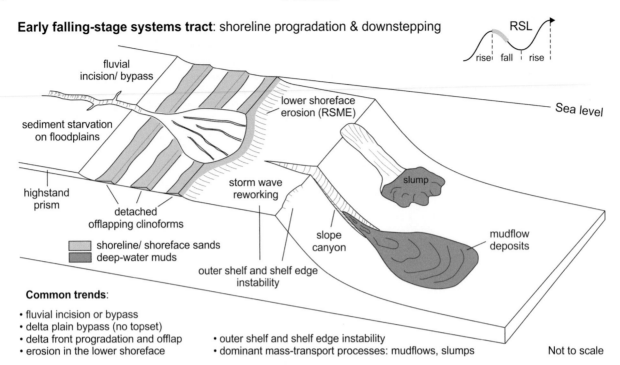

FIGURE 5.27   Depositional processes and products of the early falling-stage systems tract. Most riverborne sediment is retained on the shelf within coastal and shoreface systems. Fine grained sediment on the outer shelf is remobilized by the falling storm wave base and delivered to the deep-water environment by mudflows and slumps. Departures from these general trends are possible and reflect the variability of the tectonic and depositional settings (details in Chapters 4 and 8). Abbreviations: RSL—relative sea level; RSME—regressive surface of marine erosion.

Late falling-stage systems tract: shoreline progradation & downstepping

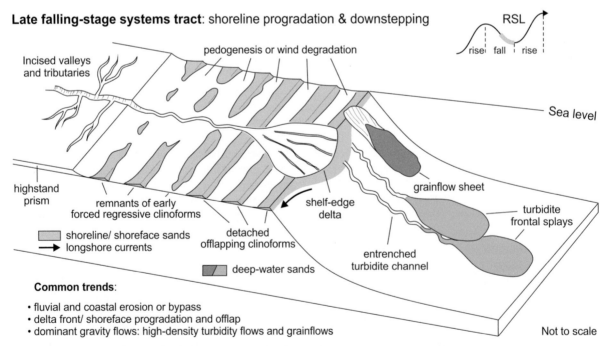

FIGURE 5.28   Depositional processes and products of the late falling-stage systems tract. Most riverborne sediment is delivered to the deep-water setting as the shoreline approaches the shelf edge. A shoreline located at the shelf edge promotes grainflows, which result from the collapse of coastal systems (Fig. 4.20). Departures from these general trends are possible and reflect the variability of the tectonic and depositional settings (details in Chapters 4 and 8). Abbreviation: RSL—relative sea level.

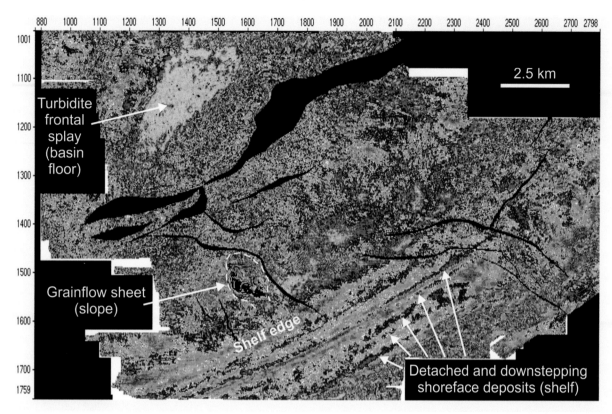

FIGURE 5.29   Seismic horizon slice (amplitude extraction) along a subaerial unconformity and its correlative conformity, showing forced regressive shoreface and gravity-flow deposits (from Catuneanu et al., 2003; image courtesy of PEMEX). Lithologies include sandstones (blue) and mudstones (orange).

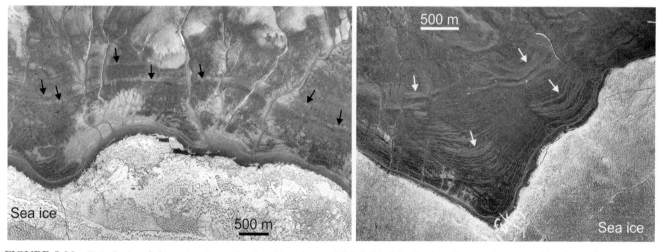

FIGURE 5.30   Detached and downstepping paleoshorelines (arrows) left behind by the forced regression caused by the Holocene glacio-isostatic rebound (Dundas Peninsula, Melville Island, Canadian Arctic Archipelago).

2019, Fig. 5.34). During early fall, when the shoreline is still far from the shelf edge, the riverborne sediment may not yet reach the shelf edge. However, the lowering of the storm wave base causes instability on the outer shelf, which triggers mass-transport processes into the deep-water environment. These mass-transport deposits include mainly the fine-grained sediment accumulated

on the outer shelf during the previous highstand normal regression, as well as during the earliest phases of forced regression, and are represented by slumps and cohesive debris flows (mudflows) (Fig. 5.27). During the late stages of relative sea-level fall, as the shoreline approaches the shelf edge the riverborne sediment can be delivered to the slope, most commonly

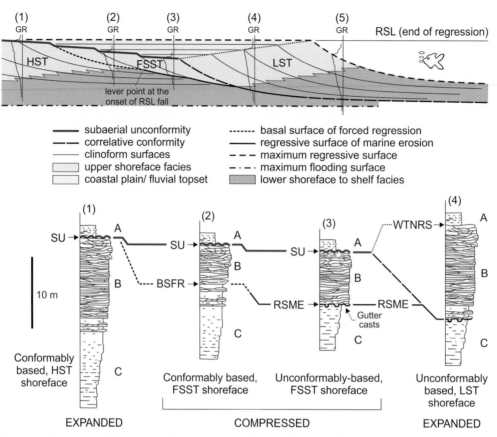

FIGURE 5.31   Compressed vs. expanded shoreface deposits in wave-dominated settings (modified after Plint, 1988; Posamentier et al., 1992b; Walker and Plint, 1992; Posamentier and Allen, 1999). Forced regressive shoreface clinoforms are "compressed" due to progradation during relative sea-level fall, which limits the amount of accommodation available for shoreface deposition. In contrast, normal regressive clinoforms are "expanded" due to progradation during relative sea-level rise, which increases the amount of accommodation available for shoreface aggradation. Log 5 has the same conformably based profile as log 1. Whether conformably or unconformably based, the forced regressive clinoforms are characteristically sharp-based (Figs. 4.4, 4.45, 4.46, and 4.47). In contrast, a normal regressive shoreface is typically conformably and gradationally based (Fig. 4.30), except for the earliest lowstand clinoform which sits on top of the most distal portion of the RSME (log 4). Fundamentally, the forced and normal regressive clinoforms differ in terms of their downstepping vs. upstepping facies trends in a basinward direction, respectively (Fig. 4.54). The "lever point" marks the updip termination of the RSME (Fig. 5.32). Facies: A—coastal plain; B—shoreface; C—shelf fines. Abbreviations: GR—gamma-ray log; HST—highstand systems tract; FSST—falling-stage systems tract; LST—lowstand systems tract; SU—subaerial unconformity; BSFR—basal surface of forced regression; RSL—relative sea level; RSME—regressive surface of marine erosion; WTNRS—within-trend normal regressive surface.

FIGURE   5.32   Wave-dominated shallow-marine succession showing the transition between conformably based (A) and unconformably based (B) forced regressive shoreface facies (Upper Cretaceous, Blackhawk Formation, Utah). The dashed line represents the basal surface of forced regression (preserved paleoseafloor at the onset of forced regression), and the solid line marks the regressive surface of marine erosion (see Figs. 5.31 and 5.33 for context).

as sandy gravity flows (i.e., turbidity flows and grain-flows; Figs. 4.20 and 5.28). The transition from mudflows to turbidites and grainflows during the falling stage leads to the coarsening-upward of the forced regressive submarine fans (Fig. 4.20; more details in Chapter 8).

The distinction between the muddy and sandy mass-transport deposits in a deep-water setting is critical in the process of petroleum exploration, prior to drilling. Slumps and mudflows typically have a high sediment-to-water ratio (i.e., high density), and therefore are most erosive and tend to move along the steepest slope gradients (Figs. 5.35 and 5.36). Any pieces of substrate ripped up by such flows are dragged along the seafloor and generate linear grooves that can be observed at seismic scales (Figs. 5.37 and 5.38). These basal grooves are typical of cohesive debris flows, and can be used as indicators of paleodip directions. The cohesive nature of slumps and mudflows generates an internal shear strength which causes the flows to "freeze" on deceleration. This explains the chaotic internal architecture of the deposits (Fig. 5.39), as well as the formation of structures such as compressional ridges and thrust faults within the lobes (Fig. 5.40). The high viscosity of slumps and mudflows also explains their tendency to accumulate at more proximal locations relative to turbidites, although hydroplaning can increase the distance of travel to hundreds of kilometers (Fig. 5.41).

The sandy gravity flows of the late forced regression tend to include the coarsest sediment of a stratigraphic cycle, and therefore the best conventional reservoirs of a deep-water sequence. The high-density turbidites become prevalent when the bathymetry at the shelf edge is less than the depth of the storm wave base, and grainflows occur when coastal systems reach the shelf edge (Catuneanu, 2019a, 2020a; Sweet et al., 2019, Figs. 4.20 and 5.28). While sandy flows are typically less dense than the mudflows (Fig. 5.42), turbidity currents and grainflows still generate erosion on the slope, expressed as entrenched channels and slope scars (Figs. 5.28, 5.43, and 5.44). Beyond the base of the slope, turbidite channels and splays become aggradational as the energy of the system declines on the lower gradients of the basin floor (Fig. 5.28). The higher the density of the turbidity flows the thicker the "traction carpet" (Fig. 5.42) and the less fine-grained sediment is available to extend the construction of levees into the basin. As a result, higher density turbidity flows tend to have shorter channels, and their frontal splays are typically larger and more proximally located than the frontal splays of low-density turbidity flows (Figs. 5.45, 5.46, and 5.47).

Levees are important elements in turbidite systems, which maintain the flow confined while the system is active (Fig. 5.44). However, deeply incised channels on steeper slopes may not form levees, if the finer grained sediment fractions of the flow are unable to reach the overbank area. Turbidite channels are typically narrower and deeper on the upper slope, due to the steeper seafloor gradients and higher flow velocities, and wider and shallower toward the toe of the slope and onto the basin floor as gradients and energy decrease (Beaubouef et al., 1999). These trends correlate with the common transition recorded by most turbidite systems from erosional on the slope to aggradational on the basin floor (Figs. 5.28, 5.34, and 5.43). However, all channels fill eventually, either with hemipelagic sediment following abandonment (i.e., "passive fill") or with turbidites while the system is still active (i.e., "active fill"). As turbidite aggradation starts in the distal, lower energy parts of the system, active fills tend to display a backstepping character as the proximal limit of the depositional area shifts gradually in an updip direction. At any given time, active turbidite systems may include areas of erosion, concurrent bypass and accumulation (e.g., lateral accretion of point bars), and turbidite aggradation, whose limits control the types of sediment that can be expected within the system (details below and in Chapter 8).

The falling-stage is typically the least developed systems tract in carbonate settings, where the fall in relative sea level exposes the carbonate platform and shuts down the carbonate factory. Where conditions are favorable for the precipitation of evaporites, the falling-stage systems tract can include up to km-scale successions of salt that accumulates as interior seas or lakes desiccate (e.g., the Aptian desiccation of the early South Atlantic, or the Messinian "crisis" of the Mediterranean Sea; Csato et al., 2013, 2015; Tedeschi et al., 2017; details in Chapter 8).

5.2.3.2 Lowstand systems tract

The lowstand systems tract is defined by a normal regressive stacking pattern (i.e., progradation and upstepping of the subaerial clinoform rollovers; Figs. 4.1 and 4.21) which follows a forced regression of the same hierarchical rank (Figs. 4.39, 5.9, 5.14, and 5.15). The lowstand normal regression occurs during the early stage of relative sea-level rise, when the rates of sedimentation exceed the rates of relative sea-level rise at the shoreline (Fig. 4.32; see discussion in Chapter 4).

The lowstand systems tract is bounded at the base by the subaerial unconformity and/or the correlative conformity (Fig. 4.39). Where the lowstand systems tract is followed by transgression, the upper boundary is represented by the maximum regressive surface reworked in part by the transgressive surface of erosion (i.e., the case of transgression accompanied by fluvial aggradation; Fig. 4.39A), or by a composite surface which includes the marine portion of the maximum regressive surface, the transgressive surface of erosion, and the

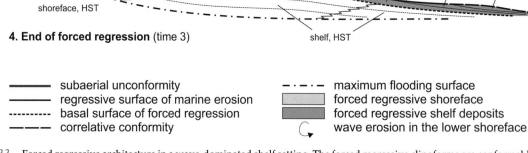

FIGURE 5.33   Forced regressive architecture in a wave-dominated shelf setting. The forced regressive clinoforms are conformably based above the basal surface of forced regression, and unconformably based where the basal surface of forced regression is reworked by the regressive surface of marine erosion. The inner shelf environment widens during forced regression, due to the combination of relative sea-level fall and shelf aggradation, and accumulates hummocky cross-stratified tempestites which aggrade during storms forming meter-scale high and hundreds of meters wide macroforms above the average concave-up seafloor profile (Catuneanu, 2003; Arnott et al., 2004). Abbreviations: HST—highstand systems tract; HCS—hummocky cross-stratification; SCS—swaley cross-stratification; FWB—fairweather wave base; SWB—storm wave base.

subaerial unconformity (i.e., the case of transgression accompanied by fluvial erosion; Fig. 4.39B). Where the lowstand systems tract is followed by forced regression, the upper boundary is represented by the subaerial unconformity and/or the basal surface of forced regression.

Lowstand systems tracts typically include a continental topset and a marine foreset and bottomset, and tend to display a concave-up shoreline trajectory (Fig. 4.51). In the most common scenario, whereby the subaerial unconformity forms during forced regression (Fig. 4.39A), the lowstand topset includes the highest

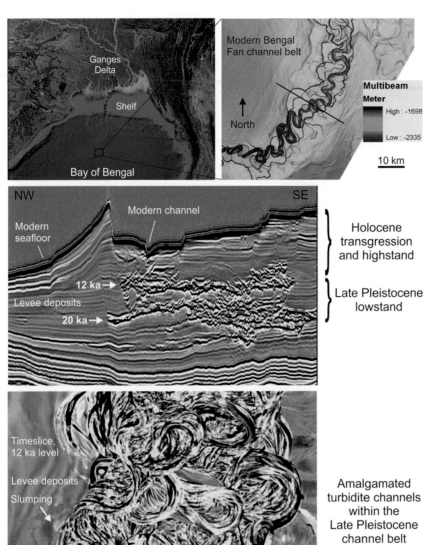

FIGURE 5.34 Lowstand shedding in a siliciclastic setting (Bay of Bengal, India; images
courtesy of Reliance Industries, India; further
details in Kolla et al., 2012). Sediment supply to
the deep-water setting was highest during the
Late Pleistocene glaciation (lowstand in relative
sea-level, when the shelf was exposed), and
decreased during the Holocene transgression
and highstand. Note the Late Pleistocene unfilled incised valley across the submerged shelf,
which still facilitates the transfer of riverborne
sediment to the present-day shelf edge. The
location of the 2D seismic line is indicated on the
top images.

energy fluvial systems of a depositional sequence. In
the less common cases in which the subaerial unconformity forms during transgression (Fig. 4.39B), the lowstand topset includes the lowest energy fluvial
systems of a depositional sequence (see discussion in
Chapter 4). In either case, since the formation of lowstand systems tracts is driven by sediment supply outpacing the rates of creation of accommodation at the
shoreline (Fig. 4.32), the depositional processes are
commonly linked to low rates of coastal progradation
and aggradation. As accommodation is generated on
both sides of the shoreline, the lowstand systems tract
is expected to include the entire suite of depositional
systems, from fluvial through to deep water (Fig. 5.48).

Coastal aggradation during lowstand normal regression commonly leads to a lowering of fluvial gradients
and energy (Fig. 5.14), which results in an overall upward decrease in grain size. The fining-upward trend
is further enhanced by the increase with time in the
rate of relative sea-level rise, which promotes the deposition and preservation of floodplain deposits. However,
if rivers remain unconfined (e.g., braided style), the lowstand topsets may be dominated by amalgamated channels overlying subaerial unconformities formed during
forced regressions, which is most common in nature
(Figs. 5.49 and 5.50). In this case, lowstand topsets accumulate on uneven landscapes sculptured by differential
erosion during relative sea-level fall (Fig. 2.81), with

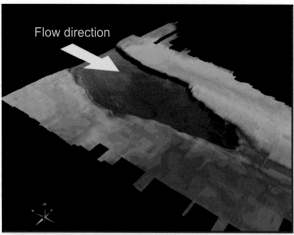

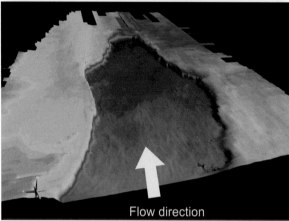

FIGURE 5.35 Time-structure map at the base of a mudflow deposit, from a 3D seismic volume (Pleistocene, Gulf of Mexico; images courtesy of Henry Posamentier). The scour is c. 30 km long, c. 12.5 km wide, and c. 240 m deep.

deposition starting within incised valleys following the onset of relative sea-level rise (Fig. 5.51). It is common that incised valleys are filled at least in part with lowstand fluvial sediments (e.g., Shanley and McCabe, 1991, 1993, 1994; Wright and Marriott, 1993; Gibling and Bird, 1994, Figs. 5.26 and 5.51), leading to the development of lowstand topsets with a discontinuous geometry and significant changes in thickness along dip and strike (Fig. 5.25).

Where forced regression does not result in valley incision, or where incised valleys are filled completely by lowstand fluvial sediments, the lowstand topset can also extend across the interfluve areas between the unincised or incised rivers of the falling stage (Fig. 5.48). There are also cases where the lowstand fluvial deposits are missing from the infill of incised valleys, due to either nondeposition or erosion during subsequent transgression. In such cases, the subaerial unconformity at the base of the incised valley is reworked by a transgressive surface of erosion, and the incised valley may be filled entirely with transgressive deposits (e.g.,

Dalrymple et al., 1992; Ainsworth and Walker, 1994). Where present, the attribution of fluvial deposits to the lowstand topset requires evidence of a process—response relationship between the aggradation of river-borne sediment and the lowstand normal regression of the shoreline (e.g., using regional cross sections of correlation or seismic lines; Kerr et al., 1999). Fig. 5.52 provides examples of sedimentary structures that document the marine influence on fluvial processes at the time of sedimentation. Fluvial deposits that accumulate independently of relative sea-level changes are better ascribed to high- or low-amalgamation systems tracts which form in upstream-controlled settings (Figs. 3.11 and 4.70; discussion below).

A more detailed discussion of the fluvial styles that can be associated with lowstand topsets is presented in Chapter 4. Irrespective of the type of river system, the aggradation of lowstand fluvial strata starts from the river mouth (i.e., the lowstand delta) and gradually extends upstream by onlapping the subaerial unconformity (Figs. 5.12 and 5.13). This trend is triggered by the increase in coastal elevation during the normal regression, which requires a continuous readjustment of the graded fluvial profile, starting from the coastline where the energy of the rivers is lowered first. The pattern of fluvial onlap widens the stratigraphic hiatus of the subaerial unconformity in a landward direction, as the overlying fluvial strata become younger toward the basin margin (Fig. 5.13). This pattern of deposition of the lowstand topset generates a prism of fluvial sediment with a wedge-shaped geometry, which tapers off upstream toward the youngest point of fluvial onlap (Fig. 5.12).

The updip extent of the lowstand topset depends on the magnitude of relative sea-level rise (i.e., the amount of coastal aggradation), topographic gradients, and sediment supply. Low-gradient shelf settings are conducive to fluvial aggradation over large areas, whereas a steep topography (e.g., in a high-gradient ramp setting, such as a fault-bounded basin margin) restricts the area of fluvial aggradation (Blum and Tornqvist, 2000). In the latter case, the subaerial unconformity may be overlain directly by transgressive fluvial strata over much of its extent (Embry, 1995; Dalrymple, 1999). Data from the US Gulf Coastal Plain indicate that the landward limit of fluvial onlap also correlates with the amount of sediment supply, and is inversely proportional to the gradient of the onlapped floodplain surface. This distance may vary significantly, from c. 40 km in the case of the steeper gradient and low-supply Nueces River to at least 300—400 km for the low-gradient and high-supply Mississippi River (Blum and Tornqvist, 2000).

In shallow-water settings, the lowstand systems tract corresponds to the last stage of shoreline progradation, with the maximum regressive surface placed at the top

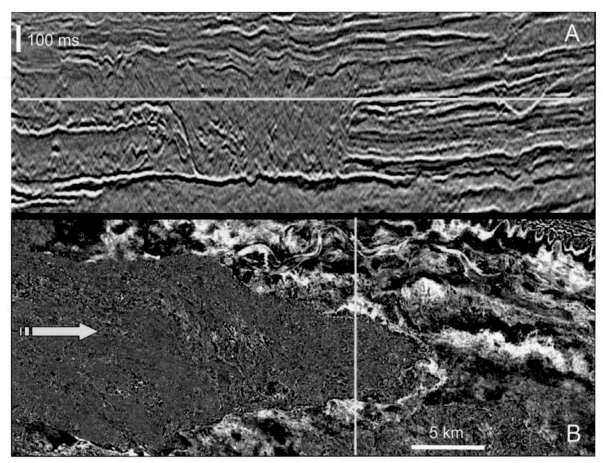

FIGURE 5.36 Transverse (A) and planar (B) sections through a Pleistocene mudflow lobe at the toe of a continental slope (Gulf of Mexico; images courtesy of Henry Posamentier). In transverse view, the lobe resembles the shape of a channel due to the substantial erosional relief; however, the flow is not channelized.

of a coarsening-upward trend (Fig. 5.13). The shoreface deposits of the lowstand systems tract are generally gradationally and conformably based, due to the upstepping of the wave base during normal regression, except for the earliest lowstand clinoform which overlies the youngest segment of the regressive surface of marine erosion (Figs. 5.14 and 5.31). Beyond the fairweather wave base, the extent of shelf facies may be limited due to the potential proximity of the shoreline to the shelf edge at the end of forced regression and/or during the lowstand normal regression (Fig. 5.12). In this case, the subtidal facies of the shoreface or shelf-edge deltas may pass directly into deep-water slope facies, which consist primarily of gravity-flow deposits (Figs. 4.6, 5.12, and 5.48).

The vertical profiles in deep-water settings do not replicate the coarsening-upward trends of the shallow-water systems, as the deep-water lowstand normal regressive deposits are typically finer grained than the underlying late forced regressive sediments (Figs. 4.20 and 5.13). This is due to the selective retention of the coarser fraction of the riverborne sediment within the

lowstand topset as fluvial aggradation resumes or accelerates following the onset of relative sea-level rise, which reduces both the volume and the caliber of the sediment that can reach the shelf edge. As a result, the turbidity flows of the lowstand are commonly of lower density than those of the falling stage (Fig. 5.47). The maximum grain size of the riverborne sediment that reaches the shelf edge is also expected to decline during the lowstand normal regression, due to the lowering with time of the fluvial gradients and related fluvial competence (Figs. 4.20 and 5.13). Grainflows may continue to occur during the lowstand normal regression, for as long as coastlines are still located at the shelf edge (Fig. 5.48).

The lower density turbidity flows of the lowstand are less erosive than the denser flows of the falling stage, but can still form entrenched channels on the slope (Fig. 5.53). Beyond the toe of the slope, turbidite systems aggrade on the basin floor, due to the loss of energy (Figs. 5.54, 5.55, and 5.56). The buoyancy of the flows can generate high-sinuosity leveed channels starting with the slope (Fig. 5.57) and continuing on the basin

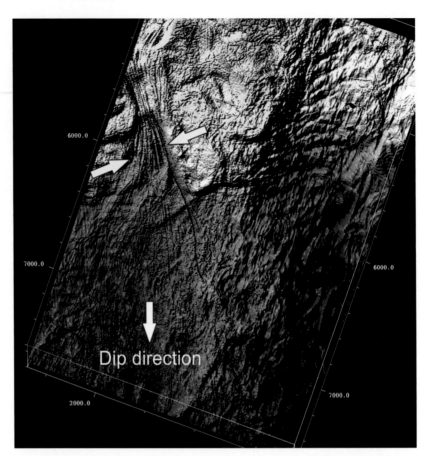

FIGURE 5.37 Basal grooves generated by mudflows, diverging in a downdip direction (seismic horizon slice; Mahanadi Basin, offshore India; image courtesy of Reliance Industries, India).

FIGURE 5.38 Seismic timeslice illustrating grooves generated by the passage of mudflows across an unconsolidated substrate (seismic amplitude domain; Pleistocene, eastern Gulf of Mexico; image courtesy of Henry Posamentier). The image is diachronous, as it includes an older turbidite channel with sediment waves in the lower part of the 3D volume.

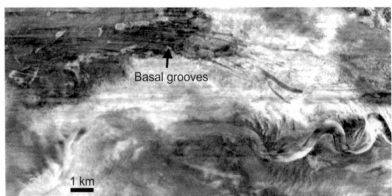

floor (Figs. 5.41, 5.54, and 5.58). Levees are unique to turbidity flows (i.e., they do not form in relation to any other mass-transport processes), and are built by the finer grained sediment fractions which are able to escape from the flow onto the overbank (Figs. 5.34, 5.44, and 5.56). The length of the leveed channels is proportional to the amount of finer grained sediment made available at the shelf edge. As a flow runs out of finer grained sediment, the height of the levees decreases in a downdip direction until it becomes insufficient to maintain the current confined (Fig. 5.59); at that point, the flow becomes unconfined, losing momentum and generating a frontal splay (Figs. 5.45, 5.46, 5.58, and 5.60). Part of the flow can also break through the levees, especially at the outer bends of the high-sinuosity channels, generating crevasse splays and sediment waves on the overbank (Figs. 5.61 and 5.62). In the absence of barriers to flow, and given sufficient supply of fine sand and

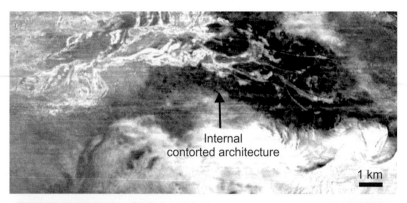

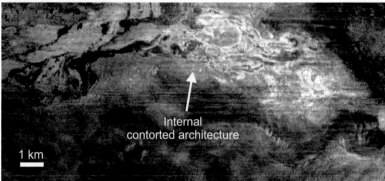

FIGURE 5.39 Internal contorted architecture of a mudflow deposit, caused by freezing on deceleration (Pleistocene, eastern Gulf of Mexico; images courtesy of Henry Posamentier). The high density of the flow prevents sediment settling from suspension and the development of stratification. Instead, mudflow deposits display a chaotic makeup, as well as compressional features (Fig. 5.40). These seismic timeslices are above the image in Fig. 5.38.

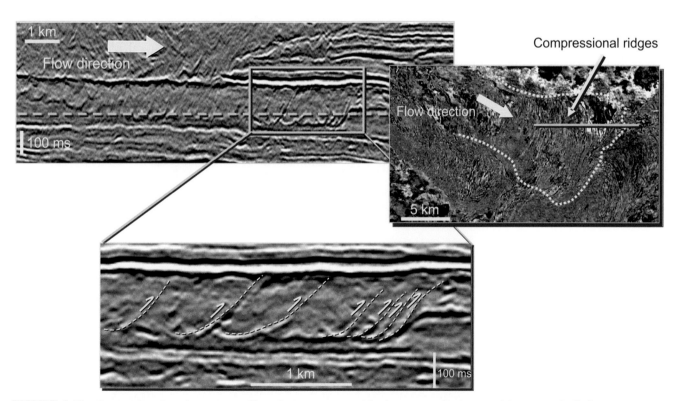

FIGURE 5.40 Compressional structures in mudflow deposits, generated by freezing on deceleration (Pleistocene, Gulf of Mexico; images courtesy of Henry Posamentier). Due to its plastic rheology caused by the cohesion between grains, the flow "freezes" when the gravitational shear becomes less than the internal shear strength of the sediment/water mixture. The frontal part of the flow freezes first, when reaching areas of lower slope gradient, while the tail part of the flow still moves. This causes the formation of internal thrust faults (see cross-sectional views), which are expressed as compressional ridges in plan view.

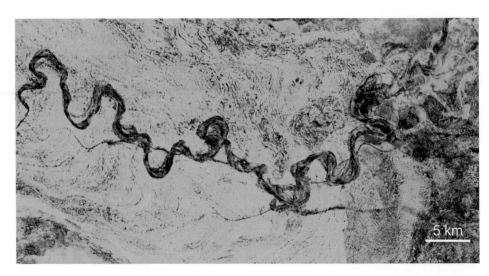

FIGURE 5.41 Channel of low-density turbidity current at the top of a mudflow deposit in a basin-floor setting (Mahanadi Basin, offshore India; spectral decomposition image, courtesy of Reliance Industries, India). Note the compressional structures of the mudflow deposit, which are typical of cohesive debris flows. The mudflow travelled over 200 km to reach this location, aided by hydroplaning.

FIGURE 5.42 Types of gravity flows. The erosive power of a gravity flow is proportional to its density and velocity. At equal velocities, denser flows are more "abrasive." Travel distances generally increase with the decrease in density, although hydroplaning can extend the mobility of impermeable mudflows. Grainflows and turbidity flows are permeable, and therefore less susceptible to hydroplaning. Grainflows are generally restricted to slope settings, whereas mudflows and turbidity flows can travel onto the basin floor. Traction carpets refer to the denser, non-turbulent lower part of turbidity flows. Grainflows are essentially stand-alone "traction carpets," equivalent to the division A of Bouma sequences. The higher the density of turbidity flows the shorter the channels and the closer the affinity to grainflows. Abbreviations: CDF—cohesive debris flows (mudflows); NDF—non-cohesive debris flows (grainflows); HDT—high-density turbidity flows; LDT—low-density turbidity flows.

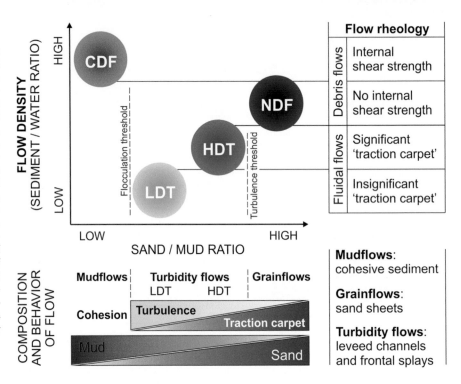

mud, low-density turbidity currents can travel farther than any other type of gravity flow, potentially for thousands of kilometers; e.g., the present-day turbidity channels fed by the Ganges Delta in the Bay of Bengal are over 2,700 km long.

The end of lowstand deposition is marked by the onset of transgression, which results in the formation of estuaries and other backstepping coastal systems. The change from progradation to retrogradation (i.e., the maximum regressive surface) is most evident in areas adjacent to the shoreline (e.g., at the top of coarsening-upward trends in shallow-water systems, or at the base of the earliest estuarine facies in coastal

systems), although criteria can be defined to identify this limit in fully fluvial and deep-water systems as well (e.g., Kerr et al., 1999; Catuneanu, 2020a; see examples in Chapter 6).

5.2.3.3 Transgressive systems tract

The transgressive systems tract is defined by a retrogradational stacking pattern, which develops in relation to the shoreline transgression (Figs. 4.1, 4.21, 5.9, 5.14, and 5.15). Transgression occurs during relative sea-level rise, when the rates of sedimentation are lower than the rates of relative sea-level rise at the shoreline (4.32; see discussion in Chapter 4).

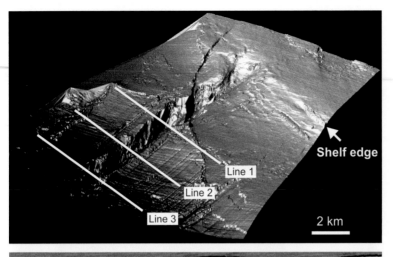

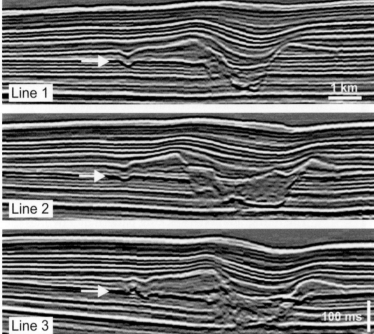

FIGURE 5.43 Entrenched channel of a high-density turbidity flow in a slope setting (Pleistocene, Gulf of Mexico; data courtesy of Henry Posamentier). The horizon slice is mapped along the basal surface of forced regression in Fig. 4.6. Note the dip-oriented grooves generated by the passage of mudflows in the earlier stages of forced regression.

The transgressive systems tract is bounded at the base by the maximum regressive surface reworked in part by the transgressive surface of erosion, and at the top by the maximum flooding surface (Fig. 4.39). Where transgression is accompanied by fluvial aggradation, the fluvial portion of the transgressive systems tract may also rest directly on top of the subaerial unconformity, beyond the updip termination of the lowstand topset (Fig. 4.39A). Most commonly, transgressive systems tracts record the highest rates of fluvial aggradation (Shanley and McCabe, 1993; Wright and Marriott, 1993; Emery and Myers, 1996; Kerr et al., 1999), leading to the lowest rates of riverborne sediment supply to the marine environment. This results in the sediment starvation of the seafloor and the formation of marine condensed sections, which are most often associated with transgression (Loutit et al., 1988).

Where transgression is accompanied by fluvial erosion (Figs. 4.39B and 4.56), the rates of riverborne sediment supply to the shoreline are the highest of the entire stratigraphic cycle. However, this does not necessarily translate into high sediment supply to the shelf edge, particularly in the case of wider shelves which can retain the sediment within the shallow-water systems. The efficiency and the mechanisms of sediment transfer from the shoreline to the shelf edge need to be assessed on a case-by-case basis, and depend on a variety of factors including the physiography of the basin (e.g., shelf width), the presence or absence of unfilled incised valleys across the shelf that could serve as

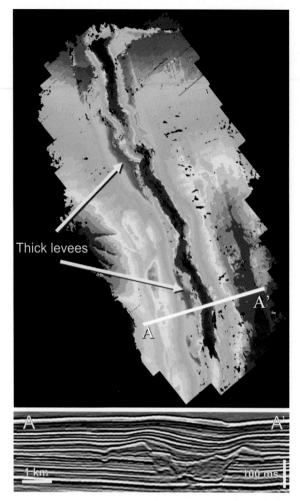

FIGURE 5.44 Turbidite leveed channel in an upper slope setting (Late Pleistocene, De Soto Canyon area, Gulf of Mexico; images courtesy of Henry Posamentier). As seen on the seismic isochron map, channel sinuosity is higher than levee sinuosity. The 2D seismic line (A−A′) reveals the wedge-shaped geometry of levee deposits. Levees are the elements that keep the flow confined during the motion of the sediment−water mixture. As levees are built by the finer grained sediment fractions of the flow, turbidity currents become unconfined when the systems run out of mud, leading to the formation of frontal splays at the end of leveed channels (Fig. 5.45). Due to the higher sand/mud ratio of the high-density turbidity flows, their frontal splays tend to be located closer to the toe of the slope (i.e., insufficient amount of mud to sustain the formation of levees over large distances). Notwithstanding other variables, such as the seascape morphology, the location of frontal splays may thus provide additional clues to evaluate the type of turbidity flows (high- vs. low-density) at syn-depositional time.

conduits of sediment transport, the magnitude of storms (i.e., depth of the storm wave base), and the tidal range. Riverborne sediment supply to the deep-water environment tends to remain lowest at the time of maximum flooding, despite the high supply to the shoreline, due to the location of the shoreline at the greatest distance from the shelf edge at the end of transgression (Fig. 4.20). In this scenario, the shoreline at the end of

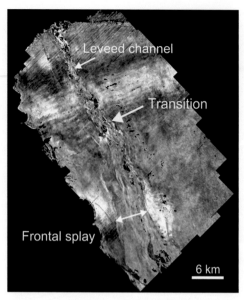

FIGURE 5.45 Transition from turbidite channel to frontal splay in a slope setting (Late Pleistocene, De Soto Canyon area, Gulf of Mexico; image courtesy of Henry Posamentier). The transition from leveed channel to frontal splay occurs where the height of the levees, which typically decreases in a downdip direction, is no longer sufficient to maintain the flow confined. This transition is placed more proximally in the case of higher density turbidity flows, due to the lower mud content of the sediment−water mixture. The leveed channel in the upper part of the image is 1.8 km wide.

transgression marks the updip limit of the transgressive systems tract (Fig. 4.39B).

The retrogradational stacking pattern that defines the transgressive systems tract is most often expressed as a fining-upward trend in both continental and marine settings, with the bulk of the sediment trapped within fluvial and/or coastal systems. Where transgression is conducive to fluvial aggradation (Fig. 4.39A), the transgressive fluvial deposits accumulate during a stage of lowering of topographic gradients and fluvial energy, at the same time with a rise of the water table relative to the topographic profile. These conditions are increasingly favorable to the accumulation of peat during the transgression, leading to the formation of the thickest and most regionally extensive coal beds at the stratigraphic position of the maximum flooding surface (Fig. 5.50). This upper boundary of the transgressive systems tract also correlates to the top of the youngest estuarine facies. In contrast, the maximum regressive surface is placed at the base of the earliest estuarine facies, and should not be confused with the facies contact between transgressive fluvial facies and the overlying estuarine strata, which develops within the transgressive systems tract (Fig. 5.14). This facies contact is diachronous, becoming progressively younger in an updip direction (more details in Chapter 6).

Estuaries are diagnostic of transgression, and act as a buffer between the riverborne sediment and the marine

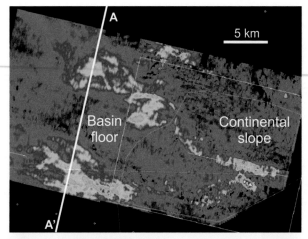

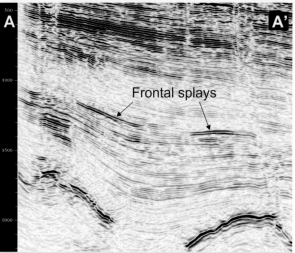

FIGURE 5.46 Frontal splays at the toe of the slope, indicative of high-density turbidity flows (images courtesy of PEMEX). The lower mud content of the high-density turbidity flows prevents the formation of levees over large distances, triggering the transition from leveed channels (confined flow) to frontal splays (unconfined flow) at more proximal locations.

environment (Fig. 5.63). The cut-off of sediment supply to the marine environment during transgression leads to the formation of condensed sections or even unconformities (e.g., hardgrounds) on the outer shelf (Loutit et al., 1988; Galloway, 1989). The low sediment supply coupled with a rapid relative sea-level rise also lead to instability at the shelf edge, particularly during the late stages of transgression (Fig. 5.64). As a result, the transgressive systems tract may consist of two distinct wedges separated by an area of non-deposition or erosion, one on the shelf (fluvial to shallow-water deposits) and one in the deep-water setting (Figs. 5.12, 5.64, and 5.65). Both wedges expand in a landward direction during transgression, generating fluvial and coastal onlap on the shelf, and marine onlap in the slope setting (i.e., the "slope apron" of Galloway, 1989, Figs. 4.6 and 5.65). Among the three types of onlap,

only the coastal onlap (i.e., shallow-water "healing phase" deposits onlapping the wave-ravinement surface; Fig. 4.55) is unequivocally diagnostic of transgression (see Chapter 4 for details).

The fluvial portion of the transgressive systems tract most commonly includes isolated channels engulfed within floodplain fines, in contrast to the more amalgamated channels of the underlying lowstand topset (Fig. 5.50; Wright and Marriott, 1993; Kerr et al., 1999). This change in the degree of channel amalgamation occurs where the rates of coastal aggradation increase following the onset of transgression, as observed in many stratigraphic studies (Shanley et al., 1992; Shanley and McCabe, 1993; Kerr et al., 1999). The increase in the rates of coastal aggradation across the maximum regressive surface parallels the increase in the rates of relative sea-level rise, as long as sufficient sediment supply is available to maintain a shoreline trajectory that is steeper than the fluvial profile. This condition is met in most documented field studies, whereby the dominance of downstream controls extends into the continental realm (i.e., typically in low-gradient settings; see discussion in Chapter 4). Coastal aggradation lowers the gradient of the fluvial graded profile, which triggers adjustments in fluvial style at the same time with an upward increase in tidal influence, palynological marine index, and the frequency of occurrence of crevasse splays during the transgression.

Transgressive deposits may form a significant portion of incised-valley fills, and may also accumulate in the interfluve areas of filled incised valleys. Where incised valleys are still unfilled following the lowstand normal regression, their downstream reaches are commonly converted into estuaries at the onset of transgression (Dalrymple et al., 1994). In such cases, the maximum regressive surface at the base of the transgressive systems tract may be preserved at the contact between the lowstand fluvial topset and the overlying estuarine facies, or it may be reworked by tidal- or wave-ravinement surfaces (Rahmani, 1988; Allen and Posamentier, 1993; Ainsworth and Walker, 1994; Breyer, 1995; Rossetti, 1998; Cotter and Driese, 1998). In this setting, the maximum regressive surface can be mapped at the abrupt change from the coarser fluvial deposits of the lowstand topset ("clean" and blocky sand on well logs) to the overlying finer grained and tidally influenced estuarine facies (Fig. 5.66; Allen and Posamentier, 1993). In contrast to the "cleaner" sands of the lowstand topset, the transgressive fluvial deposits are typically finer grained and interbedded with estuarine facies.

In coastal settings, the transgressive systems tract may include open shoreline systems (e.g., backstepping beaches, lagoon—barrier island systems, tidal flats), estuaries, and even proper deltas in the case of larger rivers (Figs. 5.67 and 5.68). The type of coastline may

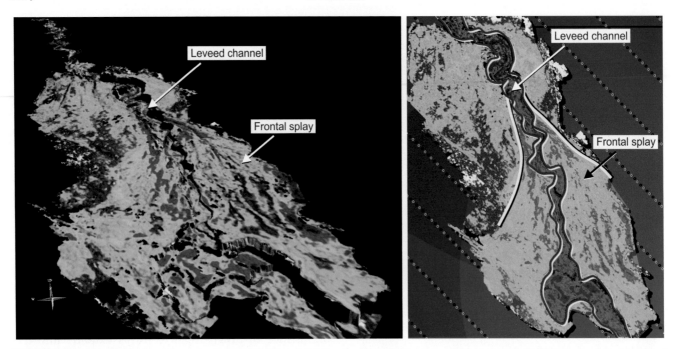

FIGURE 5.47   Turbidite channel crossing the frontal splay of a higher density turbidite system (Pleistocene, Gulf of Mexico; images courtesy of Henry Posamentier). Where unobstructed, the lower density turbidity flows travel farther into the basin than the higher density flows, due to their ability to maintain the construction of levees over larger distances. These images capture the shift from the late falling-stage frontal splays to the lowstand leveed channels (i.e., the correlative conformity in Fig. 4.6).

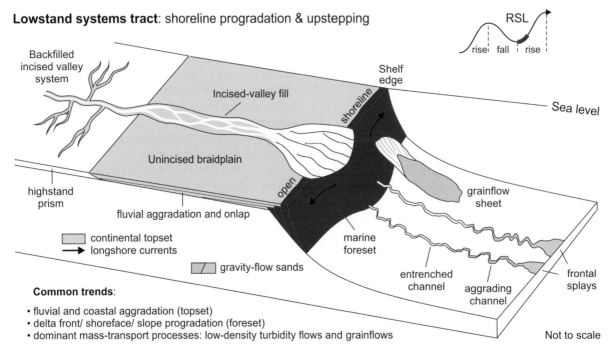

FIGURE 5.48   Depositional processes and products of the lowstand systems tract. In contrast to the falling-stage systems tract (Figs. 5.27 and 5.28), the lowstand sediment is more evenly distributed across the basin, with aggradation taking place in all depositional environments. The lowstand topset expands landward during the progradation and upstepping of the shoreline, generating fluvial onlap. The aggradation of the fluvial topset reduces sediment supply to the shelf edge; as a result, the grainflows and turbidity flows of the lowstand systems tract are finer grained than those of the late falling stage (Fig. 5.47). Abbreviation: RSL—relative sea level.

FIGURE 5.49   Outcrop examples of lowstand fluvial topsets (Upper Cretaceous, Castlegate Formation, Utah). A—Amalgamated channels, unconformably overlying forced regressive shoreface deposits (FSST: Blackhawk Formation); B, C, D, E—amalgamated channels of braided fluvial systems; F—climbing dunes indicating high sediment supply in the high-energy braided streams. Abbreviations: FSST—falling-stage systems tract; LST—lowstand systems tract; SU—subaerial unconformity.

also change from transgressive to normal regressive as a function of the shifting balance between the rates of relative sea-level rise and the rates of sedimentation along the shoreline (Fig. 5.68). The limit between coeval transgressive and normal regressive coastlines may be constrained by structural elements, or it may shift in response to variations in the rates of subsidence and sedimentation through time, leading to the formation of diachronous systems-tract boundaries (Catuneanu, 2019a). Wehr (1993) noted that "spatial variations in sedimentation rates … might locally shift the onset of

prograadation to an earlier time and delay the onset of retrogradation." Subsequent work on the changes in depositional trends along a coastline provided further details on the 3D variability of the stratigraphic framework (e.g., Martinsen and Helland-Hansen, 1995; Helland-Hansen and Martinsen, 1996; Catuneanu et al., 1998b; Madof et al., 2016; Schultz et al., 2020; Zecchin and Catuneanu, 2020; more details in Chapter 7).

The common element of all transgressive coastlines is the retrogradation of the open shoreline systems (Fig. 5.68). Within the overall transgressive setting,

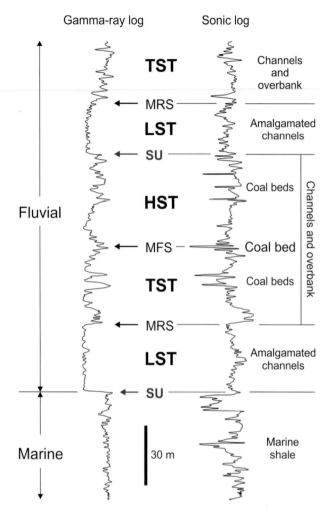

FIGURE 5.50 Systems tracts and bounding surfaces in a downstream-controlled fluvial succession (Jurassic, Sverdrup rift basin; modified after Embry and Catuneanu, 2001). The delineation of systems tracts relies on the observation of changes in fluvial style, marine influence, and water table (see text for details). Further validation is provided by correlation with the coeval shallow-water systems in a downdip direction. Abbreviations: LST—lowstand systems tract; TST—transgressive systems tract; HST—highstand systems tract; SU—subaerial unconformity; MRS—maximum regressive surface; MFS—maximum flooding surface.

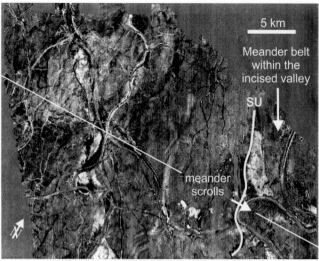

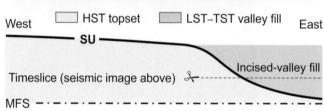

FIGURE 5.51 Amplitude extraction map (timeslice at 196 ms below the sea level) in the Malay Basin, offshore Malaysia (modified after Miall, 2002; seismic image courtesy of Andrew Miall). The image shows juxtaposed highstand and lowstand fluvial systems of Pleistocene age, separated by a subaerial unconformity that formed during an intervening stage of relative sea-level fall and fluvial incision. The incised valley cuts into the highstand topset, and is filled with lowstand and transgressive deposits. Abbreviations: HST—highstand systems tract; LST—lowstand systems tract; TST—transgressive systems tract; SU—subaerial unconformity; MFS—maximum flooding surface.

the river mouth environments may record various depositional trends, from diagnostic estuaries to non-diagnostic deltas, depending on the balance between accommodation and the riverborne sediment supply (Figs. 5.67 and 5.68). Larger rivers with high sediment load may prograde into the basin despite the transgression of the adjacent open shorelines (case B in Fig. 5.68; Figs. 5.69 and 5.70). At the opposite end of the spectrum, and more commonly, smaller rivers which do not supply enough sediment to fill all accommodation generated by relative sea-level rise are converted into estuaries. Where the transgression of the open shoreline is slower than the drowning of the river, the estuary is fully developed, which is commonly the case with incised valleys

(Dalrymple et al., 1994; case D in Fig. 5.68). Where the transgression of the open shoreline is faster than the transgression of the river, which is possible in the case with unincised channels, the estuary itself is drowned and represented only by its backstepping bayhead delta (Fig. 5.67; case C in Fig. 5.68). This is most common in lakes, where the reworking of the sediment brought by the river is minimal (Fig. 5.71).

The vertical profiles of proper (prograding) deltas and bayhead (retrograding) deltas are illustrated in Fig. 5.68. Deltas sensu stricto may form in relation to any controls on deltaic processes (i.e., river, waves, or tides), and may be found in any systems tract (i.e., they have a non-diagnostic occurrence within the transgressive systems tract). In contrast, bayhead deltas are generally restricted to wave-dominated estuaries or lakes, and form only in transgressive settings. The fundamental difference between the two types of deltas rests with the long-term depositional trends of progradation vs. retrogradation associated with the dominant direction of shoreline shift, as reflected by the overall

FIGURE 5.52 Tidal influences in lowstand fluvial systems, including cross-bedding with mud drapes (A) and sigmoidal bedding, mud drapes, and trace fossils (B) (Upper Cretaceous, Castlegate Formation, Utah). The evidence of marine influences indicates proximity to the shoreline, and that fluvial processes are at least in part controlled by relative sea-level changes; i.e., deposition within a downstream-controlled area, which justifies the usage of conventional ("lowstand" in this case) systems tract nomenclature (Figs. 3.11 and 5.8).

coarsening- and fining-upward profiles, respectively (Fig. 5.68). Both types of deltas may include higher frequency regressive—transgressive cycles, whose dominant component defines the long-term depositional trend (Fig. 5.68). The preservation of transgressive coastal systems depends on the rates of coastal aggradation relative to the rates of subsequent erosion associated with processes of tidal and/or wave ravinement.

Sediment supply to the marine environment during transgression is generally limited, and may include riverborne sediment, detritus generated by wave-ravinement processes, and intrabasinal carbonate. Sediment in the shoreface contributes to the development of backstepping coastal systems during fairweather, and it is dispersed seaward onto the shelf by storm surges and tidal currents, or as hypopycnal plumes. Sedimentation from suspension is a dominant process during transgression, due to the prevalence of the finer grained sediment, and leads to the deposition of "healing-phase" wedges which smooth out ("heal") the seafloor profile (Fig. 4.57; Posamentier and Allen, 1993). As the influx of coarser sediment to any specific location beyond the shoreface decreases during the backstepping of the shoreline, healing-phase wedges display typical fining-upward profiles. Healing-phase wedges may form both in shallow- and deep-water settings, at different scales, filling in the bathymetric lows outboard of the youngest regressive clinoforms. In shallow-water settings, healing-phase wedges are associated with coastal onlap and are generally thin, within a range of 10^0-10^1 m, often below the resolution of the seismic data. In the deep-water setting, healing-phase wedges onlap the slope (i.e., marine onlap; Fig. 4.8), and may reach thicknesses of 10^1-10^2 m (Fig. 4.6). Where above the seismic resolution, healing-phase deposits are often defined by "transparent" seismic facies due to the relatively homogeneous nature of condensed sections (Fig. 5.72).

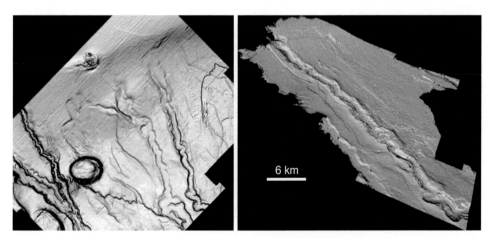

FIGURE 5.53 Entrenched channels of low-density turbidity currents in a slope setting (lowstand or early transgression; Late Pleistocene, De Soto Canyon area, Gulf of Mexico; images courtesy of Henry Posamentier). The horizon in these images was mapped along the maximum regressive surface in Fig. 4.6 within a 3D seismic volume. Note the absence of mudflow grooves at this stratigraphic level (compare with Fig. 5.43).

FIGURE 5.54   High-sinuosity aggrading channel of a low-density turbidity current in a basin floor setting (Late Pleistocene, De Soto Canyon area, Gulf of Mexico; images courtesy of Henry Posamentier). Extensive leveed channels on a basin floor are typical of low-density turbidity flows, whose mud content is high enough to sustain the formation of levees over large distances. Such basin-floor channels are age-equivalent to entrenched channels on the slope (Fig. 5.53). The shift from erosion on the slope to aggradation on the basin floor relates to changes in slope gradients and associated energy of the flow. Note the extremely low gradients that can sustain the motion of the flow. For this reason, low-density turbidity flows can travel hundreds or even thousands of kilometers across a basin floor (e.g., 2,700 km in the present-day Bay of Bengal).

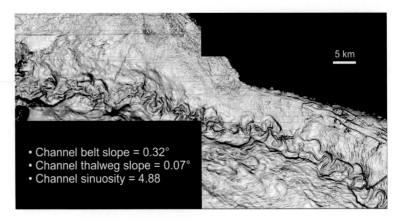

- Channel belt slope = 0.32°
- Channel thalweg slope = 0.07°
- Channel sinuosity = 4.88

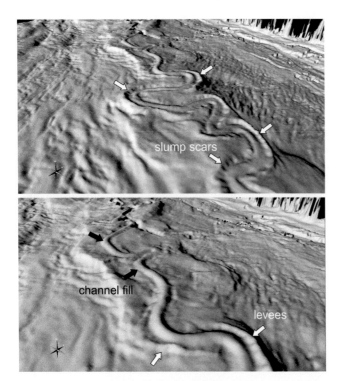

FIGURE 5.55   Morphology of a turbidite leveed channel in a basin-floor setting (details from Fig. 5.54; images courtesy of Henry Posamentier). The raised appearance of the channel fill, with a convex-up top, is caused by post-depositional differential compaction and indicates a coarser lithology within the channel relative to the overbank. Levees are better developed along the outer channel bends, and their inner margins are marked by scoop-shaped slump scars. The relief of the channel belt above the adjacent basin plain is c. 65 m. The channel fill is c. 625 m wide.

Shallow-water settings may also provide favorable conditions for the accumulation of sandy macroforms on the shelf, or for the deposition of carbonate condensed sections in areas starved of extrabasinal sediment, particularly during late stages of transgression (Figs. 5.64 and 5.73). In the process of sediment reworking by waves and tides, the coarsest clasts in the shoreface may be left behind as a transgressive lag on top of the wave-ravinement surface (Swift, 1976, Fig. 4.46). The sediment that can be transferred onto the shelf may accumulate as sand sheets or ridges, depending on the mechanism of sediment transport. Sand sheets are relatively thin, generally within a range of 1–3 m, but potentially extensive upward-fining tempestite deposits which accumulate on inner shelves following storms (e.g., Swift, 1968; Swift and Field, 1981; Belknap and Kraft, 1981; Demarest and Kraft, 1987; Kraft et al., 1987; Masterson and Eggert, 1992; Helland-Hansen et al., 1992; Eschard et al., 1993; Abbott, 1998). Shelf ridges are also sand prone, usually consisting of 5–10 m thick upward-fining successions of well-sorted, cross-bedded to bioturbated fine- to coarse-grained sediment reworked by tidal currents (Snedden et al., 1994).

The Miocene section of the Java shelf provides an example of shelf ridges on seismic data calibrated with well logs and core (Posamentier, 2002). These sand bodies are up to 17 m thick, 0.3–2.0 km wide, and more than 20 km long. They are asymmetrical, thicker along the sharp leading edge, gradually thinning toward the more irregular trailing edge (Figs. 5.74–5.78). Smaller scale sand waves may also be observed at the top of shelf ridges, oriented oblique to the long axes of the ridges and to the direction of ridge migration (Posamentier, 2002). Shelf ridges overlie transgressive wave-ravinement surfaces abandoned on the shelf following the retreat of the shoreline, and tend to be oriented parallel to the axes of structural embayments that channelize the energy of tidal currents. Shelf ribbons, which are smaller scale equivalents of shelf ridges, up to 5 m thick and 100 m wide, may also concentrate sand on a transgressive shelf (Posamentier, 2002). The scale of shelf ridges is proportional to the tidal range, and it may reach 100 km in length, 8 km in width, and 20 m in height in macrotidal settings (Fig. 5.79; Saha et al., 2016). The formation of tidal macroforms is enhanced during transgression, when a larger area of the shelf is submerged, with preservation enabled by the overlying highstand bottomsets.

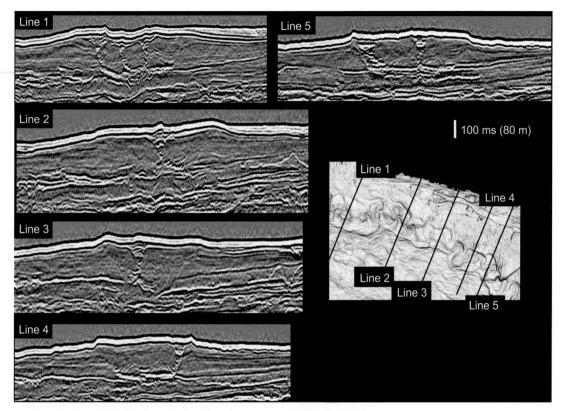

FIGURE 5.56   Transverse sections illustrating the evolution of the leveed channel in Fig. 5.54 during the Late Pleistocene—Holocene (images courtesy of Henry Posamentier). The 2D seismic lines indicate channel aggradation, as well as lateral migration with time. The turbidite channel-fill records higher seismic amplitudes relative to the adjacent finer grained levees. The channel fill is c. 625 m wide.

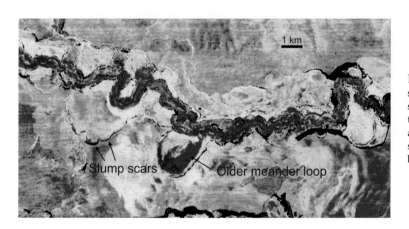

FIGURE 5.57   Turbidite leeved channel in a continental slope setting (3D seismic interval attribute map: seafloor +80 ms, maximum negative polarity; Late Pleistocene, offshore Nigeria; image courtesy of Henry Posamentier). Flow direction from right to left. The high sinuosity despite the steep slope gradient is due to the buoyancy of the flow in a subaqueous environment.

Deep-water settings also accumulate sediment during transgression, albeit in smaller amounts than other systems tracts (i.e., commonly, less than the falling-stage and lowstand systems tracts in siliciclastic settings, and less than the highstand systems tract in carbonate settings). The types of gravity-driven processes may also change during transgression, depending on the sediment available at the shelf edge, from low-density turbidity flows to slumps and mudflows (Figs. 4.20, 5.63, and 5.64; Posamentier and Kolla, 2003; Catuneanu, 2020a). However, even though the same set of principles

underlies the stratigraphic framework of all sedimentary basins, each setting is somewhat unique and may not necessarily fit the sedimentological makeup predicted by generalized templates (see details in Chapter 8). Beyond the basin-specific variability of the sedimentary record, the end of transgression is always the time when the shoreline is at the greatest distance from the shelf edge, which results in the highest degree of stratigraphic condensation in both siliciclastic and carbonate settings. As such, the maximum flooding surface is a significant and reliable stratigraphic marker in offshore

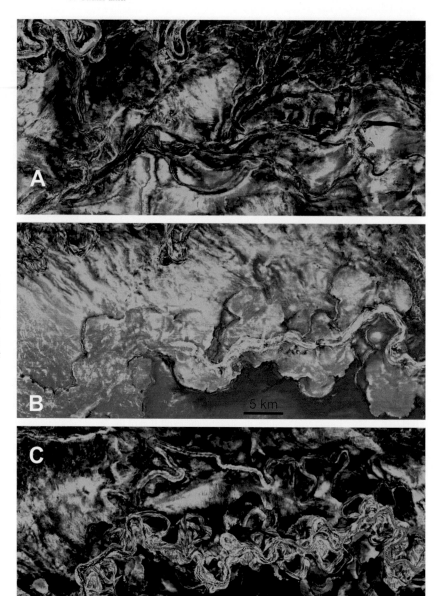

FIGURE 5.58 Elements of low-density turbidite systems in a basin-floor setting (Mahanadi Basin, offshore India; 3D seismic timeslices between 3.8–4.0 sec below sea level; spectral decomposition images, courtesy of Reliance Industries, India). A—Channel and splay elements; B—channel with slumped levees, and sediment waves; C—high-sinuosity channels.

studies, which can be identified on the basis of sedimentological, biostratigraphic, and geochemical criteria (see discussions in Chapters 2, 4, and 6).

Transgression typically results in a decrease with time in the amount and caliber of the sediment that is made available at the shelf edge, in both siliciclastic and carbonate settings. The primary reason is the retrogradation of the shoreline, which increases the distance between the riverborne (extrabasinal) sediment entry points into the basin and the shelf edge, coupled with the increase in water depth which inhibits the precipitation of intrabasinal sediment. Other factors which can reduce sediment supply to the shelf edge include the increase in the rates of relative sea-level rise and coastal aggradation at the

onset of transgression, which result both in the lowering of fluvial energy and the retention of more sediment within fluvial systems. The result is a fining-upward profile which continues the trend started at the onset of relative sea-level rise (Fig. 4.20). As the transgressive systems tract maintains the depositional trend of the underlying lowstand systems tract, with no marked changes in grain size and in the types of turbidites, the maximum regressive surface is often cryptic within the submarine fan complex (more details in Chapters 6 and 8).

Most documented field studies show that during early transgression, when the shoreline is still relatively close to the shelf edge and the bathymetry at the shelf edge is less than the storm wave base, sand is delivered

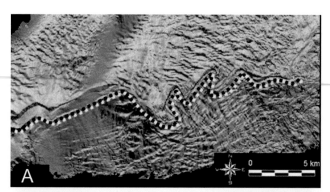

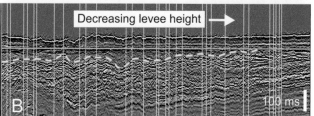

FIGURE 5.59 Seismic imaging of a turbidite channel (A) and its levee height (B) in a basin-floor setting (Late Pleistocene, Gulf of Mexico; images courtesy of Henry Posamentier). The seismic traverse follows the levee crest, and shows a decrease in levee height in a downdip direction. The datum (green line) marks the top of the levee deposits. The levee height decreases basinward as the system runs out of finer sediment, to the point where the levees are no longer able to keep the flow confined; at that point the flow disperses and a frontal splay forms at the end of the channel.

to the deep-water environment mainly by low-density turbidity flows (Figs. 4.20 and 5.63). These flows may still entrench on the slope, although the channels may be filled with time by backstepping turbidites as sediment supply (and hence the erosive power of the current) decreases. The turbidity flows aggrade on the basin floor, and may travel large distances if sufficient fine-grained sediment fractions are available to maintain the construction of levees. With time, turbidity currents wind down as the supply of riverborne sediment to the shelf edge is gradually cut off. During late transgression, when the sediment entry points into the marine basin are far from the shelf edge and the bathymetry at the shelf edge exceeds the depth of the storm wave base, the staging area for the deep-water gravity flows (i.e., the outer shelf below the storm wave base) may only accumulate fine-grained sediment from suspension. At that point, mudflows and slumps become the dominant processes of sediment mass-transport into the deep-water environment (Figs. 4.20 and 5.64). Subordinately, low-density turbidity flows may still continue where unfilled incised valley across the submerged shelves facilitate the transfer of sediment from the river mouth to the shelf edge (more details in Chapter 8).

The transgressive systems tract typically has the highest preservation potential among all systems tracts,

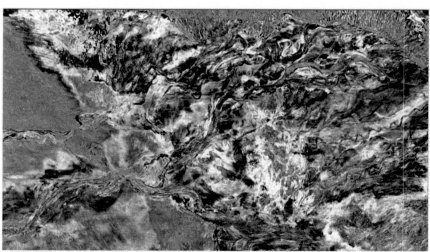

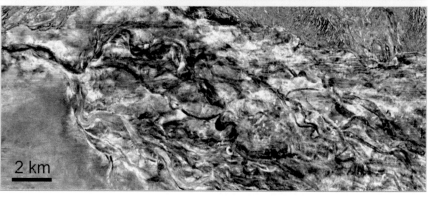

FIGURE 5.60 Examples of turbidity-flow frontal splays in a basin-floor setting (Late Pleistocene, Gulf of Mexico; seismic amplitude extraction maps, courtesy of Henry Posamentier).

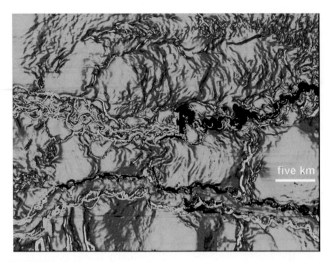

FIGURE 5.61 Seafloor reflection dip-magnitude map showing sediment waves related to "flow stripping" in the overbank area of channelized basin-floor turbidity flows (offshore Nigeria; from Posamentier and Kolla, 2003; image courtesy of Henry Posamentier).

as the subsequent highstand normal regression entails sediment aggradation in all depositional environments (Fig. 5.14). The preservation of all other systems tracts may be affected by unconformities that form during forced regressions or transgressions; i.e., subaerial unconformities at the top of highstand, falling-stage or lowstand deposits, transgressive surfaces of erosion at the top of lowstand or older deposits, and regressive surfaces of marine erosion at the top of highstand deposits (Fig. 4.39).

5.2.3.4 *Highstand systems tract*

The highstand systems tract is defined by a normal regressive stacking pattern (i.e., progradation and upstepping of the subaerial clinoform rollovers; Figs. 4.1 and 4.21) which follows a transgression of the same hierarchical rank (Figs. 4.39, 5.9, 5.14, and 5.15). The highstand normal regression occurs commonly during late stages of relative sea-level rise,

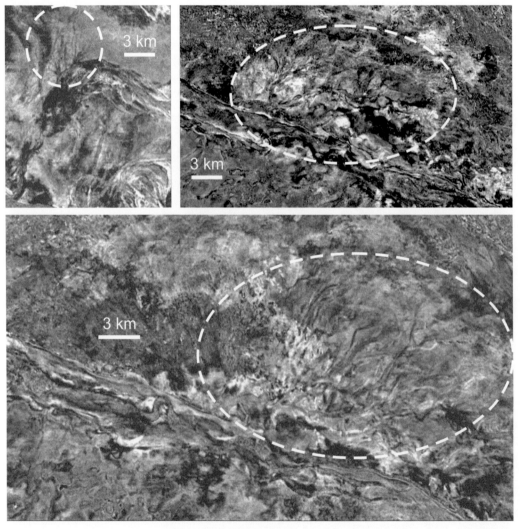

FIGURE 5.62 Examples of turbidity-flow crevasse splays in a basin-floor setting (Late Pleistocene, Gulf of Mexico; seismic amplitude extraction maps, courtesy of Henry Posamentier).

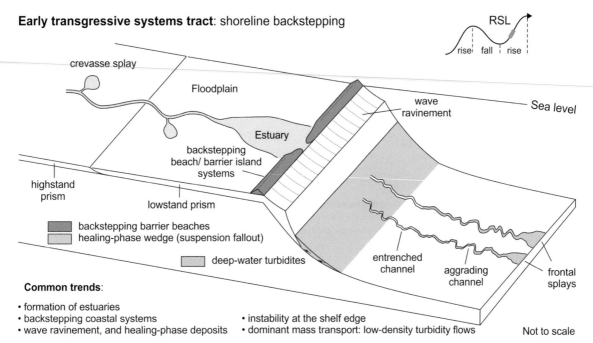

FIGURE 5.63 Depositional processes and products of the early transgressive systems tract. Transgressions are marked by the formation of estuaries, which trap riverborne sediment and further reduce sediment supply to the sea (compare with Fig. 5.48). This is in part compensated by wave-ravinement processes, which generate sediment in the shoreface by coastal erosion. For as long as the shelf edge remains above the storm wave base, sediment supply to the deep-water setting continues uninterrupted, leading commonly to low-density turbidity flows similar to the ones of the lowstand systems tract (Figs. 4.20 and 5.48). Healing-phase wedges are typical of transgression, and infill, or heal over, the bathymetric profile established at the end of regression (Fig. 4.6). Abbreviation: RSL—relative sea level.

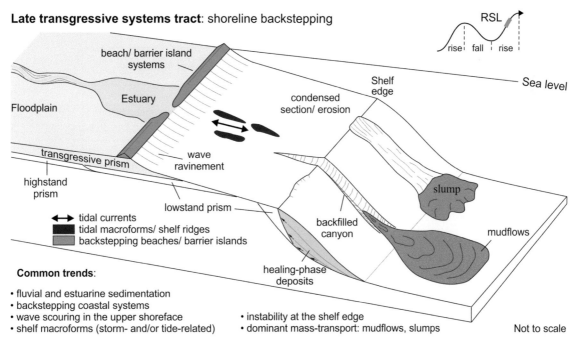

FIGURE 5.64 Depositional processes and products of the late transgressive systems tract. Late transgression refers to the stage when the bathymetry at the shelf edge exceeds the depth of the storm wave base. This results in a significant reduction of terrigenous sediment supply to the deep-water setting, although under special circumstances (e.g., the presence of submerged but yet unfilled incised valleys across the shelf; Fig. 5.34) riverborne sediment may still reach the shelf edge (details in Chapter 8). The fine-grained sediment that accumulates on the outer shelf is a source for mudflows and slumps, which dominate the late stages of transgression. Sediment reworking by storms or tides may generate shelf macroforms which are common, and have the best preservation potential, in transgressive settings. Departures from these trends (e.g., fluvial erosion instead of aggradation) are possible and reflect the variability of the tectonic and depositional settings (details in Chapters 4 and 8). Abbreviation: RSL − relative sea level.

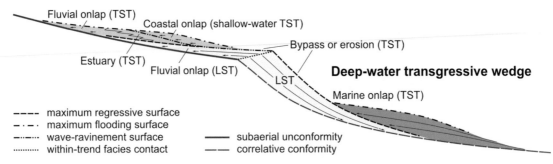

Shelf transgressive wedge

Fluvial onlap (TST)
Coastal onlap (shallow-water TST)
Bypass or erosion (TST)
Estuary (TST)
Fluvial onlap (LST)
LST
Deep-water transgressive wedge
Marine onlap (TST)

- - - - maximum regressive surface
- · - · maximum flooding surface
- ···· wave-ravinement surface
········ within-trend facies contact
———— subaerial unconformity
— — correlative conformity

FIGURE 5.65 Stratigraphic architecture of transgressive systems tracts. Transgressive systems tracts often consist of two distinct wedges separated by an area of sediment bypass or erosion around the shelf edge. The shelf wedge may or may not include fluvial deposits, depending on the gradient of the shoreline trajectory relative to the landscape gradient (details in Chapter 4).

FIGURE 5.66 Stratigraphic architecture of an incised-valley fill, based on the Gironde estuary (modified after Allen and Posamentier, 1993). This case study provides an example where all systems tracts that form during relative sea-level rise are represented in the rock record of the valley fill.

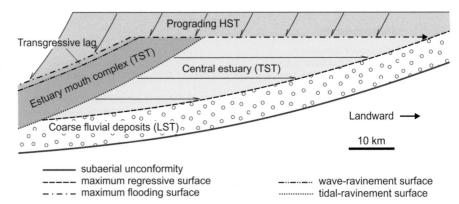

Prograding HST
Transgressive lag
Estuary mouth complex (TST)
Central estuary (TST)
Coarse fluvial deposits (LST)
Landward →
10 km

——— subaerial unconformity
- - - - maximum regressive surface
- · - · maximum flooding surface
- ··· - wave-ravinement surface
·········· tidal-ravinement surface

River-mouth environments		Conditions of formation
Deltas	Prograding deltas *(commonly larger rivers, higher sediment supply)*	Sedimentation > accommodation at the river mouth
	Retrograding ('bayhead') deltas/ incomplete (drowned) estuaries *(rivers with unincised channels)*	Accommodation > sedimentation at the river mouth. Transgression of open shoreline > drowning of the river
Estuaries	Complete estuaries *(commonly smaller rivers within incised valleys)*	Accommodation > sedimentation at the river mouth. Drowning of the river > transgression of open shoreline

FIGURE 5.67 River-mouth environments in transgressive settings. Transgressive settings are defined by backstepping open shorelines, while river mouths may range from estuaries to deltas depending on the local balance between accommodation and sedimentation. Most river mouths in transgressive settings are estuaries, but deltas may also form where the riverborne sediment supply exceeds the amount of available accommodation in front of the river. However, deltas may only be considered as part of a transgressive systems tract where the adjacent open shorelines are transgressive (Fig. 5.68). Between proper (prograding) deltas and fully developed estuaries, stand-alone retrograding ("bayhead") deltas may also occur where estuaries themselves are drowned due to the rapid transgression of the adjacent open shorelines. This is the case where channels are unincised, and the transgression of the open shoreline is faster than the rate of drowning of the river (i.e., sediment supply at the river mouth is higher than the amounts of sediment available for the construction of backstepping beaches; Fig. 5.68). Complete estuaries tend to form within incised valleys, which facilitate the drowning of rivers at rates higher than the rates of transgression of the adjacent open shorelines.

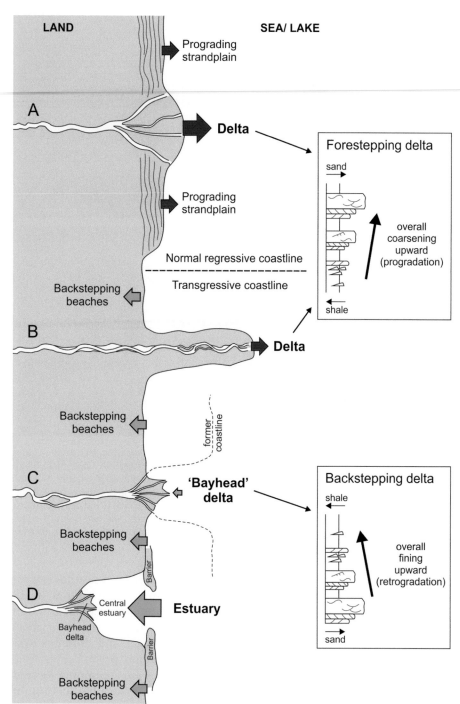

FIGURE 5.68 Types of coastal systems that may develop during relative sea-level rise. The progradational or retrogradational character of a coastal system in both river-mouth and open-shoreline settings is controlled by the local balance between the rates of accommodation and sedimentation. Normal regressive coastlines (lowstand or highstand) prograde in both river-mouth and open-shoreline settings. Transgressive coastlines are defined by the retrogradation of open shorelines, while the river mouth can be either progradational (deltas) or retrogradational (estuaries, or stand-alone bayhead deltas) (Fig. 5.67). A—Normal regressive coastline, with a delta and prograding strandplains; B—delta in a transgressive setting (Figs. 5.69 and 5.70); C—stand-alone retrogradational ("bayhead") delta in a transgressive setting (Fig. 5.71); D—fully developed estuary.

when the rates of sedimentation exceed the rates of relative sea-level rise at the shoreline (Fig. 4.32; see discussion in Chapter 4).

The highstand systems tract is bounded at the base by a maximum flooding surface with both continental and marine portions (transgression accompanied by fluvial aggradation; Fig. 4.39A), or by a marine maximum flooding surface and a subaerial unconformity (transgression accompanied by fluvial erosion; Fig. 4.39B).

Where the highstand systems tract is followed by forced regression, the upper boundary is represented by a subaerial unconformity and/or the basal surface of forced regression (Fig. 4.39). Where the highstand systems tract is followed by transgression, the upper boundary is represented by the maximum regressive surface reworked in part by the transgressive surface of erosion (transgression accompanied by fluvial aggradation), or by a composite surface which includes the marine

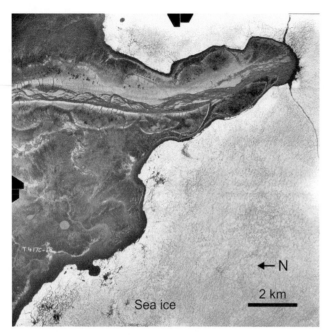

FIGURE 5.69 Aerial photograph of a river-dominated delta in a transgressive setting (case B in Fig. 5.68; Chads Point, western Sabine Peninsula, Melville Island, Canadian Arctic Archipelago; image courtesy of John England). Sediment supply at the river mouth exceeds accommodation, while the adjacent open shorelines record the opposite trend.

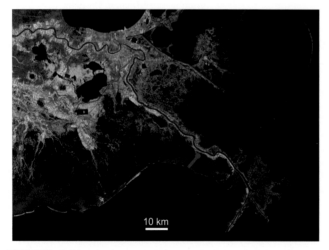

FIGURE 5.70 Satellite image of a river-dominated delta in a transgressive setting (case B in Fig. 5.68; Mississippi Delta, Louisiana; image released by the US Geological Survey National Wetlands Research Center). The deltaic progradation is driven by the high riverborne sediment supply, while the adjacent open shorelines record transgression.

portion of the maximum regressive surface, the transgressive surface of erosion, and the subaerial unconformity (transgression accompanied by fluvial erosion).

Highstand systems tracts typically include a continental topset and a marine foreset and bottomset, and ideally, display a convex-up shoreline trajectory which follows the pattern of decelerating relative sea-level

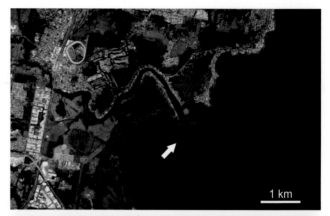

FIGURE 5.71 Backstepping bayhead deltas in lacustrine settings (images courtesy of Google Earth). A—Tuggerah Lake, New South Wales, Australia; B—Great Bear Lake, Northwest Territories, Canada. Typical of bayhead deltas, as opposed to "normal" (prograding) deltas, channels widen toward the river mouth, which is indicative of transgression. This helps to differentiate between the two types of deltas where only plan-view images are available. Vertical profiles provide further unequivocal evidence (Fig. 5.68).

rise (Fig. 4.51). Even though highstand normal regressions are conducive to sediment aggradation in all depositional environments, from continental through to deep water, the bulk of the sediment is most often sequestered in fluvial, coastal, and shoreface systems, which define the "highstand prism" (Figs. 5.80 and 5.81). Highstand shorelines are commonly located closer to the basin margins, following the flooding of the shelf during the preceding transgressions, but exceptions may occur depending on the width and gradients of the shelf, the magnitude of the transgressions, and the amount of progradation during the highstand normal regression (more details in Chapter 8). Under exceptional circumstances, the highstand shorelines may even reach the shelf edge, where shelves are narrow and/or sediment supply is extremely high (Carvajal and Steel, 2006).

In the most common scenario, where the subaerial unconformity forms during forced regression (Fig. 4.39A), the highstand topset has a low-gradient

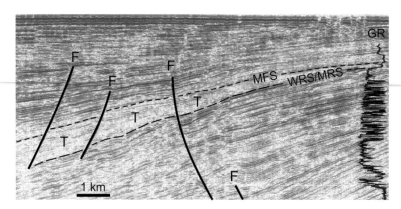

FIGURE 5.72 Dip-oriented seismic transect showing a Pliocene–Holocene succession in a shelf to upper slope setting (image courtesy of PEMEX). The seismic facies are calibrated with well data. Regionally extensive transgressive shales with a "transparent" seismic facies can be mapped across the area, and can be used as a marker for stratigraphic correlation. The shales accumulated below the storm wave base, following the abrupt water deepening and flooding of the shelf. The underlying sands accumulated above the storm wave base, in inner shelf to coastal environments. The maximum flooding surface is overlain by highstand (prograding) deposits. Abbreviations: T—transgressive shale; F—faults; WRS/MRS—wave-ravinement surface reworking a maximum regressive surface; MFS—maximum flooding surface; GR—gamma-ray log.

FIGURE 5.73 Maximum flooding surface (arrow) at the limit between coastal coal and marine limestone (Pennsylvanian Sydney Mines Formation, Sydney Basin, Nova Scotia; image courtesy of Martin Gibling). The coal bed lies within the transgressive systems tract, and it is overlain by highstand carbonates which accumulated following the maximum flooding of the basin. The maximum flooding surface merges with the transgressive wave-ravinement surface at the top of the coal bed.

profile and includes the lowest energy fluvial systems of a depositional sequence. In the less common cases where the subaerial unconformity forms during transgression (Fig. 4.39B), topographic gradients are steeper and the highstand topset includes the highest energy fluvial systems of a depositional sequence (see discussion in Chapter 4). Highstand prisms are often left stranded in the proximal (inner) parts of the shelf, following the rapid regression of the shoreline during subsequent relative sea-level fall (Figs. 4.34 and 5.17). In most cases, highstand prisms are also subject to fluvial incision during the subsequent forced regressions, as the exposed highstand clinoforms are commonly steeper than the fluvial graded profile (Figs. 5.16, 5.17, 5.27, and 5.28). These fluvial processes have long been recognized (Saucier, 1974; Leopold and Bull, 1979; Rahmani, 1988; Blum, 1991; Posamentier et al., 1992b; Allen and Posamentier, 1994) and reproduced in flume experiments (Wood et al., 1993; Koss et al., 1994).

The trends recorded by the fluvial portion of the highstand systems tract may be described in terms of (1) energy and related competence (i.e., maximum grain size that can be transported by rivers), and (2) degree of

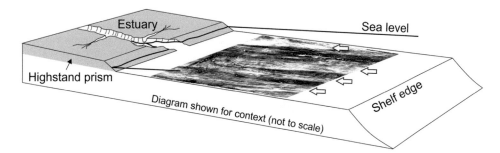

FIGURE 5.74 Tide-generated shelf ridges (white areas: high negative amplitudes on the reflection amplitude extraction map, 775 m subsea; Miocene, offshore northwest Java; image courtesy of Henry Posamentier). Ridges are hundreds of meters wide and several kilometers long, and form typically during transgression when a significant portion of the shelf is submerged. Preservation is enhanced by the subsequent aggradation of highstand bottomsets. For these reasons, such sand-prone shelf macroforms are most commonly found within transgressive systems tracts (Posamentier, 2002).

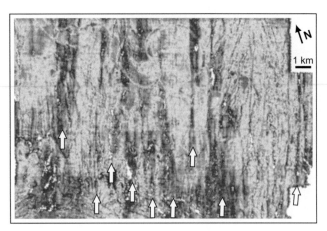

FIGURE 5.75   Tide-generated shelf ridges (red areas on the reflection amplitude extraction map, 810 m subsea; Miocene, offshore northwest Java; image courtesy of Henry Posamentier). Ridges are hundreds of meters wide and several kilometers long, and form typically during transgression when a significant portion of the shelf is submerged. Preservation is enhanced by the subsequent aggradation of highstand bottomsets. For these reasons, such sand-prone shelf macroforms are most commonly found within transgressive systems tracts (Posamentier, 2002).

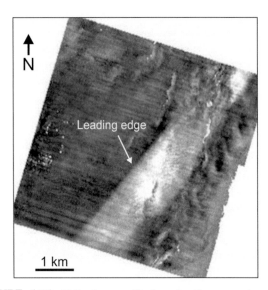

FIGURE 5.76   Reflection amplitude extraction map showing a close-up of a Miocene shelf ridge (white feature on the map: high negative amplitudes), offshore northwest Java (horizon slice 720 m subsea; image courtesy of Henry Posamentier). The leading edge of the macroform is sharper than the trailing edge (see Figs. 5.77 and 5.78 for the definition of leading and trailing edges of shelf ridges).

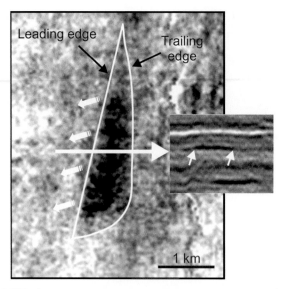

FIGURE 5.77   Morphology of a Miocene shelf ridge, offshore northwest Java, as seen on an amplitude extraction map from a horizon located 775 m subsea (images courtesy of Henry Posamentier). On the 2D seismic line, the macroform tunes within one reflection, with only the amplitude variations indicating its location between the two arrows. In plan view, the shelf ridge has an asymmetrical shape with a straight and well-defined leading edge, and a more irregular trailing edge. The direction of ridge migration is indicated by arrows on the map.

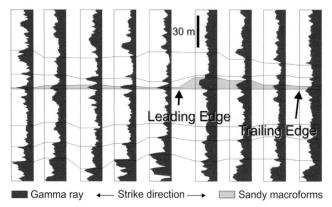

FIGURE 5.78   Well-log cross section showing the morphology of a Miocene shelf ridge located 850 m subsea, offshore northwest Java (modified after Posamentier, 2002); the cross section is 6 km long, perpendicular to the ridge). Note the asymmetrical shape of the ridge, with a thicker and better defined leading side, and a tapering trailing side. The integration of 3D seismic and well data affords the full 3D reconstruction of the shelf-ridge morphology.

channel amalgamation (i.e., overall sand/mud ratio). While the caliber of the riverborne sediment decreases with time due to the decline in fluvial gradients and energy in response to coastal aggradation, the degree of channel amalgamation may increase in response to the decrease in the rates of coastal and floodplain aggradation (Legaretta et al., 1993; Shanley and McCabe, 1993; Aitken and Flint, 1994; Emery and Myers, 1996). The latter trend follows the idealized pattern of decelerating relative sea-level rise during highstand normal regression (Fig. 4.32), which predicts a shift from dominantly aggradational to dominantly progradational depositional trends (i.e., the convex-up shoreline trajectory in Fig. 4.51). As departures from this idealized accommodation model do occur (Fig. 4.33), the predicted changes in the degree of channel amalgamation are less reproducible

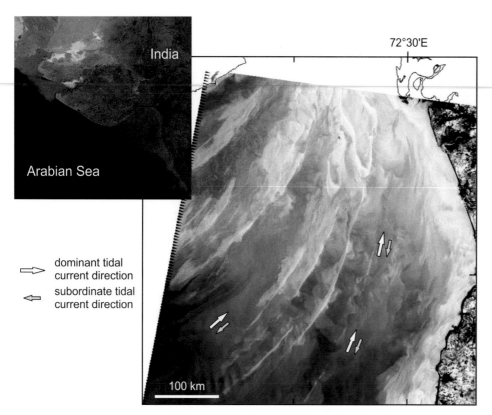

India

Arabian Sea

72°30'E

⇨ dominant tidal
   current direction
⇦ subordinate tidal
   current direction

100 km

FIGURE 5.79 Shelf tidal ridges in the outer Gulf of Cambay, India (Landsat TM image dated April 22nd, 2003; tidal processes described by Saha et al., 2016). In this macrotidal setting, individual ridges may be 100 km long, 8 km wide and 20 m high.

than the fining-upward trends that are commonly observed in field studies (Figs. 4.53 and 5.50).

Highstand topsets are usually subject to erosion during subsequent relative sea-level fall, with a higher preservation potential in the interfluve areas of incised-valley systems where pedogenesis becomes the dominant process (Wright and Marriott, 1993, Figs. 5.16, 5.27, and 5.28). Fig. 5.51 provides an example of a typical low-energy highstand fluvial system preserved in the interfluve area of a Pleistocene incised valley. The highstand channels are confined and isolated within a topset that is dominated by floodplain deposits. Both the highstand topset and the younger incised-valley fill were subsequently flooded during the Holocene transgression. The shallow-water equivalents of the highstand topset display a coarsening-upward profile related to the progradation of the coastline, and include gradationally based, expanded clinoforms, diagnostic of normal regression (Fig. 5.31). Notably, the vertical profiles of the highstand systems tract vary between different depositional systems (e.g., fluvial vs. marine), so any interpretations of log motifs or lithologs must be preceded by a study of the depositional setting.

The shallow-water clinoforms of the highstand systems tract downlap the maximum flooding surface (Figs. 1.11, 2.62, and 4.39). Depending on the scale of observation, the progradation of the shoreline can be uninterrupted, which is the case at the lowest hierarchical level within a stratigraphic framework, or it can be interrupted temporarily by higher frequency transgressions which punctuate the long-term progradational trend. The latter is typically the case with systems tracts observed at seismic scales, which only describe longer term depositional trends that can be discerned above the resolution of the seismic data. Seismic scale highstand systems tracts consist of higher frequency sequences dominated by a progradational component, in which each clinoform extends farther seaward relative to the previous one (Figs. 1.11 and 2.62). Due to the normal regressive nature of the highstand progradation, which involves aggradation as well, the component clinoforms are always attached (e.g., in contrast with the clinoforms of the falling-stage systems tract, which may also be detached; Figs. 5.29 and 5.30).

The degree of vertical overlap of the highstand clinoforms tends to be more pronounced during early highstand, when the rates of relative sea-level rise are still high following the end of transgression, and the normal regression has a strong aggradational component. In contrast, the late highstand is often characterized by a decrease in the rates of aggradation and an increase in the rates of progradation, particularly where highstand is followed by forced regression, as the relative sea-

Highstand systems tract: shoreline progradation & upstepping

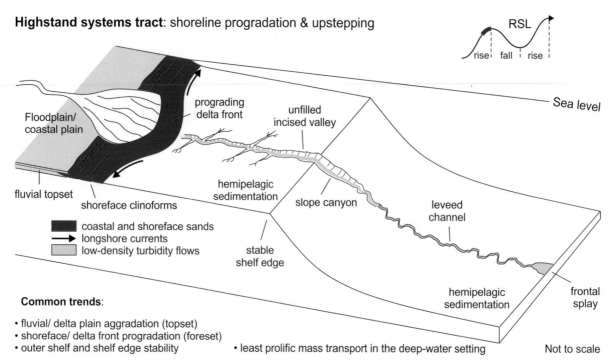

RSL

rise fall rise

Common trends:

- fluvial/ delta plain aggradation (topset)
- shoreface/ delta front progradation (foreset)
- outer shelf and shelf edge stability
- least prolific mass transport in the deep-water setting

Not to scale

FIGURE 5.80 Depositional processes and products of the highstand systems tract. The "highstand prism" includes fluvial, coastal, and shoreface deposits, and retains the majority of the riverborne sediment. As a result, terrigenous sediment supply to the shelf edge is typically low during stages of highstand, although exceptions may occur in the case of narrow shelves, or shelves dissected by unfilled incised valleys (Fig. 5.34; details in Chapter 8). The shelf edge is hydraulically stable during highstands in relative sea level, when the water depth (typically greater than the storm wave base) only varies at generally low rates. The shelf and deep-water environments are dominated by hemipelagic sedimentation, and the contribution of mass-transport processes to the deep-water setting is least significant relative to all other systems tracts. Abbreviation: RSL—relative sea level.

FIGURE 5.81 Satellite image of the Indus Delta (Pakistan), showing the aggrading and prograding alluvial and delta plains of a modern highstand topset (image courtesy of Henry Posamentier). The intertidal part of the delta plain is marked by tidal creeks. Floodplain aggradation is most active near the river, which explains the seaward encroachment of the alluvial plain along the river.

level rise decelerates to approach the point of highstand (Fig. 4.32). As a result, the thickness of the higher frequency sequences within the highstand topset, which reflects the degree of vertical overlap of the shallow-water clinoforms, decreases with time, as the balance between aggradation and progradation shifts in favor of the latter (Fig. 4.52). In this case, the highstand clinoforms become thinner with time and in a basinward direction, but with a wider geographic extent, assuming a quasi-constant sediment supply. These trends no longer apply where highstand normal regressions are followed by transgressions rather than forced regressions (Fig. 4.33).

The outer shelf is commonly starved of siliciclastic sediment during highstand, leading to a paucity of gravity flows in the deep-water environment. Exceptions may occur when highstand deltas are close to the shelf edge, or when submerged but still unfilled incised valleys facilitate the transfer of riverborne sediment from deltas to the shelf edge (Figs. 5.80 and 5.82; Catuneanu, 2020a; see details in Chapter 8). Even in the latter case, the volume of riverborne sediment that can reach the shelf edge during stages of highstand is significantly smaller than that delivered to the shelf edge during forced regressions (Sweet et al., 2019; Fig. 5.34). The opposite trends are recorded in carbonate settings, where the shallow-water carbonate factories are typically most productive during highstand normal regressions, resulting in the transfer of potentially large amounts of carbonate debris to the basinal setting (Schlager et al., 1994; Belopolsky and Droxler, 2003; Schlager, 2005; Fig. 5.83). This highstand shedding can generate seismic-scale slope clinoforms that downlap the maximum flooding surface, geometrically similar to the shallow-water clinoforms in siliciclastic settings (Fig. 5.83).

5.2.4 Systems tracts in upstream-controlled settings

In upstream-controlled settings, systems tracts form independently of relative sea/lake-level changes and shoreline shifts (Figs. 4.2 and 5.8), and reflect the combined influence of accommodation, climate, source-area uplift, and autogenic processes on sedimentation (Fig. 3.11). Upstream-controlled systems tracts can be defined in both fluvial and eolian systems. This section focuses on fluvial systems, which are more widely documented and included in sequence stratigraphic studies. A discussion of sequences and systems tracts that develop in eolian systems is provided in Chapter 8.

In upstream-controlled fluvial settings, systems tracts are defined by the dominant depositional elements (i.e., high vs. low channel-to-overbank ratio; see Chapter 4 for a discussion of stacking patterns), and can be observed at different scales (Figs. 4.2, 4.63, 4.69, and 4.70). The higher degree of channel amalgamation and the coarsest sediment are typically found at the base of depositional sequences, in relation to the higher energy fluvial systems of the stratigraphic cycle. To that extent, fluvial sequences in upstream- and downstream-controlled settings display similar fining-upward trends that reflect the decline with time in fluvial energy during the accumulation of a sequence (Figs. 2.8, 4.53, 4.65, and 5.50).

The nomenclature of "unconventional" systems tracts is no longer tied to shoreline trajectories or to the relative sea level, nor to the inferred accommodation conditions at syn-depositional time (Catuneanu, 2017). Accommodation generated by basin subsidence is, in most cases, already overfilled during the development of upstream-controlled depositional sequences and component systems tracts (Fig. 3.12). While accommodation (i.e., rates of subsidence) may still play an important role, fluvial processes, including the rates of floodplain aggradation, are also influenced by all other controls that modify the balance between sediment supply and energy flux at any location (i.e., climate, source-area uplift, and autocyclicity; Fig. 3.11). The shift from underfilled to overfilled stages in the evolution of

sedimentary basins shows that upstream controls such as climate and/or source-area uplift can sustain long-term aggradation at rates higher than the rates of subsidence. This demonstrates the variability of processes that can lead to the development of fluvial sequences and component systems tracts at different scales (see Chapter 4 for details).

The high-amalgamation (HAST) and low-amalgamation (LAST) systems tracts (Fig. 5.8; Catuneanu, 2017) were previously defined as "low-accommodation" and "high-accommodation" systems tracts, respectively, based on the assumption that accommodation is the main control on the degree of channel amalgamation (e.g., Olsen et al., 1995; Boyd et al., 2000; Arnott et al., 2002; Leckie and Boyd, 2003). However, it has become evident that accommodation alone cannot explain the development of fluvial stacking patterns across all temporal scales, particularly where the rates of accumulation of depositional elements do not match the rates of creation of accommodation (Miall, 2015). Therefore, the revised systems tract terminology emphasizes the observation of strical stacking patterns rather than the interpretation of the underlying controls (Catuneanu, 2017).

The relative contribution of high- and low-amalgamation systems tracts to the composition of sequences is variable, depending on all factors that control the degree of channel amalgamation (Fig. 4.63). Either systems tract may become dominant, or even form entire sequences. Both systems tracts are commonly documented in Phanerozoic successions, where vegetation contributes to the stabilization of channels and the preservation of overbank deposits (Figs. 4.53, 4.65, 4.69, 4.70, 4.71, and 4.72). In contrast, the lack of vegetation during the Precambrian, coupled with the different patterns of weathering, erosion, transport, and sedimentation, led to the dominance of unconfined fluvial systems which are prone to higher degrees of channel amalgamation. The scarcity of floodplain fines within Precambrian fluvial sequences may also be related to a greater eolian influence, as dust storms remove mud more efficiently from barren surfaces (Ramaekers and Catuneanu, 2004). The result is the prevalence of high-amalgamation systems tracts within Precambrian sequences (Ramaekers and Catuneanu, 2004; Eriksson and Catuneanu, 2004a; see more details on the Precambrian vs. Phanerozoic stratigraphic frameworks in Chapter 9).

The only type of sequence stratigraphic surface that is common between the downstream- and upstream-controlled settings is the subaerial unconformity, which marks the boundary of depositional sequences (Fig. 4.65). All other "conventional" sequence stratigraphic surfaces (Fig. 4.39) require a marine/lacustrine environment, and therefore are restricted to the

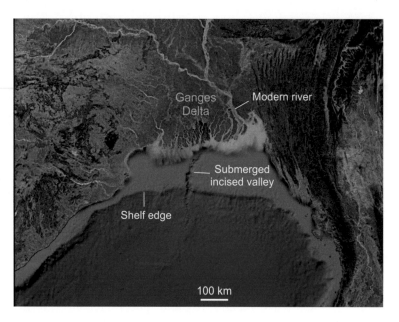

FIGURE 5.82   Satellite image of the Ganges Delta (Bangladesh), showing the location of the incised valley of the Ganges River during the Late Pleistocene lowstand when the shelf was exposed (image courtesy of Google Earth). The valley is now submerged but still unfilled, with a talweg steeper than the gradient of the shelf. This enables turbidity flows along the valley, which facilitate the transfer of riverborne sediment from the modern delta to the shelf edge, maintaining an active turbidite system in the deepwater setting of the Bay of Bengal (Fig. 5.34).

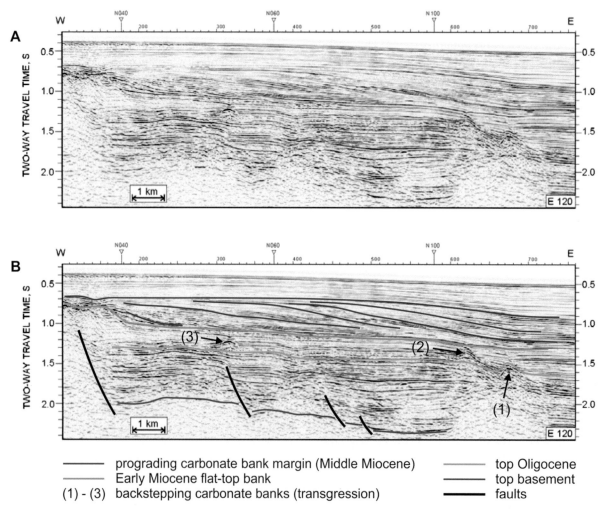

FIGURE 5.83   Highstand shedding in a carbonate setting (Maldives Islands, Indian Ocean; modified from Belopolsky and Droxler, 2003). The productivity of the carbonate platform is highest during highstand, when large amounts of carbonate debris are shed into the basin. Note the retrogradation of the carbonate banks during transgression (late Oligocene to early Miocene) and the progradation of clinoforms during the highstand normal regression (middle Miocene). The maximum flooding surface ("downlap surface") is downlapped by highstand clinoforms.

downstream-controlled settings. Within upstream-controlled depositional sequences, systems tracts are separated by a "top-HAST" (or "top-amalgamation") surface, which marks the change from a high to a low degree of channel amalgamation (Fig. 4.70). The change in the degree of channel amalgamation from one systems tract to another is not necessarily accompanied by a change in fluvial style; and, within each systems tract, the fluvial style may change between different locations within the basin (Catuneanu and Bowker, 2001; Catuneanu and Elango, 2001). Therefore, even though some types of rivers (e.g., braided) are more naturally prone to channel amalgamation, no general relationship can be established between systems tracts and specific fluvial styles (Fig. 4.63; see discussion in Chapter 4).

Both subaerial unconformities and top-HAST surfaces can be observed at different scales (i.e., hierarchical orders; Fig. 4.70), and may be diachronous, as the conditions required by their development may be met at different times at different locations. The top-HAST surface is conformable, as it marks a shift to lower energy conditions during the accumulation of a fluvial sequence, with a timing which depends on the interplay of all parameters that control the degree of channel amalgamation at any location (Fig. 4.63). Systems tracts in upstream-controlled settings may be observed over a wide range of physical and temporal scales, generally within $10^0–10^2$ m and $10^2–10^6$ yrs. As the rates of fluvial aggradation can vary over at least four orders of magnitude ($10^{-1}–10^2$ m/ka; Bridge and Leeder, 1979; Miall, 2015), the ratio between the thickness and the timescale of systems tracts can also be highly variable (e.g., 10^1 m over 10^6 yrs vs. 10^2 m over 10^5 yrs; Figs. 4.71 and 4.72).

5.2.4.1 High-amalgamation systems tract

The high-amalgamation systems tract (formerly termed "low-accommodation" systems tract) is defined by a high degree of channel amalgamation (i.e., high channel-to-overbank ratio; Figs. 4.65A, 4.69A, 4.71, 4.72, and 5.84), and it is bounded at the base by the subaerial unconformity and at the top by the top-HAST surface (4.70). The formation of high-amalgamation systems tracts is promoted by (1) low rates of floodplain aggradation; (2) unconfined fluvial channels; and (3) a high frequency of channel avulsion (Fig. 4.63; Bristow and Best, 1993; see Chapter 4 for details). To a lesser extent, confined channels may also amalgamate under conditions of low rates of floodplain aggradation and high frequency of avulsion (Shanley and McCabe, 1993; Wright and Marriott, 1993; Emery and Myers, 1996). The processes that lead to channel amalgamation depend on all factors that control fluvial sedimentation, including accommodation (i.e., rates of subsidence), climate, source-area uplift, and autocyclicity (Fig. 3.11).

High-amalgamation systems tracts commonly dominate the lower part of fluvial sequences, which accumulate under higher energy conditions (Figs. 4.53, 4.68, 4.70, 4.71, and 4.72). They typically include the coarsest sediment of fluvial sequences, which fills the erosional relief of the subaerial unconformity. As a result, the multi-storey channels of the high-amalgamation systems tracts are often found within incised valleys, overlying the subaerial unconformity and forming part or all of the valley-fills, leading to a discontinuous and irregular systems-tract development. These features make the high-amalgamation systems tract somewhat analogous to the fluvial topsets that mark the restart of continental aggradation following the formation of subaerial unconformities in downstream-controlled settings (i.e., the topset of the lowstand systems tract, where the subaerial unconformity forms during forced regression, or the topset of the highstand systems tract where the subaerial unconformity forms during transgression; Fig. 4.39). However, no genetic relationship or temporal correlation is inferred between systems tracts that form in upstream- and downstream-controlled settings.

5.2.4.2 Low-amalgamation systems tract

The low-amalgamation systems tract (formerly termed "high-accommodation" systems tract) is defined by the dominance of floodplain deposits (i.e., low channel-to-overbank ratio; Figs. 4.65B, 4.69B, 4.71, 4.72, and 5.85), and it is bounded at the base by the top-HAST surface and at the top by the subaerial unconformity (4.70). The formation of low-amalgamation systems tracts is promoted by (1) high rates of floodplain aggradation; (2) confined fluvial channels; and (3) a low frequency of channel avulsion (Fig. 4.63; Bristow and Best, 1993; see Chapter 4 for details). However, this stacking pattern may also form under conditions of low rates of floodplain aggradation and high frequency of avulsion, where channels are stable (i.e., with a high degree of confinement) and the avulsion does not undermine the preservation of floodplain deposits (i.e., a low channel-to-overbank ratio is maintained despite the frequent avulsion; Fig. 4.68). The processes that lead to a low degree of channel amalgamation depend on all factors that control fluvial sedimentation, including accommodation (i.e., rates of subsidence), climate, source-area uplift, and autocyclicity (Fig. 3.11).

Low-amalgamation systems tracts commonly dominate the upper part of fluvial sequences, which accumulate under lower energy conditions (Figs. 4.53, 4.68, 4.70, 4.71, and 4.72). The deposition of a low-amalgamation systems tract is less affected by the erosional relief of the subaerial unconformity, which is healed by the multi-storey channels of the underlying high-amalgamation systems tract, and therefore, this unit tends to have a more tabular geometry. Low-amalgamation systems tracts potentially include well-developed coal beds, which indicate a high water table coupled with a declining energy and low sediment supply at syn-depositional time (Fig. 5.86). These features, along with the prevalence of fine-grained overbank deposits, resemble the most common fluvial architecture of transgressive and highstand systems tracts in downstream-controlled settings (Figs. 4.53 and 5.50). However, no genetic relationship or temporal correlation is inferred between systems tracts that form in upstream- and downstream-controlled settings.

5.2.5 Nomenclature of systems tracts

Objectivity in the nomenclature of systems tracts requires emphasis on observational rather than interpretive criteria. As such, systems tracts in all geological settings, whether downstream- or upstream-controlled, are defined on the basis of observed stratal stacking patterns and stratigraphic relationships. All conventional sequence stratigraphic models account for the presence of unfilled accommodation (i.e., an interior seaway, or a lake) within the basin under analysis, and highlight the importance of shoreline trajectories within the downstream-controlled portion of the underfilled basin (i.e., the area where the stratigraphic architecture is influenced by changes in relative sea level; Figs. 1.10 and 3.11). Outside of the area of influence of relative sea-level changes, sequences and systems tracts in upstream-controlled settings are defined on the basis of field criteria that are independent of shoreline trajectories.

5.2.5.1 Downstream-controlled settings

In downstream-controlled settings, the timing of systems tracts and bounding surfaces is controlled by shoreline trajectories. This justifies a systems-tract nomenclature that makes reference to transgressions and regressions, with the latter being classified into "forced" and "normal," depending on the stacking pattern that accompanies the shoreline regression (i.e., downstepping vs. upstepping subaerial clinoform rollovers, respectively; Fig. 4.1). In turn, shoreline trajectories are controlled by the interplay of relative sea-level changes (i.e., accommodation) and base-level changes (i.e., sedimentation) *at the shoreline*. Both elements of this dual control are critically important, and none is a constant during the development of stratigraphic cycles. Changes in relative sea level control the geometrical trends of shoreline downstepping (i.e., forced regressions, during relative sea-level fall) and upstepping (i.e., normal regressions or transgressions, during relative sea-level rise; Figs. 4.1 and 4.23), whereas sedimentation controls the volume and internal makeup of forced regressive units during relative sea-level fall,

and the manifestation of normal regressions vs. transgressions during relative sea-level rise.

With emphasis on shoreline trajectories, transgression defines the transgressive systems tract, and forced regression defines the falling-stage systems tract. Any intervening stages of normal regression are classified into "lowstand" (i.e., a normal regression that follows a forced regression of equal hierarchical rank) and "highstand" (i.e., a normal regression that follows a transgression of equal hierarchical rank), which define the lowstand and highstand systems tracts, respectively (Fig. 5.9). It can be noted that all systems tracts are defined by local stacking patterns and stratigraphic relationships, independently of any global standards and interpretations. The reference to the falling stage, or to the lowstands and highstands in relative sea level, does not detract from the objective definition of the systems tracts. This nomenclature acknowledges the contribution of the relative sea level to the stratigraphic architecture, which can be demonstrated and quantified with stratigraphic and sedimentological data (e.g., the amounts of upstepping and downstepping of the subaerial clinoform rollovers and associated subtidal facies; Figs. 4.23, 4.40, and 4.41; Plint, 1988; Plint and Nummedal, 2000; Posamentier and Morris, 2000; Tesson et al., 2000; Zecchin and Tosi, 2014; Catuneanu, 2019a; Sweet et al., 2019).

Downplaying the importance of the relative sea level would, in fact, be counterproductive. The distinction between lowstands and highstands explains paleogeographic trends (e.g., the relative locations of lowstand and highstand shorelines, due to the intervening stages of forced regression and transgression; Fig. 5.87), the shelf capacity to retain extrabasinal sediment (i.e., lower during relative fall), the production of intrabasinal sediment (e.g., reduced carbonate factories at times of lowstand), and the consequent sediment supply to the deep-water setting (e.g., "lowstand shedding" in siliciclastic settings, and "highstand shedding" in carbonate settings; Schlager et al., 1994; Posamentier and Allen, 1999; Belopolsky and Droxler, 2003; Sweet et al., 2019; Figs. 5.34 and 5.83). Non-diagnostic variability in the development and composition of systems tracts (e.g., depositional trends in fluvial and coastal systems, variations in the production of extrabasinal and intrabasinal sediment, the overall increase in riverborne sediment supply to deep water in the case of narrow shelves or shelves dissected by unfilled incised valleys) reflects the unique tectonic and depositional settings of each sedimentary basin, and needs to be rationalized on a case-by-case basis (e.g., Catuneanu, 2019a, 2020a).

An alternative systems-tract nomenclature emphasizes the depositional trends recorded in coastal environments during the shifts of the shoreline: i.e., progradation to aggradation ("PA") instead of

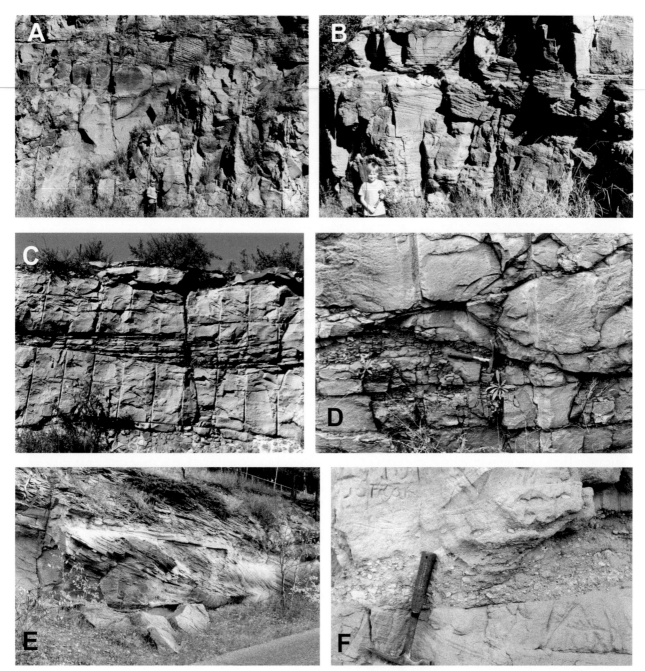

FIGURE 5.84 High-amalgamation systems tract: outcrop examples. A, B—channel fills and downstream accretion macroforms in a braided system (late Permian—early Triassic, Balfour Formation, Karoo Basin); C, D—amalgamated braided channels and downstream accretion macroforms (only small amounts of floodplain fines are preserved in this succession; Late Triassic, Molteno Formation, Karoo Basin); E, F—amalgamated braided channels with mudstone rip-up clasts derived from the erosion of the floodplain during the lateral shift of the unconfined channels (Maastrichtian, Frenchman Formation, Western Canada Basin). High-amalgamation systems tracts develop commonly in the lower (higher energy) part of depositional sequences, and include interconnected petroleum reservoirs.

lowstand normal regression; retrogradation ("R") instead of transgression; aggradation to progradation ("AP") instead of highstand normal regression; and degradation ("D") instead of forced regression (Neal and Abreu, 2009; Abreu et al., 2010). While this terminology was meant to uncouple the nomenclature of systems tracts from any reference to the relative sea level,

its proponents still assumed that depositional trends are controlled by accommodation, as "D" is attributed to relative sea-level fall (i.e., negative accommodation), and "PA," "R," and "AP" are attributed to stages of relative sea-level rise (i.e., positive accommodation). This assumption defeats the purpose of a nomenclatural change, as the terms "PA" and lowstand, as well

FIGURE 5.85   Low-amalgamation systems tract: outcrop examples (Early–Middle Triassic, Burgersdorp Formation, Karoo Basin). A—isolated channel (erosional base, fining upward) within overbank facies; B—lateral accretion macroform (point bar) in a meandering system; C—proximal crevasse splay (c. 4 m thick, sharp but conformable base, coarsening-upward) within overbank facies; D—floodplain-dominated meandering system, with isolated channels and distal crevasse splays. Low-amalgamation systems tracts develop commonly in the upper (lower energy) part of depositional sequences, and include disconnected petroleum reservoirs (i.e., isolated sandstone bodies within floodplain fines).

FIGURE 5.86   Coal beds within a low-amalgamation systems tract (Early Paleocene, Coalspur Formation; Western Canada Basin). Conditions for peat accumulation are most favorable at times dominated by floodplain aggradation, when water table is high and sediment supply is low.

as "AP" and highstand, become interchangeable (Neal and Abreu, 2009). The flaw of this assumption is that the depositional trends in coastal environments may not follow the geometrical trends of the shoreline (i.e., shoreline trajectories), as the vertical components of the two types of trend are controlled by different processes (i.e., base-level changes vs. relative sea-level changes, respectively; Fig. 4.23).

The systems-tract nomenclature based on coastal depositional trends attempts to be objective, by removing the reference to relative sea-level changes, while recognizing at the same time the importance of relative sea-level changes (i.e., the "accommodation succession method" that describes the ExxonMobil methodology; Neal and Abreu, 2009). This contradiction is rooted in the confusion between shoreline trajectories and coastal depositional trends, which triggers several model-driven errors: it overemphasizes the role of accommodation on the development of depositional trends (e.g., it assumes that subaerial unconformities form only during stages of negative accommodation, which is not always true: Fig. 4.39); it relies on ideal shoreline trajectories (i.e., concave-up vs. convex-up; Figs. 4.51 and 4.60), which may or may not be observed in the rock record; and it assumes that "A" (aggradation) requires creation of accommodation, which may or may not be the case (i.e., continental to coastal aggradation may occur during both stages of positive and negative accommodation; Figs. 4.26—4.28). The latter error hinders the distinction between forced and normal regressions, which is a significant pitfall of this approach.

The original nomenclature that makes reference to the lowstands and highstands in relative sea level remains more accurate in terms of definition of systems tracts and the separation between the deposits of negative vs. positive accommodation (see the discussion of geometrical vs. depositional trends in Chapter 4; Figs. 4.23 and 4.24). Notwithstanding the importance of sediment supply, the relative sea level has a tangible impact on the stratigraphic architecture, shallow-water processes and facies, and the transfer of sediment from the shelf to the deep-water setting (Posamentier and Morris, 2000; Zecchin and Tosi, 2014; Catuneanu and Zecchin, 2016; Catuneanu, 2019a, 2020a). The distinction between lowstand and highstand normal regressions, as well as between normal regressions and forced regressions, is critical to understand the evolution and fill of sedimentary basins. Therefore, changes in relative sea level are important to recognize and reconstruct. The identification of lowstands and highstands in relative sea level is based on local stratigraphic relationships, and not on correlations with global sea-level changes. Global cycle charts are no longer part of the sequence stratigraphic workflow and methodology.

New methods also emerge to recognize relative sea-level changes in the stratigraphic record. Geochemical proxies include the trends of change in water oxygenation levels, redox-sensitive trace metals, and the ratio between trace metals and organic carbon (Harris et al., 2013; Sano et al., 2013; Turner and Slatt, 2016; Hines et al., 2019; LaGrange et al., 2020), which can be calibrated with the changes recorded by trace-fossil assemblages (Savrda and Bottjer, 1989; MacEachern et al., 2009a,b; Dashtgard and MacEachern, 2016). Such criteria need to be adjusted to specific tectonic and depositional settings (e.g., open-circulation vs. restricted basins), and work is still in progress. Any constructive future efforts in sequence stratigraphy need to focus on refining the field criteria that afford the identification of systems tracts and bounding surfaces in the sedimentary record (Fig. 1.2).

5.2.5.2 Upstream-controlled settings

In overfilled basins, as well as in the upstream-controlled portion of underfilled basins, the definition and nomenclature of systems tracts is based on the dominant depositional elements (e.g., channels vs. floodplain deposits in fluvial settings; Figs. 4.2 and 4.70). In such cases, the relative sea level no longer plays any role in the formation of systems tracts, and the temporary base level becomes the sole control on the development of unconformity-bounded depositional sequences (see full discussions in Chapters 4 and 5). As accommodation is only one of the several controls that interplay to generate stacking patterns in fully continental settings (Fig. 3.11), any direct link between accommodation and the nomenclature of systems tracts is to be avoided. This led to a change in nomenclature from terms that make reference to inferred accommodation conditions (i.e., low- vs. high-accommodation) to terms that describe the observed stacking patterns (i.e., high- vs. low-amalgamation; Catuneanu, 2017, 2019a).

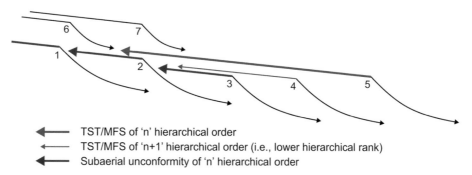

FIGURE 5.87 Stratigraphic architecture of falling-stage (timesteps 1—3), lowstand (timesteps 3—5), transgressive (timesteps 5—6), and highstand (timesteps 6—7) systems tracts, as defined by the shoreline trajectory of "n" hierarchical order. A transgression of "n+1" hierarchical order does not change the location of the shoreline (subaerial clinoform rollover) observed at the "n" scale. In contrast, a transgression of "n" hierarchical order leads to the relocation of the shoreline observed at the "n" hierarchical level. Abbreviations: TST—transgressive systems tract; MFS—maximum flooding surface.

5.3 Stratigraphic sequences

5.3.1 Definition

The definition of a "sequence" was revised and improved over time, in response to conceptual advances, the increase in the resolution of stratigraphic studies, and the need to accommodate all sequence stratigraphic approaches (Fig. 1.22). Stratal stacking patterns are at the core of the sequence stratigraphic methodology, as they provide the criteria to define all units and surfaces of sequence stratigraphy, at scales defined by the purpose of study or by the resolution of the data available. In a general sense, sequences are stratigraphic cycles of change in stratal stacking patterns, defined by the recurrence of the same type of sequence stratigraphic surface in the sedimentary record (Fig. 1.22; Catuneanu and Zecchin, 2013). This definition is inclusive of all types of stratigraphic sequences (i.e., "depositional," "genetic stratigraphic," and "transgressive-regressive"; Figs. 1.9 and 1.10). The definition of sequences is based on the observation of stratal stacking patterns, and it is independent of temporal and physical scales, stratigraphic age, and the interpreted origin of cycles (Catuneanu and Zecchin, 2013).

The delineation of depositional sequences in seismic stratigraphy relies on the observation of seismic layout geometries. In the early days of seismic stratigraphy, onlap terminations were used to interpret not only the position of subaerial unconformities (i.e., depositional sequence boundaries) but also changes in the sea level. In the absence of facies information to calibrate the seismic data, onlap terminations were assumed to indicate both the position of paleocoastlines and stages of sea-level rise (Fig. 5.88). The flaw of this assumption is that onlaps can be "marine," "coastal," or "fluvial" (Fig. 4.8), and only the coastal onlap is invariably linked to stages of relative sea-level rise. Fluvial onlap develops commonly during relative sea-level rise in relation to normal regressive topsets, but it may also form atypically during forced regressions accompanied by fluvial aggradation (see Chapter 4 for details). Similarly, marine onlap may form during either relative sea-level rise or fall (Fig. 5.89). The errors in the original assumptions led to the unnatural "saw-tooth" shape of the global cycle chart of sea-level changes, which postulates instantaneous sea-level fall (Fig. 5.88). In reality, both stages of relative sea-level rise and fall take time to occur, as recognized subsequently (Fig. 5.23).

The identification of depositional sequence boundaries on low-resolution seismic transects is generally straightforward in shelf settings where forced regressive deposits are thin and fall below the resolution of the seismic data. In this case, the sequence boundary marks an abrupt seaward shift in the location of onlap terminations, and separates packages of strata that display continuous onlap toward the basin margin (Figs. 5.88 and 5.89). However, the observation of forced regressive deposits in higher resolution studies (i.e., "shelf-perched" deposits; Posamentier and Vail, 1988; Van Wagoner et al., 1990) led to debates regarding their placement within the sequence and relative to the sequence boundary. Posamentier and Vail (1988) assigned the forced regressive strata to the lowstand systems tract, thus placing the sequence boundary at their base, whereas Van Wagoner et al. (1990) assigned the same strata to the highstand and placed the sequence boundary at their top (depositional sequence models II vs. III in Fig. 1.10). The solution to this methodological conundrum is the separation of the forced regressive deposits as a stand-alone (i.e., the "falling-stage") systems tract (depositional sequence IV in Fig. 1.10).

The separation between observation and interpretation in the definition of a "sequence" is fundamental. Sequences are often regarded as allocycles (e.g., accommodation cycles; Neal and Abreu, 2009). However, this is an oversimplification as sediment supply, which depends in part on autogenic processes, can affect the timing of formation of all types of "sequence boundary" (i.e., subaerial unconformity, maximum flooding surface, and maximum regressive surface; Catuneanu and Zecchin, 2016). The distinction between allogenic and autogenic controls on sedimentation and the development of unconformities is sometimes evident (e.g., when a relation to the relative sea level can be established), but it cannot be generalized. For example, subaerial unconformities are often assumed to be the product of negative accommodation in both downstream- and upstream-controlled settings. However, such unconformities can also form during transgression in downstream-controlled settings (i.e., interplay of positive accommodation and sedimentation; Fig. 4.39B), and in response to a variety of processes in upstream-controlled settings, some of which are unrelated to accommodation (Fig. 3.11; Catuneanu, 2019a; see discussion in Chapter 4).

5.3.2 Scale of sequences

Stratigraphic cyclicity can be observed at different scales, depending on the purpose of the study and the resolution of the data available. The range of stratigraphic scales extends from 10^2 to 10^8 yrs. Schlager (2010) also recognized stratigraphic cycles at timescales of $10^0 - 10^1$ yrs, but these periods of time become too short for the formation of geologically recognizable soil horizons as evidence of subaerial unconformities, and the accumulation of depositional elements that define depositional systems. Therefore, realistic timescales for a sequence stratigraphic analysis start with

10^2-10^3 yrs (Catuneanu, 2017, 2019a,b; Fig. 4.62). At each scale of observation (i.e., hierarchical level), sequences consist of systems tracts and component depositional systems (Fig. 4.3). A stratigraphic cycle assumes no repetition of systems tracts and sequence stratigraphic surfaces within the sequence, at the selected scale of observation; i.e., systems tracts and bounding surfaces of the hierarchical rank of the host sequence cannot occur more than once within the sequence.

Stratigraphic sequences of different scales and hierarchical ranks may develop in any type of sedimentary basin, whether connected or not to the global ocean. However, the smallest scale sequences are prone to

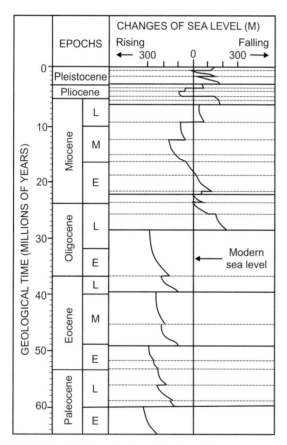

FIGURE 5.88 Global cycle chart of sea-level changes based on the interpretation of coastal onlap on seismic lines (redrafted from Vail et al., 1977b). The flaw of this reconstruction is the assumed relationship between onlap and sea level, whereby onlap was taken as evidence of sea-level rise. The "coastal onlap" in the early days of seismic stratigraphy included an undifferentiated mix of fluvial, coastal and marine onlap. Fluvial and marine onlap can form during either relative sea-level rise or fall, and all types of onlap can form during sea-level fall, where subsidence outpaces sea-level fall at the shoreline (i.e., relative sea-level rise). The error in the original assumption led to the unnatural "saw-tooth" shape of the sea-level curve, which postulates instantaneous sea-level fall. Seismic data capture the combination of multiple controls on accommodation and sedimentation. Isotope data provide the means to separate and reconstruct the record of sea-level changes (Fig. 5.23).

form in basins disconnected from the global ocean, such as lakes or interior seas, which are more susceptible to lake/sea-level changes in response to small variations in climatic conditions or local tectonism. As such, while the highest frequency sequences in pericontinental settings are commonly recorded on millennial timescales (e.g., Amorosi et al., 2005, 2009, 2017, Fig. 5.5), interior basins offer conditions for the development of full-fledged depositional sequences on centennial timescales. For example, sea-level cycles during the Holocene evolution of the Dead Sea led to the formation of 10^0-10^1 m thick depositional sequences on 700−800 yrs timescales (Bookman et al., 2004, 2006; Moran, 2020). These centennial-scale sequences correspond to full cycles of positive and negative accommodation, and include all downstream-controlled systems tracts (i.e., falling-stage, lowstand, transgressive, and highstand; Moran, 2020).

Stratigraphic cycles may be symmetrical or asymmetrical, and the corresponding sequences may include a variable number of systems tracts, up to a maximum of four in the case of downstream-controlled sequences that develop and preserve all systems tracts (Figs. 1.23, 5.8, and 4.33). Therefore, a sequence is not defined by its internal makeup, but by the recurrence of the sequence stratigraphic surface that marks its boundaries. Not all types of stratal stacking patterns (i.e., systems tracts) and bounding sequence stratigraphic surfaces may occur in the succession under analysis. For example, stratigraphic cyclicity may be defined by the repetition of transgressions and highstand normal regressions, without intervening stages of forced regression and lowstand normal regression (Figs. 5.1 and 5.5), or by the repetition of forced regressions and lowstand normal regressions, without intervening stages of transgression and highstand normal regression (Figs. 1.23, 4.33, and 5.90). Therefore, the types of recurring stacking patterns and bounding surfaces that define stratigraphic cyclicity may vary with the case study, which underlines the need for a model-independent approach to the sequence stratigraphic analysis.

Sequences of all scales may include unconformities of equal hierarchical rank (e.g., a depositional sequence may include unconformities associated with transgression; a genetic stratigraphic sequence may include a subaerial unconformity). Therefore, sequences may or may not be relatively conformable successions, but they always consist of genetically related strata that belong to the same cycle of change in stratigraphic stacking pattern (Catuneanu et al., 2009). For example, a genetic stratigraphic sequence that includes a subaerial unconformity of equal hierarchical rank cannot be described as a "relatively conformable succession" within the area of development of the unconformity, but it does consist of "genetically related strata" that belong to the same stratigraphic cycle. The subaerial unconformity

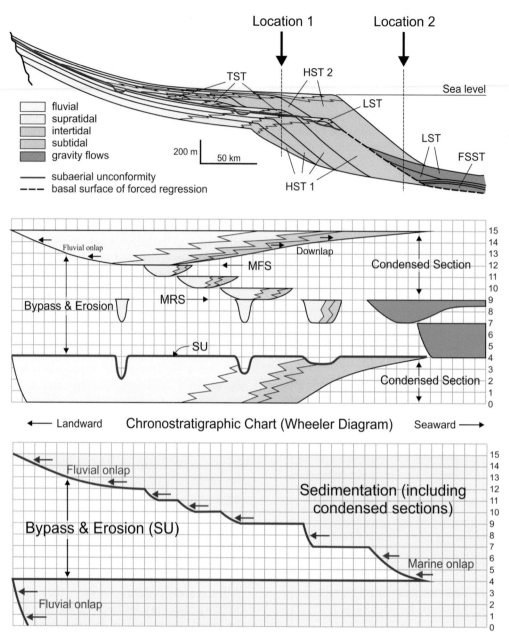

FIGURE 5.89   Stratigraphic architecture of downstream-controlled systems tracts in depth and time domains (diagrams courtesy of Henry Posamentier). In the early days of seismic stratigraphy, all types of onlap were identified as "coastal," leading to the unnatural "saw-tooth" shape of the sea-level curve which postulated instantaneous sea-level fall (Fig. 5.88).

has a smaller geographic extent than the genetic sequence hosting it, which maintains the integrity of the sequence as one stratigraphic cycle (Galloway, 1989). Beyond the termination of the unconformity, the correlative conformity preserves the continuity of sedimentation and the genetic relationship between the underlying and the overlying strata within the sequence. Unconformities with basinwide extent become sequence boundaries by default, at the scale of the underlying and overlying sequences.

Stratigraphic sequences of lower hierarchical ranks are nested within higher rank systems tracts, as illustrated by outcrop-scale sequences (10^0–10^1 m) that commonly build seismic-scale (10^1–10^2 m) systems tracts (Figs. 4.3, 4.70, and 5.1). The scale of observation is set by the resolution of the data available or by the purpose of the study (e.g., basin analysis: 10^2–10^3 m; petroleum exploration: 10^1–10^2 m; petroleum production development: 10^0–10^1 m; cyclostratigraphy of astronomical and solar radiation cycles: potentially reaching sub-meter scales). The architecture of sequences becomes increasingly complex with the increase in the scale of observation (Fig. 4.3). Beyond this general trend, there are no standards for the scale and internal makeup of sequences of any hierarchical rank; the scale, the systems tract composition, and the relative development of systems tracts within sequences vary with the tectonic, climatic, and depositional settings (e.g., Fielding et al., 2006, 2008; Martins-Neto and Catuneanu, 2010; Csato and Catuneanu, 2012). The scale and the internal makeup of sequences are basin or even sub-basin specific, reflecting the influence of local controls on accommodation and sedimentation.

High-frequency sequences (and component systems tracts and depositional systems) are commonly observed at scales of 10^0-10^1 m and 10^2-10^5 yrs (e.g., Tesson et al., 1990; 2000; Lobo et al., 2004; Amorosi et al., 2005; 2009; 2017; Bassetti et al., 2008; Nanson et al., 2013; Nixon et al., 2014; Magalhaes et al., 2015; 2020; Ainsworth et al., 2017; Pellegrini et al., 2017, 2018; Zecchin et al., 2017a,b; Catuneanu, 2019b; Moran, 2020), which defines the scope of high-resolution sequence stratigraphy (Figs. 4.62 and 5.4). The stacking pattern of high-frequency sequences defines systems tracts and component depositional systems of higher hierarchical ranks in lower resolution studies (Figs. 4.6, 4.58, 4.70, 5.1, and 5.5). Despite the nested architecture of stratigraphic cycles, the stratigraphic record is not truly fractal as sequences of different scales, as well as sequences of similar scales, may have different origins and internal composition of systems tracts (see Chapter 7 for more details).

5.3.3 Types of sequences

The concept of "sequence" evolved since the 1940s, in terms of the scale of a sequence and the nature of the sequence boundary. The original concept regarded sequences as large-scale units bounded by the most significant, basin-scale unconformities (Longwell, 1949; Sloss et al., 1949). Subsequent increases in stratigraphic resolution led to the identification of sequences at progressively smaller scales (Fig. 1.22). This required unconformities of smaller extent and magnitudes, as well as correlative conformities beyond the areas of

development of the unconformities, to be employed as sequence boundaries (Wheeler, 1964; Fig. 1.21). At the same time, the types of surfaces selected as the "sequence boundary" diversified as well, from surfaces of subaerial exposure (Sloss, 1963; Mitchum et al., 1977; Posamentier et al., 1988; Van Wagoner et al., 1988) to maximum flooding surfaces (Frazier, 1974; Galloway, 1989) and maximum regressive surfaces (Johnson and Murphy, 1984; Embry and Johannessen, 1992) (Figs. 1.9 and 1.10). In this process, conformable surfaces have become increasingly important to the delineation of sequences, to the point that some types of high-frequency sequences (e.g., genetic stratigraphic) no longer require unconformities at the sequence boundary.

Correlative conformities enabled the leap to higher resolution studies, and are now an integral part of modern sequence stratigraphy. While instrumental to the refinement of the sequence stratigraphic methodology, correlative conformities have also been a source of confusion and disagreements with respect to their timing and physical attributes in the rock record. Fig. 5.91 illustrates the position (timing of formation) of the correlative conformity (i.e., the conformable portion of the sequence boundary) in six different sequence stratigraphic models. Interesting to note is the fact that, with the exception of Galloway (1989), five out of these six models use the subaerial unconformity as the unconformable portion of the sequence boundary. Therefore, the difference between these models is most evident in the marine portions of sedimentary basins, where sequence boundaries are picked within conformable successions. As model "F" in Fig. 5.91 is an evolution of model "A," only five distinct models are left to consider, as illustrated in Fig. 1.10. All these models include correlative conformities, albeit with different names (Fig. 5.91).

Among the five sequence models in Fig. 1.10, three fall under the category of "depositional sequences," which are all anchored to the use of the subaerial unconformity as the fundamental portion of the sequence boundary. These models stem from the original

FIGURE 5.90 Stratigraphic cyclicity defined by an alternation of forced regressions (orange arrows: downstepping clinoform rollovers) and lowstand normal regressions (yellow arrows: upstepping clinoform rollovers) (Late Pleistocene, Le Castella, Calabria, southern Italy; from Zecchin and Catuneanu, 2013; modified after Zecchin et al., 2010a).

depositional sequence of seismic stratigraphy (Mitchum et al., 1977; Fig. 1.9), but differ in terms of the selection of the conformable portion of the sequence boundary and/or the definition of systems tracts (Fig. 1.10). In contrast, the "genetic stratigraphic sequence" model requires the use of the maximum flooding surface as the sequence boundary, whereas the "transgressive–regressive sequence" model is centered on the use of the maximum regressive surface as the sequence boundary (Fig. 5.92). With the exception of the "correlative conformities" that form at the onset and at the end of relative sea-level fall (i.e., the basal surface of forced regression and the correlative conformity in Fig. 4.6, respectively), whose timing is independent of sedimentation, the formation of all other types of sequence boundaries may be influenced by both accommodation (i.e., relative sea-level changes) and sedimentation (i.e., base-level changes) (see Chapter 4 for details).

None of the three types of sequences (i.e., depositional, genetic stratigraphic, and transgressive–regressive; Figs. 1.9, 1.10, and 5.92) can describe the stratigraphic cyclicity in all geological settings. In upstream-controlled settings, depositional sequences provide the suitable option since only subaerial unconformities are available to define sequence boundaries (Fig. 4.64). At the opposite end of the spectrum, sequences that form during continuous relative sea-level rise in downstream-controlled settings may be missing subaerial unconformities, in which case only genetic stratigraphic or transgressive–regressive sequences can be used to delineate stratigraphic cyclicity (Fig. 4.33). It follows that all models have limitations, as none is applicable to the entire range of case studies. The modern approach to the definition of stratigraphic sequences (Fig. 1.22) provides the flexibility to apply the sequence stratigraphic methodology in a manner that is independent of model and geological setting, and at all scales afforded by the data available. The model-independent approach to stratigraphic analysis transcends the differences between the various sequence models, and enables a data-driven methodology that is free of any predetermined model templates.

The defining features of the different types of sequences are outlined below. Ultimately, all models describe the same stratigraphic successions only using different terminology and style of delineation of sequences and component systems tracts. The contrasts between the different sequence models are explained in order to facilitate communication among practitioners embracing alternative approaches to stratigraphic analysis. This discussion should also help the less experienced practitioners to rationalize the apparently conflicting stratigraphic information published by researchers who follow different schools of thought.

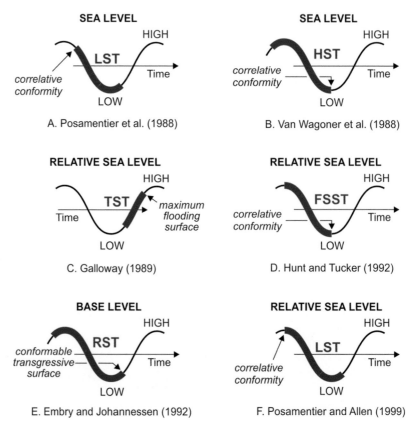

FIGURE 5.91 Timing of the conformable portion of the sequence boundary in light of different models. These "correlative conformities" correspond to different types of stratigraphic surfaces, including the "correlative conformity" in the sense of Hunt and Tucker (1992) (same age as the basinward termination of the subaerial unconformity), the "basal surface of forced regression" (i.e., the "correlative conformity" in the sense of Posamentier and Allen, 1999; older than the basinward termination of the subaerial unconformity), the "maximum regressive surface" (i.e., the "conformable transgressive surface" of Embry and Johannessen, 1992; younger than the basinward termination of the subaerial unconformity), and the "maximum flooding surface." Abbreviations: FSST—falling-stage systems tract; LST—lowstand systems tract; TST—transgressive systems tract; HST—highstand systems tract.

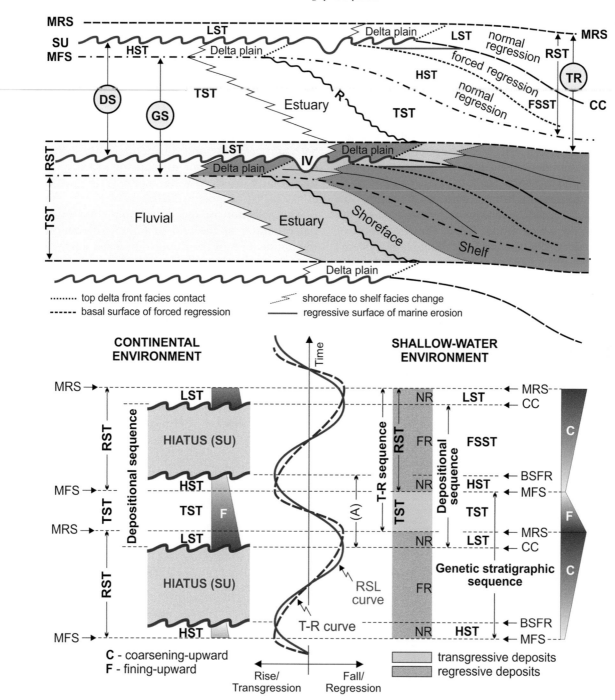

FIGURE 5.92 Sequences, systems tracts and bounding surfaces in downstream-controlled settings. Departures from these general trends may be encountered as a function of local conditions of accommodation and sedimentation (Fig. 4.33; details in Chapter 4). Abbreviations: SU—subaerial unconformity; CC—correlative conformity in the sense of Hunt and Tucker (1992); BSFR—basal surface of forced regression (correlative conformity in the sense of Posamentier and Allen, 1999); MRS—maximum regressive surface; MFS—maximum flooding surface; R—wave-ravinement surface; IV—incised valley; (A)—positive accommodation; NR—normal regression; FR—forced regression; LST—lowstand systems tract; TST—transgressive systems tract; HST—highstand systems tract; FSST—falling-stage systems tract; RST—regressive systems tract; DS—depositional sequence; GS—genetic stratigraphic sequence; TR—transgressive-regressive sequence.

5.3.3.1 Depositional sequences

The depositional sequence is a stratigraphic sequence bounded by subaerial unconformities or their correlative conformities (Mitchum, 1977). Different types of depositional sequences have been defined (Fig. 1.9), which can be classified into two groups as a function of the timing of the marine portion of the sequence boundary (i.e., the "correlative conformity"; Fig. 1.10):

one group considers the correlative conformity as the paleoseafloor at the onset of forced regression (i.e., the correlative conformity *sensu* Posamentier et al., 1988; herein referred to as the "basal surface of forced regression"); and another group considers the correlative conformity as the paleoseafloor at the end of forced regression (i.e., the correlative conformity *sensu* Van Wagoner et al., 1988; herein referred to as the "correlative conformity"). This difference of opinion was underlain by the assignment of the forced regressive deposits to the lowstand systems tract by Posamentier et al. (1988), but to the highstand systems tract by Van Wagoner et al. (1988). Subsequent refinements of the depositional sequence model led to the separation of forced regressive deposits as the falling-stage systems tract, and the placement of the correlative conformity at the end of forced regression (Figs. 1.10, 5.92, and 5.93).

The most recent depositional sequence model (Hunt and Tucker, 1992) presents the advantage of separating and assigning forced and normal regressions to different systems tracts. This distinction is necessary in light of the different sediment dispersal patterns associated with the two types of regressions, which is significant on several grounds, including the distribution of petroleum plays (Posamentier et al., 1992b; Posamentier and Morris, 2000). The proposal of two distinct "correlative conformities" in the 1980s led to confusion, and

ultimately a choice had to be made with respect to the nomenclature of the two sequence stratigraphic surfaces. Since the paleoseafloor at the end of forced regression is the surface that correlates with the basinward termination of the subaerial unconformity, the term "correlative conformity" was retained for this surface, while the base of forced regressive strata was termed the "basal surface of forced regression" (Figs. 4.6, 5.13, 5.92, and 5.93). Beyond this nomenclatural choice, both surfaces are important to map for practical purposes, particularly in deep-water settings (Fig. 4.20; Sweet et al., 2019; Catuneanu, 2020a).

The selection of the correlative conformity as the paleoseafloor at the end of forced regression affords the delineation of the depositional sequence boundary as a throughgoing surface which is intercepted only once at any location (Fig. 4.6). In contrast, the subaerial unconformity and the basal surface of forced regression may be intercepted at different stratigraphic levels within the area of forced regression, where they are separated by forced regressive deposits (Figs. 4.6, 5.12, 5.13, and 5.14). This poses a methodological challenge to the delineation of the sequence boundary, although a distinction can be made between the proximal portion of the subaerial unconformity at the top of the highstand topset (i.e., the "master" sequence boundary; Posamentier and Allen, 1999) and the distal

FIGURE 5.93   Delineation of the depositional sequence boundary in fluvial to shallow-water settings, according to different models. Depositional sequence II refers to Posamentier et al. (1988). Depositional sequences III and IV refer to Van Wagoner et al. (1988) and Hunt and Tucker (1992), respectively (Fig. 1.10). The correlative conformity of depositional sequence II is truncated by the subaerial unconformity at the location of the shoreline at the onset of forced regression, whereas the correlative conformity of depositional sequences III and IV meets the basinward termination of the subaerial unconformity at the location of the shoreline at the end of forced regression; hence, the nomenclatural change of CC1 to the "basal surface of forced regression"). Abbreviations: HST—highstand systems tract; RSL—relative sea level; (1) to (8)—prograding clinoforms; SU—subaerial unconformity; CC1—correlative conformity in the sense of Posamentier et al. (1988); CC2—correlative conformity in the sense of Van Wagoner et al. (1988) and Hunt and Tucker (1992); RSME—regressive surface of marine erosion.

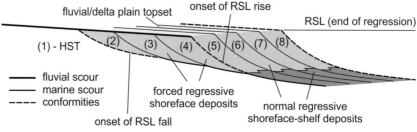

A. Lowstand systems tract (*sensu* Posamentier et al., 1988):

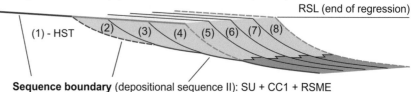

B. Sequence boundary (*sensu* Posamentier et al., 1988):

Sequence boundary (depositional sequence II): SU + CC1 + RSME

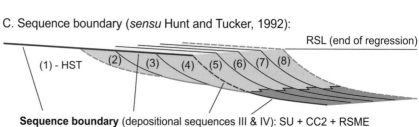

C. Sequence boundary (*sensu* Hunt and Tucker, 1992):

Sequence boundary (depositional sequences III & IV): SU + CC2 + RSME

portion of the unconformity at the top of the forced regressive clinoforms. However, the latter portion of the unconformity is also significant in terms of the development of stratigraphic petroleum plays (e.g., in relation to incised valleys or forced regressive coastal and shallow-water systems), so it is common practice to map the subaerial unconformity as one physical surface across its entire area of development (Figs. 4.6, 5.12, 5.13, and 5.14).

Regardless of the depositional sequence model employed, the definition of the sequence boundary as the "subaerial unconformity and its correlative conformity" is oversimplified. In practice, there is a high probability that at least part of the correlative conformity that forms in shallow-water settings is reworked and replaced by the regressive surface of marine erosion (Fig. 5.93). As such, the regressive surface of marine erosion becomes part of the depositional sequence boundary, whether the correlative conformity is picked at the base or at the top of forced regressive deposits (Fig. 5.93; more examples in Chapter 6). Attempts were also made to classify depositional sequence boundaries into a "type 1" vs. a "type 2," depending on the extent and the magnitude of erosion associated with the subaerial unconformities; i.e., widespread erosion in the case of "type 1" sequence boundaries vs. minimal erosion in the case of "type 2" sequence boundaries (Vail et al., 1984; Posamentier and Vail, 1988). However, since between these end members any situations are possible across a continuum of stratigraphic scales, the usage of the "type 1" and "type 2" terminology was eventually discarded (Posamentier and Allen, 1999; see the discussion of systems tracts above, for more details).

The concept of depositional sequence applies to both downstream- and upstream-controlled settings, where subaerial unconformities may form. All types of depositional sequences, as originally defined, assume full cycles of change in accommodation and relate the sequence boundary to stages of negative accommodation (Vail, 1987; Posamentier and Vail, 1988; Van Wagoner et al., 1988; Hunt and Tucker, 1992; Duval et al., 1998; Leckie and Boyd, 2003; Neal and Abreu, 2009). However, subaerial unconformities in downstream-controlled settings may also form during stages of relative sea/lake-level rise and transgression (Figs. 4.28 and 4.39B). Similarly, depositional sequences in upstream-controlled settings may correspond to tectonic cycles, climate cycles, or cycles of autogenic migration of alluvial channel belts, with or without changes in accommodation (see Chapter 4 for details). Therefore, the interpretation of the underlying controls that are responsible for the formation of depositional sequences needs to be separated from the methodological workflow which relies on the observation of stratal stacking patterns.

5.3.3.2 Genetic stratigraphic sequences

The genetic stratigraphic sequence is a stratigraphic sequence bounded by maximum flooding surfaces, which can be traced on either side of the coastline in downstream-controlled settings (Figs. 5.50 and 5.91; Galloway, 1989). One argument for this choice of sequence boundary is that the main changes in the paleogeographic distribution of depositional systems and depocenters occur during times of maximum shoreline transgression (Galloway, 1989). As maximum flooding surfaces form during stages of positive accommodation, the formation of genetic stratigraphic sequences does not require stages of negative accommodation. At any scale of observation, a genetic stratigraphic sequence corresponds to a regressive—transgressive cycle, which may occur during a full cycle of change in accommodation or during a stage of positive accommodation (Fig. 4.33). In the latter case, the genetic stratigraphic sequence does not include falling-stage and lowstand systems tracts, nor any sequence stratigraphic surface that is exclusively associated with forced regression (i.e., the basal surface of forced regression, the regressive surface of marine erosion, and the correlative conformity; Fig. 4.39).

Genetic stratigraphic sequences may include subaerial unconformities of equal hierarchical rank (Frazier, 1974; Galloway, 1989). In this case, the genetic stratigraphic sequence cannot be described as a "relatively conformable succession" within the area of development of the subaerial unconformity, although it does consist of "genetically related strata" which belong to the same stratigraphic cycle of change in stratal stacking patterns; therefore, satisfying the definition of a stratigraphic sequence (Fig. 1.22; Galloway, 1989; Catuneanu et al., 2009, 2011). The internal subaerial unconformity has a smaller extent than the host genetic sequence, as it otherwise becomes a depositional sequence boundary by default; this maintains the integrity of the genetic sequence as one stratigraphic cycle. Beyond the termination of the subaerial unconformity, the correlative conformity preserves the continuity of sedimentation and the genetic relationship between the underlying and the overlying strata within the sequence.

The genetic stratigraphic sequence approach does not rely on the development and recognition of subaerial unconformities and correlative conformities, which is an advantage where stratigraphic cyclicity develops within conformable successions, or where the lack of seismic data prevents the observation of stratal terminations that facilitate the identification of subaerial unconformities and correlative conformities. Instead, the physical record of transgression provides "readily recognized regionally correlative, easily and accurately datable, and robust sequence boundaries" (Galloway, 1989). Maximum flooding surfaces are among the most reliable and easily identifiable stratigraphic markers on well logs, particularly in shallow-water settings, as well as by means of biostratigraphic and geochemical data in all depositional systems (more details in Chapters 2 and 6). They also typically provide the best reference surfaces for correlation and flattening of well-log cross sections and seismic volumes (Fig. 1.11). For these reasons, it is common practice to start a stratigraphic analysis with a survey of maximum flooding surfaces, no matter the model of choice of the practitioner. The concept of genetic stratigraphic sequence applies to downstream-controlled settings, where maximum flooding surfaces may form.

5.3.3.3 Transgressive–regressive sequences

The transgressive–regressive (T-R) sequence is a stratigraphic sequence bounded by maximum regressive surfaces (Johnson and Murphy, 1984). Maximum regressive surfaces can be traced on either side of the coastline in downstream-controlled settings, and present the advantage of being readily identifiable, particularly in fluvial to shallow-water systems, both in outcrop and in the subsurface (Figs. 4.6 and 5.50). In its original definition, the formation of a T-R sequence does not require negative accommodation. As maximum regressive surfaces of any hierarchical rank may form during stages of positive accommodation, T-R sequences may be generated either during full cycles of change in accommodation or during periods of positive accommodation. In the latter case, the T-R sequence consists solely of transgressive and highstand systems tracts (Fig. 4.33). The T-R sequence boundary is also termed "transgressive surface," where emphasis is placed on the onset of transgression rather than the end of regression (details in Chapter 6).

A proposal to modify the original definition of the T-R sequence was made to include the subaerial unconformity as the continental portion of the sequence boundary (Embry and Johannessen, 1992). The intention was to combine the strength of the depositional sequence model (i.e., the use of the most prominent unconformities as part of the sequence boundary) with the strength of the original T-R sequence model (i.e., the mappability of maximum regressive surfaces in shallow-water settings). However, as the maximum regressive surface is most commonly younger than the subaerial unconformity, the marine portion of the maximum regressive surface may not meet with the basinward termination of the subaerial unconformity (Embry and Johannessen, 1992; Embry, 1995; Fig. 4.6). In this case, the maximum regressive surface and the subaerial unconformity occur at different stratigraphic levels in one vertical section (e.g., they can be intercepted at two different depths in a well), separated by the topset of the lowstand systems tract (Figs. 4.6, 4.39A,

and 5.92). Therefore, the original definition of Johnson and Murphy (1984) is still recommended, as it provides a foolproof approach.

It is noteworthy that in the 1990s, when Embry and Johannessen (1992) proposed their amendment to the definition of a T-R sequence, the criteria available to identify correlative conformities were largely restricted to seismic data (e.g., the oldest and the youngest clinoform surfaces associated with offlap; Fig. 4.6). The lack of criteria to pinpoint correlative conformities in outcrop and borehole studies was a weakness of the depositional sequence model, which the use of maximum regressive surfaces was meant to rectify. The difficulty in recognizing correlative conformities in outcrops and boreholes also led to the proposal of a "regressive systems tract" which amalgamates all forced and normal regressive deposits of a stratigraphic cycle (Figs. 1.10, 5.91, and 5.92). These amendments did not receive widespread acceptance, due to their conceptual and practical limitations; i.e., the lack of temporal and physical correlation between the continental and marine portions of the sequence boundary, and the low resolution (and hence limited utility) of the regressive systems tract.

A temporal correlation between the subaerial unconformity and the maximum regressive surface may only be achieved where the lowstand systems tract is missing, which is possible under special circumstances (e.g., in fault-bounded basins where accommodation is created abruptly at the onset of relative sea-level rise by the reactivation of faults; Martins-Neto and Catuneanu, 2010). The physical connection between the two surfaces may be made by the transgressive surface of erosion, if wave scouring during transgression removes the entire lowstand topset (Fig. 5.14). This may only happen where the thickness of the lowstand topset is less than the depth of the fairweather wave base, which is the maximum amount of erosion that can be attributed to wave-ravinement processes (Demarest and Kraft, 1987). Thicker lowstand topsets are preserved in part during transgression, preventing the subaerial unconformity and the maximum regressive surface from forming a throughgoing sequence boundary (Figs. 4.6 and 5.50). This is the norm in most sedimentary basins that include shelf-slope systems (Plint and Nummedal, 2000; Cathro et al., 2003; Hampson and Storms, 2003; Ainsworth, 2005; Schlager, 2005; Dominguez et al., 2020).

More recent developments in high-resolution sequence stratigraphy led to the definition of additional criteria that afford the identification of correlative conformities in core and outcrops (e.g., MacEachern et al., 2012; Catuneanu, 2020a,c; details and examples in Chapter 6). This enhances the applicability of depositional sequences, and enables the separation between forced and normal regressions which is fundamental to the efforts of petroleum exploration. For example, the correlative conformity in the sense of Hunt and Tucker (1992) (Figs. 4.6, 5.91, and 5.93) is arguably the most significant sequence stratigraphic surface in the deep-water setting in terms of locating the best petroleum reservoirs (Fig. 4.20), but it is overlooked within the undifferentiated regressive systems tract of Embry and Johannessen (1992). This limits the utility of the T-R sequence model, which is outdated by the higher resolution approach to the subdivision of the stratigraphic framework (Figs. 4.6, 5.92, and 5.93).

The original T-R sequence model of Johnson and Murphy (1984) remains the most dependable and logical approach to the delineation of transgressive−regressive cycles in both continental and marine settings, and its usage is justified where the maximum regressive surface is the only mappable or the most prominent type of sequence boundary within the studied section. The concept of T-R sequence applies to downstream-controlled settings, where maximum regressive surfaces may form. In most cases, T-R sequences and genetic stratigraphic sequences provide alternative approaches that are equally applicable, as both models require the formation of maximum regressive and maximum flooding surfaces (Fig. 1.11). The type of stratigraphic sequence that is most pertinent to a case study (i.e., depositional, genetic stratigraphic, or T-R) ultimately depends on the formation and mappability of its bounding surfaces at the selected scale of observation. All three types of sequences may be observed at different scales, and their applicability at any specific scale depends on the geological setting and the types and resolution of the data available (more details in Chapter 7).

5.4 Parasequences

The sequence stratigraphic framework records a nested architecture of stratigraphic cycles that can be observed at different scales, depending on the purpose of the study and the resolution of the data available. Much discussion and controversy surrounded the classification and nomenclature of these cycles, with opinions ranging from the use of the "sequence" concept at all stratigraphic scales (Vail et al., 1977b; Posamentier and Allen, 1999) to the use of different terms at different stratigraphic scales (Van Wagoner et al., 1990; Mitchum and Van Wagoner, 1991; Sprague et al., 2003; Neal and Abreu, 2009). In the latter approach, the "sequence" is a "relatively conformable" succession relative to which smaller cycles (parasequences) and larger cycles (composite sequences, megasequences) have been defined. This nomenclatural debacle has implications for the methodology, and so it is important to resolve.

Perhaps the most contentious type of stratal unit in the scale-variant classification system is the

"parasequence," due to its allostratigraphic rather than sequence stratigraphic affinity (Catuneanu and Zecchin, 2020). Problems with the parasequence concept have been pointed out in numerous studies (e.g., Krapez, 1996; Posamentier and Allen, 1999; Strasser et al., 1999; Catuneanu, 2006; Zecchin, 2007, 2010; Catuneanu et al., 2009; 2010, 2011; Miall, 2010). Parasequences are bounded by facies contacts (i.e., "flooding surfaces") which may or may not coincide with sequence stratigraphic surfaces, and are usually assumed as shallowing-upward units with only minor or no transgressive deposits (Fig. 5.94). Most authors consider parasequences as units developed without intervening stages of relative sea-level fall, despite the fact that they may pass laterally into units of the same rank that record full cycles of relative sea-level change (Zecchin, 2010). This is the case in tectonically active basins where subsidence and uplift can occur at the same time along the shoreline of an interior seaway, leading to the coeval formation of depositional sequences and parasequences (e.g., Catuneanu et al., 2002: forelands; Gawthorpe et al., 2003: half-graben rift basins). There is also evidence of significant variability in the composition of parasequences, which may consist of successions dominated by either shallowing- or deepening-upward trends (e.g. Kidwell, 1997; Saul et al., 1999; Di Celma et al., 2005; Zecchin, 2005, 2007; Di Celma and Cantalamessa, 2007; Spence and Tucker, 2007; Amorosi et al., 2017; Bruno et al., 2017; Zecchin et al., 2017a,b; Fig. 5.95).

In spite of the progress made by the publication of formal guidelines for sequence stratigraphy (Catuneanu et al., 2011), confusion still persists with respect to a number of key issues, including the scale of sequences and the difference between high-frequency sequences and parasequences. Some of these confusions are rooted in the historical development of the method, and stem from the scales of observation imposed by the resolution of the data that were used to define the concepts (e.g., in the context of seismic stratigraphy in the 1970s, the scale of sequences, systems tracts, and depositional systems had to exceed, by default, the vertical resolution of seismic data). This section revisits the reasons for this nomenclatural conundrum, the nature of parasequences as stratigraphic units, and the solution for a standard nomenclature that is in line with the modern principles and realities of sequence stratigraphy.

5.4.1 Definition

The parasequence is a succession of genetically related beds and bedsets bounded by flooding surfaces (Fig. 5.94; Van Wagoner et al., 1988, 1990). A flooding surface is a facies contact that marks an abrupt increase in water depth and, consequently, an abrupt shift to

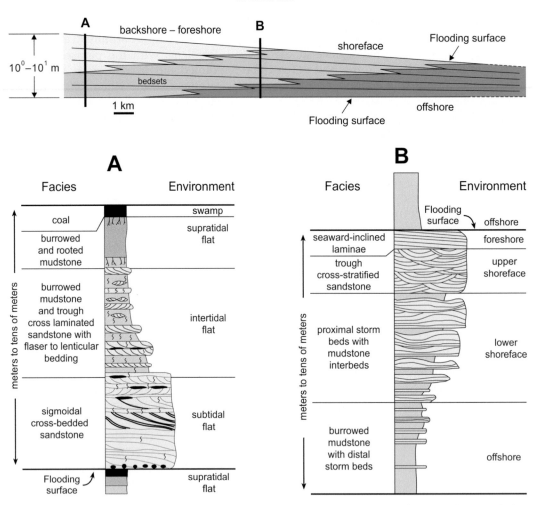

FIGURE 5.94   Schematic cross section of a parasequence along the depositional dip, and two vertical sections showing ideal parasequences in proximal (A) and distal (B) locations (vertical sections courtesy of Steven Holland; modified after Van Wagoner et al., 1990).

relatively more distal facies on top (Fig. 5.94). Such lithological discontinuities are often represented by sandstone-to-shale contacts in siliciclastic systems or by limestone-to-shale contacts in carbonate systems. In the latter case, the flooding surface is also known as a "drowning unconformity" (Schlager, 1989). Parasequences may include both transgressive and regressive deposits (Arnott, 1995), with the latter commonly forming the bulk of the unit. For this reason, the

emphasis in the original definition was placed on the regressive portion of the parasequence, which was identified as "an upward-shallowing succession of facies bounded by marine flooding surfaces" (Fig. 5.94; Van Wagoner et al., 1988, 1990).

Parasequences are commonly identified with the coarsening-upward prograding lobes in coastal to shallow-water settings, where the deposition of each lobe is terminated by events of abrupt water deepening

FIGURE 5.95   Variability in the internal architecture of parasequences (modified after Zecchin, 2007). Abbreviations: MRS—maximum regressive surface; RS—ravinement surface; MFS—maximum flooding surface; DLS—downlap surface; R—regressive; T—transgressive.

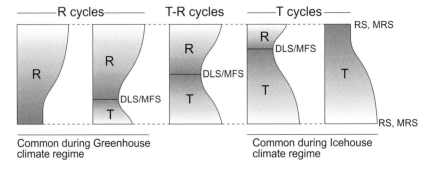

(Figs. 5.96 and 5.97). Such transgressive—regressive cycles may be observed in the context of longer term trends of coastal progradation or retrogradation, in which case they are part of larger scale systems tracts defined by the stacking pattern of parasequences (Figs. 4.29, 4.34, and 5.5). This approach is the byproduct of low-resolution seismic stratigraphy, in which the vertical seismic resolution (range of 10^1 m) defines the limit between the scales of sequences and parasequences. Subsequent developments in high-resolution sequence stratigraphy led to the recognition of sequences at parasequence scales, which initiated the debates over the usage of the "sequence" vs. "parasequence" nomenclature in high-resolution studies (Posamentier and James, 1993; Arnott, 1995; Kamola and Van Wagoner, 1995; Posamentier and Allen, 1999; Embry, 2005). Overall, the parasequence concept brought more confusion than benefits, as discussed below.

The "parasequence" of Van Wagoner et al. (1988) occupies a specific niche in the architecture of the sedimentary record, between sedimentological units (beds and bedsets; Campbell, 1967) and the systems tracts of seismic stratigraphy (Brown and Fisher, 1977). In this view, parasequences, which consist of beds and bedsets, represent the building blocks of seismic-scale systems tracts, and the smallest stratigraphic units at any location. The sedimentological makeup of parasequences may be described in terms of beds and bedsets (Campbell, 1967) or in terms of facies and facies successions (Walker, 1992). More important for stratigraphic analysis is the identification of parasequence boundaries, which may provide the means to subdivide stratigraphic successions into genetically related packages of strata separated by sharp facies contacts (Fig. 5.94). Parasequences may consist of variable facies successions, depending on depositional setting and the location within the basin, with the component facies accumulated in the order prescribed by Walther's Law (Figs. 5.94, 5.96, and 5.98).

The concept of parasequence applies to coastal and shallow-water settings, where flooding surfaces may form (Posamentier and Allen, 1999). Landward,

flooding surfaces may be traced within coastal plain settings, using facies analysis and coal-bed stratigraphy (Fig. 5.94; Van Wagoner et al., 1990; Ketzer et al., 2003a,b; Amorosi et al., 2005; Pattison, 2019), but do not extend beyond the zone of marine influence. Seaward, flooding surfaces gradually lose their identity as the increase in water depth eventually prevents the formation of contrasting facies that could be related to episodes of abrupt water deepening (Pattison, 2019; Fig. 5.99). Similar lithological contacts in deep-water settings (e.g., shale on sandstone, beyond the shelf edge) may be linked to the evolution of gravity flows, such as the avulsion of turbidite channels, rather than to episodes of water deepening. For these reasons, the proper use of the parasequence concept is restricted to coastal and shallow-water settings. Even in these settings, the identification of flooding surfaces requires evidence of transgression in order to avoid confusion with facies contacts generated by the abandonment of deltaic lobes as a result of delta-lobe switching without water deepening (Colombera and Mountney, 2020).

The utility of the parasequence concept in sequence stratigraphy is further impeded by the equivocal stratigraphic significance of the "flooding surface," which may coincide with transgressive ravinement surfaces, maximum regressive surfaces, maximum flooding surfaces, or various types of within-trend facies contacts (Fig. 5.100). Where the flooding surface coincides with a sequence stratigraphic surface (e.g., transgressive ravinement, maximum regressive, or maximum flooding), the sequence stratigraphic nomenclature conveys the precise significance of the contact. Only the abrupt facies contacts at the top of onlapping shell beds or transgressive lags (Fig. 5.101), as well as other within-trend flooding surfaces that may form as a result of sediment starvation within transgressive systems tracts (Fig. 5.100B) are flooding surfaces that do not overlap with any sequence stratigraphic surface. Whether coincident or not with a sequence stratigraphic surface, all types of flooding surfaces are allostratigraphic contacts (i.e., lithological discontinuities; Catuneanu and Zecchin, 2020).

FIGURE 5.96 Coarsening-upward parasequences in a shallow-water setting (Upper Cretaceous, Woodside Canyon, Utah). Parasequences are commonly dominated by progradational trends, meters to tens of meters thick, but exceptions may occur in terms of scales and internal makeup. More important to the definition of parasequences, flooding surfaces mark abrupt increases in water depth (arrows). In this example, parasequences are c. 10 m thick, and flooding surfaces coincide with maximum regressive surfaces.

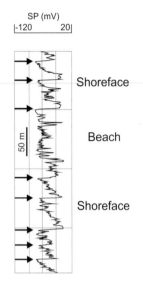

SP (mV)
|-120 20|

Shoreface

Beach

Shoreface

50 m

FIGURE 5.97 Well-log examples of parasequences in a shallow-water setting (Pliocene, Gulf of Mexico; well-log courtesy of PEMEX). Parasequence boundaries (i.e., flooding surfaces) are marked by arrows. In this example, parasequences are dominantly progradational ("R" type in Fig. 5.95). Depending on their timing (i.e., onset, during, or end of transgression), flooding surfaces may coincide with maximum regressive surfaces, within-trend facies contacts, or maximum flooding surfaces. Therefore, parasequences may correspond to different types of stratigraphic units observed at high-resolution scales.

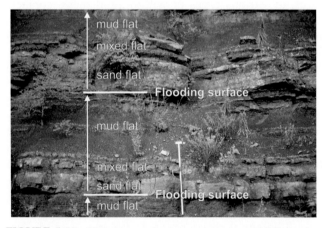

FIGURE 5.98 Fining-upward parasequences in a tidal flat setting (Ordovician Juniata Formation, Germany Valley, West Virginia; example courtesy of Steven Holland). Jacob staff is 1.5 m. In this example, flooding surfaces coincide with transgressive surfaces of erosion which replace maximum regressive surfaces.

Depending on the stratigraphic significance of the flooding surface, parasequences may be depositional sequences (where flooding surfaces are represented by wave-ravinement surfaces that rework subaerial unconformities; Fig. 5.100C), transgressive–regressive sequences (where flooding surfaces are maximum regressive surfaces; Fig. 5.100A), genetic stratigraphic sequences (where flooding surfaces are maximum flooding surface; Fig. 5.100D), or allostratigraphic units

(where flooding surfaces are within-trend facies contacts within transgressive systems tracts; Fig. 5.100B). Further confusion was introduced by the usage of the term to describe small-scale sequences developed during specific time intervals; e.g., in the sequence stratigraphic scheme of Krapez (1996), parasequences are equated with "fourth-order" sequences formed during 90,000–400,000 yrs time intervals. This approach was criticized by Posamentier and Allen (1999) who proposed a return to the original and scale-free definition of the parasequence.

The origin of the "parasequence" can be traced back to the concept of "paracycle" of relative sea level, defined as "the interval of time occupied by one regional or global relative rise and stillstand of sea level, followed by another relative rise, with no intervening relative fall" (Vail et al., 1977b). The corresponding stratal unit was termed "parasequence" (Van Wagoner, 1985, 1988), introduced as "a relatively conformable succession of genetically related beds and bedsets bounded by marine flooding surfaces and their correlative surfaces." This formulation emulates the definition of a "sequence" as "a relatively conformable succession of genetically related strata bounded by unconformities or their correlative conformities," coined in the context of seismic stratigraphy (Mitchum, 1977). However, significant differences between the two types of units were implied in terms of relative scales, origins, and bounding surfaces.

Important to the classification of stratigraphic cycles, the scales of sequences and parasequences at any location were inferred to be mutually exclusive, with parasequences being the building blocks of sequences and component systems tracts (Van Wagoner et al., 1988, 1990). Sequences were envisaged to represent full cycles of relative sea-level rise and fall, whereas parasequences were assumed to form during relative sea-level rise. The inferred link between the paracycle and the relative sea level implied an allogenic origin and widespread development of parasequences. However, it is now known that several allogenic and autogenic controls can interplay to generate parasequences, including eustasy, tectonism, compaction-driven subsidence, and autogenic changes in sediment supply (Fig. 3.8). Parasequences driven by autogenic processes may have a more limited extent, but their field expression is similar to that of allogenic cycles (Fig. 5.102). The interplay of allogenic and autogenic processes has been documented at multiple stratigraphic scales, starting with the smallest "parasequence" scales (Fig. 3.1). For this reason, the sequence stratigraphic methodology is now decoupled from the interpretation of underlying controls (Catuneanu, 2019a, 2020b).

The definitions of both sequences and parasequences make reference to "relatively conformable" and

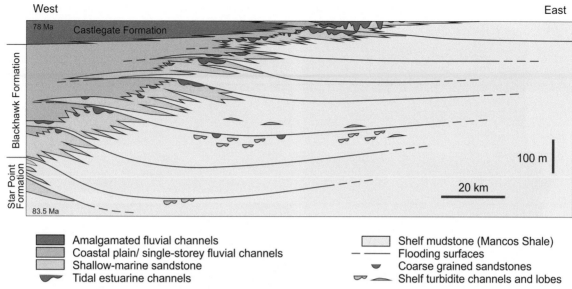

FIGURE 5.99 Dip-oriented cross section of the Book Cliffs region, from Helper Utah (West) to western Colorado (East) (modified after Pattison, 2019). At the low-resolution scale of this cross section, flooding surfaces are merged with maximum regressive, transgressive ravinement, and maximum flooding surfaces. Some of the shallow-water sandstones immediately underlying the flooding surfaces are forced regressive, in which case the contacts also include subaerial unconformities and stages of relative sea-level fall. At scales of 10^5 yrs, these "parasequences" are much more complex than the successions of beds and bedsets envisaged by Van Wagoner et al. (1990), and are better described as "depositional," "transgressive—regressive," or "genetic stratigraphic" sequences, depending on the precise surface that marks the boundary. Flooding surfaces loose their identity (physical expression as facies discontinuities) in the fluvial and basinal settings. However, maximum flooding surfaces observed at the same scales can be traced farther inland and basinward, based on sedimentological criteria (e.g., abundance of tidal structures and the development of coal beds in fluvial deposits, and marine progradational vs. retrogradational trends; Shanley et al., 1992; Shanley and McCabe, 1994; Fanti and Catuneanu, 2010), biostratigraphic criteria (e.g., palynological marine index in fluvial deposits, and abundance of microfossils in the deep-water setting; Helenes et al., 1998; Gutierrez Paredes et al., 2017), and geochemical criteria (e.g., cross calibration of organic and inorganic proxies; Dong et al., 2018; Harris et al., 2018; LaGrange et al., 2020). Therefore, genetic stratigraphic sequences observed at the scale of parasequences provide a superior alternative for correlation and the subdivision of the stratigraphic record.

"genetically related" packages of strata, implying that any interruptions in deposition during the accumulation of either type of unit are not significant enough to breach Walther's Law. Abrupt facies shifts that violate Walther's Law are expected at parasequence boundaries (i.e., flooding surfaces, at the contact between coastal or shallow-water facies below and deeper water facies above) and at the unconformable portions of sequence boundaries. However, if the scales of sequences and parasequences are mutually exclusive, and the latter are nested within the former, sequences could no longer be "relatively conformable." A solution to this inconsistency is the notion that "relatively conformable successions" can be observed at different scales, depending on the resolution of the stratigraphic study (Catuneanu, 2019b). In this case, the scale of a "relatively conformable succession" cannot be used as a reproducible reference for the classification of stratigraphic cycles.

The parasequence concept triggered confusion and controversy in sequence stratigraphy, due to its traits that are unlike the features of a sequence stratigraphic unit, including the restricted applicability to coastal and shallow-water settings, and the allostratigraphic rather than sequence stratigraphic nature of its

bounding surfaces. Attempts have been made to fix the concept by modifying the original definition to include all regional meter-scale cycles, whether or not bounded by flooding surfaces (Spence and Tucker, 2007; Tucker and Garland, 2010). However, this revised definition introduces even more ambiguity because it leaves the parasequence boundary unspecified, which creates confusion between different types of units that may develop at similar scales (e.g., Zecchin, 2007; Zecchin and Catuneanu, 2013; Catuneanu and Zecchin, 2013, 2020). Some of these meter-scale units may be genuine parasequences, whereas others are sequences of different kinds (Fig. 5.103). Since stratal units are defined by specific bounding surfaces, the concept of parasequence needs to be restricted to a unit bounded by flooding surfaces, in agreement with the original definition (Van Wagoner et al., 1988, 1990).

5.4.2 Scale of parasequences

The introduction of parasequences as the building blocks of seismic-scale systems tracts implied a two-tier system of stratigraphic scales, whereby parasequences and sequences, as well as "flooding" and "maximum

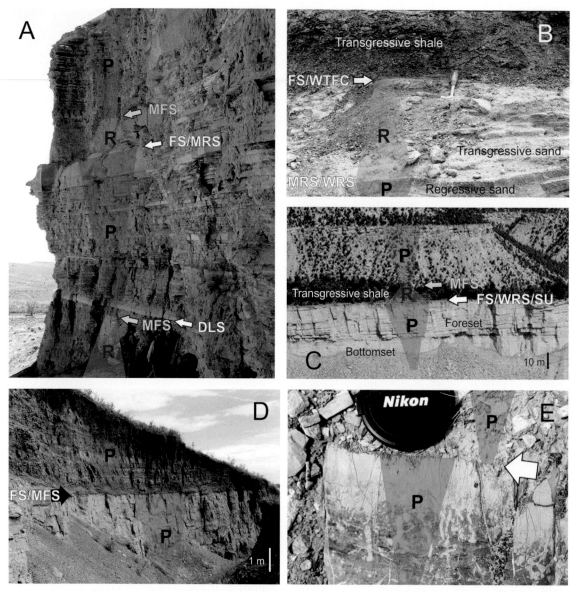

FIGURE 5.100   Stratigraphic meaning of flooding surfaces. Flooding surfaces are allostratigraphic contacts with variable meaning in sequence stratigraphy: A—flooding surface coincident with a maximum regressive surface (Upper Cretaceous, Western Interior, Utah); B—flooding surface represented by a within-trend facies contact (Upper Cretaceous, Western Interior, Alberta); C—flooding surface coincident with a transgressive ravinement surface that replaces a subaerial unconformity (Upper Cretaceous, Western Interior, Utah); D—flooding surface coincident with a maximum flooding surface (Upper Cretaceous, Western Interior, Alberta); E—flooding surface coincident with a maximum flooding surface, marked by bioturbation as a result of sediment starvation during transgression (Mississippian, Western Canada Basin, Alberta). Prograding deposits within parasequences can be normal regressive (A, B, D, and E) or forced regressive (C). In the latter case, parasequences accumulate during relative sea-level fall, and the flooding surface incorporates the hiatus of the subaerial unconformity (Fig. 4.47). Abbreviations: FS—flooding surface; SU—subaerial unconformity; MRS—maximum regressive surface; WRS—wave-ravinement surface; WTFC—within-trend facies contact; MFS—maximum flooding surface; DLS—downlap surface; P—progradation; R—retrogradation.

flooding" surfaces, would form at different hierarchical levels. In actuality, the difference between the two types of units is not a matter of scale, but a matter of definition of their bounding surfaces (i.e., lithological discontinuities that may develop during transgressions vs. surfaces that mark the end of transgressions, with or without a lithological expression; Figs. 5.100 and 5.104). Furthermore, there are more than two scales in stratigraphy, meaning that two types of surfaces

(flooding vs. maximum flooding) would not be enough to describe and differentiate, with a scale-dependent nomenclature, the multiple scales in the stratigraphic record (Fig. 5.105). The simplest and most objective solution is to keep the nomenclature independent of scale and the resolution of the data available, by applying the definitions consistently at all stratigraphic scales (e.g., see maximum flooding surfaces of different hierarchical ranks in Figs. 5.3, 5.104, and 5.105).

FIGURE 5.101 Transgressive lag and onlapping shell bed formed by wave reworking in the shoreface during shoreline transgression, bounded at the base by the wave-ravinement surface (WRS) and at the top by a within-trend flooding surface (FS). The FS marks an abrupt facies shift to the overlying deeper water deposits (Eocene, Talara Basin, Peru).

The concept of parasequence is commonly applied at scales of 10^0-10^1 m and 10^2-10^5 yrs, which coincide with the scales of high-resolution sequence stratigraphy. In contrast, the sequences of seismic stratigraphy are typically recognized at scales of 10^1-10^2 m and 10^5-10^6 yrs; (Vail et al., 1991, Duval et al., 1998; Schlager, 2010; Catuneanu, 2019a,b). The observation of sequences at seismic scales led to the proposal of a scale-variant hierarchy system which postulates orderly patterns in the sedimentary record (i.e., bedsets < parasequences < sequences; Van Wagoner et al., 1990; Sprague et al., 2003; Neal and Abreu, 2009; Abreu et al., 2010). However, this scheme does not provide a reproducible standard,

as parasequences and depositional sequences of equal hierarchical ranks can coincide (e.g., in the case of orbital cycles: Fig. 5.106; Strasser et al., 1999; Fielding et al., 2008; Tucker et al., 2009), or may form side by side in tectonically active basins (Catuneanu et al., 2002; Gawthorpe et al., 2003; Zecchin, 2010). As noted by Schlager (2010), "data on sequences of 10^3-10^7 years duration, the interval most relevant to practical application of sequence stratigraphy, do not conform well to the ordered-hierarchy model. Particularly unsatisfactory is the notion that the building blocks of classical sequences (approximate domain 10^5-10^6 years) are parasequences bounded by flooding surfaces (Van Wagoner et al., 1990; Duval et al., 1998)."

The early hypotheses about the origins and relative scales of sequences and parasequences proved to be contentious (Posamentier and Allen, 1999; Catuneanu, 2006; Catuneanu et al., 2009, 2011; Miall, 2010; Schlager, 2010). Both parasequence boundaries (i.e., flooding surfaces) and sequence boundaries (e.g., subaerial unconformities in the case of depositional sequences, or maximum flooding surfaces in the case of genetic stratigraphic sequences) can form at the same stratigraphic scales, in relation to the same cycles of relative sea-level change. Accommodation cycles are recorded at all scales, starting from the sedimentological scales of tidal cycles, and exposure surfaces are as common as flooding surfaces in the rock record (Vail et al., 1991; Schlager, 2004, 2010; Sattler et al., 2005; Fig. 5.106). Moreover, every transgression that leads to the formation of a flooding surface ends with a maximum flooding observed at the scale of that transgression. Therefore, the distinction between sequences and parasequences is not based on scale or accommodation conditions at syn-depositional time, but on the nature of their

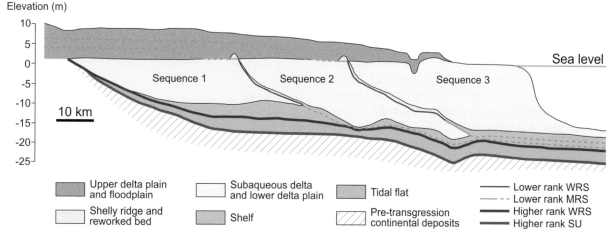

FIGURE 5.102 Dip-section through the Yellow River delta, which prograded after the post-glacial Holocene transgression (from Zecchin and Catuneanu, 2020; modified after Xue, 1993). This part of the delta consists of three high-frequency sequences generated by autocyclic delta-lobe switching followed by water deepening and transgression. Each transgression was accompanied by wave erosion and a shoreline retreat of more than 10 km. The retrogradation of shelf facies across the wave ravinement or maximum regressive surfaces defines episodes of flooding, and therefore the three sequences also qualify as parasequences.

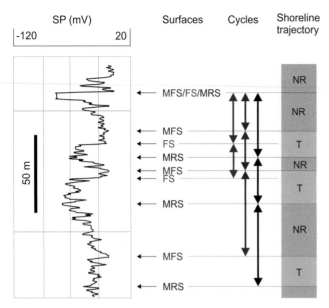

FIGURE 5.103 Stratigraphic cycles in a shallow-water setting: parasequences (red arrows), genetic stratigraphic sequences (blue arrows), transgressive—regressive sequences (black arrows) (from Catuneanu, 2019a). Sequences and parasequences can co-exist at the same stratigraphic scales. Flooding surfaces are allostratigraphic contacts which may or may not coincide with systems tract boundaries. Abbreviations: T—transgression; NR—normal regression; MRS—maximum regressive surface; MFS—maximum flooding surface; FS—flooding surface.

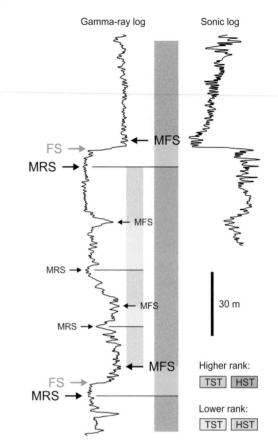

FIGURE 5.104 Stratigraphic cyclicity observed at two different scales in a siliciclastic shallow-water rift setting (Jurassic, Sverdrup Basin, Canada; modified after Embry and Catuneanu, 2001). In this example, flooding surfaces are allostratigraphic contacts within transgressive systems tracts. Sequences can be observed at the parasequence scale, as well as at smaller scales. Abbreviations: MRS—maximum regressive surface; FS—flooding surface; MFS—maximum flooding surface; TST—transgressive systems tract; HST—highstand systems tract.

bounding surfaces. Flooding surfaces, as well as all types of sequence stratigraphic surfaces, can form at multiple stratigraphic scales (hierarchical levels) (Figs. 5.104 and 5.105).

The application of the parasequence concept in the sense of Van Wagoner et al. (1988, 1990) requires the use of flooding surfaces of the lowest hierarchical rank as parasequence boundaries, to ensure that parasequences consist only of beds and bedsets (Fig. 5.94). However, there is a high degree of subjectivity in the application of the concept, which muddles the stratigraphic meaning of "parasequences" as delineated by different authors. Stratal units referred to as parasequences are highly variable in terms of timescales and internal makeup, ranging from the smallest sedimentary cycles bounded by flooding surfaces (i.e., typically observed at scales of 10^2-10^3 yrs; Amorosi et al., 2009, 2017; Pellegrini et al., 2017) to much larger units that encompass entire regressive—transgressive shelf-transit cycles (i.e., observed at scales of 10^4-10^5 yrs; Mitchum and Van Wagoner, 1991; Ainsworth et al., 2018). In the former approach, the parasequence consists of beds and bedsets (i.e., sedimentological cycles), as per the original definition. In the latter approach, the "parasequence" is much more complex than a succession of beds and bedsets, as it records internal stratigraphic cyclicity over multiple scales. In this case, the parasequence concept is used

beyond the original meaning of Van Wagoner et al. (1990), to describe petroleum reservoirs irrespective of their sequence stratigraphic significance (Ainsworth et al., 2018; Colombera and Mountney, 2020), which only deepens the nomenclatural confusion.

5.4.3 Parasequence architecture

Parasequence boundaries form during transgression, when "flooding" occurs. The abrupt water-deepening episodes are typically short-lived events that punctuate longer term trends of coastal progradation or retrogradation. The result is stepped progradation, marked by a set of forestepping parasequences (e.g., Posamentier et al., 1992b; Amorosi et al., 2005, 2017), or stepped retrogradation, marked by a set of backstepping parasequences (e.g., Martinsen and Christensen, 1992; Bruno et al., 2017). Forestepping parasequences may be either upstepping or downstepping, depending on the overall

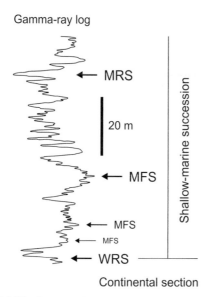

FIGURE 5.105 Stratigraphic cyclicity in a siliciclastic shallow-water rift setting (Jurassic, Sverdrup Basin, Canada; modified from Embry and Catuneanu, 2001). Note the development of maximum flooding surfaces at three scales of observation (hierarchical levels). At each scale, transgressions may or may not result in the formation of flooding surfaces, but they always end with maximum flooding surfaces. Where they do form, flooding surfaces are restricted to coastal and shallow-water settings, whereas maximum flooding surfaces of the same hierarchical ranks extend beyond the areas of development of flooding surfaces, in both updip and downdip directions. For these reasons, maximum flooding surfaces are more reliable for correlations and the construction of stratigraphic frameworks. Abbreviations: WRS—wave-ravinement surface; MFS—maximum flooding surface; MRS—maximum regressive surface.

accommodation conditions during progradation (i.e., positive or negative, respectively). Backstepping parasequences are always upstepping, as being associated with positive accommodation and transgression. Upstepping parasequences consist mainly of normal regressive and/or transgressive deposits, whereas downstepping parasequences may consist primarily of forced regressive, along with other types of deposits. However, in all cases, positive accommodation and transgression are required at the time of formation of the parasequence boundary. The stacking pattern of parasequences describes longer term normal regressions (Fig. 5.5), forced regressions (Fig. 4.34), or transgressions (Fig. 4.29), which define systems tracts of higher hierarchical rank.

In the case of stepped progradation, the regressive trend (either "normal" or "forced") is interrupted by higher frequency transgressions that lead to short-term changes in coastal depositional environments during the formation of flooding surfaces (e.g., short-term estuaries that interrupt temporarily the longer term deltaic progradation; Fig. 5.1). The architecture of stepped progradation is typically defined by asymmetrical forestepping parasequences dominated by regressive deposits (Figs. 4.34 and 5.5). Each high-frequency transgression *may* be accompanied by the formation of a flooding surface (if the diagnostic lithological discontinuity develops), but it always starts with a maximum regressive surface, possibly reworked by a wave-ravinement surface, and ends with a maximum flooding surface (Fig. 5.103). Where transgressions and water

FIGURE 5.106 Nested stratigraphic cycles in peritidal carbonates, generated by orbital forcing (Triassic, The Dolomites, Italy). In this example, the stratigraphic cycles satisfy the definition of both depositional sequences and parasequences. Abbreviations: FS/SU—flooding surface (FS) superimposed on an exposure surface (subaerial unconformity, SU).

◄ higher rank FS/SU   ⇐ lower rank FS/SU

deepening are gradual, flooding surfaces may not form in the shallow-water systems that accumulate downdip of the wave-ravinement surface. For this reason, transgressive–regressive or genetic stratigraphic sequences of the same hierarchical rank with parasequences provide a more reliable alternative for correlation, both within and beyond the confines of coastal and shallow-water systems (Catuneanu et al., 2009, 2011; Zecchin and Catuneanu, 2013). The stepped progradation of the Po coastal plain during the Middle to Late Holocene (Amorosi et al., 2005, 2017) documents forestepping parasequences of 10^0 m and c. 1000 yrs scales, delineated by brief stages of transgression that punctuated the longer term trend of progradation. In this case, each transgression resulted in a change in coastal environment (e.g., a centennial-scale estuary interrupting the millennial-scale deltaic progradation), as well as in the formation of one flooding surface and one maximum flooding surface. Where these surfaces coincide, parasequences become genetic stratigraphic sequences. Changes in the direction of shoreline shift and associated coastal environment afford the definition of depositional systems, systems tracts, and sequences at parasequence scale.

In the case of stepped retrogradation, the transgressive trend is punctuated by episodes of abrupt water deepening that lead to the formation of flooding surfaces. The architecture of stepped retrogradation is defined by a series of backstepping parasequences, each of which is dominated by regression (Fig. 4.29) or transgression (Fig. 5.5), depending on the local conditions of accommodation and sedimentation. In either case, transgression is required at the parasequence boundary. Where the backstepping parasequences include regressive deposits, the formation of parasequences is accompanied by changes in coastal environments (e.g., estuaries during times of flooding, replaced by deltas during the regressive phases). In this case, each transgression results in the formation of one flooding surface and one maximum flooding surface, which may or may not coincide; either way, genetic stratigraphic sequences (and component systems tracts and depositional systems) can be defined at the parasequence scale. Where regression is suppressed during the formation of parasequences, there are no changes in the types of coastal environments during the transgression, but only variations in the rates of shoreline backstepping (e.g., a stepwise retreat of estuaries, with higher rates during the formation of flooding surfaces). In this case, several flooding surfaces may form during the transgression, but only one maximum flooding surface at the end of it (Fig. 5.5). Such parasequences are expressed as bedsets within the transgressive systems tract, bounded by allostratigraphic facies contacts which mark no change in stratal stacking pattern. The stepped retrogradation of the Po coastal plain during the Early Holocene provides an example of backstepping

parasequences of 10^0 m and c. 1,000 yrs scales, built by transgressive deposits (Bruno et al., 2017). Without changes in the type of shoreline trajectory, these parasequences are bedsets within the lowest rank transgressive systems tract and component depositional systems.

5.4.4 Sequences vs. parasequences

Stratigraphic cyclicity in downstream-controlled settings, whether leading to the formation of sequences or parasequences, is always linked to changes in the type of shoreline trajectory. Criteria to differentiate stratigraphic cycles (sequences, parasequences) from sedimentological cycles (bedsets) that form without changes in shoreline trajectory have been provided by Zecchin et al. (2017a,b). Sequences are invariably stratigraphic cycles, as they always involve changes in the type of shoreline trajectory no matter what sequence stratigraphic surface is selected as the sequence boundary (subaerial unconformity, maximum flooding surface, or maximum regressive surface). In contrast, parasequences may correspond to stratigraphic cycles, where they include regressive deposits, or to bedsets where they consist only of transgressive deposits. In either case, flooding surfaces are facies discontinuities (allostratigraphic contacts) which may or may not coincide with sequence and systems tract boundaries (sequence stratigraphic surfaces).

Sequences are subdivided into systems tracts, which are stral units that can be mapped from continental through to deep-water settings on the basis of specific stacking patterns. In coastal to shallow-water settings, where parasequences may form, the stacking patterns that are diagnostic to the definition and identification of systems tracts are linked to the trajectory of subaerial clinoform rollovers (i.e., shoreline trajectories: progradation with upstepping, progradation with downstepping, and retrogradation; Fig. 4.1). Systems tract boundaries are surfaces of sequence stratigraphy, irrespective of their physical expression and conformable or unconformable character. The attribute they all have in common is that they mark a change in stratal stacking pattern; e.g., a maximum flooding surface is at the limit between retrogradational strata below and progradational strata above, even though it may be lithologically cryptic within a conformable succession. The same types of sequence stratigraphic surfaces can be observed at different scales; e.g., maximum flooding surfaces of different hierarchical ranks form in relation to transgressions of different magnitudes (Fig. 5.105). Where a transgression results in the formation of a flooding surface, the maximum flooding surface that forms at the end of the transgression is invariably more extensive, both in updip and downdip directions, therefore providing better means for regional correlation.

The assumption that the stratigraphic framework is organized according to an orderly pattern in which parasequences form at smaller scales and include only transgressive and highstand strata, and sequences form at larger scales and include all systems tracts (Van Wagoner et al., 1988, 1990; Duval et al., 1998; Sprague et al., 2003; Neal and Abreu, 2009; Abreu et al., 2010) is an idealization that stems from the premises and scales of seismic stratigraphy. High-resolution studies show that sequences and parasequences are not mutually exclusive in terms of scales and composition (e.g., Schlager, 2005, p. 96; Figs. 5.103 and 5.107). For example, high-frequency sequences controlled by orbital forcing may develop at scales equal to, or smaller than those of many parasequences (e.g., Strasser et al., 1999; Fielding et al., 2008; Tucker et al., 2009; Zecchin et al., 2010b; Catuneanu et al., 2011; Csato et al., 2014, Fig. 5.106). Other allogenic or autogenic processes can also generate sequences at parasequence scales, or even sequences that are nested within larger scales parasequences (Fig. 5.104).

The distinction between sequences and parasequences is not based on scale, composition, or underlying controls, but on the nature of their bounding surfaces (Figs. 5.103 and 5.108). Irrespective of the sequence stratigraphic model ("depositional," "transgressive—regressive," or "genetic stratigraphic"), the sequence boundary always has a precise genetic meaning and separates different types of stratal stacking patterns that are diagnostic to the definition of systems tracts (Fig. 4.1). In contrast, flooding surfaces do not mark a change in stratal stacking pattern, unless they coincide with a maximum regressive or a maximum flooding surface (Fig. 5.103). Where flooding surfaces are facies contacts within transgressive deposits, i.e., within-trend flooding surfaces (Catuneanu, 2006) or local flooding surfaces (Abbott and Carter, 1994; Zecchin and Catuneanu, 2013), they only have allostratigraphic significance. Some types of flooding surfaces imply a specific mechanism of formation (e.g., "local" flooding surfaces at the base of backlap shell beds relate to sediment starvation on the shelf toward the end of transgression;

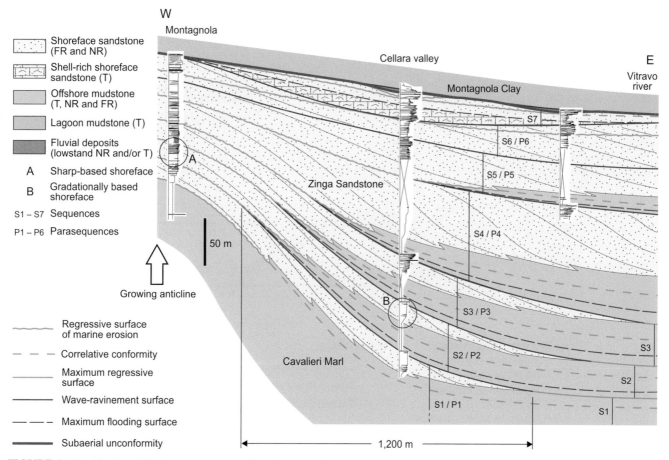

FIGURE 5.107 Stratigraphic sequences generated by glacio-eustatic cycles during the lower Pliocene (Crotone Basin, southern Italy; modified after Catuneanu and Zecchin, 2013; and Zecchin et al., 2003). Sequences amalgamate in an updip direction due to the syn-depositional growth of a salt-cored anticline. Sequence boundaries can be mapped across the study area despite the lateral changes of facies. In this example, wave-ravinement surfaces also qualify as "flooding surfaces," as they mark abrupt shifts from proximal to overlying distal facies. Due to the limited extent of flooding surfaces, sequences that develop at parasequence scales extend across larger areas, and therefore provide a better alternative for regional correlation.

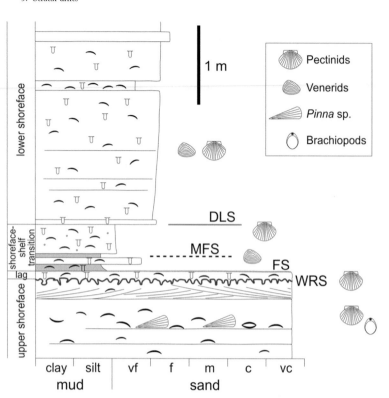

FIGURE 5.108 Detailed section illustrating Gelasian shallow-marine deposits of the Crotone Basin (southern Italy) across the limit between two high-frequency sequences separated by a wave-ravinement surface replacing a maximum regressive surface (transgressive—regressive sequences), or by a maximum flooding surface (genetic stratigraphic sequences) (from Catuneanu and Zecchin, 2020). In an allostratigraphic approach, the cycle boundary is represented by the flooding surface, at a point of marked facies change and water deepening, which does not correspond to any sequence stratigraphic surface. Abbreviations: DLS—downlap surface; FS—flooding surface; MFS—maximum flooding surface; WRS—wave-ravinement surface.

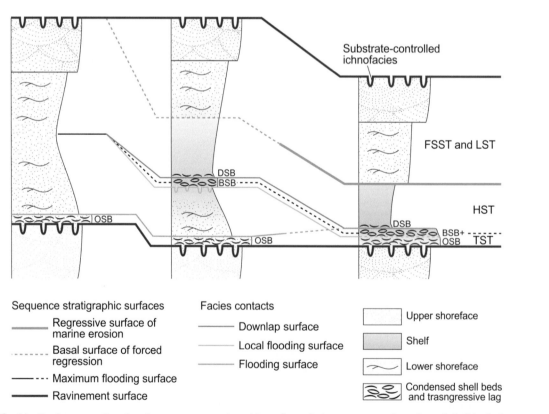

FIGURE 5.109 Idealized cross section showing sequence stratigraphic surfaces, facies contacts, and condensed shell beds that are commonly found in high-frequency sequences (modified after Zecchin, 2007). Cycle boundaries may be defined by sequence stratigraphic surfaces or by prominent facies contacts that mark abrupt water deepening (flooding surfaces). The correlation of flooding surfaces may result in the cross-cutting of sequence stratigraphic surfaces and facies contacts. Abbreviations: BSB—backlap shell bed; DSB—downlap shell bed; FSST—falling-stage systems tract; HST—highstand systems tract; LST—lowstand systems tract; OSB—onlap shell bed (with transgressive lag); TST—transgressive systems tract.

Fig. 5.109; Abbott and Carter, 1994; Zecchin and Catuneanu, 2013). However, if a flooding surface is not coincident with a sequence stratigraphic surface, it remains a mere lithological contact within the sequence stratigraphic framework (Figs. 5.108 and 5.109).

Both sequences and parasequences may include normal regressive, transgressive, and forced regressive deposits, and hence are able to form under variable accommodation conditions (Fig. 5.100; Catuneanu et al., 2011; Catuneanu and Zecchin, 2020). While the timing of sequence boundaries (either relative sea-level fall or rise) and parasequence boundaries (always during relative sea-level rise) may be offset relative to one another, the scales of sequences and parasequences may be defined by the same cycles of relative sea-level change (Figs. 5.106 and 5.107). Even without falls in relative sea level, both sequence and parasequence boundaries may form in relation to the same stages of positive accommodation and transgression. This is the norm with maximum regressive surfaces (T-R sequence boundaries), flooding surfaces (parasequence boundaries), and maximum flooding surfaces (genetic stratigraphic sequence boundaries), and it is also the case with subaerial unconformities that form during transgression (depositional sequence boundaries; Leckie, 1994; Catuneanu and Zecchin, 2016).

Sequences and parasequences of equal hierarchical ranks offer different alternatives for correlation and the delineation of stratigraphic cycles (Strasser et al., 1999; Fielding et al., 2008; Tucker et al., 2009; Catuneanu et al., 2011; Csato et al., 2014; Fig. 5.103). However, parasequences have a smaller extent than sequences and systems tracts, being restricted to coastal and shallow-water settings where flooding surfaces may form and can be demonstrated; and, flooding surfaces may not even form in shallow-water settings where conformable successions accumulate during gradual transgressions and water deepening. These limitations prevent the dependable use of the parasequence concept in the methodological workflow of sequence stratigraphy. Sequences that develop at parasequence scales provide a more reliable alternative for correlation, both within and outside of the coastal and shallow-water settings, rendering parasequences obsolete (see Chapter 9 for more details on the methodology).

Stratigraphic surfaces

Not all stratigraphic surfaces in the sedimentary record are surfaces of sequence stratigraphy. A sequence stratigraphic surface is a type of stratigraphic contact that can serve, at least in part, as a systems tract boundary (Catuneanu et al., 2009, 2011). As systems tract boundaries, sequence stratigraphic surfaces mark changes in stratal stacking pattern between the units below and above the contact (Fig. 4.39). This defining attribute separates a sequence stratigraphic surface from any other type of stratigraphic contact, and provides the basis for sequence stratigraphic correlation (Fig. 1.6). Sequence stratigraphic surfaces define the framework of sequences and component systems tracts. Within this framework, lithological changes inside systems tracts define "within-trend" facies contacts, which have physical expression but do not mark a change in stratal stacking pattern (Fig. 6.1). Both types of stratigraphic surfaces are important in stratigraphic studies. However, the workflow of sequence stratigraphy requires the identification of sequence stratigraphic surfaces first, followed by the placement of facies contacts within the framework of systems tracts and bounding sequence stratigraphic surfaces (Fig. 1.13).

Within-trend facies contacts bound stratal units defined by lithological criteria, and provide the basis for lithostratigraphic or allostratigraphic correlations. Such stratal units may develop across systems tract boundaries, and are significant for petroleum reservoir studies, particularly at the scale of high-frequency sequences (Zecchin and Catuneanu, 2013, 2015). For example, condensed sections are stratal units that may include both transgressive and highstand sediment, bounded by flooding surfaces at the base (within transgressive systems tracts) and downlap surfaces at the top (at the limit between clinoforms and outer shelf fines within highstand systems tracts) (Figs. 6.2 and 6.3; Zecchin and Catuneanu, 2013). In this example, the maximum flooding surface (i.e., the systems tract boundary) lies within the condensed section (Figs. 6.2 and 6.3, Posamentier and Allen, 1999). Within-trend facies contacts have been defined in all systems tracts,

in relation to normal regressions (within-trend normal regressive surfaces), forced regressions (within-trend forced regressive surfaces), and transgressions (within-trend flooding surfaces) (Catuneanu, 2006).

Sequence stratigraphic surfaces may or may not have a lithological expression. Conformable sequence stratigraphic surfaces are prone to be lithologically "cryptic" (e.g., maximum flooding surfaces within marine condensed sections), as the change in stratal stacking pattern, rather than grain size, is the defining attribute. In the case of maximum flooding surfaces, the overlying highstand systems tract includes part of the condensed section, as well as the clinoforms that display the typical prograding and upstepping stacking pattern. The limit between the condensed section and the overlying clinoforms is a within-trend facies contact that may have a stronger physical expression than the maximum flooding surface (i.e., the downlap surface in Fig. 6.3; Zecchin and Catuneanu, 2013). Nevertheless, and notwithstanding the importance of the within-trend facies contact as the limit between conventional petroleum reservoirs (clinoforms) and potential source rocks (outer shelf fines), the maximum flooding surface is more significant for regional correlation and the construction of the sequence stratigraphic framework, and it displays a lower degree of diachroneity (Fig. 6.2).

The identification of stratigraphic surfaces in the sedimentary record depends on the types of data available, and it is enhanced by the integration of independent datasets. For example, a maximum flooding surface that is lithologically cryptic in outcrop or core may be mapped on 2D seismic lines as a "downlap surface" (Fig. 6.4). This is the case where the highstand outer shelf fines fall below the seismic resolution, resulting in the tuning of the maximum flooding surface with the downlap surface at the base of highstand clinoforms into a single seismic reflection (Fig. 6.3). Generally, where two or more stratigraphic surfaces are superimposed, it is common to use the name of the youngest one to define the composite contact ("downlap surface" in this example), as it is the last surface which leaves the

Surfaces of Sequence Stratigraphy

Relative sea-level fall	Relative sea-level rise
• Subaerial unconformity [1]	• Subaerial unconformity [2]
• Correlative conformity *	• Maximum regressive surface
• Basal surface of forced regression **	• Maximum flooding surface
• Regressive surface of marine erosion	• Ravinement surfaces (transgressive)

Within-trend Facies Contacts

Regression	Transgression
• Within-trend NR surface • Within-trend FR surface	• Flooding surfaces within transgressive deposits

Sequence stratigraphic surfaces serve, at least in part, as systems tract boundaries. This is the attribute that separates them from any other type of stratigraphic surface.

Within-trend facies contacts are lithological discontinuities within systems tracts. They are surfaces of litho- or allostratigraphy which can be rationalized within the sequence stratigraphic framework.

FIGURE 6.1 Types of stratigraphic surfaces (modified after Embry and Catuneanu, 2001, 2002). Notes: * in the sense of Hunt and Tucker (1992); ** correlative conformity in the sense of Posamentier et al. (1988); [1] where the shoreline trajectory during forced regression is steeper than the landscape gradient; [2] where the landscape gradient is steeper than the shoreline trajectory during transgression (Fig. 4.39). Other types of within-trend facies contacts may also form in relation to sedimentological processes (see text for details). Abbreviations: NR—normal regressive; FR—forced regressive.

defining imprint on the contact. This principle also applies to composite unconformities, where two or more surfaces coalesce into a single physical contact due to nondeposition or erosion. In such cases, the unconformity is defined by the attributes of the youngest surface of the composite contact, which is the one preserved in the sedimentary record (Fig. 6.5); e.g., a subaerial unconformity reworked by waves during subsequent transgression becomes a transgressive wave-ravinement surface (Fig. 4.46).

The correct identification of stratigraphic surfaces is critical to the construction of the sequence stratigraphic framework, and it is based on criteria that involve the nature of the stratigraphic contact (conformable or unconformable), the nature of the sedimentary facies below and above the contact, and the stratal stacking patterns below and above the contact (Fig. 6.6). Application of these criteria requires a preliminary analysis of the tectonic and depositional settings, in order to place the data in the proper context (see the workflow of sequence stratigraphy in Chapter 9). The approach to the analysis of the tectonic and depositional settings depends on the types of data available and the scale of the stratigraphic study (e.g., seismic- vs. outcrop-based studies). Each type of data affords unique insights, and mutual calibration leads to the most reliable results (details in Chapter 2). The regional context for smaller scale studies is commonly provided by 2D seismic transects which constrain the key elements of the basin fill (e.g., structural elements, shelf edge, clinoforms, stratal terminations, paleoshorelines; Figs. 4.7 and 4.8).

6.1 Surfaces of sequence stratigraphy

Seven types of sequence stratigraphic surfaces have been defined in the context of "conventional" sequence stratigraphy, in relation to shoreline trajectories (Figs. 4.39, 5.92, 6.1, and 6.7). With the exception of the subaerial unconformity, which may form in both upstream- and downstream-controlled settings, all other "conventional" sequence stratigraphic surfaces require a marine/lacustrine environment to form, and therefore, are restricted to downstream-controlled settings. Only four of these surfaces (i.e., the basal surface of forced regression, the correlative conformity, the maximum regressive surfaces, and the maximum flooding surface) extend into the deep-water system. Therefore, the most

FIGURE 6.2 Sequence stratigraphic surfaces and facies contacts in a siliciclastic shelf setting (modified after Zecchin and Catuneanu, 2013). The SU, CC, MRS, WRS, MFS, BSFR, and RSME are surfaces of sequence stratigraphy, as they invariably serve, at least in part, as systems tract boundaries. The LFS and DLS are examples of within-trend facies contacts (see text for details). Abbreviations: SU—subaerial unconformity; CC—correlative conformity; MRS—maximum regressive surface; WRS—wave-ravinement surface; MFS—maximum flooding surface; BSFR—basal surface of forced regression; RSME—regressive surface of marine erosion; LFS—local flooding surface; DLS—downlap surface; LST—lowstand systems tract; TST—transgressive systems tract; HST—highstand systems tract; FSST—falling-stage systems tract.

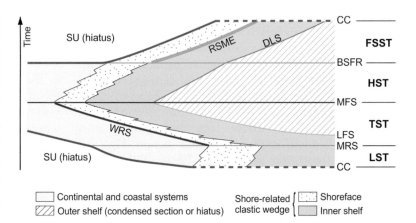

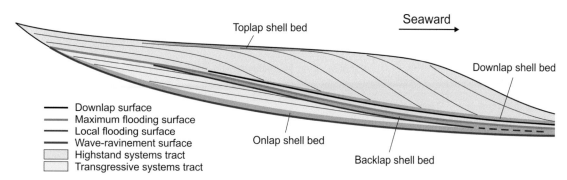

FIGURE 6.3 Types of shell beds within the sequence stratigraphic framework of a clastic shelf (modified after Zecchin, 2007). Shell beds accumulate under conditions of high wave energy and low sediment supply, commonly during transgressions and highstands. Condensed shell beds typically concentrate at the base and at the top of transgressive deposits (onlap and backlap shell beds, respectively), and at the base and at the top of the highstand clinoforms (downlap and toplap shell beds, respectively). The downlap surface may be separated from the maximum flooding surface by a condensed section of outer shelf fines which accumulate below the storm wave base. Where the condensed section falls below the seismic resolution, the maximum flooding surface can be mapped as a "downlap surface" on seismic lines (i.e., the maximum flooding and the downlap surface tune into one seismic reflection; Fig. 6.4).

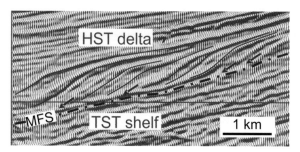

FIGURE 6.4 Seismic expression of a maximum flooding surface in a shallow-water setting (Cretaceous, offshore South Africa; modified after Brown et al., 1995). The maximum flooding surface is a "downlap surface," as it is downlapped by the highstand clinoforms.

stratigraphic surfaces may form exclusively in upstream-controlled settings, independently of relative sea-level changes and shoreline trajectories (e.g., the top-amalgamation surface in fluvial systems; Fig. 4.70).

Among all surfaces of sequence stratigraphy, three are always unconformable; i.e., the subaerial unconformity, the transgressive ravinement surfaces (tidal- and wave-ravinement), and the regressive surface of marine erosion (Fig. 6.5). All other "conventional" sequence stratigraphic surfaces can be conformable, and serve as "correlative conformities" in light of one model or another (Fig. 1.10). The maximum flooding and maximum regressive surfaces have an unambiguous nomenclature. The "correlative conformities" of the depositional sequence models, however, required a nomenclatural adjustment in order to avoid the usage of the same term for two distinct surfaces (Figs. 1.10 and 6.7). The correlative conformity of Posamentier and Vail (1988) and Posamentier et al. (1988) is truncated

complete sequence stratigraphic framework, which potentially includes the entire set of "conventional" sequence stratigraphic surfaces, is found in the continental to shallow-water systems of downstream-controlled settings (Fig. 4.39). Other sequence

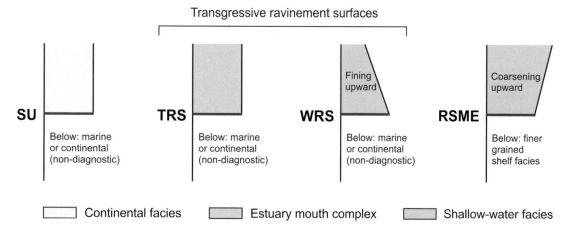

FIGURE 6.5 Unconformable surfaces of sequence stratigraphy. The identification of the different types of unconformities is based on the nature of the overlying strata. In the case of composite unconformities, the overlying strata are diagnostic of the youngest surface of the composite contact, which overprints the attributes of the older surface(s). For this reason, a composite unconformity bears the name of the younger surface, which is the one preserved in the sedimentary record. Abbreviations: SU—subaerial unconformity; TRS—tidal-ravinement surface; WRS—wave-ravinement surface; RSME—regressive surface of marine erosion.

Stratigraphic surface	Nature of contact	Facies		Stacking patterns [1]		Temporal attributes
		below	above	below	above	
Subaerial unconformity	Erosion or nondeposition	Variable	Continental	NR or FR	NR or T	Diachronous, dip and strike
Basal surface of forced regression [2]	Conformable or scoured	Marine (c-u on shelf)	Marine, c-u	NR	FR	Diachronous along strike
Correlative conformity [3]	Conformable	Marine, c-u	Marine (c-u on shelf)	FR	NR	Diachronous along strike
Maximum regressive surface	Conformable	Variable [4]	Variable (where marine, f-u)	NR	T	Diachronous along strike
Maximum flooding surface	Conformable or hiatal	Variable (where marine, f-u)	Variable (where marine, c-u)	T	NR	Diachronous along strike
Transgressive wave ravinement	Erosional	Variable (where marine, c-u)	Marine, f-u (healing phase)	NR or T	T	Diachronous, dip and strike
Transgressive tidal ravinement	Erosional	Variable (where marine, c-u)	Estuary mouth complex	NR or T	T	Diachronous, dip and strike
Regressive surface of marine erosion	Erosional	Shelf, c-u	Shoreface, c-u	NR or FR	FR	Diachronous, dip and strike
Within-trend NR surface	Conformable	Delta front or beach	Delta plain or coastal plain	NR	NR	Diachronous, dip and strike
Within-trend FR surface [5]	Conformable	Prodelta	Delta front (river dominated)	FR	FR	Diachronous, dip and strike
Within-trend flooding surface	Conformable or erosional	Variable	Marine, f-u	T	T	Diachronous, dip and strike

FIGURE 6.6   Attributes of stratigraphic surfaces in downstream-controlled settings (modified after Catuneanu, 2002, 2003, 2006; Embry and Catuneanu, 2002). Notes: (1)—where all systems tracts are preserved; (2)—correlative conformity in the sense of Posamentier et al. (1988); (3)—correlative conformity in the sense of Hunt and Tucker (1992); (4)—where marine, coarsening-upward in shallow water and fining-upward in deep water; (5)— this facies contact develops in supply-dominated settings, instead of the regressive surface of marine erosion (Fig. 4.4). Abbreviations: c-u—coarsening-upward; f-u—fining-upward; NR—normal regression; FR—forced regression; T—transgression.

FIGURE 6.7 Timing of sequence stratigraphic surfaces relative to the main events of a relative sea-level cycle (modified after Catuneanu et al., 1998a; Embry and Catuneanu, 2002). Transgressive ravinement surfaces include wave- and tidal-ravinement surfaces, which are superimposed in open shoreline settings but separated by the estuary-mouth complex in river-mouth settings. Notes: [1] where the shoreline trajectory during forced regression is steeper than the landscape gradient; subaerial unconformities may also form during transgression where the landscape gradient is steeper than the shoreline trajectory (Fig. 4.39); (−A)—negative accommodation.

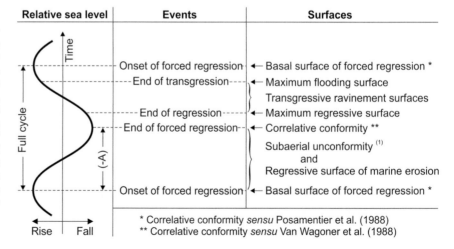

by the subaerial unconformity at the location of the shoreline at the onset of forced regression, whereas the correlative conformity of Van Wagoner et al. (1988, 1990) and Hunt and Tucker (1992) meets the basinward termination of the subaerial unconformity at the location of the shoreline at the end of forced regression (Fig. 5.93). The former designation of the "correlative conformity" has the advantage of a stronger physical expression in the deep-water setting, but it is neither conformable at the base of gravity-flow deposits nor does it correlate with the termination of the subaerial unconformity; hence, the nomenclatural change to the "basal surface of forced regression" (Fig. 6.7).

Wherever sediment supply plays a role in the timing of sequence stratigraphic surfaces, their origin may relate to any combination of allogenic and autogenic processes (Catuneanu and Zecchin, 2013). In downstream-controlled settings, this is the case with all sequence stratigraphic surfaces that form in relation to shoreline transgression (i.e., maximum regressive surfaces, transgressive surfaces of erosion, maximum flooding surfaces, and some subaerial unconformities; Fig. 4.39). Only surfaces associated with forced regression (i.e., the basal surface of forced regression, the regressive surface of marine erosion, and the correlative conformity; Fig. 4.39) have their timing controlled exclusively by relative sea-level changes, and therefore, by allogenic factors (Fig. 6.8). The origin of sequence stratigraphic surfaces in upstream-controlled settings is also variable, with influence from both allogenic and autogenic processes, as it may relate to tectonism, climate, or the autogenic migration of alluvial channel belts (details in Chapter 4). All sequence stratigraphic

Sequence stratigraphic surface	Control	
	Allogenic	Autogenic
Subaerial unconformity	√	√
Maximum regressive surface	√	√
Maximum flooding surface	√	√
Transgressive ravinement surface	√	√
Basal surface of forced regression	√	
Regressive surface of marine erosion	√	
Correlative conformity	√	

√ where the subaerial unconformity forms during transgression

FIGURE 6.8 Controls on the formation of sequence stratigraphic surfaces in downstream-controlled settings. Allogenic processes are commonly involved in the formation of all sequence stratigraphic surfaces. Autogenic processes can also play a role where sediment supply influences the formation of a surface. This is typically the case with maximum regressive surfaces, transgressive surfaces of erosion, and maximum flooding surfaces, but also with subaerial unconformities that form during transgression (Fig. 4.39B).

surfaces are potentially diachronous, with various degrees of diachroneity along dip and/or strike directions (details in Chapter 7).

6.1.1 Subaerial unconformity

The subaerial unconformity is a hiatal surface that forms in continental environments as a result of fluvial erosion or bypass, pedogenesis, wind degradation, or karstification (Fig. 6.9). The usage of subaerial unconformities as sequence boundaries was pioneered in the 1940s (Sloss et al., 1949), and consolidated in the 1960s (Sloss, 1963, Wheeler, 1964) and 1970s (Mitchum, 1977). Subaerial unconformities provide natural candidates to delineate depositional sequences, especially in low-resolution stratigraphic studies, since the largest gaps in the sedimentary record are typically related to periods of subaerial exposure (Sloss, 1963). A subaerial unconformity may be reworked and replaced subsequently by other types of unconformities (e.g., wave-ravinement surfaces; Figs. 4.46, 6.10, and 6.11), but the bulk of the hiatus remains associated with the period of subaerial exposure. As a rule of thumb, any time the hiatus of a subaqueous unconformity such as a wave-ravinement surface or a regressive surface of marine erosion can be detected by means of biostratigraphy in a shelf setting, a preceding subaerial unconformity can be inferred (Figs. 2.9 and 6.11).

Subaerial unconformities may form in both downstream- and upstream-controlled settings, under variable accommodation conditions. More important than accommodation, changes in the temporary base level, which account for all controls on the balance between sedimentation and erosion at any location, play the key role in the formation of unconformities (see Chapter 3 for details). Subaerial unconformities may form not only during stages of negative accommodation, as generally assumed, but also during stages of positive accommodation and transgression or in relation to processes of autogenic migration of alluvial channels belts that are independent of accommodation (Miall, 2015; Catuneanu and Zecchin, 2016; see Chapter 4 for details). The link between subaerial unconformities and stages of negative accommodation is evident in carbonate settings, but it is less constrained in siliciclastic settings. Alternative terms include "lowstand unconformity" (Schlager, 1992) and "regressive surface of fluvial erosion" (Plint and Nummedal, 2000). However, the term "subaerial unconformity" is preferred because it does not link, nor restrict, the formation of this surface to stages of lowstand or regression.

In downstream-controlled settings, subaerial unconformities may form during forced regressions (negative accommodation) or transgressions (positive accommodation) (Fig. 4.39), or both (Zecchin et al., 2022). Where

FIGURE 6.9 Subaerial unconformity (arrow) at the limit between the floodplain-dominated fluvial succession of the Las Cabras Formation and the overlying channel-dominated fluvial system of the lower Potrerillos Formation (Triassic, Cuyana Basin, Argentina). Diagnostic for the subaerial unconformity is the preservation of continental deposits on top.

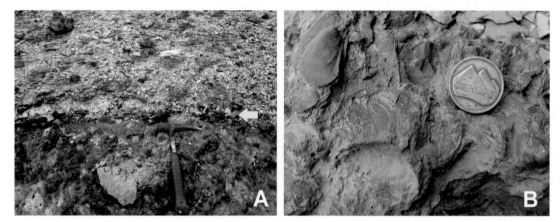

FIGURE 6.10 Outcrop example of a transgressive wave-ravinement surface which replaces a subaerial unconformity (Lower Cenomanian, Bahariya Formation, Western Desert, Egypt). A—The wave-ravinement surface (arrow) separates an iron-rich paleosol horizon (ferricrete) from the overlying glauconitic marine deposits. The formation of ferricrete is attributed to the *in-situ* alteration of marine glauconite under subaerial conditions (i.e., a paleo-seafloor exposed during relative sea-level fall; El-Sharkawi and Al-Awadi, 1981; Catuneanu et al., 2006). The contact was subsequently modified by transgression, and it is preserved as a wave-ravinement surface with diagnostic transgressive marine deposits on top (Fig. 6.5); B—Onlap shell bed on top of the wave-ravinement surface.

FIGURE 6.11 Subaerial unconformity reworked and replaced by a wave-ravinement surface during subsequent transgression. The composite unconformity (arrow) has a 6 My hiatus, and marks the limit between "clean" Middle Eocene highstand limestone (A) and "dirty" Early Oligocene transgressive limestone interbedded with glauconitic silts and marls (B) (Kutch Basin, India; from Catuneanu and Dave, 2017).

a subaerial unconformity forms during forced regression, it is commonly found at the top of the highstand topset and of the falling-stage foreset (Figs. 4.6 and 4.39A). However, in cases where the exposure is associated with significant erosion (e.g., leading to the formation of incised valleys; Fig. 6.12), the underlying highstand and falling-stage systems tracts may not be preserved and the subaerial unconformity may truncate any older systems tracts (Fig. 6.13). Under particular circumstances, where the landscape profile is steeper than the shoreline trajectory, the subaerial unconformity may also form during periods of positive accommodation and transgression (Figs. 4.28, 4.39B, and 4.56). Subaerial unconformities that form during transgressions are age-equivalent to the coastal onlap, as they develop during the retrogradation of the shallow-water "healing-phase" deposits that onlap the transgressive wave-ravinement surface (Fig. 4.55).

In upstream-controlled settings, subaerial unconformities form independently of relative sea-level changes

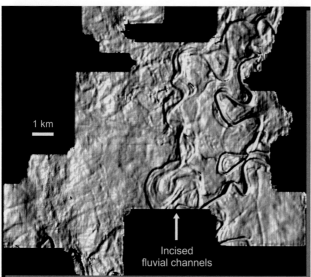

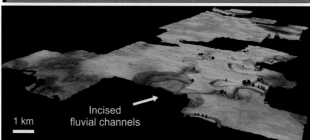

FIGURE 6.12 Subaerial unconformity at the limit between Devonian carbonates and the overlying Early Cretaceous fluvial deposits of the Mannville Group, Western Canada Basin (images courtesy of Henry Posamentier). The 3D seismic images (illuminated horizon above and time-structure map below) capture the morphology of a first-order sequence boundary (i.e., a shift in tectonic setting from divergent continental margin to foreland basin), with high-sinuosity fluvial channels incised into the underlying carbonates.

and shoreline trajectories, with a timing controlled by climate cycles, source-area tectonism, or autogenic migration of alluvial channel belts (details in Chapter 4). For example, incised valleys cut during interglacial periods of ice melting, isostatic rebound, and increased fluvial discharge in upstream-controlled settings (Fig. 3.21) are commonly coeval with stages of eustatic rise and fluvial aggradation in downstream-controlled settings (Blum, 1994, Fig. 3.20). Such subaerial unconformities are temporally offset, or even out-of-phase with those that form in relation to shoreline trajectories (Fig. 3.20). The subaerial unconformities in upstream-controlled fluvial settings are typically erosional at the base of amalgamated channels, and mark an abrupt shift from lower energy fluvial systems below to higher energy fluvial systems above. The hiatus of the subaerial unconformity corresponds to a stage of increase in fluvial energy (e.g., increase in gradients or fluvial discharge) accompanied by erosion. Most commonly, the stacking patterns change from isolated to amalgamated channels across the subaerial unconformity (Fig. 4.70).

The physical expression of subaerial unconformities is variable, from sharp erosional surfaces at the base of incised valleys (Fig. 6.14) to paleosols in interfluve areas (Fig. 6.15), deflation surfaces in eolian settings (Figs. 3.18, 3.19, and 6.16), and karst surfaces in carbonate settings (Fig. 6.17). Karstification in carbonate settings is most effective where dissolution by meteoric water and drainage systems is promoted by a humid climate during the period of subaerial exposure. In the absence of dissolution, due to insufficient meteoric water and drainage under more arid climatic conditions, pedogenesis at the top of the exposed highstand topset may become the dominant process (Fig. 4.52). Paleosols in both siliciclastic and carbonate settings may have variable expressions and physical properties, from higher porosity and lower acoustic velocity (e.g., due to the presence of root traces, more common where pedogenesis occurs under humid climatic conditions; Fig. 6.18) to higher density and higher acoustic velocity (e.g., in the case of calcretes, silcretes, or ferricretes which involve chemical precipitation under arid climatic conditions; Fig. 6.15A). In siliciclastic settings with riverborne sediment supply, the subaerial unconformity commonly marks a sharp increase in grain size due to the abrupt shift to more proximal and/or higher energy systems on top, in both downstream- and upstream-controlled areas (Figs. 4.45, 4.70, 6.9, and 6.19).

The stratigraphic hiatus associated with the subaerial unconformity is commonly variable across its area of development, due to differential fluvial incision or eolian degradation, and the gradual shift of the areas of erosion and sedimentation along dip or strike directions (Fig. 3.19). As a result, subaerial unconformities

FIGURE 6.13 Subaerial unconformity (*arrow*) at the contact between (A) transgressive tidal flat deposits marked by the *Glossifungites* Ichnofacies with *Gyrolithes*-like burrows and (B) the overlying lowstand braided stream deposits (Plio-Pleistocene, Algarve Basin, Portugal; image courtesy of Mirian Menegazzo). Fluvial erosion removed the highstand and the marine transgressive deposits, and generated a composite unconformity which includes the wave-ravinement surface (top of the *Glossifungites* firmground), the maximum flooding surface, and the subaerial unconformity. The fluvial deposits on top preserve the defining attribute of the subaerial unconformity.

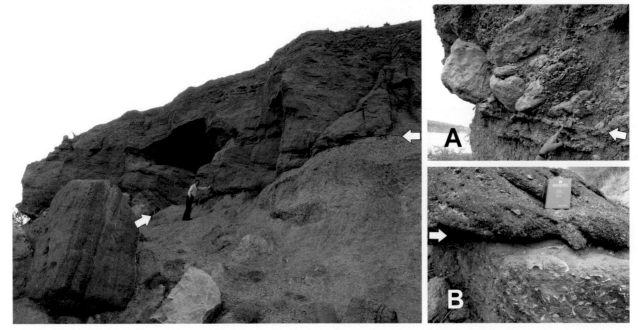

FIGURE 6.14 Subaerial unconformity (arrows) at the base of an incised valley cut into Early Miocene (Burdigalian) carbonates (Kutch Basin, India). The carbonate platform was uplifted during the hard collision of India with Asia in the Middle Miocene, which led to a change in sedimentation regime from carbonates to siliciclastics derived from the emerging Himalayas. The incised valley is filled with coarse-grained braided fluvial deposits. Note the meter-scale carbonate rip-up clasts at the base of the incised-valley fill (A), and the sharp contact between fluvial and carbonate facies (B).

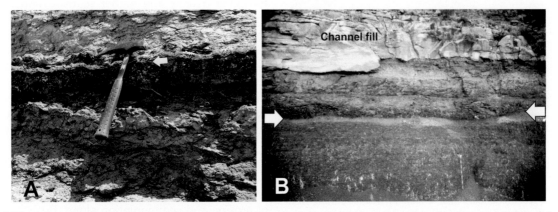

FIGURE 6.15 Subaerial unconformities (arrows) at the top of paleosol horizons. In both examples, the paleosols are underlain and overlain by fluvial overbank facies. A—Ferruginous paleosol (ferricrete) with plant roots (Lower Cenomanian, Bahariya Formation, Western Desert, Egypt; from Catuneanu et al., 2006); B—paleosol with rootlets (Early-Middle Triassic, Burgersdorp Formation, Karoo Basin, South Africa; the scale is 1.4 m long).

FIGURE 6.16 Deflationary "supersurfaces" (i.e., sub-aerial unconformities in an eolian system; white arrows) delineating a 20-m thick eolian sequence (Permian Cedar Mesa Sandstone, Squaw Butte, SE Utah; image courtesy of Nigel Mountney). Note the rhizoliths associated with the sequence boundaries, and the point of pinch-out of a wet interdune pond (black arrow).

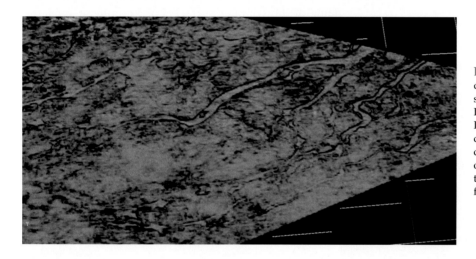

FIGURE 6.17 Karst surface with circular dissolution features, overlain by a fluvial system (Middle Miocene, Saurashtra Shelf, India; image courtesy of Reliance Industries, India). The karst surface is a subaerial un-conformity generated by the exposure of the carbonate platform during the hard collision of India with Asia. The fluvial systems mark the onset of siliciclastic sedimentation following the initiation of the Himalayas.

FIGURE 6.18 Subaerial unconformity (arrow) at the contact between (A) a paleosol horizon (protosol in the classification of Mack et al., 1993) and (B) the overlying fluvial system (Upper Cretaceous, Adamantina Formation, Bauru Basin, Brazil; image courtesy of Mirian Menegazzo). The paleosol has a higher porosity due to the root traces, and hence a lower acoustic velocity on sonic logs. The fluvial deposits on top preserve the defining attribute of the subaerial unconformity.

in both downstream- and upstream-controlled settings can develop a high degree of diachroneity, as the conditions required by their formation may be met at different times in different locations. For example, subaerial unconformities that form during forced regressions extend basinward with time until the end of relative sea-level

fall, becoming younger in a downdip direction (Figs. 5.13 and 5.15). As sedimentation resumes after forced regression, the gradual expansion of the fluvial sedimentation area toward the basin margin during relative sea-level rise (i.e., the development of fluvial onlap) further increases the magnitude of the hiatus in

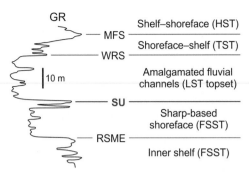

FIGURE 6.19 Stratigraphic architecture in a rift setting (Lower Cretaceous, Brooks-Mackenzie Basin, Arctic Canada; modified after Millar, 2021). Note the contrast in grain size between the sharp-based (forced regressive) shoreface and the overlying lowstand fluvial deposits, due to the abrupt shift from distal to proximal facies across the subaerial unconformity. Abbreviations: GR—gamma-ray log; RSME—regressive surface of marine erosion; SU—subaerial unconformity; WRS—wave-ravinement surface; FSST—falling-stage systems tract; LST—lowstand systems tract; TST—transgressive systems tract; HST—highstand systems tract.

an updip direction (Figs. 5.13 and 5.15). The mechanics of formation of subaerial unconformities during forced regressions are illustrated in Figs. 4.42A and 4.44. Note that the subaerial unconformity not only expands basinward as the seafloor is exposed by the falling relative sea level, but at the same time it also expands in landward direction due to the upstream migration of fluvial knickpoints (Figs. 4.42 and 4.43).

The identification of a subaerial unconformity in the rock record requires the preservation of continental deposits (fluvial or eolian) on top (Figs. 6.5, 6.19, and 6.20). The facies preserved below a subaerial unconformity may range from continental to marine, and therefore, they are not diagnostic to the identification of this surface (Figs. 6.6 and 6.20). In downstream-controlled settings, subaerial unconformities may be subsequently reworked and replaced by younger erosional surfaces, in which case the composite unconformity takes the name of the younger surface. For example, a subaerial

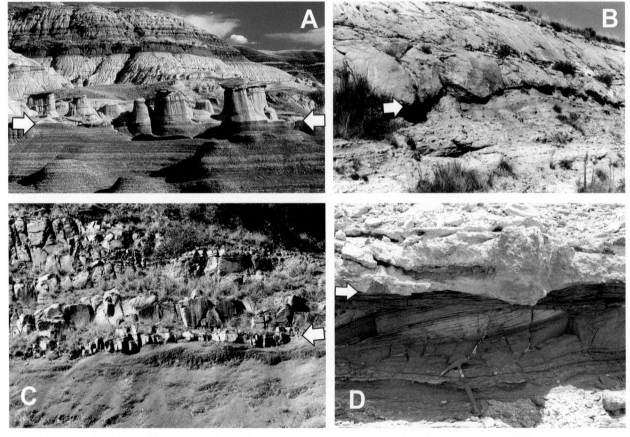

FIGURE 6.20 Outcrop examples of subaerial unconformities (arrows). A—Contact between shallow-marine shales (Bearpaw Formation) and the overlying incised-valley-fill fluvial sandstones (Late Cretaceous, Horseshoe Canyon Formation, Western Canada Basin; facies interpretations from Ainsworth, 1994). B—Contact between the Bamboesberg and Indwe members of the Molteno Formation (Late Triassic, Dordrecht region, Karoo Basin). C—Contact between the Balfour Formation and the overlying Katberg Formation (Early Triassic, Nico Malan Pass, Karoo Basin). The successions in B and C are fully fluvial and upstream-controlled, with abrupt increases in grain size and energy levels across the contacts. D—Contact between swaley cross-stratified shoreface and the overlying fluvial deposits (Lower Cenomanian, Bahariya Formation, Western Desert, Egypt).

unconformity that is reworked by waves during subsequent transgression is replaced by a wave-ravinement surface (Figs. 4.46, 6.10, and 6.11). In this case, the deposits on top of the composite unconformity are transgressive marine, diagnostic of the younger surface (i.e., the subaerial unconformity is not preserved as a physical surface, and its temporal attributes are transferred to the wave-ravinement surface; Fig. 6.5). Conversely, subaerial unconformities can also rework and replace older surfaces, in which case the composite contact is defined by the continental deposits on top (Fig. 6.13).

6.1.2 Basal surface of forced regression

The basal surface of forced regression (Hunt and Tucker, 1992) is a sequence stratigraphic surface that marks a change in stratal stacking pattern from normal regression (below) to forced regression (above) (Figs. 5.12, 5.92, 6.6, and 6.7). Most commonly, the underlying normal regression is highstand, in the case of sequences that include a transgressive systems tract (Fig. 4.39), but it can also be lowstand in the case of sequences that consist only of falling-stage and lowstand systems tracts (Figs. 1.23 and 4.33). In either case, the basal surface of forced regression marks the onset of forced regression and the base of forced regressive deposits; therefore, the criteria that afford the distinction between forced and normal regressive deposits (e.g., Posamentier and Morris, 2000; Catuneanu and Zecchin, 2016, Figs. 4.6, 4.36, 4.40, and 4.41) are critical to the identification of this surface (see details in Chapter 4). The basal surface of forced regression corresponds to the "correlative conformity" in the sense of Posamentier et al. (1988) (i.e., the depositional sequence II model in Fig. 1.10).

Field examples that illustrate the physical expression of the basal surface of forced regression have been documented on different datasets, from seismic to core and outcrop (Figs. 4.6, 6.21, and 6.22). Where the subaerial unconformity forms during forced regression (Fig. 4.39A), the basal surface of forced regression is a marine surface (i.e., the paleoseafloor at the onset of forced regression) truncated at the top by the subaerial unconformity (Fig. 4.6). Where the subaerial unconformity forms during transgression, the basal surface of forced regression may extend into both marine and continental settings (Fig. 4.39B). In shallow-water settings, it is common that the basal surface of forced regression is reworked in part by the regressive surface of marine erosion (Figs. 5.33 and 6.23). The degree of preservation of the basal surface of forced regression depends on the gradient of the shoreline trajectory during forced regression relative to the seafloor gradient at the onset of forced regression (Fig. 4.37). Steeper seafloor gradients, such as in ramp settings, afford a higher degree of

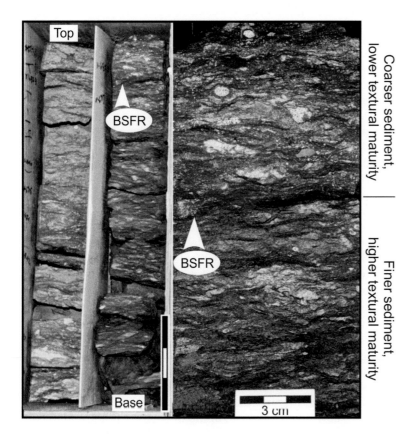

FIGURE 6.21 Lithological expression of the basal surface of forced regression (BSFR) in a siliciclastic shallow-water setting (Early Cretaceous, Viking Formation, Western Canada Basin; from MacEachern et al., 2012). Where preserved from wave scouring, the BSFR is a conformable paleo-seafloor that marks an increase in sediment supply to the sea following the onset of forced regression, due to the decreased capacity of fluvial systems to retain riverborne sediment during relative sea-level fall. As a result, the forced regressive clinoforms include coarser and less sorted sediment than the underlying highstand clinoforms.

FIGURE 6.22 Lithological expression of the basal surface of forced regression (BSFR) in a mixed carbonate-siliciclastic shallow-water setting (Aptian, Araripe Basin, Brazil). Facies: A—limestone, horizontally laminated (outer shelf); B—marls, horizontally stratified (inner shelf); C— shoreface clinoforms downlapping the regressive surface of marine erosion (RSME); D—structureless shoreface sandstone, concretionary toward the top; E—transgressive silts and shale. The carbonate factory is shut down by siliciclastic sediment influx during forced regression. As a result, limestones are replaced by marls in the inner shelf, and by sandstone in the shoreface. The BSFR marks the limit between the highstand limestone and the overlying falling-stage sediments that record an increase in the rates of progradation and in the amount of terrigenous influx to the lake. Abbreviations: WRS—wave-ravinement surface; HST—highstand systems tract; FSST—falling-stage systems tract; TST—transgressive systems tract.

preservation of the basal surface of forced regression from subsequent wave scouring (Figs. 4.37A and 6.24).

In seismic stratigraphic terms, the basal surface of forced regression is the oldest clinoform surface associated with offlap (Fig. 4.6). However, while offlap is common in the stratigraphic record (Fig. 4.34), it is not always preserved, in which case other criteria need to be employed in order to recognize forced regression (Fig. 4.36). Where high-magnitude falls in relative sea level are accompanied by significant erosion across the entire shelf, the subaerial unconformity may obliterate any evidence of shallow-water forced regressive clinoforms. In this case, the preservation of offlap and of the basal surface of forced regression is only possible downdip of the location of the shelf edge at the onset of forced regression (Fig. 4.6). For lower magnitude falls in relative sea level, or where the exposure of the shelf is accompanied by less erosion, the shallow-water forced regressive clinoforms may be preserved, allowing the observation of offlap in the shelf setting as well (Fig. 4.34). However, this does not guarantee the preservation of the basal surface of forced regression, which

may still be replaced by the regressive surface of marine erosion (Fig. 4.37).

In low-gradient shelf settings where the basal surface of forced regression is more susceptible to wave scouring, its most proximal portion is still preserved as a conformable clinoform surface at the base of the earliest forced regressive clinoform (Figs. 5.32 and 5.33). This is due to the combination of depositional and erosional processes that maintain the wave equilibrium profile of the seafloor during relative sea-level fall (Fig. 4.44; Bruun, 1962; Plint, 1988; Dominguez and Wanless, 1991; see Chapter 4 for details). The location of the lever point that marks the limit between the areas of sedimentation and erosion during the earliest stage of forced regression (i.e., the contact between the basal surface of forced regression and the regressive surface of marine erosion; Fig. 5.33) depends on the balance between sediment supply and wave energy, shifting seaward with the increase in sediment supply and landward with the increase in wave energy. The forced regressive wave-dominated shoreface is conformably based (atypical) updip of the lever point, and unconformably based (typical) downdip of the lever point

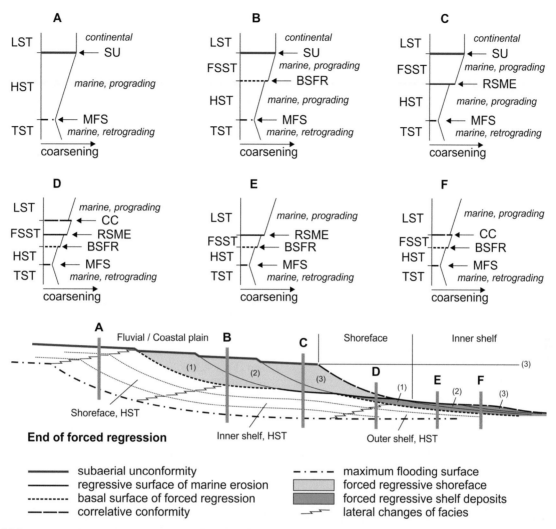

FIGURE 6.23 Stratigraphic architecture in a wave-dominated shallow-water setting, where all systems tracts are formed and preserved. The BSFR is the seafloor at the onset of forced regression, which is reworked in part by the RSME. Where the BSFR is preserved, both the BSFR and the RSME are present in the same vertical section (e.g., location D), separated by forced regressive shelf deposits. Abbreviations: TST—transgressive systems tract; HST—highstand systems tract; FSST—falling-stage systems tract; LST—lowstand systems tract; SU—subaerial unconformity; CC—correlative conformity; BSFR—basal surface of forced regression; RSME—regressive surface of marine erosion; MFS—maximum flooding surface.

(Figs. 5.31, 5.32, and 6.23). Nevertheless, all forced regressive clinoforms are downstepping during the progradation of the shoreline and display a reduced thickness relative to the normal regressive clinoforms (Figs. 4.30 and 5.31; details in Chapter 4).

The change from relative sea-level rise to relative sea-level fall which marks the timing of the basal surface of forced regression (Figs. 4.32 and 6.7) triggers an increase in riverborne sediment supply to the shoreline, as well as instability on the outer shelf due to the lowering of the storm wave base (details in Chapters 4 and 5). These changes in the sedimentation regimes lead to a sharp increase in grain size across the basal surface of forced regression in shallow-water settings (Figs. 6.21, 6.22, 6.24, and 6.25), and to a renewed cycle of mass-transport processes in deep-water settings (Figs. 4.20 and 6.26; details in Chapter 8). The basal surface of forced regression is one of the four sequence

stratigraphic surfaces (i.e., systems tract boundaries) that extend into the deep-water setting (Catuneanu, 2020a). Beyond the shelf edge, the basal surface of forced regression is no longer reworked by the regressive surface of marine erosion, which is restricted to shallow-water settings, but it is commonly scoured by mass-transport processes (e.g., slumping) and gravity flows (e.g., mudflows) which accompany the fall in relative sea level (Fig. 4.20; details in Chapter 8).

6.1.3 Correlative conformity

The correlative conformity (Van Wagoner et al., 1988, 1990; Hunt and Tucker, 1992, Fig. 1.10) is a sequence stratigraphic surface that marks a change in stratal stacking pattern from forced regression (below) to lowstand normal regression (above) (Figs. 4.32, 4.39, 6.6, and 6.7). This type of sequence stratigraphic surface

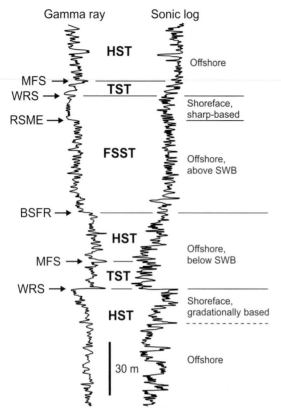

FIGURE 6.24   Well-log expression of the basal surface of forced regression in a rift setting (Jurassic, Sverdrup Basin; modified after Embry and Catuneanu, 2001). The steeper seafloor gradients in ramp settings enable the deposition and preservation of thicker offshore deposits between the basal surface of forced regression and the regressive surface of marine erosion (Fig. 4.37). The basal surface of forced regression is marked by an increase in grain size from the highstand to the falling-stage systems tracts, due to the decrease in the capacity of rivers to retain sediment following the onset of relative sea-level fall (Fig. 6.21). Lowstand systems tracts tend to be poorly developed or missing in rift basins, where accommodation is generated abruptly by the reactivation of faults (Fig. 4.33). Abbreviations: FSST—falling-stage systems tract; TST—transgressive systems tract; HST—highstand systems tract; WRS—wave-ravinement surface; MFS—maximum flooding surface; BSFR—basal surface of forced regression; RSME—regressive surface of marine erosion; SWB—storm wave base.

was first recognized by Wheeler (1964, p. 606), as a "continuity" surface that correlates with the downdip termination of a subaerial unconformity. The continuity surface of Wheeler (1964) was renamed as the

"correlative conformity" in the 1970s, when it was incorporated in the revised definition of a sequence (Mitchum, 1977, Fig. 1.22). Only subaerial unconformities that form during forced regression have correlative conformities *sensu stricto* (Fig. 4.39A). In the case of subaerial unconformities that form during transgression, their correlative conformable surfaces are represented by maximum regressive surfaces, and the physical connection between subaerial unconformities and their correlative maximum regressive surfaces is made by the transgressive surface of erosion (Fig. 4.39B). However, in a broader sense, the term "correlative conformity" can be retained for the surface that marks the end of forced regression, irrespective of the timing of the subaerial unconformity, for nomenclatural consistency and simplicity (Fig. 4.39).

Where the subaerial unconformity forms during forced regression (Fig. 4.39A), the correlative conformity is a marine surface (i.e., the paleoseafloor at the end of forced regression) which connects physically with the downdip termination of the subaerial unconformity at the location of the shoreline at the end of forced regression (Figs. 4.6, 5.13, and 6.26). This scenario is most common in the stratigraphic record. Where the subaerial unconformity forms during transgression (Fig. 4.39B), the "correlative conformity" becomes a conformable surface which extends within both marine and continental realms, without a physical and temporal relationship with the subaerial unconformity. In either case, the correlative conformity marks the end of forced regression and the beginning of lowstand normal regression (Fig. 4.39); therefore, the criteria that afford the distinction between forced and normal regressive deposits (e.g., Posamentier and Morris, 2000; Catuneanu and Zecchin, 2016, Figs. 4.6, 4.36, 4.40, and 4.41) are critical to the identification of this surface (see details in Chapter 4).

Field examples that illustrate the physical expression of the correlative conformity have been documented on different datasets, from seismic to core and outcrop (Figs. 4.6, 5.10, 6.27, and 6.28). In seismic stratigraphic terms, the correlative conformity is the youngest clinoform surface associated with offlap (Fig. 4.6). Where evidence of offlap is missing, other criteria that can be used to recognize forced regression include the

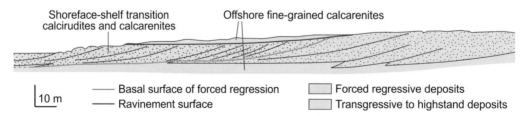

FIGURE 6.25   Stratigraphic architecture of the Pliocene "Calcarenite di Gravina" in Materna, southern Italy (from Zecchin and Catuneanu, 2013; modified after Pomar and Tropeano, 2001). The basal surface of forced regression marks an increase in sediment supply to the sea and a steepening of the prograding clinoforms from the highstand to the falling-stage systems tracts.

FIGURE 6.26 Stratigraphic framework of a mixed carbonate-siliciclastic shelf-slope system (Triassic, The Dolomites, Italy). Shelf facies: HST—subtidal platform carbonates; LST—intertidal to supratidal platform carbonates; TST—transgressive shales. Slope facies: A (HST) and C (LST)—carbonate grainflows; B (FSST)—mixed carbonate-siliciclastic debris flows. The BSFR marks the change from relative sea-level rise (carbonate production on the platform) to relative sea-level fall (carbonate factory shut down). The CC marks the change from relative sea-level fall to relative sea-level rise (carbonate factory switched on again). Abbreviations: SU—subaerial unconformity; BSFR—basal surface of forced regression; CC—correlative conformity; WRS/MRS—wave-ravinement surface reworking a maximum regressive surface; MFS—maximum flooding surface; HST—highstand systems tract; FSST—falling-stage systems tract; LST—lowstand systems tract; TST—transgressive systems tract.

downstepping and the compressed nature of clinoforms, and the stratigraphic foreshortening in shelf settings in a downdip direction (Figs. 4.30, 4.36, 4.40, 4.41, and 5.31; more details in Chapter 4). In sedimentological terms, the change from relative sea-level fall to relative sea-level rise which marks the timing of the correlative conformity (Figs. 4.32 and 6.7) triggers a decrease in the volume and caliber of the riverborne sediment supply to the shoreline. This results in a decrease in grain size across the correlative conformity, which can be observed in both shallow- and deep-water settings (Figs. 6.23, 6.27, and 6.28; details in Chapters 5 and 8).

Among the two candidates for the conformable portion of the depositional sequence boundary (Figs. 1.10, 5.93, and 6.7), the correlative conformity has a higher preservation potential because its formation is followed by normal regression, when aggradation is the prevalent depositional trend in all depositional environments (Figs. 4.39, 5.12, 5.13, and 5.14). This is in contrast to the basal surface of forced regression, which

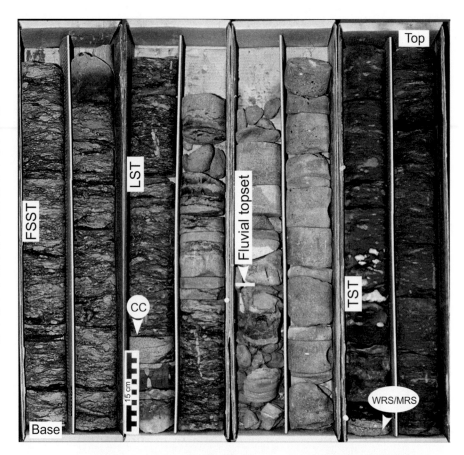

FIGURE 6.27 Lithological expression of the correlative conformity in a siliciclastic shallow-water setting (Early Cretaceous, Viking Formation, Western Canada Basin; from MacEachern et al., 2012). The correlative conformity marks a decrease in grain size from the falling-stage to the overlying lowstand clinoforms, due to the increased capacity of rivers to retain sediment following the onset of relative sea-level rise. Both falling-stage and lowstand clinoforms display coarsening-upward trends. The lowstand foreset is overlain by the fluvial topset of the LST. Abbreviations: CC—correlative conformity; WRS/MRS—wave-ravinement surface reworking a maximum regressive surface; FSST—falling-stage systems tract; LST—lowstand systems tract; TST—transgressive systems tract.

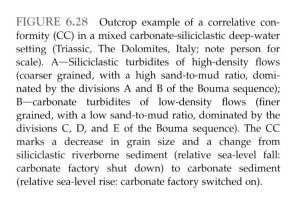

FIGURE 6.28 Outcrop example of a correlative conformity (CC) in a mixed carbonate-siliciclastic deep-water setting (Triassic, The Dolomites, Italy; note person for scale). A—Siliciclastic turbidites of high-density flows (coarser grained, with a high sand-to-mud ratio, dominated by the divisions A and B of the Bouma sequence); B—carbonate turbidites of low-density flows (finer grained, with a low sand-to-mud ratio, dominated by the divisions C, D, and E of the Bouma sequence). The CC marks a decrease in grain size and a change from siliciclastic riverborne sediment (relative sea-level fall: carbonate factory shut down) to carbonate sediment (relative sea-level rise: carbonate factory switched on).

is more likely to be scoured by waves in shallow-water shelf settings (Figs. 4.37 and 5.33) or by mass-transport processes and gravity flows in deep-water settings (Fig. 4.20; details in Chapter 8). For this and other reasons discussed earlier in the book, including the connection with the basinward termination of the subaerial unconformity (Figs. 4.6, 5.12, 5.13, 5.14, and 5.93), the correlative conformity as defined herein is the most common choice for the conformable portion of the depositional sequence boundary. Nonetheless, both surfaces are important components of the sequence stratigraphic framework, and are critical for the prediction of sedimentary facies in the process of petroleum exploration, particularly in deep-water settings (details in Chapter 8).

6.1.4 Maximum regressive surface

The maximum regressive surface is a sequence stratigraphic surface that separates regressive strata below from transgressive strata above (Helland-Hansen and Martinsen, 1996, Figs. 4.32, 6.6, and 6.7). The formation and usage of this surface are restricted to downstream-controlled settings, where shoreline shifts provide the reference to define regressive and transgressive stacking patterns. Most commonly, the underlying deposits display a normal regressive stacking pattern (Figs. 4.6 and 4.39). The normal regression may be lowstand, in the case of sequences that include a falling-stage systems tract (i.e., a stage of negative accommodation during the formation of the sequence; Figs. 4.6 and 4.39), or highstand, in the case of sequences that consist only of transgressive and highstand systems tracts without intervening stages of negative accommodation (i.e., sequences that form during continuous relative sea-level rise, as a result of variations in the rates of accommodation and sedimentation at the shoreline; Fig. 4.33) (e.g., Csato and Catuneanu, 2012, 2014).

Where sequences record both stages of positive and negative accommodation, but the lowstand systems tract is missing, the maximum regressive surface coincides with the correlative conformity at the end of forced regression (e.g., in extensional settings where accommodation at the onset of relative sea-level rise is generated rapidly by the reactivation of faults; Martins-Neto and Catuneanu, 2010). Therefore, maximum regressive surfaces may top any type of regressive deposit (i.e., lowstand normal regressive, highstand normal regressive, or forced regressive), so the actual stratigraphic relationship in each study needs to be determined on a case-by-case basis. However, in all cases, the maximum regressive surface marks the end of a regression and the beginning of subsequent transgression. Sequences which are missing transgressive systems tracts do not include maximum regressive surfaces or any other surfaces associated with transgression, such as transgressive surfaces of erosion and

maximum flooding surfaces. This is the case with sequences which consist only of falling-stage and lowstand systems tracts, where transgressions are suppressed by sediment supply at the scale of the observed sequences (Fig. 4.33).

The maximum regressive surface is the depositional surface at the end of regression, which includes the paleoseafloor at the end of coastal progradation and its correlative depositional surface within the continental setting (Figs. 6.29 and 6.30). At least part of the continental portion of the maximum regressive surface is invariably reworked and replaced by the transgressive surface of erosion during subsequent transgression (Figs. 4.39 and 6.29). The marine portion of the maximum regressive surface has a better preservation potential, as it is typically onlapped by transgressive marine or lacustrine "healing-phase" deposits which accumulate from suspension in shallow- and deep-water environments (i.e., coastal and marine onlap, respectively; Figs. 4.6, 4.8, and 6.31). Within its area of preservation, the maximum regressive surface is most commonly conformable, as the environmental energy typically declines following the onset of transgression in both continental and marine settings. Where the seafloor is starved of sediment during transgression, the maximum regressive and maximum flooding surfaces may merge into a composite unconformity expressed as a firmground or a hardground, depending on the degrees of seafloor erosion or cementation (Loutit et al., 1988; Galloway, 1989; Wilson and Palmer, 1992; Vinn and Wilson, 2010).

The physical expression of the maximum regressive surface depends on the depositional setting. In fluvial settings, the onset of transgression is most often marked by a decrease in fluvial energy and the degree of channel amalgamation (Figs. 6.29 and 6.30; see discussion in Chapters 4 and 5). This may correspond to a change from braided to meandering fluvial styles (Kerr et al., 1999; Ye and Kerr, 2000, Fig. 6.30), although instances of lowstand topsets consisting of meandering systems have also been documented (Miall, 2000; Posamentier, 2001). In the latter case, the position of the maximum regressive surface may be indicated by the least amount of tidal influences coupled with the lowest palynological marine index at any location. Following the pattern of fluvial onlap recorded by the underlying lowstand topset, the fluvial portion of the maximum regressive surface may also onlap the subaerial unconformity (Figs. 4.39A, 5.12–5.14). The location of the updip termination of the maximum regressive surface depends on the basin physiography (landscape gradients), the duration of the lowstand normal regression, and the rates of fluvial aggradation during the lowstand normal regression (see Chapter 4 for details).

In coastal settings, the maximum regressive surface is placed at the base of the earliest estuarine deposits, which mark the onset of transgression (Fig. 6.30). The contact between estuarine and underlying fluvial facies becomes progressively younger in an updip direction, diverging from the maximum regressive surface (Fig. 6.30). Therefore, it is important to distinguish between the base of estuarine deposits that is placed at the lowest stratigraphic level and the younger facies contact that develops within the transgressive systems tract. This distinction may be made on the basis of sedimentological criteria, as the maximum regressive surface is largely overlain by central estuary facies, whereas the within-trend facies contact sits mostly at the base of bayhead deltas in wave-dominated settings or estuary channels in tide-dominated settings (Fig. 6.30). The base of the transgressive deposits can be further constrained by correlation with the coeval shallow-water systems, where the change from progradation to retrogradation is typically more evident (Fig. 6.31).

In shallow-water settings, the maximum regressive surface is placed at the top of coarsening-upward siliciclastic facies (Figs. 6.29, 6.31−6.33), or at the top of prograding carbonates that are overlain by upward-deepening backstepping facies (Figs. 6.26, 6.34, and 6.35). The identification of maximum regressive surfaces in deep-water settings is more difficult, within successions that record a gradual reduction in the amount of

terrigenous sediment supply to the slope and basin floor environments (Fig. 4.20). The maximum regressive surface may be placed at the top of the youngest grain-flow deposits of a stratigraphic cycle, in cases where the coastline reaches the shelf edge at the end of regression, which requires the construction of composite profiles that integrate sedimentological and stratigraphic data from larger areas within a basin (Catuneanu, 2020a; see discussion in Chapter 8). In seismic stratigraphic terms, the maximum regressive surface is the youngest clino-form surface which is onlapped by transgressive "healing-phase" strata (coastal or marine onlap; Fig. 4.6).

Depending on the rates of subsequent transgression, as well as on the location within the basin, the maximum regressive surface may or may not be associated with a lithological contrast. Where transgression is slow and sediment supply to the shoreline continues to be high following the onset of transgression, the surface that marks the end of progradation may be lithologically cryptic (Figs. 5.104 and 6.33). In such cases, the position of the maximum regressive surface is constrained by the change in stacking patterns between the strata below and above, which is often evident from the breaks in slope gradients that can be observed in outcrops

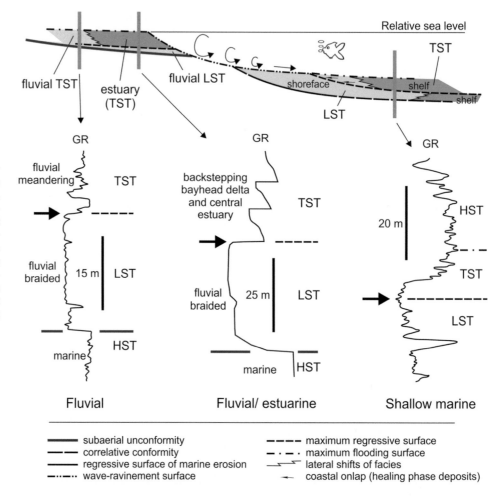

FIGURE 6.29 Well-log expression of the maximum regressive surface (arrows) in fluvial, coastal, and shallow-water settings. Log examples from Kerr et al. (1999) (left and center), and Embry and Catuneanu, 2001 (right). See Fig. 6.6 for a summary of attributes of the maximum regressive surface. Abbreviations: GR—gamma ray; LST—lowstand systems tract; TST—transgressive systems tract; HST—highstand systems tract.

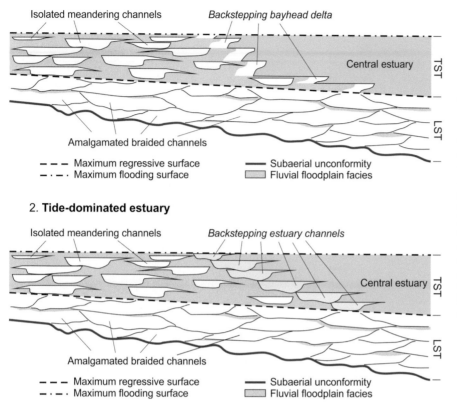

1. Wave-dominated estuary

Isolated meandering channels

Backstepping bayhead delta

Central estuary

TST

Amalgamated braided channels

LST

- – – – Maximum regressive surface
- – · – · Maximum flooding surface
- —— Subaerial unconformity
- ▨ Fluvial floodplain facies

2. Tide-dominated estuary

Isolated meandering channels

Backstepping estuary channels

Central estuary

TST

Amalgamated braided channels

LST

- – – – Maximum regressive surface
- – · – · Maximum flooding surface
- —— Subaerial unconformity
- ▨ Fluvial floodplain facies

FIGURE 6.30 Dip-oriented cross sections through fluvial to estuarine successions in wave- and tide-dominated settings (modified after Kerr et al., 1999). The lowstand systems tract (LST) consists of higher energy fluvial systems (lowstand topset) on top of the subaerial unconformity. The transgressive systems tract (TST) includes lower energy fluvial and estuarine facies. The maximum regressive surface may be traced at the base of the earliest transgressive deposits. Beyond the landward limit of the estuary at the onset of transgression, the shift from estuarine to coeval fluvial strata is a diachronous facies contact within the TST, which may be traced at the base of backstepping bayhead deltas (in wave-dominated settings) or at the base of backstepping estuary channels (in tide-dominated settings).

FIGURE 6.31 Outcrop examples of wave-ravinement (WRS), maximum flooding (MFS), and maximum regressive (MRS) surfaces (Triassic, Potrerillos—Cacheuta formations, Cuyana lacustrine rift basin, Argentina). The WRS is unconformable (wave scouring above the fairweather wave base), with transgressive "healing-phase" deposits on top. The MRS and MFS may be conformable (deposition below the fairweather wave base). In this example, the WRS and MRS coincide with flooding surfaces (episodes of abrupt water deepening). The two MFS surfaces are lithologically cryptic within condensed sections. This cyclicity describes the internal architecture of a larger scale (higher hierarchical rank) transgressive systems tract.

(Fig. 6.33). Where transgressions are fast and accompanied by abrupt changes in sediment supply to the seafloor, maximum regressive surfaces may record sharp lithological shifts, such as from sandstone to mudstone in siliciclastic settings (Fig. 6.31 and 6.32) or from limestone to shale in carbonate settings (i.e., "drowning unconformities"; Fig. 6.34). In this case, maximum regressive surfaces may coincide with allostratigraphic contacts such as flooding surfaces (Figs. 5.100, 5.103, 6.32, and 6.34; more details in Chapter 5).

FIGURE 6.32   Outcrop examples of maximum regressive surfaces in shallow-water settings, in which the tops of the coarsening-upward trends are marked by fully lithified layers (barriers to flow: A and C) or concretions (B) due to the preferential fluid migration and cementation along the maximum regressive surface after burial. This is the case where the onset of transgression is accompanied by an abrupt cut-off of sediment supply to the shallow-water environments, leading to a sharp permeability change across the maximum regressive surface. A—Late Campanian, Bearpaw Formation, Western Canada Basin; B—Late Permian, Waterford Formation, Ecca Group, Karoo Basin; C—top of the Late Campanian Kipp Member of the Bearpaw Formation, Western Canada Basin, exposed by the weathering and erosion of the overlying (and more recessive) transgressive deposits.

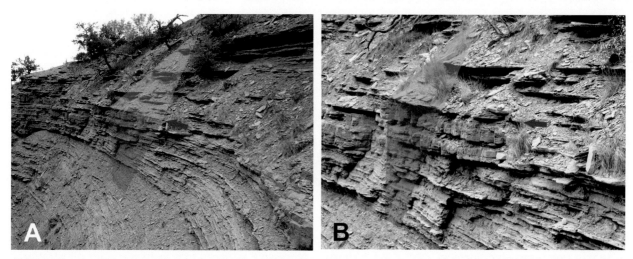

FIGURE 6.33   Maximum regressive surface (*arrows*) in a conformable shallow-water succession, at the limit between coarsening-upward (red) and fining-upward (blue) trends (Campanian, Panther Tongue, Utah; image B: detail from A). In this example, the shift from progradation to retrogradation is more gradual (compare with Fig. 6.32), although still marked by a break in slope gradients which indicates textural changes in grain size across the contact.

FIGURE 6.34 Maximum regressive surface in a mixed carbonate-siliciclastic deep-water setting (Triassic, The Dolomites, Italy). In this example, the maximum regressive surface coincides with a drowning surface generated by rapid transgression and water deepening. Facies: A—slope calcirudites; B—transgressive shales; C—slope calcarenites. Abbreviations: LST—lowstand systems tract; TST—transgressive systems tract; HST—highstand systems tract; MRS—maximum regressive surface; MFS—maximum flooding surface.

The stacking patterns which afford the identification of maximum regressive surfaces in the sedimentary record (i.e., progradation vs. retrogradation) are defined on the basis of physical criteria (e.g., coarsening- vs. fining-upward trends), which are often referred to in terms of bathymetric trends (i.e., shallowing- vs. deepening-upward). However, the two types of trends are not necessarily interchangeable. The seafloor at the end of regression may not correspond to the peak of shallowest water, especially in offshore areas beyond the shoreface where subsidence rates may be higher and sedimentation rates may be lower than those recorded at the shoreline. As a result, the peak of shallowest water may be within the underlying regressive

deposits, with the maximum regressive surface forming within a deepening-upward succession (see discussion in Chapter 7). For this reason, it is preferable to make reference to observed physical features (e.g., grading trends in siliciclastic settings) rather than inferred bathymetric trends (shallowing- vs. deepening-upward). The two types of trends usually coincide in areas adjacent to the shoreline, such as the shoreface, but are likely to diverge offshore.

Maximum regressive surfaces may be observed at different scales, in relation to regressions of different hierarchical ranks (Fig. 5.104). At each hierarchical level, the change from progradation to retrogradation occurs during relative sea-level rise, when accommodation

FIGURE 6.35   Maximum regressive (MRS) and within-trend normal regressive (WTNRS) surfaces in a carbonate setting (Aptian, Maestrat Basin, Spain; modified after Bover-Arnal et al., 2009; image courtesy of Telm Bover-Arnal). A—Slope facies (lowstand foreset); B—platform facies (lowstand topset); C—backstepping platform (transgressive systems tract). The MRS is isochronous along dip, whereas the WTNRS becomes younger in the direction of progradation.

starts to outpace the rates of sedimentation at the shoreline (Fig. 4.32; details in Chapter 4). As both subsidence and sedimentation rates may vary along the shoreline, the maximum regressive surface may develop a high degree of diachroneity in a strike direction (details in Chapter 7). The maximum regressive surface is also known as the "transgressive surface" (Posamentier and Vail, 1988). The term "maximum regressive surface" is recommended where emphasis is placed on the end of regression (e.g., top of a coarsening-upward trend in a shallow-water system); the term "transgressive surface" is recommended where emphasis is placed on the onset of transgression (e.g., the base of the oldest facies of an estuarine system). The two terms are interchangeable, as the end of regression and the beginning of transgression are coincident at stratigraphic timescales. However, depending on the nature of the facies that are observed (i.e., regressive or transgressive), either term may become most suitable within the context of a particular study.

6.1.5 Maximum flooding surface

The maximum flooding surface is a sequence stratigraphic surface that separates transgressive strata below from highstand normal regressive strata above (Frazier, 1974; Posamentier et al., 1988; Van Wagoner et al., 1988; Galloway, 1989, Figs. 4.32, 6.6, and 6.7). The formation and usage of this surface are restricted to downstream-controlled settings, where shoreline shifts provide the reference to define transgressive and highstand normal regressive stacking patterns. The maximum flooding surface forms during relative sea-level rise, when the sedimentation rates start to outpace the rates of

accommodation at the shoreline (Fig. 4.32; details in Chapter 4). As both subsidence and sedimentation rates may vary along the shoreline, the timing of the change from retrogradation to progradation may record a high degree of diachroneity in a strike direction (details in Chapter 7). Maximum flooding surfaces may be observed at different scales, in relation to transgressions of different magnitudes (Figs. 5.104 and 5.105). However, not every sequence includes a maximum flooding surface, but only those which incorporate transgressive systems tracts (Fig. 4.33).

The maximum flooding surface is the depositional surface at the end of transgression, which includes the paleoseafloor at the end of coastal retrogradation and its correlative depositional surface within the continental setting (Fig. 6.36). The latter portion of the maximum flooding surface may form where transgression is accompanied by fluvial or eolian aggradation (Fig. 4.39A), which is the case in most settings. In the less common scenario in which transgression is accompanied by fluvial erosion or eolian degradation (Fig. 4.39B), the maximum flooding surface is restricted to the marine (or lacustrine) portion of the sedimentary basin. Within its area of formation, the maximum flooding surface is commonly conformable, as aggradation continues during the highstand normal regression in all depositional environments. However, sediment starvation coupled with instability caused by a rapid increase in water depth may lead to nondeposition or even scouring of the seafloor during transgression, particularly in outer shelf and upper slope settings (Loutit et al., 1988; Galloway, 1989, Fig. 5.65). In such cases, the maximum flooding surface is unconformable and potentially merged with the maximum regressive surface (Fig. 5.100D and E).

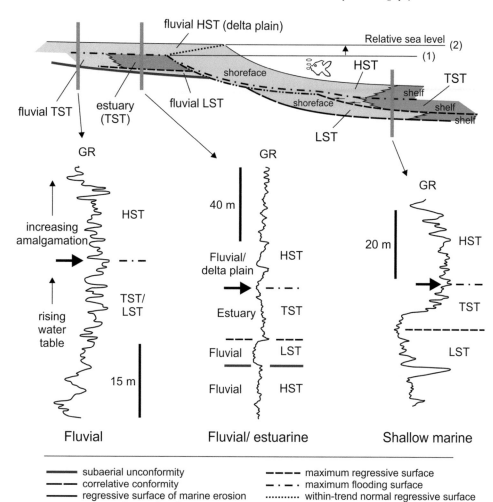

FIGURE 6.36 Well-log expression of the maximum flooding surface (arrows) in fluvial, coastal, and shallow-water settings. Log examples from the Wapiti Formation (Late Cretaceous, Western Canada Basin: left and center), and Embry and Catuneanu (2001) (Jurassic, Sverdrup Basin: right). See Fig. 6.6 for a summary of attributes of the maximum flooding surface. Abbreviations: GR—gamma ray; LST—lowstand systems tract; TST—transgressive systems tract; HST—highstand systems tract.

Maximum flooding surfaces are among the most useful stratigraphic markers for correlation and the subdivision of sedimentary successions, as they form within easily mappable and regionally extensive condensed sections (Frazier, 1974; Galloway, 1989). They also typically provide the best reference horizons (datums) for flattening well-log cross-sections or seismic volumes, at the top of transgressive deposits which "heal" pre-existing lows or irregular bathymetric profiles (Fig. 4.6). Condensed sections often exhibit a transparent facies on seismic data, due to their lithological homogeneity dominated by hemipelagic sediment, and a high gamma-ray response caused by increased concentrations of organic matter and radioactive elements (Fig. 5.72). However, organic matter may also concentrate in isolated depozones that are subject to restricted water circulation and limited terrigenous sediment influx, such as in some paralic environments, at times which may not correspond to the maximum transgression of the shoreline. Therefore, a comprehensive analysis of the tectonic and depositional settings is required in order to place the data in the proper context (see the workflow of sequence stratigraphy in Chapter 9).

Condensed sections associated with stages of maximum flooding may contain glauconite, siderite, and other carbonates or biochemical precipitates which may exhibit a wide range of log motifs (Posamentier and Allen, 1999). For these reasons, well-log data must be integrated with all other available datasets (e.g., biostratigraphic and geochemical; see Chapter 2 for details), as well as with the observation of the regional stratal stacking patterns, for reliable interpretations. Processes of chemical and biochemical precipitation at the seafloor are typically enhanced during the late stages of transgression, when the amount of terrigenous sediment supply diminishes, potentially leading to the early cementation of the seafloor and the formation of hardgrounds. The peak of the transgression may also be marked by pelagic carbonates in the distal areas of siliciclastic sediment-starved sedimentary basins (Hart, 2015), as well as by the highest degree of stratigraphic condensation (Fig. 4.20). The latter criterion has become a dependable tool in stratigraphic studies, with reproducible results (Gutierrez Paredes et al., 2017). Within conformable successions, the interval containing the maximum flooding surface can be further constrained with micropaleontological data (e.g., the abundance, diversity, and % fragmentation of benthic foraminifera, as well as the ratio between distal and proximal benthic species; Zecchin et al., 2021).

The physical expression of maximum flooding surfaces varies with the depositional setting, as well as with the effects of transgression on sediment supply to different areas within the basin, and ranges from lithologically cryptic to lithological discontinuities. In spite of this variability, all maximum flooding surfaces can be identified on the basis of the same criterion which relies on the observation of stratal stacking patterns (Fig. 6.6). The continental portion of the maximum flooding surface is commonly associated with the highest water table relative to the topographic profile, and therefore, it may be marked by the development of regional coal beds (e.g., Gastaldo et al., 1993; Wright and Marriott, 1993; Hamilton and Tadros, 1994; Shanley and McCabe, 1994; Bohacs and Suter, 1997; Tibert and Gibling, 1999; Holz et al., 2002; Fanti and Catuneanu, 2010, Figs. 5.50 and 5.73). It is also common to record the most frequent occurrence of crevasse splays within low-energy fluvial systems at the time of maximum flooding, due to the higher water table, as well as the highest palynological marine index and amount of tidal influences at any location.

In coastal settings, the maximum flooding surface is placed at the top of the youngest estuarine facies, marking the change to sedimentation in a deltaic environment (Figs. 5.92, 6.36, and 6.37). The backstepping of estuaries extends the area of tidal influence toward the basin margin, providing criteria for the identification of transgressive systems tracts and their bounding surfaces in fluvial deposits based on changes in the abundance of tide-related physical and biogenic structures, including sigmoidal, wavy and lenticular bedding, paired mud/silt drapes, shrinkage cracks, multiple reactivation surfaces, inclined heterolithic strata, bidirectional cross-beds, and brackish to marine trace fossils (Shanley et al., 1992). Tidal influences, including tidal-current reversals, often extend for tens of kilometers upstream from the shoreline (Shanley et al., 1992), and, depending on river discharge, gradients, and tidal range, may reach as far as 130 km (Allen and Posamentier, 1993) or even over 200 km inland from the river mouth (Miall, 1997). At any location, the highstand fluvial deposits overlying a maximum flooding surface record a decline in the abundance of tidal structures, and a gradual increase in the degree of channel amalgamation as the rates of flood-plain aggradation decrease toward the end of relative sea-level rise (Fig. 6.36, Wright and Marriott, 1993; Shanley and McCabe, 1993; Emery and Myers, 1996).

In marine settings, maximum flooding surfaces may be lithologically cryptic inside condensed sections where sedimentation continues during transgression (Carter et al., 1998; Posamentier and Allen, 1999; Catuneanu, 2006, Figs. 5.10 and 6.31), or marked by lithological discontinuities where sediment supply ceases or decreases significantly during transgression (Fig. 5.100D and E). Within successions which preserve at least in part transgressive systems tracts, maximum flooding surfaces are placed at the top of fining-

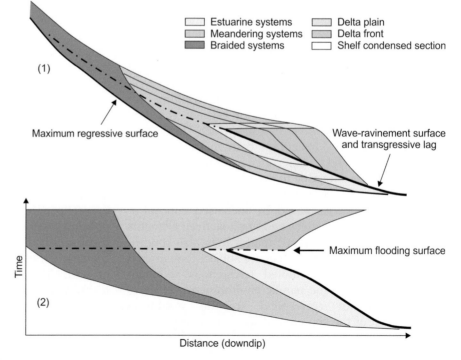

FIGURE 6.37 Stratigraphic architecture of a fluvial to shallow-water succession, in depth (1) and time (2) domains (Upper Cretaceous, southern Utah; modified after Shanley et al., 1992). Note the decline in fluvial energy during relative sea-level rise (transgression and highstand in this example), leading to the landward migration of the limit between braided and meandering fluvial systems, and the fining-upward trends that are commonly observed in the fluvial portions of systems tracts.

Estuarine systems
Meandering systems
Braided systems
Delta plain
Delta front
Shelf condensed section

(1)

Maximum regressive surface

Wave-ravinement surface and transgressive lag

Time

(2)

Maximum flooding surface

Distance (downdip)

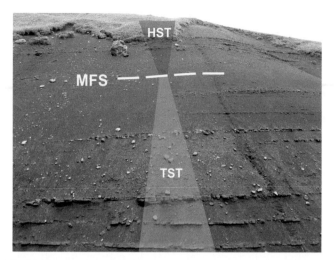

FIGURE 6.38 Maximum flooding surface (MFS) in a mixed carbonate-siliciclastic deep-water setting (Triassic, The Dolomites, Italy). The MFS corresponds to a time of minimum influx of platform sediment to the deep-water setting, leading to the highest degree of stratigraphic condensation. The frequency of turbidity flows decreases during transgression and increases during subsequent highstand normal regression. Abbreviations: TST—transgressive systems tract; HST—highstand systems tract.

upward (retrogradational) trends. This is true in both siliciclastic and carbonate settings, and from shallow- to deep-water environments. In carbonate settings, the maximum flooding surface may have a lithological expression even in conformable successions, as the change from transgressive to highstand systems tracts is often marked by a shift from shales to limestones, as the carbonate factory is switched on again following drowning (Figs. 6.26 and 6.34). In deep-water settings, the maximum flooding surface marks the end of a stage of waning of the frequency and density of turbidity flows, as sediment supply decreases and the degree of stratigraphic condensation increases during transgression (Figs. 4.20 and 6.38).

The mappability of maximum flooding surfaces depends on the types of data available, as well as on the depositional setting and the nature of the stratigraphic contact. In conformable successions where the end of retrogradation is not marked by a lithological discontinuity, maximum flooding surfaces are difficult to pinpoint in outcrop or core within the monotonous facies of condensed sections. This is commonly the case with conformable successions in siliciclastic shallow-water settings, as well as in all deep-water settings which are not starved of detrital (whether carbonate or siliciclastic) sediment (Figs. 5.10, 6.31, and 6.38). In such instances, well logs calibrated with grain-size analysis in core or rock cuttings may help to pinpoint the position of maximum flooding surfaces (Figs. 5.104 and 5.105). Notably, the top of a fining-upward retrogradational trend in a marine succession does not necessarily correspond to the peak of deepest water, especially in offshore areas beyond the

shoreface (see Chapter 7 for details). Therefore, as discussed above in the case of maximum regressive surfaces, the stacking patterns which afford the identification of maximum flooding surfaces need to be described in terms of observed physical trends (fining- vs. coarsening-upward) rather than inferred bathymetric trends (deepening- vs. shallowing-upward).

At seismic scales, the identification of maximum flooding surfaces is based on the observation of the large-scale stratigraphic architecture, seismic reflection terminations (in this case, downlap at the base of highstand clinoforms), and seismic facies (Figs. 5.72 and 6.4). The information derived from the seismic data is typically sufficient for reconnaissance studies in frontier basins, but it can be further constrained by calibration with well logs, lithologs, and biostratigraphic data in the more advanced stages of basin exploration. Notably, "downlap surfaces" interpreted as maximum flooding surfaces are typically observed on low-resolution seismic transects, where the bottomsets of highstand clinoforms fall below the seismic resolution (Figs. 4.8 and 6.4). In high-resolution studies, a sedimentological downlap surface may be defined within the highstand systems tract, at the limit between the hemipelagic condensed section, within which the maximum flooding surface resides, and the overlying highstand clinoforms (Zecchin and Catuneanu, 2013, Fig. 6.2). This within-trend facies contact may have a stronger physical expression than the maximum flooding surface, but it is diachronous, with a rate that is inversely proportional to the rate of shoreline progradation. Despite its potentially cryptic appearance, the maximum flooding surface remains more important for regional correlation.

The termination of maximum flooding surfaces in an updip direction is typically described by onlap against the underlying landscape or seascape, which can be "fluvial" where transgression is accompanied by fluvial aggradation (Fig. 5.65), "coastal" where transgression is accompanied by fluvial erosion (Fig. 4.39B), or "marine" in deep-water settings where the transgressive systems tract is missing around the shelf edge (Figs. 4.8, 5.12, and 5.65). Maximum flooding surfaces may downlap older paleoseafloors in a downdip direction at the end of the areas of deposition of transgressive sediments in shallow- and deep-water environments (Fig. 5.65). Some of these downlap terminations observed on seismic lines may be only apparent (i.e., seismic artifacts), as the thickness of the transgressive systems tract falls below the seismic resolution in distal areas characterized by lower rates of sedimentation. In all depositional settings, maximum flooding surfaces have a high preservation potential, being overlain by aggrading and prograding highstand normal regressive deposits (Figs. 4.39 and 5.14).

Alternative terms include "final transgressive surface" (Nummedal et al., 1993) and "maximum transgressive surface" (Helland-Hansen and Martinsen, 1996). The term "maximum flooding surface" is still recommended, as it is strongly entrenched in the literature and has historical priority. It is important to avoid confusion between maximum flooding surfaces and flooding surfaces. The distinction is not a matter of scale (e.g., major vs. minor transgressions, as there are more than two scales in stratigraphy), but a matter of definition: a maximum flooding surface marks a change in stratal stacking pattern (i.e., it is a sequence stratigraphic surface), and it may or may not be associated with a lithological contrast; a flooding surface is a lithological discontinuity (i.e., an allostratigraphic surface), which may or may not mark a change in stratal stacking pattern. Both flooding surfaces and maximum flooding surfaces may form at the same scale of observation (i.e., hierarchical rank), in relation to the same transgressions (Figs. 5.100, 5.103, and 5.106; details in Chapter 5).

6.1.6 Transgressive surface of erosion

The transgressive surface of erosion (Posamentier and Vail, 1988) is an unconformity that forms during transgression by means of wave scouring (i.e., "wave-

ravinement surface"; Swift, 1975) or tidal scouring (i.e., "tidal-ravinement surface"; Allen and Posamentier, 1993) in subtidal to intertidal environments (Fig. 6.6). The two types of transgressive ravinement surfaces merge into a single contact in open-shoreline settings between the river mouths, where backstepping beaches are reworked by both waves and tides during transgression; in this case, the composite surface is described as a wave-ravinement, since waves in the subtidal environment leave the last imprint on the composite contact. In river-mouth settings, the tidal- and wave-ravinement surfaces form at the same time in different environments, within the estuary and the shoreface, respectively (Figs. 6.39 and 6.40); in this case, the two transgressive ravinement surfaces may be preserved as distinct contacts separated by the estuary-mouth complex (i.e., with the tidal-ravinement below, and the wave-ravinement above; Posamentier and Allen, 1999; Catuneanu, 2006, Fig. 5.66).

Transgressive ravinement surfaces young in an updip direction (Nummedal and Swift, 1987), with rates inversely proportional to the rates of shoreline transgression. The oldest portion of any transgressive ravinement surface invariably reworks part of the maximum

FIGURE 6.39 Tidal- and wave-ravinement surfaces in a wave-dominated estuarine setting (modified after Dalrymple et al., 1992; Reinson, 1992; Zaitlin et al., 1994; Shanmugam et al., 2000). Both tidal- and wave-ravinement surfaces expand landward during transgression, truncating central estuary and estuary-mouth complex facies, respectively. Where both ravinement surfaces are preserved, the wave-ravinement is stratigraphically higher at any location, and separated from the tidal-ravinement by the estuary-mouth complex (Fig. 5.66). Abbreviations: TRS—tidal-ravinement surface; WRS—wave-ravinement surface.

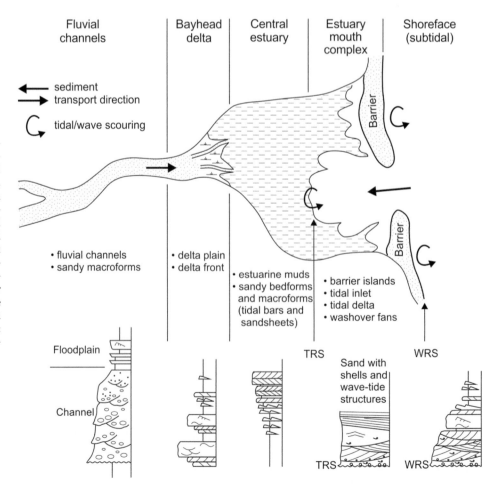

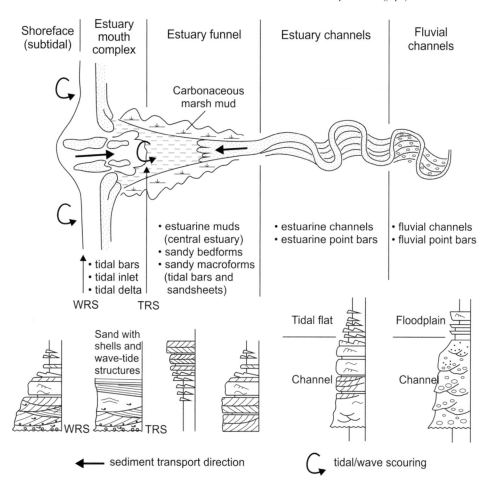

FIGURE 6.40 Tidal- and wave-ravinement surfaces in a tide-dominated estuarine setting (modified after Allen, 1991; Dalrymple et al., 1992; Allen and Posamentier, 1993; Shanmugam et al., 2000). Both tidal- and wave-ravinement surfaces expand landward during transgression, truncating central estuary and estuary-mouth complex facies, respectively. Where both ravinement surfaces are preserved, the wave-ravinement is stratigraphically higher at any location, and separated from the tidal-ravinement by the estuary-mouth complex (Fig. 5.66). Abbreviations: TRS—tidal-ravinement surface; WRS—wave-ravinement surface.

regressive surface, thus becoming a systems tract boundary (Catuneanu et al., 2011, Figs. 4.39, 5.14, and 6.41); this makes the transgressive surface of erosion a surface of sequence stratigraphy. Transgressive surfaces of erosion may also rework and replace other sequence stratigraphic surfaces, including subaerial unconformities (e.g., where lowstand topsets are eroded during transgression; Figs. 4.46, 5.14, and 6.11). The amount of erosion associated with transgressive ravinement processes is proportional to the energy of waves and tides along the transgressive coastline (i.e., the depth of the wave base and the magnitude of the tidal range). In most cases, the erosion is limited to the depth of the fair-weather wave base, with an average of 10-20 m of section being removed by ravinement processes during transgression (Demarest and Kraft, 1987; Abbott, 1998). This amount can increase in areas of exceptionally high energy (e.g., c. 40 m along the Canterbury Plains of New Zealand; Leckie, 1994), but it can also be negligible where the substrate is indurated by pedogenic or other processes prior to the transgression (Fig. 6.10). Coal beds can also limit the amount of ravinement erosion, due to their resilient nature, and as a result some transgressive ravinement surfaces are found directly on top of xylic substrates (Fig. 5.73).

Depending on the scale of observation, and on the character of the transgression within the study area (i.e., uninterrupted or punctuated by higher frequency regressions; Figs. 4.6, 4.29, and 6.41), the transgressive surface of erosion may be a single through-going surface or a composite of multiple lower rank surfaces. Where transgression is punctuated by stages of short-term progradation, the long-term backstepping of the shoreline is accompanied by the formation of multiple transgressive ravinement surfaces of lower hierarchical rank separated by the prograding facies (Fig. 6.41A2 and B2). In this case, the short-term transgressions and regressions define high-frequency genetic stratigraphic sequences within the larger scale transgressive systems tract, and the higher rank ravinement surface becomes a composite contact that can be traced at the base of all lower rank highstand clinoforms (Fig. 6.41A2 and B2). Where the transgression is uninterrupted, the transgressive surface of erosion forms a single through-going physical surface at the hierarchical level of the associated transgression (Fig. 6.41A1 and B1).

Diagnostic to the transgressive surface of erosion is the presence of estuary-mouth complex (in the case of tidal-ravinement surfaces) or marine "healing-phase" (in the case of wave-ravinement surfaces) deposits on

A. Transgression with fluvial aggradation

1. Uninterrupted transgression punctuated by flooding events

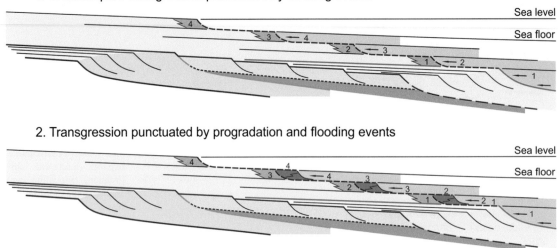

2. Transgression punctuated by progradation and flooding events

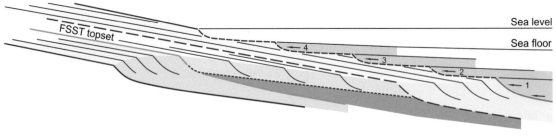

B. Transgression with fluvial erosion

1. Uninterrupted transgression punctuated by flooding events

2. Transgression punctuated by progradation and flooding events

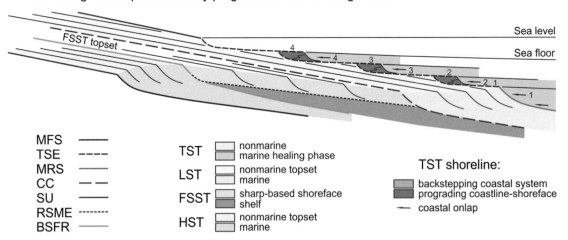

MFS	——
TSE	----
MRS	——
CC	— —
SU	——
RSME	········
BSFR	——

TST	nonmarine
	marine healing phase
LST	nonmarine topset
	marine
FSST	sharp-based shoreface
	shelf
HST	nonmarine topset
	marine

TST shoreline:

backstepping coastal system
prograding coastline-shoreface
← coastal onlap

FIGURE 6.41 Evolution of coastal systems during stepped transgression punctuated by flooding events. A — Shoreline trajectory steeper than the fluvial profile: transgressive fluvial and coastal systems aggrade and may be preserved below the transgressive surface of erosion (TSE). B — Fluvial profile steeper than the shoreline trajectory: the TSE reworks the subaerial unconformity and no transgressive fluvial and coastal systems are preserved below the TSE. Both A and B: the flooding events may or may not be followed by short-term progradation; in the former case, the TSE is a composite of multiple lower rank TSEs, and the backstepping parasequences are high-frequency genetic stratigraphic sequences; in the latter case, the TSE is a single physical surface, and the backstepping parasequences are bedsets within the transgressive systems tract. The TSEs in this diagram are wave-ravinement surfaces. Where the backstepping coastal system is an estuary, tidal-ravinement surfaces may also form at the base of the estuary mouth complex. Abbreviations: BSFR — basal surface of forced regression; RSME — regressive surface of marine erosion; SU — subaerial unconformity; CC — correlative conformity; MRS — maximum regressive surface; MFS — maximum flooding surface; HST — highstand systems tract; FSST — falling-stage systems tract; LST — lowstand systems tract; TST — transgressive systems tract.

1. Open-shoreline setting

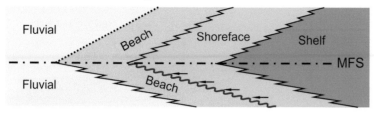

Disconnect between wave-ravinement and within-trend normal regressive surface

2. River-mouth setting

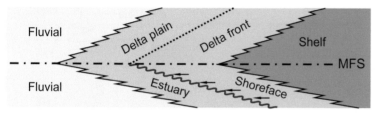

Connection between wave-ravinement and within-trend normal regressive surface

– · – · maximum flooding surface (MFS)   ⌇ facies changes
∿∿∿ transgressive wave-ravinement surface   ← coastal onlap
·········· within-trend normal regressive surface: topset-foreset contact

FIGURE 6.42 Stratigraphic architecture of depositional systems in transgressive and highstand open-shoreline and river-mouth settings. The within-trend normal regressive surface (limit between topset and foreset) may or may not connect with the updip end of the wave-ravinement surface, depending on the type of coastline. In open-shoreline settings, beaches are part of the highstand foreset (Fig. 4.50B), whereas in river-mouth settings delta plains are part of the highstand topset (Fig. 4.50A).

top (Posamentier and Allen, 1993, 1999; Catuneanu, 2006, Figs. 6.6 and 6.5). Depending on the frequency of the transgressive–regressive cycles, the diagnostic deposits on top can be very thin, potentially reduced to transgressive lags or onlapping shell beds toward the proximal end of the ravinement surfaces (Figs. 5.2, 5.5, 6.41A2 and B2). Transgressive ravinement surfaces, transgressive lags and shell beds, and maximum flooding surfaces of equal hierarchical ranks merge together toward the location of the shoreline at the end of the transgression, in which case the composite contact is defined by the diagnostic attributes of the younger surface (i.e., the maximum flooding; Figs. 4.39 and 6.6). The facies preserved below a transgressive surface of erosion may range from continental to marine, and therefore are not diagnostic to the identification of the surface (Fig. 6.5).

Transgressive ravinement surfaces provide favorable conditions for the formation of substrate-controlled ichnofacies, as they are always scoured and overlain by marginal-marine to shallow-marine facies. Depending on the amount of tidal and/or wave scouring, as well as on the nature of facies that are subject to erosion, the transgressive ravinement surfaces may be marked by firmgrounds (*Glossifungites* Ichnofacies; Figs. 2.47, 2.48), hardgrounds (*Trypanites* Ichnofacies; Fig. 2.49), or woodgrounds (*Teredolites* Ichnofacies; Fig. 2.50). The colonization of these substrates occurs during or immediately after a ravinement surface is cut, and before the deposition of the overlying transgressive facies (MacEachern et al., 1992, 2012). Numerous case studies documenting the ichnology of transgressive ravinement surfaces have

been published from both modern settings and ancient successions (e.g., MacEachern et al., 1992, 1999, 2012; Taylor and Gawthorpe, 1993; Pemberton and MacEachern, 1995; Ghibaudo et al., 1996; Krawinkel and Seyfried, 1996; Pemberton et al., 2001; Gingras et al., 2004; Buatois and Mangano, 2011; Knaust and Bromley, 2012).

6.1.6.1 Wave-ravinement surface

The wave-ravinement surface forms by wave scouring in the shoreface as the seafloor profile is maintained in equilibrium with the wave energy during shoreline transgression (Bruun, 1962; Swift et al., 1972; Swift, 1975; Dominguez and Wanless, 1991, Fig. 4.55). The resulting unconformity is placed at the top of coastal systems in both open-shoreline and river-mouth settings (Fig. 6.42). Where coastal systems are not preserved following the wave erosion, the wave-ravinement surface may be placed at the top of any type of depositional system, from continental to marine (Figs. 6.5, 6.31, and 6.43). Irrespective of the nature of the underlying depositional system, the wave-ravinement surface is always onlapped by transgressive shallow-water deposits which heal the bathymetric profile of the seafloor (i.e., coastal onlap; Figs. 4.8, 5.12, 5.65, 6.5, and 6.6).

Within shallow-water successions, both wave-ravinement and maximum regressive surfaces mark a similar change from coarsening-upward to overlying fining-upward trends. However, the former surfaces are unconformable, and tend to be marked by sharper contacts, whereas the latter are conformable, with a smoother transition from prograding to retrograding trends (Fig. 6.44). Wave-ravinement surfaces are

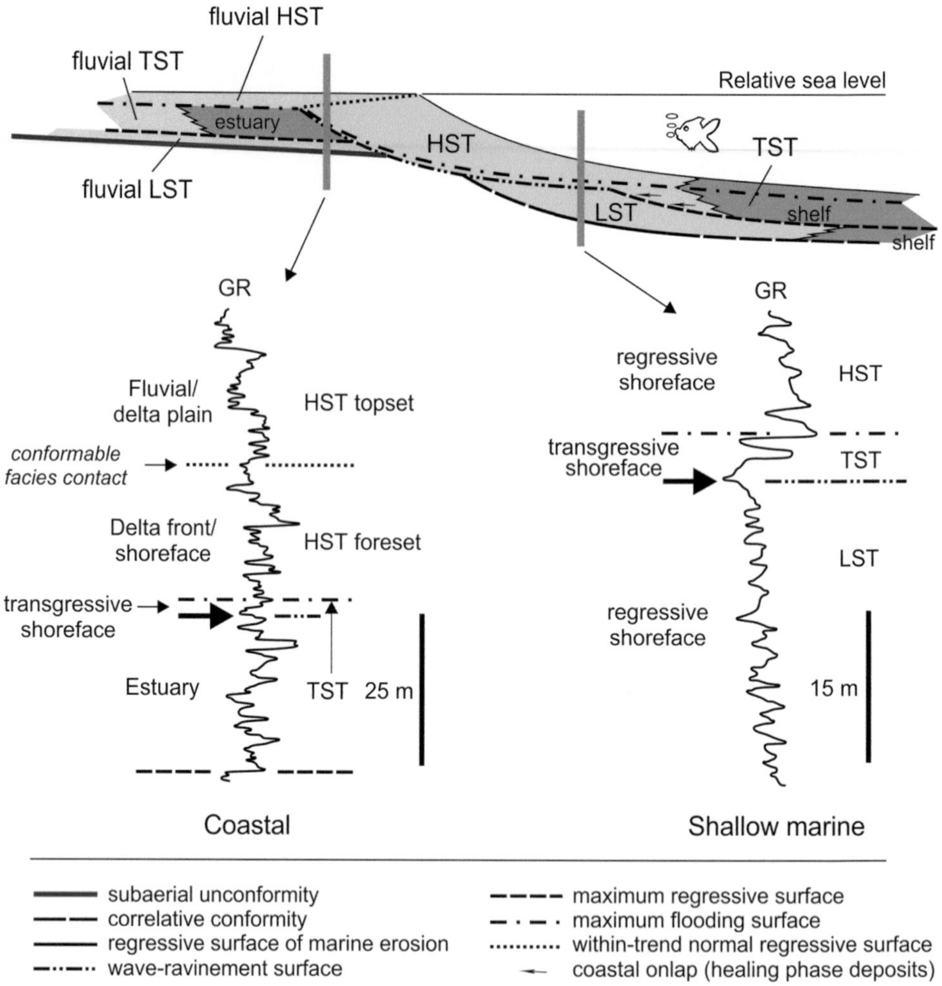

FIGURE 6.43   Well-log expression of the transgressive wave-ravinement surface (arrows) in coastal and shallow-water settings. Log examples from the Bearpaw Formation (Late Cretaceous, Western Canada Basin: left) and Embry and Catuneanu (2001) (Jurassic, Sverdrup Basin: right). See Fig. 6.6 for a summary of attributes the wave-ravinement surface. Abbreviations: GR—gamma ray; LST—lowstand systems tract; TST—transgressive systems tract; FSST—falling-stage systems tract; HST—highstand systems tract.

commonly overlain by cm- to dm-thick transgressive lags and/or concentrations of onlapping shell beds (Figs. 4.46, 5.101, 6.3, 6.37, 6.45, 6.46, and 6.47; Kidwell, 1991a,b; Zecchin and Catuneanu, 2013). However, the thickness of transgressive lags and shell beds can be highly variable, depending on the nature of the substrate that is being eroded and the energy of the waves, from insignificant to several meters in extreme cases (Catuneanu and Biddulph, 2001, Fig. 6.48). The transgressive lags are often marked by low gamma-ray deflections on well logs, which allow inferring the unconformable nature of the contact (Figs. 6.43 and 6.44).

6.1.6.2 Tidal-ravinement surface

The tidal-ravinement surface is a scour generated by tidal erosion in coastal environments during shoreline transgression (Allen and Posamentier, 1993). While this scouring may occur in all tidal settings, the tidal-ravinement is typically preserved as a distinct surface within estuarine systems, where the estuary-mouth complex protects it from subsequent wave erosion in the shoreface. The formation of tidal-ravinement

surfaces within estuaries involves the coalescence of channel scours in the process of landward migration of tidal channels (Dalrymple et al., 1992; Kitazawa and Murakoshi, 2016, Figs. 6.39 and 6.40). The resulting unconformity is placed at the top of central estuary facies, or any older depositional systems where central estuary sediments are not preserved, but always at the base of estuary-mouth complex deposits (Figs. 6.39, 6.40, and 6.49). The latter provides the diagnostic criterion for the identification of this surface in the sedimentary record (Figs. 6.5 and 6.6). The preservation of a tidal-ravinement surface requires that the amount of aggradation of the estuary-mouth complex exceeds the amount of subsequent wave-ravinement erosion; otherwise, the wave-ravinement surface reworks the tidal-ravinement surface, and the two contacts become superimposed.

The formation of tidal-ravinement surfaces requires tidal action, and therefore connection to the global ocean. Interior basins devoid of tides do not provide the necessary conditions to generate these scour surfaces. Moreover, open shoreline settings in pericontinental basins that experience a tidal range are not prone

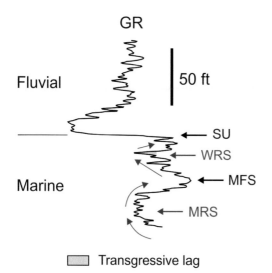

FIGURE 6.44 Well-log expression of wave-ravinement vs. maximum regressive surfaces in shallow-water successions (Jurassic, Sverdrup Basin, Canada; modified after Embry and Catuneanu, 2001). Wave-ravinement surfaces are typified by sharp changes from coarsening- to fining-upward trends, due to their unconformable nature. In contrast, maximum regressive surfaces are conformable and display a smoother shift from coarsening- to fining-upward trends. In this example, the two surfaces belong to two different stratigraphic cycles, with the well intercepting the maximum regressive surface of the lower cycle (location downdip from the shoreline at the onset of transgression) and the wave-ravinement surface of the upper cycle (location updip from the shoreline at the onset of transgression). Abbreviations: GR—gamma-ray log; MRS—maximum regressive surface; MFS—maximum flooding surface; WRS—wave-ravinement surface; SU—subaerial unconformity.

to preserve tidal-ravinement surfaces, as subsequent wave erosion during the transgression overprints the evidence of tidal scouring. Consequently, the wave-ravinement surface is usually the type of transgressive ravinement that is documented in the majority of studies. Irrespective of setting, the wave-ravinement surface has a wider spread development at the base of all fully marine (or lacustrine) transgressive deposits, both in front of estuaries and along open shorelines, and it is therefore accounted for as the primary transgressive surface of erosion (Fig. 4.39). Where the estuary-mouth complex is preserved in the rock record, the wave-ravinement surface is always intercepted at a higher stratigraphic level than the tidal-ravinement surface, as the shoreface shifts on top of the older estuary during the transgression (Figs. 5.66 and 6.50).

Tidal-ravinement surfaces are most commonly documented within incised-valley fills, where estuarine systems are common and the syn-depositional tidal range may be enhanced. The Gironde estuary in France provides an example of a mixed tide- and wave-influenced coastal setting, where the fill of the incised valley preserves the full succession of lowstand fluvial, transgressive estuarine, and highstand deltaic systems (Fig. 5.66; see Allen and Posamentier, 1993, for core photographs). In this case study, the tidal-ravinement surface is preserved at the contact between central estuary and the overlying estuary-mouth facies. In coastal settings characterized by rapid transgression following

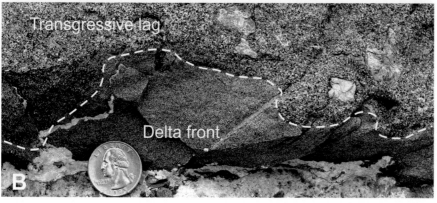

FIGURE 6.45 Wave-ravinement surface (WRS) at the contact between forced regressive clinoforms and the overlying transgressive healing-phase deposits (Campanian, Panther Tongue Formation, Utah). B—detail from A.

FIGURE 6.46 Wave-ravinement surface (arrow) at the contact between coal-bearing fluvial facies and the overlying onlap shell beds (oyster coquinas) and transgressive shales (Late Cretaceous, contact between the Dinosaur Park and Bearpaw formations, Western Canada Basin). In this example, the fluvial deposits are also transgressive (Hamblin, 1997), and therefore this portion of the wave-ravinement surface develops within a transgressive systems tract.

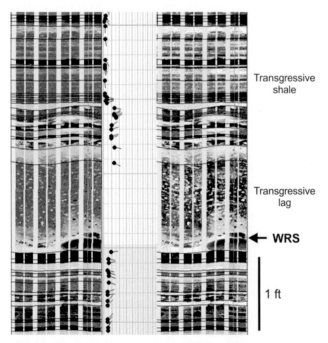

FIGURE 6.47 Wave-ravinement surface (WRS) on a microresistivity (image) log (Turonian, Cauvery Basin, India; image courtesy of Reliance Industries, India). Note the erosional nature of the WRS, and the structureless composition of the transgressive lag as indicated by the dipmeter data.

the onset of relative sea-level rise, and/or high tidal range, the lowstand fluvial deposits and the fine-grained central estuary facies may not be preserved in the rock record. In such cases, the tidal-ravinement surface reworks the subaerial unconformity, and the

underlying highstand facies may range from fluvial to shallow-marine (Figs. 6.49 and 6.50).

6.1.7 Regressive surface of marine erosion

The regressive surface of marine erosion is an unconformity cut by waves in the lower shoreface in response to the lowering of the wave base during relative sea/lake-level fall (Plint, 1988). Alternative terms include "regressive ravinement surface" (Galloway, 2001) and "regressive wave ravinement" (Galloway, 2004) (Fig. 6.6). As the relative sea level falls and the shoreline is forced to regress, the regressive surface of marine erosion expands in a downdip direction, thus becoming younger basinward with the rate of shoreline regression (Figs. 5.33 and 6.6). The oldest portion of the regressive surface of marine erosion invariably reworks part of the basal surface of forced regression, thus becoming a systems tract boundary (Figs. 4.39, 5.31, and 5.33); for this reason, the regressive surface of marine erosion is a surface of sequence stratigraphy. The extent to which the basal surface of forced regression is reworked by waves during relative sea-level fall depends on the gradient of the shoreline trajectory during forced regression relative to the seafloor gradient at the onset of forced regression (Fig. 4.37). However, since the regressive surface of marine erosion is restricted to shallow-water settings, the basal surface of forced regression is always preserved in deeper water environments (Fig. 4.37).

The amount of scouring associated with the formation of the regressive surface of marine erosion is commonly in a range of meters (Plint, 1991). Seaward from the fairweather wave base, erosion is replaced by bypass and eventually by sediment aggradation in the deeper water environments (Plint, 1991, Figs. 5.33 and 6.23). Enhanced wave energy during storms generates gutter casts at the base of the forced regressive shoreface, which may be observed along the regressive surface of marine erosion (Fig. 6.51). In areas of higher sediment supply, the wave scouring during relative sea-level fall may be suppressed and replaced by seafloor aggradation, leading to a rapid foreshortening of the shallow-water clinoforms. This is the case with river-dominated deltas, where sediment supply exceeds the capacity of waves and tides to rework and redistribute the sediment (Fig. 4.44B). As a result, the shift from prodelta to delta front is conformable, without an erosional contact (Fig. 4.46). However, cases of a conformable transition from bottomsets to foresets during forced regression are atypical, and, with the exception of river-dominated deltas, the majority of forced regressive shallow-water settings afford the formation of the regressive surface of marine erosion in lower shoreface or lower delta front environments.

The inner shelf is generally an area of sediment bypass during forced regression, although meter-thick

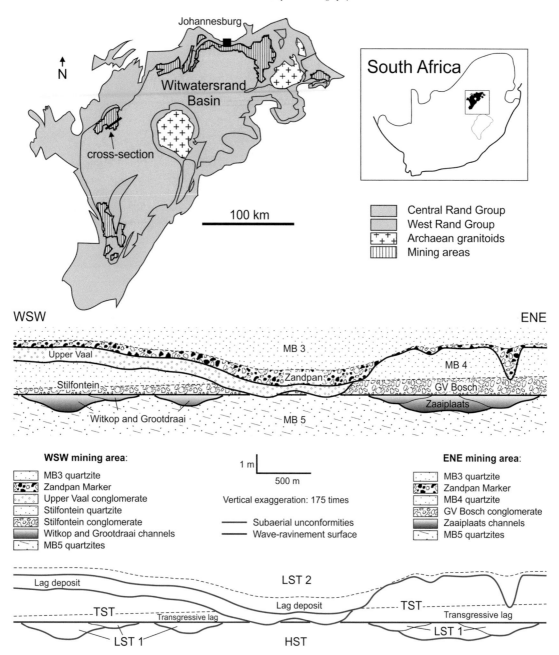

FIGURE 6.48 Gold placers ("reefs": Zandpan, Upper Vaal, GV Bosch, and basal Stilfontein) in a sequence stratigraphic framework (modified after Catuneanu and Biddulph, 2001; see map for the location of the cross section). In this example, the gold placers are lag deposits associated with subaerial unconformities and the wave-ravinement surface. Abbreviations: LST—lowstand systems tract; TST—transgressive systems tract; HST—highstand systems tract.

hummocky cross-stratified tempestites may still accumulate above the storm wave base (Plint, 1991). However, the preservation potential of these tempestites is relatively low because, as the relative sea level falls, the wave-scoured lower shoreface shifts across the former inner shelf, and as a result the hummocky cross-stratified beds are truncated by the regressive surface of marine erosion (Figs. 5.33 and 6.23). Beyond the storm wave base, the outer shelf may record continuous aggradation, providing that the fall in relative sea level

does not expose the entire shelf (Plint, 1991). Given the low preservation potential of forced regressive inner shelf facies, it is common to find swaley cross-stratified upper to middle shoreface deposits directly overlying falling-stage outer shelf (Plint and Nummedal, 2000) or even highstand outer shelf fines where the forced regressive shelf deposits are missing (Figs. 6.23, 6.51, and 6.52). Whether or not forced regressive shelf deposits are preserved, the regressive surface of marine erosion is always placed at the sharp contact between prograding

FIGURE 6.49 Tidal-ravinement surface (TRS) at the contact between highstand shelf deposits and a transgressive estuary-mouth complex (Lower Cretaceous, Ft. Collins and Horsetooth members of the Muddy Formation, Colorado; image courtesy of Henry Posamentier). In this example, the tidal-ravinement surface reworks a subaerial unconformity, and neither lowstand nor central estuary deposits are preserved following the tidal scouring.

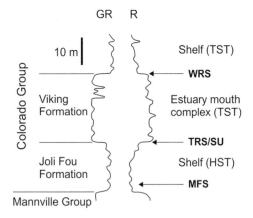

FIGURE 6.50 Well-log expression of a tidal-ravinement surface at the contact between highstand shelf deposits below and transgressive estuary-mouth sands above (Albian, Colorado Group, Crystal Field, Alberta). The estuary-mouth complex is truncated by a wave-ravinement surface during the backstepping of the shoreline. The tidal-ravinement surface reworks the subaerial unconformity at the base of the incised valley which hosts these transgressive deposits. Abbreviations: HST—highstand systems tract; TST—transgressive systems tract; MFS—maximum flooding surface; SU—subaerial unconformity; TRS—tidal-ravinement surface; WRS—wave-ravinement surface.

FIGURE 6.51 Regressive surface of marine erosion at the contact between forced regressive shoreface sands (above) and outer shelf fines (below) (Late Cretaceous, Marshybank Formation, Alberta; image courtesy of A.G. Plint). The sharp-based shoreface deposits have large, shore-normal gutter casts at their base (arrows), formed by scouring during storms.

shelf and shoreface facies, and corresponds to a missing section that incorporates at least the lower shoreface (Figs. 5.33, 6.33, 6.51, and 6.52).

Diagnostic to the regressive surface of marine erosion is the presence of sharp-based forced regressive shoreface deposits on top (Figs. 4.41, 4.45, 6.5, 6.6, 6.22, 6.51–6.53). The forced regressive wave-dominated shoreface clinoforms are typically unconformably based, with the exception of the earliest clinoform above the preserved (most proximal) segment of the basal surface of forced regression (i.e., updip of the landward termination of the regressive surface of marine erosion; Figs. 5.31, 5.32, 6.23). Whether unconformably or conformably based, all forced regressive shoreface clinoforms in wave-dominated settings are downstepping and

compressed, with a thickness constrained by the depth of the fairweather wave base (details in Chapter 4). Where forced regression is followed by lowstand normal regression, the youngest forced regressive clinoform is overlain by the first upstepping clinoform of the normal regression, which increases the overall thickness of the shoreface (Fig. 6.52). However, the forced and normal regressive shoreface deposits can still be separated, as the onset of relative sea-level rise is marked by a decline in riverborne sediment supply to the shoreline, which results in a decrease in grain size across the correlative conformity (Figs. 6.27 and 6.52).

The youngest segment of the regressive surface of marine erosion is overlain by the first clinoform of the lowstand systems tract, which therefore becomes

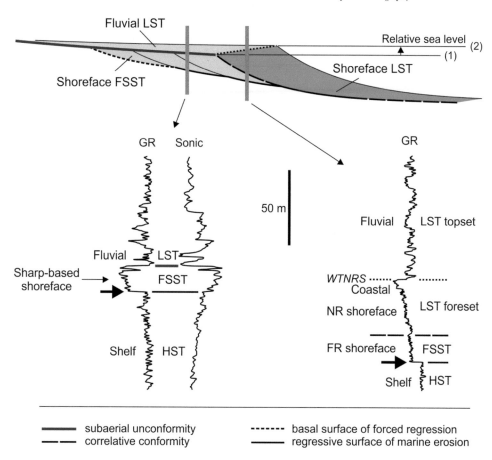

FIGURE 6.52 Well-log expression of the regressive surface of marine erosion (arrows). Log examples from the Cardium Formation (left) and the Lea Park Formation (right) (Late Cretaceous, Western Canada Basin). See Fig. 6.6 for a summary of attributes of the regressive surface of marine erosion. Note the decrease in grain size across the correlative conformity, at the limit between forced regressive (FR) and normal regressive (NR) shoreface deposits. Abbreviations: GR—gamma ray log; HST—highstand systems tract; FSST—falling-stage systems tract; LST—lowstand systems tract; WTNRS—within-trend normal regressive surface.

FIGURE 6.53 Regressive surface of marine erosion (arrow) at the limit between shelf fines accumulated below the fairweather wave base and the overlying "sharp-based" forced regressive shoreface sands (Triassic, upper Potrerillos Formation, Cuyana Basin, Argentina). Diagnostic for the regressive surface of marine erosion are the "sharp-based" prograding shoreface deposits on top.

and thicker than the forced regressive clinoforms that are supplied with sediment by the same river systems. This is due to the progradation of the normal regressive shoreface during relative sea-level rise, which promotes aggradation on both sides of the shoreline and enhances the trapping of the coarser sediments within the fluvial topset (details in Chapter 4). Notwithstanding the atypical development of the earliest forced regressive clinoform (conformably based) and of the earliest lowstand normal regressive clinoform (unconformably based), the regressive surface of marine erosion is always downlapped by the forced regressive shoreface clinoforms (Fig. 4.34), and marks an abrupt shift to finer grained shelf facies below, which may belong to either the falling-stage or the highstand systems tracts (Figs. 4.39, 6.23, 6.24, 6.51, and 6.52).

The regressive surface of marine erosion has a strong physical expression in the sedimentary record due to the facies contrast across the contact, even though both the underlying and the overlying deposits are part of a regressive succession (Figs. 6.6, 6.23, 6.24, and 6.52). The process of wave scouring during forced regression leads to the exhumation of semi-lithified marine sediments, resulting in the formation of firmgrounds colonized by the *Glossifungites* Ichnofacies tracemakers (MacEachern et al., 1992, 2012; Chaplin, 1996; Buatois et al., 2002). Such firmgrounds may

unconformably based (Figs. 6.23E and 6.52). This is atypical of normal regression, and can be differentiated from a forced regressive shoreface by the upstepping shoreline trajectory which is diagnostic of relative sea-level rise. Normal regressive clinoforms are also finer grained

separate deposits with contrasting ichnofacies, due to the abrupt shift in environmental conditions that prevailed during the deposition of the juxtaposed facies across the contact. MacEachern et al. (1992) and Buatois et al. (2002) provide case studies where the regressive surface of marine erosion, marked by the *Glossifungites* Ichnofacies, separates finer-grained shelf deposits with *Cruziana* Ichnofacies from overlying shoreface sands with a *Skolithos* assemblage. The basinward extent of the forced regressive *Glossifungites* firmground is limited to the area affected by fairweather wave erosion, beyond which the stratigraphic hiatus collapses, being replaced by the correlative conformity (Figs. 6.23 and 6.52).

Due to its diachronous formation during the entire stage of relative sea-level fall, the regressive surface of marine erosion merges with the basal surface of forced regression in a landward direction, and with the correlative conformity in a basinward direction (Figs. 4.39, 5.13, 5.31, 6.23, and 6.52). As a result, the regressive surface of marine erosion may be placed at the base (Fig. 6.23C), within (Fig. 6.23D), or at the top (Fig. 6.23E) of the falling-stage systems tract. For this reason, it has been recognized that this unconformity "is neither a logical nor practical surface at which to place the sequence boundary" (Plint and Nummedal, 2000). The regressive wave ravinement is to some extent the counterpart of the transgressive wave ravinement, which is also diachronous in a dip direction, merging with the maximum regressive surface basinward and with the maximum flooding surface landward (Fig. 4.39). However, the two wave-ravinement surfaces differ in terms of timing (i.e., during relative sea-level fall vs. transgression; Figs. 5.13, 6.7), dominant place of scouring (i.e., lower vs. upper shoreface; Figs. 4.44 and 4.55), and younging direction (i.e., seaward and landward; Figs. 4.39 and 5.13).

6.1.8 Other surfaces of sequence stratigraphy

In addition to the seven "conventional" sequence stratigraphic surfaces discussed above, other systems-tract boundaries can be defined exclusively in upstream-controlled settings, within fully continental sequences which form independently of relative sea-level changes. Unconventional systems tracts and bounding surfaces reflect depositional conditions controlled and modified by tectonism, climate, and autogenic processes (Fig. 3.11), which generate the stratal stacking patterns that can be observed in fluvial and eolian systems. Both fluvial and eolian sequences in upstream-controlled settings are bounded by subaerial unconformities, which are the only sequence stratigraphic surfaces that can be found in both upstream- and downstream-controlled settings. Irrespective of depositional environment and the specific underlying control(s), all upstream-controlled sequences correspond to stratigraphic cycles of rise and fall in the temporary base level, which can be observed at different scales (Figs. 1.19 and 3.22; see Chapter 3 for details on the base level).

6.1.8.1 Top-amalgamation surface

Fluvial sequences in upstream-controlled settings can be subdivided into two "unconventional" systems tracts defined by the degree of channel amalgamation (i.e., high- vs. low-amalgamation systems tracts; details in Chapters 4 and 5), which are separated by the top-amalgamation surface (Fig. 4.70). As the name indicates, the "top-amalgamation" surface marks a change in stacking pattern from amalgamated channels below to a floodplain-dominated succession above. Such changes in stacking patterns occur during stages of decline in fluvial energy which lead to the accumulation of depositional sequences as relatively conformable successions at the scale of the observed sequences (details in Chapters 5 and 7). As a result, the top-amalgamation surface develops as a conformable contact at the limit between the higher energy amalgamated channels below and the lower energy floodplain-dominated succession above. The change in the degree of channel amalgamation depends on multiple factors (Fig. 4.63) and may occur at different times in different areas within the basin; therefore, the top-amalgamation surface can be diachronous along both dip and strike directions. The top-amalgamation surface can also be observed at different scales, in relation to stratigraphic cycles of different magnitudes (Fig. 4.70).

The physical expression of top-amalgamation surfaces may be similar to that of maximum regressive surfaces in the fluvial portion of downstream-controlled settings (Figs. 6.29, 6.30, and 5.50); however, no correlation is implied between the two surfaces, as the timing of upstream- and downstream-controlled sequences depends on different controls and therefore can be out of phase (Fig. 3.20).

6.1.8.2 Maximum expansion level

Eolian sequences in upstream-controlled settings can be subdivided into constructional and contractional systems tracts, which correspond to stages of growth and decline of eolian ergs (Fig. 6.54; Galloway and Hobday, 1996; Catuneanu, 2003). Specific facies depend on the nature of the eolian systems, which can be classified into dry, wet, or stabilized, largely as a function of the position of the water table relative to the surface profile (Loope, 1985; Kocurek and Nielson, 1986; Chan and Kocurek, 1988; Kocurek, 1988, 1991, 1998; Clemmensen and Hegner, 1991; Kocurek and Havholm, 1993; Crabaugh and Kocurek, 1993; Clemmensen et al., 1994; Galloway and Hobday, 1996; Mountney and Jagger, 2004; Mountney, 2006, 2012; Rodriguez-Lopez et al.,

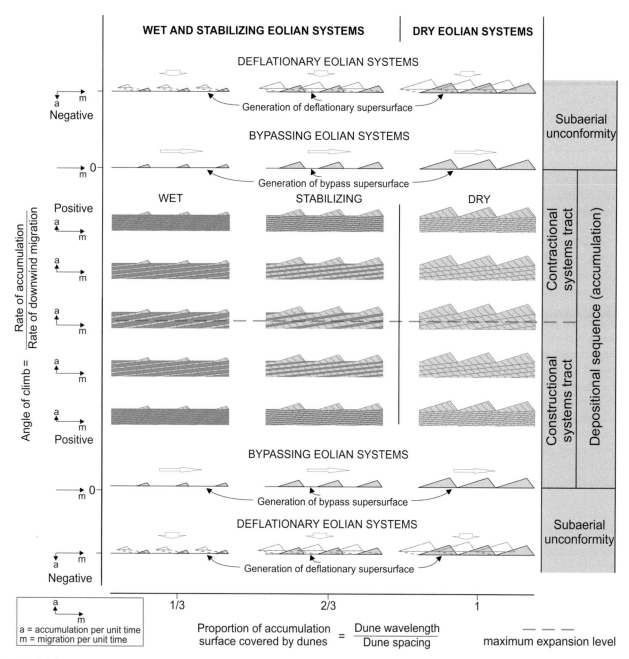

FIGURE 6.54   Stratigraphic architecture of eolian sequences in upstream-controlled settings (modified after Mountney, 2012, and Rodriguez-Lopez et al., 2014). The maximum expansion level separates a stage of growth of the eolian system from a receding stage that culminates with the formation of a sequence-bounding "supersurface." The rates of sediment accumulation are highest at the maximum expansion level (see text for details).

2012, 2014). The development of eolian sequences requires sediment accumulation, which can be sustained by a rise in the temporary base level. The position of the temporary base level is proxied by the elevation of the interdune areas, which is constrained by the water table in wet and stabilizing systems, and by the balance between sediment supply and the wind capacity to transport the sediment in dry systems (see Chapter 3 for details on the base level). Stages of base-level fall result in the formation of sequence-bounding unconformities (i.e., "supersurfaces" in Fig. 6.54).

Stratal stacking patterns in eolian systems are defined by the trends of change in the thickness of the dune–interdune cycles and the associated angle of climb of the interdune migration surfaces (Fig. 6.54). During the growth stage, the constructional systems tract records an increase in the rates of base-level rise (i.e., rates of sediment accumulation), which results in a corresponding increase in the thickness of the dune–interdune cycles and of the angle of climb of the interdune migration surfaces. In contrast, the contractional systems tract records a decrease in the rates of base-level rise and sediment accumulation, which results in a decrease with time in the thickness of the dune—interdune cycles and of the angle of climb of the interdune migration surfaces (Fig. 6.54). These generalized patterns may be obscured by the higher frequency changes in base level that are superimposed on the longer term trends (Fig. 1.19). Therefore, a hierarchy of sequence stratigraphic elements may apply to any particular eolian system, just as in all other depositional settings, with sequences, systems tracts, and bounding surfaces being observed at different scales (details on hierarchy in Chapter 7).

The boundary between the constructional and contractional systems tracts within depositional sequences is represented by the maximum expansion level, which is placed at the core of the stratigraphic interval that includes the thickest dune—interdune cycles, with the steepest angle of climb of the interdune migration surfaces. This level corresponds to the time of the highest rate of base-level rise during the accumulation of a sequence (Fig. 6.54), when the erg reaches its peak of development. The maximum expansion can be attained at different times in different areas, and therefore it can be described as a 3D and diachronous stratigraphic level, rather than a discrete physical surface, which crosses the higher diachroneity dune—interdune facies contacts (Fig. 6.54). For practical purposes (e.g., to constrain fluid flow pathways within heterogeneous reservoirs and aquifers, or to describe the architecture of eolian facies), most studies rely on the identification of sedimentological surfaces that develop within sequences, such as the interdune migration surfaces, superimposition surfaces, and reactivation surfaces (details in Chapter 8).

FIGURE 6.55 Within-trend normal regressive surface (WTNRS) in a carbonate setting (Triassic, The Dolomites, Italy). A—Slope calcarenites (highstand foreset); B—platform carbonates (highstand topset). Note the oblique stratification of the slope clinoforms, in contrast with the horizontal stratification of the platform-top carbonates. The WTNRS is diachronous, younging in the downdip direction with a rate inversely proportional to the rate of progradation.

6.2 Within-trend facies contacts

Within-trend facies contacts refer to surfaces of lithostratigraphy, allostratigraphy, or sedimentology which develop within systems tracts at the limit between contrasting facies, as a result of shifts in paleoenvironmental conditions at syn-depositional time. These are not sequence stratigraphic surfaces as they do not serve as systems tract boundaries, but have a strong physical expression and are important to map for practical purposes (e.g., the delineation of water or hydrocarbon reservoirs) and to describe the internal composition of systems tracts (Fig. 6.1). Within-trend facies contacts are best rationalized within the genetic context of the sequence stratigraphic framework. Therefore, in the workflow of sequence stratigraphy, within-trend facies contacts are only dealt with after the framework of sequence stratigraphic surfaces has been constructed.

Some within-trend facies contacts develop in relation to specific types of shoreline trajectories (i.e., the within-trend normal regressive, forced regressive, and flooding surfaces), whereas others are generated by processes that can operate during the accumulation of different systems tracts (e.g., downlap surfaces and surf diastems, which may form during forced and normal regressions; and bedset boundaries, which may form in relation to any type of shoreline trajectory). A discussion of the most prominent types of within-trend facies contacts follows below.

6.2.1 Within-trend normal regressive surface

The within-trend normal regressive surface is a conformable or scoured facies contact that develops during normal regressions at the limit between the topsets and the foresets of lowstand and highstand systems tracts (Figs. 4.49, 4.50, and 6.6). The topset may include fluvial, coastal plain, delta plain, or peritidal carbonate facies, whereas the underlying foreset may include beach, shoreface, delta front, or slope carbonate facies (Figs. 6.42 and 6.55). Intertidal facies may be part of the topset in river-mouth and carbonate settings (e.g., lower delta plain, peritidal carbonates), or part of the coarsening-upward foreset clinoforms in open-shoreline siliciclastic settings (beach sands; Figs. 1.11, 6.42, 6.56). This surface has a strong physical expression, commonly marked by an abrupt shift from coarser to finer sediments in siliciclastic settings (Figs. 6.52 and 6.56), or from detrital slope facies to *in-situ* platform facies in carbonate settings (Figs. 6.35 and 6.55). However, the mere observation of lithological contrasts (e.g., mud on sand) is not sufficient to identify a facies contact as a within-trend normal regressive surface, as other contacts, such as flooding surfaces, may exhibit a similar juxtaposition of facies. Other key observations include

FIGURE 6.56 Within-trend normal regressive surfaces in siliciclastic settings, at the contact between foresets and topsets: A—contact between beach sands and overlying coal-bearing fluvial deposits (Early Maastrichtian, base of Horseshoe Canyon Formation, Western Canada Basin); B—distributary channel that scours locally the contact in image A; C—top of beach sands exposed by the erosion of the overlying (and more recessive) fluvial floodplain fines (Late Permian, contact between the Ecca and Beaufort groups, Karoo Basin); D—contact between delta front and delta plain deposits (Late Cretaceous, Ferron Delta, Utah).

depositional, geometrical, and bathymetric trends, in order to demonstrate the normal regressive nature of the succession. Conformable within-trend normal regressive surfaces may be scoured locally by channels that belong to the overlying topset (e.g., wells 6 and 7 in Fig. 1.11; Fig. 6.56B). Where amalgamation occurs within the topset, the within-trend normal regressive surface becomes a composite scour at the base of the coalesced distributary channels.

The development of conformable facies contacts at the base of continental topsets demonstrates that the base of continental sections is not necessarily unconformable, as commonly inferred. The conformable vs. unconformable nature of the marine-to-continental facies shifts needs to be assessed on the basis of local stratigraphic relationships, including evidence of missing sections and the degree of change in sediment composition across the contact (Fig. 5.50). In contrast to the base of marine deposits, which is always unconformable (i.e., the transgressive wave-ravinement surface; Fig. 1.11), the base of continental deposits can be either conformable (e.g., the within-trend normal regressive surface; Fig. 6.42) or unconformable (i.e., the subaerial unconformity;

Figs. 4.45 and 6.19). In the case of conformable shifts from the foreset to the topset of a prograding system, the contact may be marked by a decrease in grain size from beach/delta front to the overlying floodplain/delta plain deposits; however, the distributary channels of the continental topset are filled with the same type of sediment that builds the clinoforms (Fig. 6.56). In the case of unconformable contacts at the base of continental deposits, the systems below and above the unconformity are of different ages, and record an abrupt shift to more proximal and typically coarser facies on top of the subaerial unconformity (Figs. 4.45 and 6.19).

The extent of within-trend normal regressive surfaces along dip directions is highly variable, in a range of 10^0-10^3 km, depending on the duration and rates of the normal regression. In spite of its strong physical expression and possible regional extent, the within-trend normal regressive surface has little value for chronostratigraphic correlations, as it is highly diachronous, due to the relatively low rates of normal regression (e.g., an average diachroneity of 1 My per 100 km during the regression of the Bearpaw Seaway from the Western Canada Basin; Fig. 6.57). However, this surface remains

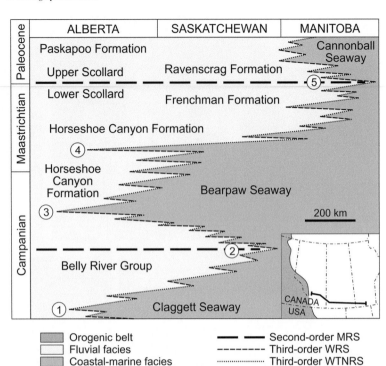

FIGURE 6.57 Second- and third-order cycles of the first-order Western Canada Foreland Basin (location on inset map; modified after Catuneanu et al., 2000). Biostratigraphic dating based on ammonite zonation: 1—*Baculites mclearni*: maximum flooding surface of the Claggett second-order transgression; 2—*Baculites scotti*: maximum regressive surface of the Claggett second-order regression (c. 76 Ma); 3—*Baculites compressus*: maximum flooding surface of the Bearpaw second-order transgression (c. 74 Ma); 4—*Baculites grandis*: maximum flooding surface of the third-order Drumheller Marine Tongue; 5—maximum regressive surface of the Bearpaw second-order regression (K-T boundary: c. 65 Ma). Abbreviations: MRS—maximum regressive surface; WRS—wave-ravinement surface; WTNRS—within-trend normal regressive surface.

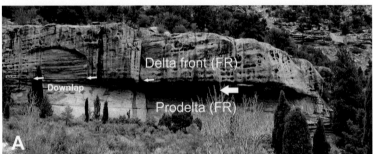

FIGURE 6.58 Within-trend forced regressive surfaces in river-dominated delta settings, at the contact between prodelta bottomsets and delta front foresets (Late Cretaceous, Panther Tongue, Utah; note person for scale in image A). This facies contact is conformable and diachronous. The rate of diachroneity of within-trend forced regressive surfaces is lower than that of within-trend normal regressive surfaces (Figs. 6.55 and 6.56), as forced regressions are typically faster than normal regressions. A and C: dip-oriented transects; B: strike-oriented transect. Abbreviations: FR—forced regressive; WTFRS—within-trend forced regressive surface; WRS—transgressive wave-ravinement surface; SU—subaerial unconformity.

A. **High-resolution study** (e.g., based on outcrop or well data)

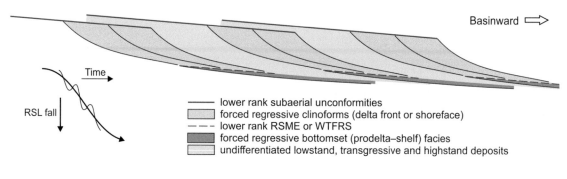

Basinward ⟹

Time

RSL fall

——— lower rank subaerial unconformities
▢ forced regressive clinoforms (delta front or shoreface)
– – – lower rank RSME or WTFRS
▮ forced regressive bottomset (prodelta–shelf) facies
▢ undifferentiated lowstand, transgressive and highstand deposits

B. **Low-resolution study** (e.g., based on seismic data)

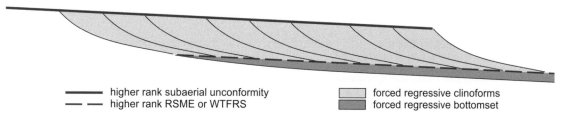

——— higher rank subaerial unconformity
– — higher rank RSME or WTFRS

▢ forced regressive clinoforms
▮ forced regressive bottomset

FIGURE 6.59   Forced regressions observed at different stratigraphic scales. The regressive surface of marine erosion (RSME) forms in energy-dominated settings (wave- or tide-dominated), whereas the within-trend forced regressive surface (WTFRS) forms in supply-dominated settings (river-dominated deltas). Higher rank surfaces are composites of discrete lower rank contacts, and describe overall trends in lower resolution studies. RSL—relative sea level.

important for the definition of litho- and allostrati-graphic units, as well as to describe the internal architecture of lowstand and highstand systems tracts. As with all other types of stratigraphic contacts, the within-trend normal regressive surface can be observed at different scales in relation to normal regressions of different hierarchical ranks. At larger scales of observation, the limit between the topset and the foreset of a normal regressive unit is a composite of lower rank surfaces which upstep in a downdip direction, as the long-term progradation is interrupted by higher frequency transgressions (Figs. 4.59 and 6.57).

6.2.2 Within-trend forced regressive surface

The within-trend forced regressive surface is a conformable facies contact that develops during forced regressions at the limit between the foreset (delta front) and the bottomset (prodelta) of river-dominated deltas (Figs. 6.6 and 6.58). This type of facies contact does not develop in wave- or tide-dominated settings, where the regressive surface of marine erosion forms instead in both river-mouth and open shoreline settings (Figs. 4.44 and 5.33). As with all other facies contacts, the within-trend forced regressive surface is diachronous, younging basinward with a rate that is inversely proportional to the rate of the forced regression. However, the degree of diachroneity of this surface is typically less than that of the within-trend normal regressive surface, as forced regressions are commonly

faster than normal regressions (details in Chapter 4). Within-trend forced regressive surfaces can also be observed at different stratigraphic scales, with the higher rank surfaces being composites of lower rank contacts that form in relation to higher frequency cycles during a longer term fall in relative sea level (Fig. 6.59).

The within-trend forced regressive surface does not have an equivalent in normal regressive settings, where the shift from prodelta to delta front, or from shelf to

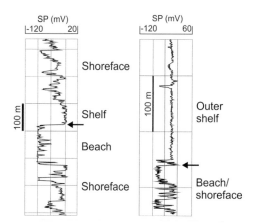

FIGURE 6.60   Flooding surfaces (arrows) at the sharp contact between shoreface or beach sands, and the overlying shelf fines. In these examples, the flooding surfaces are represented by wave-ravinement surfaces. The transgressive portion of the overlying shelf fines may be very thin, with the maximum flooding surfaces being near or even coincident with the flooding surfaces.

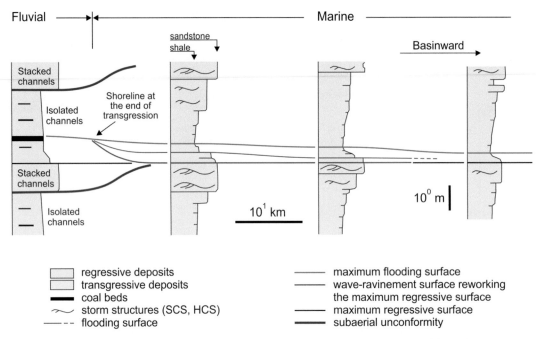

FIGURE 6.61 Stratigraphic architecture at parasequence scales, along a dip-oriented transect. The workflow of sequence stratigraphy starts with facies analysis and the identification of stratigraphic surfaces in vertical sections, followed by correlation across the study area. In this example, the flooding surface is a within-trend facies contact. Depending on the local conditions of sedimentation, flooding surfaces may also coincide with maximum regressive surfaces, wave-ravinement surfaces, or even maximum flooding surfaces in areas of sediment starvation during transgression (Fig. 5.100). Therefore, the precise stratigraphic meaning of a flooding surface needs to be determined on a case-by-case basis. Abbreviations: SCS—swaley cross-stratification; HSC—hummocky cross-stratification.

shoreface, is more gradual (Fig. 4.50). The higher rates of forced regression enable a sharper shift of facies from bottomsets to foresets across the regressive surface of marine erosion in energy-dominated settings (Figs. 4.45, 6.51) or across the within-trend forced regressive surface in supply-dominated settings (Figs. 4.4 and 6.58). However, even though the latter surface appears to be a distinct facies contact between prodelta and delta front facies, a closer look reveals that the change occurs within a narrow transition zone rather than across a discrete surface. As such, the within-trend forced regressive surface combines a relatively thin transition interval (e.g., c. 4—5 m in Fig. 4.47) into a single mappable horizon in lower resolution studies. Therefore, the forced regressive delta front in a river-dominated setting is sharp- but conformably based, as no erosion of the bottomset occurs during progradation (Fig. 4.54; compare the log motifs in Figs. 4.47 and 6.52).

The conformably based delta front clinoforms that overlie the within-trend forced regressive surface differ from the forced regressive delta front or shoreface deposits in energy-dominated settings, as the latter are bounded at the base by an unconformity (i.e., the regressive surface of marine erosion; Figs. 6.51 and 6.52) and have a compressed thickness constrained by the depth of the fairweather wave base (details in Chapter 4). In contrast, a forced regressive river-dominated delta front can be thicker, as the toe of the clinoforms can reach

below the fairweather wave base (Figs. 4.44B and 4.47). This is because the scale of the clinoforms in supply-dominated settings is controlled by the amounts of sediment influx and accommodation available, rather than the energy of waves or tides which is insufficient to rework and redistribute the incoming sediment. While both forced and normal regressive clinoforms in conformable successions can expand to thicknesses greater than the depth of the fairweather wave base, they can still be separated as the former are sharp based and accompanied by the downstepping of the shoreline and adjacent foreset facies (Fig. 4.54).

6.2.3 Within-trend flooding surface

The flooding surface is a facies contact across which there is evidence of an abrupt increase in water depth (Van Wagoner, 1995). Flooding surfaces form during shoreline transgression, and are commonly documented in coastal to shallow-water settings where abrupt water deepening can be demonstrated (Figs. 5.94 and 6.60). Owing to their specific mode of formation, flooding surfaces are always overlain by transgressive marine or lacustrine deposits, while the underlying facies can vary from continental through to marine or lacustrine (Fig. 6.6). Surfaces that satisfy the definition of a flooding surface can have multiple stratigraphic meanings, from sequence stratigraphic surfaces to surfaces of

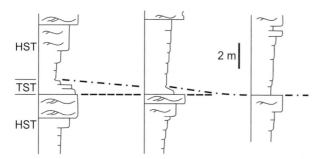

1. Sequence stratigraphic approach to correlation

--- maximum regressive surface
-·- maximum flooding surface

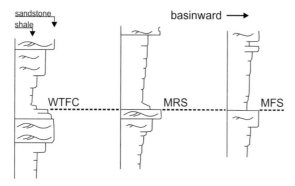

2. Allostratigraphic approach to correlation

----- flooding surface (lithologic discontinuity)
～ storm-generated structures (HCS, SCS)

FIGURE 6.62 Sequence stratigraphic vs. allostratigraphic approaches to correlation in shallow-water settings (Turonian, Cardium Formation, Western Canada Basin). Sequence stratigraphic surfaces mark changes in stratal stacking patterns, whereas allostratigraphic surfaces are lithological discontinuities. Where the transgressive deposits are missing, the MRS and MFS collapse into one omission surface (e.g., a hardground), which bears the name of the younger surface (i.e., the MFS). The MRS may also be reworked in part by the wave-ravinement surface during transgression (Fig. 6.61). Note the change in the sequence stratigraphic meaning of the flooding surface along the dip-oriented transect, depending on the stratigraphic position of the lithological discontinuity at each location. Abbreviations: WTFC—within-trend facies contact; MRS—maximum regressive surface; MFS—maximum flooding surface; TST—transgressive systems tract; HST—highstand systems tract; HCS—hummocky cross-stratification; SCS—swaley cross-stratification.

lithostratigraphy or allostratigraphy. Where flooding surfaces coincide with sequence stratigraphic surfaces (i.e., maximum regressive, wave-ravinement, or maximum flooding; Fig. 5.100), the sequence stratigraphic nomenclature is preferred in order to convey the unequivocal meaning of the surface. In all other cases, flooding surfaces that form within the marine/lacustrine portion of transgressive systems tracts can be referred to as "within-trend flooding surfaces" (Figs. 5.100B and 5.104).

Flooding surfaces do not mark a change in stratal stacking pattern, unless they coincide with a maximum regressive or a maximum flooding surface (Fig. 5.103).

Where flooding surfaces are facies contacts within transgressive marine deposits, they only have allostratigraphic meaning as lithological discontinuities that are in breach of Walther's Law. Some types of flooding surfaces imply a specific mechanism of formation (e.g., "local" flooding surfaces at the base of backlap shell beds relate to sediment starvation on the shelf toward the end of transgression; Figs. 5.109, 6.3; Abbott and Carter, 1994; Zecchin and Catuneanu, 2013), although in most cases only generic episodes of abrupt water deepening are implied, no matter the processes involved (e.g., rapid sea/lake-level rise or tectonic subsidence, with or without a break in sedimentation). In either case, within-trend flooding surfaces are lithological contacts of lesser stratigraphic significance than the transgressive systems tract boundaries that are observed at the same hierarchical level (Fig. 6.61; see Chapter 5 for details).

The increase in water depth that results in the formation of a flooding surface is accompanied by the inundation of previously dry land, as well as by a decrease in sediment supply to the distal areas of the shelf. These variations in depositional conditions can cause a through-going flooding surface to change its stratigraphic significance in a dip direction, from a wave-ravinement surface to a within-trend facies contact, maximum regressive surface, and maximum flooding surface (Fig. 6.62). Therefore, a flooding surface is not simply a maximum flooding surface observed at smaller scales (i.e., a maximum flooding surface of lower hierarchical rank), which is a common confusion, but a different type of contact altogether (i.e., a lithological discontinuity with or without a change in stacking pattern, as opposed to a change in stacking pattern with or without a lithological expression, respectively). Flooding surfaces may be observed at the stratigraphic scales of maximum flooding surfaces that end the transgressions associated with the episodes of water deepening, or even at larger scales than maximum flooding surfaces of lower hierarchical ranks (Fig. 5.104; Catuneanu and Zecchin, 2020).

Within-trend flooding surfaces may be conformable, where sedimentation continues during their formation (Fig. 5.100B), or may be associated with a break in sedimentation due to a lack of sediment supply (nondeposition), sediment bypass, or erosion. In the latter case, the "omission" surface may be represented by firmgrounds or hardgrounds demarcated by substrate-controlled ichnofacies (Fig. 6.63). The shift to deeper water conditions above the contact usually triggers an increase in faunal abundance and ichnodiversity following the flooding event (Pemberton et al., 2001), as well as a sharp increase in the bioturbation index (Siggerud and Steel, 1999). Water deepening also leads to an increase in hydrostatic pressure at the seafloor (water loading), which may

FIGURE 6.63   Within-trend flooding surface (arrow) at the contact between transgressive lower shoreface sands (Coniacian, Bad Heart Formation) and the overlying transgressive outer shelf shales (Santonian, Puskwaskau Formation) (Western Canada Basin; images courtesy of Andrew Mumpy). The flooding surface marks an episode of abrupt water deepening which led to sediment starvation and pervasive bioturbation. The contact displays no lag deposits or any other evidence of scouring. Images A–D show the hardground underlying the flooding surface; images E and F show the fabric of the substrate-controlled ichnofacies.

further enhance compaction and the firmness of the substrate, and hence the formation of substrate-controlled ichnofacies (Snedden, 1991). In the absence of transgressive sediments on the outer shelf, the flooding surface may be draped by "backlap shell beds" which concentrate remains of benthic fauna that lived in deeper water relative to the fauna that contributes to the formation of onlap shell beds (Fig. 6.3; Kidwell, 1991a,b; Naish and Kamp, 1997; Kondo et al., 1998; Di Celma et al., 2005; Zecchin and Catuneanu, 2013).

Whether conformable or unconformable, within-trend flooding surfaces are better suited for allostratigraphic correlations, which rely on lithological discontinuities (Fig. 6.62). However, their identification remains important to describe the internal architecture and the sedimentological makeup of transgressive systems tracts. Within-trend flooding surfaces are typically more diachronous and less extensive than the systems tract boundaries of the same hierarchical rank (Fig. 6.61), and therefore they are best rationalized after the construction

FIGURE 6.64 Sequence stratigraphic surfaces and within-trend facies contacts in a siliciclastic shallow-water setting (Late Cretaceous, Woodside Canyon, Utah; modified after Zecchin et al., 2017a, and Catuneanu and Zecchin, 2020). Downlap surfaces mark the limit between shore-related clinoforms (shoreface and inner shelf deposits) and the underlying outer shelf condensed sections. The clinoforms consist of bedsets bounded by scour surfaces (green lines: base of tempestites), with each bedset recording an upward decrease in energy levels and the degree of amalgamation of event beds. Abbreviations: P—progradational trend; R—retrogradational trend; MFS—maximum flooding surface; MRS—maximum regressive surface; DLS—downlap surface.

and in the context of the sequence stratigraphic framework. For this reason, surfaces of sequence stratigraphic significance need to be mapped first, even though they may lack the lithological expression of allostratigraphic contacts. More details on flooding surfaces are provided in the discussion of parasequences in Chapter 5.

6.2.4 Other within-trend facies contacts

In contrast to the within-trend facies contacts described above, which form in relation to specific types of shoreline trajectories, the facies contacts below may be generated by sedimentological processes in a manner that is independent of relative sea-level changes and shoreline trajectories. However, these facies contacts are still important as they separate sediment bodies

with distinct lithologies, therefore providing the means to describe the internal architecture of systems tracts. Downlap surfaces and surf diastems form during normal regressions and forced regressions, and may therefore be observed within falling-stage, lowstand and highstand systems tracts. Bedsets have a wider range of occurrence, as they constitute the sedimentological building blocks of all systems tracts.

6.2.4.1 Downlap surface

The downlap surface is a facies contact between the riverborne sediment of clastic clinoforms (shoreface to inner shelf deposits) and the underlying hemipelagic condensed sections that accumulate below the storm wave base in outer shelf settings (Figs. 6.2, 6.64; Zecchin and Catuneanu, 2013). Downlap surfaces may form during both forced and normal regressions, when coastal progradation occurs (Fig. 6.2). As progradation takes time, downlap surfaces are typically diachronous, with a rate of diachroneity that is inversely proportional to the rate of shoreline regression (Fig. 6.2). Downlap surfaces may be overlain by shell beds that concentrate as a result of storm reworking at the toe of the clinoforms, near the storm wave base where the rates of sedimentation are lower (i.e., the "downlap shell beds" of Kidwell, 1991a,b; Fig. 6.3). The downlap shell beds tend to be more common in sand-rich basins than in mud-rich basins, due to the increased production of shells coupled with the transport provided by storm surges in higher energy settings (Kondo et al., 1998).

The deposition of hemipelagic sediment during shoreline regression includes contribution from terrigenous mud transported offshore as hypopycnal plumes. Depending on the lateral variations in sediment supply and subsidence, downlap surfaces may record a significant degree of diachroneity not only along dip, but also along strike directions (Carter et al., 1998). The condensed section that accumulates beyond the toe of the clinoforms on the outer shelf separates the downlap surface from the maximum flooding surface below (Figs. 6.2 and 6.3). Where the thickness of the condensed section is less than the vertical seismic resolution, the maximum flooding surface can be mapped as a "downlap surface" on seismic lines (i.e., the maximum flooding and the downlap surface tune into one seismic reflection; Fig. 6.4). Where the condensed section is missing due to nondeposition or erosion, the downlap surface and the maximum flooding surface merge into one unconformable contact (Fig. 6.64).

6.2.4.2 Surf diastem

The surf diastem is a facies contact generated by the seaward migration of longshore troughs and rip channels during coastal progradation, and separates lower shoreface facies below from trough cross-bedded upper

FIGURE 6.65 Surf diastem separating lower shoreface from upper shoreface deposits in the middle Pleistocene Cutro terrace, southern Italy (from Zecchin and Catuneanu, 2013; modified after Zecchin et al., 2011).

shoreface deposits above (Zhang et al., 1997; Swift et al., 2003; Clifton, 2006; Zecchin and Catuneanu, 2013, Fig. 6.65). This facies contact may develop in any regressive systems tract (i.e., falling-stage, lowstand, and highstand), as it forms independently of relative sea-level changes, during both forced and normal regressions (Clifton, 2006). Therefore, the surf diastem should not be confused with the regressive surface of marine erosion, which is the product of relative sea-level fall. Where the lower shoreface deposits are missing, and the upper shoreface facies overlie directly shelf sediments within a falling-stage systems tract (Fig. 4.45), then the surf diastem coincides with the regressive surface of marine erosion (Zecchin and Catuneanu, 2013).

6.2.4.3 Bedset boundaries

Bedsets are commonly meter-scale stratal units that are distinctive in terms of the orientation, texture, and composition of the component beds (Campbell, 1967). Bedsets correspond to sedimentological cycles within the smallest scale stratigraphic units, and therefore they

can be viewed as the sedimentological building blocks of the lowest rank systems tracts (Fig. 3.24). The formation of bedsets and bedset boundaries relates to fluctuations in sediment supply and energy conditions within a depositional environment, without changes in systems tract and component depositional systems. In coastal settings, bedsets may be linked to individual beach ridges, with bedset boundaries reflecting minor reorganizations of the shoreline, such as in response to storm events, during uninterrupted progradation or retrogradation (Hampson et al., 2008; Somme et al., 2008). The stacking pattern of bedsets describes depositional and geometrical trends that define systems tracts at stratigraphic scales (Enge et al., 2010; Zecchin and Catuneanu, 2013, Figs. 3.24 and 4.62).

Bedset boundaries are represented by nondepositional or erosional discontinuities, which are typically most distinctive within lower shoreface and shelf deposits, becoming cryptic in both landward and seaward directions (Fig. 6.66; Hampson, 2000; Somme et al., 2008; Zecchin and Catuneanu, 2013).

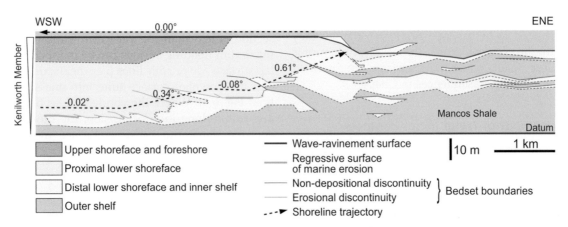

FIGURE 6.66 Sequence stratigraphic surfaces and bedset boundaries within the Kenilworth Member of the Blackhawk Formation (Late Cretaceous, Utah; modified after Hampson, 2000, and Helland-Hansen and Hampson, 2009). Note changes between downstepping and upstepping shoreline trajectories during the progradation of this wave-dominated deltaic system.

The nondepositional vs. erosional nature of bedset boundaries depends to a large extent on the pattern of energy changes during the deposition of bedsets. Nondepositional discontinuities record an abrupt decrease in the thickness and the frequency of storm-generated event beds above the contact, as well as an increase in bioturbation (Hampson, 2000). This is the case where bedsets accumulate during periods of increasing energy levels, as indicated by an upward increase in the thickness and the degree of amalgamation of event beds (Fig. 6.67). In contrast, erosional discontinuities record an abrupt increase in bed amalgamation and grain size across the contact, and may be marked by gutter casts and the *Glossifungites* Ichnofacies (Hampson, 2000; Zecchin and Catuneanu, 2013). In this case, bedsets accumulate during periods of decline in energy, as indicated by the decrease with time in the thickness and the degree of amalgamation of event beds (Fig. 6.64).

With the unprecedented increase in the resolution of stratigraphic studies in recent years, refined criteria are needed to differentiate between stratigraphic and sedimentological cycles (i.e., high-frequency sequences vs. bedsets; Zecchin et al., 2017a,b; see details in Chapter 7), and implicitly between low-rank stratigraphic surfaces and bedset boundaries. Sedimentological cycles describe the internal architecture of depositional systems within the lowest rank systems tracts, whereas stratigraphic cycles involve changes in systems tracts and component depositional systems (Figs. 3.24 and 4.62; details in Chapters 4 and 7). As a result, bedset boundaries have a more limited areal extent, generally restricted to the confines of individual depositional systems (Fig. 6.66). The variations in sediment supply and/

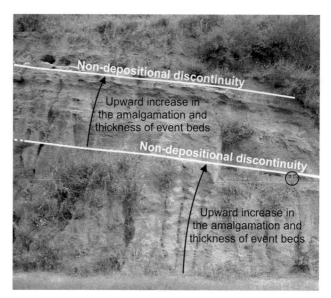

FIGURE 6.67 Bedsets bounded by non-depositional discontinuities showing an upward increase in the degree of amalgamation of event beds, in the shoreface—shelf deposits of the Gelasian Strongoli Sandstone (Crotone Basin, southern Italy; from Zecchin and Catuneanu, 2013; note hammer for scale). The bedset boundaries record a sharp decline in energy from tempestites to fairweather deposits; hence, the non-depositional nature of discontinuities, in contrast with erosional discontinuities which mark abrupt increases in energy levels across bedset boundaries (see text for details).

or energy flux that generate most sedimentological cycles may be linked to changes in climate, relative sea level, and shoreline location over sub-stratigraphic scales of $\leq 10^2$ yrs, shorter than the time required to form a depositional system (Fig. 4.62; details in Chapters 4, 5, and 7).

CHAPTER

7

Sequence stratigraphic framework

The sequence stratigraphic framework records a nested architecture of stratigraphic cycles that can be observed at different scales, depending on the purpose of the study and the resolution of the data available. At each scale of observation (i.e., hierarchical level), stratigraphic cycles define sequences (Fig. 1.22), which consist of systems tracts and component depositional systems (see Chapter 5 for the definition and scales of sequences, systems tracts, and depositional systems). With the refinements in sequence stratigraphic methodology, whereby high-frequency sequences replace parasequences in the sequence stratigraphic workflow, the building blocks of the sequence stratigraphic framework include sequences, systems tracts, and depositional systems at all hierarchical levels. This provides consistency in methodology and nomenclature at all stratigraphic scales, irrespective of geological setting and the types and resolution of the data available, which is key to the standard application of sequence stratigraphy.

Despite the nested architecture of stratigraphic cycles, the sequence stratigraphic framework is not truly "fractal," as sequences of different scales, as well as sequences of similar scales, may differ in terms of underlying controls and internal composition of systems tracts. As a general principle, no correlation can be assumed between scale, origin, and the systems-tract composition of sequences. For example, sequences of similar scales and origins may differ in terms of systems-tract composition (e.g., sequences driven by Milankovitch cycles may include or exclude normal regressions, depending on the availability of sediment during relative sea-level rise; Figs. 4.6 vs. 4.34), and sequences of different scales and origins may happen to be similar in terms of systems-tract composition (Figs. 7.1–7.5). The scales, the origins, and the systems-tract composition of sequences are basin specific, reflecting the local conditions of accommodation and sedimentation. Therefore, the observation of stratal stacking patterns, at scales selected by the practitioner or imposed by the resolution of the data available, overrides any model assumptions in the process of constructing a sequence stratigraphic framework.

The basin-specific nature of the sedimentary record in terms of architecture, scales, underlying controls, timing, and composition of sequences prevents the definition of a reproducible pattern that could describe stratigraphic frameworks worldwide. At any location, the systems tracts of the lowest rank sequences are the smallest stratigraphic units of a sequence stratigraphic framework (Figs. 4.3 and 4.62). The scale and the degree of sedimentological complexity of the lowest rank systems tracts and component depositional systems at any location depend on the geological setting (i.e., local accommodation and sedimentation conditions), and the time involved in the development of the defining stacking pattern (e.g., 10^2–10^3 yrs vs. 10^4–10^5 yrs). Therefore, the sedimentological makeup of stratigraphic units cannot be predicted or depicted by standard templates. Similarly, no orderly pattern can be defined to describe a standard architecture of stratigraphic cycles worldwide. As a safe norm, every study needs to start from the premise that stratigraphic frameworks are basin or even sub-basin specific, reflecting the particular conditions of each geological setting.

7.1 Scale in sequence stratigraphy

Scale is a key issue in sequence stratigraphy, with implications for methodology, nomenclature, and practical applications. The scale of observation may be selected by the practitioner according to the purpose of study, or it may be constrained by local parameters, including the geological setting (e.g., local subsidence rates and sediment supply) and the resolution of the data available (e.g., seismic vs. well data). There are no temporal or physical standards for the scale of any type of unit or surface of sequence stratigraphy; the same types of stratal units and bounding surfaces may record variable scales, both between and within sedimentary basins. For this reason, the methodology and nomenclature must remain independent of scale and all local variables, for a consistent application of sequence stratigraphy across the entire range of geological settings, stratigraphic scales, and types of data available.

FIGURE 7.1 Sequence hierarchy in the Northwest Basin of Argentina (data courtesy of GEOMAP Argentina; modified after Hernandez et al., 2008). Not to scale. Dominant depositional systems: [1] continental; [2] lacustrine-marine; [3] lacustrine. The first-order rift sequence records a duration of 75 My and a thickness of 10^3 m. The second-order sequences developed over timescales of 10^6–10^7 yrs, with thicknesses of 10^2–10^3 m. The Yacoraite Formation (73.5–64 Ma, c. 250 m thick) is well exposed and affords the observation of 3rd-order (10^6 yrs and 10^1 m scales), 4th-order (10^5 yrs and c. 10 m scales), and 5th-order (10^4 yrs and 1–3 m scales) sequences. Fifth-order sequences can be mapped over distances of 10^1 km; all higher rank sequences can be mapped across the entire basin (10^2 km scale). Stratigraphic cyclicity was controlled mainly by the interplay of tectonism and climate. The role of tectonism becomes increasingly dominant at larger scales, and it is evident at the 3rd, 2nd, and 1st hierarchical orders. The role of climate is evident at smaller scales (5th, 4th, and 3rd orders). Abbreviations: WRS—wave-ravinement surface; MFS—maximum flooding surface; LST—lowstand systems tract; TST—transgressive systems tract; HST—highstand systems tract.

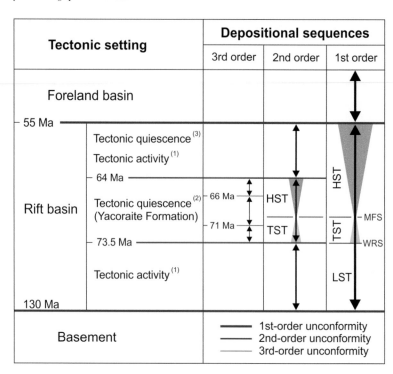

Seismic stratigraphy in the 1970s introduced by default a minimum scale for sequences, systems tracts, and depositional systems, which had to exceed the vertical seismic resolution. As a result, the building blocks of the seismic stratigraphic framework are commonly observed at scales of 10^1–10^2 m. The perception that sequences and their component systems tracts and depositional systems develop typically at these scales is an artifact of seismic resolution, but it dominated stratigraphic thinking for decades. The integration of seismic data with information from wells and outcrops led to the emergence of "high-resolution" sequence stratigraphy which tackles scales of 10^0–10^1 m and 10^2–10^5 yrs (Figs. 4.62 and 5.4). The gradual increase in stratigraphic resolution is reflected in the revisions to the definition of a "sequence," which highlight the fact that as the scale of observation decreases, the magnitude and areal extent of unconformities decrease as well, and conformable surfaces become increasingly important to delineate sequences (Fig. 1.22).

The reality of sequences, systems tracts, and depositional systems at sub-seismic scales has become evident with the advances in high-resolution sequence stratigraphy (e.g., Tesson et al., 1990, 2000; Lobo et al., 2004; Amorosi et al., 2005, 2009, 2017; Bassetti et al., 2008; Catuneanu and Zecchin, 2013; Nanson et al., 2013; Zecchin and Catuneanu, 2013, 2015, 2017; Csato et al., 2014; Nixon et al., 2014; Magalhaes et al., 2015; Zecchin et al., 2015, 2017a,b; Ainsworth et al., 2017, 2018; Pellegrini et al., 2017, 2018; Catuneanu, 2019b). The improvements in stratigraphic resolution also demonstrated that unconformities may form over a wide range of scales, both below and above the seismic resolution, and therefore, unconformity-bounded units are not restricted to the scales of seismic stratigraphy (e.g., Miall, 2015; Strasser, 2016, 2018). In fact, it is now clear that most commonly, the building blocks of a seismic stratigraphic framework are higher frequency sequences that develop at sub-seismic scales; e.g., seismic-scale systems tracts (10^1–10^2 m) consist typically of outcrop-scale sequences (10^0–10^1 m), which are different from and should not be confused with parasequences (see Chapter 5 for details).

As the building blocks of systems tracts, depositional systems can be observed at different scales. The formation of depositional systems, and implicitly of systems tracts, requires typically minimum timescales of 10^2 yrs, and it may be sustained for as long as the defining environments are maintained as dominant sediment fairways (see Chapter 5 for details). Within the transit area of a shoreline, where changes in depositional environment are most frequent, only the lowest rank depositional systems consist strictly of process-related facies accumulated in specific environments; these depositional systems *sensu stricto* develop commonly at scales below the resolution of seismic stratigraphy (Fig. 5.2). At larger scales (higher hierarchical ranks), depositional systems *sensu lato* reflect dominant depositional trends in lower resolution studies, but may record higher frequency changes in depositional environment (Figs. 5.1, 5.3, and 5.5). The distinction between depositional systems *sensu stricto* and *sensu lato* becomes less meaningful outside of the shoreline transit area, where stratigraphic cyclicity may develop without changes in depositional environment.

FIGURE 7.2 Stratigraphic cyclicity within the Yacoraite succession of inter-bedded lacustrine limestones and shales (Northwest Basin, Argentina; see Fig. 7.1 for context). The unconformities originated as subaerial (with evidence of exposure: karstification, desiccation, pedogenesis), now preserved as wave-ravinement surfaces. Abbreviation: MFS—maximum flooding surface.

7.1.1 Stratigraphic resolution

The trend in the development of sequence stratigraphy was to gradually improve the resolution of stratigraphic studies, by applying the method to increasingly smaller scales of observation. This involved refinements to the definition of a "sequence" and the recognition of sequences at progressively smaller scales (Fig. 1.22). Following this trend, the resolution of sequence stratigraphy improved from $10^2–10^3$ m in the 1940s–1960s (i.e., scales relevant to continent-wide correlations; Sloss et al., 1949; Sloss, 1963), to $10^1–10^2$ m in the 1970s (i.e., scales relevant to petroleum exploration in seismic stratigraphy; Payton, 1977), and eventually to $10^0–10^1$ m with the advent of high-resolution sequence stratigraphy (i.e., scales relevant to reservoir studies and petroleum production development; e.g., Zecchin and Catuneanu, 2013, 2015; Magalhaes et al., 2015, 2020, Figs. 1.22 and 5.4). A resolution below 10^0 m is also possible, especially in carbonate systems which are more sensitive to small changes in environmental conditions (e.g., Mawson and Tucker, 2009).

The revised definition of a "sequence" as "a relatively conformable succession of genetically related strata bounded by unconformities or their correlative conformities" in the 1970s (Fig. 1.22, Mitchum, 1977) introduced new criteria for the identification of sequences: (1) sequences are relatively conformable successions, with no resolvable internal unconformities; (2) sequences consist of strata that are genetically related; and (3) unconformities in the stratigraphic record delineate sequence boundaries. None of these attributes are tied to any particular scale, because (1) relatively conformable successions can be observed at different scales, depending on the resolution of the stratigraphic study; (2) genetically related strata that belong to the same stratigraphic cycle can also be observed at different scales; and (3) unconformities may form over a wide range of temporal and physical scales, both below and above the seismic resolution (Fig. 7.6). Relatively conformable successions consist of strata that are genetically related, but the opposite is not always true, as genetically related strata (e.g., a genetic stratigraphic sequence) may or may not be relatively conformable (Catuneanu et al., 2009).

In the context of seismic stratigraphy, only unconformities that can be detected with seismic data can be used to delineate sequences. In this case, relatively conformable successions are observed at scales that exceed the vertical seismic resolution, most commonly in a range of $10^1–10^2$ m (Figs. 4.6 and 7.7). However, unconformity-bounded units can also be identified at sub-seismic scales with higher resolution datasets (Fig. 5.2). Therefore, relatively conformable successions can be observed at different scales, both below and above the seismic resolution, depending on the resolution of the data available (Fig. 7.6). If only the lowest rank (smallest magnitude) unconformities were considered as sequence boundaries, most if not all relatively conformable successions *sensu stricto* would be found at sub-seismic scales. This means that sequences of seismic stratigraphy are only "relatively conformable" in the context of the lower resolution data that are used to delineate them (Fig. 7.6).

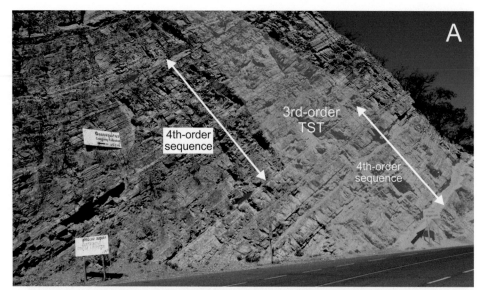

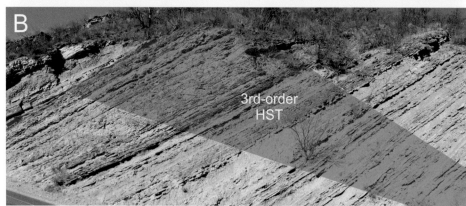

FIGURE 7.3 Third-order systems tracts (10^1 m scale) within the Yacoraite Formation, defined by trends of change in bathymetry and the limestone-to-shale ratio (Northwest Basin, Argentina; see Fig. 7.1 for context). A—Red arrows indicate fourth-order unconformities; B—arrows indicate an angular third-order unconformity (depositional sequence boundary). Abbreviations: TST—transgressive systems tract; HST—highstand systems tract.

Despite the variability in the scale of relatively conformable successions, the "sequence" of seismic stratigraphy was used as a reference for the classification of stratal units, relative to which smaller and larger scale units bearing different names were subsequently defined (e.g., parasequence < systems tract < sequence < sequence set < composite sequence < composite sequence set < megasequence; Sprague et al., 2003; Neal and Abreu, 2009; Abreu et al., 2010). This led to nomenclatural inconsistency, as the resolution of the data available for stratigraphic analysis may vary with the case study (e.g., "systems tracts" and "sequences" identified with a low-resolution dataset become "sequence sets" and "composite sequences," or even "composite sequence sets" and "megasequences" in higher resolution studies). Therefore, a nomenclatural system that is tied to the resolution of the data available is arbitrary in terms of the stratigraphic meaning of the terminology used to define units at different scales. Further confusion was caused by the usage of alternative terminology to designate "groups of superposed sequences," such as "supersequences" (Mitchum et al., 1977).

The "classic" seismic sequences observed at scales of 10^1-10^2 m (Duval et al., 1998) commonly correspond to the 10^6 yrs "third-order" stratigraphic cycles of Vail et al. (1977b). However, the founders of seismic stratigraphy also indicated that the scale of sequences can vary within a huge range, potentially from mm to km (Mitchum et al., 1977). Scale is therefore not a reliable factor for the classification of stratigraphic cycles, and even less so, the resolution of the data available which is an arbitrary variable that changes with the case study. The confusion caused by the definition of different types of units at different scales on the basis of the resolution of the data available (e.g., using the "relatively conformable" seismic-scale sequences with no resolvable internal unconformities as the anchor for the classification) can be eliminated with a scale-independent approach to methodology and nomenclature. At the same time, this also simplifies the sequence stratigraphic workflow, and ensures objectivity and consistency in methodology and nomenclature despite all local variables. The variability in the scale of unconformities and of the "relatively conformable" successions that they separate is discussed below.

FIGURE 7.4   Fourth-order systems tracts within the Yacoraite Formation, defined by trends of change in bathymetry and the limestone-to-shale ratio over ≤10 m scales (Northwest Basin, Argentina; see Fig. 7.1 for context). Abbreviations: MFS—maximum flooding surface; WRS/SU—wave-ravinement surface reworking and replacing a subaerial unconformity; HST—highstand systems tract; TST—transgressive systems tract.

7.1.1.1 Unconformities vs. diastems

The recognition of unconformities in the rock record (Steno, 1669; Hutton, 1788, 1795; Jameson, 1805; De la Beche, 1830; Lyell, 1830; Grabau, 1905; Blackwelder, 1909; Willis, 1910; see Chapter 1 for details) provided the means to subdivide the sedimentary record into "relatively conformable" successions separated by breaks in deposition. This revolutionized sedimentary geology and the approach to stratigraphic correlation, leading the way eventually to the emergence of sequence stratigraphy. As a general term for a hiatal surface, an unconformity indicates missing time in the geological record, of whatever cause and magnitude, and with or without erosion (Fig. 1.18).

Hiatal surfaces form at all temporal and physical scales, ranging in relevance from the scope of sedimentology to the scope of stratigraphy. The wide variability in the magnitude of depositional gaps prompted attempts to differentiate between "major" and "minor" breaks in the sedimentary record. This classification was pioneered by Barrell (1917) who introduced the concept of diastem as a "minor" break relative to the resolution of biostratigraphy. According to Barrel (1917), diastems are "breaks... too brief to give a clue by means of a faunal or floral change" (Barrell, 1917, p. 748); otherwise, in terms of the missing time, diastems may approach 10^6-10^7 years in duration: "Diastems range of all values from seasonal cessations of sedimentation to those which approach geological epochs in duration" (Barrell, 1917, p. 748).

The original definition of a diastem was intended to separate the "minor" hiatuses that fall below the

FIGURE 7.5   Fifth-order sequences and component systems tracts within the Yacoraite Formation, defined by trends of change in bathymetry and the limestone-to-shale ratio over 1–3 m scales (Northwest Basin, Argentina; see Fig. 7.1 for context). Abbreviations: MFS—maximum flooding surface; WRS/SU—wave-ravinement surface reworking and replacing a subaerial unconformity; HST—highstand systems tract; TST—transgressive systems tract.

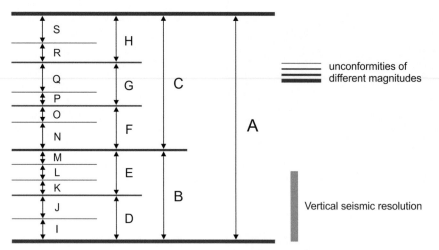

FIGURE 7.6 Relatively conformable successions (A—S) observed at different scales. The scale of a relatively conformable succession depends on the resolution of the stratigraphic study (i.e., the resolution of the data available or the selected scale of observation). With the vertical seismic resolution indicated in this diagram, only sequences A, B, and C can be detected by means of seismic stratigraphy. Relatively conformable successions at seismic scales consist of nested unconformity-bounded sequences at sub-seismic scales. Only sequences B and C are "relatively conformable" *sensu stricto* at the seismic scale (i.e., with no resolvable internal unconformities). A relatively conformable succession *sensu lato* is a stratal unit whose internal unconformities are negligible relative to the scale of the unit and of its bounding unconformities. Internal unconformities of lower hierarchical ranks are negligible in the sense that they do not break the continuity in the paleogeographic evolution observed at the scale of the host unit, thus maintaining the relatively conformable character of the succession (see text for details).

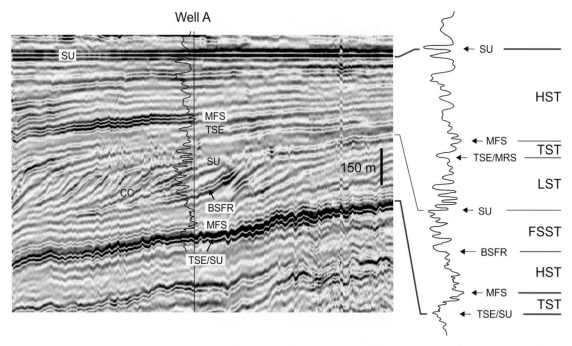

FIGURE 7.7 Sequence stratigraphic framework of the Vaca Muerta (foreset and bottomset)—Quintuco (topset) shelf-slope system, observed at the seismic scale (data courtesy of YPF Argentina). The seismic data are calibrated with well data (gamma-ray log shown). Thicker lines indicate surfaces of higher hierarchical rank. The seismic-scale systems tracts consist of higher frequency sequences at sub-seismic scales (e.g., the FSST consists of three lower rank sequences in Well A). Abbreviations: TSE/SU—transgressive surface of erosion (TSE) reworking and replacing a subaerial unconformity (SU); MFS—maximum flooding surface; BSFR—basal surface of forced regression; MRS—maximum regressive surface; TST—transgressive systems tract; HST—highstand systems tract; FSST—falling-stage systems tract; LST—lowstand systems tract.

biostratigraphic resolution from the "major" breaks that record abrupt changes in faunal or floral content, to which the term "unconformity" would apply. However, stratigraphic resolution only provides an arbitrary

standard to separate diastems from unconformities. As such, a diastem may become an unconformity as stratigraphic resolution improves and affords the detection of a hiatus; similarly, a Precambrian diastem may be equivalent to a Phanerozoic unconformity, as strati- graphic resolution coarsens with the increase in strati- graphic age. Physical criteria too are insufficient to quantify the magnitude of a hiatus, as the degree of paleogeographic change and associated facies shift across a contact are not necessarily proportional to the stratigraphic significance of the contact (e.g., regional flooding surfaces associated with high-frequency trans- gressions may generate prominent facies contacts, but of low stratigraphic rank).

Subsequent to Barrell's (1917) work, improvements in the resolution of biostratigraphy, as well as the develop- ment of radiochronology and other time-measuring methods, enhanced the perception that the term "dia- stem" should be restricted to *brief* gaps in deposition, although what is "brief" remains unquantified to the present day. According to Neuendorf et al. (2005), an un- conformity assumes "a substantial break or gap in the geologic record ... a change that caused deposition to cease for a considerable span of time. An unconformity is of longer duration than a diastem." In contrast, "Dia- stems are not ordinarily susceptible of individual mea- surement, even qualitatively, because the lost intervals are too short" (Neuendorf et al., 2005). In this approach, an unconformity implies a measurable hiatus that is above the resolution of the available age-dating methods (e.g., biostratigraphy, radiochronology), whereas a dia- stem implies an undeterminable hiatus that falls below the resolution of the age-dating techniques. Therefore, the timescale of what is "substantial" (i.e., unconfor- mity) vs. "brief" (i.e., diastem) can only be evaluated

relative to the resolution of the available age-dating methods, but it remains ambiguous in absolute terms, as resolution varies with the method (e.g., biostratig- raphy vs. radiochronology), data available (e.g., marine vs. continental fossils), and stratigraphic age (e.g., Pre- cambrian vs. Phanerozoic).

The modern definition of a diastem indicates "a depo- sitional break of lesser magnitude than a paraconform- ity, or a paraconformity of very small time value" (Neuendorf et al., 2005). The equivocal distinction be- tween the two types of surfaces underlies the fact that there is no natural limit between "major" and "minor" hiatuses, as unconformities and diastems "grade into each other" (Barrell, 1917, p. 794; Fig. 7.8); therefore, the selection of any specific limit is bound to be arbitrary. One solution to this nomenclatural conundrum is to restrict the usage of diastems to sedimentology (i.e., bed or bedset boundaries). Alternatively, the concept of diastem can be eliminated from sedimentary geology altogether. Ultimately, "every bedding plane is, in effect, an unconformity" (Ager, 1993).

Discarding the concept of diastem from stratigraphy not only eliminates ambiguity in the designation of strati- graphic discontinuities, but it provides nomenclatural con- sistency at all hierarchical levels. This becomes important in light of the ongoing trend of gradually increasing the res- olution of stratigraphic studies, which implies the defini- tion of stratigraphic sequences at scales below the resolution of age-dating techniques (Catuneanu, 2019b, Fig. 1.22). This approach eliminates subjectivity in the nomenclatural designation of stratigraphic breaks, and

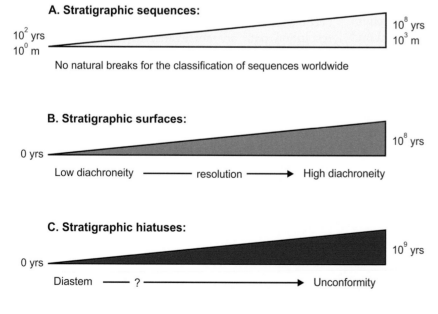

FIGURE 7.8 Scales in sequence stratigraphy. Stratigraphic sequences, bounding surfaces, and hiatuses may develop across a continuum of physical and temporal scales. A. There are no natural breaks for the classification of strati- graphic sequences worldwide. However, natural breaks may occur in individual basins, reflecting the local conditions of accommodation and sedimentation. B. All sequence stratigraphic sur- faces are potentially diachronous. The degree of diachroneity ranges from "low" (diachroneity below the resolution of the available age data: the surface may be approximated with a time line) to "high" (diachroneity measurable with the available age data). C. There are no standards to separate diastems from unconformities. A diastem (i.e., a "minor" unconformity) is a hiatus below the resolution of the available age data. As strati- graphic resolution varies with age (e.g., Precam- brian vs. Phanerozoic) and depositional setting (e.g., types of fossils available), a Precambrian diastem may be equivalent to a Phanerozoic unconformity. The concept of diastem may be abandoned, or restricted to sub-stratigraphic scales (see text for details).

is consistent with the scale-independent definition of un-conformities as stratigraphic discontinuities of any magnitude: "any interruption in sedimentation, what-ever its cause or length, usually a manifestation of nonde-position and accompanying erosion; an unconformity" (Neuendorf et al., 2005, Fig. 1.18).

7.1.1.2 *Relatively conformable successions*

The scale of sequences as "relatively conformable" successions is linked to the scale of the unconformities selected as sequence boundaries (Fig. 7.6). Sequences bounded by continental-scale unconformities may reach thicknesses of 10^2-10^3 m (Longwell, 1949; Sloss et al., 1949; Sloss, 1963); relative to this stratigraphic resolu-tion, internal unconformities of smaller magnitudes and areal extent are negligible. With the increase in the resolution of stratigraphic studies, unconformities of smaller magnitude and areal extent may be selected as sequence boundaries, defining "relatively conformable" sequences at seismic scales of 10^1-10^2 m, or at sub-seismic scales of 10^0-10^1 m in the context of high-resolution sequence stratigraphy (Fig. 5.4). Notably, as the scale of observation decreases, the magnitude and the areal extent of unconformities decrease as well, and conformable surfaces become increasingly important to delineate sequences (Catu-neanu, 2017, 2019a,b).

In the context of seismic stratigraphy, the definition of a sequence as a "relatively conformable succession" (Mitchum, 1977, Fig. 1.22) inadvertently linked the scale of a sequence to the resolution of the seismic data. The subsequent definition of other types of strati-graphic cycles at smaller and larger scales (e.g., "parase-quences" below the scale of sequences, and "composite sequences" and "megasequences" above the scale of sequences; Van Wagoner et al., 1988, 1990; Mitchum and Van Wagoner, 1991; Sprague et al., 2003; Neal and Abreu, 2009; Abreu et al., 2010) led to nomenclatural inconsistency, since the scale of the reference unit (i.e., the "relatively conformable" sequence) varies with the resolution of the data available (Fig. 7.6). Sequences do not occupy any specific niche within a framework of nested stratigraphic cycles, and can be observed at all stratigraphic scales, depending on the geological setting (i.e., local conditions of accommoda-tion and sedimentation), the resolution of the data avail-able (e.g., seismic vs. borehole or outcrop data), and the scope of the study (e.g., petroleum exploration vs. production development).

Sequences of any scale may include unconformities of equal and lower hierarchical ranks. Therefore, se-quences may or may not be "relatively conformable suc-cessions," but they always consist of "genetically related strata" that belong to the same cycle of change in strati-graphic stacking pattern (Catuneanu et al., 2011). Inter-nal unconformities of lower hierarchical ranks are

negligible relative to the scale of the host sequence, thus maintaining the relatively conformable character of the succession (Fig. 7.6). Internal unconformities of equal hierarchical rank may break the relatively conformable character of a sequence within the area of development of the unconformity (e.g., a subaerial un-conformity within a genetic stratigraphic sequence may mark an abrupt shift in depositional systems at the scale of the host sequence). However, beyond the termination of the unconformities, correlative confor-mities preserve the genetic relationship between the un-derlying and the overlying strata within the sequence, thus maintaining the integrity of the sequence as one stratigraphic cycle (Frazier, 1974; Galloway, 1989). Un-conformities with basinwide extent become sequence boundaries by default, as they break the genetic relation-ship between the strata below and above at any location, at the scale of the underlying and overlying sequences (Catuneanu, 2019a,b).

The perception of what constitutes a "relatively conformable" sequence within a hierarchical framework of "parasequences" < "sequences" < "composite sequences" < "megasequences" is subjective even where high-resolution data are available. For example, shoreline shelf-transit cycles of 10^4-10^5 yrs designated as "parasequences" (e.g., Ainsworth et al., 2018, and ref-erences therein) are "sequences" or even "composite se-quences" for authors who define parasequences at 10^2-10^3 yrs timescales (e.g., Amorosi et al., 2005, 2009, 2017; Bassetti et al., 2008; Mawson and Tucker, 2009; Pel-legrini et al., 2017, 2018). Therefore, the standards used by different authors to define the scale of "relatively conformable" sequences are inconsistent, with a tempo-ral variance of at least one or two orders of magnitude. The scales of "relatively conformable successions" become even more difficult to constrain where high-resolution data are not available (e.g., in all seismic-based case studies). The solution to this problem is an approach to hierarchy, and implicitly to methodology and nomenclature, which is independent of scale and data resolution.

A hierarchy system that is anchored to the resolution of the data available is superfluous, as it promotes a complex but volatile nomenclature that changes with the dataset (e.g., a "sequence" defined with low-resolution data may become a "composite sequence" or even a "megasequence" in higher resolution studies, which strips this terminology of stratigraphic meaning). In this context, the argument that the concept of "sys-tems tract" should only be applied at one scale within a framework of nested stratigraphic cycles (i.e., at the scale of "relatively conformable" sequences; Neal and Abreu, 2009) is flawed because the scale of sequences is tied to data resolution, and hence it is variable. Se-quences of any scale may include unconformities whose identification depends on the resolution of the data available. Internal unconformities that are not resolvable with a low-resolution dataset become bounding surfaces for smaller scale sequences in higher resolution studies (Fig. 7.6). If high-resolution data were available in every study, "relatively conformable" sequences may only be found at sub-seismic scales, which would render seismic stratigraphy obsolete.

There is, however, a solution to "save" seismic stratigraphy. In a broader view, the scale of "relatively conformable" successions is set by the scale of observation rather than the resolution of the data available (Catuneanu, 2019b, Fig. 7.6). In this approach, "relatively conformable" successions *sensu lato* can be observed at all stratigraphic scales, as stratal units whose internal unconformities are negligible relative to the scale of the unit and of its bounding unconformities (Fig. 7.6). At the largest stratigraphic scales, first-order basin-fill sequences are relatively conformable successions in the sense that the internal unconformities are negligible relative to the scale of the sequence (i.e., they do not break the tectonic significance of the first-order sequence and the continuity in the paleogeographic evolution observed at the basin scale; Fig. 5.3). This scale-independent approach to the classification of stratigraphic cycles expands the application of Mitchum's (1977) definition of a "sequence" to all stratigraphic scales, independently of data resolution.

7.1.2 Sedimentological vs. stratigraphic scales

The unprecedented increase in the resolution of stratigraphic studies, which is now approaching scales of 10^0 m and 10^2 yrs, requires refined criteria to define the limit between the scales of sedimentology *sensu stricto* and stratigraphy, and to discriminate between sedimentological and stratigraphic cycles (Fig. 4.62; Zecchin et al., 2017a,b, Catuneanu, 2019b). The observation of depositional systems at different scales, depending on the resolution of the study, defines the overlap between the scopes of sedimentology and stratigraphy. At each scale of observation, depositional systems are the largest units of sedimentology and the building blocks of systems tracts in stratigraphy; i.e., the end point of a sedimentological study and the starting point of a stratigraphic study. Therefore, depositional systems provide the link between the scopes of sedimentology and stratigraphy at each hierarchical level.

The scale and the internal makeup of systems tracts and component depositional systems depend on the local conditions of accommodation and sedimentation, the lifespan of the depositional environments that accumulate the sediment, and the resolution of the study. Systems tracts and depositional systems of the lowest hierarchical rank consist solely of sedimentological cycles (i.e., beds and bedsets; Fig. 5.2). All other systems tracts and component depositional systems of higher

hierarchical ranks consist of higher frequency (lower rank) stratigraphic cycles (i.e., sequences; Figs. 5.1 and 5.5). The scale of the lowest rank systems tracts at any location defines the highest resolution that can be achieved with a stratigraphic study, and the limit between stratigraphy and sedimentology *sensu stricto* (Fig. 4.62).

The nested depositional cycles that build the sedimentary record can be observed from sedimentological to stratigraphic scales (Figs. 3.24, 3.25, 4.3, and 4.62). At sedimentological scales, beds and bedsets form in response to fluctuations in sediment supply and/or environmental energy, without changes in depositional system and systems tract. At stratigraphic scales, sequences form in response to cycles of change in accommodation and/or sedimentation conditions that involve changes in coastal systems and associated systems tracts in downstream-controlled settings, or in the dominant depositional elements and associated systems tracts in upstream-controlled settings (Catuneanu, 2019a,b). The stacking pattern of sedimentological cycles (i.e., beds and bedsets) defines the lowest rank systems tracts, which are the smallest units of sequence stratigraphy at any location. The stacking pattern of stratigraphic cycles (i.e., sequences) defines higher rank systems tracts in lower resolution studies (Figs. 3.24, 4.3, and 4.62).

In downstream-controlled settings, changes in the location of the shoreline can be observed at both stratigraphic and sedimentological scales. Shoreline shifts at stratigraphic scales are defined by the trajectory of subaerial clinoform rollovers, which is diagnostic to the definition of systems tracts (Fig. 3.25). Cycles of change in stratigraphic shoreline trajectory involve changes in coastal systems and systems tracts, and result in the formation of sequences (Fig. 5.5). Shoreline shifts at sedimentological scales occur within coastal environments (e.g., in relation to tidal or fairweather–storm cycles), without changes in coastal system and systems tract, and result in the formation of beds and bedsets within the lowest rank systems tracts and component depositional systems (Fig. 3.25). Shoreline shifts of all scales may be accompanied by changes in accommodation. At stratigraphic scales, relative sea-level changes are measured with reference to the mean sea level. Changes in relative sea level at sedimentological scales are measured with reference to the actual sea level (e.g., tidal lowstand vs. highstand, or storm vs. fairweather sea level). Sedimentological shoreline shifts across an intertidal environment are geometrically analogous to the stratigraphic shoreline shifts across a shelf, with transit areas located updip of their respective subaerial clinoform rollovers (Figs. 3.25, 4.58, and 4.60; Catuneanu, 2019a,b).

The development of both sedimentological and stratigraphic cycles may involve a combination of allogenic and autogenic processes, and may be accompanied by changes in accommodation, energy flux, sediment supply, and shoreline trajectory (Figs. 3.24 and 4.62). Both types of sedimentary cycles can form across wide ranges of temporal and physical scales, which overlap for timespans of 10^2-10^4 yrs and thicknesses of 10^0 m (i.e., bedsets have been documented at scales of $10^{-1}-10^0$ m and 10^0-10^4 yrs, although most would form during periods of time of $\leq 10^2$ yrs, while high-frequency sequences may span scales of 10^0-10^1 m and 10^2-10^5 yrs; Hampson et al., 2008, Zecchin et al., 2017a,b, 2019, Catuneanu, 2019a, Fig. 5.4). However, at any location, the largest bedsets are always nested within the lowest rank systems tracts and component depositional systems, and therefore, are order(s) of magnitude smaller than the high-frequency sequence that hosts them (Zecchin et al., 2017a,b, 2019, Catuneanu, 2019a).

The distinction between sedimentological and stratigraphic cycles is perhaps most challenging in the deep-water setting, where both types of cycle may develop within one depositional system, without any significant interruptions in sedimentation (Fig. 7.9). The internal makeup of deep-water systems is shaped by the interplay of several independent processes, including gravity-driven mass failures, contour currents, and suspension fallout. While all types of deposit are important to evaluate in hydrocarbon reservoir studies, not all are equally relevant to the construction of the sequence stratigraphic framework. Specifically, the sequence stratigraphic framework describes the cyclicity recorded by gravity-driven processes, which is linked to shoreline trajectories and can be used to the designation of systems tracts. For this reason, sedimentological work must precede the sequence

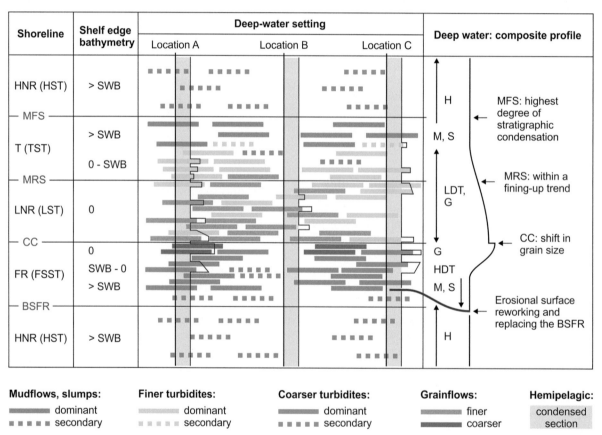

FIGURE 7.9 Stratigraphic cyclicity in the deep-water setting during a shoreline shelf-transit cycle (modified from Catuneanu, 2019b). The composite profile describes the relative chronology of the dominant gravity-driven processes in the case of wide and low-gradient siliciclastic shelves that are subject to high-magnitude changes in relative sea level (i.e., shelves exposed at times of lowstand, and submerged at times of highstand). The sedimentological makeup of the composite profile (e.g., grain size and the types of gravity flows) may be modified by basin-specific factors. The stratigraphic criteria that define the deep-water sequence stratigraphic framework relate to the *trends of change* in gravity flows during the shoreline transit cycles on the shelf, and are independent of the sedimentological variability of the composite profile. Note the difference in terms of timing, frequency and grading trends between the stratigraphic cycle defined by the composite profile and the sedimentological cycles at specific locations. Abbreviations: HNR—highstand normal regression; FR—forced regression; LNR—lowstand normal regression; T—transgression; HST—highstand systems tract; FSST—falling-stage systems tract; LST—lowstand systems tract; TST—transgressive systems tract; BSFR—basal surface of forced regression; CC—correlative conformity; MRS—maximum regressive surface; MFS—maximum flooding surface; SWB—storm wave base; H—hemipelagic sediment; M—mudflows; S—slumps; HDT—high-density turbidity flows; G—grainflows; LDT—low-density turbidity flows.

stratigraphic analysis, in order to separate the depositional elements that originate from processes with different degrees of relevance to sequence stratigraphy. The stratigraphic interpretation of the deep-water system may be validated further by correlation with the stratigraphic cycles on the shelf, wherever possible.

The application of sequence stratigraphy to the deepwater setting relies on the construction of composite profiles that illustrate the relative chronology of the different types of gravity-driven processes at basinal or sub-basinal scales (Fig. 7.9). Stratigraphic cyclicity relevant to the definition of sequences and systems tracts is described by the composite rather than local profiles. The place of accumulation of individual gravity-driven deposits depends on the location of sediment entry points along the shelf edge, the types of gravity-driven processes, and the seafloor morphology (e.g., structural elements, salt tectonics, or mud diapirism). The allocyclic or autocyclic shifts of deep-water depositional elements (e.g., the lateral shifts of leveed channels) further enhance the offset between local trends and the regional composite profile in terms of timing and frequency of cycles, timing of coarsening- and finingupward trends, and timing of the coarsest sediment (Fig. 7.9). The sedimentological cycles defined by local trends must not be confused with the stratigraphic cycles defined by regional composite profiles. Notwithstanding these caveats, the stratigraphic architecture of basin floors that lack any significant relief can be remarkably consistent (e.g., van der Merwe et al., 2010, Fig. 5.7; see Chapter 8 for details).

The internal makeup of systems tracts and component depositional systems varies with the scale of observation, from sedimentological cycles to stratigraphic cycles (Figs. 3.24 and 4.3). At sub-stratigraphic scales, the beds and bedsets of the lowest rank systems tracts can be classified into several hierarchical orders that describe depositional units relevant to sedimentology, with various degrees of internal complexity (e.g., Miall, 1996; Gardner et al., 2003; Ainsworth et al., 2017). At stratigraphic scales, higher rank systems tracts are defined by the stacking pattern of lower rank sequences (Figs. 4.3, 4.70, 5.1, 5.5, and 7.1–7.5). For example, a transgressive systems tract defined by long-term retrogradation may include prograding systems, as well as maximum flooding surfaces, of lower hierarchical ranks (i.e., higher frequency transgressions and regressions during the long-term retrogradational trend; Figs. 4.29, 5.5, 6.31, and 6.41). Similarly, long-term prograding systems may also include higher frequency transgressions and regressions, which define lower rank sequences and component systems tracts and depositional systems (Figs. 5.1 and 5.5).

The nested architecture of depositional cycles resembles a fractal pattern (Schlager, 2004), in which the internal makeup of sedimentological and stratigraphic units becomes increasingly complex with the increase in the scale of observation. However, the stratigraphic architecture is not truly fractal, because sequences of different scales may have different underlying controls (Fig. 3.1) and systems-tract composition (i.e., different combinations and/or relative development of systems tracts; Schlager, 2010; Catuneanu, 2019b, Fig. 1.23). Similarly, the architecture of the sedimentary record at sedimentological scales does not follow a fractal pattern, as different types of units form at different hierarchical levels (e.g., bedforms vs. macroforms; Miall, 1996). Moreover, the different types of sedimentological units, as well as their architecture, are not necessarily reproducible from one basin to another, due to differences in the parameters that control sedimentation, including accommodation, sediment supply, and environmental energy. The construction of sedimentological to stratigraphic frameworks needs to be performed on a caseby-case basis, honoring the local data and the reality of each basin.

7.1.2.1 Sedimentological cycles

Sedimentological cycles (i.e., beds and bedsets; Fig. 3.24) are the building blocks of the lowest rank systems tracts and component depositional systems. The minimum timescale that affords the development of a depositional system, and implicitly of a systems tract, is typically in a range of 10^2 yrs (details in Chapter 5). Therefore, sedimentary cycles on timescales of $< 10^2$ yrs are most commonly of sedimentological nature. Where depositional systems take longer time to develop, the formation of bedsets may be extended up to timescales of 10^4 yrs (Hampson et al., 2008), although cyclicity within the range of orbital forcing (10^4 yrs), and even on millennial timescales (10^3 yrs), is more commonly stratigraphic in nature (Csato et al., 2014; Strasser, 2018, Fig. 5.106). As opposed to stratigraphic cycles, which extend across larger areas of systemstract development, beds and bedsets are commonly restricted to the confines of specific depositional environments (Fig. 6.66; details in Chapter 6). The approaches to the classification of sedimentological and stratigraphic cycles are also based on the observation of different types of stratal stacking patterns (Figs. 3.24 and 4.62).

Sedimentological cycles may display a nested architecture defined by cyclic changes in sedimentological stacking patterns observed at different scales (Figs. 3.24 and 4.62). Sedimentological stacking patterns describe the stratal architecture of a depositional system at sub-stratigraphic scales (e.g., the pattern of change in the degree of amalgamation of storm beds, without changes in depositional system and systems tract; Figs. 6.64 and 6.67). At each sedimentological scale of observation, bedsets correspond to cycles of change in the sedimentological stacking pattern, which are generated by fluctuations in sediment supply and/or energy conditions within a depositional environment without changes in the stratigraphic stacking pattern; e.g., cycles of change in the frequency of storm events during continuous progradation or retrogradation of coastal systems, without changes in the type of shoreline trajectory (Figs. 6.64 and 6.67). The architecture of beds and bedsets delineates the stratigraphic stacking patterns that define the lowest rank systems tracts (see Chapter 6 for more details on bedsets and bedset boundaries).

Beds and bedsets can be classified into several hierarchical orders that describe sedimentological units with various degrees of internal complexity. Classifications of the sedimentological units that build the stratigraphic framework have been proposed for specific depositional settings (e.g., and , for coastal to shallow-water settings). While realistic for the areas which afforded the collection of the supporting data, the orderly patterns proposed by each classification system do not apply to all cases, and departures from the model predictions can be expected as a function of local accommodation and sedimentation conditions, as well as the time available to build a depositional system (e.g., 10^2-10^3 yrs vs. 10^4-10^5 yrs or longer timescales). The interplay of these variables determines the sedimentological makeup and the degree of sedimentological complexity of stratigraphic units at any location.

7.1.2.2 Stratigraphic cycles

Stratigraphic cycles define sequences in the sedimentary record (Fig. 1.22). Sequences and their component systems tracts and depositional systems can be observed at all stratigraphic scales (Figs. 4.3, 4.70, 5.1, 5.2, 5.3, and 7.1–7.5), starting with timescales of 10^2-10^3 yrs (Amorosi et al., 2009, 2017; Catuneanu, 2019b, Fig. 4.62). At each scale of observation, stratigraphic cyclicity refers to cycles of change in the stratigraphic stacking pattern, which involve changes in systems tracts and component depositional systems. Stratigraphic stacking patterns describe the stratal architecture of systems tracts, which is linked to specific types of shoreline trajectories in downstream-controlled settings, and to the dominant depositional elements in upstream-controlled settings (Figs. 3.24, 4.62; see Chapter 4 for details). Unless otherwise specified, the default usage of "stratal stacking patterns" throughout the book is in a stratigraphic sense.

The sequence stratigraphic framework is scale invariant in the sense that the same stratigraphic stacking patterns can be observed at different scales (Schlager, 2004, 2010; Catuneanu, 2019b). This enables a consistent application of sequence stratigraphy, irrespective of the resolution of the data available and the scale of sequences that may develop at particular

locations. At each stratigraphic scale (i.e., hierarchical level), the building blocks of the sequence stratigraphic framework are represented by sequences and their component systems tracts and depositional systems (Figs. 4.3, 4.70, 5.1, 5.3, and 5.5). Despite the nested architecture of stratigraphic cycles, the stratigraphic framework is not fractal as sequences of different scales, as well as sequences of similar scales, may differ in terms of underlying controls and the internal composition of systems tracts. This variability reflects the local conditions of accommodation and sedimentation, and therefore it is basin or even sub-basin specific. The classification of stratigraphic cycles that define the nested architecture of the stratigraphic framework is a key aspect of sequence stratigraphy, with implications for methodology and nomenclature, as discussed below.

7.2 Hierarchy in sequence stratigraphy

Hierarchy refers to the classification of stratal units and bounding surfaces on the basis of their absolute or relative scales. The classification of stratigraphic elements that develop at different scales is necessary in order to define the relationship between the nested cycles of the sequence stratigraphic framework. Several hierarchy systems have been proposed since the 1970s, based on criteria that emphasize different attributes of sequences, from their temporal scales (e.g., Vail et al., 1977b, 1991; Krapez, 1996, 1997) to their physical features (e.g., Embry, 1995) or their "relatively conformable" character (e.g., Mitchum and Van Wagoner, 1991; Sprague et al., 2003; Abreu et al., 2010). None of these hierarchy systems has received universal acceptance or validation by data in all depositional and tectonic settings.

The approach to the classification of stratigraphic cycles impacts the way sequence stratigraphy is applied, from a scale-dependent to a scale-independent methodology and nomenclature. In a scale-dependent approach to stratigraphic methodology and nomenclature, different types of units are defined at different scales. In a scale-independent approach, the same types of units are recognized at all stratigraphic scales, albeit of different hierarchical ranks. These contrasting approaches present the practitioner with different alternatives to the construction of a sequence stratigraphic framework, whose merits and pitfalls are discussed below. Beyond personal preferences, a robust and objective approach to sequence stratigraphic methodology and nomenclature requires consistency across the entire range of case studies, despite the variability of the stratigraphic record. This requirement underlies the standard approach to sequence stratigraphy, which relies on core principles that are independent of model and any local variables such as the geological setting and the types and resolution of the data available.

Hierarchy systems typically require a reference (i.e., an "anchor"), relative to which smaller or larger units can be defined. Classifications that employ a nomenclature based on hierarchical orders typically use the "first order" units and bounding surfaces as the anchor for the hierarchy system. In sedimentology, "first order" designates the smallest units and bounding surfaces that develop at bedform scales, and the hierarchical ranks scale up from bedforms to macroforms, architectural elements, and depositional systems (e.g., Miall, 1996). In stratigraphy, "first order" designates the largest genetic units and bounding surfaces of a sedimentary basin, and the hierarchical ranks scale down to progressively smaller stratigraphic cycles, as far as permitted by the resolution of the data available (Figs. 4.3, 5.4, and 7.10; e.g., Vail et al., 1977b, 1991; Embry, 1995; Posamentier and Allen, 1999; Catuneanu, 2006).

The contrast between sedimentological and stratigraphic approaches reflects the different types of data that are commonly used to classify sedimentary cycles. Sedimentological classifications are based on the observation of facies (e.g., in outcrops, core, or modern environments), whereas the reconstruction of the larger scale stratigraphic architecture requires more regional data, such as seismic profiles. The latter approach is particularly evident in frontier basins where only seismic data are available (e.g., offshore basins prior to drilling). The construction of stratigraphic frameworks in mature basins, where all types of data are available, can be approached from both ends, either starting from sedimentological units (e.g., beds and bedsets) and scaling up or starting from the basin fill and scaling down. Anchors for a hierarchy system can also be selected at intermediate scales (e.g., shoreline shelf-transit cycles, or "relatively conformable" sequences), in which case the designation of hierarchical ranks is more logically based on names (e.g., parasequence, sequence, composite sequence, etc.) rather than numerical orders (i.e., a reference selected at an intermediate scale cannot be of "first" order).

Irrespective of approach, the underlying reality is that stratigraphic frameworks are basin specific (e.g., potentially nothing in common between an interior basin and a coeval continental margin), reflecting the unique accommodation and sedimentation conditions of each geological setting. Therefore, a system of stratigraphic classification that can be applied universally must be independent of any features that are specific to a tectonic or depositional setting. For example, the shelf-transit cycle of a shoreline at Milankovitch scales of 10^4–10^5 yrs (Burgess and Hovius, 1998; Porebski and Steel, 2006; Ainsworth et al., 2018) may provide a reference for the definition of smaller and larger units in shelf settings dominated by climate cycles, but may have no relevance or expression in overfilled basins or ramp settings controlled tectonic cycles (Catuneanu and Elango, 2001; Martins-Neto and Catuneanu, 2010). Given the uniqueness of each sedimentary basin, stratigraphic features observed at intermediate scales fail to provide a reproducible anchor for the classification of stratigraphic cycles worldwide. The only anchor for stratigraphic classification that can be used universally in all geological settings, and in both frontier and mature basins (i.e., irrespective of tectonic setting, depositional setting, and the types and resolution of the data available), is the "first-order" sedimentary basin cycle, relative to which stratigraphic cycles of lower hierarchical ranks can be defined in the context of each basin. In this case, sequences of any hierarchical order are basin specific in terms of origin, timing, scales, and internal makeup.

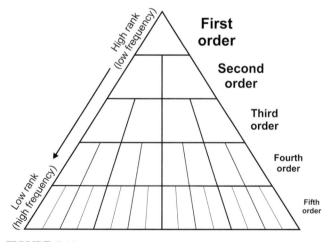

FIGURE 7.10 Concept of hierarchy, illustrating the nested architecture of the stratigraphic framework. First-order sequences can be subdivided into second-order and lower rank sequences. "First-order" may be referred to as "high order," with reference to the position in the hierarchy pyramid, or as "low order" with reference to the numerical value. To avoid confusion, it is preferable to make reference to the higher vs. lower hierarchical *ranks* (i.e., "first-order" is the highest hierarchical rank within a sequence stratigraphic framework). The frequency of occurrence of sequence boundaries in the sedimentary record typically increases with the decrease in the hierarchical rank, but time is not the only attribute used in the classification of stratigraphic cycles (see text for details).

7.2.1 Criteria for stratigraphic classification

Historically, both temporal and physical criteria have been employed to devise hierarchy systems for the classification of stratigraphic cycles, with evidence derived from Phanerozoic case studies (e.g., Vail et al., 1977b, 1991; Embry, 1995). The time-based hierarchy is based on the frequency of occurrence of sequence boundaries in the sedimentary record (i.e., the duration of stratigraphic cycles), and emphasizes eustasy as the main driving force behind stratigraphic cyclicity. In turn, eustatic changes have been attributed primarily to a combination of plate tectonic and orbital controls (Fig. 3.4). As eustasy is global in nature, the philosophy behind this hierarchy system led to the construction of global cycle charts (Vail et al., 1977b; Haq et al., 1987, 1988) coupled with the assumption of global correlations of the "geochronologic units" defined by the global cycles (Vail et al., 1977b). The usage of global cycle charts as a reference for the classification and dating of local stratigraphic cycles was placed under intense scrutiny, and it was eventually excluded from the sequence stratigraphic workflow (Miall, 1992, 1997; Posamentier and Allen, 1999; Catuneanu, 2006). The reason is that sedimentary basins form and evolve in response to tectonic processes that operate on regional to continental scales, so stratigraphic cycles around the world are unlikely to be synchronous (e.g., Miall, 2000).

The hierarchy system based on cycle duration poses two challenges to the practicing geologist: (1) from a practical perspective, *time control* is always required to designate hierarchical orders; and (2) from a theoretical perspective, the controls on stratigraphic cyclicity are assumed to comply with the law of *uniformitarianism* throughout the Earth's history. Despite these issues, the time-based hierarchy was eventually expanded to the Precambrian as well, with orders of cyclicity defined by the following durations: c. 364 My for first-order cycles; 22–45 My for second-order cycles; 1–11 My for third-order cycles; and 90–400 kyr for fourth-order cycles (Krapez, 1996, 1997). Each of these hierarchical orders was linked to particular tectonic (and to a lesser extent climatic) controls whose periodicities were assumed to be more or less constant during geological time. For example, the 364 My duration of first-order cycles was calculated based on the assumption that nine equal-period global tectonic (Wilson) cycles of

supercontinent assembly and breakup occurred during the 3500–224 Ma interval (Krapez, 1993, 1996). The need for a hierarchy system based on cycle duration stems from the argument that "There are no physical criteria with which to judge the rank of a sequence boundary. Therefore, sequence rank is assessed from interpretations of the origin of the strata contained between the key surfaces, and of the period of the processes that formed these strata" (Krapez, 1997).

Stratigraphic uniformitarianism assumes that the controls on cyclicity remained unchanged during the Earth's evolution, allowing equal periodicity for stratigraphic cycles of the same hierarchical order, irrespective of age. This is difficult to establish solely from the study of the Phanerozoic, whose window into the past is too narrow to demonstrate whether the nature and periodicity of processes controlling stratigraphic cyclicity were constant throughout geological time. Work on Precambrian basins shows that the mechanisms controlling the formation and evolution of sedimentary basins, for the greater part of geological time, were far more diverse and erratic in terms of origins and rates than originally inferred from the study of the Phanerozoic record (Catuneanu and Eriksson, 1999, 2002; Eriksson et al., 2004, 2005a,b, 2013; Catuneanu et al., 2005, 2012; Bumby et al., 2012, Fig. 7.11). Similar conclusions have been reached by studies of orbital (Milankovitch scale) processes, which demonstrated that the periods of precession and obliquity have changed significantly over time due to the evolution of the Earth–Moon system (Lambeck, 1980; Walker and Zahnle, 1986; Algeo and Wilkinson, 1988; Berger and Loutre, 1994). This means that *time* is largely irrelevant for the designation of a universally applicable hierarchy system.

The inadequacy of a time-based hierarchy system is illustrated by the contradictions between the supporters of this approach to stratigraphic classification. For example, a "second-order" cycle is assigned a duration

FIGURE 7.11 Contrasts between Precambrian and Phanerozoic in terms of aspects relevant to sequence stratigraphy (from Catuneanu et al., 2005).

Attributes	Precambrian	Phanerozoic
Time span	c. 88% of Earth's history	c. 12% of Earth's history
Facies preservation	Relatively poor (due to post-depositional tectonics, diagenesis, metamorphism)	Relatively good
Time control	Relatively poor (based on marker beds and lower resolution radiochronology)	Relatively good (marker beds, as well as several age-dating techniques)
Basin-forming mechanisms	Competing plume tectonics and plate tectonics, more erratic regime	Plate tectonics, more stable regime

of 10–100 My by Vail et al. (1977b), 3–50 My by Mitchum and Van Wagoner (1991), and 22–45 My by Krapez (1996); a "third-order" cycle has a periodicity of 1–10 My in the view of Vail et al. (1977b), 0.5–3 My in the hierarchy of Mitchum and Van Wagoner (1991), and 1–11 My according to Krapez (1996). This inconsistency stems from the fact that several competing sequence-forming mechanisms (e.g., regional tectonism, global eustasy, and climate changes), each operating over different timescales, interplay to generate the stratigraphic record. The periodicity recorded by cycles above the Milankovitch band is typically specific to each sedimentary basin, reflecting the local conditions of accommodation and sedimentation. Even for cycles within the Milankovitch band, there is increasing evidence that other processes which operate within the same temporal range, such as intraplate stress fluctuations and local tectonism, may distort and obscure the stratigraphic response to orbital forcing (Peper and Cloetingh, 1995; Wilkinson et al., 2003; Strasser, 2018). Therefore, attempts to generalize a relationship between sequence scales and specific controls beyond the confines of individual sedimentary basins are problematic and unrealistic.

Physical criteria for the classification of stratigraphic cycles provide an alternative to temporal scales, and emphasize observable features irrespective of the temporal duration of cycles. Such observations may focus on the attributes of sequences (e.g., their "relatively conformable" character; Mitchum, 1977; Sprague et al., 2003; Neal and Abreu, 2009; Abreu et al., 2010, Fig. 7.6), or on the attributes of sequence boundaries (Embry, 1995, Fig. 7.12). The latter approach relies on six attributes to determine the hierarchical order of sequence boundaries, and implicitly of the sequences delineated by them: the areal extent over which the sequence boundary can be recognized; the areal extent of the unconformable portion of the boundary; the degree of deformation that strata underlying the unconformable portion of the boundary underwent during the boundary formation; the magnitude of the deepening of the sea and flooding of the basin margin as indicated by the facies and extent of the transgressive strata overlying the boundary; the degree of change of the sedimentary regime across the boundary; and the degree of change of the tectonic setting of the basin and surrounding areas across the boundary (Embry, 1995, Fig. 7.12).

The emphasis on the physical attributes of sequence boundaries helps to classify surfaces on the basis of their relative stratigraphic significance, and provides guidelines to the classification of sequences as well: a sequence cannot contain within it a sequence boundary of equal or greater hierarchical rank than the rank of its lowest rank boundary (e.g., second-order sequence cannot contain a first-order boundary within it, but can include third- and lower rank boundaries); and the order of a sequence is equal to the order of its lowest rank boundary (Embry, 1995, Fig. 7.12). Two pitfalls of this approach to classification have been discussed by Miall (1997, p. 330–331). One is that it implies tectonic control in sequence generation. Sequences generated by glacio-eustasy, such as the Late Paleozoic cyclothems of North America and those of Late Cenozoic age on modern continental margins, would be of first order on the basis of their areal extent, but of a lower order on the basis of the nature of their bounding surfaces. The second problem is that this classification requires

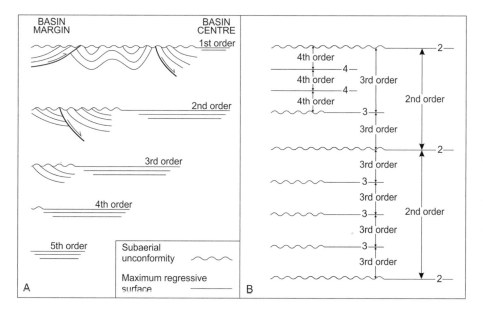

FIGURE 7.12 Hierarchy system of stratigraphic classification in tectonically active basins, based on the Jurassic Sverdrup rift basin in northern Canada (modified after Embry, 1993, 1995). A—Physical characteristics that differentiate bounding surfaces of different hierarchical orders. B—Principles of classification: a sequence cannot include a sequence boundary of equal or higher hierarchical rank; the order of a sequence is equal to the order of its lowest rank boundary.

good preservation of the basin margin in order to assess the areal extent of the unconformable portion of the boundary or the degree of deformation across the boundary. For these reasons, the applicability of this hierarchy system is limited to tectonically active and well-preserved basins, such as the rift basin from which the supporting data were derived.

As discussed by Carter et al. (1991), and confirmed by subsequent field and statistical studies, a hierarchy of sequences does exist in the rock record; however, the distinctiveness of these sequences in terms of duration, periodicity, and physical scales is only approximate, at best. The temporal and physical scales of stratigraphic sequences do not define mutually exclusive ranges, and any model-selected boundaries between such ranges (hierarchical classes) become arbitrary. In spite of the lack of universal standards that could define a hierarchy system, Carter et al. (1991) note that "… some orders of sequence do indeed embrace lower orders, e.g., the major first-order thermo-tectonic cycle that incorporates the complete sedimentary history of the Canterbury Basin … includes examples of second, third, fourth, and probably fifth-order sequences." Nested architectures of stratigraphic sequences can be expected in any sedimentary basin, with temporal and physical scales that define basin-specific patterns. The degree of correlation between basin-specific sequences and global cycles depends on the dominant controls at syn-depositional time (e.g., local tectonism vs. eustasy), and needs to be assessed on a case-by-case basis following the construction of sequence stratigraphic frameworks.

7.2.2 Absolute vs. relative stratigraphic scales

The classification of stratigraphic cycles may be approached from two different perspectives, one based on the absolute scales of cycles (i.e., the temporal duration of cycles), and another based on the relative scales of cycles (i.e., the scales of cycles relative to each other, irrespective of temporal durations). Both approaches have been used to develop hierarchy systems for sequence stratigraphic units and bounding surfaces, inciting debates and controversy with respect to the best practice. Furthermore, both systems were subject to divided opinions, such as the actual duration of cycles attributed to specific hierarchical ranks (e.g., 1–10 My vs. 0.5–3 My to define "third-order" cycles; Vail et al., 1977b, 1991, Figs. 3.4 and 7.13), or the scale-variant vs. scale-invariant nomenclature of stratal units and bounding surfaces that develop at different hierarchical levels within the nested architecture of stratigraphic cycles (Schlager, 2010).

Classifications based on absolute scales have been applied since the dawn of modern sequence

Hierarchical order	Duration (My)
First order	50 +
Second order	3-50
Third order	0.5-3
Fourth order	0.08-0.5
Fifth order	0.03-0.08
Sixth order	0.01-0.03

FIGURE 7.13 Hierarchy system based on the duration of stratigraphic cycles (modified after Vail et al., 1991).

stratigraphy (e.g., Vail et al., 1977b, 1991; Krapez, 1996; Duval et al., 1998). Even though each of the several controls on stratigraphic cyclicity may record natural periodicities (Miall, 2010, Fig. 3.1), this approach is both impractical and artificial, as shown by two significant pitfalls: (1) age data are not always available, and the duration of stratigraphic cycles becomes increasingly difficult to measure in older successions with the loss of stratigraphic resolution; and (2) the scales of stratigraphic cycles are basin specific, reflecting the interplay of multiple local and global controls. The scales of any specific natural processes may or may not be expressed in a local stratigraphic framework, as several processes with different natural periodicities interplay to generate a unique architecture. This is true even in the case of orbital (Milankovitch) cycles, which are among the most precise timekeepers in the sedimentary record, as the orbital signal may be too weak to be recorded or it may be distorted or overprinted by other local or regional processes (Wilkinson et al., 2003; Strasser, 2018).

Data analyses from a wide range of geological settings and stratigraphic ages indicate that the scale of stratigraphic cycles worldwide is statistically random across a continuum of time spans and physical dimensions (Carter et al., 1991; Drummond and Wilkinson, 1996, Fig. 7.8). No natural breaks can be replicated in all sedimentary basins, so any arbitrary limits (e.g., 1 My between third- and fourth-order cycles; Vail et al., 1977b, Fig. 3.4) are bound to be artificial. Numerous case studies demonstrate the variability of the stratigraphic record across the entire spectrum of scales (e.g., Posamentier et al., 1992a,b; Thorne, 1995; Fouke et al., 2000; D'Argenio, 2001; Van Wagoner et al., 2003; Schlager, 2004, 2010; Miall, 2010; Csato et al., 2014). This is further enhanced by the fact that the rates and periodicity of natural processes changed over time, from the Precambrian to the Phanerozoic (e.g., Eriksson et al., 2004, 2005a,b, 2013; Catuneanu et al., 2005, 2012; Bekker et al., 2010; details in Chapter 9). Ultimately, stratigraphic frameworks are basin specific in terms of scales and architecture, reflecting unique combinations of local and global controls on accommodation and sedimentation.

Classifications based on relative scales require a reference unit ("anchor") relative to which larger and/or smaller units can be defined, irrespective of their absolute temporal and physical scales. The "relatively conformable" sequence of seismic stratigraphy (Mitchum, 1977, Fig. 1.22) introduced a reference relative to which smaller units (parasequences and systems tracts) and larger units (sequence sets, composite sequences, composite sequence sets, and megasequences) have been defined (Sprague et al., 2003). The drawback of this approach is that "relatively conformable" sequences can be observed at different scales, depending on the resolution of the data available; e.g., "relatively conformable" sequences observed at seismic scales are typically unconformable at the sub-seismic scales of high-resolution sequence stratigraphy (Fig. 7.6). In this context, an increase in the resolution of a stratigraphic study results in a decrease in the scale of "relatively conformable" sequences, triggering changes to the nomenclature of all previously defined units; e.g., "sequences" become "composite sequences" or "megasequences," and "systems tracts" become "sequence sets" or "composite sequence sets."

It follows that the "relatively conformable" sequence does not provide an objective and reproducible anchor for the classification of stratigraphic cycles, as it leads to nomenclatural inconsistency between sequence stratigraphic frameworks constructed with different datasets in different basins or sub-basins of the same sedimentary basin. A standard system of stratigraphic classification must remain independent of local variables, to ensure consistency across the entire range of case studies. The solution is to select an anchor that is independent of the types and resolution of the data available; i.e., the basin-fill cycle, which is the largest genetic unit of any stratigraphic framework. Within the basin fill, the nested architecture of the component cycles may be unique to each sedimentary basin in terms of temporal and physical scales, and so it needs to be defined on a case-by-case basis. Criteria to discriminate between the relative stratigraphic significance of cycles of different hierarchical ranks have been proposed for specific settings (e.g., Embry, 1995, for rift basins; Fig. 7.12), as well as in a broader sense for all tectonic settings (Catuneanu, 2006, p. 330–334; Fig. 7.14).

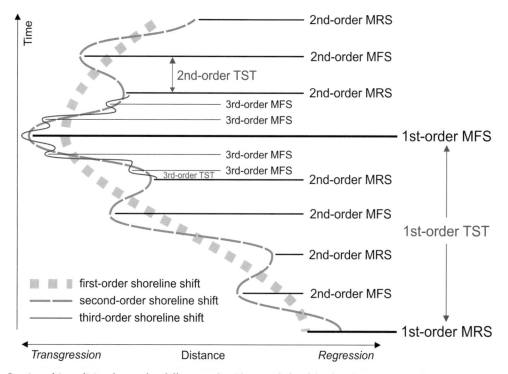

FIGURE 7.14 Stratigraphic cyclicity observed at different scales. The true shifts of the shoreline are recorded at the lowest hierarchical level. At higher hierarchical levels, shoreline shifts represent longer term trends in lower resolution studies. In a general sense, the classification of sequences reflects the relative magnitude of stratigraphic cycles within the nested architecture of the basin fill, generated by base-level changes in upstream-controlled settings and by the interplay of base level and relative sea-level changes in downstream-controlled settings. Abbreviations: MRS—maximum regressive surface; MFS—maximum flooding surface; TST—transgressive systems tract.

7.2.3 Approaches to stratigraphic nomenclature

In terms of nomenclature of hierarchical ranks, one approach is to assign numerical orders to stratigraphic cycles of different scales (e.g., Vail et al., 1977b, 1991; Embry, 1995; Krapez, 1996; Posamentier and Allen, 1999; Catuneanu, 2003, 2006), although the criteria employed to define the different orders of cyclicity may vary from temporal to physical standards. In this approach, the nomenclature of stratal units and bounding surfaces is consistent at all scales, while their relative stratigraphic significance is indicated by hierarchical orders (Figs. 4.3 and 4.70). For example, maximum flooding surfaces can be observed at multiple scales, in relation to transgressions of different magnitudes (Figs. 5.1, 5.3, and 5.5). A third-order maximum flooding surface is more significant than a fourth-order maximum flooding surface, but the name of the surface remains the same at all scales of observation (i.e., hierarchical ranks). The same is true for sequences, systems tracts, and depositional systems: e.g., a third-order transgression is more significant than a fourth-order transgression, but estuaries and transgressive systems tracts can be observed at both scales (Figs. 5.1 and 5.5). Therefore, this approach to nomenclature is independent of scale.

An alternative approach is to assign different names to cycles that develop at different scales, using the "relatively conformable" sequence as a reference for the hierarchy system (e.g., parasequences < sequences < composite sequences < megasequences; Van Wagoner et al., 1990; Mitchum and Van Wagoner, 1991; Sprague et al., 2003; Neal and Abreu, 2009). This approach leads to ambiguity in the nomenclatural designation of stratal units, as the scale of a "relatively conformable" sequence depends on the resolution of the data available, and hence, it is variable (Fig. 7.6). As a result, the complex terminology promoted by this hierarchy system is not only confusing but also inconsistent between basins or sub-basins where different types of data are available (e.g., a seismic-scale "sequence" in an area with lower resolution data may be equivalent to a "composite sequence" or even a "megasequence" in another area where well data are available). Another significant drawback of this hierarchy system is that the classification focuses only on stratal units, but it overlooks the nomenclature of bounding surfaces (e.g., maximum flooding surfaces; Fig. 7.14) that develop at different scales, in relation to stratigraphic cycles of different magnitudes.

7.2.4 Orderly vs. variable stratigraphic patterns

The definition of a "relatively conformable" sequence as an anchor to hierarchy, with different types of units

observed at different scales (Mitchum, 1977; Van Wagoner et al., 1990; Mitchum and Van Wagoner, 1991), implies that the stratigraphic record is organized according to an orderly pattern in which each type of unit has a specific internal makeup (i.e., parasequences are the building blocks of systems tracts and sequences; sequences are the building blocks of sequence sets and composite sequences; and composite sequences are the building blocks of composite sequence sets and megasequences; Sprague et al., 2003; Neal and Abreu, 2009; Abreu et al., 2010). This model also ties the stratigraphic architecture to specific accommodation conditions, by linking processes of coastal aggradation to stages of positive accommodation, and the formation of sequence boundaries to stages of negative accommodation (Neal and Abreu, 2009; Abreu et al., 2010). The flaws of these assertions have been pointed out by independent data-based and numerical modeling studies (e.g., see Catuneanu, 2019a, for a summary).

An underlying assumption of the orderly-pattern model is that the "standard" sequences of seismic stratigraphy (statistically within the third-order range of cyclicity, by both temporal and physical standards: 10^5-10^6 yrs and 10^1-10^2 m; Vail et al., 1991, Duval et al., 1998; Schlager, 2010, Figs. 4.3 and 5.4) form during full cycles of accommodation and include lowstand systems tracts, whereas any lower rank stratigraphic cycles (i.e., parasequences) form during periods of positive accommodation, therefore missing lowstand deposits (Van Wagoner et al., 1990; Duval et al., 1998). In this approach, hierarchical ranks are tied both to scale and internal makeup, predicting that "standard" seismic-scale sequences include all systems tracts, whereas their building blocks at sub-seismic scales (i.e., parasequences) consist only of transgressive and highstand deposits. The pitfall of this model is that accommodation cycles are recorded at all scales, starting from the sedimentological scales of tidal cycles, and exposure surfaces are as common as flooding surfaces in the rock record (Vail et al., 1991; Schlager, 2004, 2010; Sattler et al., 2005, Fig. 5.106). Therefore, parasequences may consist of any types of deposits, and the scales of sequences and parasequences are not mutually exclusive; i.e., the two types of units are merely different alternatives for the definition of stratigraphic cycles at the scales of high-resolution sequence stratigraphy (Catuneanu et al., 2010, 2011, Fig. 5.103).

As summarized by Schlager (2010), "data on sequences of 10^3-10^7 years duration, the interval most relevant to practical application of sequence stratigraphy, do not conform well to the ordered-hierarchy model. Particularly unsatisfactory is the notion that the building blocks of classical sequences (approximate domain 10^5-10^6 years) are parasequences bounded by flooding surfaces (Van Wagoner et al., 1990; Duval et al., 1998)." Several pitfalls of the parasequence concept prevent its dependable usage as the building block of sequences and component systems tracts, as originally intended (Catuneanu and Zecchin, 2020; details in Chapter 5). Parasequences have become obsolete with the advent of high-resolution sequence stratigraphy, as sequences that develop at parasequence scales provide a better and more reliable alternative for stratigraphic correlation. As a result, parasequences are no longer part of the methodological workflow of sequence stratigraphy (details in Chapters 5, 9, and 10). Full-fledged depositional sequences that may include all systems tracts have been documented starting from the smallest stratigraphic scales of 10^0 m and 10^2 yrs (details in Chapter 5).

The conceptual and nomenclatural pitfalls of the orderly-pattern model call for a more realistic approach to stratigraphic classification that is independent of data resolution and unaffected by the natural variability of the sedimentary record. Sequences of all scales can include all or any combinations of systems tracts (e.g., Posamentier et al., 1992a,b; Catuneanu et al., 2011; Csato and Catuneanu, 2012, 2014; Moran, 2020, Figs. 1.23, 4.33). This variability requires the construction of stratigraphic frameworks on the basis of local data and the unbiased observation of stratal stacking patterns, free of any model assumptions (Catuneanu, 2019a,b). The model-independent approach to the classification of stratigraphic cycles is presented below.

7.2.5 Basin-specific stratigraphic frameworks

Statistical surveys show that sequences in sedimentary basins worldwide are not organized into discrete classes of temporal or physical scales, but are rather part of a stratigraphic continuum (Carter et al., 1991; Drummond and Wilkinson, 1996, Fig. 7.8). This variability is the result of the complex interplay of multiple local and global controls on accommodation and sedimentation, which can alter or override the orderly patterns that may be expected from the natural periodicity of any specific controls on stratigraphic cyclicity (Fig. 3.1). As a result, the durations and thicknesses of sequences worldwide have log-normal distributions that lack significant modes (Drummond and Wilkinson, 1996). Similar conclusions were reached by Algeo and Wilkinson (1988), as well as Peper and Cloetingh (1995), who demonstrated that the periodicities of stratigraphic cycles have random distributions relative to any specific sequence-forming mechanism. Therefore, the definition of any temporal or physical standards for a universal classification of stratigraphic sequences in all tectonic and depositional settings becomes unrealistic, as it would merely provide an arbitrary subdivision of a stratigraphic continuum (Drummond and Wilkinson, 1996; Schlager, 2010; Catuneanu, 2017).

Stratigraphic cyclicity is basin or even sub-basin specific. Sequences in different sedimentary basins should not be expected to correlate, by default, since even those basins that are related genetically can form at different times in different places (e.g., in relation to the diachronous rifting and opening of new oceans, or in relation to the diachronous accretion of allochthonous terrains along convergent plate margins). As a result, different types of basins may coexist (e.g., rift basins and divergent continental margins, or backarc basins and foreland systems, etc.), each controlled by distinct subsidence mechanisms. Every sedimentary basin displays a potentially unique architecture of nested stratigraphic cycles, reflecting the importance of local controls on accommodation and sedimentation. The stratigraphic framework of a basin reflects its unique evolution, and may differ from the stratigraphic frameworks of other basins in terms of timing and duration of cycles, the physical scales of sequences, and the underlying controls. Stratigraphic frameworks may also differ between sub-basins of the same sedimentary basin, as a result of changing accommodation and sedimentation conditions across sub-basin boundaries (e.g., Catuneanu, 2004a,b, 2019a,c; Miall et al., 2008; Schultz et al., 2020).

The natural variability of the stratigraphic record indicates that the classification of sequences and bounding surfaces is best approached separately for each sedimentary basin. Stratigraphic frameworks are basin-specific and need to be constructed on the basis of local data rather than information extrapolated from other basins, such as global standards or reference cycle charts. Subsequently, global correlations can still be tested once local frameworks are in place. Within each sedimentary basin, the nested architecture of sequences defines their relative stratigraphic significance (Figs. 4.3, 4.70, 5.1, 5.5, and 7.1—7.6). In downstream-controlled settings, the scale of stratigraphic cycles reflects the scale of observation of coastal depositional systems (Figs. 5.1 and 5.5), and the cyclicity of shoreline trajectories is expressed in the stratigraphic architecture on both sides of the coastline (Figs. 5.7 and 5.50). In upstream-controlled settings, the scale of stratigraphic cycles reflects the scale of observation of depositional-element associations (e.g., channel- vs. floodplain-dominated in fluvial settings) within the same depositional system (Fig. 4.70).

Hierarchical orders may be assigned within the context of each sedimentary basin, starting with the "first-order" basin fill as a reference (Figs. 7.1—7.5). The sedimentary basin cycle defines the largest genetic unit of a stratigraphic framework, and therefore provides a reference that is independent of the types and resolution of the data available. The acquisition of higher resolution data in more advanced studies affords the recognition of smaller scale units and unconformities, without affecting the definition and nomenclature of the larger scale framework that is already established with lower resolution data. A first-order sequence is the fill of a sedimentary basin that accumulated in a specific tectonic setting (i.e., with accommodation controlled by a related set of subsidence mechanisms; Fig. 7.1). In the case of polyphase basins, first-order sequence boundaries mark changes in the type of basin (e.g., from rift to passive margin, from passive margin to foreland, etc.), which are the most significant changes that can be observed within a sedimentary succession irrespective of the timespan between two such consecutive events (Figs. 5.3, 7.1, and 7.15). For example, the post-Gondwana basins in Colombia include a backarc stage (c. 160—65 Ma) followed by a foreland stage (c. 65 Ma to the present day; Fig. 5.3). These stages should not be lumped together into one first-order sequence simply because they add up to a convenient duration (e.g., in excess of 100 My), nor should they be regarded as second-order sequences because their durations fit an arbitrary time span (e.g., 10—100 My; Vail et al., 1977b). Irrespective of their temporal and physical scales, each stage corresponds to a first-order sequence defined by a specific tectonic setting, and the limit between them is a first-order sequence boundary because it marks a change in the tectonic setting.

At the first-order level of stratigraphic cyclicity, the emphasis on the tectonic setting and changes thereof is universally applicable, as the formation and classification of all types of sedimentary basins are tied to tectonic criteria (Busby and Ingersoll, 1995; Einsele, 2000; Miall, 2000; Allen and Allen, 2013; Allen et al., 2015). First-order sequences can be subdivided into lower rank sequences according to the nested architecture that defines their stratigraphic relationships (Figs. 4.3, 4.70, 5.1, 6.57, 7.1—7.5, and 7.14). In this approach, hierarchical orders have no time or thickness connotations, but only a relative stratigraphic significance in relation to each other within the context of each basin (Fig. 7.14). Where regional data are insufficient to observe the entire basin fill, sequences of different scales within the stratigraphic interval of interest can be referred to in relative terms (e.g., higher vs. lower frequency sequences of lower vs. higher hierarchical ranks, respectively), without the designation of hierarchical orders (Fig. 7.16). The specification of numerical orders is not required for sequence stratigraphic analysis, as the same methodology applies to all scales of observation (Catuneanu, 2017, 2019a, 2020c).

Stratigraphic cycles of different hierarchical ranks are independent of each other in terms of systems-tract composition, which is determined by the accommodation and sedimentation conditions that controlled the formation of stratal stacking patterns at each stratigraphic scale. The relative ranking of sequences of different scales is defined by their stacking patterns, as lower rank sequences are nested within higher rank systems tracts (Figs. 4.3, 4.70, 5.1, 5.5, 5.87, 7.1—7.5, and 7.16). In turn, the high-frequency sequences that build the higher rank systems tracts consist of their own systems tracts, which are independent of the higher rank systems tract in which they are

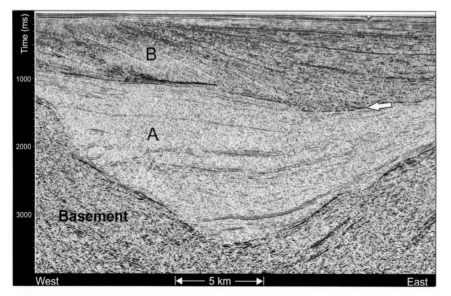

FIGURE 7.15   First-order sequence boundary (arrows) at the limit between a rift sequence (A) and the overlying "passive" margin sequence (B). Core data indicate that the sequence boundary is represented by a wave-ravinement surface at the contact between fluvial deposits below and marine shales above (Krishna-Godavari Basin, offshore India; data courtesy of Reliance Industries, India).

A. Architecture of lower frequency (higher rank) depositional sequences and systems tracts

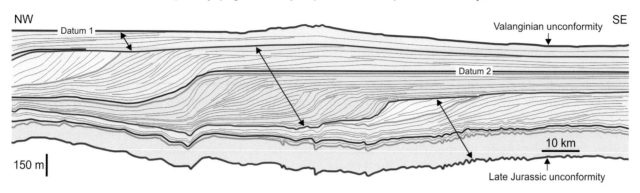

B. Architecture of higher frequency (lower rank) depositional sequences and systems tracts

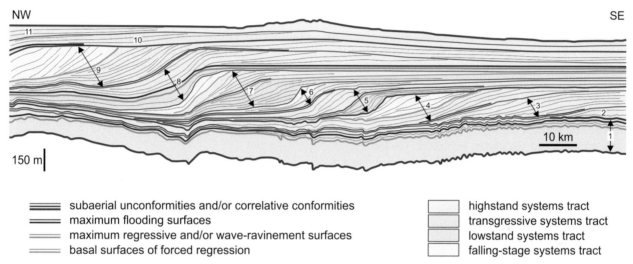

≡≡≡ subaerial unconformities and/or correlative conformities
≡≡≡ maximum flooding surfaces
≡≡≡ maximum regressive and/or wave-ravinement surfaces
≡≡≡ basal surfaces of forced regression

▢ highstand systems tract
▢ transgressive systems tract
▢ lowstand systems tract
▢ falling-stage systems tract

FIGURE 7.16   Sequence stratigraphic framework of the Vaca Muerta—Quintuco system (Late Jurassic—Early Cretaceous, Neuquén Basin, Argentina), based on seismic data calibrated with well data (modified after Dominguez et al., 2016; and Dominguez and Catuneanu, 2017). Line thicknesses are proportional to the hierarchical rank of surfaces. The trajectory of the shelf edge defines the first-order shoreline trajectory. At any smaller scales, the shoreline shifted within transit areas across the shelf. Changes in relative sea level modified the shelf capacity to retain extrabasinal sediment, the production of intrabasinal sediment, and the sediment supply to the shelf edge, as shown by the petrographic composition of systems tracts from the shelf to the basin floor. The distinction between lowstands and highstands in relative sea level, as well as the recognition of forced regressions and transgressions is important to explain sedimentological variations and the distribution of organic carbon within the basin (Dominguez et al., 2020).

nested. For example, a higher rank lowstand systems tract may include lower rank sequences comprised of lowstand, transgressive, highstand and falling-stage systems tracts (Fig. 7.16); similarly, a higher rank falling-stage systems tract may include lower rank sequences comprised of transgressive, highstand and falling-stage systems tracts (Fig. 4.25), etc.

The physical criteria that can be used to subdivide first-order basin fills into lower rank sequences vary with the tectonic and depositional settings, and may include the degree of deformation or tectonic tilt associated with the formation of unconformities (Figs. 2.10, 7.12); the areal extent of sequence boundaries (Fig. 7.12); and the impact of the cycle on

paleogeography and depositional systems within the basin (Fig. 6.57), irrespective of the duration of cycles (Embry, 1995; Neal et al., 1999; Catuneanu, 2006). However, not all criteria are applicable in every sedimentary basin. For example, tectonically "passive" basins may not undergo deformation or tilt during sedimentation, at any stratigraphic scale; and the degree of change in paleogeography and depositional systems is not always proportional to the scale of the stratigraphic cycle (Fig. 4.61). Ultimately, the classification of sequences of different hierarchical ranks reflects the relative magnitude of stratigraphic cycles within the nested architecture of the basin fill, related to base-level cycles in upstream-controlled settings and shoreline transit cycles in downstream-controlled settings, observed at different scales (Figs. 4.70, 5.1, 5.5, 5.87, 7.1–7.5, 7.14, and 7.16). This flexible approach can be applied to any sedimentary basin, regardless of stratigraphic age, geological setting, availability of time control, or degree of preservation of the basin margins.

As an example, the first-order foreland sequence of the Western Canada Sedimentary Basin (Catuneanu, 2019c) may be subdivided into several second-order sequences that reflect major cycles of basin-scale transgressions and regressions of the Western Interior seaway. Each of these second-order sequences is punctuated by third-order transgressive-regressive cycles, and so on (Fig. 6.57). The origin of these stratigraphic cycles has been attributed to either eustatic (e.g., Plint, 1991) or tectonic controls (e.g., Catuneanu et al., 2000), and, even in the case of the latter, no deformation is recorded across sequence boundaries due to the nature of the dominant flexural tectonics. The synsedimentary basin margin of the Western Canada foreland system is not preserved, due to a combination of orogenic front progradation (c. 165 km during the Late Cretaceous—Paleocene; Price, 1994) and post-orogenic uplift and erosion that removed c. 3 km of stratigraphic section along the basin margin during the post-Paleocene isostatic rebound (Issler et al., 1999; Khidir and Catuneanu, 2005). Under these circumstances, the orders of cyclicity are not defined by temporal durations or tectonic attributes, but by the relative magnitude of stratigraphic cycles within the nested architecture of the basin fill.

The possibility that stratigraphic frameworks may correlate from one basin to another in response to global controls on accommodation or sedimentation may not be excluded, particularly in the case of tectonically "passive" settings, but also, it cannot be generalized. The importance of local controls on accommodation and sedimentation is demonstrated by the 3D variability of the sequence stratigraphic framework in terms of the coeval deposition of different systems tracts and the development of diachronous sequence stratigraphic surfaces (Catuneanu, 2006, 2019a; Schultz et al., 2020; Zecchin and Catuneanu, 2020). For this reason, hierarchical orders are only meaningful within the context of the sedimentary basin in which they are defined, and their scales and significance may change from one basin to another. For example, "third-order" sequences may occur in any sedimentary basin, but they may differ from one basin to another in terms of timing, temporal and physical scales, systems-tract composition, internal sedimentological makeup, and underlying controls. The "third-order" connotation is only meaningful within the context of the sedimentary basin in which it was defined, relative to the lower and higher rank stratigraphic cycles within the same basin.

Statistical data indicate broad trends in the stratigraphic record, whereby most stratigraphic sequences develop at scales of 10^3 m and 10^6–10^8 yrs (1st order), 10^2–10^3 m and 10^5–10^7 yrs (2nd order), 10^1–10^2 m and 10^4–10^6 yrs (3rd order), 10^0–10^1 m and 10^3–10^5 yrs (4th order), and $\leq 10^0$ m and 10^2–10^4 yrs (5th order and lower ranks) (Figs. 4.3, 5.4, and 7.1–7.5). However, sedimentary basins need to be examined on a case-by-case basis, and exceptions from statistical trends should be considered as a safe norm. There are no standards for the scale and internal makeup of any type of sequence stratigraphic unit, or for the lowest hierarchical rank (e.g., 4th vs. 5th order) that should be expected in a sedimentary basin. The record of the higher frequency stratigraphic cycles is strongly influenced by the tectonic and depositional settings, being typically more evident in basins dominated by chemical or biochemical sedimentation. Siliciclastic settings commonly afford the development of at least four orders of stratigraphic cyclicity, whereas carbonate settings, which are more susceptible to smaller variations in environmental conditions, can host a more complex stratigraphic architecture with a greater number of hierarchical levels (Mawson and Tucker, 2009; Catuneanu, 2019b).

7.3 Time in sequence stratigraphy

Sequence stratigraphy is not chronostratigraphy, as all sequence stratigraphic surfaces are potentially time-transgressive, with various degrees of diachroneity (Figs. 6.6 and 7.17). This implies that different systems tracts can accumulate at the same time within a sedimentary basin. Whether or not the diachronous character of a stratigraphic surface can be demonstrated depends on the resolution of the available age-data relative to the amount of diachroneity. The criteria employed to define a sequence stratigraphic surface can also influence its degree of diachroneity. For example, maximum regressive surfaces have been referred to as surfaces that mark the end of coastal progradation (i.e., definition based on stacking patterns) or surfaces that mark the top of shallowing-upward trends (i.e., definition based on bathymetric criteria). While these definitions are interchangeable in areas close to the shoreline (e.g., within a shoreface), they point to different surfaces offshore, which diverge in a downdip direction. This section clarifies the difference between physical and bathymetric criteria in the definition of surfaces that form in subaqueous environments, the correct approach to sequence stratigraphic analysis, and the implications of diachroneity for sequence stratigraphic correlation within a 3D framework.

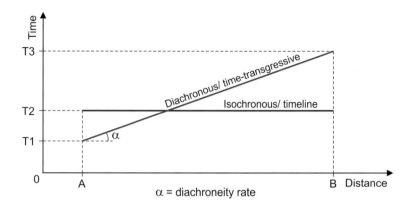

FIGURE 7.17 Time attributes of stratigraphic surfaces. A diachronous surface is "time-transgressive"; i.e., of varying ages in different areas, cutting across time planes (Bates and Jackson, 1987). An isochronous surface is a timeline, forming everywhere at the same time.

Some of the early assumptions in sequence stratigraphy have been debunked, notably the idea that sequence stratigraphic surfaces are timelines of chronostratigraphic significance. The assumptions made in the early days of seismic and sequence stratigraphy, in the 1970s and the 1980s, overemphasized the role of eustasy in the generation of sequences and bounding surfaces. The result was a model of globally synchronous stratigraphic cyclicity, which postulated that sequences worldwide define geochronologic units generated by global changes in sea level (Vail et al., 1977b, 1991; Haq et al., 1987, 1988). The timing of these global cycles, as depicted by the global cycle chart, was assumed to provide a beacon of geological time in the stratigraphic record, even more precise than that availed by local biostratigraphic or radiometric data. The global eustasy model oversimplified the true time-stratigraphic relationships, as it underrated the importance of local parameters on stratigraphic cyclicity, including tectonism and sedimentation, whose rates may vary both along dip and strike directions. As a result, none of the stratigraphic surfaces discussed in this book are truly isochronous within a 3D framework (Fig. 7.17).

While the time-transgressive character of stratigraphic surfaces varies across a continuum of scales, the degree of diachroneity can be described in qualitative terms ("low" vs. "high"), in relation to the resolution of the available age-dating methods: a "low diachroneity" refers to a difference in age that falls below the resolution of the available dating techniques (i.e., undetectable by means of biostratigraphy, radiochronology, magnetostratigraphy, or any other methods available); a "high diachroneity" refers to a more significant difference in age between the different portions of a stratigraphic surface, which can be detected and measured with the available dating techniques. The sections below explore the factors that impact the temporal attributes of stratigraphic surfaces, the reasons that hampered the analysis of time-stratigraphic relationships in the past, and the degrees of diachroneity of the various stratigraphic surfaces.

7.3.1 Reference curve for stratigraphic surfaces

The timing of all sequence stratigraphic surfaces and systems tracts is defined relative to a reference curve that describes a full cycle of sea-level, relative sea-level, or base-level changes, depending on the model that is being employed (Fig. 5.91). For example, the original correlative conformity of Posamentier et al. (1988) was considered to form during early sea-level fall (Fig. 5.91), which was later revised to the onset of sea-level fall (Posamentier et al., 1992b) or the onset of relative sea-level fall (Posamentier and Allen, 1999). The correlative conformity of Hunt and Tucker (1992) is taken at the end of relative sea-level fall (Fig. 5.91), and so on. Irrespective of the conceptual approach, each model shows *one reference curve* relative to which all surfaces and systems tracts are defined. These reference curves are typically generic, with an unspecified position within the sedimentary basin. The variable that is represented by the reference curve (i.e., sea level, relative sea level, or base level) makes a difference to the timing and meaning of the stratigraphic cycles that are defined. It is now understood that relative sea-level changes, along with sedimentation, are relevant to the definition of sequences and component systems tracts in downstream-controlled settings (Fig. 4.32), while base-level changes are relevant to the definition of stratigraphic cycles in upstream-controlled settings (details in Chapters 3, 4, and 5).

The generic nature of the reference curve originates from the early seismic and sequence stratigraphic models of the mid-1970s to the late 1980s, which were based on the assumption that eustasy is the main driving force behind sequence formation at all stratigraphic scales (Fig. 3.4). Since eustasy is global in nature, there was no need to specify a location for the reference curve within the basin under analysis. Subsequent realization that tectonism is as important as eustasy in controlling stratigraphic cyclicity led to the replacement of the eustatic curve with the curve of relative sea-level changes (Fig. 5.91). The next step was to specify the position of the reference curve of relative sea-level changes within the basin, since subsidence is typically variable along both dip and strike directions. Fig. 7.18 shows a dip-oriented transect through an extensional basin in which subsidence rates increase in a downdip direction (Fig. 7.19). As a result, different locations

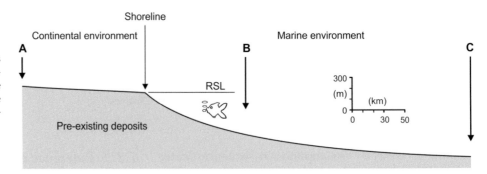

FIGURE 7.18 Dip-oriented cross section through a hypothetical extensional basin. Locations A, B, and C are characterized by different subsidence rates, as illustrated in Fig. 7.19. Abbreviation: RSL—relative sea level.

FIGURE 7.19 Changes in sea level, subsidence, and relative sea level along the profile in Fig. 7.18, during a period of time of 1.5 My (data courtesy of Henry Posamentier). Changes in sea level, subsidence, relative sea level, and cumulative relative sea level are shown for 100,000 yrs timesteps. The curve of sea-level changes is the same for all three locations in Fig. 7.18. Subsidence rates increase basinward, and are considered to be stable during the 1.5 My time interval: $0\,m/10^5$ yrs for location A, $5\,m/10^5$ yrs for location B, and $10\,m/10^5$ yrs for location C. Eustasy and subsidence control the change in relative sea level (Δ RSL) for each timestep. The cumulative relative sea-level change (Σ RSL) is calculated in the last column of the table. Key: * (x 10^5 yrs), # $m/10^5$ yrs, + m.

	Time*	Δ Eustasy#	− Δ Subsidence#	= Δ RSL#	Σ RSL+
Location A	0 to 1	30	0	30	30
	1 to 2	20	0	20	50
	2 to 3	10	0	10	60
	3 to 4	0	0	0	60
	4 to 5	−20	0	−20	40
	5 to 6	−30	0	−30	10
	6 to 7	−20	0	−20	−10
	7 to 8	0	0	0	−10
	8 to 9	10	0	10	0
	9 to 10	20	0	20	20
	10 to 11	30	0	30	50
	11 to 12	20	0	20	70
	12 to 13	10	0	10	80
	13 to 14	0	0	0	80
	14 to 15	−5	0	−5	75
Location B	0 to 1	30	−5	35	35
	1 to 2	20	−5	25	60
	2 to 3	10	−5	15	75
	3 to 4	0	−5	5	80
	4 to 5	−20	−5	−15	65
	5 to 6	−30	−5	−25	40
	6 to 7	−20	−5	−15	25
	7 to 8	0	−5	5	30
	8 to 9	10	−5	15	45
	9 to 10	20	−5	25	70
	10 to 11	30	−5	35	105
	11 to 12	20	−5	25	130
	12 to 13	10	−5	15	145
	13 to 14	0	−5	5	150
	14 to 15	−5	−5	0	150
Location C	0 to 1	30	−10	40	40
	1 to 2	20	−10	30	70
	2 to 3	10	−10	20	90
	3 to 4	0	−10	10	100
	4 to 5	−20	−10	−10	90
	5 to 6	−30	−10	−20	70
	6 to 7	−20	−10	−10	60
	7 to 8	0	−10	10	70
	8 to 9	10	−10	20	90
	9 to 10	20	−10	30	120
	10 to 11	30	−10	40	160
	11 to 12	20	−10	30	190
	12 to 13	10	−10	20	210
	13 to 14	0	−10	10	220
	14 to 15	−5	−10	5	225

experience different histories of relative sea-level changes, which vary in terms of magnitudes and timing of lowstands and highstands (Fig. 7.20). None of these location-specific curves of relative sea-level changes is representative for the stratigraphic cyclicity that develops within the basin, and so none can serve as a reference for the timing of stratigraphic surfaces.

The formation of sequence stratigraphic surfaces in downstream-controlled settings is timed by changes in the type of shoreline trajectory (Fig. 4.32). Therefore, the reference curve of relative sea-level changes (Figs. 1.10, 5.92, and 6.7) describes changes in accommodation *at the shoreline*, wherever the transient shoreline is located within the basin at any given time. The interplay of accommodation and sedimentation at the shoreline controls the timing of the four events that mark changes in stratal stacking pattern and the formation of systems tract boundaries, at any specific site along the shoreline:

the onset of forced regression, the end of forced regression, the end of lowstand normal regression, and the end of transgression (Figs. 1.10 and 6.7). Away from the shoreline, changes in relative sea level can differ significantly from those recorded at the shoreline; e.g., a forced regression may be coeval with relative sea-level rise offshore, due to variations in the subsidence rates, which is the scenario envisaged by Vail et al. (1984) for the formation of "type 2" sequence boundaries (see Chapter 5 for details). Subsidence also varies in a strike direction, modifying the magnitudes and timing of relative sea-level lowstands and highstands along the shoreline. Therefore, each dip-oriented transect has its own reference curve that describes the timing of stratigraphic surfaces.

A simple numerical simulation demonstrates the difference between the location-specific relative sea-level curves (Fig. 7.20) and the reference curve for one dip-

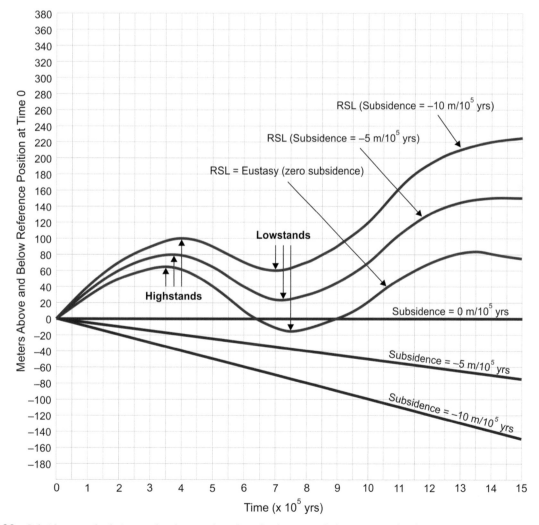

FIGURE 7.20 Subsidence and relative sea-level curves based on the data provided in Fig. 7.19, for the 1.5 My time interval (data courtesy of Henry Posamentier). For location A (Figs. 7.18 and 7.19), where subsidence is zero, the sea level curve coincides with the relative sea-level curve. For locations B and C (Figs. 7.18 and 7.19), the relative sea-level curves account for the combined effects of eustasy and subsidence.

oriented transect (Figs. 7.18 and 7.19). Starting from the initial profile in Fig. 7.18, and based on the sea level and subsidence data in Fig. 7.19, the stratigraphic architecture is built for 15 timesteps, as illustrated in Figs. 7.21–7.24. For simplicity, sediment supply is maintained constant by assigning the same volume of sediment to all 15 depositional wedges. The final stratigraphic architecture and facies relationships are shown in Fig. 7.25. Changes in accommodation at the shoreline during the 15 timesteps were calculated by interpolation between profiles A and B, or B and C, depending on the location of the shoreline at each timestep. The types of shoreline trajectories are indicated on the cross sections, and the differential subsidence was taken into account by tilting the profile accordingly, from one timestep to the next. The resultant reference curve of relative sea-level changes is shown in Fig. 7.26. The interplay between sedimentation and *this* curve of relative sea-level changes controls the

timing of all systems tracts and bounding surfaces along the dip-oriented transect in Fig. 7.25.

7.3.2 Stratal stacking patterns vs. bathymetric trends

Sequence stratigraphic surfaces in marine (or lacustrine) settings are paleoseafloors which correspond to specific events or stages during the development of sequences (Figs. 5.92 and 6.7). The identification of these surfaces in the sedimentary record is based on physical criteria; i.e., the observation of stratal stacking patterns, whose timing is linked to the evolution of the shoreline (Figs. 1.10, 4.32, 5.92, 6.7, and 7.26). For example, a maximum flooding surface is placed at the limit between retrograding strata below and prograding strata above, which marks the top of a fining-upward trend in shallow-water systems (Fig. 6.6). The application of such physical criteria does not require knowledge of

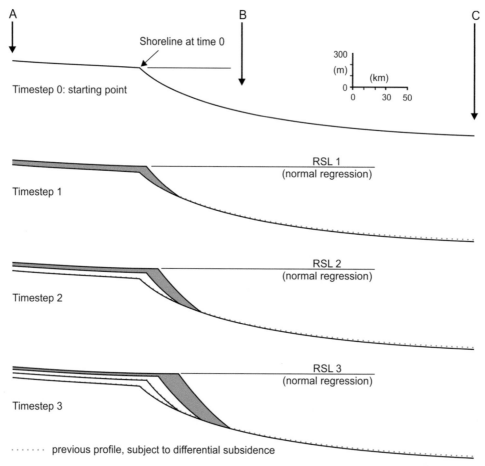

FIGURE 7.21 Forward modeling (timesteps 1–3) based on the data provided in Figs. 7.18–7.20 (see text for details). The amount of accommodation generated at the shoreline decreases with time, triggering an increase in the rates of progradation. Accommodation is calculated by interpolation between the relative sea-level curves at locations A and B (Fig. 7.20), according to the position of the shoreline during each timestep. Each cross section is tilted relative to the previous one according to the rates of differential subsidence. All facies prograde in relation to the basinward shift of the shoreline.

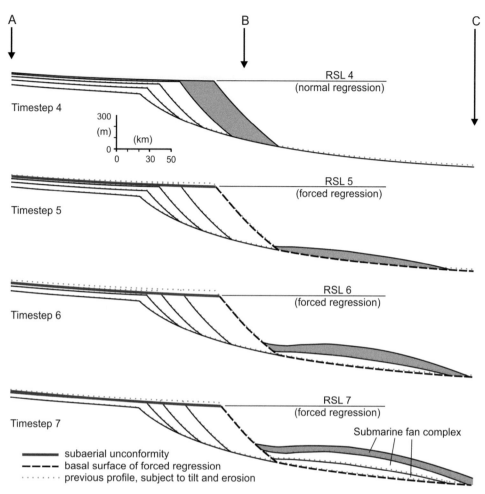

FIGURE 7.22 Forward modeling (timesteps 4–7) based on the data provided in Figs. 7.18–7.20 (see text for details). The top of the normal regressive deposits is subject to erosion during the forced regression of the shoreline, resulting in the formation of the subaerial unconformity. The amount of erosion at the shoreline is calculated by interpolation between the relative sea-level curves at locations A and B (Fig. 7.20), according to the position of the shoreline during each timestep. Each cross section is tilted relative to the previous one according to the rates of differential subsidence. No offlapping geometries are preserved in areas affected by erosion.

the stratigraphic age of the succession or of the water depth at the time of sedimentation. A common but misguided practice is to substitute physical criteria with bathymetric criteria; e.g., by defining the maximum flooding surface as the peak of deepest water instead of the top of a fining-upward trend. While interchangeable in areas above the fairweather wave base, the physical and bathymetric criteria lead to different stratigraphic levels offshore.

The offset between physical and bathymetric criteria is due to the fact that the former reflect the interplay of accommodation and sedimentation *at the shoreline*, whereas the latter account for the rates of accommodation and sedimentation at each individual location. As accommodation and sedimentation vary along dip and strike directions, the bathymetric trends of water deepening and shallowing are also location specific, and so they may not correlate with the events recorded at the

shoreline (Fig. 6.7). As such, the finest sediment does not necessarily accumulate at the time of deepest water, and similarly, the coarsest sediment does not necessarily correspond to the peak of shallowest water. This is most evident in siliciclastic settings, where sediment grading is linked to the transgressive and regressive shifts of the shoreline (i.e., fining- vs. coarsening-upward trends, respectively; Fig. 5.92). To demonstrate these points, the model in Figs. 7.18–7.26 was used to calculate changes in bathymetry at two locations during the development of the stratigraphic architecture (Fig. 7.27). At each location and timestep, water depth is quantified by the vertical distance between the sea level and the seafloor.

The model results in Fig. 7.27 indicate that water-depth changes at different locations may not correlate to each other or to the timing of shoreline transgressions and regressions. Under the assumption that subsidence

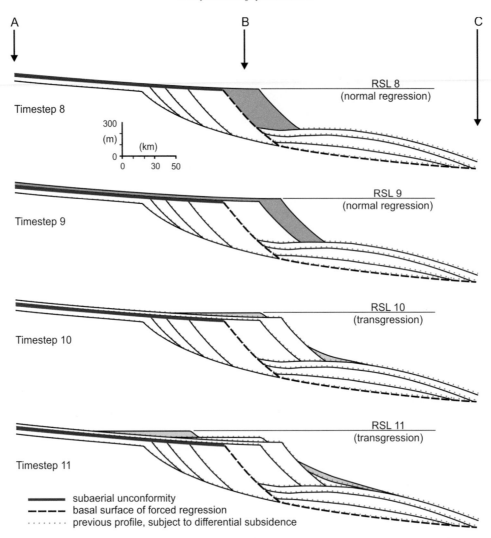

FIGURE 7.23   Forward modeling (timesteps 8–11) based on the data provided in Figs. 7.18–7.20 (see text for details). The change from normal regression (timesteps 8 and 9) to transgression (timesteps 10 and 11) depends on the interplay of sedimentation and accommodation at the shoreline. Since sediment supply is not quantified in this exercise, the transgression was arbitrarily selected to start with timestep 10, because of the accelerated rates of relative sea-level rise in Fig. 7.20. Accommodation at the shoreline is calculated by interpolation between the relative sea-level curves at locations A, B, and C (Fig. 7.20), according to the position of the shoreline during each timestep. Each cross section is tilted relative to the previous one according to the rates of differential subsidence. The rates of progradation and retrogradation change with time, and reflect the interplay of accommodation and sedimentation at the shoreline.

rates increase, and sedimentation rates decrease, in a basinward direction (Figs. 7.18, 7.19, and 7.20), the transgression of the shoreline is coeval with water deepening in the basin, whereas regressions may occur at the same time with water shallowing or deepening offshore (Fig. 7.27). This conclusion is supported by field studies which show that the maximum water depth, as inferred from benthic foraminifera, often occurs within highstand systems tracts (Naish and Kamp, 1997; T. Naish, pers. comm., 1998). Similar conclusions were reached by Vecsei and Duringer (2003), who demonstrated that the "maximum depth interval" within the Middle Triassic marine succession of the Germanic Basin is younger than the maximum flooding surface, and

occurs within the highstand systems tract (Fig. 7.28). In this study, the difference in age between the maximum flooding surface and the maximum water-depth interval is attributed to variations in sedimentation rates between the basin margin and the basin center, which is in agreement with the predictions of numerical models (e.g., Catuneanu et al., 1998a).

The integration of sedimentological and micropaleontological data in high-resolution studies provides a powerful tool to constrain the position of both sequence stratigraphic and bathymetric surfaces in the sedimentary record (Zecchin et al., 2021). Foraminiferal data help to discriminate stratal stacking patterns from bathymetric trends, and to clarify the relationship

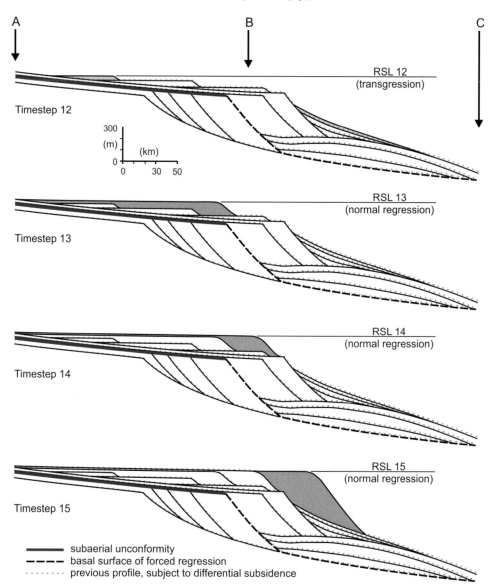

FIGURE 7.24 Forward modeling (timesteps 12—15) based on the data provided in Figs. 7.18—7.20 (see text for details). The change from transgression (timestep 12) to normal regression (timesteps 13—15) depends on the interplay of sedimentation and accommodation at the shoreline. Since sediment supply is not quantified in this exercise, the normal regression was arbitrarily selected to start with timestep 13, because of the lower rates of relative sea-level rise in Fig. 7.20. Accommodation at the shoreline is calculated by interpolation between the relative sea-level curves at locations A, B, and C (Fig. 7.20), according to the position of the shoreline during each timestep. Each cross section is tilted relative to the previous one according to the rates of differential subsidence. The rates of progradation and retrogradation change with time, and reflect the interplay of accommodation and sedimentation at the shoreline.

between the physical stratigraphic architecture and the environmental conditions at syn-depositional time. Parameters such as the % fragmentation of benthic foraminifera and the ratio between distal and proximal benthic species are indicators of shoreline shifts and constrain the position of sequence stratigraphic surfaces, including those that are lithologically cryptic (e.g., maximum flooding surfaces within condensed sections). In contrast, the ratio between planktonic and benthic foraminifera is indicative of water-mass variations, and therefore can be used as a proxy for bathymetric

changes (Zecchin et al., 2021, Figs. 2.28 and 2.29). In the case of the Plio-Pleistocene succession of the Crotone Basin, Italy, the surface of maximum water depth is placed above the stratigraphic level of the maximum flooding surface, within the highstand systems tract, and the offset between the two surfaces increases in a basinward direction (Zecchin et al., 2021, Figs. 2.28 and 2.29).

The model assumptions in Figs. 7.18—7.27 typify extensional settings with depocenters located away from the basin margins. Other settings display different

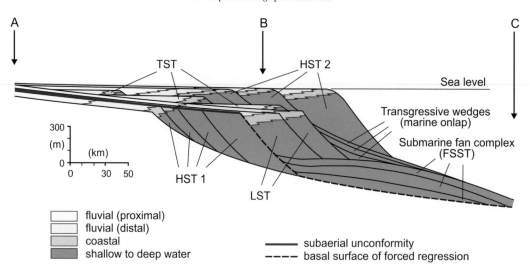

FIGURE 7.25 Stratigraphic architecture of the succession modeled in Figs. 7.21–7.24 (data courtesy of Henry Posamentier). Abbreviations: LST—lowstand systems tract; TST—transgressive systems tract; HST—highstand systems tract; FSST—falling-stage systems tract.

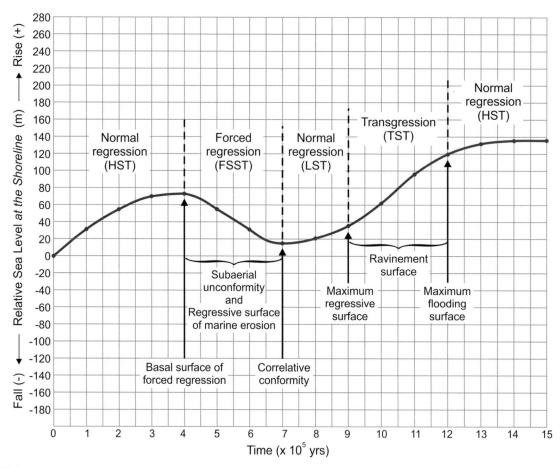

FIGURE 7.26 Reference curve of relative sea-level changes for the stratigraphic succession in Fig. 7.25. This curve indicates changes in accommodation *at the shoreline*, wherever the shoreline is located at any given time. It is the interplay of this reference curve and sedimentation that controls the timing of all systems tracts and bounding surfaces.

accommodation and sedimentation patterns, which also lead to a lack of correlation between shoreline trajectories and the offshore bathymetric trends. For example, subsidence and sedimentation rates in foreland basins decrease in a downdip direction, in which case the deposition of fining-upward transgressive deposits may occur in deepening water in areas adjacent to the shoreline but under shallowing water conditions in distal

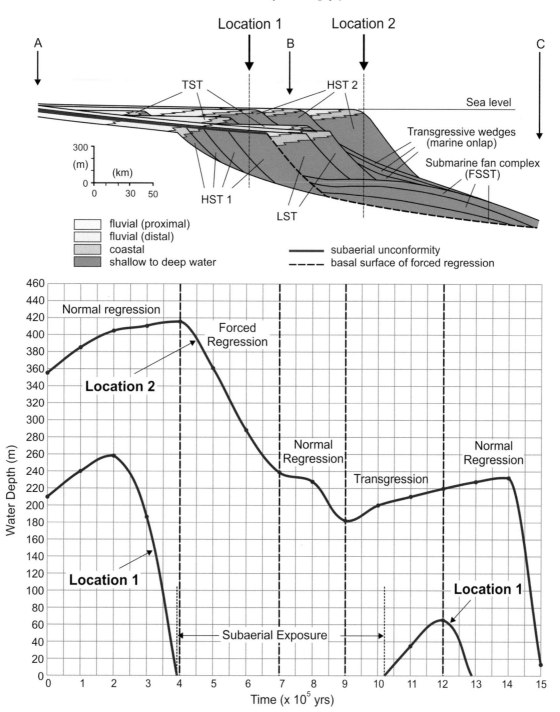

FIGURE 7.27 Water-depth changes at locations 1 and 2 on the cross section (see text for details). Abbreviations: LST—lowstand systems tract; TST—transgressive systems tract; HST—highstand systems tract; FSST—falling-stage systems tract.

areas (Catuneanu, 2004b, 2019c). Carbonate platforms on continental shelves may also record bathymetric "anomalies," such as the growth of distal barrier reefs in balance with the rise in relative sea level during slow transgressions (i.e., no change in water depth during a local "keep-up" phase), although the shoreline is transgressive and the rest of the shelf may experience water deepening (details in Chapter 8). These examples show that stacking patterns and water-depth changes are not always in a linear relationship, as commonly assumed, and that deviations from the "norm" can occur in any geological setting.

The assumed relationship between bathymetric and grading trends is modified by changes in sediment supply and environmental energy, which in turn depend on multiple factors including shoreline shifts, changes in

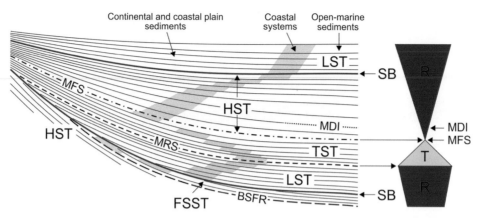

FIGURE 7.28 Sequence stratigraphic architecture in an intracratonic low-gradient ramp setting (modified after Vecsei and Duringer, 2003). The maximum water depth (MDI) occurs within the highstand systems tract, above the maximum flooding surface, due to differences in the rates of subsidence and sedimentation between the basin margin and the depocenter (see text for details). Abbreviations: HST—highstand systems tract; FSST—falling-stage systems tract; LST—lowstand systems tract; TST—transgressive systems tract; SB—depositional sequence boundary; BSFR—basal surface of forced regression; MRS—maximum regressive surface; MFS—maximum flooding surface; MDI—maximum depth interval; R—regression and coarsening upward; T—transgression and fining upward.

FIGURE 7.29 Progradation of a coarsening-upward succession into a basin that is subject to differential subsidence. At any given time, the threshold "T" is placed where subsidence and sedimentation are in balance. Landward from T, sedimentation outpaces subsidence and the water is shallowing. Seaward from T, accommodation is generated faster than it is consumed by sedimentation, and therefore the water is deepening. The succession that accumulates in the deepening water is still coarsening-upward, reflecting the progradation of the shoreline. This is the case of the Mahakam delta in Indonesia (sedimentation > subsidence nearshore, and subsidence > sedimentation offshore).

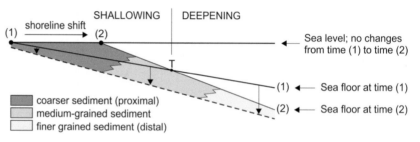

seafloor gradients, and the mechanisms of sediment transport (e.g., traction currents or gravity-driven processes). For example, deeper water is generally correlated with low depositional energy and fine-grained sediments. However, depending on the seafloor gradients and sediment supply, gravity flows may distort this relationship, leading to increases in depositional energy and grain size in a manner that is independent of water depth. Differential subsidence in tectonically active basins may lead to the progradation of coarsening-upward clinoforms in deepening water (Fig. 7.29). In this example, the inverse grading is the result of shoreline progradation, while the water deepening reflects the change in the balance between the rates of subsidence and sedimentation in a down-dip direction (Fig. 7.29). In this case, the seafloor steepens with time, and the contribution of gravity flows to the transport of sediment becomes increasingly important, regardless of the bathymetric trends (Figs. 7.30 and 7.31). The water deepening at the toe of the clinoforms (e.g., in a lower delta front to prodelta

FIGURE 7.30 Sandy turbidites (divisions A and B of the Bouma sequence) accumulated in a lower delta front setting (Upper Cretaceous, river-dominated Ferron delta, Utah). Conventional views account for a linear relationship between water-depth, depositional energy, and grain size. Gravity flows alter this relationship, and enable the progradation of coarser sediments into deeper (or deepening) water.

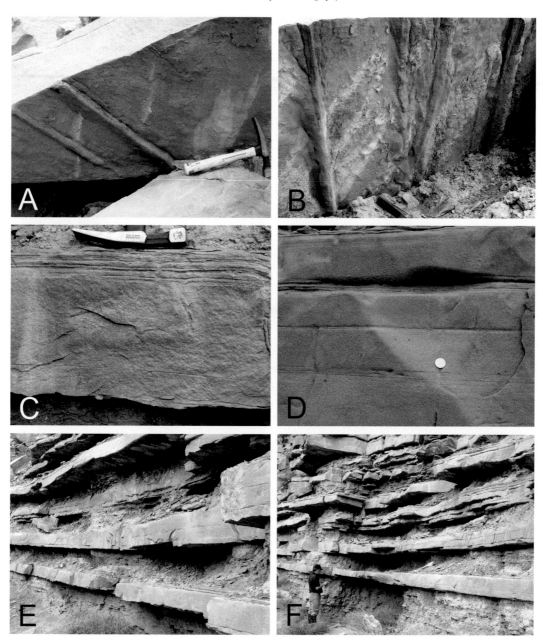

FIGURE 7.31   Sandy turbidites accumulated in the lower delta front of a river-dominated delta (Upper Cretaceous, Panther Tongue, Gentle Wash Canyon, Utah). With deposition dominated by gravity flows, no direct relationship may be established between water depth and grain size, or between water-depth changes and grading trends. Instead, grain size correlates with depositional energy, which if often independent of bathymetry. A, B—sole marks at the base of turbidite rhythms; C, D—sandy turbidites, dominated by the divisions A and B of the Bouma sequence; E, F—turbidite cycles in the distal portion of the deltaic system.

area) does not prevent the progradation of coarser sediment, because the depositional energy is increasing in response to the change in sediment supply and slope gradients.

Similar coarsening-upward trends in deepening water may also be generated without changes in seafloor gradients, with accommodation created exclusively by sea level rise (Fig. 7.32). In this case, the progradation of coarsening-upward clinoforms in deepening water

is not accompanied by an increase in the participation of gravity flows with time, but it only reflects the increase in sediment supply as a result of shoreline regression (Fig. 7.32). However, gravity flows may still facilitate the sediment transport to the toe of the clinoforms (Figs. 7.30 and 7.31). In the absence of differential subsidence, the progradation of a coarsening-upward succession in deepening water requires the rates of topset aggradation to exceed the rates of aggradation of the

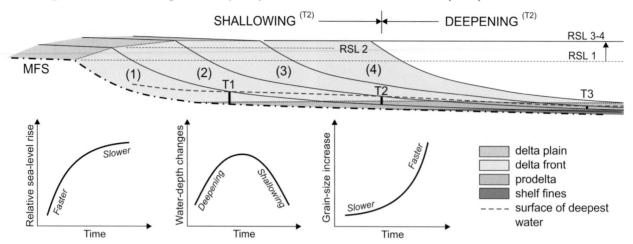

1. **Highstand normal regression** (HST): the rates of relative sea-level (RSL) rise decrease with time

- Time 4 (highstand): shallowing along the entire profile, in the absence of differential subsidence
- Most of the maximum flooding surface (MFS) forms in deepening water
- The surface corresponding to the peak of deepest water is diachronous and forms *within* the HST

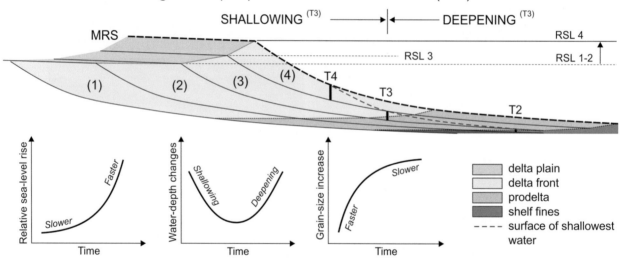

2. **Lowstand normal regression** (LST): the rates of relative sea-level (RSL) rise increase with time

- Time 1 (lowstand): shallowing along the entire profile, in the absence of differential subsidence
- Most of the maximum regressive surface (MRS) forms in deepening water
- The surface corresponding to the peak of shallowest water is diachronous and forms *within* the LST

FIGURE 7.32 Relative sea level, water depth, and grading trends during highstand and lowstand normal regressions, in the absence of differential subsidence. Similar deltas, with the lower delta front prograding into deepening water, were presented by Berg (1982) and Bhattacharya and Walker (1992), accounting for deltaic bottomsets thinner than the topsets. The threshold "T" marks the point of balance between accommodation and sedimentation at each timestep, where no change in water depth is recorded. The higher the rates of topset aggradation, the closer the threshold T is to the shoreline. The maximum flooding and maximum regressive surfaces form in deepening water, and connect with the surfaces of deepest and shallowest water at the shoreline. The inverse grading of the clinoforms reflects the increase in sediment supply associated with the progradation of the shoreline, irrespective of the bathymetric trends at the toe of the clinoforms. The identification of sequence stratigraphic surfaces relies on physical rather than bathymetric criteria (see text for details).

bottomset (Fig. 7.32). This condition is not necessary in basins that are subject to differential subsidence, where progradation in deepening water may occur even without coastal aggradation (Fig. 7.29). The two scenarios of clinoform progradation under conditions of

water deepening in the bottomset area are summarized in Fig. 7.33. It can be concluded that water depth is not a control on grain size and depositional energy, but merely a consequence of the interplay of accommodation and sedimentation at every location.

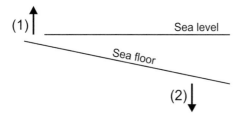

FIGURE 7.33 Controls on water deepening during the progradation of coarsening-upward clinoforms. These controls may operate independently or in conjunction: (1)—coastal aggradation, at rates higher than the rates of aggradation of the bottomset (i.e., topset > bottomset); (2)—differential subsidence, with rates increasing basinwards (i.e., accommodation > sedimentation in the bottomset area). The higher the rates of aggradation at the shoreline, or the more pronounced the pattern of differential subsidence, the more evident the offset between grading and bathymetric trends (i.e., the point of balance between accommodation and sedimentation—threshold "T" in Fig. 7.32—is closer to the shoreline).

The offset between the timing of grading vs. bathymetric trends typically increases basinwards, as the rates of accommodation and sedimentation become increasingly different from those recorded at the shoreline (Fig. 7.34). The physical criteria that afford the identification of sequence stratigraphic surfaces take precedence over the bathymetric trends which can be measured independently with fossil data. Variations in subsidence rates also modify the timing and the rates of relative sea-level changes across a sedimentary basin (Fig. 7.20); however, these variations do not affect the timing of systems tract boundaries. For example, the two types of correlative conformities in Fig. 1.10 mark reversals in the direction of relative sea-level change *at the shoreline*, but may form during rising relative sea level offshore (Fig. 7.34). Similarly, the maximum regressive and maximum flooding surfaces mark changes in bathymetric trends *near the shoreline*, but may form under deepening-water conditions offshore (Fig. 7.34). Surfaces of shallowest and deepest water are lithologically undeterminable within prograding systems, but may still be identified based on paleobathymetric data derived from body fossils (e.g., benthic foraminifera) and trace fossils (e.g., Pekar and Kominz, 2001). Such surfaces meet the systems tract boundaries at the shoreline, but diverge from them in an offshore direction (Fig. 7.34).

7.3.3 Diachroneity of stratigraphic surfaces

All sequence stratigraphic surfaces are potentially diachronous along dip and/or strike directions (Fig. 6.6). In downstream-controlled settings, the four event-related surfaces (Figs. 1.10 and 6.7) may be conformable, whereas the three stage-related surfaces (i.e., the subaerial unconformity, the regressive surface of marine erosion, and the transgressive surface of erosion; Fig. 6.7) are invariably unconformable (see Chapter 6 for details). The event-related surfaces represent the landscape and seascape profiles at specific moments in time (i.e., "events") when changes from one

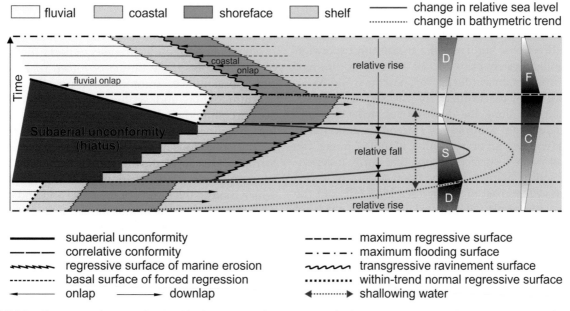

FIGURE 7.34 Changes in relative sea level and bathymetric conditions where subsidence rates increase, and sedimentation rates decrease, in a basinward direction (see Catuneanu et al., 1998a; for full numerical calculations). In such basins, forced regressions (i.e., relative sea-level fall at the shoreline) are coeval with relative sea-level rise offshore; water shallowing in the shoreface is coeval with water deepening offshore; and maximum regressive and maximum flooding surfaces form in deepening water. Abbreviations: S—shallowing; D—deepening; C—coarsening-upward; F—fining-upward.

type of shoreline trajectory to another occur in coastal environments. These events lead to changes in stratal stacking patterns, which define systems tract boundaries. The event-related surfaces are isochronous in a dip direction, as each dip-oriented transect intersects the shoreline at one location with a unique timing of the four events; e.g., along each dip-oriented transect there is only one moment in time when the shoreline reaches its landwardmost position, which marks the timing of the maximum flooding surface (Figs. 4.32 and 7.34). However, the timing of the four events and of their related surfaces changes in a strike direction due to variations in the rates of accommodation and sedimentation along the shoreline (e.g., Catuneanu et al., 1998a).

It follows that the event-related surfaces are isochronous along dip but diachronous along strike, with a timing linked to the evolution of the shoreline. Variations in the rates of accommodation and sedimentation away from the shoreline are only important in controlling the syn-depositional bathymetric trends and the thickness of systems tracts, but without influencing the timing of their boundaries (Fig. 7.34). In contrast to the event-related surfaces, the stage-related unconformities are diachronous along both dip and strike directions (Figs. 5.92 and 7.34). The along-dip diachroneity of unconformities in downstream-controlled settings is due to their formation in relation to shoreline shifts, during forced regressions and transgressions (Figs. 4.39 and 7.34). As the shoreline takes time to regress or transgress, these unconformities are always diachronous along dip, with rates inversely proportional to the rates of shoreline shift. In upstream-controlled settings, the top-amalgamation surface is also potentially diachronous, as topographic gradients, fluvial styles, and the degrees of channel amalgamation may change both along strike and dip directions. Similarly, the maximum expansion level in eolian systems can also be attained at different times in different areas (details in Chapters 6 and 8).

The degree of diachroneity of any sequence stratigraphic surface along the depositional strike depends on its underlying controls (i.e., accommodation and sedimentation vs. accommodation alone). Among the four event-related surfaces, those that form independently of sedimentation (i.e., the basal surface of forced regression at the onset of relative sea-level fall, and the correlative conformity at the end of relative sea-level fall) are typically closer to time lines. In contrast, maximum regressive and maximum flooding surfaces are more diachronous, as their timing depends both on accommodation and sedimentation. Similarly, the degree of diachroneity of the three stage-related unconformities also varies along strike, with those that form during forced regression (and hence independently of sedimentation: i.e., the regressive surface of marine erosion, and some subaerial unconformities;

Fig. 4.39A) recording lower degrees of diachroneity than those that form during transgression (and hence in part dependent on sedimentation: i.e., the transgressive surface of erosion, and some subaerial unconformities; Fig. 4.39B). The following is a brief summary of the time attributes of stratigraphic surfaces in downstream-controlled settings (Figs. 6.6, 7.35).

7.3.3.1 Subaerial unconformities

Subaerial unconformities form most commonly during forced regressions (Figs. 4.6, 4.36, 4.39A, 5.92, and 6.7), but, under particular circumstances, they may also form during transgressions (Figs. 4.28, 4.39B; see Chapter 4 for details). All subaerial unconformities are diachronous along dip and strike directions (Figs. 6.6 and 7.35). Along dip, the degree of diachroneity is inversely proportional to the rate of shoreline shift (Fig. 7.35). Additional time-transgressive processes that operate in a dip direction and contribute to the along-dip diachroneity are the upstream migration of fluvial knickpoints following the erosion at the river mouth (Figs. 4.42 and 4.43), and the pattern of fluvial onlap following the resumption of fluvial aggradation (Fig. 7.34). The former refers to the fact that fluvial incision may continue to propagate upstream even after the cessation of erosion at the river mouth, although the migration rates of fluvial knickpoints are generally high enough to produce nearly isochronous unconformities at geological timescales (Posamentier, 2001). Fluvial onlap, on the other hand, widens the hiatus associated with the subaerial unconformity in an updip direction, as the area of fluvial sedimentation expands with time toward the basin margin, and can generate a much more significant time difference along the unconformity (Fig. 7.34; details in Chapter 5).

The along-strike diachroneity reflects the variability of the controls on the timing of forced regressions and transgressions (i.e., accommodation vs. the interplay of accommodation and sedimentation, respectively). As both accommodation and sedimentation can vary along the shoreline, all subaerial unconformities are potentially diachronous in a strike direction. Subaerial unconformities that form during forced regressions are only susceptible to strike variations in accommodation, which modify the timing of the onset and end of relative sea-level fall (i.e., the age of the basal surface of forced regression and of the correlative conformity; Figs. 4.32 and 6.7) along the shoreline. As a result, the stratigraphic hiatuses of these unconformities can vary in terms of timing and duration in a strike direction. Subaerial unconformities that form during transgressions are susceptible to variations in both accommodation and sedimentation along the shoreline, and therefore record higher degrees of diachroneity in a strike direction. The timing and duration of the stratigraphic hiatuses of

Stratigraphic surface			Diachroneity along dip	Diachroneity along strike
Sequence stratigraphic surface	Event-significant	Correlative conformity	None	Variations in the rates of accommodation
		Basal surface of forced regression		
		Maximum regressive surface	None	Variations in the rates of sedimentation and accommodation
		Maximum flooding surface		
	Stage-significant	Subaerial unconformity	Inversely proportional to the rates of shoreline shift	Variations in the rates of sedimentation and/or accommodation
		Transgressive ravinement surfaces		Variations in the rates of sedimentation and accommodation
		Regressive surface of marine erosion		Variations in the rates of accommodation
Within-trend facies contact		Within-trend normal regressive surface	Inversely proportional to the rates of shoreline shift	Variations in the rates of sedimentation and accommodation
		Within-trend flooding surface		
		Within-trend forced regressive surface		Variations in the rates of accommodation

FIGURE 7.35   Time attributes of stratigraphic surfaces in downstream-controlled settings (see text for details).

these unconformities vary with the age of the onset and the end of transgressions (i.e., the age of the maximum regressive and maximum flooding surfaces; Figs. 4.32 and 6.7) along the shoreline.

The subaerial unconformity was traditionally perceived as a "time barrier" (Winter and Brink, 1991; Embry, 2001) based on the assumption that timelines do not cross this surface; i.e., all strata below the unconformity are older than the strata above it. However, the formation of diachronous unconformities which cross timelines is also possible, in relation to the migration of uplifted areas (e.g., Cohen, 1982; Johnson, 1991; Crampton and Allen, 1995; Catuneanu, 2004b) as well as in relation to shoreline shifts. The latter scenario was documented in the case of the Brazos River of the Texas Gulf Coast (Sylvia and Galloway, 2001; Galloway and Sylvia, 2002), where the diachronous development of the subaerial unconformity during the late Pleistocene glaciation was accompanied by coeval deposition below and above the unconformity (Fig. 7.36). As the shoreline shifted basinwards during forced regression, expanding the areal extent of the subaerial unconformity, the updip limit of the downstream-controlled area shifted accordingly, allowing the deposition of upstream-controlled fluvial sediments on top of the

older portion of the unconformity. Those fluvial sediments above the unconformity (timesteps 1, 2, and 3 in Fig. 7.36) are contemporaneous with the forced regressive shallow-water deposits that are truncated by the same physical surface (Fig. 7.36). Therefore, timelines cross the subaerial unconformity, as the early fluvial sediments above are older than the late forced regressive deposits below.

7.3.3.2 Correlative conformities

The basal surface of forced regression and the correlative conformity (both labeled as "correlative conformities" in light of one model or another; Fig. 1.10), share similar temporal attributes along dip and strike directions (Fig. 7.35). The timing of these surfaces is marked by the onset and the end of relative sea-level fall *at the shoreline*, respectively (Figs. 1.10, 4.32, and 6.7). As the age of these events is unique for each dip-oriented transect (Fig. 7.26), the two surfaces are timelines in a dip direction. However, since subsidence rates may vary along the shoreline, the timing of the two events of the relative sea-level cycle may also change in a strike direction. The degree of this strike diachroneity depends on the type of sedimentary basin and its subsidence mechanisms; it is considered "low" in the

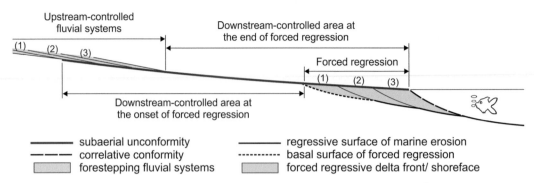

FIGURE 7.36   Diachronous development of a subaerial unconformity during relative sea-level fall (modified after Sylvia and Galloway, 2001). As the area of influence of relative sea-level change follows the regression of the shoreline, the fluvial strata that accumulate on top of the proximal portion of the unconformity are age-equivalent to the shoreface deposits truncated by the distal portion of the same unconformity. Therefore, the unconformity is not a "time barrier," as it crosses timelines while it youngs in a basinward direction.

case of tectonically "passive" basins (e.g., divergent continental margins or intracratonic basins), but it may become significantly higher in basins that are more active tectonically (Posamentier and Allen, 1999). The timing of the two "correlative conformities" is independent of sedimentation, as well as of the relative sea-level changes away from the shoreline (Fig. 7.34).

7.3.3.3 Maximum regressive and maximum flooding surfaces

The maximum regressive and maximum flooding surfaces are also similar in terms of their underlying controls and temporal attributes along dip and strike directions (Fig. 7.35). The timing of these surfaces depends on the interplay of sedimentation and accommodation *at the shoreline* (Fig. 4.32), and it is not affected by the variations in the rates of sedimentation and accommodation offshore (Fig. 7.34). At any location along the shoreline, the age of maximum regressive and maximum flooding surfaces is marked by the onset and the end of transgressions, respectively (Figs. 1.10, 4.32, and 6.7). As the timing of these events is unique for each dip-oriented transect (Fig. 7.26), the two surfaces are timelines in a dip direction; i.e., there is only one point in time when the shoreline changes its direction of shift along a dip-oriented transect. However, a strike diachroneity is recorded in relation to the variations in accommodation and sedimentation along the shoreline. The dependence of maximum regressive and maximum flooding surfaces on sedimentation makes them more diachronous along strike than the two "correlative conformities," possibly within the range of biostratigraphic resolution (Gill and Cobban, 1973).

7.3.3.4 Transgressive ravinement surfaces

The transgressive ravinement surfaces (Figs. 6.6, 6.7) are unconformities with a diachronous character along both dip and strike directions (Fig. 7.35). These surfaces form during the entire stage of transgression, from the location of the shoreline at the onset of transgression to the location of the shoreline at the end of transgression; therefore, they make the physical connection between the maximum regressive and the maximum flooding surfaces across the transgressive systems tract (Figs. 4.39 and 6.41). The transgressive ravinement surfaces become invariably younger in an updip direction, with a degree of diachroneity that depends on the duration and the rates of transgression; e.g., prolonged and slow transgressions lead to higher degrees of diachroneity. Along strike, the timing of these surfaces is affected by variations in the rates of accommodation and sedimentation along the shoreline, which modify the age of the onset and of the end of transgression at any location. Therefore, the transgressive ravinement surfaces can cross timelines along both dip and strike directions, with the direction of the highest degree of diachroneity depending on the specific conditions of each sedimentary basin.

7.3.3.5 Regressive surfaces of marine erosion

The regressive surface of marine erosion is also a stage-related unconformity (Fig. 6.7) which is diachronous along both dip and strike directions (Fig. 7.35). This surface forms during the fall in relative sea level at the shoreline (Fig. 4.44), between the shoreline locations at the onset and at the end of forced regression; therefore, it makes the physical connection between the basal surface of forced regression and the correlative conformity across the falling-stage systems tract (Fig. 4.39). Owing to its development during forced regression, the regressive surface of marine erosion becomes invariably younger in a downdip direction, with a rate that is inversely proportional to the rate of shoreline shift. Along strike, the timing of the onset and of the end of forced regression is modified by variations in the rates of subsidence along the shoreline, which confer the regressive surface of marine erosion a diachronous character. The amount of strike diachroneity depends on the degree of differential subsidence, which is typically higher in tectonically active basins. However, as the timing of this unconformity is independent of sedimentation, its strike diachroneity is lower than that of other stage-related unconformities that develop within the same basin, such as the transgressive ravinement surfaces (Fig. 7.35).

7.3.3.6 Within-trend facies contacts

Within-trend facies contacts are not sequence stratigraphic surfaces, yet are important to describe facies relationships within systems tracts (details in Chapter 6). Each type of shoreline trajectory and associated systems tract affords the development of specific within-trend facies contacts, the most prominent of which include the within-trend normal regressive surface (formed during lowstand or highstand normal regressions; Figs. 4.49, 6.56, and 6.6), the within-trend forced regressive surface (formed during forced regressions in supply-dominated settings; Figs. 6.6 and 6.58), and the within-trend flooding surface (formed during transgressions; Figs. 5.100B, 5.104, 6.6, and 6.63). All these facies contacts are diachronous along both dip and strike directions (Fig. 7.35). Along dip, the rates of diachroneity are inversely proportional to the rates of shoreline shift, with the facies contacts that form during normal regressions being commonly the most diachronous, as normal regressions are typically slower than forced regressions and transgressions (details in Chapters 4 and 6). Along strike, the rates of diachroneity reflect the variability recorded by the controls on the different types of shoreline trajectories, namely accommodation and sedimentation in the case of normal regressions and transgressions, and accommodation in the case of forced regressions (Fig. 7.35; details in Chapter 4).

7.3.4 Three-dimensional stratigraphic architecture

The diachronous development of sequence stratigraphic surfaces implies that different types of systems tracts can accumulate at the same time within a sedimentary basin (Martinsen and Helland-Hansen, 1995; Catuneanu et al., 1998a; Madof et al., 2016; Schultz et al., 2020; Zecchin and Catuneanu, 2020, Fig. 7.37). The 3D variability of the stratigraphic architecture is the result of changes in the rates of accommodation and sedimentation along dip and strike directions, which affect the scales of stratal units and the timing of their bounding surfaces (Wehr, 1993; Martinsen and Helland-Hansen, 1995; Catuneanu et al., 1999, 2002; Helland-Hansen and Hampson, 2009). In particular, variations in accommodation and sedimentation along a shoreline are responsible for the coeval deposition of different "conventional" systems tracts in downstream-controlled settings (Posamentier and Allen, 1999; Catuneanu, 2006; Csato and Catuneanu, 2014; Catuneanu and Zecchin, 2013, 2016). In upstream-controlled settings, "unconventional" systems tracts and bounding surfaces may also be diachronous due to variations in the sedimentation conditions in all directions across the basin (details in Chapters 4, 5, and 6).

A common occurrence of stratigraphic variability is the coeval progradation and retrogradation along a shoreline, as documented both in the geological record (e.g., along the proximal shore of the Western Interior seaway; Gill and Cobban, 1973) and in the modern environment (Figs. 7.38 and 7.39). Changes in the type of shoreline trajectory along a coastline reflect strike variations in accommodation and sedimentation. Consequently, different systems tracts may form at the same time in juxtaposition, separated by structural elements (e.g., active faults that limit sub-basins with different

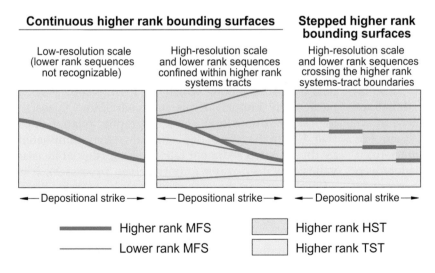

Continuous higher rank bounding surfaces

Low-resolution scale (lower rank sequences not recognizable)

High-resolution scale and lower rank sequences confined within higher rank systems tracts

Stepped higher rank bounding surfaces

High-resolution scale and lower rank sequences crossing the higher rank systems-tract boundaries

←— Depositional strike —→   ←— Depositional strike —→   ←— Depositional strike —→

— Higher rank MFS   ▢ Higher rank HST
— Lower rank MFS   ▢ Higher rank TST

FIGURE 7.37 Stratigraphic relationships between higher and lower rank systems-tract boundaries (from Zecchin and Catuneanu, 2020). For simplicity, only maximum flooding surfaces and transgressive and highstand systems tracts are shown. Seismic-scale systems tracts consist of sequences observed at higher resolution (outcrop or well-log) scales. Therefore, a "systems tract" observed at seismic scales becomes a "sequence set" or even a "composite sequence set" at sub-seismic scales. This nomenclatural inconsistency led to the demise of the scale-variant systems of stratigraphic classification. The definition of stratigraphic units is now independent of scale and data resolution. Abbreviations: HST—highstand systems tract; MFS—maximum flooding surface; TST—transgressive systems tract.

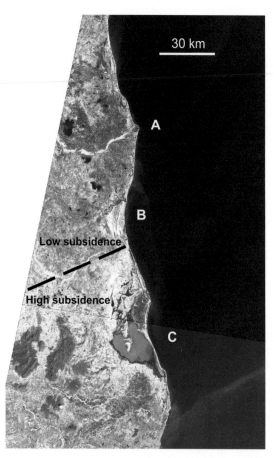

FIGURE 7.38 Active fault separating an area of low subsidence and coastal progradation (A—delta; B—prograding strandplain) from an area of high subsidence and coastal backstepping (C—transgressive lagoon-barrier island system) (Pennar Basin, eastern India).

subsidence rates; Fig. 7.38), sediment entry points (e.g., deltas; Fig. 7.39), or merely the points of balance between accommodation and sedimentation along the coastline. More evident in the latter case, the shift in the location of those points of balance along the coastline leads to the development of diachronous sequence stratigraphic surfaces, which may be expressed as single physical surfaces (Ito and O'Hara, 1994) or as composites of higher frequency surfaces (Plint and Nummedal, 2000; Magalhaes et al., 2015; Schultz et al., 2020; Zecchin and Catuneanu, 2020, Fig. 7.37).

At larger scales, contrasting trajectories may also be observed between the proximal and distal shorelines of an interior seaway, due to regional variations in accommodation and sedimentation across the basin. For example, retroarc foreland systems are characterized by out-of-phase flexural tectonics between the foredeep and the forebulge flexural provinces, which result in "reciprocal" stratigraphic frameworks across the flexural hingelines (e.g., Catuneanu et al., 1999). In such settings, tectonism exerts a fundamental control on the development of proximal vs. distal reciprocal

stratigraphies, which refer to out-of-phase depositional sequences and subaerial unconformities in overfilled forelands, and the coeval deposition of different conventional systems tracts along the dip direction of underfilled forelands (e.g., summaries in Catuneanu et al., 1999; Catuneanu, 2019c). Contrasting stratigraphic frameworks between different sub-basins of a sedimentary basin may also develop in any other tectonic setting, in relation to variations in accommodation and sedimentation conditions across sub-basin boundaries (e.g., Gawthorpe et al., 2003, who described coeval subsidence and uplift in rift basins).

The 3D variability of the sequence stratigraphic framework does not change the methodological workflow (Figs. 1.2 and 7.40), which typically starts with the analysis of 1D (e.g., outcrop or borehole) or 2D (e.g., seismic) data, nor does it change the principles of interpretation of 1D and 2D datasets. In fact, the criteria that lead to the identification of the elements of the sequence stratigraphic framework are best demonstrated with 1D and 2D datasets, which afford the observation of stratal stacking patterns that are diagnostic to the definition of systems tracts and bounding surfaces (e.g., Figs. 4.7 and 4.8). This is the reason why the sequence stratigraphic methodology is typically illustrated with 1D and 2D profiles (e.g., Fig. 4.39), which depict the stratigraphic architecture of particular areas and along specific directions within the 3D framework. Once sufficient 1D and 2D datasets have been collected, the full 3D architecture can be assembled by integrating the entire information available (e.g., the construction of fence diagrams from a network of cross sections oriented in different directions; Eriksson et al., 2001; Fanti and Catuneanu, 2009; Zubalich et al., 2021).

The strategy of selection of the most relevant 2D sections within a 3D dataset depends on the depositional setting under analysis. In the case of coastal to shallow-water systems, it is advisable to start the analysis with dip-oriented 2D profiles, which afford the observation of clinoforms, stratal terminations, and stacking patterns (Fig. 1.11), followed by the integration of the 2D profiles into the 3D stratigraphic architecture. The construction of the 3D stratigraphic framework also benefits from the integration of all other types of data available, among which the fossil record plays a particularly important role in constraining paleoecological conditions and age-stratigraphic relationships (Fig. 1.1). In the case of depositional systems dominated by dip-oriented sediment fairways (e.g., fluvial incised valleys, or deep-water gravity-driven processes), it is advisable to start the analysis with strike-oriented lines, which afford the best visualization of the sediment-transport systems (Fig. 5.25). Where 3D seismic volumes are available, the 2D transects are best complemented by the observation of depositional elements in

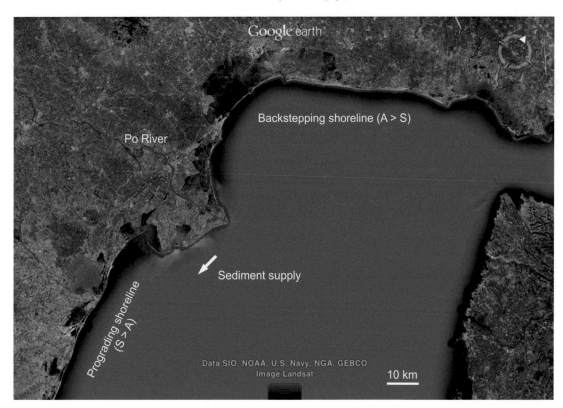

FIGURE 7.39 Coeval transgression and regression along the coastline of the Adriatic Sea (Italy), driven by differences in sediment supply. The sediment from the Po River is directed to the south by longshore currents, leading to coastal progradation (c. 5 km within the last 2,000 yrs), while sediment starvation to the north leads to coastal retrogradation. Abbreviations: S—sedimentation rate; A—accommodation rate (in this case, rate of relative sea-level rise).

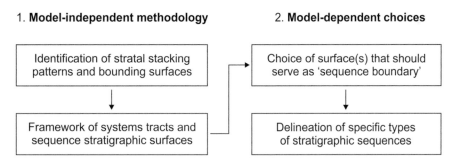

FIGURE 7.40 Workflow of sequence stratigraphy (modified after Catuneanu et al., 2011). Stratal stacking patterns provide the basis for the definition of all units and surfaces of sequence stratigraphy, at scales defined by the purpose of study or constrained by the resolution of the data available. Beyond the framework of systems tracts and bounding surfaces, the selection of the "sequence boundary" becomes a model-dependent choice, often guided by the mappability of the different types of sequence stratigraphic surfaces that are present within the study area.

plan view (i.e., seismic geomorphology on horizon slices; Posamentier, 2003, 2004; Davies et al., 2007; details in Chapter 2).

The 3D variability of the stratigraphic architecture can be observed at different scales. A sequence stratigraphic framework constructed at a specific scale of observation (i.e., as constrained by the scope of the study and the resolution of the data available) reflects the physical and temporal relationships of stratal units and bounding surfaces that develop at that particular scale. The degree of diachroneity of the stratigraphic framework may vary with the scale of observation (i.e., hierarchical level; Schultz et al., 2020), depending on the conditions that govern the development of stratal stacking patterns at each stratigraphic scale. Where the coeval systems tracts consist of higher frequency sequences, the systems-tract boundaries may be either stepped or continuous surfaces, depending on the areal development of the stratal units observed at different scales. If the high-frequency sequences cross the higher rank systems-tract boundaries, the higher rank surfaces display a stepped geometry along the depositional strike. In contrast, if the high-frequency sequences are confined to the areas of development of the higher rank systems tracts, the higher rank surfaces form as continuous physical contacts (Zecchin and Catuneanu, 2020, Fig. 7.37).

The 3D nature of the stratigraphic architecture demonstrates that sequence stratigraphic frameworks are basin-specific, reflecting the importance of local controls on accommodation and sedimentation. This reiterates the fact that the methodology needs to be decoupled from global standards, and be based on data acquired from within the basin under analysis. For example, the identification of lowstand and highstand systems tracts relies on local stratigraphic relationships (i.e., normal regressions that follow forced regressions or transgressions, respectively; details in Chapters 4 and 5), and not on correlations with the global sea level. The 3D basin-specific variability of the stratigraphic framework, which entails the coeval deposition of different types of systems tracts, explains why global cycle charts are no longer part of the sequence stratigraphic workflow and methodology. Beyond the immediate purpose of basin-specific applications, sequence stratigraphy still provides a benchmark to assess the extent of global controls on sedimentation, as global correlations can be tested following the construction of local stratigraphic frameworks.

Variability of stratigraphic sequences

The same principles underlie the application of sequence stratigraphy in all geological settings. The sequence stratigraphic methodology and nomenclature are independent of any local controls on the stratigraphic framework, such as the tectonic and depositional settings, or the climatic regimes at the time of sedimentation. Unless otherwise specified, the concepts presented in this book are by default illustrated with examples from siliciclastic settings. However, the geometry of sequences, as well as their internal systems-tract composition and sedimentological makeup are highly variable as a function of the syn-depositional conditions of accommodation and sedimentation (Fig. 4.33). Differences in the types of subsidence, depositional elements, and basin physiography modify the physical attributes of the stratigraphic framework, and require versatile sedimentological skills to identify sequence stratigraphic units and bounding surfaces in various settings. Beyond the 3D aspects of the stratigraphic architecture discussed in Chapter 7, this chapter describes the variability of stratigraphic sequences that form in distinct tectonic and depositional settings, and under different climatic conditions.

8.1 Variability with the tectonic setting

The stratigraphic variability introduced by the tectonic setting has become evident following the publication of numerous case studies in different types of sedimentary basins: divergent continental margins (e.g., Posamentier et al., 1988; Simpson and Eriksson, 1990; Boyd et al., 1993; Donovan, 1993; Contreras et al., 2010; Yang and Escalona, 2011; Safronova et al., 2014; Hampson and Premwichein, 2017; Kress et al., 2021); intracratonic basins (e.g., Jackson et al., 1990; Lindsay et al., 1993; Vecsei and Duringer, 2003; Husinec, 2016); grabens and rifts (e.g., Embry, 1993, 1995; Gawthorpe et al., 1994; Davies and Gibling, 2003; Martins-Neto and Catuneanu, 2010; Ichaso et al., 2016; Maravelis et al., 2017); convergent continental margins (e.g., Takashima et al., 2004; Buchs et al., 2015; Maravelis

et al., 2016; Mora et al., 2018); backarc basins (e.g., Salazar et al., 2016; Dominguez et al., 2020); foreland systems (e.g., Devlin et al., 1993; Posamentier and Allen, 1993; Hart and Plint, 1993; Plint et al., 1993; Catuneanu et al., 1997a,b, 1998b, 1999, 2000, 2002; Donaldson et al., 1998, 1999; Giles et al., 1999; Miller and Eriksson, 2000; Bordy and Catuneanu, 2001, 2002a,b; Catuneanu, 2004a,b, 2019c, Miall et al., 2008; Li et al., 2014; Avarjani et al., 2015; Menegazzo et al., 2016; Breckenridge et al., 2019); and pull-apart basins (e.g., Ryang and Chough, 1997; Brister et al., 2002; Wu et al., 2009).

Each type of tectonic setting is unique in terms of subsidence mechanisms (e.g., extensional, thermal, or flexural), sediment supply and dispersal patterns, location of depocenters, and basin physiography. The variability imposed by the tectonic setting to the stratigraphic architecture is one of the reasons for the proposal of different models (Figs. 1.9 and 1.10), as each model is based on conclusions drawn from data collected in different types of sedimentary basins (e.g., "passive" margins, forelands, or rifts). Contrasts between the stratigraphic frameworks that develop in different tectonic settings include the scales, the architecture, and the systems-tract and sedimentological composition of sequences. All types of sedimentary basins are tectonically active, since all require a tectonic mechanism to form, although they are generally classified into "passive" vs. "active" with reference to the rates of subsidence and the variability in the rates of subsidence with time (i.e., low vs. high, in relative terms). For example, divergent continental margins are often referred to as "passive" margins because they are dominated by thermal subsidence which operates with relatively low and constant rates, although they still record tectonic activity and the majority of accommodation is owed to subsidence.

For the purpose of basin analysis, the classification of sedimentary basins is based on plate-tectonic criteria (i.e., type of underlying crust, type of plate margin associated with the formation of the basin, and distance relative to the plate margin; Busby and Ingersoll, 1995; Einsele, 2000; Miall, 2000; Allen and Allen, 2013), and

341

so it recognizes the role of tectonism in the formation of all types of sedimentary basins. For the purpose of stratigraphic analysis, the fact that the methodology is independent of the local controls on accommodation and sedimentation makes it unnecessary to customize the workflow to the tectonic setting, even though the mechanisms and the patterns of subsidence have an evident impact on the stratigraphic framework. In this context, the broad classification of sedimentary basins into "passive" vs. "active" is sufficient to describe tectonic settings in terms of the relative contributions of tectonism and sea-level changes to stratigraphic cyclicity; i.e., tectonically "passive" settings are more susceptible to the influence of sea-level changes on the development of sequences, whereas the "active" settings are dominated by tectonic controls. These underlying conditions affect the areal extent, the composition, and the degree of 3D temporal and physical variability of stratigraphic sequences.

Tectonism during the evolution of a sedimentary basin can also influence the architecture of the basin fill, including the development of ramps vs. shelf-slope systems (Fig. 8.1). Sedimentary basins with a high tectonic activity (and instability) are prone to the formation of ramps or of poorly developed shelf-slope systems with narrower shelves (e.g., the rift-fill geometry in Figs. 7.15 and 8.2). In contrast, tectonically "passive" basins can sustain the long-term progradation of slope clinoforms, given sufficient sediment supply, resulting in the development of well-defined shelf-slope systems with wider shelves (Figs. 8.2—8.4).

Ramps and shelf-slope systems may also coexist within a sedimentary basin, due to variations along strike in tectonic activity and sediment supply (Dominguez et al., 2020), with ramps typically preceding the development of shelf-slope systems at any location (Fig. 7.16). Physiography makes a difference to the modes of sediment transport and the types of depositional elements that may be found within a sedimentary basin. For example, sediment dispersal across a shelf is most commonly related to traction currents (e.g., rivers, longshore currents, storm currents, tidal currents), whereas ramps are more prone to gravity-driven processes.

8.1.1 Sequences in tectonically "passive" basins

Tectonically "passive" basins are typically located in stable areas remote from plate margins and the influence of tectonic processes associated with plate margins. Examples include divergent ("passive") continental margins and intracratonic sag basins, which are located within the interior of tectonic plates and continents, respectively. Backarc basins may also display similar features, including the long-term progradation of clinoforms which results in the formation of shelf-slope systems (Figs. 7.16 and 8.4). The scale of the clinoforms may range from 10^1 m in the case of shallow intracratonic sags to 10^2 m in backarc basins and 10^3 m along mature divergent continental margins, depending on the thickness and composition of the underlying crust (i.e., the thinner

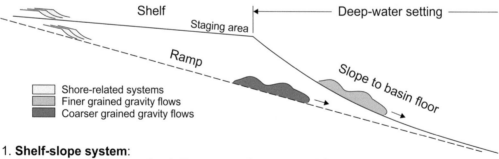

1. **Shelf-slope system:**
   • Shallow-water clinoforms on the shelf: coarsest sediment at the MRS in a vertical section
   • Deep-water gravity flows beyond the shelf edge: coarsest sediment at the CC (end of forced regression)

2. **Ramp:**
   • Area with shallow-water clinoforms: coarsest sediment at the MRS in a vertical section
   • Downdip of the shallow-water clinoforms: equivalent to the deep-water setting of a shelf-slope system

FIGURE 8.1   Shelf-slope vs. ramp systems in siliciclastic settings. Shelf-slope systems may develop in a variety of tectonic settings with low-gradient basin margins (e.g., divergent continental margins, but also some convergent margins, backarc basins, and forelands). Ramps are more typical of fault-bounded basins with steeper gradient basin margins, which promote a coarser clastic sediment supply. Downdip of the shore-related clinoforms, the ramp setting is similar to the deep-water setting of a shelf-slope system in terms of the dominant processes of sediment transport and deposition, except that the gravity flows tend to include coarser sediment. In the absence of gravity flows (e.g., not enough gradient or sediment supply), the ramp beyond the shallow-water clinoforms becomes equivalent to a shelf in terms of processes of sediment transport and deposition (e.g., storm currents, sedimentation from suspension). Abbreviations: MRS—maximum regressive surface; CC—correlative conformity.

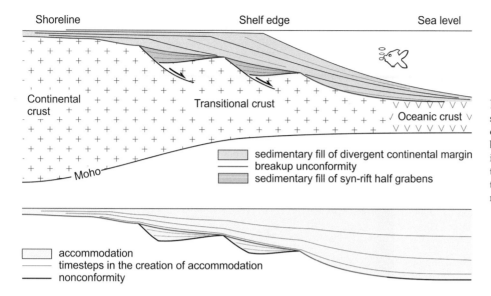

FIGURE 8.2 Subsidence patterns and stratigraphic architecture in a divergent continental margin setting. Below the breakup unconformity, subsidence rates increase toward the boundary faults, typically in a landward direction. Above the breakup unconformity, subsidence rates increase in a basinward direction.

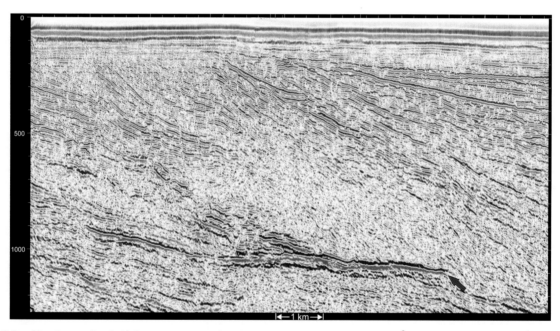

FIGURE 8.3 Clinoforms of a shelf-slope system in a divergent continental margin setting (10^3 m scale), downlapping the breakup unconformity (red arrow: first-order sequence boundary at the top of the underlying rift sequence) (Krishna—Godavari Basin, offshore India; data courtesy of Reliance Industries). Vertical scale: two-way travel time (milliseconds). Well data indicate that the breakup unconformity is represented by a wave-ravinement surface reworking a subaerial unconformity, at the contact between fluvial sediments and the overlying shallow-marine sediments of the early "passive" margin (Fig. 7.15).

and the denser the crust, the higher the subsidence rates). Subsidence rates in tectonically "passive" basins increase in a basinward direction, resulting in the distal location of depocenters and the divergence of time horizons toward the depocenters (Fig. 8.2). The limited tectonic activity during the formation and evolution of these basins restricts the uplift of new source areas for extrabasinal sediment, and maintains generally low gradients along the basin margins. These conditions afford any style of sedimentation, from siliciclastic to intrabasinal biochemical and chemical precipitates (e.g., carbonates and evaporites; Csato et al., 2013, 2015; Dominguez et al., 2020; Reijenstein et al., 2020).

Stratigraphic sequences that accumulate within tectonically "passive" basins may be strongly influenced by sea-level changes, leading to the development of laterally extensive units with tabular geometries and a limited 3D variability. Such sequences have a greater chance to correlate with stratigraphic cycles in other tectonically "passive" basins, although this assumption

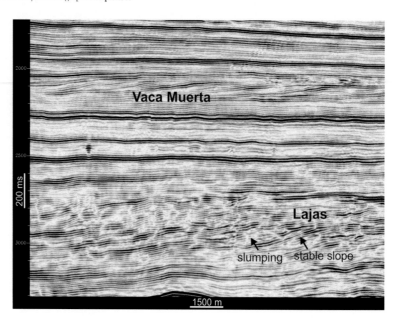

FIGURE 8.4 Clinoforms of the Lajas and Vaca Muerta—Quintuco shelf-slope systems (10^2 m scales) in a backarc setting (Late Jurassic—Early Cretaceous, Neuquén Basin, Argentina; data courtesy of YPF Argentina). The Lajas clinoforms are steeper (coarser sediment, higher angle of repose), with cyclic changes between slope stability (periods of lowstand and sand delivery to the slope) and instability (periods of transgression and highstand, when only mud was available to the slope), which is typical of siliciclastic systems. In contrast, the Vaca Muerta—Quintuco clinoforms are finer grained (lower angle of repose), with a carbonate platform on the shelf. The smaller scale of the Vaca Muerta—Quintuco clinoforms indicates a shallower bathymetry in the later stages of basin evolution.

can never be the starting point of a sequence stratigraphic study (see the discussion on workflow and methodology in Chapter 9). The possibility of interbasinal correlations can still be tested after the construction of basin-specific stratigraphic frameworks based on the acquisition of local data. The low and relatively constant rates of subsidence that typify tectonically "passive" basins are also conducive to the development of all systems tracts within sequences, including lowstand systems tracts which are often missing in tectonically "active" basins such as rifts (discussion below). The development of all systems tracts in tectonically "passive" settings is facilitated by the fact that the rates of accommodation and sedimentation commonly vary within overlapping ranges, allowing the occurrence of all types of shoreline trajectories (i.e., lowstand and highstand normal regressions, forced regressions, and transgressions; Fig. 4.32).

8.1.2 Sequences in tectonically "active" basins

Sedimentary basins in tectonically "active" settings form and evolve under the direct control of tectonic processes, including extension (grabens and rifts), compression (convergent margins, orogenic belts), orogenic loading (foreland systems), and strike-slip motion (pull-apart basins). In such settings, tectonism commonly outruns the influence of sea-level changes on stratigraphic cyclicity, leading to the deposition of strongly asymmetrical sequences with a significant 3D temporal and physical variability. Subsidence rates typically increase toward the basin margins where the tectonic activity is more intense (e.g., in fault-bounded basins, or in foreland systems); as a result, depocenters

occupy proximal locations, and sequences thicken and time horizons diverge in an updip direction (e.g., the rift fills in Figs. 7.15 and 8.2, or the foreland fills in Figs. 8.5—8.7). Tectonic instability prevents the extensive progradation of clinoforms, due to the frequent collapse of unstable slopes, and promotes the development of ramps (Figs. 7.15 and 8.1). Shelf-slope systems may still form, but typically with narrow shelves, as in the case of forearc basins along convergent continental margins (e.g., Dickinson, 1995; Fildani et al., 2008; Hessler and Sharman, 2018; Orme and Surpless, 2019).

Tectonically "active" basins are subject to rapid changes in the rates of accommodation and sedimentation, which may promote or suppress the formation of specific systems tracts. The reactivation of basin-margin faults in grabens and rifts marks episodic pulses of extension when accommodation is created rapidly, followed by longer periods of tectonic quiescence when sedimentation consumes the available space. The abrupt increases in accommodation at the onset of new depositional cycles trigger transgressions and prevent the development of lowstand systems tracts, as sedimentation is outpaced by accommodation. Following a tectonic pulse, sediment supply fills part or all of the available accommodation, generating highstand normal regressions. These cyclic changes in the balance between accommodation and sedimentation in fault-bounded basins result in the development of incomplete sequences dominated by transgressive and highstand systems tracts, with sequence boundaries overprinted by flooding surfaces (Martins-Neto and Catuneanu, 2010, Fig. 8.8). This is in contrast to the complete sequences, in terms of the component systems tracts, which are commonly observed in "passive"-margin (Posamentier

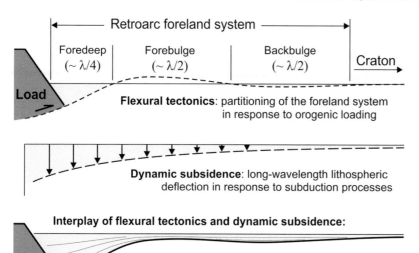

FIGURE 8.5 Subsidence patterns and stratigraphic architecture in a retroarc foreland setting (modified after Catuneanu, 2019c). Subsidence rates generally increase, and timelines diverge, toward the orogenic load. Flexural tectonics and dynamic subsidence operate over different timescales, with variable rates and magnitudes (see Catuneanu, 2019c, for details).

et al., 1988) or foreland settings (Van Wagoner et al., 1988, 1990).

Foreland systems record the interplay of orogen-driven flexural tectonics, subduction-driven dynamic subsidence and sedimentation, which generates a strong 3D stratigraphic variability along both dip and strike directions (Figs. 8.5–8.7). Orogenic cycles of loading (thrusting) and unloading (rebound) result in a reciprocal flexural response between the foredeep and the forebulge of a foreland system, leading to the development of out-of-phase stratigraphic patterns on time-scales of $0.5-10^{0}$ My (Catuneanu et al., 1997a, 1999, 2000, 2002; Catuneanu and Sweet, 1999). The strike variability in orogenic loading adds a third dimension to the temporal and physical variability of foreland sequences (Fig. 8.7). The interplay of tectonism and sedimentation during the lifespan of a foreland system controls the shift from underfilled to overfilled accommodation conditions, which defines the first-order foreland cycle (Catuneanu, 2019c, Fig. 8.6). Tectonism is the dominant control on stratigraphic cyclicity at all timescales ≥0.5 My, below which sea-level changes become a possible contender at the Milankovitch scales of orbital cycles (Varban and Plint, 2008). Despite the 3D complexity of foreland systems, foreland sequences are prone to include all four systems tracts, as the rates of flexural and dynamic subsidence are commonly within the range of the sedimentation rates.

Tectonic activity during the formation and evolution of sedimentary basins described herein is prone to the uplift of new source areas for extrabasinal sediment and/or to the development of steep-gradient basin margins, which promote a siliciclastic style of sedimentation. Depending on the location, elevation, and connectivity to the global ocean, deposition may occur in a variety of environments ranging from fully continental through to deep water (Hempton and Dunne, 1984; Leeder and Gawthorpe, 1987; Frostick and Steel, 1993a,b; Nottvedt et al., 1995; Catuneanu and Eriksson, 1999, 2002; Barka et al., 2000; Scherer et al., 2007; Folkestad and Satur, 2008; Gurbuz, 2010; Mumpy and Catuneanu, 2019; Csato et al., 2021). Tectonically "active" basins are also most unpredictable and diverse in terms of the systems-tract composition of sequences; e.g., high sediment supply may suppress transgressions, resulting in the development of sequences consisting only of falling-stage and lowstand systems tracts; high rates of subsidence at the onset of relative sea-level rise can prevent the accumulation of lowstand systems tracts; and continuous relative sea-level rise denies the formation of falling-stage and lowstand systems tracts (Csato and Catuneanu, 2012, 2014, Fig. 4.33). However, the sequence stratigraphic methodology remains the same irrespective of the variability imposed by the tectonic setting.

8.2 Variability with the depositional setting

The sedimentological composition of sequences and systems tracts, as well as the relative development of systems tracts within sequences, varies with the depositional setting. Depositional settings may be broadly classified into continental (or nonmarine; beyond the reach of marine flooding), coastal (or marginal marine; intermittently flooded by marine water), and marine

I. Underfilled phase: marine environment within the foreland system

Stage 1: Flexural uplift > dynamic subsidence

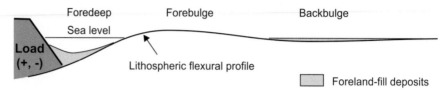

Stage 2: Dynamic subsidence > flexural uplift

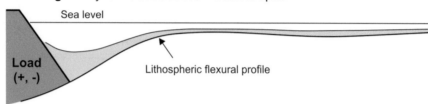

FIGURE 8.6 Patterns of sedimentation in a retroarc foreland setting, as a function of the interplay between accommodation and sedimentation (from Catuneanu, 2019c). Retroarc foreland systems typically start with an underfilled phase, in which accommodation is created faster than the rates of sedimentation, and end with an overfilled phase of continental sedimentation (see Catuneanu, 2004b, for a review of case studies).

II. Overfilled phase: fluvial environment across the foreland system

Stage 3: Flexural uplift > dynamic subsidence

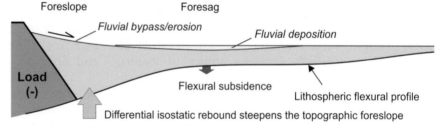

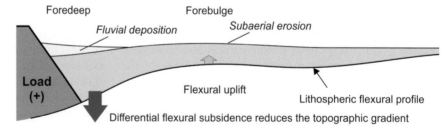

(permanently covered by marine water) (Figs. 4.10 and 4.11). The same classification applies to lacustrine settings, whereby the coastline separates continental from lacustrine environments, and the relative lake level replaces the relative sea level. The subenvironments that encompass the transition from continental to marine environments are illustrated in Fig. 4.11. Along the coastline, the river-mouth environments (i.e., extrabasinal sediment entry points into the marine basin: deltas or estuaries) are separated by stretches of open shoreline where prograding strandplains, backstepping beaches,

or lagoon-barrier island systems may develop. The glacial environment is not included in the classification scheme in Fig. 4.10 because it is climatically controlled and may overlap with any continental, coastal, or marine/lacustrine setting.

The progradation of open shorelines, particularly under normal regressive conditions, can lead to the development of relatively flat and broad coastal plains, extending up to 10^2 km inland from the shoreline (Fig. 4.12). The body of sediment that accumulates in this process (i.e., the coastal prism in Fig. 4.12)

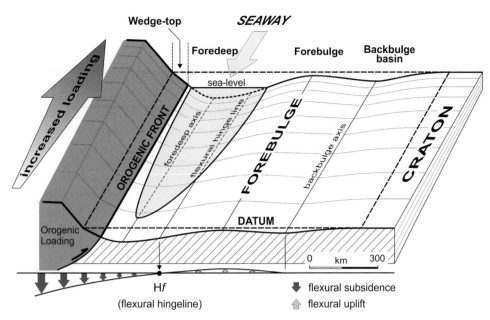

FIGURE 8.7 Strike variability in orogenic loading, and the resulting tilt of the foreland system (modified after Catuneanu, 2004b, 2019c). The datum represents the top of available accommodation generated by basin-forming mechanisms (Fig. 8.5).

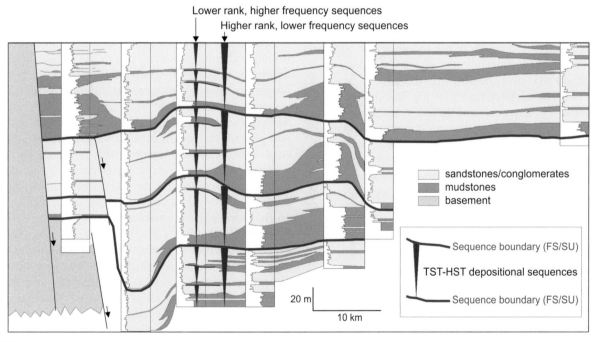

FIGURE 8.8 Well-log cross section of correlation in the Viking Graben, North Sea (modified after Frostick and Steel, 1993b). Depositional sequences display coarsening-upward profiles, and are initiated by episodes of abrupt water deepening triggered by the reactivation of basin-margin faults. In this case, depositional sequence boundaries amalgamate subaerial unconformities and transgressive ravinement surfaces, and also fit the lithological description of flooding surfaces. Abbreviations: TST—transgressive systems tract; HST—highstand systems tract; FS—flooding surface; SU—subaerial unconformity.

describes the architecture of lowstand and highstand systems tracts in shelf settings. Particularly in the case of the latter, the development of flat and wide coastal plains is favored by the low energy of the highstand rivers, which most commonly are the lowest energy rivers of an entire stratigraphic cycle (see discussion in Chapters 4 and 5). The low-relief of the highstand coastal plains is the reason why fluvial incision in downstream-controlled settings is commonly recorded during stages of relative sea-level fall, when the trajectory of the shoreline becomes steeper than the gradient of the highstand coastal plain

(Fig. 4.42A). The steepening of the fluvial profile across the exposed seafloor leads to an increase in fluvial energy, and hence erosion. The resulting incised valleys often inherit the sinuous morphology of the highstand rivers, which can be maintained even if fluvial gradients and energy subsequently increase, as rivers get trapped within valleys or canyons (Figs. 5.18, 5.20, and 5.21).

Coastal environments are critically important in sequence stratigraphy, as the timing of systems tracts and bounding surfaces in downstream-controlled settings is linked to shoreline trajectories. A distinction needs to be made between the geometrical trends that define shoreline trajectories (i.e., the upstepping or downstepping of subaerial clinoform rollovers) and the depositional trends of fluvial to coastal aggradation or erosion that accompany the shifts of the shoreline (Figs. 4.23, 4.26, and 4.28). The two types of trends are controlled by different processes and can be out of phase, and only the former are diagnostic to the definition of systems tracts and bounding surfaces (see discussion in Chapter 4). The development of sequence stratigraphic concepts started with the study of the transition zone between marine and nonmarine environments, where the relationship between facies shifts and stratigraphic cyclicity is most evident. From the shoreline, the application of sequence stratigraphy was gradually expanded in both landward and basinward directions, until a comprehensive understanding of the stacking patterns that define the sequence stratigraphic framework was reached, from fully continental to deep-water systems.

A reality that is commonly overlooked is that coastlines can change their transgressive vs. regressive character along strike, due to variations in subsidence and sedimentation rates (Fig. 4.11). This results in the coeval deposition of different systems tracts and the diachronous development of sequence stratigraphic surfaces, which should safely be regarded as the norm rather than the exception (Schultz et al., 2020; Zecchin and Catuneanu, 2020). Therefore, the architecture and age relationships of depositional systems and systems tracts that can be demonstrated on a particular 2D dip-oriented transect are not necessarily representative for other locations along the shoreline. The 3D variability of the stratigraphic framework has been demonstrated with biostratigraphic data for several decades (e.g., Gill and Cobban, 1973), but it has not been seriously considered in sequence stratigraphy until the 1990s (e.g., Martinsen and Helland-Hansen, 1995; Catuneanu et al., 1998a; Posamentier and Allen, 1999). Full details on the 3D variability of the stratigraphic architecture are presented in Chapter 7.

The relationship between the vertical and lateral changes of facies and associated depositional environments is described by Walther's Law (Fig. 1.5). This is a fundamental principle of stratigraphy, which affords the prediction of lateral changes of facies based on the observation of vertical profiles in outcrops and wells. Vertical changes in litho- and biofacies have long been used to reconstruct paleogeography and, with the aid of Walther's Law, to predict lateral shifts of depositional environments. However, such interpretations are only valid within conformable successions. Vertical changes across unconformities potentially reflect shifts of facies between successions that are genetically unrelated, and therefore such changes cannot be used to reconstruct the paleogeography at any specific time. A delta is a good illustration of Walther's Law. The deltaic system includes prodelta, delta front, and delta plain facies, "which occur side by side in that order and the products of which occur together in the same order in vertical succession" (Miall, 1990). Beyond the scale of a depositional system, Walther's Law enables predictions of lateral changes of facies within systems tracts, at all hierarchical levels (Figs. 1.16, 5.2, and 5.3), as long as the successions are "relatively conformable" at the scale of the stratigraphic study (Fig. 7.6).

8.2.1 Sequences in continental settings

Sequences in continental settings develop within fluvial and eolian systems, under the influence of either upstream or downstream controls (Fig. 3.11). Subaerial unconformities are natural candidates to delineate sequences in continental settings, since they typically have a strong physical expression (Figs. 4.65 and 6.16), and provide the means to subdivide the stratigraphic record into relatively conformable successions at the selected scales of observation (Figs. 4.70 and 6.54). In downstream-controlled settings, continental deposits are part of "conventional" sequence stratigraphic frameworks, forming the proximal portion of shore-related systems tracts (Fig. 4.39). In upstream-controlled settings, continental deposits build stand-alone sequences and systems tracts which describe the architecture of "unconventional" sequence stratigraphic frameworks, with a timing that is independent of shoreline trajectories. Stratigraphic cyclicity in downstream- and upstream-controlled settings forms in response to different factors (i.e., the interplay of relative sea-level changes and base-level changes vs. base-level changes alone, respectively; details in Chapters 3 and 4), and therefore the two stratigraphic frameworks can be out of phase (Fig. 3.20).

8.2.1.1 Sequences in fluvial settings

Fluvial systems are the default deposits used to illustrate the composition of sequences and systems tracts in continental settings throughout this book. A comprehensive discussion of fluvial processes and stacking patterns is provided in Chapters 4 and 5. The same types of fluvial stacking patterns can be observed in both downstream- and upstream-controlled settings, as they form in response to a common set of underlying processes: the degree of channel confinement, the rate of floodplain aggradation, and the frequency of avulsion (Fig. 4.63). The combination of these processes controls the degree of channel amalgamation, which is usually highest at the base of depositional sequences (Fig. 8.9). This trend accompanies the overall decline in fluvial energy and decrease in sediment caliber that are typically recorded during the accumulation of a fluvial sequence (Figs. 4.53, 4.68, 5.50, and 8.9; details in Chapter 4). The degree of channel amalgamation defines the high- and low-amalgamation systems tracts in upstream-controlled settings (Figs. 4.69 and 4.70), which are often similar to the lowstand vs. the transgressive and high-stand systems tracts in downstream-controlled settings (Figs. 5.50, 6.30, and 8.9).

Despite these general trends, fluvial sequences may display a significant variability as a function of the tectonic setting and the syn-depositional climatic regimes,

which both can modify fluvial styles and the types of depositional elements that comprise systems tracts. Fluvial sequences and systems tracts in both downstream- and upstream-controlled areas may form under high or low accommodation conditions, depending on the subsidence patterns at syn-depositional time. For example, the foredeep of a foreland system may qualify as a high-accommodation setting during times of orogenic loading, whereas the backbulge area is a low-accommodation setting due to the much lower subsidence rates recorded at the same time (Leckie and Boyd, 2003; Catuneanu, 2004b, 2019c). However, each of these settings can afford the development of all systems tracts, albeit with different thicknesses. Therefore, a distinction needs to be made between the degree of channel amalgamation, which defines fluvial stacking patterns, and the accommodation conditions at syn-depositional time. For this reason, the nomenclature of systems tracts in upstream-controlled settings is now decoupled from accommodation, since the degree of channel amalgamation depends on several factors, some of which unrelated to

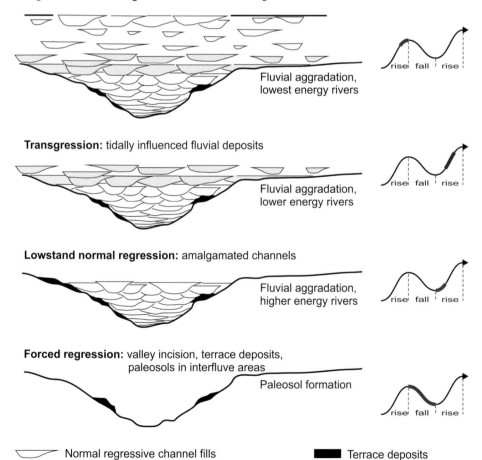

Highstand normal regression: isolated to amalgamated channels

Fluvial aggradation, lowest energy rivers

Transgression: tidally influenced fluvial deposits

Fluvial aggradation, lower energy rivers

Lowstand normal regression: amalgamated channels

Fluvial aggradation, higher energy rivers

Forced regression: valley incision, terrace deposits, paleosols in interfluve areas

Paleosol formation

Normal regressive channel fills
Tidally influenced (transgressive) channel fills
Terrace deposits
Subaerial unconformities

FIGURE 8.9 Stratigraphic architecture of a fluvial depositional sequence in downstream-controlled settings where subaerial unconformities form during forced regressions (modified after Shanley and McCabe, 1993). Subsequent erosion lowers the preservation potential of the highstand amalgamated channels. The degree of channel confinement tends to increase upward, in response to the decrease in topographic gradients and fluvial energy during the accumulation of the sequence. While departures from this model may occur, the trends depicted herein are commonly observed in the stratigraphic record. Similar stacking patterns form in upstream-controlled settings: high-amalgamation systems tracts, similar to the lowstand amalgamated channels, and low-amalgamation systems tracts similar to the floodplain-dominated transgressive and highstand fluvial deposits. However, no time equivalence is implied between the downstream- and upstream-controlled stratigraphic frameworks.

Physical attributes	Tectonic setting	
	Low rates of subsidence	High rates of subsidence
Unconformities	Common, closely spaced	Rare, widely spaced
Incised valleys	Multiple and compound	Rare
Stratigraphic sections	Thin, HST deposits commonly truncated	Thick, with preservation of all systems tracts
Channel sandbodies	More amalgamation, potentially isolated near MFS	Amalgamated near SU, isolated near MFS
Floodplain fines	Rare, potentially present near MFS	Common and abundant
Tidal deposits	Most abundant near MFS	Most abundant near MFS
Underlying topography	Enhanced control	Weak control
Coal seams	Commonly absent; compound where present	Abundant, thicker, and simpler (fewer hiatuses)
Paleosols	Well-developed, multiple and compound	Thinner and widely spaced, more organic-rich

FIGURE 8.10 Contrasts in the stratigraphic architecture of low- vs. high-subsidence settings (modified after Leckie and Boyd, 2003). Despite these differences, both types of settings afford the formation of all "conventional" and "unconventional" systems tracts in downstream- and upstream-controlled areas, respectively.

accommodation (Fig. 4.63; Catuneanu, 2017; details in Chapters 4 and 5).

The main differences between fluvial sequences that develop in areas of low vs. high subsidence are summarized in Fig. 8.10. Despite these differences, both settings afford the formation of all types of systems tracts, including all "conventional" systems tracts in downstream-controlled areas and the high- and low-amalgamation systems tracts in upstream-controlled areas. For example, upstream-controlled fluvial sequences in high-subsidence settings may include both high- and low-amalgamation systems tracts, and not only the latter as commonly assumed (Fig. 4.71). The same is true for fluvial sequences in low-subsidence settings, whereby both types of stacking patterns may develop, and not only amalgamated channels as generally expected (i.e., floodplain-dominated successions may also accumulate where channels are confined and the frequency of avulsion is low, or where the rates of floodplain aggradation exceed the rates of subsidence; details in Chapters 4 and 5). Most physical attributes in Fig. 8.10 are common between downstream- and upstream-controlled sequences, with the exception of the types of systems tracts ("conventional" vs. "unconventional") and the presence vs. the absence of tidal influences, respectively.

The cyclicity recorded in fluvial settings, whether downstream- or upstream-controlled, may develop in response to tectonic cycles that modify the topographic gradients, and hence the energy of fluvial systems; climate cycles that modify the balance between sediment supply and the transport capacity of rivers; and sea-level cycles that change the elevation of river mouths, which, along with sediment supply and subsidence patterns, affect depositional processes and fluvial styles (Miall, 1991; Schumm, 1993; Holbrook et al., 2006; Catuneanu et al., 2011). Tectonic cycles have been documented most commonly at scales of ≥ 0.5 My, typically leading to changes in topographic gradients, fluvial styles, and dip directions across subaerial unconformities (Figs. 2.10 and 4.64). Climate and sea-level cycles are most evident at the 10^4-10^5 yrs scales of orbital forcing, with the greatest influence on the stratigraphic architecture during Icehouse periods (Gibling et al., 2005; Fielding et al., 2008). Superimposed on these allogenic controls, autogenic switching of alluvial channel belts on timescales of 10^3-10^5 yrs can further modify sediment supply to any particular area, and therefore the timing of fluvial aggradation or erosion at any location (Fig. 3.1).

Separating the controls on the development of fluvial sequences and systems tracts is a difficult task, which can be tackled on a case-by-case basis. For example, the effect of climate on fluvial processes of aggradation and degradation becomes evident in tectonically-stable inland regions, such as along the cratonic margins of foreland systems that are beyond the influence of sea-level fluctuations (Gibling et al., 2005). The unconformity-bounded late Quaternary fluvial sequences in the southern Gangetic Plains provide an example where changes in monsoonal precipitation triggered by climate shifts generated cycles of floodplain aggradation and degradation over timescales of 10^4 yrs. In this case, sequences record periods when floodplains were inundated and experienced sustained aggradation, whereas declining flood frequency on parts of the interfluves resulted in low-relief degradation surfaces and badland ravines, as well as local soil development (Gibling et al., 2005). Irrespective of the relative contributions of the underlying controls on fluvial processes, the deposition of fluvial sequences and the formation of sequence-bounding subaerial unconformities reflect cycles of base-level rise and fall, respectively, which can be observed at different scales (Fig. 4.70; details in Chapters 3–5).

8.2.1.2 Sequences in eolian settings

Depositional sequences in eolian settings correspond to cycles of expansion and contraction in the evolution of ergs, with sequence-bounding unconformities representing periods of cessation of sedimentation (i.e., the constructional, contractional, and deflationary stages in Fig. 6.54). This stratigraphic cyclicity is driven by fluctuations in the temporary base level, which may occur at different scales in response to variations in the position of the water table, the amount of sediment supply, and the wind regime. In turn, these variations reflect the interplay of climate and tectonism, as well as the influence of sea-level changes on the water table where the erg extends into downstream-controlled areas. Climate controls the wind regime, as well as the humidity levels (arid vs. humid conditions) which affect the water table, the growth of vegetation, and the efficiency of weathering and related sediment supply. Tectonism controls changes in elevation, the rates of subsidence or uplift at syn-depositional time, the amount of accommodation that affords the long-term preservation of eolian sequences, and it may also exert an influence on the position of the water table.

Ergs are dominated by dunefields, but may also include sand sheets particularly along the erg margins and at the base and at the top of eolian sequences, in relation to the higher energy flow regimes that lead to the deflation of ergs. The position of the water table relative to the surface profile is a key element in the classification of eolian systems into dry, wet, and stabilized (Kocurek, 1998; Rodriguez-Lopez et al., 2014). Dry systems (e.g., the Namib Sand Sea, Namibia; the Algodones Dunes, SE California) are typified by dry interdunes and noncohesive sand above a deep water table which has no effect on sedimentation. Processes in such systems are controlled by the interplay of sediment supply and wind regime. Wet systems (e.g., the White Sands dunefield, New Mexico; the Liwa Oasis, United Arab Emirates; the Bahariya Oasis, Western Desert, Egypt)

are defined by wet or damp interdune flats in which the water table or its capillary fringe intersects the surface profile. Processes in such systems are primarily controlled by fluctuations in the water table. Stabilizing systems (e.g., the Kalahari Desert, Botswana; the Thar Desert, Rajasthan, India; the Negev-Sinai erg, Israel-Egypt) are defined by interdune areas which are at least in part stabilized by vegetation or cemented crusts. These systems too are to a large extent controlled by fluctuations in the water table. Notably, the interdune areas play a key role in the definition of all types of eolian systems, and their elevation is a proxy for the position of the temporary base level (see Chapter 3 for details on the base level).

Sedimentary cyclicity in eolian systems develops at both stratigraphic and sedimentological scales. Stratigraphic cycles are defined by rises and falls in the temporary base level, which result in the formation of sequences and sequence boundaries, respectively (Fig. 6.54). At sub-stratigraphic scales, the migration of dunes or draas generates sedimentological cycles of various hierarchical ranks (Fig. 8.11). At the largest sedimentological scales, the dune–interdune cycles are bounded by interdune migration surfaces. The thickness of these cycles is proportional to the rates of sediment accumulation (i.e., rates of base-level rise) whose increase also tends to steepen the angle of climb of the interdune migration surfaces (Fig. 6.54). The trends of change in the thickness of dune–interdune cycles and the associated angle of climb of the interdune migration surfaces define stratal stacking patterns and corresponding systems tracts in eolian settings (Fig. 6.54). Eolian sequences are bounded by "supersurfaces" generated by processes of wind deflation or sediment bypass (Kocurek, 1988, Figs. 6.16 and 6.54). Supersurfaces are often expressed as paraconformities, where the lack of tectonic tilt during deflation or bypass prevents the formation of angular unconformities. The process of wind erosion also promotes the formation of flat deflation surfaces which lack the erosional relief that defines disconformities (Figs. 1.18 and 6.16).

The amount of base-level rise during the accumulation of a sequence is quantified by the relative increase in the elevation of interdune areas, which is driven by the rise in the water table relative to the surface profile in wet and stabilized systems, and by sediment supply exceeding the transport capacity of the wind in dry systems. The constructional stage of a stratigraphic cycle is characterized by accelerating growth (i.e., an increase with time in the rates of base-level rise), which results in a corresponding increase in the thickness of the dune–interdune cycles up to the maximum expansion level (Fig. 6.54). The contractional stage of a stratigraphic cycle records a decelerating growth (i.e., a decrease with time in the rates of base-level rise), with

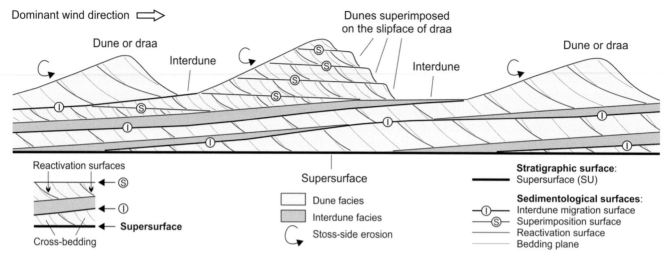

FIGURE 8.11   Architecture of stratigraphic and sedimentological surfaces in eolian systems (modified after Abrantes et al., 2020). Super-surfaces are bypass or deflationary sequence-bounding unconformities. Within sequences, the interdune migration surfaces are the most extensive and the highest rank sedimentological surfaces, with a key role in the definition of stacking patterns in eolian systems. Superimposition surfaces form in relation to the oblique migration of dunes over the slipface of draas, cutting across reactivation surfaces and cross-bedding sets. Reactivation surfaces involve the scouring of cross-beds caused by short-term increases in wind velocity, in a direction that is generally concordant with the orientation of cross-bedding; these scours develop at a lower angle than the cross-beds, due to the higher flow regime at the time of their formation, truncating the strata below and being downlapped by the overlying cross-beds (see text for details). Abbreviation: SU—subaerial unconformity.

a consequent reduction in the thickness of the dune—interdune cycles until the rates of accumulation become nil (sediment bypass) or negative (wind erosion) (Fig. 6.54). Stratigraphic cyclicity can be observed at different scales, depending on the magnitude of base-level cycles (Fig. 1.19). At each stratigraphic scale, the limit between the constructional and contractional systems tracts is the maximum expansion level, which marks the highest rates of sequence development at any location (Fig. 6.54; see discussion in Chapter 6).

The maximum expansion level is placed at the core of the stratigraphic interval that includes the thickest dune—interdune cycles. This level does not have a physical expression, as it crosses the interdune migration surfaces and the associated dune—interdune facies contacts, and its placement requires a study of the stacking patterns of the dune—interdune cycles observed at sequence scales (Fig. 6.54). Physical surfaces in eolian systems include supersurfaces at stratigraphic scales, and, in the order of decreasing hierarchical ranks, interdune migration surfaces, superimposition surfaces, reactivation surfaces, and bedding planes at sedimentological scales (Figs. 6.16, 8.11, and 8.12; Brookfield, 1977, 1992; Kocurek, 1981, 1988, 1996; Collinson, 1986; Havholm and Kocurek, 1994; Galloway and Hobday, 1996; Mountney and Jagger, 2004; Abrantes et al., 2020). Each type of surface terminates against surfaces of higher hierarchical ranks: bedding planes terminate against reactivation surfaces or any other higher rank surfaces; reactivation surfaces terminate against

superimposition surfaces or any other higher rank surfaces; superimposition surfaces terminate against interdune migration surfaces or the associated interdune facies; and interdune migration surfaces terminate against supersurfaces (Fig. 8.11). All these surfaces are diachronous, with higher degrees of diachroneity recorded typically at sedimentological scales.

Supersurfaces may be associated with variable stratigraphic hiatuses, depending on the hierarchical rank of the unconformity, although the amount of time missing is generally difficult to measure in eolian sands. It is estimated that supersurfaces associated with closely spaced tree-size rhizoliths may take 10^4–10^5 yrs to form (Loope, 1985), which falls within the range of Milankovitch-scale orbital cycles. As a rule of thumb, deflationary supersurfaces are of higher hierarchical ranks than the bypass supersurfaces, as the latter correspond to brief stages of balance between accumulation and erosion which cannot be maintained for long periods of time. As the base level is a dynamic surface, always in motion, the bypass supersurfaces either mark stages of brief cessation of erg development followed by a resumption of deposition (i.e., high-frequency/low-rank sequence boundaries) or are transient toward longer stages of accumulation or deflation and therefore part of higher rank sequence boundaries (Fig. 6.54). Sequences and sequence-bounding deflationary supersurfaces have been attributed to climate cycles driven by orbital forcing with periodicities of ±20–400 kyr (Fig. 2.15, Loope, 1985; Clemmensen et al., 1994; Mountney, 2006; Jordan and

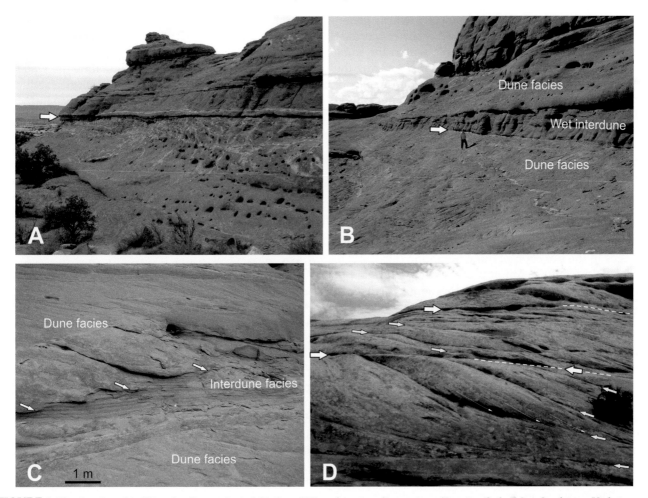

FIGURE 8.12 Stratigraphic (A) and sedimentological (B, C, and D) surfaces in eolian systems (Permian Cedar Mesa Sandstone, Utah; images A, B, and C courtesy of Nigel Mountney; image D courtesy of Brad Macurda). A. Supersurface (arrow) marked by rhizoliths; B. Interdune migration surface (arrow); C. Dune migration surface; note the intertongued relationship between the dune and interdune facies (arrows), as the dune migration was contemporaneous with sedimentation in the interdune environment; D. Superimposition (white arrows) and reactivation (yellow arrows) surfaces; reactivation surfaces are quasi-concordant with the cross-beds, and terminate against superimposition surfaces. Note the downlap of bedding planes against the underlying reactivation surface (see text for details).

Mountney, 2010, 2012; Rodriguez-Lopez et al., 2012, 2014), although tectonic subsidence-uplift cycles on timescales of 10^5-10^6 yrs can also control the formation of continental sequences (Catuneanu and Bowker, 2001; Catuneanu and Elango, 2001).

The long-term preservation of eolian sequences in the rock record is largely controlled by the amount of accommodation generated by subsidence below the elevation of the outflow ridge by the time the basin becomes overfilled (Fig. 3.12). Similar to what has been documented in fluvial successions (Miall, 2015, 2016), the amount of time that is represented in eolian sequences is thought to be significantly less than the duration of hiatuses associated with the sequence boundaries (Loope, 1985). It thus appears that the preserved eolian sequences likely account for only a fraction of the geological time over which the eolian systems were active (Rodriguez-Lopez et al., 2014). This is in line with

previous findings that the sedimentary record is highly fragmentary, "more gap than record" (Ager, 1981, 1993), particularly in continental settings (Miall, 2015). As pointed out by Bailey and Smith (2010), "such stratigraphic records are better viewed as the outcome of temporary cessation of the erosion and redistribution of sediment." As such, the preserved record captures evidence of a limited set of formative processes, and therefore it is biased toward those processes (Rodriguez-Lopez et al., 2014).

The formative processes that are responsible for the sedimentological makeup of stratigraphic sequences include the migration of dunes or draas separated by interdune flats, which results in the formation of dune—interdune cycles and interdune migration surfaces; the migration of dunes across the lee side of draas, resulting in the formation of superimposition surfaces at the limit between different dunes; the scouring of cross-beds due to short-term increases in wind velocity, which generates reactivation surfaces that are generally concordant with the orientation of the cross-beds within the same dune; and the lee-face accretion of sediment during the migration of dunes, which results in the formation of bedding planes (Figs. 8.11 and 8.12). Superimposition surfaces form without interdune flats between the migrating dunes; therefore, these surfaces are not associated with interdune facies but only with dune facies on both sides of the contact, and truncate the lower rank surfaces of the underlying dune (i.e., reactivation surfaces and bedding planes; Figs. 8.11 and 8.12D). At smaller intradune scales, reactivation surfaces rework older cross-beds and are downlapped by the younger bedding planes of the same cross-bedding set, without changes in the direction of dune migration (Fig. 8.12D).

The field identification of process-related surfaces in Fig. 8.11 may be challenging to the non-specialists in eolian sedimentology, which prompted the proposal of an alternative classification based solely on the observation of physical relationships of strata, devoid of process connotations (Hasiotis et al., 2021). The latter approach follows the principles of stratal hierarchy pioneered by Campbell (1967), but extending the classification of stratal units and bounding surfaces from the scales of laminae, beds and bedsets to the scales of stratigraphic sequences. However, an understanding of the processes that generate the different types of units and bounding surfaces is still needed in order to separate between elements of sedimentology and stratigraphy (Figs. 6.54 and 8.11). The distinction between sedimentological and stratigraphic elements becomes important when selecting surfaces that can be used to delineate stratigraphic sequences, vs. surfaces that describe the sedimentological architecture of sequences and systems tracts. Therefore, a correspondence between the process-based hierarchy of eolian surfaces (Brookfield, 1977, 1992; Kocurek, 1981, 1988, 1996; Mountney and Jagger, 2004; Abrantes et al., 2020) and the hierarchy based on geometrical criteria is still inferred (Hasiotis et al., 2021).

The geographic extent of eolian systems depends not only on the vertical rates of growth but also on the development of adjacent depositional environments which may enlarge or restrict the area available for erg deposition. Therefore, the rates of sedimentation which define the stages of construction and contraction in Fig. 6.54 may or may not correlate with the trends of lateral expansion and contraction of the eolian system. For example, interglacial ergs tend to contract areally due to the expansion of marine, marsh, lacustrine, and fluvial environments (Clemmensen and Hegner, 1991; Clemmensen et al., 1994; Galloway and Hobday, 1996). At the same time, the rise of the water table, as well as the more humid conditions that promote weathering and the production of clastic sediment, are conducive

to higher rates of aggradation in all eolian systems (wet, dry, or stabilized). Conversely, glaciations allow eolian systems to expand laterally as the adjacent environments contract, while at the same time the water table falls and the production of clastic sediment decreases, with the exception of glaciogenic silt (Muhs and Bettis, 2003, Fig. 3.5). Therefore, the vertical and horizontal aspects of eolian sequences may be out of phase, with the vertical trends providing the key attributes for the definition of eolian stacking patterns and systems tracts (Fig. 6.54). This approach also presents the advantage of not being affected by the degree of preservation of the erg margins, which may be limited for older sequences.

8.2.2 Sequences in coastal to shallow-water settings

Sequences in coastal to shallow-water settings develop under the influence of downstream controls (Fig. 3.11), and therefore consist of conventional systems tracts defined in relation to shoreline trajectories (Figs. 4.39 and 5.15). Coastal and shallow-water systems are best represented in shelf settings, where they contribute to the topset of shelf-slope systems (Fig. 4.60). Sedimentation within shallow seas can be highly variable, from siliciclastic to chemical or biochemical, depending on the tectonic setting (e.g., leading or not to enhanced extrabasinal sediment supply) and the climatic conditions at syn-depositional time (e.g., conducive or not to chemical or biochemical precipitation). Siliciclastic deposits are exemplified as default systems throughout the book, unless otherwise specified. However, the natural complexity of coastal to shallow-water systems is important to consider, as it explains the variability of sequences in terms of sedimentological makeup, relative development of systems tracts, and the patterns of sediment distribution along dip and strike directions. Fluvial incision during stages of relative sea-level fall may also affect the preservation of highstand deposits and further complicate the stratigraphic architecture (Fig. 8.13).

At the limit between continental and marine or lacustrine environments, coastlines play a key role in the architecture of the downstream-controlled sequence stratigraphic framework, as they record the timing of formation of systems tracts and bounding surfaces (Fig. 5.92). Irrespective of the extrabasinal or intrabasinal origin of the sediment, it is the interplay of accommodation and sedimentation *in coastal environments* that controls shoreline trajectories and changes thereof (Fig. 4.32). In a paleogeographic context, the location of the coastline at any timestep controls the areal extent of unconformities (e.g., the downdip limit of subaerial unconformities, or the updip limit of transgressive

FIGURE 8.13 Stratigraphic architecture in a coastal to shallow-water setting (Upper Miocene, Cape Bon, northern Tunisia; image courtesy of Serge Ferry). A—Transgressive succession of shoreface bedded tempestites to offshore clays; IVF—incised valley fills (estuarine megarippled sandstones); RS—wave-ravinement surface; SU—subaerial unconformity. The subaerial unconformity separates two depositional sequences which consist exclusively of transgressive deposits at this location.

surfaces of erosion), which is important for a number of reasons, including the exploration for mineral deposits associated with unconformities (e.g., placers). Downdip from the coastline, the shallow-water systems in shelf settings extend as far as the shelf edge, which marks the limit between shallow- and deep-water environments (Fig. 4.60). The downdip extent of shallow-water systems is more difficult to pinpoint in ramp settings, where processes of sediment transport beyond the toe of coastal clinoforms may show affinity to either shallow- or deep-water environments, depending on the seafloor gradients and sediment supply (Fig. 8.1).

8.2.2.1 Sequences in siliciclastic settings

Stratigraphic sequences in siliciclastic shallow-water settings are commonly asymmetrical, dominated by regressive deposits, and prevalent in tectonically active basins which generally receive more extrabasinal sediment. The relatively poor representation of transgressive deposits is due primarily to the retention of riverborne sediment within backstepping coastal systems such as estuaries, which act as a buffer between the continental and marine environments. As a result, the seafloor is usually starved during transgressions, leading to the formation of "healing-phase" condensed sections (Fig. 4.57) or even unconformities (Loutit et al., 1988; Galloway, 1989). Other factors that may lead to the poor development of shallow-water transgressive deposits are the high rates and short durations that characterize most transgressions driven by tectonic and climatic controls (Figs. 4.34, 5.23, and 8.8). Despite these trends, sandy macroforms ("shelf ridges") may still form during transgressions, with sediment sourced from the tidal reworking of underlying regressive coastal systems (Figs. 5.64, 5.74–5.79; Posamentier, 2002; Saha et al., 2016). The formation of shelf ridges is enhanced during transgressions, due to the larger extent of shallow-water environments, with preservation owed to the overlying highstand sediments.

Siliciclastic sequences in tectonically active basins tend to be wedge shaped, thickening toward the active margin where the subsidence rates are highest (Figs. 7.15 and 8.5). Siliciclastic sequences may also develop in tectonically "passive" basins, although potentially mixed with chemical or biochemical sediments (Fig. 7.16). Among the four systems tracts of the conventional sequence stratigraphic framework, highstands tend to be dominant in tectonically active basins (Fig. 8.8), whereas both normal regressive systems tracts (lowstand and highstand) are well represented in tectonically "passive" basins (Fig. 7.16). These differences in stratigraphic architecture reflect to a large extent the patterns of basin subsidence (Fig. 4.33). Apart from normal regressive and transgressive deposits, siliciclastic sequences may also include falling-stage systems tracts in any sedimentary basin, whether tectonically "active" or "passive," where forced regressions are recorded. Falling-stage systems tracts may become dominant in the case of strongly asymmetrical cycles such as those that develop in response to glacial–interglacial stages during Icehouse periods (Figs. 4.34, 5.23).

Coastal to shallow-water environments are shaped by the interaction between sediment supply and the energy of the sediment-transport agents, which results in the partitioning of this area into several subenvironments with distinct morphological and sedimentological characteristics (Figs. 4.11 and 8.14). Coastlines in siliciclastic settings include river-mouth environments, which represent sediment entry points into the marine basin, as well as open shorelines in the areas between river mouths (Figs. 4.10 and 4.11). Most of the siliciclastic sediment is terrigenous in origin, transported from source areas to the receiving basin by water (rivers, runoff) or wind. Additional sediment supply may originate from coastal erosion (Figs. 4.55 and 4.56), as well as marine erosion, most commonly by processes of wave scouring in the shoreface (e.g., Figs. 4.44, 4.55, and 4.57). It can be noted that the river-mouth and shoreface environments collect most of the incoming sediment, therefore acting

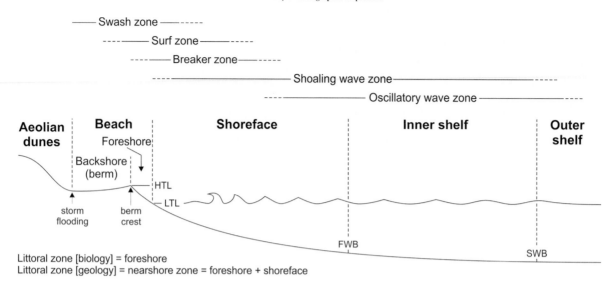

FIGURE 8.14   Partitioning of coastal and shallow-water environments into subenvironments separated by the high tide level (HTL), low tide level (LTL), fairweather wave base (FWB), and storm wave base (SWB) (modified after Walker and Plint, 1992, and Reading and Collinson, 1996). The effect of wave energy on the seafloor increases toward the coastline, generating a concave-up equilibrium profile. Preservation of this equilibrium profile during transgressions and forced regressions explains the formation of wave-ravinement surfaces and regressive surfaces of marine erosion, respectively (Figs. 4.44 and 4.55).

as staging areas before the sediment is dispersed to other depozones by various processes that operate within the coastal to shallow-water environments.

Processes of sediment transport in shallow water relate to several factors, including tides and fairweather waves, episodically enhanced by storms; sediment-gravity flows and even slumps, where the seafloor is steep enough to produce the necessary gravitational shear; and hypopycnal plumes of suspended sediment. Tides affect shorelines by rising and lowering the sea level, thus shifting the site of wave action and generating currents that can move large volumes of water back and forth across the intertidal area. In tide-dominated settings, tidal currents may become sufficiently strong to transport and rework sediment across the shelf (Figs. 5.64, 5.74–5.79). Fairweather waves give rise to a range of traction currents within the shoreface, which may be directed onshore or oblique to the shore, parallel to the shore (longshore currents), and offshore (rip currents). Storms increase the intensity

of fairweather processes, and result in the erosion of coastal systems (Fig. 8.15) and the movement of sediment beyond the intertidal and shoreface areas, into the supratidal and inner shelf environments. In the latter case, the storm currents result in the deposition of tempestites as far offshore as the storm wave base.

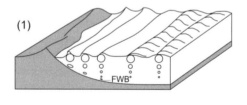

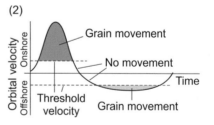

FIGURE 8.16   The fairweather "littoral energy fence" (modified after Swift and Thorne, 1991). (1) Wave transformation with depth and distance to the shoreline. The orbital diameter decreases with depth and becomes asymmetrical as it nears the seafloor and frictional drag increases. This frictional drag results in a to-and-fro motion of the sediment on the seafloor. (2) The effects on sediment movement during the passage of a shoaling wave. The onshore stroke of the wave as the crest passes carries more sediment than the offshore stroke associated with the passage of the trough. As a result, the sediment in the shoreface is pushed landward where it contributes to the construction of beaches. The opposite occurs during storms, when beaches are eroded and the sediment is transported seaward by storm currents, resulting in the deposition of tempestites (Fig. 8.15).

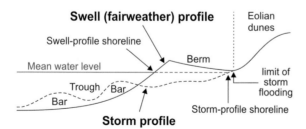

FIGURE 8.15   The beach cycle, between a fairweather profile, when a berm is constructed due to the landward sediment transport in the shoreface, and a storm profile, when the beach is eroded and the sediment is redistributed to the shoreface and shelf (modified after Komar, 1976).

The siliciclastic sediment in the shoreface does not pass easily out onto the shelf because the onshore stroke of the passing fairweather waves is stronger than the offshore stroke, resulting in a net sediment transport toward the land (Fig. 8.16). This asymmetrical orbital motion of the waves keeps any sediment coarser than mud within the shoreface and intertidal area, preventing the transfer of sediment onto the shelf during fairweather ("littoral energy fence"; Fig. 8.16). In addition to this "energy fence," the fairweather rip currents die out in the lower shoreface, which also helps to retain the sediment within the shoreface. The "energy fence" can be bypassed by (1) storm currents, which enable the redistribution of coastal and shoreface sediment onto the inner shelf; (2) tidal currents, which may surpass the influence of waves; and (3) gravity flows, which operate independently of the location of the wave base. In shelf settings, gravity flows and even small-scale slumps are most common in the steeper delta fronts of river-dominated deltas (Figs. 7.30 and 7.31), and also enable the transfer of sediment from deltas to the shelf edge where the shelf is dissected by unfilled incised valleys with sufficiently steep thalwegs (Fig. 5.34). Hypopycnal plumes provide an additional mechanism to transport fine-grained riverborne sediment out onto the shelf as a buoyant suspended load. Depending on the prevailing winds, waves and currents, hypopycnal plumes may carry fines far from the river mouth, even beyond the shelf edge, until sediments flocculate and settle down from suspension onto the seafloor.

At any location within the coastal to shallow-water environments, the seafloor gradient reflects the balance between sediment supply and the combined energy of all sediment-transport agents. Due to the loose nature of the sediment, the seafloor profile in siliciclastic shallow-water settings tends to be smooth, generally becoming steeper toward the coast where the sediment is coarser (i.e., with a steeper angle of repose) and the energy of waves and currents is higher (Fig. 8.14). This explains the concave-up profile of the shoreface–intertidal zone (Figs. 8.14 and 8.16), which represents the temporary base level at any point in time (Fig. 3.22). The tendency to maintain this graded seafloor profile through time explains the formation of wave-ravinement surfaces and regressive surfaces of marine erosion during transgressions and forced regressions, respectively (Figs. 4.44 and 4.55). The rigid frameworks that accumulate in shallow-water settings dominated by chemical or biochemical precipitation prevent the formation of the ideal concave-up profiles that typify the siliciclastic environments. The ragged seafloor profiles encountered in carbonate settings reflect a more pronounced 3D morphology of the temporary base level, which is influenced by the variable degree of lithification of the sediment across the seafloor. However, the principles that govern the formation of shoreline trajectories, systems tracts, and bounding surfaces remain the same in all depositional settings.

8.2.2.2 Sequences in carbonate settings

The sequence stratigraphic methodology is independent of tectonic and depositional settings, and therefore it applies to carbonate systems in the same way as it does to siliciclastics. The key differences relate to the intrabasinal origin of the carbonate sediment and the processes by which the sediment is generated. In turn, these differences affect the types of sedimentary facies that comprise systems tracts, and the development of systems tracts in the different subenvironments within the basin. Pure carbonate settings receive little or no riverborne clastic influx, and produce sediment by means of chemical or biochemical precipitation, most commonly within shallow-water "carbonate factories." The productivity of carbonate factories depends on several factors which are controlled by climate, geological setting, and stratigraphic age: water temperature, bathymetry, illumination, salinity, acidity, oxygenation, and turbidity; amounts of clastic influx and nutrients; the surface area of the "factory"; the rates of relative sea-level changes; and the types of carbonate-producing organisms (Tucker and Wright, 1990; Coniglio and Dix, 1992; James and Kendall, 1992; Jones and Desrochers, 1992; Pratt et al., 1992; Schlager, 2005; James and Jones, 2015). The main types of carbonate settings are illustrated in Fig. 8.17.

The *in situ* production of carbonate sediment, much of which is linked to depth-dependent photosynthetic activity, establishes a close tie between carbonate production and water depth, with the highest rates of generation near the sea level. The depth-dependent character of most *in situ* carbonate facies triggers a high response of carbonate systems to changes, as well as to the rates of change, in relative sea level. In that respect, carbonate productivity and facies change more significantly than their siliciclastic counterparts between the lowstands and the highstands in relative sea level. Additionally, carbonate sediments often have a biochemical origin and so are influenced by the chemistry of the water from which they precipitate. This means that carbonate facies can also be used to determine changes in paleoecology and related changes in

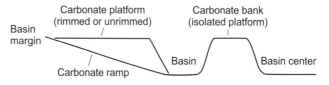

FIGURE 8.17 Carbonate settings (modified after James and Kendall, 1992).

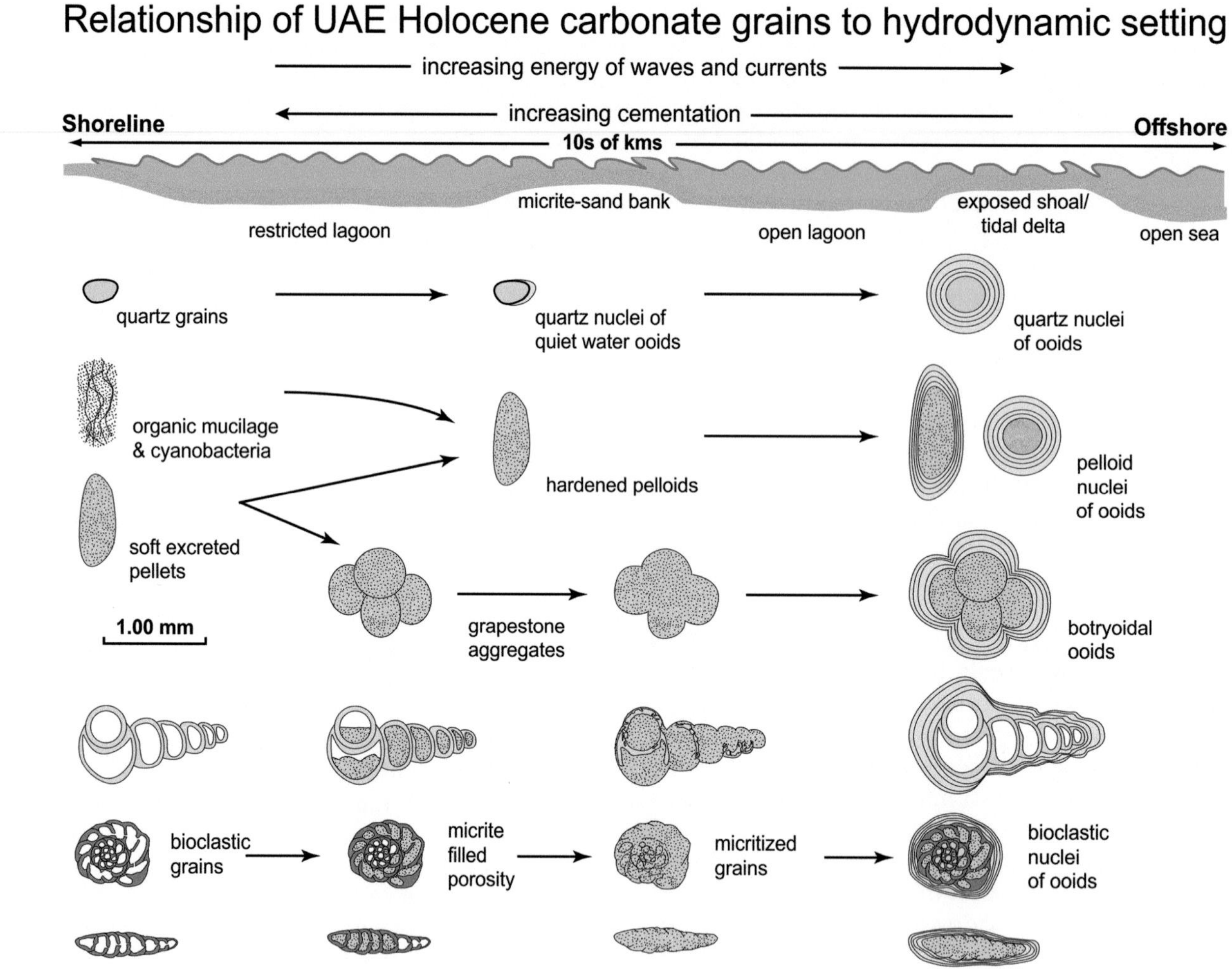

Relationship of UAE Holocene carbonate grains to hydrodynamic setting

————— increasing energy of waves and currents —————→

Shoreline   ←————— increasing cementation —————   **Offshore**

←————— 10s of kms —————→

micrite-sand bank

restricted lagoon   open lagoon   exposed shoal/ tidal delta   open sea

quartz grains → quartz nuclei of quiet water ooids → quartz nuclei of ooids

organic mucilage & cyanobacteria

soft excreted pellets

hardened pelloids →   pelloid nuclei of ooids

1.00 mm

grapestone aggregates →   botryoidal ooids

bioclastic grains →   micrite filled porosity →   micritized grains →   bioclastic nuclei of ooids

FIGURE 8.18   Relationship of carbonate grains to cementation and the energy of waves and currents, based on the Holocene carbonate ramp of the United Arab Emirates (see text for details). This scenario is also analogous to epeiric sea settings (from Catuneanu et al., 2011; diagram courtesy of Christopher Kendall).

paleogeography, including the development of isolation or access to the open sea (e.g., restricted conditions in low-energy lagoons promote penecontemporaneous cementation but prevent the formation of cortical layers, whereas the opposite is true for higher energy settings connected to the open sea; Fig. 8.18).

Beyond the differences in the origin of the sediment, all concepts that enable the application of sequence stratigraphy, including *accommodation* (i.e., relative sea-level changes), *sedimentation* (i.e., base-level changes), and the types of shoreline trajectories, systems tracts, and bounding surfaces that result from the interplay of accommodation and sedimentation, remain the same for carbonate settings. Below the sea level, which marks the top of the available accommodation, the seafloor is shaped by competing processes of sediment generation and reworking. Following the precipitation of carbonates, sediment reworking and redistribution within the basin may occur as a result of mechanical erosion by waves and currents,

and bioerosion. The bulk of this sediment is generated within the shallow-water carbonate factory, and part of it may be remobilized and transported to the deeper portions of the basin by storm currents and gravity flows (Hine et al., 1981, 1992; Tucker et al., 1993; Spence and Tucker, 1997; Ineson and Surlyk, 2000). What makes carbonates unique is the variety of processes that generate sediment with different degrees of consistency, from loose grains to rigid frameworks. Penecontemporaneous cementation also stabilizes the seafloor and restricts sediment mobility.

The morphology of the seafloor depends on the type and consistency of the sediment, which may vary with the location. In the case of loose carbonate sediment (e.g., oolitic, peloidal, bioclastic), the seafloor displays the concave-up profile that is observed in siliciclastic settings, which reflects the gradual changes in grain size and energy levels with water depth (Fig. 8.14). In the case of coherent sediment, the seafloor has a more irregular 3D morphology that reflects the variability of the ecosystem and the degree of sediment consistency, although the water depth at any location is still controlled by the same interplay between carbonate production and erosion. In both cases, processes of sediment production and reworking compete toward a state of equilibrium which is attained at the *temporary base level* (Fig. 3.22; details in Chapter 3). As the type and consistency of the sediment, as well as the environmental energy, change with time and location, the temporary base level is a dynamic 3D surface which is constantly adjusting to new equilibrium conditions. Proposals were made to distinguish between "physical" and "ecological" accommodation based on the dominant depositional process (loose vs. coherent sediment; Pomar and Kendall, 2008); however, these are aspects of sedimentation rather than accommodation, as the latter is always "physical" irrespective of the depositional setting (Fig. 3.22; details in Chapter 3).

Examples of carbonate settings with strong 3D seafloor morphology include unrimmed (open-shelf) platforms with large-skeleton metazoans and microbial structures like stromatolites; rimmed platforms with barrier reefs; and ramps with fringe reefs and mud-mounds generated below the photic zone (MacNeil and Jones, 2006; James et al., 2010; James and Jones, 2015). The top of active biogenic frameworks is kept in check by the energy of waves and currents, which limits the growth of carbonate buildups. Any carbonate sediment above the temporary base level is transported by waves and currents, and contributes to the development of coastal systems during fairweather or it may be shed onto the slope during storms. The higher the productivity of the carbonate factory the more sediment is delivered to the deeper basin as clastic carbonate, potentially generating slope clinoforms that resemble the morphology of a siliciclastic foreset (Figs. 5.83 and 6.35). While the relative sea level is only one of several factors which influence carbonate production, it has been observed that the highest productivity is most commonly recorded during times of highstand, with a consequent high delivery of carbonate detritus to the deeper basin (i.e., "highstand shedding"; Droxler and Schlager, 1985; Eberli et al., 1994; Andresen et al., 2003; Belopolsky and Droxler, 2003; Schlager, 2005, Fig. 5.83).

The response of carbonate systems to relative sea-level changes is illustrated in Fig. 8.10. Fundamentally, changes in relative sea level have the opposite effect in carbonate vs. siliciclastic settings in terms of the patterns of sediment dispersal across a basin. As shown by studies of the sedimentation rates during the late Quaternary relative sea-level cycles in various low- and high-latitude continental margin settings (Droxler and Schlager, 1985; Schlager, 1992, 2005; Sweet et al., 2019), deep-water siliciclastic deposits accumulate most rapidly during lowstands in relative sea level, when terrigenous sediment is delivered most efficiently across subaerially exposed continental shelves to the shelf edge ("lowstand shedding"; Fig. 5.34). At the same time, carbonate production can be reduced substantially during times of lowstand, and instead, the rates of aggradation of deep-water carbonate deposits are highest during relative sea-level highstands, when the carbonate factory on the shelf is most productive ("highstand shedding"; Figs. 5.83 and 8.20). This opposite response of carbonate and clastic systems to relative sea-level changes is a consequence of the intra- vs. extrabasinal origin of the sediment, respectively. However, the response of carbonate systems to relative sea-level changes also depends on the type of carbonate setting (e.g., carbonate ramps vs. platforms and banks), as discussed below.

In contrast to the siliciclastic systems, which can aggrade in all depositional settings as long as sediment is available, carbonates are much more susceptible to water depth and environmental conditions in general. The carbonate factory can be shut down by various factors, including subaerial exposure, drowning, excess of clastic influx, or lack of nutrients. When the relative sea level rises, whether or not carbonate production increases in tandem will depend on the rate at which the new space is added relative to the ability of the carbonate factory to keep up (Fig. 5.83). When carbonate accumulation cannot keep up with the rate of generation of new accommodation, water depth progressively increases and carbonate systems struggle to catch up with the rise in relative sea level (case 4 in Fig. 8.19); in this case, condensed sections or even marine hardgrounds may develop. This decrease in productivity may not vary linearly with respect to the rate of relative sea-level rise, as other environmental factors also play a role (e.g., transgressive deposits in warm water are potentially thicker than those that accumulate in cold water or in siliciclastic settings; Kerans et al., 1995; Kerans and Loucks, 2002). However, rapid water deepening may also cause the carbonate factory to "give up" and drown, irrespective of other environmental controls (Schlager, 1989; case 6 in Fig. 8.19; Fig. 8.21).

Coastlines and carbonate factories backstep during transgressions in response to the increase in water depth. Despite this general trend, barrier reefs along rimmed platform margins may still be able to aggrade and prograde, if the transgression is slow enough to prevent drowning (case 4 in Fig. 8.19). This underlines the point that it is the shift of the shoreline rather than the shift of the shelf edge which matters to the definition of systems tracts and bounding surfaces (details in Chapter 4). Where drowning occurs, carbonate factories decrease in size during transgression, becoming restricted to increasingly smaller areas adjacent to the

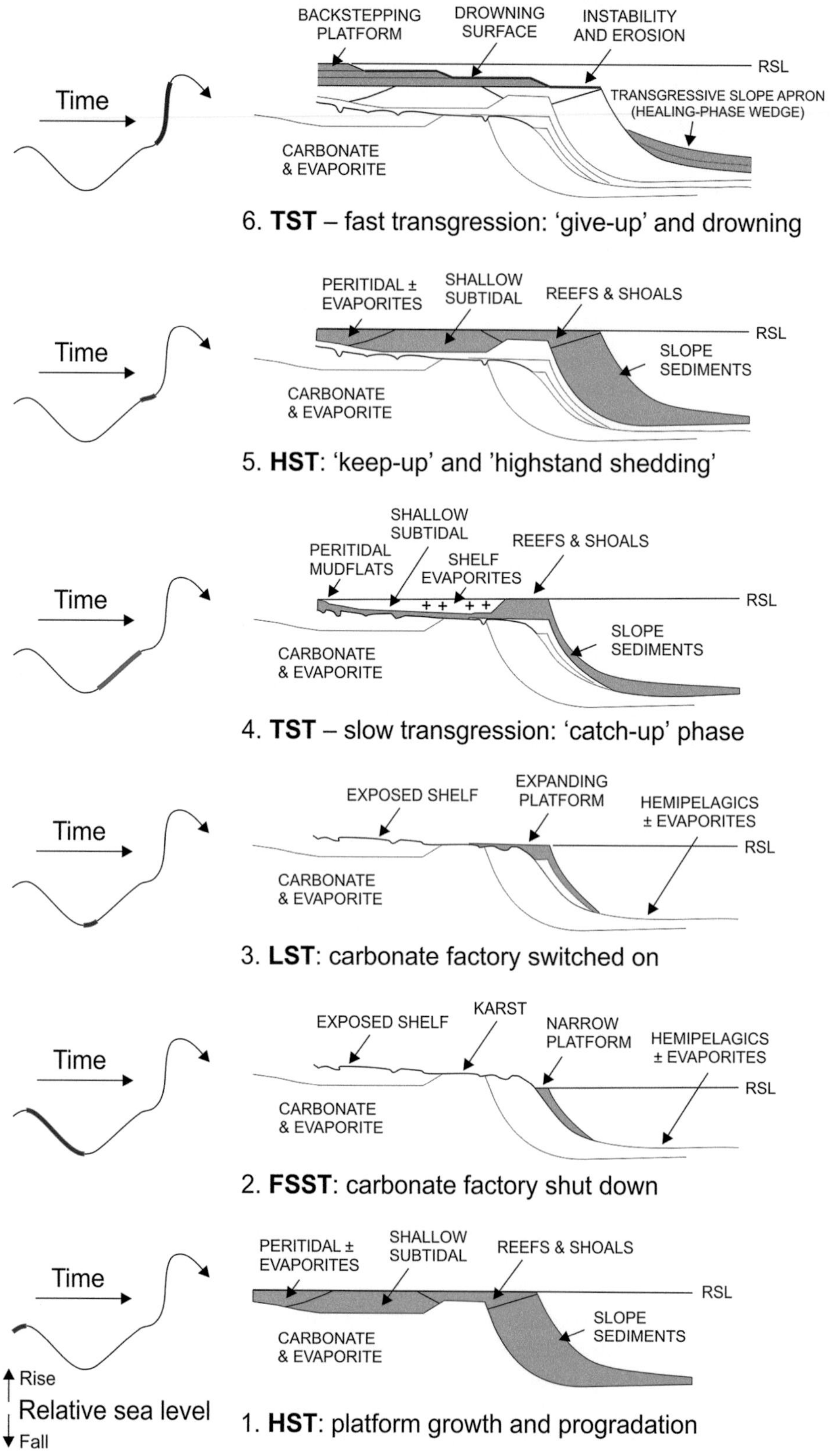

FIGURE 8.19 Evolution of a carbonate platform during various stages of relative sea-level change (modified after James and Kendal, 1992; Jones and Desrochers, 1992; and Schlager, 1992). Coastlines and carbonate factories prograde during normal and forced regressions, and backstep during transgressions. However, barrier reefs along platform margins may prograde at any time, including during shoreline transgression. This shows that it is the shift of the shoreline rather than of the shelf edge that matters to the definition of systems tracts and bounding surfaces. The stages indicated in this diagram may occur in any succession; e.g., stages 1–4 may repeat without the formation of drowning surfaces. Where drowning surfaces do form, due to rapid increases in water depth, they are typically diachronous, younging in a landward direction, and overlain by hemipelagic sediment following the shutdown (drowning) of the carbonate factory.

shoreline (case 6 in Fig. 8.19). In this case, the backstepping stacking pattern is most evident, and limestones are overlain abruptly by transgressive fines. This lithological contact was termed "drowning unconformity"

(Schlager, 1989, 1992), which is a flooding surface in a carbonate setting (Fig. 8.21). As transgression and backstepping take time, the drowning unconformity is diachronous, younging in an updip direction (Figs. 8.19

FIGURE 8.20 Highstand systems tract in a carbonate shelf-slope setting (Triassic, The Dolomites, Italy). Topset (platform top): *in situ* peritidal facies; foreset (slope): allochthonous calcirudites and calcarenites; bottomset (basin floor): calcilutites. Carbonate slopes are often steeper than the clinoforms of siliciclastic shelf-slope systems, due to the rigid nature of carbonate frameworks or to the typically coarser carbonate detritus.

FIGURE 8.21 Drowning unconformity (DU) at the top of backstepping platform carbonates (Triassic, The Dolomites, Italy). The drowning unconformity is a type of flooding surface which forms as a result of abrupt water deepening in a carbonate setting. Following the drowning, the limestone is overlain by hemipelagic transgressive sediment (more recessive, not preserved in this example). During drowning, carbonate factories decrease in size and become increasingly restricted to smaller areas adjacent to the backstepping shoreline. As transgression takes time, the drowning unconformity is diachronous, younging in the updip direction.

and 8.21). Where transgression and water deepening are gradual, drowning unconformities do not form; instead, the transgressive succession may be represented by "dirty" limestones interbedded with shales, in which the percentage of shale increases upward to the maximum flooding surface (Fig. 8.22).

Where the rates of carbonate sedimentation keep pace with the rates of accommodation added, water depth is

FIGURE 8.22 Stratigraphic architecture of a carbonate platform (Triassic, The Dolomites, Italy). HST: massive ("clean") limestone, subtidal facies; LST: intertidal to supratidal facies; TST: "dirty" shelf limestone interbedded with shale, with the shale content increasing upsection toward the MFS. Abbreviations: HST—highstand systems tract; LST—lowstand systems tract; TST—transgressive systems tract; SU—subaerial unconformity; WRS/MRS—wave-ravinement surface reworking a maximum regressive surface; MFS—maximum flooding surface.

maintained or reduced, and aggradation may be accompanied by progradation if sediment production exceeds the space available to retain it. The extent to which progradation occurs depends on the amount of excess sediment, which can contribute both to the progradation of the coastline and to the lateral build out of the margin (Fig. 5.83). Both lowstand and highstand normal regressions are times when sedimentation exceeds accommodation, which happens most commonly when the rates of relative sea-level rise are low (Fig. 8.19). However, highstands are prone to generate more sediment, due to the larger size of the carbonate factory following

FIGURE 8.23 Sedimentological contrasts between highstand, lowstand, and transgressive systems tracts in a carbonate platform setting (Triassic, The Dolomites, Italy). The highstand consists of massive ("clean") limestone accumulated in a subtidal to shelf environment. The lowstand consists of interbedded limestone and continental fines, indicative of an intertidal to supratidal environment. Transgressive shales cap the regressive deposits. Abbreviations: HST—highstand systems tract; LST—lowstand systems tract; TST—transgressive systems tract; SU—subaerial unconformity; MRS—maximum regressive surface.

transgression. At any location, the highstand and lowstand systems tracts differ in terms of their faunal and sedimentological character, due to the abrupt basinward shift of facies across the subaerial unconformity (Figs. 6.26 and 8.23). The faunal and sedimentological character of transgressive deposits can also differ from those of the preceding lowstands or the following highstands, generally with higher contributions from hemipelagic sediment and evaporites (Figs. 8.19 and 8.23). Highstand systems tracts always follow transgressions, and may be followed either by exposure (i.e., relative sea-level fall; Figs. 8.19 and 8.24) or by transgression (i.e., an increase in the rates of relative sea-level rise; Figs. 8.19 and 8.25).

The lowering of the wave base during relative sea-level fall leads to increased reworking of the seafloor; as a result, the falling-stage systems tract may include coarser carbonate detritus than the underlying highstand and overlying lowstand systems tracts (Kunzmann et al., 2020), including slope aprons of limestone breccias (Eberli et al., 1994; Spence and Tucker, 1997; Playton and Kerans, 2002). In this case, the basal surface of forced regression and the correlative conformity are placed at the base and at the top of the coarser clastic carbonates, respectively (Fig. 8.26). A fall in relative sea level leads to the shut down of the carbonate factory in the exposed region, while carbonate production may still continue basinward of the subaerial unconformity (Fig. 8.19). Subaerial exposure may result in karstification, particularly under humid climatic conditions where rivers and/or meteoric water produce dissolution features (Fig. 6.17), or in the development of calcrete horizons in arid and semiarid regions. While stages of relative sea-level fall generate subaerial unconformities in all carbonate settings, the carbonate production is less affected in ramp settings than it is on carbonate platforms and banks (Fig. 8.27). This is because ramps do

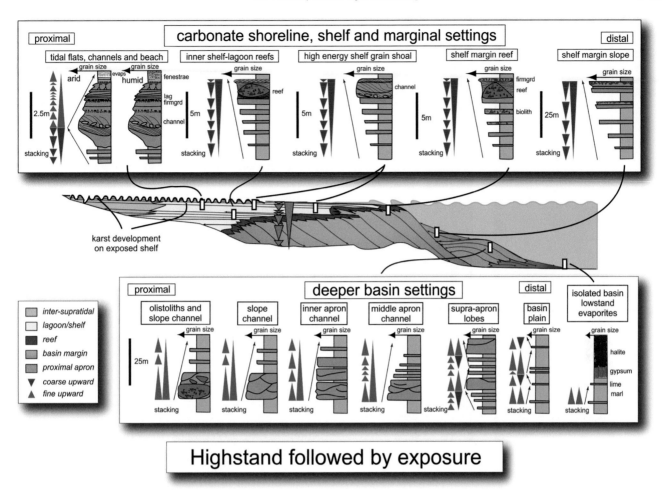

FIGURE 8.24 Highstand carbonate stacking in a platform to basinal setting, followed by exposure and the development of a subaerial unconformity (karst surface) (modified after Catuneanu et al., 2011; diagram courtesy of Christopher Kendall).

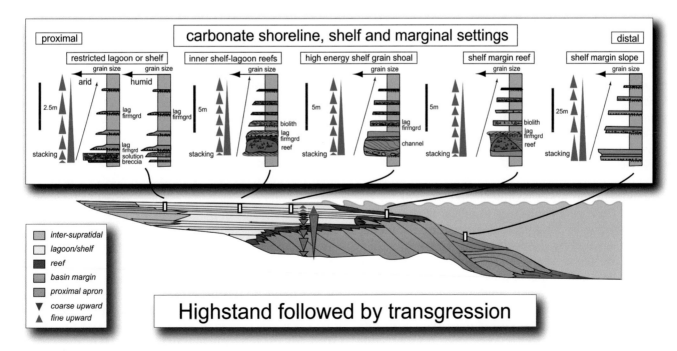

FIGURE 8.25 Highstand carbonate stacking in a platform to basinal setting, followed by transgression and backstepping of the carbonate factory (modified after Catuneanu et al., 2011; diagram courtesy of Christopher Kendall).

FIGURE 8.26 Basal surface of forced regression (BSFR) and correlative conformity (CC) in a Proterozoic carbonate succession (McArthur Basin, Australia; images courtesy of Marcus Kunzmann; from Kunzmann et al., 2020). The lowering of the wave base during relative sea-level fall led to increased reworking of the seafloor; as a result, the falling-stage systems tract includes coarser carbonate detritus relative to the underlying and the overlying highstand and lowstand systems tracts. In this case, the BSFR and the CC are placed at the base and at the top of the coarser clastic carbonates, respectively. Abbreviations: HST—highstand systems tract; FSST—falling-stage systems tract; LST—lowstand systems tract.

not record abrupt increases in water depth beyond the downdip limit of the carbonate factory, so the area of carbonate production can shift basinward during relative sea-level fall.

Even though accommodation and sedimentation in carbonate settings can be measured independently of each other, just as in siliciclastic settings, changes in accommodation do affect sedimentation in terms of the ability of the carbonate factory to adjust to new bathymetric conditions. This process—response relationship is often mediated by the seafloor physiography, which controls the areas available for biota to thrive. For example, the aggradation of lowstand topsets may start around the highstand reefal buildups before expanding to other areas. During transgressions that are slow enough to prevent the drowning of the carbonate factory (stage 4 in Fig. 8.19), barrier reefs tend to grow along platform margins, where nutrients are made available

by upwelling currents related to deep-water circulation. The physiography of carbonate settings (e.g., shelf-slope systems vs. ramps; Fig. 8.17) also influences the relative development of systems tracts, as relative sea-level changes have a much more profound effect on low-gradient carbonate factories (e.g., at the top of platforms and banks; Fig. 8.17) than on sloping ramps where the location of the "factory" can adjust to changes in relative sea level. As a result, the difference between lowstands and highstands is much more evident in the case of carbonate platforms and banks (Figs. 6.26, 8.19, 8.23) than it is in the case of carbonate ramps (Fig. 8.27).

The relationship between accommodation and sedimentation in carbonate settings is complex, and also modulated by changes in environmental conditions that accompany the fluctuations in relative sea level. For example, stages of relative sea-level rise, especially in tectonically "passive" basins that support the development of carbonate systems, are often accompanied by increases in temperature which favor the precipitation of carbonates. At the same time, global warming and sea-level rise may also trigger changes in water acidity, circulation, and nutrients, which all affect the productivity of the carbonate factory. A lag time may also occur before the carbonate-producing organisms can adjust and respond to changes in accommodation (Schlager, 1981; Enos, 1991; Tipper, 1997), although this adjustment may only require calendar timescales and therefore can be neglected at stratigraphic scales (Denton, 2000). This illustrates the point made throughout this book that observations need to be separated from interpretations; the methodology that affords the identification of systems tracts based on the observation of stratal stacking patterns is independent of the interpretation (i.e., modeling) of the underlying controls (Fig. 3.3; see discussion in Chapter 9).

8.2.2.3 Sequences in mixed siliciclastic-carbonate settings

Sequences in mixed siliciclastic-carbonate settings include both terrigenous sediment and *in situ* carbonate produced inorganically or by organisms, and therefore require at least in part favorable conditions for carbonate production. Terrigenous influx into a carbonate setting may change the style of sedimentation to siliciclastic, either locally or regionally, as the rates of clastic deposition can be orders of magnitude higher than the rates of carbonate precipitation. Therefore, even though carbonate production may continue after the onset of clastic influx, the amount of carbonate in the sediment may be too small to change the overall siliciclastic nature of the deposit. Terrigenous material also has a detrimental effect on carbonate production, as it affects the carbonate-secreting organisms in several ways. Turbidity caused by fine-grained material in suspension affects organisms by reducing light penetration, driving

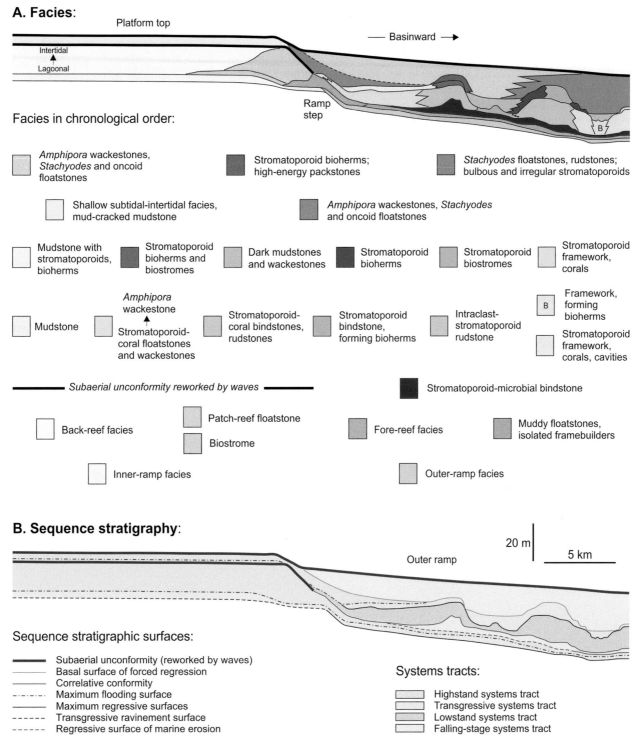

A. Facies:

Platform top

— Basinward →

Intertidal

Lagoonal

Ramp step

Facies in chronological order:

Amphipora wackestones, Stachyodes and oncoid floatstones

Stromatoporoid bioherms; high-energy packstones

Stachyodes floatstones, rudstones; bulbous and irregular stromatoporoids

Shallow subtidal-intertidal facies, mud-cracked mudstone

Amphipora wackestones, Stachyodes and oncoid floatstones

Mudstone with stromatoporoids, bioherms

Stromatoporoid bioherms and biostromes

Dark mudstones and wackestones

Stromatoporoid bioherms

Stromatoporoid biostromes

Stromatoporoid framework, corals

Mudstone

Amphipora wackestone

Stromatoporoid-coral floatstones and wackestones

Stromatoporoid-coral bindstones, rudstones

Stromatoporoid bindstone, forming bioherms

Intraclast-stromatoporoid rudstone

B Framework, forming bioherms

Stromatoporoid framework, corals, cavities

Subaerial unconformity reworked by waves

Stromatoporoid-microbial bindstone

Back-reef facies

Patch-reef floatstone

Biostrome

Fore-reef facies

Muddy floatstones, isolated framebuilders

Inner-ramp facies

Outer-ramp facies

B. Sequence stratigraphy:

20 m

5 km

Outer ramp

Sequence stratigraphic surfaces:

Subaerial unconformity (reworked by waves)
Basal surface of forced regression
Correlative conformity
Maximum flooding surface
Maximum regressive surfaces
Transgressive ravinement surface
Regressive surface of marine erosion

Systems tracts:

Highstand systems tract
Transgressive systems tract
Lowstand systems tract
Falling-stage systems tract

FIGURE 8.27　Stratigraphic architecture of a Late Devonian reef system that developed on a gently sloping epicontinental ramp (Alexandra Formation, Western Canada Sedimentary Basin; modified after MacNeil and Jones, 2006). Vertical exaggeration is 250 times.

animals and plants to live at shallower depths, while clogging up their feeding systems. Sudden influxes of terrigenous mud and sand can smother and bury organisms, and the high nutrient levels that often accompany terrigenous input can lead to intense microbial activity while being harmful to the many oligotrophic organisms which produce carbonate skeletal material. However, if a balance is reached between siliciclastics and carbonates, a depositional setting may display mixed features that typify both settings.

Carbonate-producing organisms have made many adaptations to clastic influx, and so there are many examples today where such animals can still thrive in areas of terrigenous sedimentation. For example, this can lead to the formation of patch reefs on clastic shelves (e.g., the inner part of the Queensland shelf, inboard from the Great Barrier Reef; Larcombe et al., 2001), or to reefs able to grow on abandoned delta-front bars in areas dominated by terrigenous mud (e.g., the Mahakam Delta in Indonesia; Wilson and Lockier, 2002). In the Red Sea, coral reefs are in close proximity to wadi-fan delta systems which periodically shed clastic sediment over the reefs (Tucker, 2003). There are also numerous examples in the geological record of reefal carbonates associated with clastics (e.g., Santisteban and Taberner, 1988; Braga et al., 1990). In some cases, carbonate and siliciclastic deposits alternate in a vertical succession, each dominating specific systems tracts, whereas in other cases they may coexist side by side within the same systems tracts due to lateral changes in the dominant style of sedimentation. These mixed systems occur where carbonate platforms or ramps are attached to terrigenous source areas or where there is an axial supply of siliciclastic sediment through longshore drift or contour currents.

The timing and location of siliciclastic input into the basin are variable, depending on multiple factors including sea-level changes, tectonism (e.g., fault reactivation in fault-bounded basins, advancing thrust sheets in orogen-related basins), autogenic shifts of alluvial channel belts, and climate. For this reason, no single model can describe all mixed siliciclastic-carbonate sequences worldwide. Numerous variations can be identified, involving vertical changes, lateral changes, or both (Fig. 8.28; Zecchin and Catuneanu, 2017). Vertical changes between systems tracts involve most commonly transgressive—highstand carbonates

and falling-stage—lowstand siliciclastics, as exemplified by the Quaternary deposits of the Caribbean area where clastic input to the deep sea was much greater during glacial times and carbonate production was higher during the interglacials when platforms were flooded (Schlager et al., 1994). Ancient examples are documented from the Cambrian of northern Greenland (Ineson and Surlyk, 2000), Devonian of the Canning Basin (Southgate et al., 1993; Playford et al., 2009), the Permian of Texas (e.g., Saller et al., 1989), and the Tertiary of northeast Australia and Gulf of Papua (Davies et al., 1989). This scenario defines the notion of "reciprocal sedimentation," which involves the "lowstand shedding" of clastics and the "highstand shedding" of carbonates (Van Siclen, 1958; Wilson, 1967).

Departures from the "reciprocal sedimentation" model are common, including cases where transgressive carbonates are overlain by highstand clastics. In this case, siliciclastic sediment starvation of the seafloor during transgressions enables the deposition of carbonates, whereas the subsequent progradation of highstand deltas shuts down the carbonate factory. This scenario requires availability of riverborne sediment, as documented in the mid-Carboniferous of North America (e.g., Soreghan, 1997; Rankey et al., 1999; Miller and Eriksson, 2000; Smith and Read, 2001) and Western Europe (Tucker et al., 2009). Conversely, the transgressive deposits can be dominantly siliciclastic, while the highstands record the deposition of carbonates (Fig. 6.26). Examples of sequences which include basinal to outer shelf transgressive mudstones grading upward into shallow-water highstand limestones have been documented within the Cambrian "Grand Cycles" of North America and China (Aitken, 1978; Meng et al.,

FIGURE 8.28 Variability of mixed siliciclastic—carbonate sequences between fully siliciclastic and fully carbonate systems (from Zecchin and Catuneanu, 2017). Carbonates are generally more common in the TST—HST (A) or, less frequently, in the FSST—LST (B). Abbreviations: TST—transgressive systems tract; HST—highstand systems tract; FSST—falling-stage systems tract; LST—lowstand systems tract; TSE/MRS—transgressive surface of erosion replacing a maximum regressive surface; MFS—maximum flooding surface; RSME/BSFR—regressive surface of marine erosion replacing the basal surface of forced regression; CC—correlative conformity; MRS—maximum regressive surface.

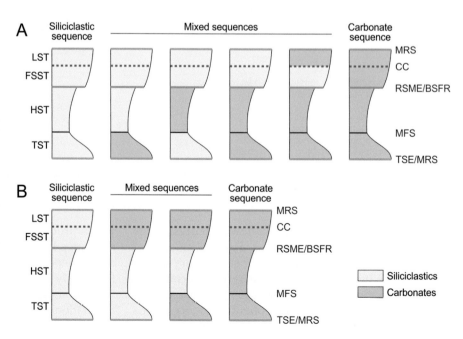

1997; Glumac and Walker, 2000), the Carboniferous strata of South Wales (Burchette and Wright, 1992), the Triassic Muschelkalk of Germany (Aigner, 1984), and the Cretaceous of the central and western Pyrenees (Simo, 1989; Lenoble and Canerot, 1993) and Austrian Alps (Sanders and Hofling, 2000).

Mixed siliciclastic—carbonate sequences with transgressive limestones accumulated during the developing glaciation in Gondwana, in the transition interval between the Lower Carboniferous carbonate sequences and the Upper Carboniferous clastic sequences (Fig. 8.29; Tucker et al., 2009). The transgressive limestones of the mid-Carboniferous "Yoredale cycles" consist of bedded bioclastic packstones and wackestones with local coral biostromes, underlain by a transgressive lag up to 0.5 m thick of sand- or mud-rich bioturbated bioclastic packstone, with the clastic material derived from reworking of the deposits below. The "Yoredale" sequences vary in thickness from 5 to 70 m, with the carbonate part up to 30 m thick. It is estimated that deposition occurred in a low-energy epeiric-type platform setting with water depths of 10—40 m. Following the accumulation of transgressive limestones and highstand clastics, the shelf was subject to exposure and fluvial incision during relative sea-level fall. The lowstand fills of the incised valleys complete a full cycle of sedimentation. The Yoredale sequences are attributed

to glacio-eustatic changes as a result of orbital forcing at the scale of 100 kyr eccentricity cycles, with 400 kyr modulation (Davydov et al., 2010). However, tectonic overprinting can also locally mask the orbital-forcing signal (Wilkinson et al., 2003).

Less commonly, the highest amount of carbonate may be included in the falling-stage systems tract, while the other systems tracts are hybrid or siliciclastic, and less developed (Soreghan, 1997; Zecchin et al., 2010a, 2016; Massari and D'Alessandro, 2012, Figs. 8.30 and 8.31). The Gelasian succession in Capodarso, Italy, consists of hybrid transgressive facies, dominantly siliciclastic highstand mudstones and siltstones, and mostly carbonate falling-stage deposits (Massari and Chiocci, 2006; Massari and D'Alessandro, 2012, Fig. 8.30). In this case, carbonate production was enabled by a dry climate during glacio-eustatic sea-level fall, leading to the accumulation of 10—20 m thick wedges of packstones, grainstones, and rudstones (Massari and D'Alessandro, 2012, Fig. 8.30). Similar carbonate-rich forced regressive clinoforms have been documented in the late Pleistocene Le Castella terrace in southern Italy, overlying much thinner transgressive and highstand hybrid deposits (Zecchin et al., 2010a, 2016, Fig. 8.31). Besides *in situ* carbonate facies, falling-stage systems tracts may also include breccias and other clastic carbonates produced by the collapse of platform margins, but this

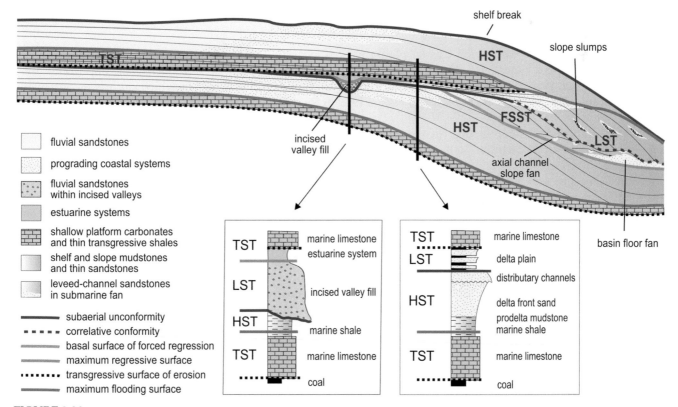

FIGURE 8.29 Stratigraphic architecture of the mid-Carboniferous Yoredale sequences in Northern England (modified after Tucker et al., 2009). Abbreviations: TST—transgressive systems tract; HST—highstand systems tract; FSST—falling-stage systems tract; LST—lowstand systems tract.

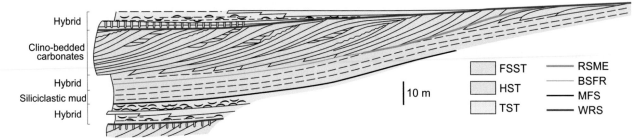

FIGURE 8.30   Architecture of the mixed siliciclastic-carbonate Gelasian sequence at Capodarso, Sicily, Italy (modified after Massari and D'Alessandro, 2012). Note the dominance of siliciclastic mudstone in the HST and of carbonates in the clinoforms of the FSST. In this case, carbonate production was favored by dry climate during glacio-eustatic sea-level fall. Abbreviations: BSFR—basal surface of forced regression; FSST—falling-stage systems tract; HST—highstand systems tract; MFS—maximum flooding surface; WRS—wave-ravinement surface; RSME—regressive surface of marine erosion; TST—transgressive systems tract.

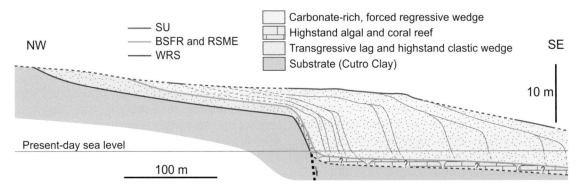

FIGURE 8.31   Dip section through the late Pleistocene Le Castella terrace, southern Italy (modified after Zecchin et al., 2010a, 2016). The Le Castella terrace includes transgressive and highstand systems tracts in the lower part, and a carbonate-rich forced regressive wedge that prograded seaward of a fault step. The highstand deposits consist of distal reefs and an attached nearshore clastic wedge. Abbreviations: BSFR—basal surface of forced regression; WRS—wave-ravinement surface RSME—regressive surface of marine erosion; SU—subaerial unconformity.

is more commonly the case with classic platforms that include highstand limestones (Eberli et al., 1994; Spence and Tucker, 1997; Ineson and Surlyk, 2000; Playton and Kerans, 2002; Kunzmann et al., 2020, Figs. 8.19 and 6.26).

Lateral variations in the dominant style of sedimentation are most commonly recorded within transgressive and highstand systems tracts, both along dip and strike directions (Zecchin and Catuneanu, 2017). Examples include the concurrent deposition of proximal siliciclastics and distal carbonates, or *vice versa*, as well as changes along strike that reflect variations in the amount of terrigenous sediment delivered to different areas along a shoreline. The latter depend on the location of river mouths (terrigenous sediment entry points) and the patterns of sediment dispersal within the basin (e.g., the direction and intensity of longshore currents). Autogenic processes of fluvial avulsion and delta switching modify the location of areas dominated by siliciclastic sedimentation, leading to the development of diachronous facies contacts within or between systems tracts. At any given time, the carbonates occupy the areas of lower terrigenous influx between the zones of

siliciclastic sedimentation (Sanders and Hofling, 2000). The lateral shifts of the siliciclastic depozones may lead to "*in situ* mixing," in which calcareous organisms grow on siliciclastic substrates (Mount, 1984). Facies mixing can also occur at the limit between coeval clastics and carbonates, resulting in hybrid deposits with diffuse lateral facies boundaries (Mount, 1984).

Shelves that exhibit lateral variations along dip may host proximal clastics and distal carbonates, as observed in the Triassic of British Columbia (Zonneveld et al., 2001), the Paleogene of North Carolina (Coffey and Read, 2007), the Neogene of Florida (McNeill et al., 2004), the Quaternary of the Crotone Basin, Italy (Nalin and Massari, 2009; Zecchin et al., 2009b, 2010a, 2011; Zecchin and Caffau, 2011), and modern environments (e.g., Purdy and Gischler, 2003; Brandano and Civitelli, 2007) (Figs. 8.31—8.33), although the distal carbonates are ultimately expected to pass into marls or siliciclastic mud toward the outer shelf and slope settings once the water depth exceeds the zone of optimal carbonate production (Coffey and Read, 2007; McNeill et al., 2012). The opposite trend of proximal carbonates and distal clastics has also been documented, in the Holocene of

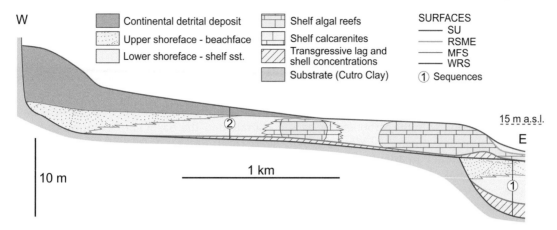

FIGURE 8.32   Dip section through the late Pleistocene Capo Colonna Terrace, southern Italy (modified after Zecchin et al., 2010b, 2016). Note the prevalence of siliciclastic deposits in proximal locations and of carbonate deposits in distal locations in sequence 2. The transgressive deposits include a mollusc-rich assemblage (OSB) above the WRS, overlain by a bryozoan-bearing accumulation (BSB) which straddles the MFS. The highstand deposits are represented by coralline algal patch reefs that pass laterally into calcarenites (Zecchin and Caffau, 2011). Abbreviations: WRS—wave-ravinement surface; MFS—maximum flooding surface; RSME—regressive surface of marine erosion; SU—subaerial unconformity; OSB—onlap shell bed; BSB—backlap shell bed.

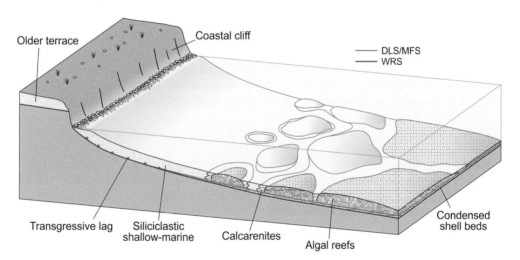

FIGURE 8.33   Depositional systems of the late Pleistocene Capo Colonna terrace (southern Italy) during highstand time (modified after Zecchin and Caffau, 2011). Sedimentation was dominated by siliciclastics in proximal settings and by carbonates offshore. The calcarenites generated by the erosion of algal reefs are locally interbedded with siliciclastic sandstones. Abbreviations: WRS—wave-ravinement surface; MFS—maximum flooding surface; DLS—downlap surface.

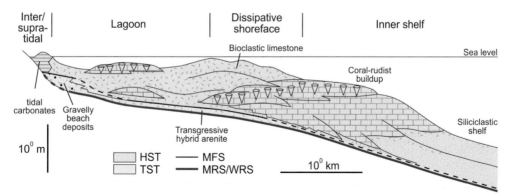

FIGURE 8.34   Mixed siliciclastic-carbonate shallow-water systems in the late Cretaceous succession of the Northern Calcareous Alps, Austria (modified from Sanders and Höfling, 2000). Note the dominance of highstand carbonates in the proximal areas of the basin, and the transition to siliciclastics offshore. Abbreviations: TST—transgressive systems tract; HST—highstand systems tract; MRS—maximum regressive surface; WRS—wave-ravinement surface; MFS—maximum flooding surface.

the Persian Gulf (Park, 2011) and the Late Cretaceous of the Austrian Alps (Sanders and Hofling, 2000) (Fig. 8.34). The former trend is more common, particularly in basins with terrigenous sediment influx which retain most of the riverborne sediment in coastal and shoreface systems. Beyond the fairweather wave base, storm currents may transport the sediment farther into the inner shelf, where mixing with the carbonate facies is most likely (i.e., the "punctuate mixing" of Mount, 1984).

Vertical and lateral changes in the composition of mixed sequences can be linked to a variety of controls, including climate, sea-level changes, local tectonism, and autogenic shifts in sediment supply, which may operate in tandem or independently. Local tectonism and autocyclicity may lead to both vertical and lateral facies changes, at basinal or sub-basinal scales. The clastics-to-carbonates facies contacts are often time-transgressive, and may or may not coincide with the systems tract boundaries. In the latter case, the within-trend facies contacts are more diachronous than the systems tract boundaries. For example, clastic sedimentation lasts longer in deltaic environments than in other areas where carbonate production can start earlier, resulting in age variations of the facies contacts along strike. Changes in climate and sea level trigger larger scale shifts in the sedimentation regimes, at basinal or even global scales. In some cases, these two controls may be out of phase, such as when climate shifts lead to the accumulation of carbonates during drier stages and siliciclastics during humid stages, irrespective of the coeval changes in sea level (Massari and D'Alessandro, 2012; Schwarz et al., 2016, Figs. 8.30 and 8.31). However, most field examples indicate a correlation between climate and sea-level changes, which is most evident in relation to glacial—interglacial cycles.

Mixed carbonate—clastic sequences are best developed at times of high amplitude sea-level change, and since these are typical of icehouse conditions, they are well-represented in the Quaternary and Permo—Carboniferous. During these times the high-frequency sea-level fluctuations were on the scale of 10s of meters and this was sufficient to bring terrigenous sediment into the depositional environment during a sea-level fall. The Permo—Carboniferous was also a time of global sea-level lowstand, when more of the continental landmasses were exposed for weathering, erosion, and denudation than at other times. The late Paleozoic was also a period of supercontinent assembly, when mountain ranges were forming, being uplifted and eroded (i.e., providing new sources of extrabasinal sediment). In contrast, during greenhouse times (e.g., lower- and mid-Paleozoic and Mesozoic), low amplitude sea-level changes were the norm and extensive carbonate platforms and clastic shelf seas were typical. Most shallow-marine continental-shelf sequences of these periods are composed of single, uniform lithofacies (all clastic or all carbonate), with mixed sequences being relatively rare.

8.2.2.4 Sequences in mixed carbonate-evaporite settings

Mixed carbonate-evaporite sequences develop where climate and the basin configuration provide the necessary conditions for the partial or total evaporation of water. Salinity is the key factor which controls the types of minerals that can precipitate at any given time, from carbonates to gypsum, halite and bittern salts as water concentration increases (Schreiber and Hsu, 1980). In turn, salinity reflects the balance between inflow and evaporation, which depends primarily on climate and tectonic setting (Schreiber and Hsu, 1980). Precipitation of evaporites can occur in a range of shallow- to deep-water environments at times of increased aridity (Warren, 2006). The majority of the world's thickest and most extensive evaporite successions were precipitated in basins separated from the global ocean by some barrier, placed in intracratonic, extensional or compressional settings. Examples include the Cambrian Siberian, Silurian-Devonian Michigan, Devonian Williston-Elk Point, Carboniferous (Pennsylvanian) Paradox, Permian Zechstein and Delaware, Aptian South Atlantic, and late Miocene (Messinian) Mediterranean and Paratethyan basins (Schreiber and Hsu, 1980; Warren, 2006; Csato et al., 2013, 2015; Brandao et al., 2020). In most of these, carbonate platforms developed around the basin margins when there was free circulation within the basins and open connection with the ocean (e.g., Csato et al., 2021).

Evaporites precipitate commonly in sabkhas and hypersaline lagoons, often extremely extensive, behind carbonate rims (Fig. 8.35). The evaporites within the basin centers, where the thickest successions occur (i.e., the saline giants), precipitate at times of reduced connection with, or complete isolation from, the open ocean, through the operation of a sill or barrier. Once sea level has dropped below the height of the barrier to cut the basin off totally, then the basin may draw down very quickly through evaporation. Some water can seep into the basin through marginal permeable rocks, and there may be some reflux of hypersaline brines out of the basin as well. In a similar way, global sea level may only need to rise a few meters to overtop the barrier and flood the basin. Much greater and more rapid water-level changes can thus be expected within such silled basins compared with those normally considered for siliciclastics and carbonates deposited on passive margins and in other completely open basins. In addition, water and brine levels within an isolated saline basin may vary independently of global sea level, responding especially to climate changes and local tectonism (e.g., the present-day Dead Sea).

Carbonate Mound

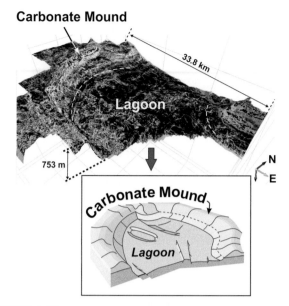

FIGURE 8.35 Lagoon identified with 3D seismic data along the northern margin of the Precaspian Basin (Upper Devonian—Tournaisian, Kazakhstan; from Csato et al., 2021). The lagoon formed by the reactivation of basement structures in the Devonian, and the marginal highs became sites of carbonate mound formation in Famennian—Tournaisian. The lagoon was subsequently filled with mixed sediments during the lower Visean (see Csato et al., 2021 for details).

The sequence stratigraphy of carbonate-evaporite systems has been explored and refined over the last three decades (Tucker, 1991; Sarg, 2001; Warren, 2006; Catuneanu et al., 2011). The origin of sequences relates to cycles of salinity change, but other ecological factors such as temperature and humidity are also important in controlling facies and faunal cyclicity in carbonate-evaporite settings (Pratt and Haidl, 2008). Prior to the onset of evaporite precipitation, a saline basin is typically connected to the global ocean, with the sea level above the sill height. This leads to the development of carbonate platforms or rims around the basin, under the generally ideal hot and dry climate, where there is limited siliciclastic input (Figs. 8.36A and 8.37A). The subsequent fall in relative sea level may result in incomplete or complete drawdown, depending on the sea-level position relative to the elevation of the outflow barrier. In the case of incomplete drawdown (i.e., sea level still above the sill height; Fig. 8.36), the salinity cannot increase enough to trigger the precipitation of halite and bittern salts, and the sequence is dominated by gypsum and carbonates. In the case of complete drawdown (i.e., sea level below the sill height; Fig. 8.37), evaporation can result in the precipitation of the entire series of salts, including the most soluble ones (i.e., halite and bittern salts).

A. Highstand open-basin carbonates

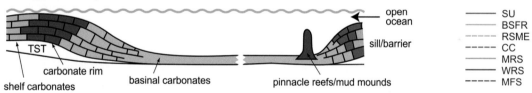

B. Falling-stage marginal gypsum wedge

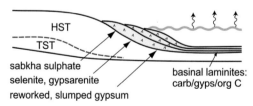

C. Lowstand sabkha/evaporitic lagoons

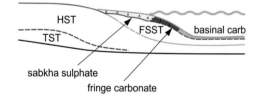

D. Transgressive carbonates

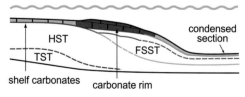

E. Highstand sabkha and carbonates

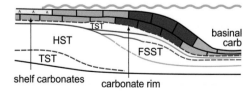

FIGURE 8.36 Stratigraphic architecture of carbonate—evaporite intracratonic basins with incomplete drawdown, leading to basin-margin gypsum wedges (modified after Tucker, 1991). Abbreviations: TST—transgressive systems tract; HST—highstand systems tract; FSST—falling-stage systems tract; MFS—maximum flooding surface; WRS—wave-ravinement surface; MRS—maximum regressive surface; CC—correlative conformity; RSME—regressive surface of marine erosion; BSFR—basal surface of forced regression; SU—subaerial unconformity.

A. Highstand open-basin carbonates

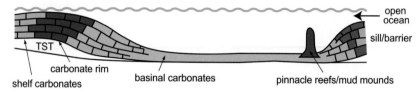

B. Falling-stage gypsum wedge, expanding salt pans

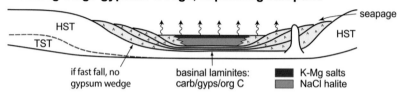

FIGURE 8.37 Stratigraphic architecture of carbonate–evaporite intracratonic basins with complete drawdown, leading to basin-fill halite (modified after Tucker, 1991). Abbreviations: TST—transgressive systems tract; HST—highstand systems tract; FSST—falling-stage systems tract; LST—lowstand systems tract. See Fig. 8.36 for the key to sequence stratigraphic surfaces.

C. Lowstand marginal carb-gypsum, shrinking halite zone

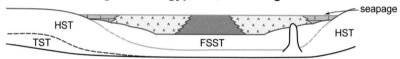

D. Transgressive gypsum and expanding carbonates

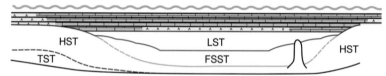

E. Highstand gypsum and shrinking carbonates

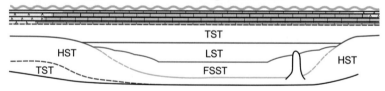

Systems tracts in carbonate-evaporite settings are defined by the same stacking-pattern criteria that are used in all other depositional settings (i.e., the balance between accommodation and sedimentation). Accommodation is modified by changes in relative sea level, and sedimentation refers to the rates of precipitation of evaporites. A rise in relative sea level can be driven by subsidence or eustatic rise, where the saline basin is connected to the global ocean, or by inflow outpacing evaporation in enclosed basins. Conversely, a fall in relative sea level can be driven by uplift or eustatic fall in open basins, or by evaporation outpacing inflow in isolated basins. In all cases, an increase in water concentration is required for evaporite precipitation. As long as salinity exceeds the solubility of evaporites, precipitation will occur whether the relative sea level is rising or falling. However, the order in which evaporite minerals precipitate reflects the trends of salinity changes, which correlate with relative sea-level changes. A rise in relative sea level (inflow exceeding evaporation) leads to dilution, in which case minerals precipitate in the order of decreasing solubility (e.g., gypsum following halite, or carbonate following gypsum). A fall in relative sea level (evaporation exceeding inflow) leads to an increase in salinity, which results in the precipitation of increasingly soluble minerals with time (e.g., halite after gypsum, or bittern salts after halite) (Figs. 8.36 and 8.37).

While the same types of systems tracts may form in a variety of settings which may experience complete or incomplete drawdown, their mineralogical composition reflects the local salinity conditions which can be modified by the degree of basin isolation, rates of evaporation, rates of fresh or seawater inflow, and influx of hydrothermal fluids. The latter may alter the relationship between salinity and relative sea-level changes in terms of the predicted mineralogical trends of systems tracts. However, when mineralogy is combined with stratal architecture, systems tracts can still be identified reliably. Figs. 8.36 and 8.37 describe the most common composition of systems tracts in mixed carbonate-evaporite settings, although departures from these trends may be encountered due to the variability in salinity levels between different basins. Following the highstand carbonates which precede evaporite precipitation in most mixed settings (Figs. 8.36A and 8.37A), the falling stage is typically the most prolific time for chemical deposition in both open and isolated saline basins. Open-basin settings are dominated by the less soluble components in the gypsum-carbonates range, whereas enclosed basins can accumulate more soluble salts (halite and K-Mg salts) (Figs. 8.36B and 8.37B).

As water concentration increases during relative sea-level fall, the falling-stage systems tract typically includes the most soluble salts of an evaporite or mixed carbonate-evaporite sequence, which expand areally with time at the expense of the less soluble minerals (Figs. 8.36B and 8.37B). Gypsum is a common occurrence along the basin margins, as well as within the basin on local topographic highs (Slowakiewicz and Mikolajewski, 2009). Resedimentation of gypsum into deeper water by storms, slope failure and turbidity currents generates graded beds, slumps, slides and breccias, and contribute toward the progradation of the gypsum platforms (Fig. 8.38). Within the basin center, fine-grained gypsum can precipitate interlaminated with carbonate and organic matter, reflecting seasonal variations in precipitation and plankton blooms (Fig. 8.36B). Water circulation becomes increasingly

FIGURE 8.38 Gypsum facies in saline basins (photos courtesy of Charlotte Schreiber). A—Shallow-water clastic gypsarenite (Messinian, Eraclea Minoa, Sicily); B—oscillation ripples in gypsarenite (Messinian, Salemi, Sicily); C—gypsum turbidites with graded bedding and flame structures; D—gypsum block in basin center, embedded in marls (note primary gypsum beds separated by darker microbial carbonates; Messinian, Sutera, Sicily).

restricted during relative sea-level fall, leading to salinity stratification of the water mass and enhanced preservation of organic matter. As the drawdown continues, halite may precipitate on the basin floor across increasingly larger areas (Fig. 8.37B). The deposition of the falling-stage systems tract may end with potash salts in cases of complete desiccation (Fig. 8.37B).

The onset of relative sea-level rise marks the beginning of a new stage of positive (generation of) accommodation (water replenishment), driven by the inflow of seawater from the adjacent ocean or fresh water from surface run-off. While evaporation may continue, dilution typically occurs, expanding the area of deposition of the less soluble minerals at the expense of the more soluble ones (e.g., carbonates vs. gypsum, or gypsum vs. halite; Figs. 8.36C and 8.37C), which reverses the trends observed in the underlying falling-stage systems tract. In open-basin settings, the rates of precipitation may be in balance or even exceed the rates of relative sea-level rise, leading to the progradation of the shoreline (Fig. 8.36C). In enclosed basins, deposition within salt flats may continue prior to the transgression, as long as the rates of precipitation keep pace with the rates of relative sea-level rise (Fig. 8.37C). This stage of progradation (sedimentation outpacing accommodation in open basins) or salt-flat aggradation (sedimentation in balance with accommodation in enclosed basins), following forced regression and preceding transgression, defines the lowstand systems tract of mixed carbonate-evaporite sequences.

Transgression leads to the submergence of the formerly exposed carbonate platforms, typically with water inflow from the open ocean, as a result of which carbonate deposition resumes with a retrogradational stacking pattern (Figs. 8.36D and 8.37D). The transgressive systems tract may still include evaporites accumulated in backstepping sabkhas and hypersaline lagoons, as well as in shelf settings behind barrier reefs (Figs. 8.19 and 8.37D). Where transgression follows the formation of lowstand salt flats in enclosed basins, the initial shallow sea may still be hypersaline due to the dissolution of salts by the incoming seawater. As a result, the earliest transgressive deposits may include gypsum in the depocenter, before the water salinity decreases within the range of carbonate precipitation (Fig. 8.37D). During transgression, the inflow exceeds evaporation and the rates of sedimentation, leading to water deepening and dilution with time, as reflected by the shift in depositional trends from more soluble (e.g., gypsum) to less soluble (e.g., carbonate) minerals. Water typically reaches the normal salinity level of the connecting ocean by the end of the transgression, a trend which continues into the subsequent highstand.

Highstand deposition occurs after transgression, when the basin is fully connected to the open ocean.

The normal salinity of the seawater is conducive to the precipitation of carbonates, which describe a normal regressive stacking pattern of progradation and upstepping (Figs. 8.36E and 8.37E). Under the ideal arid-climate conditions, sabkhas and hypersaline lagoons may still develop in coastal settings, leading commonly to the precipitation of gypsum (Figs. 8.36E and 8.37E). The relative sea-level still rises during the highstand normal regression, but at low rates that allow sedimentation to keep up and even exceed the rates of accommodation in areas adjacent to the shoreline. In the case of formerly enclosed basins that lack any significant differential subsidence, the low-gradient topography inherited from the underlying lowstand salt flats may be maintained during the subsequent transgression and highstand (Fig. 8.37E). This enables the establishment of potentially extensive peri- or epicontinental shallow seas in which deposition occurs under shallowing-water conditions, conducive to increasingly limited water circulation. As a result, the highstand carbonates may show a more restricted fauna and an upward increase in the occurrence of evaporites (Figs. 8.36E and 8.37E).

The physical expression of sequence stratigraphic surfaces depends on the nature of the facies that are in contact across the systems tract boundaries (Figs. 8.36, 8.37, 8.39, 8.40). Among the most prominent surfaces, the subaerial unconformity can be traced at the top of the exposed highstand carbonates and falling-stage evaporites. The latter consist of downstepping gypsum wedges in open-basin settings, as well as hypersaline salt flats in enclosed and desiccated basins. The basal surface of forced regression also marks a major facies change from the highstand carbonates to the falling-stage evaporites in both open-basin and isolated-basin settings. The correlative conformity may not form in enclosed basins that are subject to complete desiccation during the falling stage, where depocenters become exposed, but it can be mapped at the limit between the mixed falling-stage basinal laminites and the overlying lowstand basinal carbonates in open basins. The proximal parts of the subaerial unconformity and the maximum regressive surface are typically reworked by the wave-ravinement surface during subsequent transgression in open basins (Fig. 8.36), while the maximum regressive surface has an even lower preservation potential in enclosed basins (Fig. 8.37).

Mixed carbonate-evaporite sequences may develop at different scales, in relation to flooding-desiccation cycles of different magnitudes. The placement of the sequence boundary varies with the model (Fig. 1.10). Maximum flooding surfaces have a good preservation potential and may be used to define genetic stratigraphic sequences, but their physical expression is often subtle within one lithology (Figs. 8.36 and 8.37). Subaerial

FIGURE 8.39 Basal surface of forced regression (arrows) in an evaporite setting (Upper Messinian, near Basilicoi section, Crotone Basin, Calabria, southern Italy; image courtesy of Laurent Gindre). A—Diatomites interbedded with marlstones (highstand systems tract); B—massive bedded gypsum (falling-stage systems tract).

FIGURE 8.40 Transgressive wave-ravinement surface (arrow) formed during the marine re-flooding of the Mediterranean Sea after the paroxysm of the Messinian salinity crisis (latest Miocene, Los Molinos Pass, Sorbas Basin, SE Spain; image courtesy of Jean-Pierre Suc). A—Uppermost Messinian gypsum; B—reworked algal reef carbonates formed along the shore of the lagoon where the gypsum accumulated (Clauzon et al., 2015; Roveri et al., 2020).

unconformities and their correlative conformities delineate depositional sequences, and are particularly relevant in mixed carbonate-evaporite settings where they form while the precipitation of evaporites is most effective. During the falling stage, the marginal carbonate platforms are exposed to subaerial erosion, karstification, or pedogenesis, and, potentially, dolomitization. Fluvial incision and siliciclastic influx may occur if there is hinterland topography, although under an arid climate the amount of sediment supply is typically low due to the lesser efficiency of weathering and erosion. The basal surface of forced regression is another option for the sequence boundary, although its development is restricted to the areal extent of the falling-stage

systems tract (Figs. 8.36 and 8.37). Irrespective of model, the methodology remains consistent and based on the identification of stratal stacking patterns and changes thereof in the sedimentary record (details in Chapter 9).

8.2.3 Sequences in deep-water settings

The deep-water setting is the last frontier in sequence stratigraphy, due to the generally sparser data available and the difficulty to observe processes in modern environments. The stratigraphic architecture and sedimentological makeup of deep-water systems reflects the interplay of several allogenic and autogenic processes (Fig. 3.1), and the co-existence of multiple sediment sources and modes of sediment transport. Sedimentation in deep-water settings relates to gravity-driven processes and pelagic/hemipelagic fallout in both enclosed and oceanic basins, as well as to contour currents in basins connected to the global ocean. Despite this complexity, reproducible field criteria can be defined to identify the elements of the deep-water sequence stratigraphic framework, in a manner that is independent of any basin-specific (non-diagnostic) variability (Catuneanu, 2020a). Accommodation and sedimentation on the shelf, which depend on relative sea-level changes, the production of extrabasinal and intrabasinal sediment, and the physiography of the basin, define the "dual control" on shoreline trajectories and sediment supply to the shelf edge. Both elements of this "dual control" contribute in discernable ways to the architecture and makeup of the stratigraphic record, and can be separated based on stratigraphic and sedimentological criteria (Catuneanu, 2019a; details in Chapters 3 and 4).

The sequence stratigraphy of deep-water systems is underlain by the genetic relationship between the stratigraphic shoreline-transit cycles on the shelf and the corresponding cycles of change in gravity flows in the deep-water setting (Posamentier and Kolla, 2003; De Gasperi and Catuneanu, 2014; Catuneanu, 2020a, Fig. 4.20). The correlation of shelf and deep-water sequences may be hampered by processes of erosion or nondeposition at the shelf edge or on the slope, but can be demonstrated where age data are available (e.g., by means of biostratigraphy; Gutierrez Paredes et al., 2017). However, the preservation of a continuous sedimentation record from the shelf to the deep-water setting is not required for the construction of a deep-water sequence stratigraphic framework (van der Merwe et al., 2010). The distinction between "shelf" and "deep-water" systems applies to shelf-slope settings, which include a shelf edge. However, this distinction can also be extrapolated to ramp settings, which are devoid of a shelf edge, whereby the shoreline transit area is equivalent to the shelf, and the distal area dominated by sediment-gravity flows defines the deep-water setting (Fig. 8.1). Therefore, in a general sense, the limit between shallow- and deep-water environments is defined by the dominant modes of sediment transport, such as the extent to which gravity controls sedimentary processes, rather than bathymetric or physiographic criteria.

Sedimentation cycles in the deep-water environment can have different origins (e.g., changes in the location of sediment entry points along the shelf edge, changes in the location of depositional elements on the sea floor, changes in relative sea level, and/or changes in sediment supply), and not all have stratigraphic significance. The meaning of sedimentary cyclicity in the deep-water setting is difficult to clarify solely from the

analysis of vertical profiles (e.g., individual outcrops or wells). Some of the sedimentary cycles observed at specific locations are stratigraphic (i.e., delineated by the recurrence of sequence stratigraphic surfaces), whereas others reflect lateral shifts of depositional elements without a stratigraphic significance (i.e., sedimentological cycles). The separation between sedimentological and stratigraphic cycles requires the reconstruction of regional depositional trends within a 3D stratigraphic framework. Seismic data are critical for this purpose, as they provide the means to clarify the relative chronology of different types of gravity flows at regional scales (Fig. 8.41). In addition to the grids of 2D seismic lines, 3D seismic data also afford the plan-view visualization of depositional elements (Figs. 2.76, 5.41, and 5.58), providing further constraints to discriminate between the various types of gravity-driven processes (Figs. 5.42 and 8.42).

The deep-water setting is unique in the sense that the scales of sedimentological and stratigraphic cycles at any location are not mutually exclusive. In all other settings, sedimentological cycles are typically nested within the smallest scale stratigraphic cycles (Fig. 3.24; Catuneanu, 2019b). The difference relates to the patterns of sediment dispersal and deposition in the deep-water environment, whereby areas away from the paths of gravity-driven transport can accumulate lithologically monotonous pelagites or contourites for periods of time encompassing multiple stratigraphic cycles (Fig. 8.41). At such locations, the facies that contribute to the definition of sequences and systems tracts may be missing, and the scale of sedimentological cycles defined by the recurrence of the same types of depositional elements can exceed the scale of stratigraphic cycles. This can deepen the confusion with respect to the sedimentological vs. stratigraphic nature of sedimentary cycles observed at specific locations. In this case, clarifying the origin of sedimentary cycles requires correlation with other sections or integration with geochemical data that can reveal stratigraphic patterns within apparently monotonous fine-grained successions (Harris et al., 2013, 2018; LaGrange et al., 2020).

8.2.3.1 Controls on stratigraphic cyclicity

Stratigraphic cyclicity in the deep-water setting is controlled by the interplay of accommodation and sedimentation on the shelf, which modifies the sediment supply to the shelf edge. The timing of systems tracts and bounding surfaces depends on the balance between relative sea-level changes and sedimentation in coastal environments, which controls shoreline trajectories and the delivery of extrabasinal and intrabasinal sediment to the deep-water environment (Figs. 4.1 and 4.20). Deposition or erosion in the deep-water setting is constrained by several factors, including the mechanisms

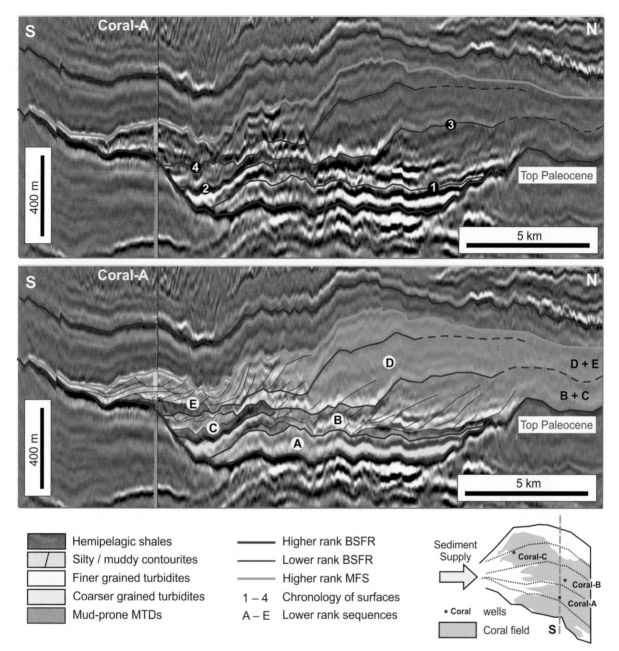

FIGURE 8.41 Stratigraphic framework of the Eocene Coral field in northern Mozambique (modified after Fonnesu et al., 2020). Depositional sequences consisting of mixed gravitational—contourite systems can be observed at two stratigraphic scales. The base of each lower rank sequence marks renewed sediment supply from the shelf to the deep-water setting. Sequences record cyclic changes in the types of gravity flows, from mud-prone MTDs and coarser turbidites attributable to the falling-stage systems tract, to finer grained turbidites that indicate a decline in sediment supply to the deep-water setting. The submarine fan complex is flanked by contourite drift mounds to the north, and the mixed system migrates to the south during the deposition of the higher rank sequence. Abbreviations: MTD—mass-transport deposits; BSFR—basal surface of forced regression; MFS—maximum flooding surface.

of sediment transport (e.g., types of gravity-driven processes; Figs. 5.42 and 8.42) and the physiography of the basin (e.g., seafloor gradients, or barriers to flow). In basins connected to the global ocean, contour currents may also scour the seafloor or accumulate sediment in relation to or independently of relative sea-level changes and shoreline trajectories. In the latter case, the

deep-water unconformities and contourites are unrelated to the sequence stratigraphic framework (i.e., they do not provide criteria for the definition of systems tracts and bounding surfaces; Fig. 8.43). However, if changes in relative sea level modify the paths, energy and sediment load of contour currents, the different products of contour current activity can be rationalized

1. Lithified or semi-lithified sediment:

Type of transport	Diagnostic features
Rock falls: lithified sediment, small scale Slides: lithified sediment, large scale	Original stratification preserved and undeformed
Slumps: semi-lithified sediment, small or large scale	Original stratification preserved and deformed; commonly muddy

2. Gravity flows (loose sediment):

Type of flow		Clast-support mechanism
Fluidal flows	Turbidity flow *Liquefied flow* [1] *Fluidized flow* [1]	Water turbulence Pore-water pressure Water escape
Debris flows	Grainflow [2] (non-cohesive debris flow)	Grain collision (sands)
	Mudflow (cohesive debris flow)	Grain cohesion (clays)

[1] secondary flows, within the 'traction carpet' of turbidity flows
[2] stand-alone flows, or within the 'traction carpet' of turbidity flows

FIGURE 8.42　Classification of gravity-driven processes in deep-water settings (modified after Reading, 1996).

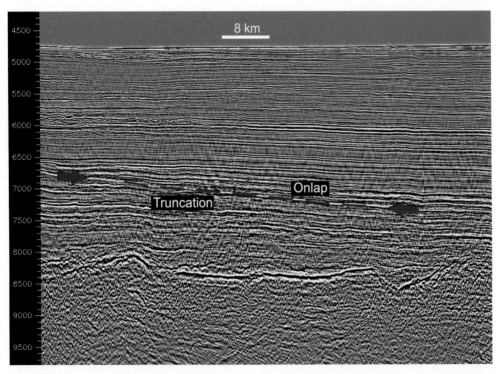

FIGURE 8.43　Regional unconformity cut by deep-water contour currents in a basin floor setting (Bay of Bengal, offshore India; image courtesy of Reliance Industries, India). Where erosion is triggered by tilting of continental margins, contourite-bounding unconformities may coincide with sequence stratigraphic surfaces (e.g., a basal surface of forced regression at the onset of continental uplift and forced regression). Where erosion relates only to changes in the pattern of deep-water circulation, the unconformity that marks the new seafloor equilibrium profile may be unrelated to accommodation and shoreline trajectories.

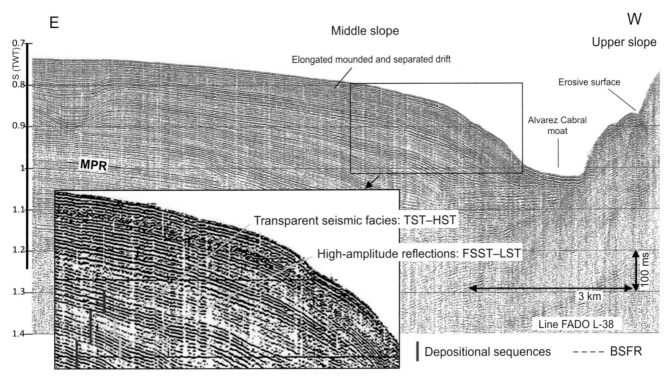

FIGURE 8.44 Stratigraphic cycles in Quaternary contourites on the continental slope of the Gulf of Cadiz (modified after Llave et al., 2001, and Mestdagh et al., 2019). Changes in seismic facies are attributed to fluctuations in the strength of the Mediterranean outflow water, which in turn are linked to relative sea-level changes. Systems tracts are interpreted based on the correlation of the slope contourites with the sequences identified on the shelf, as afforded by high-resolution seismic data calibrated with IODP (International Ocean Discovery Program) well data. The MPR marks a change from an obliquity-controlled periodicity prior to c. 900–920 ka (41 kyr glacial–interglacial cycles during the Upper Pliocene—Middle Pleistocene) to a dominant eccentricity-driven periodicity (100 kyr cycles) during the Middle—Late Pleistocene (Llave et al., 2001). Siliciclastic sediment supply to contourite systems is typically higher during forced regressions and lowstands, and lower during transgressions and highstands (Mestdagh et al., 2019). Abbreviations: MPR—Middle Pleistocene Revolution; TST—transgressive systems tract; HST—highstand systems tract; FSST—falling-stage systems tract; LST—lowstand systems tract; BSFR—basal surface of forced regression.

within the sequence stratigraphic framework (Fig. 8.44; Llave et al., 2001, 2006; Brackenridge et al., 2011; Mestdagh et al., 2019). Contourites can also be integrated in the sequence stratigraphic framework by correlation with areas of accumulation of gravity-flow deposits, where the delineation of sequences is more evident (Fig. 8.41).

The distinction between accommodation (i.e., relative sea-level changes) and sedimentation (i.e., base-level changes) is important in coastal environments, where their interplay generates the three types of shoreline trajectories that define systems tracts and bounding surfaces (Figs. 4.1 and 4.32; details in Chapters 3 and 4). This distinction becomes less meaningful away from the coastline, where the rates of accommodation and sedimentation no longer influence shoreline trajectories. Both accommodation and sedimentation record a 3D variability across the basin, with rates that change along dip and strike directions; however, only the rates of accommodation and sedimentation *at the shoreline* are relevant to the timing of systems tracts and bounding surfaces, including those in deep-water settings. For example, the Plio-Pleistocene stratigraphic sequences

of the Mississippi fan complex include falling-stage systems tracts that formed during stages of glaciation and forced regression on the shelf (i.e., relative sea-level fall), while the deep-water setting, where deposition occurred, experienced higher rates of subsidence and consequent relative sea-level rise (Weimer, 1990; Weimer and Dixon, 1994; Madof et al., 2019; Catuneanu, 2020a).

Sediment supply to the shoreline depends on the production and distribution of extrabasinal and/or intrabasinal sediment. The production of extrabasinal sediment is typically unrelated to changes in relative sea level, reflecting the rates of weathering and erosion of extrabasinal source areas. The production of intrabasinal sediment depends in part on the relative sea level (e.g., Fig. 8.19), as well as on other factors such as water temperature and chemistry (e.g., Fig. 8.36). Variations in sediment supply to the shoreline can modify the stacking patterns that define systems tracts, but only in terms of the manifestation of normal regressions vs. transgressions during stages of relative sea-level rise (Figs. 4.1 and 4.32). Forced regressions remain controlled solely by stages of relative sea-level fall, and play a key role in the definition of stratigraphic cyclicity in deep-water settings (Fig. 4.20; Posamentier and Kolla, 2003; van der Merwe et al., 2010; De Gasperi and Catuneanu, 2014; Catuneanu, 2020a).

While the relative sea level is only one element of the "dual control" on the stratigraphic architecture, changes thereof affect the transfer efficiency of riverborne sediment to the deep-water setting (higher during relative sea-level fall), the shelf capacity to retain extrabasinal sediment (lower during relative sea-level fall), and the production of intrabasinal sediment (lower during relative sea-level fall in carbonate settings). Therefore, relative sea-level changes, and the distinction between lowstands and highstands in relative sea level, remain critically important to understanding the patterns of sediment distribution across a basin, including sediment supply to the deep-water setting. Any basin-specific variability in the type and amount of sediment supply to the shelf edge (e.g., siliciclastic vs. carbonate settings; the increase in riverborne sediment supply to deep water in the case of narrow shelves or shelves dissected by unfilled incised valleys; Sinclair and Tomasso, 2002; Clift, 2006; Covault et al., 2007; Covault and Graham, 2010; Sweet and Blum, 2016; Catuneanu, 2020a; Sweet, 2020) is not diagnostic to the definition of systems tracts, and needs to be rationalized on a case-by-case basis.

8.2.3.2 Stratigraphic vs. sedimentological cycles

The nested depositional cycles that build the sedimentary record can be observed from sedimentological to stratigraphic scales (Fig. 3.24). Within the transit area of a shoreline (i.e., typically on a shelf), where changes in depositional environment are most frequent, sedimentological cycles (beds, bedsets) are always nested within the lowest rank systems tracts and component depositional systems. In this case, the scales of stratigraphic and sedimentological cycles at any location are mutually exclusive: sedimentological cycles form without changes in depositional system and systems tract, and their stacking patterns define the lowest rank systems tracts; in contrast, stratigraphic cycles (i.e., sequences) involve changes in depositional systems and systems tracts, and their stacking patterns define systems tracts of higher hierarchical ranks (Catuneanu, 2019b).

Outside of the shoreline transit area (i.e., in fully continental settings and in deep-water settings), stratigraphic cyclicity may develop without changes in depositional environment. In this case, establishing the relationship between sedimentological and stratigraphic cycles depends on the ability to recognize sequences and component systems tracts. In fully continental settings, the development of subaerial

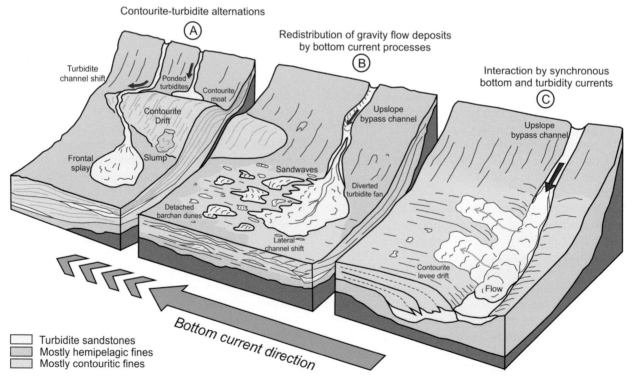

FIGURE 8.45 Interplay of gravity flows and contour currents in mixed deep-water systems (modified after Fonnesu et al., 2020). Contour currents may alternate with gravity flows (A), redistribute previously accumulated gravity-flow deposits (B), or interact with gravity flows at syndepositional time (C). While all types of deposits are important to evaluate in hydrocarbon reservoir studies, only gravity flows provide the criteria for the identification of systems tracts (see text for details).

unconformities affords the delineation of depositional sequences. In this case, the distinction between sedimentological and stratigraphic cycles is straightforward, and bedsets are always nested within the lowest rank sequences. The two types of sedimentary cycles become more difficult to separate in deepwater settings, where both sedimentological and stratigraphic units may develop without significant interruptions in sedimentation (Fig. 7.9). In this case, sedimentological cycles defined by the recurrence of same-type depositional elements can be either smaller or larger than the stratigraphic cycles that define sequences; i.e., the scales of sedimentological and stratigraphic cycles at specific locations are no longer mutually exclusive.

The sedimentological makeup of deep-water systems depends on sediment supply (extrabasinal vs. intrabasinal), as well as on the interaction and relative contributions of several independent processes of sediment transport and deposition, including gravity-driven mass failures, contour currents, and suspension fallout (Fig. 8.45). While all types of deposits are important to evaluate in hydrocarbon reservoir studies, not all are equally relevant to the construction of the sequence stratigraphic framework. Specifically, the sequence stratigraphic framework describes the cyclicity recorded by gravity-driven processes, which is linked to shoreline trajectories and can be used to the designation of systems tracts (Figs. 4.20 and 8.41). For this reason, sedimentological work must precede the stratigraphic analysis, in order to separate the depositional elements that originate from processes with different degrees of relevance to sequence stratigraphy.

The distinction between stratigraphic cycles and sedimentological cycles in deep-water settings is difficult based only on vertical profiles (e.g., well data). The identification of stratigraphic trends requires an integration of seismic and sedimentological data from multiple locations, in order to constrain the relative chronology of the different types of gravity-driven processes at regional scales (Figs. 7.9 and 8.41). Without this regional perspective, the sedimentological vs. stratigraphic meaning of sedimentary cycles observed at specific locations can be equivocal. In areas with frequent gravity flows, sedimentological cycles are usually nested within stratigraphic cycles (Fig. 7.9). However, in areas of continuous sedimentation dominated by non-diagnostic processes (i.e., hemipelagic fallout, contour currents), sedimentological cycles defined by the recurrence of same-type depositional elements (e.g., mudflow deposits) can also exceed the scale of stratigraphic cycles.

8.2.3.3 Sequence stratigraphic framework

The same types of systems tracts and bounding surfaces can form in all downstream-controlled tectonic and depositional settings, and at all stratigraphic scales (Catuneanu, 2019a,b). The deep-water setting is no exception, with a stratigraphic architecture that includes falling-stage, lowstand, transgressive, and highstand systems tracts (Fig. 5.7). The formation of these systems tracts and of their bounding sequence stratigraphic surfaces is linked to shoreline trajectories and changes thereof on the shelf, which modify the sediment supply to the shelf edge during the shoreline shelf-transit cycles (Fig. 4.20). The dominant and subordinate types of gravity flows that contribute to the development of deepwater sequences during the various stages of a relative sea-level cycle are summarized in Fig. 8.46, based on outcrop and subsurface studies (Posamentier and Kolla, 2003; van der Merwe et al., 2010; De Gasperi and Catuneanu, 2014; Gutierrez Paredes et al., 2017; Sweet et al., 2019; Sweet, 2020). However, departures from the common trends may be recorded as a function of the magnitude of relative sea-level changes, as well as basin-specific variables such as physiography, subsidence patterns, and sediment supply (e.g., narrower and steeper shelves enable the delivery of coarser siliciclastic sediment to the shelf edge, thus increasing the contribution of high-density turbidity flows and grainflows to the deep-water system; Sinclair and Tomasso, 2002; Clift, 2006; Covault and Graham, 2010; Sweet and Blum, 2016; Catuneanu, 2020a; Sweet, 2020).

The composition and the relative development of systems tracts in deep water depend to a large extent on the depositional setting on the shelf. In siliciclastic settings, the highest sediment supply to the shelf edge is commonly delivered at the end of forced regression (i.e., "lowstand shedding"; Posamentier and Allen, 1999; Catuneanu, 2006; Sweet et al., 2019, Fig. 5.34), while the lowest sediment supply is typically associated with the maximum flooding surface (e.g., Gutierrez Paredes et al., 2017, Figs. 4.20 and 6.38). In carbonate settings, sediment delivery to the shelf edge follows different trends, with the highest supply during times of highstand, when the carbonate factory is usually most productive (i.e., "highstand shedding"; Schlager et al., 1994, Fig. 5.83). These "reciprocal sedimentation" patterns, which involve the "lowstand shedding" of clastics and the "highstand shedding" of carbonates, have been observed before the emergence of sequence stratigraphy (Van Siclen, 1958; Wilson, 1967). Notwithstanding these commonly observed trends, the deep-water systems need to be analyzed case-by-case, in order to separate any non-diagnostic (basin specific) variability from the diagnostic trends that define the sequence stratigraphic framework.

The relative location of the main components of a deep-water sequence is illustrated in Fig. 8.47, based on the rheology and travel distances of the different types of gravity flows (Figs. 5.42 and 8.42) Debris flows

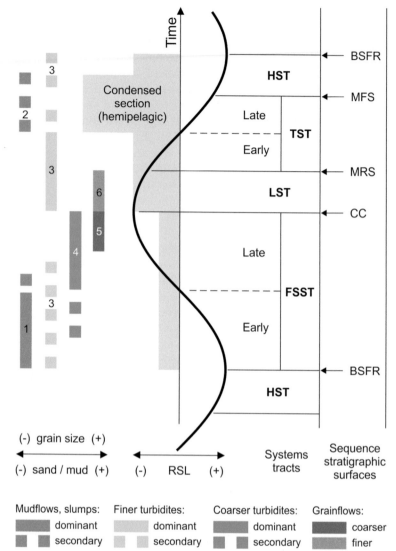

FIGURE 8.46 Commonly observed trends of change in deep-water gravity flows during a shoreline transit cycle on the shelf (modified from De Gasperi and Catuneanu, 2014). Diagnostic for the identification of systems tracts and bounding surfaces are the trends of change rather than the actual types of gravity flows, as the latter depend on several basin-specific variables (see text for details). Conducive factors: 1—muds on the outer shelf, and lowering of the storm wave base; 2—muds on the outer shelf, and instability at the shelf edge; 3—mixed mud and sand at the shelf edge, lower sand-to-mud ratio; 4—mixed sand and mud at the shelf edge, higher sand-to-mud ratio; 5—coastal systems at the shelf edge, coarser sediment; 6—coastal systems at the shelf edge, finer sediment. Abbreviations: FSST—falling-stage systems tract; LST—lowstand systems tract; TST—transgressive systems tract; HST—highstand systems tract; BSFR—basal surface of forced regression; CC—correlative conformity; MRS—maximum regressive surface; MFS—maximum flooding surface; RSL—relative sea level.

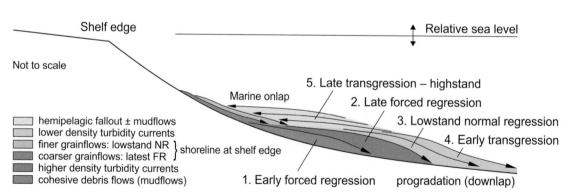

FIGURE 8.47 Dominant processes that contribute to the formation of a submarine fan complex during a relative sea-level cycle. Grainflows travel the shortest distances, as the sand sheets are unconfined (no levees) and permeable (no ability to trap water underneath for hydroplaning), with a steep angle of repose. At the opposite end of the spectrum, the low-density turbidity currents travel the largest distances (e.g., over 2,700 km in the present-day Bay of Bengal). Departures from these common trends reflect the magnitude of relative sea-level changes, and basin-specific variables including physiography, subsidence patterns, and sediment supply (see text for details).

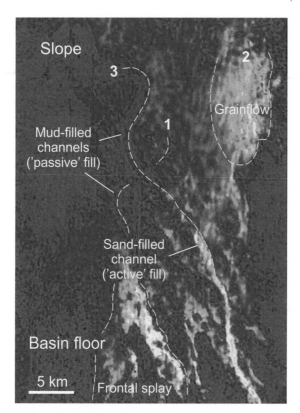

FIGURE 8.48 Relative chronology of gravity flows in a slope to basin floor setting (3D seismic horizon slice, amplitude extraction; Bay of Bengal; image courtesy of Reliance Industries, India). 1—High-density turbidity flow, with entrenched channel on the slope and a frontal splay at the toe of the slope (late falling stage); 2—grainflow on the slope (late falling-stage—lowstand); 3—low-density turbidity flow, with entrenched channel on the slope and aggrading channel on the basin floor (lowstand—transgression). Entrenched channels were filled after abandonment, with suspension sediment ("passive" fill). Aggrading channels were filled with turbidites while the system was active ("active" fill).

travel shorter distances, due to the inherent resistance to motion (i.e., internal shear strength) of mudflows, and the dissipative nature of grainflows which require steep gradients to move. Grainflow sand sheets have steep angles of repose and contribute to the progradation of slope clinoforms, whereas the denser and less permeable mudflows can reach the basin floor by

hydroplaning (Fig. 5.41). In contrast, fluidal flows (i.e., turbidity currents and associated flows; Fig. 8.42) are channelized, and hence able to maintain momentum for a longer time, reaching farther into the basin. In particular, the lower density turbidity currents are the lightest and least erosive types of gravity flows, which can sustain the formation of levees for larger distances; as a result, they can travel beyond the frontal splays of the higher density turbidity flows (Figs. 5.47 and 8.48), potentially for thousands of kilometers (e.g., over 2,700 km in the present-day Bay of Bengal). Fig. 8.49 exemplifies the trends shown in Fig. 8.47, which assume unobstructed flows. Obstacles can divert the flows or shorten travel distances, modifying the general trends. More details on the characteristics of gravity flows that contribute to the formation of systems tracts are presented in Chapter 5.

The model in Fig. 8.47 describes a complete sequence that includes all systems tracts and the full range of gravity flows. Deviations from the ideal sequence may occur in terms of the development (e.g., icehouse sequences are dominated by falling-stage deposits) and composition (types of sediment and gravity flows) of systems tracts, as well as the location of depositional elements on the seafloor. The latter adds a significant degree of complexity to the architecture of the deep-water stratigraphic framework, due to the lateral shifts of gravity flows as they follow the paths of least resistance across a constantly evolving bathymetric profile (Figs. 7.9, 8.41, and 8.48). Gravity flows may be confined to pre-existing canyons or channel belts (Figs. 5.34, 8.50, and 8.51) or may establish new paths in unconfined areas of the seafloor (Figs. 5.44 and 5.45), with the timing of the two styles of sediment transport being in or out of phase (Fisher et al., 2021). The course of gravity flows also depends on the location of the feeder systems, such as unfilled incised valleys on the shelf (Fig. 5.34), shelf-edge deltas (Figs. 8.52 and 8.53), or other areas of instability at the shelf edge (Figs. 5.28 and 5.48). The variability in the place of accumulation of depositional elements during a stratigraphic cycle, along with the subsequent reworking

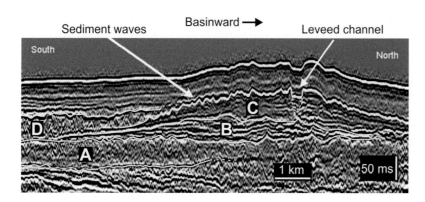

FIGURE 8.49 Seismic transect showing the common elements of a submarine fan complex in a basin-floor setting (Pleistocene, Gulf of Mexico; modified after Posamentier and Kolla, 2003). A—mudflow deposits (chaotic internal facies) of the early falling-stage systems tract; B—frontal-splay turbidites (parallel reflections, convex-up top) of the late falling-stage systems tract; C—leveed channel of the lowstand systems tract or early transgressive systems tract; D—mudflow deposits of the late transgressive systems tract. Note the progradation of gravity-flow deposits into the basin from A to C, and the retrogradation from C to D (compare with Fig. 8.47).

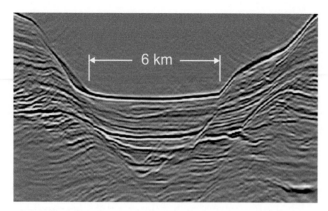

FIGURE 8.50 Submarine canyon filled in part with fine-grained sediment slumped from the flanks, following the canyon abandonment (Gulf of Mexico; image courtesy of Henry Posamentier).

caused by slumping or contour currents, poses the greatest challenge to stratigraphic correlations in deep-water settings. This challenge is best addressed by integrating multiple datasets, including seismic, biostratigraphic, and geochemical.

Among the seven surfaces of the "conventional" sequence stratigraphy (Fig. 6.7), only four extend into the deep-water environment: the basal surface of forced regression, the correlative conformity, the maximum regressive surface, and the maximum flooding surface (Figs. 4.20, 8.46, and 8.54). The basal surface of forced regression and the correlative conformity, which mark changes in the direction of shift of the relative sea level,

tend to have the strongest physical expression in the deep-water setting (Figs. 4.20, 7.9, and 8.41). Field data indicate that the basal surface of forced regression marks an increase in siliciclastic sediment supply and/or a decrease in carbonate sediment supply to the shelf edge (Fig. 6.26). In contrast, the correlative conformity marks a decrease in siliciclastic sediment supply and/or an increase in carbonate sediment supply to the shelf edge (Fig. 6.28). These trends relate to the shelf capacity to retain riverborne sediment and to produce carbonate sediment, both lowered during relative sea-level fall. Seismic data also reveal that the basal surface of forced regression is commonly scoured by the passage of early falling-stage mudflows (Fig. 8.54A), whereas the correlative conformity marks a change from coarser to finer grained turbidites (Fig. 8.54B) (more details in Chapter 5). The physical expression of these surfaces affords their identification on both dip- and strike-oriented seismic lines (Figs. 4.6 and 8.41).

The maximum regressive surface and the maximum flooding surface tend to be lithologically cryptic, as they develop during relative sea-level rise within fining-upward and condensed intervals, respectively (Figs. 4.20 and 7.9). Seismic data indicate no discernable difference between the lowstand and early transgressive turbidity flows (Figs. 5.48, 5.53, 5.56, and 5.63), and a paucity or absence of mudflows within the lowstand— early transgressive interval that encompasses the maximum regressive surface (Fig. 8.54C). Similarly, the plan-view expression of the maximum flooding surface

FIGURE 8.51 Seismic imaging of the modern seafloor (A) and cross-sectional view (B: location indicated in image A) of the Mississippi canyon (Gulf of Mexico; images courtesy of Henry Posamentier). The 2D seismic line shows the complex nature of the canyon fill, which recorded multiple cut-and-fill stages related to the activity of gravity flows. The arrow in image A shows the present-day direction of gravity flows. The canyon fill is c. 720 m thick.

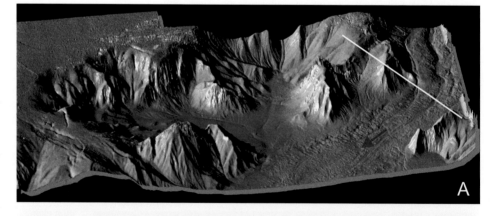

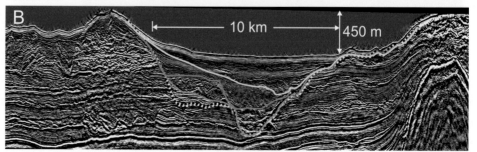

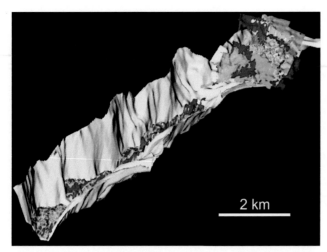

FIGURE 8.52 Shelf-edge delta and entrenched turbidity channel on the upper continental slope (Late Pleistocene, De Soto Canyon area, Gulf of Mexico; image courtesy of Henry Posamentier). The delta is c. 2 km wide at the shelf edge, and the depth of channel incision is c. 275 m.

is often undistinguishable from that of other paleoseafloors within condensed sections that span the underlying late transgressive and the overlying highstand deposits (Fig. 8.54D). For these reasons, the identification of maximum regressive and maximum flooding surfaces in deep-water settings is best approached starting with the analysis of dip-oriented seismic lines, particularly where the correlation of deep- and shallow-water systems is possible (Figs. 2.51 and 4.6). Once the systems tracts are identified based on 2D stratal relationships, the bounding sequence stratigraphic surfaces can be imaged in plan view where 3D seismic data are available (Fig. 8.54).

Despite the cryptic expression of maximum regressive and maximum flooding surfaces in deep-water settings, sedimentological, biostratigraphic, and geochemical criteria may still be defined to identify these surfaces within lithologically monotonous successions (Donovan et al., 2015; Turner et al., 2015, 2016; Gutierrez Paredes et al., 2017; Dong et al., 2018; Harris et al., 2018; Catuneanu, 2019a, LaGrange et al., 2020). In sedimentological terms, the maximum regressive surface approximates the end of grainflow deposition (e.g., collapse of beach sands at the shelf edge; Figs. 5.48, 7.9, and 8.46), where siliciclastic shelves become entirely continental by the end of progradation. In carbonate settings, the maximum regressive surface may coincide with a drowning surface where the subsequent transgression and water deepening are rapid (Fig. 6.34). The maximum flooding surface typically marks the time of the lowest sediment supply to deep water in both siliciclastic and carbonate settings (Figs. 4.20, 6.38, 7.9, 8.46), which often results in the highest abundance of microfossils within the condensed sections (Fillon, 2007; Gutierrez Paredes et al., 2017).

Geochemical criteria are particularly useful in monotonous mudrock intervals where the lack of lithofacies diversity reduces the effectiveness of conventional facies analysis. Both organic and inorganic geochemistry have relevance to sequence stratigraphy and can be used to identify systems tracts and bounding surfaces in fine-grained successions, and to reconstruct the history of relative sea-level changes at syn-depositional time (e.g., Slingerland et al., 1996; Harris et al., 2013, 2018; Dong et al., 2018; LaGrange et al., 2020; see Chapter 2 for details). The most reliable application of

Axial section

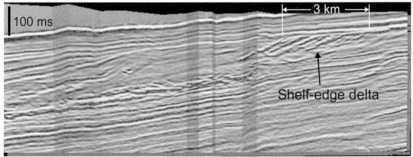

Planar section   Transverse section

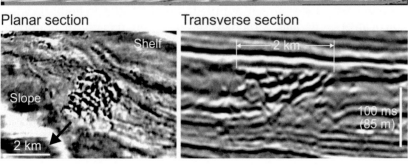

FIGURE 8.53 Shelf edge delta observed in axial, planar, and transverse views (Late Pleistocene, De Soto Canyon area, Gulf of Mexico; images courtesy of Henry Posamentier). The arrow in the planar section indicates the direction of progradation. The delta is c. 2 km wide and 3 km long.

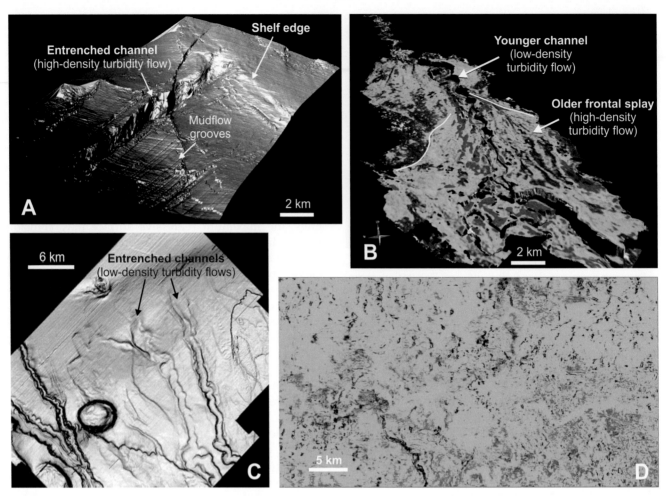

FIGURE 8.54   Plan-view expression of stratigraphic surfaces in deep-water settings (3D seismic horizon slices; images A, B, and C courtesy of Henry Posamentier; image D courtesy of Reliance Industries, India). A—basal surface of forced regression (slope setting, Gulf of Mexico); B—correlative conformity (change from high-density to low-density turbidity flows; slope setting, Gulf of Mexico); C—maximum regressive surface (slope setting, Gulf of Mexico); D—maximum flooding surface (within a condensed section, with fluid-escape faults; basin floor setting, Bay of Bengal). The stratigraphic position of surfaces A, B, and C is indicated in Fig. 4.6.

geochemistry is for the identification of maximum flooding and maximum regressive surfaces, based on TOC values and elemental proxies for terrigenous influx, grain size, and biogenic silica (Fig. 2.51; LaGrange et al., 2020). The basal surface of forced regression and the correlative conformity are less evident on geochemical datasets, but can be identified on the basis of other criteria (e.g., petrography of framework grains, or the geometry of seismic reflections; Figs. 4.6, 6.26, and 6.28). The integration of independent datasets provides the most effective approach to the identification of all elements of the sequence stratigraphic framework, and geochemical proxies are most reliable when calibrated with sedimentological, ichnological, and biostratigraphic data. Geochemical proxies also need to be adjusted to stratigraphic age, tectonic setting, and climatic regime, which modify surface processes and the chemistry and biology of sedimentary basins (details in Chapters 2 and 9).

The timing of the basal surface of forced regression and of the correlative conformity is independent of sediment supply; i.e., these surfaces mark the onset and the end of relative sea-level fall, respectively (Figs. 6.7 and 8.46). In contrast, the maximum regressive surface and the maximum flooding surface form during relative sea-level rise, with a timing controlled in part by the sediment supply to the shoreline (Figs. 4.32 and 8.46). Relative sea-level changes modify the sediment supply to the shelf edge in both siliciclastic and carbonate settings, and remain an important component of the "dual control" on the stratigraphic architecture. The relationship between the relative sea level and the timing of sequence stratigraphic surfaces is evident in all depositional settings, from carbonate or mixed carbonate-siliciclastic (Figs. 6.26 and 6.28) to fully siliciclastic (Figs. 4.6 and 5.7). This relationship, as well as the identification of deep-water systems tracts and bounding surfaces, can be further validated by the correlation of deep-water systems with the shallow-water systems on the shelf, wherever enabled by stratal preservation and the availability of data (Figs. 2.5 and 6.26).

8.2.3.4 Stratigraphic scales in deep-water settings

The sequence stratigraphic framework of deep-water systems records a nested architecture of stratigraphic cycles (i.e., sequences) that can be observed at different scales (Weimer, 1990; Weimer and Dixon, 1994; Gardner et al., 2003, 2009; Catuneanu, 2019b, Figs. 8.41 and 8.55). There are no temporal or physical standards for the scale of sequences that develop at different hierarchical levels. Scales are basin specific, reflecting the interplay of local and global controls on accommodation and sedimentation (Fig. 3.1). With the exception of orbital forcing, the periodicity of all other controls on stratigraphic cyclicity can vary unpredictably. Even in the case of orbital cycles, whose periodicities are most regular, the process–response correlation with stratigraphic sequences is still tentative because the orbital signal may be distorted or overprinted by local or regional processes (Strasser, 2018). Despite the nested architecture of stratigraphic cycles, the sequence stratigraphic framework is not truly fractal as sequences of different scales may differ in terms of underlying controls and internal composition of systems tracts.

The timescales of stratigraphic cycles in the deep-water setting reflect the cyclicity of shoreline shifts on the shelf, within transit areas located updip of the shoreline trajectories of immediately higher hierarchical ranks (Fig. 4.60). At each scale of observation, the width of the shoreline transit area depends on the gradient of the shelf, the magnitude of relative sea-level changes, and the rates of sedimentation at the shoreline. Shoreline transit cycles can be observed at different scales (i.e., hierarchical levels), from 10^2–10^3 yrs (e.g., short-term variations in the balance between accommodation and sedimentation at the shoreline; Amorosi et al., 2017) to 10^4–10^5 yrs (e.g., glacial–interglacial cycles related to orbital forcing, which are most evident during icehouse regimes; Burgess and Hovius, 1998; Porebski and Steel, 2006; Ainsworth et al., 2017, 2018) and 10^6–10^7 yrs (e.g., in the case of long-term climate changes associated with icehouse–greenhouse cycles; Fig. 3.1). The Mississippi fan complex in the Gulf of Mexico provides an example of nested stratigraphic cycles related to forced regressions of different magnitudes during the Plio-Pleistocene icehouse (Figs. 8.55 and 8.56). In this example, sequences of all hierarchical ranks are dominated by the falling-stage systems tract, which is a typical signature of icehouse regimes (Weimer, 1990; Weimer and Dixon, 1994; Tesson et al., 2000; Fielding et al., 2006, 2008; Isbell et al., 2008; Zecchin et al., 2015; Sweet et al., 2019).

At each scale of observation, the growth and the progradation of slope clinoforms take place in incremental steps, reflecting the episodic delivery of sediment to the shelf edge (e.g., "highstand shedding" in the case of carbonates or "lowstand shedding" in the case of siliciclastic systems; Schlager et al., 1994; Posamentier and Allen, 1999; Sweet et al., 2019; Fisher et al., 2021). These cyclic fluctuations in sediment supply to the shelf edge are linked to the shoreline shelf-transit cycles, which are controlled by all factors that modify accommodation and sedimentation on the shelf. The location of the shoreline transit areas that develop at different stratigraphic scales also affects the bathymetry at the shelf edge during the shoreline transit cycles, influencing further the type of sediment that can reach the shelf edge (Fig. 4.20). At the "first-order" scale of shelf-slope systems, changes in sediment supply to the shelf edge relate to the second-order and lower rank stratigraphic cycles of shoreline shifts across the shelf (Fig. 4.60; Catuneanu, 2019a,b). The scales of deep-water sequences that develop in relation to the shoreline shelf-transit cycles may range from sub-seismic (10^0–10^1 m) to seismic (10^1–10^2 m), and their identification depends on the resolution of the data available.

8.2.3.5 Conclusions

Sedimentation in deep-water settings is the result of several independent processes of sediment transport and deposition, including gravity flows and suspension fallout in both enclosed and oceanic basins, as well as contour currents in basins connected to the global ocean (Figs. 8.41 and 8.45). While all types of deposits are important to the sedimentological makeup of deep-water systems, only the trends of change in gravity flows during the shoreline shelf-transit cycles provide dependable criteria for the construction of the sequence stratigraphic framework in all settings (Figs. 4.20, 8.41, and 8.46). The actual types of gravity flows which contribute to the formation of systems tracts are non-diagnostic, as they depend on basin-specific parameters that control sediment supply to the shelf edge during the shoreline transit cycles: the physiography of the shelf (width, seafloor gradients, the presence or absence of unfilled incised valleys across the shelf); the magnitude and rates of relative sea-level changes at the shoreline; the rates of sedimentation at the shoreline; and the extrabasinal vs. intrabasinal origin of the sediment. Contour currents further modify the sedimentological makeup of deep-water sequences in oceanic basins (Figs. 8.41 and 8.45), and contribute to the identification of systems tracts and bounding surfaces where variations in their sediment composition are tied to relative sea-level changes and shoreline shelf-transit cycles (Fig. 8.44). However, contourites and related unconformities may also form in

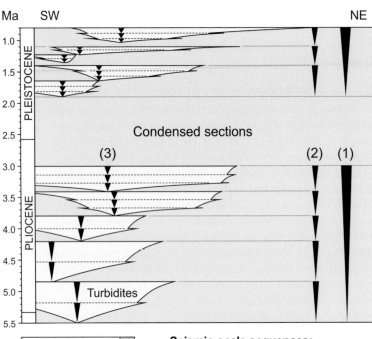

FIGURE 8.55 Nested architecture of stratigraphic cycles in the deep-water setting of the Gulf of Mexico (see map for location; modified after Weimer, 1990; Weimer and Dixon, 1994; Madof et al., 2019; Catuneanu, 2019b). Sequences of all scales record an increase in sediment supply during deposition, driven by forced regressions of corresponding magnitudes that dominated the Plio-Pleistocene icehouse. In this example, sequences of all hierarchical ranks are dominated by the falling-stage systems tract, which is a typical signature of icehouse regimes. Coeval with the forced regressions on the shelf (i.e., stages of relative sea-level fall) the deep-water setting was subject to higher rates of subsidence and consequent relative sea-level rise. Only the relative sea-level changes *at the shoreline* are relevant to the timing of sequences and systems tracts (details in Chapter 5).

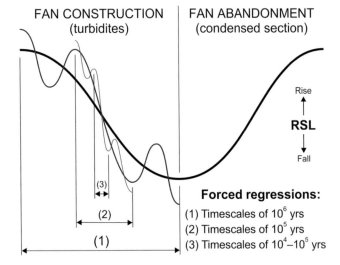

FIGURE 8.56 Relative sea-level changes (RSL) on a continental shelf, generating nested sequences such as those observed in the Mississippi submarine fan complex (Fig. 8.55). The submarine fan complex was constructed during forced regressions of three different scales. Intervening stages of relative sea-level rise resulted in fan abandonment at the corresponding scales of observation. Coeval with the relative sea-level cycles on the shelf, the deep-water setting was subject to relative sea-level rise due to the higher rates of subsidence of the basin floor. The timing of deep-water sequences is linked to the relative sea-level changes on the shelf, which control the sediment supply to the shelf edge (see text for details).

relation to oceanographic processes decoupled from relative sea-level changes and shoreline transit cycles on the shelf, in which case they do not provide criteria for the identification of systems tracts and bounding surfaces (Fig. 8.43).

Stratigraphic cyclicity in the deep-water setting reflects the interplay of accommodation and sedimentation on the shelf, which controls shoreline trajectories, sediment supply to the shelf edge, and the timing of deep-water sequences, systems tracts, and bounding surfaces. Changes in the types, volume, and composition of gravity flows during the shoreline shelf-transit cycles define stratigraphic trends that are diagnostic to the identification of systems tracts and bounding surfaces. Forced regressions invariably lower the shelf capacity to retain extrabasinal sediment and to generate intrabasinal carbonate sediment, thus modifying the sediment supply to the shelf edge, and are fundamental to the delineation of deep-water depositional sequences (Fig. 8.55). Notably, forced regressions are always driven by the fall in relative sea level on the shelf, independently of sedimentation (Figs. 5.7, 6.26, 6.28, and 8.55). Therefore, the relative sea level plays a key role in the stratigraphic architecture of deep-water systems, and the distinction between lowstands and highstands remains critical to understanding the distribution of extrabasinal and intrabasinal sediment across a sedimentary basin. Within the stratigraphic framework, the non-diagnostic variability in the sedimentological makeup of systems tracts reflects the unique tectonic and depositional settings of each sedimentary basin and needs to be rationalized on a case-by-case basis. The distinction between stratigraphic trends and the non-diagnostic sedimentological variability is key to the construction of the sequence stratigraphic framework.

The application of sequence stratigraphy to the deep-water setting relies on the construction of composite profiles that describe the relative chronology of the different types of gravity flows at regional scales (Fig. 7.9). This is typically accomplished with seismic data (Figs. 5.47, 8.41, and 8.49), which can be calibrated with well data (e.g., biostratigraphy; Gutierrez Paredes et al., 2017). The cyclicity that is relevant to the delineation of sequences is defined by the composite rather than local profiles. The place of accumulation of gravity-flow deposits depends on the location of the sediment entry points along the shelf edge, the types of gravity flows, and the seafloor morphology (e.g., the location of depocenters controlled by structural elements, salt tectonics, or mud diapirism). The allocyclic and autocyclic shifts of deep-water depositional elements (e.g., the lateral shifts of leveed channels) further enhance the offset between local trends and the regional composite profile in terms of timing and frequency of cycles, timing of coarsening- and fining-upward trends, and timing of coarsest sediment (Fig. 7.9). The sedimentological cycles defined by local trends must not be confused with the stratigraphic cycles defined by regional trends. Notwithstanding these caveats, the stratigraphic architecture of basin floors that lack any significant relief can be remarkably consistent (van der Merwe et al., 2010, Fig. 5.7).

8.3 Variability with the climatic regime

The influence of climate on the stratigraphic architecture has been explored in numerous studies (Bartek et al., 1991, 1997; Kidwell, 1997; Naish and Kamp, 1997; Saul et al., 1999; Fielding et al., 2001, 2006, 2008; Naish et al., 2001; Cantalamessa et al., 2007; Di Celma and Cantalamessa, 2007; Massari and D'Alessandro, 2012; Zecchin et al., 2015; Schwarz et al., 2016; Csato et al., 2021). Climate changes operate over various timescales, from long-term Greenhouse—Icehouse cycles (10^6–10^7 yrs) to shorter term glacial—interglacial cycles related to orbital forcing (10^4–10^5 yrs) or solar radiation (10^3 yrs and under; Fig. 3.1). At each scale of observation, climate changes trigger sea-level fluctuations as well, with magnitudes that reflect the scale of the cycle

and the background climatic regime. For example, orbital cycles during Permian Icehouse periods generated sea-level changes of 70–80 m, as opposed to only 20–30 m during intervening Greenhouse stages (Fielding et al., 2008; details in Chapter 3). The magnitude of sea-level changes also varied among the different icehouses, reaching 100–150 m during the Pleistocene on timescales of 100 kyr (Bard et al., 1990; Raymo, 1997). The global rise in sea level within the last 20,000 yrs since the Last Glacial Maximum of the Plio-Pleistocene Icehouse is estimated to c. 135 m (Lambeck et al., 2014). These changes in sea level and the underlying climatic conditions affect both the architecture and the sedimentological composition of sequences and component systems tracts.

The influence of climate on sedimentation has been documented from basinal to global scales. The Quaternary deposits of the Caribbean Sea reveal a clastic input during glacial times (lower temperatures and sea-level fall) and carbonate production during the interglacials (higher temperatures and sea-level rise) when platforms were flooded (Schlager et al., 1994). While sea-level changes and the underlying climatic conditions operate most often in tandem, they may also be out of phase, in which case climate usually takes the dominant role (e.g., carbonate accumulation during stages of sea-level fall, due to the drier climatic conditions; Massari and D'Alessandro, 2012; Schwarz et al., 2016, Figs. 8.30 and 8.31). Most commonly, the more significant changes in the sedimentation regimes between carbonates and siliciclastics occurred during Icehouse periods, when the magnitudes of sea-level changes were higher, such as during the Quaternary and the Permo—Carboniferous glaciations. Significant contrasts in the sedimentation regimes may also occur at the same time between

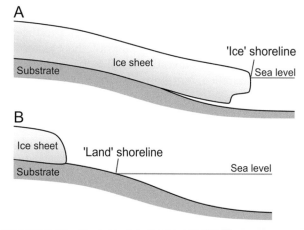

FIGURE 8.57 The concepts of "ice" and "land" shorelines (from Zecchin et al., 2015). A—The "ice" shoreline is represented by the water—ice contact, when ice sheets occupy the shelf; B—the "land" shoreline, represented by the water—land contact, is found when ice retreats inland.

high-latitude glaciated settings and the lower latitude non-glaciated settings, due to the presence or absence of ice shelves and related "ice shorelines," respectively (Fig. 8.57; Zecchin et al., 2015).

8.3.1 Sequences in non-glaciated settings

Non-glaciated settings are illustrated by default throughout the book, and are exemplified in all standard sequence stratigraphic models. Fundamentally, non-glaciated settings are defined by the presence of land shorelines, whereby ice does not directly modify the amount of accommodation available in coastal and marine environments (Fig. 8.57). In this case, the stratigraphic architecture is controlled by the interplay of relative sea-level changes and sediment supply at the shoreline, as accounted for in the conventional sequence stratigraphy (details in Chapter 4). The presence of ice shelves, however, can affect significantly the stratigraphic architecture, as well as the timing and sedimentology of systems tracts (Zecchin et al., 2015). This is due to the fact that ice shelves modify the nature of the shoreline, from land–water to the ice–water contact, which can migrate independently of relative sea-level changes and sediment supply. Glaciated settings also promote the deposition of glacigenic and glacimarine facies, which are distinctly different from the sedimentary facies that are typically observed in non-glaciated environments. Notably, the ice-free (non-glaciated) settings, for which the classic approach to sequence stratigraphy was developed, can be regarded as an end member of a range of stratigraphic scenarios, with the ice-permanent (glaciated) settings defining the opposite end member (Fig. 8.58; discussion below).

8.3.2 Sequences in glaciated settings

Glaciated settings are defined by the development of ice shelves connected to landmasses, which produce ice shorelines (Fig. 8.57). As remnants of the Quaternary icehouse, ice shelves are still found today at high latitudes in Antarctica, Northern Canada, Greenland, and the Russian Arctic, but were more prevalent at other times during the Phanerozoic in relation to the early Paleozoic, late Paleozoic, and Cenozoic glaciations (Semtner and Klitzsch, 1994; Visser, 1997; Naish et al., 2001, 2009; Pazos, 2002; Isbell et al., 2008; McKay et al., 2009; Fielding et al., 2011; Passchier et al., 2011; Bart and De Santis, 2012). A significant departure from the usual norms that govern the formation of stratigraphic sequences in non-glaciated settings is that ice exerts an additional control on accommodation and sedimentation, as a result of which the timing of systems tracts and bounding surfaces becomes at least in part independent of relative sea-level changes and sediment supply. Instead, the sequence stratigraphic framework forms in relation to the advance and retreat of the ice shelves, which control the regressions and transgressions of the ice shorelines, respectively (Powell and Cooper, 2002; Dunbar et al., 2008; Zecchin et al., 2015).

The growth and decline of ice shelves relates to glacial and interglacial periods observed most commonly at Milankovitch and sub-Milankovitch scales (10^3–10^5 yrs; Raymo, 1997; Naish et al., 2001, 2008, 2009; Isbell et al., 2008; Zecchin et al., 2010b, 2011; Csato et al., 2014, 2021). Isostasy plays a particularly important role in glaciated settings, due to the changes in the volume of ice during the glacial–interglacial cycles. The effects of glacio-isostasy on the stratigraphic architecture have been documented throughout geologic time, from the Precambrian to the Holocene (Boulton, 1990; Dyke and Peltier, 2000; Eriksson et al., 2005a,b, 2006, 2013; Naish et al., 2008; Nixon et al., 2014; Nutz et al., 2015). Notably, the rates of isostatic rebound during deglaciation may be either higher (e.g., in polar areas following the last glacial maximum; Boulton, 1990; Dyke and Peltier, 2000; Nixon et al., 2014; Nutz et al., 2015) or lower than the rates of glacio-eustatic sea-level rise (e.g., during the late Oligocene deglacial stage in the Ross Sea; Naish et al., 2008). Similarly, ice loading (isostatic subsidence) during glaciation combined with the local tectonic subsidence or uplift may either outpace or be outpaced by the rates of glacio-eustatic sea-level fall. As a result, both transgressions and regressions of ice

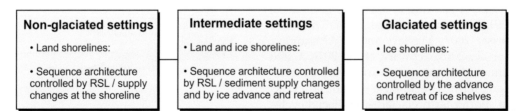

FIGURE 8.58 Range of climate-controlled settings, with end members represented by non-glaciated (ice-free) and glaciated (ice-permanent) settings (modified after Zecchin et al., 2015). Glaciated settings are characterized by glacigenic, glacimarine, and hemipelagic sedimentation, in contrast to the shoreface and shelf deposits that accumulate in shallow-water non-glaciated settings. Between these end members, sequences may form in response to a combination of land-shoreline trajectories and ice-shoreline shifts (see text for details). RSL—relative sea level.

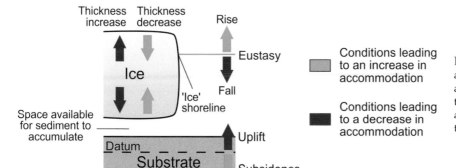

FIGURE 8.59 In the presence of ice sheets, accommodation depends not only on eustasy and tectonism but also on variations in ice thickness: an increase in ice thickness reduces accommodation, whereas a decrease in ice thickness generates accommodation.

shorelines may occur during stages of relative sea-level rise or fall, depending on the interplay between glacio-isostasy, eustatic sea-level changes, and local tectonic subsidence or uplift.

The relationship between accommodation and relative sea-level changes in glaciated settings is offset by the presence of ice, which modifies the amount of accommodation below the ice shelf (Fig. 8.59). As a result, changes in accommodation reflect not only changes in relative sea level but also variations in ice thickness: accommodation is generated by a decrease in ice thickness, and it is reduced by an increase in ice thickness (Fig. 8.59). The impact of ice-thickness variations on accommodation is greater than that of isostasy, as the rates of ice-thickness changes are higher than the rates of isostatic subsidence or rebound (Zecchin et al., 2015). For example, isostatic rebound due to decreasing ice load may lead to relative sea-level fall during the retreat of the ice shoreline (i.e., transgression; Nutz et al., 2015), while the decrease in the thickness of ice generates accommodation. Therefore, shoreline shifts in glaciated settings still correlate with changes in accommodation, while being potentially out-of-phase with the changes in relative sea level: glaciations (i.e., increase in ice thickness > isostatic subsidence, coupled with eustatic fall) lead to negative accommodation and forced regressions, while the relative sea level may rise or fall; and deglaciations (i.e., decrease in ice thickness > isostatic rebound, coupled with eustatic rise) lead to positive accommodation and transgressions, while the relative sea level may rise or fall (Fig. 8.60) (Zecchin et al., 2015).

The stratigraphic architecture of fully glaciated settings with permanent ice shelves is dominated by the falling-stage and transgressive systems tracts, which form during stages of glaciation (advance and thickening of ice shelves) and deglaciation (retreat and thinning of ice shelves), respectively. As a result, the basal surface of forced regression and the regressive surface of marine erosion are replaced by the glacial surface of erosion, the correlative conformity coincides with the maximum regressive surface, and the transgressive surface of erosion is replaced by the glacial retreat surface (Fig. 8.60). Moreover, the subaerial unconformity does not form as long as landmasses are covered with ice. Therefore, the sequence stratigraphic framework of fully glaciated settings is simpler than that of non-glaciated settings, with the two erosional surfaces that form during the advance and retreat of ice sheets playing a major role in the architecture and preservation of glacigenic deposits (Bjarnadóttir et al., 2013, 2014; Rüther et al., 2013, Fig. 8.60). A more complex stratigraphic architecture develops where landmasses get exposed during the glacial cycles, in which case lowstand and highstand systems tracts may also form in relation to land shorelines, along with the complete set of sequence stratigraphic surfaces.

Depending on the magnitude of climate changes during the glacial—interglacial cycles, as well as on the background climatic regime (Icehouse vs. Greenhouse) the landmass may or may not be exposed in the process of advance and retreat of ice shelves. Where ice shelves melt and the ice sheets retreat across the land, conventional (land) shorelines are reactivated, and the stratigraphic architecture is once again linked to shoreline trajectories controlled by the interplay of relative sea-level changes and sediment supply at the shoreline. Between the ice-free and ice-permanent end members, hybrid sequences may form in relation to a combination of ice and land shorelines, during the same stratigraphic cycle (Dunbar et al., 2008; Zecchin et al., 2015). Glacial and non-glacial conditions can alternate at specific locations, but may also occur at the same time along strike due to climate variations within a sedimentary basin. As a result, sequences controlled by glacial advance and retreat may pass laterally into sequences controlled by the trajectory of land shorelines, which can generate a significant 3D stratigraphic variability (Zecchin et al., 2015). This 3D framework can be further complicated by lateral variations in the timing of ice-shoreline and land-shoreline shifts, leading to the development of diachronous systems tracts and bounding surfaces along strike (details in Chapter 7).

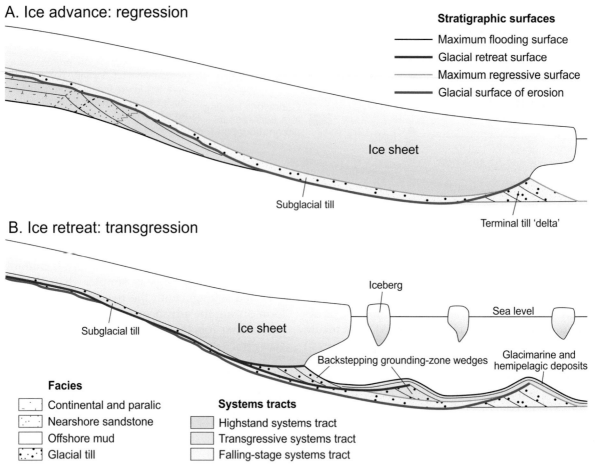

A. Ice advance: regression

Stratigraphic surfaces
— Maximum flooding surface
— Glacial retreat surface
— Maximum regressive surface
— Glacial surface of erosion

Ice sheet

Subglacial till

Terminal till 'delta'

B. Ice retreat: transgression

Iceberg

Sea level

Subglacial till

Ice sheet

Backstepping grounding-zone wedges

Glacimarine and hemipelagic deposits

Facies
- Continental and paralic
- Nearshore sandstone
- Offshore mud
- Glacial till

Systems tracts
- Highstand systems tract
- Transgressive systems tract
- Falling-stage systems tract

FIGURE 8.60   Stratigraphic architecture of glaciated settings during the advance and retreat of ice shelves (modified after Zecchin et al., 2015). Both the advance and retreat of ice result in the development of erosional surfaces at the base of the ice sheets, overlain by subglacial till; i.e., the glacial surface of erosion and the glacial retreat surface, respectively (Powell and Cooper, 2002; Dunbar et al., 2008). The falling-stage systems tract includes prograding terminal till "deltas," while the transgressive systems tract incorporates backstepping grounding-zone wedges. The lowstand and highstand systems tracts develop where land shorelines are exposed.

9

Discussion

9.1 Architecture of the stratigraphic record

Stratigraphic cyclicity in downstream-controlled settings is driven by the interplay of relative sea-level changes (accommodation) and base-level changes (sedimentation) at the shoreline (Figs. 5.1 and 5.5; details in Chapter 3). The distinction between accommodation and sedimentation becomes less meaningful in upstream-controlled settings, where stratigraphic cyclicity is controlled by base-level changes (Fig. 1.19). However, base-level changes reflect the interplay of all controls on sedimentation, including accommodation (Fig. 3.11), and may correspond to tectonic cycles, climate cycles, or cycles of autogenic migration of alluvial channel belts which generate depositional sequences and bounding unconformities (Catuneanu and Elango, 2001; Hajek et al., 2010; Hofmann et al., 2011; Miall, 2015; Catuneanu, 2017, 2019a,b; Fig. 4.70). Notably, sedimentation exceeds accommodation in overfilled settings, so it is not restricted to the amount of space made available by subsidence or sea-level rise. Processes of sedimentation or erosion are ultimately controlled by the balance between sediment supply and the energy of the sediment-transport agents at any location (details in Chapter 3).

9.1.1 Scale-independent stacking patterns

Stratal stacking patterns are scale invariant in both downstream- and upstream-controlled settings (Figs. 4.70, 5.1, 5.5, and 7.16). This provides consistency with respect to the criteria that afford the identification of sequences and systems tracts at any scales required by a stratigraphic study, whether selected by the practitioner according to the purpose of study or imposed by the resolution of the data available. The formation of the same types of stratal stacking patterns at all stratigraphic scales enables a standard application of sequence stratigraphy in a manner that is independent of the resolution imparted by the local geological setting or the types of data available. Full-fledged sequences and systems tracts may form from centennial scales (Moran, 2020, Fig. 9.1) to the 10^8 yrs scales of sedimentary-basin fills (Fig. 7.8; details in Chapters 5 and 7). The methodology remains consistent across the entire spectrum of stratigraphic scales, from the scales of seismic stratigraphy to the scales of high-resolution sequence stratigraphy (Figs. 4.6 and 5.2).

Clinoforms play a fundamental role in the definition of the sequence stratigraphic framework in downstream-controlled settings. As noted by Schlager (2010), "Thorne (1995) developed a quantitative model of the triad of topset-foreset-bottomset, a critical element in sequence architecture. He argued that this architecture remains virtually invariant for vertical scales of $10^0 - 10^3$ m. Van Wagoner et al. (2003) showed that the plan-view patterns of sediment accumulations fed by a point source, such as deltas and turbidite fans, remain invariant on scales of about 14 orders of magnitude in area." Indeed, clinoforms can be observed from the $10^0 - 10^1$ m scales of deltas (i.e., single depositional systems; Figs. 4.22A,B, 6.56D,

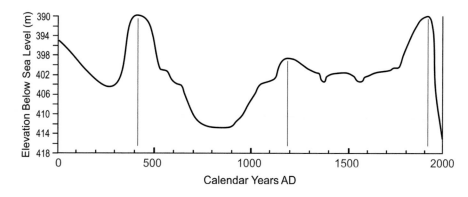

FIGURE 9.1 Centennial-scale sea-level cycles in the Dead Sea over the last 2,000 yrs (modified after Bookman et al., 2006). These changes in sea level led to the deposition of $10^0 - 10^1$ m thick unconformity-bounded sequences and component systems tracts (Moran, 2020). Basins isolated from the global ocean are most susceptible to short-term changes in sea level, and prone to the development of centennial-scale sequences and systems tracts (details in Chapters 5 and 7).

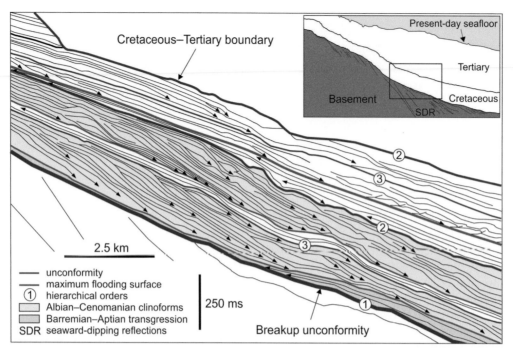

FIGURE 9.2 Dip-oriented transect in the area of the present-day continental slope, showing Albian and Cenomanian clinoforms in the early stages of development of the South Atlantic margin (offshore Argentina; tracing of seismic reflections courtesy of Pedro Kress). The development of the continental margin started from a ramp setting in the Barremian–Aptian, which evolved into a shelf-slope system by a combination of depositional and structural processes (see Kress et al., 2021 for details). The scale of the shelf-slope system increased from $\leq 10^2$ m in the Albian to 10^3 m at the present time, following the subsidence of the oceanic crust.

and 6.58) to the 10^2–10^3 m scales of shelf-slope systems (i.e., associations of depositional systems, from continental to deep water; Fig. 4.6). The scale of the clinoforms reflects the paleobathymetry of the host basin, commonly reaching 10^2 m in epicontinental settings (Fig. 8.4) and 10^3 m in pericontinental settings (Fig. 8.3). The scale of the clinoforms may also change during the evolution of a shelf-slope system, typically increasing in the early stages of basin development (Fig. 9.2), and decreasing in the later stages as the basin fills up with sediment (Fig. 8.4).

Diagnostic to the definition of systems tracts and bounding sequence stratigraphic surfaces is the trajectory of subaerial, rather than subaqueous, clinoform rollovers (e.g., it is the backstepping of the shoreline, and not of the shelf edge, that defines a transgressive systems tract or a maximum flooding surface; Catuneanu, 2019a). Notably, the shelf edge may also become a subaerial clinoform rollover during times of high-magnitude relative sea-level fall, when shelves are exposed and the shoreline coincides with the shelf edge (Figs. 4.6 and 8.53). The trajectory of the shelf edge is the first-order shoreline trajectory observed at the basin scale (Catuneanu, 2019a,b). At any scale of observation, the limit between the topset and the foreset of a prograding system represents the trajectory of maximum regressive shorelines of immediately lower hierarchical rank (Figs. 4.58 and 4.60). This is true from the scale of deltas (in which case the limit between topset and foreset

corresponds to the low tide level: i.e., the tidal lowstand; Fig. 3.25) to the scale of shelf-slope systems (in which case the limit between topset and foreset corresponds to the shelf edge: i.e., the first-order shoreline trajectory, which connects the lowstand shorelines of the second-order sequences; Fig. 4.60; Catuneanu, 2019a,b).

At any scale of observation, the transit area of the shoreline is located updip from the shoreline trajectory of higher hierarchical rank (Figs. 4.58 and 4.60). The updip extent of the transit area depends on topographic gradients, sediment supply, and the magnitude of relative sea-level changes. The shifts of the shoreline within a transit area reflect higher frequency (lower rank) transgressions and regressions within a higher rank topset (Figs. 4.6, 4.58, 4.60, and 5.87). The transit time of the shoreline across a transit area varies with the scale of observation, from timescales as short as tidal cycles in the case of intertidal settings to 10^4–10^5 yrs in the case of continental shelves (Burgess and Hovius, 1998; Porebski and Steel, 2006). At each scale of observation, the growth and the progradation of clinoforms take place in incremental steps, reflecting the episodic availability of sediment to the rollover points (e.g., highstand shedding in the case of carbonates vs. lowstand shedding in the case of siliciclastic systems). At the scale of shelf-slope systems, these changes in sediment supply to the shelf edge relate to the second-order and lower rank cycles of shoreline shift across the shelf (i.e., shoreline shelf-transit cycles; Figs. 4.58, 4.60, and 7.9).

The sedimentary record of a shoreline transit area depends on the balance between the competing processes of aggradation and erosion in continental vs. marine environments. For example, the topset of a third-order prograding system, typically interpreted as "continental" at the third-order scale of observation, may include fourth-order transgressions, which affect the preservation of continental deposits within the third-order topset (Fig. 4.58). Most commonly, the topset preserves both continental and marine systems, and it may be dominated by the former. In other cases, however, where erosion during transgression outpaces the amount of continental aggradation, the third-order topset may become entirely marine (i.e., a stack of fourth-order transgressive "healing-phase" deposits separated by fourth-order wave-ravinement surfaces). The same is true for first-order topsets (e.g., the continental shelf of a divergent continental margin) whose dominant depositional system (continental vs. marine) depends on the balance between the competing processes of continental aggradation and subsequent transgressive ravinement erosion (Figs. 4.6 and 4.60).

In upstream-controlled settings, the formation of the same types of stratal stacking patterns at different scales reflects the dominance of the defining depositional elements (e.g., channels vs. floodplains in fluvial systems) observed through different lenses of stratigraphic resolution (Fig. 4.70). The timescales that afford the development of "unconventional" stacking patterns range widely from 10^2–10^5 yrs (the scales of high-resolution sequence stratigraphy; Fig. 4.71) to 10^6 yrs and longer (Fig. 4.72). As it is also the case with downstream-controlled settings, the formation of stratal stacking patterns in upstream-controlled settings cannot be linked, by default, to any specific control on sequence development. As a result, the nomenclature of "unconventional" stacking patterns and corresponding systems tracts no longer makes reference to accommodation, which was assumed *a priori* as the dominant control on sedimentation (details in Chapter 4). Any combination of upstream-controls, including accommodation, source-area tectonism, climate, and autogenic processes can lead to the development of "unconventional" stacking patterns under specific circumstances (Fig. 3.11).

9.1.2 Classification of stratigraphic cycles

A scale-variant approach to the classification of stratigraphic cycles involves the definition of different types of units at different scales. In a scale-invariant approach, the same types of units are recognized at all stratigraphic scales. The scale-variant approach was introduced with the definition of a "sequence" as a "relatively conformable succession" (Mitchum, 1977), relative to which smaller units (parasequences) and larger units (composite sequences, megasequences) have been defined (Van

Wagoner et al., 1988, 1990; Mitchum and Van Wagoner, 1991; Sprague et al., 2003; Neal and Abreu, 2009; Abreu et al., 2010). The pitfall of this approach is that the scale of a "relatively conformable succession" depends on the resolution of the data available, as well as on the interpreter, so it does not provide a reproducible "anchor" for a hierarchy system. For example, a "sequence" defined with a low-resolution dataset (e.g., 2D seismic lines) may be equivalent to a "composite sequence" in an area where higher resolution data (e.g., well logs) are available (Fig. 7.6). This invalidates the scale-variant nomenclatural system, which is inconsistent with respect to the stratigraphic meaning of the various types of units in the classification.

A "patch" to this scale-variant hierarchy system would be to define the anchor (i.e., the "relatively conformable" sequence) as the lowest rank stratigraphic sequence, whose systems tracts and component depositional systems consist only of sedimentological cycles; this would bring more consistency to the stratigraphic meaning of the reference unit, as the smallest stratigraphic cycle at any location. However, even this more objective approach faces several problems: it would require high-resolution data in every case study, which are not always available; inconsistency would still remain in terms of the absolute scales of the lowest rank stratigraphic sequences in different areas, which depend on the local conditions of accommodation and sedimentation (i.e., specific tectonic and depositional settings); and, last but not least, there is significant subjectivity in the perception of what constitutes a "relatively conformable" sequence, even where high-resolution data are available (e.g., shoreline shelf-transit cycles of 10^4–10^5 yrs are variably referred to as sequences or parasequences, as the nomenclatural confusion continues; Fig. 4.34; details in Chapter 7).

While the range of physical scales of high-frequency sequences is relatively narrow (i.e., 10^0–10^1 m in most reported case studies: Amorosi et al., 2005, 2009, 2017; Mawson and Tucker, 2009; Nanson et al., 2013; Nixon et al., 2014; Magalhaes et al., 2015; Ainsworth et al., 2017; Pellegrini et al., 2017, 2018; Zecchin et al., 2017a,b), the temporal scales can vary widely from 10^2 to 10^5 yrs, indicating either different origins for these sequences or the fact that at least the longer term ones are not of the lowest hierarchical rank. The latter point is illustrated by the inconsistency in the identification of the lowest rank stratigraphic cycles in different studies: e.g., "parasequences" of 10^4–10^5 yrs timescales (Ainsworth et al., 2018, and references therein) become "sequences" or even "composite sequences" if the resolution of the stratigraphic study increases to 10^2–10^3 yrs timescales (Amorosi et al., 2005, 2009, 2017; Bassetti et al., 2008; Mawson and Tucker, 2009; Bruno et al., 2017; Pellegrini et al., 2017, 2018; Zecchin et al.,

2017a,b; Fig. 5.5). Hence, the standards used by different authors to define the scale of "relatively conformable" sequences are inconsistent, even where high-resolution data are available. The definition of an objective "anchor" at small stratigraphic scales becomes impossible where high-resolution data are not available (e.g., in all seismic-based studies).

The value of the "relatively conformable" character as a defining attribute of sequences is further compromised by the fact that sequences may not even be relatively conformable successions. Sequences of any scale may include unconformities of equal hierarchical rank (e.g., depositional sequences may include unconformities associated with transgression; genetic stratigraphic sequences may include subaerial unconformities), which break the relatively conformable character of the sequence within the area of development of the unconformity (details in Chapter 7). Therefore, sequences may or may not be "relatively conformable" successions, but they always consist of "genetically related" strata that belong to the same stratigraphic cycle of change in stacking patterns (Galloway, 1989; Catuneanu et al., 2011). In a most general sense, what all types of sequences have in common is not their relatively conformable character, but their nature as stratigraphic cycles defined by the recurrence of the same types of sequence stratigraphic surfaces in the sedimentary record (Figs. 1.10 and 1.22). The development of the same types of sequence stratigraphic surfaces at different hierarchical levels gives rise to the nested architecture of stratigraphic cycles in which sequences and component systems tracts are defined by the same criteria at all stratigraphic scales (Figs. 4.70, 5.1, 5.3, 5.5, 5.105, 5.106, 6.57, and 7.16).

The nested architecture of stratigraphic cycles can be observed in all depositional settings, from fully continental (Fig. 4.70) to deep water (Fig. 8.55). With the unprecedented increase in stratigraphic resolution (i.e., by three orders of magnitude since the era of low-resolution seismic stratigraphy, from 10^5-10^6 yrs to 10^2-10^3 yrs timescales), the nomenclature prescribed by the scale-variant hierarchy (i.e., parasequence < sequence < composite sequence < megasequence; Sprague et al., 2003; Neal and Abreu, 2009) has become insufficient to cover the entire spectrum of stratigraphic scales, and units larger than "megasequences" would need to be defined. This would complicate even more an already unnecessarily complex terminology. The solution is an approach to nomenclature that is independent of scale and the resolution of the data available. Within the nested architecture of sequences, the "first-order" basin fill provides the most objective anchor for classification, with first-order sequence boundaries marking changes in the tectonic setting (Figs. 5.3 and 7.1). Where data are insufficient to observe the entire basin fill, the ranking

of sequences that develop at different scales can be referred to in relative terms (Figs. 4.70, 5.106, and 7.16). The full details of the model-independent hierarchy system are presented in Chapter 7.

9.2 Sequence stratigraphy and geological time

Sequence stratigraphy can be applied to strata of all ages, at the resolution afforded by the degree of preservation of the sedimentary record (Fig. 9.3). At syndepositional time, preservation tends to increase from the continental to the marine environments of a sedimentary basin (Fig. 5.15), with the most complete record found typically in basin-floor settings on oceanic crust. Through time, however, sediments deposited on oceanic crust and in continental environments show a decline in preservation with increasing age, due to the recycling associated with crustal consumption, deformation, and erosion that accompany the assembly and breakup of supercontinents (Peters and Husson, 2017). In contrast, the preservation of shallow-water systems does not exhibit a steady decline with increasing age, but only cyclic fluctuations associated with the Wilson cycles (Peters and Husson, 2017). This explains why c. 75% of the preserved sedimentary record consists of sediments accumulated in and adjacent to shallow seas on continental crust (Peters and Husson, 2017), which proves to be the setting with the highest preservation potential in the long term.

Several long-term trends during Earth's history shaped the evolution of sedimentation systems, including the decline in mantle heat, the increase in surface oxidation, the decrease in the planetary rotation rate, and the decrease in the gravitational effects of the Moon (Bekker et al., 2010; Eriksson et al., 2013). It is also possible that the subsidence which controlled the evolution of Proterozoic and Phanerozoic basins may had been suppressed before 2.5 Ga by the thermal uplift driven by the higher concentration of radioactive isotopes in the Archean mantle (Rosas and Korenaga, 2021), leading to a switch in the overall trends of relative sea-level changes from fall to rise. Superimposed on the long-term trends, shorter term fluctuations in paleoclimate and the atmosphere—ocean chemistry were also recorded at multiple timescales, with consequences on the rates of weathering, production of extrabasinal and intrabasinal sediment, and the evolution of ecosystems. The change in the rates and periodicity of natural processes over time, along with the interplay of various local and global controls at any given time, explains the randomness of stratigraphic cycles across a continuum of temporal and physical scales (Carter et al., 1991; Drummond and Wilkinson, 1996, Fig. 7.8; details in Chapter 7).

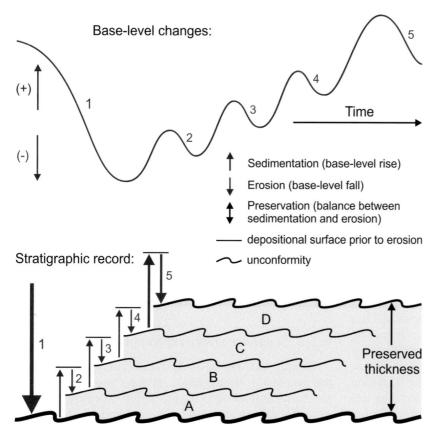

Base-level changes:

Stratigraphic record:

↑ Sedimentation (base-level rise)

↓ Erosion (base-level fall)

↕ Preservation (balance between sedimentation and erosion)

—— depositional surface prior to erosion

〜 unconformity

FIGURE 9.3 Sedimentation, erosion, and preservation. Sedimentary basins may be under-filled or overfilled; therefore, the rates of sedimentation are typically different from the rates of accommodation at any time during the evolution of a basin. Rates of sedimentation higher than the rates of accommodation lead to the shift from underfilled to overfilled basins. The rates of preservation are lower than the rates of sedimentation of the preserved units (A–D). The stacking pattern of the preserved units defines the sequence stratigraphic architecture.

The changing eco-systems through time, which is particularly evident in carbonate settings, as well as the changing rates and timescales of physical processes from the Precambrian to the Phanerozoic (Eriksson et al., 1998, 2004, 2005a,b, 2006, 2013; Catuneanu et al., 2005, 2012; Sarkar et al., 2005) raise the question of whether the sequence stratigraphic methodology depends on, or needs to be adjusted to, geological time. The variability of the stratigraphic architecture is a norm that reflects both the evolution of sedimentation conditions through time and the contrasts between different geological settings that accumulate sediment at the same time. While specific stratigraphic patterns may emerge in particular settings, any individual style of stratigraphic architecture remains unique to the set of underlying conditions that afford its development. For example, shoreline shelf-transit sequences driven by 10^4-10^5 yrs Milankovitch cycles may provide recognizable reference units on continental shelves (e.g., Burgess and Hovius, 1998; Porebski and Steel, 2006; Ainsworth et al., 2018), but may lack relevance in other settings (e.g., underfilled fault-bounded basins with no shelves, or overfilled basins controlled by tectonic cycles). Therefore, no standard pattern can be defined to describe the stratigraphic architecture in all geological settings, and even less so throughout geological time.

9.2.1 Precambrian vs. Phanerozoic sequence stratigraphy

The sequence stratigraphic methodology is independent of geological setting and the age of strata under analysis. The same workflow and set of core principles apply to all stratigraphic successions, from the Precambrian to the Phanerozoic. The study of Precambrian basins is often hampered by poorer stratal preservation and by a general lack of time control. However, where sedimentary facies are well preserved, the lack of time control may be compensated by a comprehensive reconstruction of facies architecture and relationships coupled with paleocurrent data (e.g., Catuneanu and Biddulph, 2001). The latter are particularly important in tectonically active basins, where abrupt shifts in paleoflow directions afford the identification of tectonic events and associated sequence-bounding unconformities even in successions that are relatively homogeneous from a lithological standpoint (Ramaekers and Catuneanu, 2004). A similar workflow can be applied to Phanerozoic successions, where changes in paleoflow directions afford the delineation of depositional sequences, with or without biostratigraphic support (e.g., Catuneanu and Bowker, 2001; Catuneanu and Elango, 2001).

The concepts of sequence stratigraphy have been developed from the study of the Phanerozoic record, which accounts for only c. 12% of Earth's history (Fig. 7.11). This relatively narrow window into the geological past that is offered by the Phanerozoic record is insufficient to afford a thorough understanding of the full range of mechanisms that controlled accommodation and stratigraphic cyclicity throughout geological time (Eriksson et al., 2005a,b). The inclusion of the Precambrian into the time window of study brought significant insights into the selection of criteria that are most relevant to the classification of stratigraphic sequences, and helped resolve debates generated from the sole study of the Phanerozoic record. Therefore, in spite of the limitations imposed by data availability and quality, the Precambrian offers not only challenges but also the opportunity to observe Earth's processes at broader scales and gain insights into key issues of sequence stratigraphy for which the time span of the Phanerozoic is simply too short to afford meaningful conclusions and generalizations.

The nature of basin-forming mechanisms has changed during Earth's evolution, from competing plume and plate tectonics in the Precambrian to a more stable plate-tectonic regime in the Phanerozoic (Fig. 7.11; Eriksson and Catuneanu, 2004b; Eriksson et al., 2005a,b). Sedimentary basins related to plume tectonics are prone to a dominantly nonmarine setting because the net amount of thermal uplift generally exceeds the amount of subsidence created by extension above the ascending plume. As plume tectonics was more prevalent in the Precambrian, the high- and low-amalgamation systems tracts of overfilled basins become more common with increasing stratigraphic age. In contrast, basins related to plate tectonics are dominated by subsidence, and are prone to be transgressed by the global ocean or interior seaways. Classical sequence concepts apply to such underfilled settings, where falling-stage, lowstand, transgressive, and highstand systems tracts may be recognized in relation to particular types of shoreline trajectories. Even such subsidence-dominated basins, however, may reach an overfilled state under high sediment supply conditions, in which case the recognition of fully continental systems tracts (e.g., high- vs. low-amalgamation in fluvial settings) becomes the only option for the sequence stratigraphic analysis. Overfilled plate tectonics-related basins have been documented both in the Precambrian and Phanerozoic records (e.g., Boyd et al., 1999; Zaitlin et al., 2000, 2002; Catuneanu, 2001; Wadsworth et al., 2002, 2003; Leckie and Boyd, 2003; Eriksson and Catuneanu, 2004a; Ramaekers and Catuneanu, 2004).

Arguably the most important contribution of Precambrian research to sequence stratigraphy is the definition of criteria for the classification of stratigraphic cycles in the rock record. There is increasing evidence that the tectonic regimes which controlled the formation and evolution of sedimentary basins in the more distant geological past were much more diverse and erratic than previously inferred from the study of the Phanerozoic (Fig. 7.11; Eriksson et al., 2005a,b). In this context, time becomes largely irrelevant to the classification of stratigraphic sequences, and it is rather the record of changes in the tectonic setting that provides the key criterion for the subdivision of the rock record into basin-fill successions separated by first-order sequence boundaries (Fig. 5.3). These first-order basin fills are in turn subdivided into second-order and lower rank sequences that reflect shifts in the balance between accommodation and sedimentation observed at increasingly smaller scales, irrespective of the timespan between two consecutive same-order events (Fig. 6.57). Sequences identified in a particular basin are not expected to correlate to the first-order and lower rank sequences of other basins, which may have different timing and durations (details in Chapter 7).

9.2.2 Sequence stratigraphy of the Precambrian

Unique features of the Precambrian, with impact on the sequence stratigraphic architecture, include the mechanisms leading to the formation of sedimentary basins, the atmosphere and ocean chemistry, the abundance of microbial mats, and the lack of continental vegetation. These features confer the Precambrian record a stratigraphic architecture that differs from that of the Phanerozoic in several significant ways. Therefore, the extrapolation of principles developed from the study of the Phanerozoic to the entire geological record based on the law of uniformitarianism proves to be inadequate (Eriksson et al., 2004, 2008, 2013; Catuneanu et al., 2005; Catuneanu and Eriksson, 2007). Notwithstanding the practical limitations imposed by the poorer time control and stratal preservation that hamper Precambrian research, there are still Precambrian basins where sequence stratigraphy can be applied at the highest resolution (e.g., Catuneanu and Biddulph, 2001; Sarkar et al., 2005; Magalhaes et al., 2015; Kunzmann et al., 2020). Several key aspects of the Precambrian sequence stratigraphic architecture that are in non-uniformitarian contrast with the Phanerozoic record are outlined below.

The stratigraphy and evolution of the Kaapvaal craton of South Africa during the c. 3.0—2.0 Ga (Late Archean—Early Proterozoic) interval provide a good example to illustrate the mechanisms controlling the formation of sedimentary basins in the Precambrian (Eriksson et al., 2005b). The c. 1 Gy record of Kaapvaal evolution was marked by a combination of plate tectonic and plume tectonic regimes, whose relative importance determined the type of tectonic setting and sedimentary basin at any given time. The shifting balance between these two main controls on accommodation resulted in a succession of discrete basins, starting with the

Witwatersrand (accommodation provided by subduction-related tectonic loading), followed by the Ventersdorp (accommodation generated by thermal uplift-induced extensional subsidence), and lastly by the Transvaal (accommodation created by extensional and subsequent thermal subsidence). The end of the Transvaal cycle was marked by a plume-related event that led to the emplacement of the Bushveld igneous complex. The sedimentary fill of each of these three basins relates to a specific tectonic setting and represents an unconformity-bounded first-order depositional sequence.

The temporal duration of the Kaapvaal first-order cycles varied greatly with the type of tectonic setting, from c. 5 My in the case of the plume-related Ventersdorp thermal cycle, to >600 My in the case of the extensional Transvaal Basin. This first-order cyclicity was independent of the Wilson-type cycles of supercontinent assembly and breakup, and it is rather the result of the interplay between plate tectonics (i.e., the Witwatersrand and Transvaal basins) and plume tectonics (i.e., the Ventersdorp Basin). Noteworthy, the first-order cycles controlled by plate tectonics (i.e., on timescales of 10^2 My) lasted about two orders of magnitude longer than the plume tectonics cycles (i.e., within a range of 10^0 My). Each of the Kaapvaal first-order cycles can be subdivided into second-order cycles with variable time durations, from c. 1 My in the case of the Ventersdorp Basin to c. 100 My in the case of the Witwatersrand and Transvaal basins (Catuneanu and Eriksson, 1999; Catuneanu, 2001). The study of the Kaapvaal craton indicates that time is irrelevant to the classification of stratigraphic cycles. This reinforces the conclusion that hierarchical ranks only reflect the relative magnitude of stratigraphic cycles within the nested architecture of a basin fill, rather than any temporal or physical standards. Stratigraphic frameworks are basin specific in terms of the temporal and physical scales of sequences, due to the influence of local controls on accommodation and sedimentation (details in Chapter 7).

A significant control on the stratigraphic architecture of Precambrian successions is the production of extrabasinal and intrabasinal sediment, which depends in part on the uplift of provenances around the basins, as well as on the atmosphere and ocean chemistry. Two independent lines of work indicate that Precambrian tectonism was more active than in the Phanerozoic, leading to more dynamic and frequent uplift. One line of evidence is provided by the study of tectonic regimes, which indicates that the higher heat flows of the Precambrian promoted uplift associated with plume tectonics and higher rates of plate drift (Eriksson et al., 2007, 2013). This active tectonism likely resulted in the development of ragged, immature landscapes with relatively steep topographic gradients. This conclusion is supported by the second line of research, which indicates that the most common fluvial styles in the Precambrian

relate to high-energy alluvial fans, braided rivers, and ephemeral sheetflood rivers (Long, 2011, 2019; Eriksson et al., 2013). The generally steeper gradients in the Precambrian have also been confirmed with paleohydrological data, which show Precambrian rivers flowing on paleoslopes steeper than those of their Phanerozoic and modern counterparts (Eriksson et al., 2008). Besides paleoslopes, the prevalence of unconfined rivers in the Precambrian was further enhanced by the lack of continental vegetation, which, in modern systems, contributes to bank stability (Miall, 1996; Long, 2011, 2019; Eriksson et al., 2013). The evidence of prevalent uplift supports the conclusion that the Precambrian was characterized by higher rates of weathering, erosion, and extrabasinal sediment supply.

Sediment supply in the Precambrian can also be assessed from geochemical proxies that describe the atmosphere and ocean chemistry (Nesbitt and Young, 1982; Veizer et al., 1989; Strauss, 1993; Trendall, 2002; Shields, 2007; Bekker et al., 2010; Poulton et al., 2010; Young, 2013; Satkoski et al., 2017). The general view is that the rates of weathering were higher, due to carbonic acids in rainfall (Corcoran and Mueller, 2002, 2004), although the absolute rates are subject to debate. This is in agreement with the higher rates of extrabasinal sediment supply inferred from the study of tectonic regimes, fluvial styles, and paleohydrological data discussed above. The atmospheric conditions conducive to higher rates of weathering and erosion relate to the ocean chemistry, as ocean water was oversaturated with respect to $CaCO_3$ (Sweet and Knoll, 1989; Sumner and Grotzinger, 2004; Eriksson et al., 2013), a condition which only reappeared in the Phanerozoic following major extinction events (Grotzinger and Knoll, 1999; Eriksson et al., 2013). Ocean chemistry also controlled the production of intrabasinal sediment, which included carbonates related largely to evaporative processes, forming chemical precipitates influenced by microbial processes (Eriksson et al., 2013); abiogenic chert, where evaporation in shallow-water settings concentrated silica (Maliva et al., 1989, 2005; Zentmyer et al., 2011); and iron formations, generally associated with magmatic activity and anoxic conditions (Bekker et al., 2010). Precambrian carbonates are almost exclusively dolomite, due to the low sulfate concentrations in the seawater (Warren, 2000; Machel, 2004; Zentmyer et al., 2011).

The higher sediment supply available to the Precambrian basins promoted normal regressions over transgressions during stages of relative sea-level rise in underfilled basins, thus influencing the stratigraphic architecture and the relative development of systems tracts. For this reason, systems tracts defined by a normal regressive stacking pattern tend to be more prevalent and better developed than transgressive systems tracts in the Precambrian record (e.g., Sarkar et al., 2005; Catuneanu, 2007). Furthermore, the prolific growth of microbial mats in the absence of grazing metazoans enhanced the organic binding of sediments, leading to a greater preservation potential of Precambrian coastal to shallow-water systems compared to their Phanerozoic counterparts (Parizot et al., 2005; Eriksson et al., 2007, 2010, 2012; Schieber et al., 2007). It is apparent that the balance between stages of relative sea-level fall and rise also changed during the Precambrian, with the former being more prevalent before 2.5 Ga, due to the thermal uplift caused by the higher concentration of radioactive isotopes in the Archean mantle (Rosas and Korenaga, 2021). As a result, Archean sequences may have been dominated by forced regressions and the development of subaerial unconformities across exposed landmasses in the form of volcanic islands, resurfaced seamounts, or oceanic plateaus (Rosas and Korenaga, 2021).

In tectonically active basins, the most important breaks in the rock record are commonly associated with stages of tectonic reorganization that led to changes in tilt direction across unconformities. In such cases, paleocurrent data are key to the identification of tectonic events and associated depositional sequence boundaries. This tool is particularly useful in overfilled Precambrian basins where the dominance of unconfined rivers led to fewer changes in fluvial styles and a relative homogeneity of fluvial facies. For example, the Early Proterozoic Athabasca Basin in Canada includes several second-order depositional sequences, each one defined by a unique fluvial drainage system and abrupt shifts in paleocurrent directions across the sequence-bounding unconformities (Ramaekers and Catuneanu, 2004). In this basin, paleocurrent data provide the best evidence to constrain regional correlations and to compensate for the lack of time control in a succession that is relatively homogeneous from the standpoint of fluvial facies. A similar usage of paleocurrent data can be applied to overfilled Phanerozoic basins, although the wider range of fluvial styles in the Phanerozoic provides additional tools to delineate depositional sequences (e.g., Catuneanu and Bowker, 2001; Catuneanu and Elango, 2001).

9.3 Methodology and nomenclature

The evolution of sequence stratigraphy involved advances in the understanding of the array of possible controls on sequence development, improvements in stratigraphic resolution, revisions to the definition and classification of sequences, and ultimately the emergence of a standard approach to methodology and nomenclature. The methodology improved significantly since the 1970s, from a model-driven approach underlain by assumptions regarding the dominant role of eustasy on sequence development, with consequent

assertions of global correlations, to a data-driven approach that promotes the use of local data and unbiased geological reasoning. The latter approach affords realistic constructions of local stratigraphic frameworks, which prove to be highly variable, not only from one sedimentary basin to another but also between sub-basins of the same sedimentary basin. The construction of basin-specific stratigraphic frameworks is a major breakthrough and departure from the early models.

The approach to sequence stratigraphic methodology and nomenclature hinges on the principles that underlie the hierarchy systems for the classification of stratigraphic cycles. In an idealized view, the stratigraphic record is organized according to an orderly pattern in which hierarchical ranks are defined not only by absolute or relative scales but also by the internal composition of stratigraphic cycles (e.g., including or excluding lowstand systems tracts; Van Wagoner et al., 1990; Duval et al., 1998). However, the stratigraphic record proves to be variable, with a basin-specific architecture that reflects the complex interplay of multiple local and global controls on accommodation and sedimentation (Schlager, 2010; Catuneanu, 2019a,b; details in Chapter 7). Such variable patterns require the construction of stratigraphic frameworks on the basis of local data, independently of any global standards and model predictions (Catuneanu, 2019a). The question of regional or global correlations can still be tested at the scales afforded by the resolution of the data available once the local stratigraphic frameworks are in place.

Despite the variability of the stratigraphic architecture, only a few stratigraphic stacking patterns are diagnostic to the definition of systems tracts, which can be recognized at all stratigraphic scales (details in Chapter 4 and 5). This streamlines the sequence stratigraphic workflow, whose objective is to identify and separate the diagnostic features from the non-diagnostic variability, irrespective of the resolution of the stratigraphic study (Fig. 1.2). The emphasis on the observation of the diagnostic stacking patterns affords a standard application of sequence stratigraphy in a manner that is independent of model, the resolution of the data available, and the interpretation of the underlying controls on sequence development (Catuneanu, 2019a). The sequence stratigraphic workflow requires methodological and nomenclatural consistency at all stratigraphic scales, for an objective approach that is free of subjective choices and arbitrary variables. A scale-independent approach to methodology and nomenclature is key to the standard application of sequence stratigraphy across the entire range of geological settings, stratigraphic scales, and types of data available.

The construction of the sequence stratigraphic framework can be approached from either end of the spectrum of stratigraphic scales, depending on the types of data available (e.g., seismic vs. outcrop data). At seismic scales, the methodology relies on the observation of seismic-reflection geometries and terminations, which define the stratal stacking patterns (Figs. 4.7 and 4.8). The large-scale framework defined by the architecture of seismic reflections can be subsequently refined with the identification of smaller units following the acquisition of higher resolution well data. At sub-seismic scales, the workflow can follow the opposite course, starting with facies analysis at sedimentological scales, followed by the identification of increasingly complex stratal units at stratigraphic scales. However, the methodology remains consistent in the sense that the same stacking patterns can be observed at all stratigraphic scales, even though their expression varies with the dataset; e.g., a prograding system in shallow-water settings can be expressed as a set of oblique reflections (clinoform surfaces) on a dip-oriented seismic transect, or as a shallowing-upward facies succession in vertical sections (well logs, cores, outcrops). Where both high- and low-resolution datasets are available, the results can be integrated and cross calibrated.

A number of developments arose since the publication of the first edition of this book in 2006, some more beneficial than others to the definition and clarification of principles and practices in sequence stratigraphy. Undoubtedly the most significant development was the endorsement of sequence stratigraphy by the International Commission on Stratigraphy in 2011, along with the publication of formal guidelines on methodology and nomenclature (Catuneanu et al., 2011). This concluded decades of work toward the definition of a standard methodology, following the unsuccessful attempts of two previous task groups led by A. Salvador (1995−2003) and A.F. Embry (2004−2007). Standardization was enabled by a model-independent approach to the identification of stratal units and bounding surfaces, based on the observation of stacking patterns, decoupled from the interpretation and testing of the underlying controls (Fig. 1.2). Therefore, it is important to separate methodology from modeling in sequence stratigraphy. A standard methodology does not prevent future developments in the field of stratigraphic modeling.

9.3.1 From seismic to sub-seismic scales

Sequences of seismic stratigraphy are typically observed at scales of 10^1−10^2 m and 10^5−10^6 yrs (Payton, 1977; Vail et al., 1991, Duval et al., 1998; Schlager, 2010). The use of seismic data in developing the methodology imposed by default a minimum scale on the concepts of sequence, systems tract and depositional system, which had to exceed the vertical seismic resolution (i.e., in a range of 10^1 m in the 1970s). Extending the

applications of sequence stratigraphy from the seismic scales of petroleum exploration to the sub-seismic scales of production development required an increase in stratigraphic resolution by observing stratal units at the smaller scales afforded by well logs, core, and outcrops. The high-resolution sequence stratigraphy started with the definition of parasequences as the building blocks of seismic-scale systems tracts (Van Wagoner, 1985; Van Wagoner et al., 1988, 1990).

The value of the parasequence concept, as viewed by its proponents and followers, is two-fold. First, the parasequence is used as the building block of systems tracts, according to the scale-variant system of classification of stratigraphic cycles, which requires parasequences to be below the scale of sequences at any location (Van Wagoner et al., 1988, 1990; Sprague et al., 2003; Neal and Abreu, 2009; Abreu et al., 2010). Second, the parasequence boundaries are facies contacts which violate Walther's Law (Fig. 5.94); this promotes the use of flooding surfaces for the subdivision of stratigraphic successions into stratal units composed of genetically related beds and bedsets. For this reason, it has been common practice to use flooding surfaces as the starting point for stratigraphic analyses in coastal to shallow-water settings where flooding surfaces may form (e.g., Bhattacharya, 1993; Frostick and Steel, 1993b; Pattison, 1995, 2019, Fig. 9.4).

Several limitations of the parasequence concept prevent its dependable use in the methodological workflow of sequence stratigraphy. Notably, parasequences have a smaller extent than systems tracts, typically restricted to coastal and shallow-water settings where flooding surfaces may form and can be demonstrated (Figs. 5.99 and 5.107); and, flooding surfaces may not even form in shallow-water settings where conformable successions accumulate during gradual transgressions and water deepening. Therefore, systems tracts do not necessarily consist of stacked parasequences. Moreover, the orderly patterns postulated by the scale-variant system of classification of sedimentary cycles (bedsets < parasequences < sequences; Van Wagoner et al., 1990) fail to provide a reproducible standard, as parasequences may form at both bedset and sequence scales, depending on the local conditions of accommodation and sedimentation (Figs. 5.5, 5.103, 5.104, and 6.41). More significant to the usefulness of the different types of units for stratigraphic correlation are the nature and mappability of their bounding surfaces.

The attribute that all flooding surfaces have in common is that they mark sharp shifts of facies triggered by abrupt water deepening, which violate Walther's Law (Fig. 5.94). Beyond this definition, the stratigraphic meaning of flooding surfaces is variable, from facies contacts within transgressive deposits (i.e., within-trend facies contacts; Fig. 9.5; Bruno et al., 2017) to different

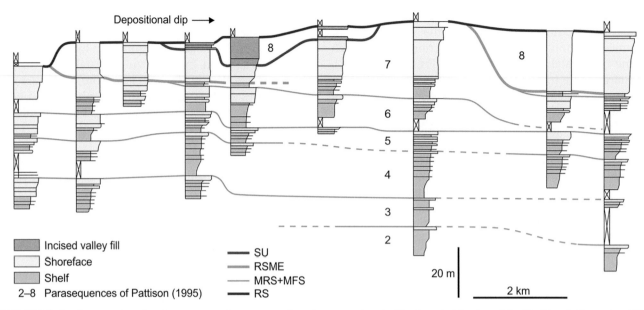

FIGURE 9.4   Cross section showing meter- to decameter-scale units composing the Kenilworth Member of the Blackhawk Formation, USA (modified after Pattison, 1995). The succession was originally described in terms of parasequences bounded by flooding surfaces, here indicated as MRSs merged with MFSs. Abbreviations: MFS—maximum flooding surface; MRS—maximum regressive surface; RS—ravinement surface; RSME—regressive surface of marine erosion; SU—subaerial unconformity.

FIGURE 9.5   Stratigraphic surfaces in a siliciclastic shallow-water setting (Upper Cretaceous Eagle Sandstone, Western Interior Basin, Billings, Montana). The FS has a strong physical expression in shallow-water systems, but the MFS can be mapped over larger areas within the basin. The contrast in acoustic impedance between shelf fines (facies B) and shoreface sands (facies A) generates clinoform "surfaces" on seismic lines. The shelf fines include transgressive facies below the MFS, and highstand bottomset facies above the MFS. Abbreviations: FS—flooding surface; MFS—maximum flooding surface; DLS—downlap surface.

types of sequence stratigraphic surfaces (maximum regressive, transgressive ravinement, or maximum flooding; Figs. 5.100, 5.103, 5.107, 5.109, and 9.4; Arnott, 1995; Pattison, 1995; Martins-Neto and Catuneanu, 2010; Zecchin et al., 2017a). All flooding surfaces with lithological expression, whether coincident or not with

sequence stratigraphic surfaces, are allostratigraphic contacts (Fig. 6.62). The physical expression of flooding surfaces as lithological discontinuities is part of their appeal as markers for mapping and correlation (Fig. 5.94). Criteria have also been defined to extend the mappability of flooding surfaces into distal, fine-grained shelf settings (Bohacs et al., 2014; Birgenheier et al., 2017; Borcovsky et al., 2017; Knapp et al., 2019).

Flooding surfaces in fine-grained shelf settings are the distal expression of the lithological discontinuities that develop in environments closer to the coastline. The distal portions of flooding surfaces can be subtle and difficult to recognize, or be defined by an abrupt decrease of silt and sand interlaminations and/or by shell beds and burrowed layers highlighting temporal hiatuses (Birgenheier et al., 2017; Borcovsky et al., 2017) or by carbonate cementation and abrupt or gradual shifts to more organic-rich deposits (Knapp et al., 2019). As the transgressive deposits wedge out basinward, these contacts tend to merge with maximum regressive and maximum flooding surfaces in a downdip direction (Fig. 6.62). Flooding surfaces in mudstone successions lose their allostratigraphic identity and usefulness as easily identifiable lithological contacts.

The precise stratigraphic meaning of a flooding surface at any location needs to be determined on a case-by-case basis (Figs. 5.100, 5.103, 5.107, 5.108, 5.109, and 6.62). Where transgressions are fast, which may be the case in tectonically active basins (e.g., rifts, forelands) or under icehouse climatic conditions, a few different surfaces (e.g., maximum regressive, wave-ravinement, within-trend flooding, and maximum flooding) can be close enough to each other in the stratigraphic section to be merged into one "flooding surface" on stratigraphic cross sections (e.g., Bhattacharya, 1993, Frostick and Steel, 1993b; Pattison, 1995, Figs. 5.99 and 9.4). This reflects the low resolution of the logs shown on the cross sections, whereby thin meter-scale intervals may "tune" into single stratigraphic horizons. A closer look, however, affords the distinction between the different types of surfaces (Figs. 5.108, 5.109, and 9.5).

In some cases, two or more surfaces can collapse into one physical contact, where the strata that should separate them are missing. Flooding surfaces may coincide with wave-ravinement surfaces, where the onlapping shell beds and/or transgressive lags are negligible or absent; with maximum regressive surfaces, where rapid transgression leads to an abrupt decrease in sediment supply to the shelf; or even with maximum flooding surfaces, where the transgressive deposits are altogether missing on the shelf (Figs. 5.100 and 5.103). Only the abrupt facies shifts to deeper water deposits at the top of the shell beds and/or transgressive lags that drape wave-ravinement surfaces, or higher up within the transgressive deposits, can be designated as stand-alone "flooding surfaces" that do not overlap with a sequence stratigraphic surface (Figs. 5.100B, 5.108, 5.109, and 6.61).

Where flooding surfaces coincide with sequence stratigraphic surfaces (maximum regressive, transgressive ravinement, or maximum flooding), the sequence stratigraphic nomenclature provides the proper genetic designation of the contact. Where flooding surfaces are within-trend facies contacts, they invariably have a smaller extent, and therefore a secondary importance, relative to the same-rank sequence stratigraphic surfaces that delineate sequences and systems tracts (Fig. 6.61). At the same scale of observation, systems tract boundaries carry more weight for regional correlation than any allostratigraphic contacts within the systems tracts. No matter how "minor" is the transgression that affords the formation of a flooding surface, it still starts from a surface of maximum regression and ends with a surface of maximum flooding observed at the scale (hierarchical rank) of that transgression. The definition of sequence stratigraphic surfaces is independent of scale, and any type of surface, including the "maximum flooding," can be observed at multiple scales (hierarchical orders; Fig. 5.105).

At the smallest stratigraphic scales, the methodological workflow of sequence stratigraphy starts with the facies analysis of vertical sections (e.g., well logs, cores, or outcrop sections), leading to the identification of stratal stacking patterns and of the various types of stratigraphic surfaces (Fig. 6.62). Some of these surfaces can be lithologically cryptic and in agreement with Walther's Law (e.g., a maximum flooding surface at the limit between retrogradational and progradation trends in a conformable succession), whereas others may have lithological expression and may be in breach of Walther's Law (e.g., flooding surfaces at the contact between strata accumulated under markedly different bathymetric conditions). The relative stratigraphic importance of the different types of surfaces is not dictated by their physical expression, as for example a "cryptic" maximum flooding surface is more significant for regional correlation than a flooding surface with strong physical expression, even at the same scale of observation (Fig. 6.61). All types of surfaces are important to record on the 1D logs, and their identification depends on several field criteria, including the depositional systems that are juxtaposed across the contact (Fig. 6.6).

Different measured sections can include different types of stratigraphic surfaces. For example, a section within a fully marine and conformable succession may include a maximum regressive surface, whereas at the same stratigraphic level the maximum regressive surface can be reworked and replaced by a wave-ravinement surface in an updip direction (Figs. 5.107 and 6.61). Similarly, flooding surfaces do not extend across the entire area of development of transgressive systems tracts, and therefore they may not be present in every measured section (Figs. 5.107 and 6.61). It follows that the methodology cannot rely on the presence of any specific type of surface; the only consistent aspect of the methodology at sub-seismic scales is the facies analysis that unravels the (potentially unique) makeup of each stratigraphic section in terms of stacking patterns and stratigraphic contacts. The next step in the workflow is the construction of cross sections of correlation in order to constrain the distribution of systems tracts and of all types of stratigraphic surfaces that exist within the study area (Figs. 5.107 and 6.61).

With the exception of basin-wide unconformities (i.e., sequence boundaries in the sense of Sloss, 1963), all stratigraphic contacts that violate Walther's Law have a limited extent and pass laterally into conformable surfaces, albeit with different amounts of missing time depending on the type of contact and the scale of observation (hierarchical rank). While important where present, these contacts are not necessarily the most significant features of the stratigraphic record, and their meaning needs to be rationalized within the context of stratal stacking patterns that define the sequence stratigraphic framework. For example, a flooding surface within a transgressive systems tract may have a stronger physical expression than the systems tract boundaries, and it may also be the only surface that violates Walther's Law in that stratigraphic section. However, the systems tract boundaries (i.e., maximum regressive and maximum flooding surfaces in this example) are more extensive and remain more important for regional correlation and for the definition of the sequence stratigraphic framework (Fig. 6.61).

Parasequences have become obsolete as sequences that develop at parasequence scales provide a more reliable alternative for correlation, both within and outside of the coastal and shallow-water settings (Catuneanu and Zecchin, 2020; details in Chapter 5). Transgressive systems-tract boundaries of the same hierarchical rank with flooding surfaces (i.e., maximum regressive and maximum flooding surfaces; Fig. 6.61) are invariably more extensive than any facies contacts that may develop during transgression. Flooding surfaces remain relevant to the description of facies relationships, but their stratigraphic meaning needs to be assessed on a case-by-case basis. The use of sequences and systems tracts in high-resolution studies provides consistency in methodology and nomenclature at all stratigraphic scales, irrespective of geological setting and the types and resolution of the data available.

9.3.2 Methodology vs. modeling in sequence stratigraphy

The development of sequence stratigraphy involved progress in perfecting the methodological analysis of outcrop and subsurface data, along with advances in the numerical simulation of the sedimentary response to the various allogenic and autogenic controls on sequence development. While intertwined in the early days of seismic and sequence stratigraphy, and particularly in the 1970s and the 1980s when global-cycle charts were constructed and embedded in the method (Vail, 1975; Vail et al., 1977b; Haq et al., 1987), it eventually became evident that observations and interpretations need to be separated for an objective approach to data analysis (e.g., Miall, 1992, 1994). This requires a clear distinction between methodology and modeling in sequence stratigraphy. The sequence stratigraphic methodology is a data-driven approach that relies on the observation of stratal stacking patterns and stratigraphic relationships, in a manner that is independent of the interpretation of the underlying controls. In contrast, stratigraphic modeling deals with the interpretation and testing of the possible controls on sequence development. The two lines of work describe different aspects of stratigraphic research, and need to be conducted independently of each other (Fig. 1.2).

The proliferation of numerical simulations in stratigraphic research within the last decade led to a "modeling revolution" whereby virtual reality has become the latest fashion in stratigraphy (Catuneanu, 2020b). Forward modeling, in particular, involves a series of assumptions with respect to the mode of sediment transport (e.g., geometric, diffusive, or process-based) and the relative contributions of the tested controls on sequence development (e.g., eustasy, subsidence, sediment supply). The outcome of the model depends on the selection of input parameters, and multiple combinations of input parameters can lead to similar results. Uncalibrated modeling can "demonstrate" any stratigraphic scenario, whether realistic or not (see discussion in Catuneanu and Zecchin, 2016). Furthermore, not all modeled scenarios are equally common in nature. For example, numerical models can predict fluvial floodplain aggradation during relative sea-level fall, but actual examples are rare (Holbrook and Bhattacharya, 2012). Therefore, what is being modeled may or may not happen in nature, and calibration with real data remains essential to validate modeling results and to draw meaningful conclusions.

Confusions regarding the role of numerical modeling in sequence stratigraphy can lead to significant setbacks. Advances in computing power enhanced the ability to perform forward modeling and test the results of various combinations of controls on sedimentation. While this may provide useful insights under realistic boundary conditions, uncalibrated simulations and overreaching conclusions can be misleading and even counterproductive. A dangerous extreme is to think that virtual reality (i.e., the outcome of numerical modeling) is more meaningful than the reality described by actual data. Software enthusiasts seem to assume so when "demonstrating" stratigraphic scenarios that have little in common with the real world, based on unrealistic selections of input parameters. Proposals that numerical simulations should become part of the workflow of sequence stratigraphy are not only impractical, but in fact counterproductive as they bring confusion between modeling and methodology. Muddling the distinction between the two lines of stratigraphic research undermines the progress made in the development of sequence stratigraphy as a data-driven methodology which relies on observations rather than model-driven assumptions (Fig. 1.2).

The separation of modeling and methodology is an important "first amendment" in sequence stratigraphy. Mixing the methodology with the interpretation of controls on sequence development was a pitfall since the inception of sequence stratigraphy, which took decades of work to correct. The early models favored eustasy as the dominant control, which led to the assumption of global correlations in the 1970s and the 1980s, while the role of sediment supply on par with accommodation was only fully recognized starting with the 1990s (i.e., the "dual control" of Schlager, 1993). Between these end members, any combinations are possible. Linking the methodology to any particular control on sequence development is ultimately misleading, as it is *always a combination* of controls that defines the stratigraphic architecture. For this reason, the methodology needs to remain neutral with respect to the interpretation of underlying controls, and any reference to a specific control (e.g., "tectono-sequence stratigraphy") should be avoided.

The latest trend in numerical modeling is the shift from an overemphasis on accommodation to an overemphasis on sediment supply. This leads to an extreme view whereby all aspects of the stratigraphic architecture can be explained by variations in sediment influx or even solely by autocyclicity. While possible in the virtual world of uncalibrated numerical models, the overemphasis on sediment supply loses touch with reality and brings more confusion than clarification. On practical grounds, this conceptual setback undermines the predictive power of sequence stratigraphy, by downplaying the role of accommodation as a control on key aspects of the stratigraphic architecture (e.g., downstepping vs. upstepping shoreline trajectories, and their effects on facies development, distribution, and relationships; Figs. 4.40, 4.41, 6.26, 7.16, and 8.55; Catuneanu, 2020a,b). This results in a critical loss of insight into the rationale of sediment distribution across a basin, which is detrimental to any practical applications. While any extreme views promoted by uncalibrated numerical models are equally deceptive, the methodology remains grounded on field data and the model-independent observation of stratal stacking patterns and stratigraphic relationships.

Models decoupled from field data can reach the point of challenging facts and common sense. Failure to separate realistic from unrealistic results promotes a "chaos theory" whereby the reality is overruled by artificial model predictions. As R.P. Feynman once described such fashionable but misguided "revolutions" in science, the distortion of reality with uncalibrated models remains an empty claim "... that some obvious and correct fact, accepted and checked for years, is false" (Gleick, 1992). Field data lend order to chaos, by providing the criteria to identify the elements of the sequence stratigraphic framework. Uncalibrated models muddle the distinction between the "dual controls" on the stratigraphic architecture, leading to bias toward randomness over predictability in terms of the process—response relationships that can be expected in the sedimentary record. In real case studies, both components of the "dual control," including the changes in relative sea level at different stages of sequence development, can be measured and demonstrated with sedimentological and stratigraphic data (Figs. 4.40, 4.41, and 5.9; Plint, 1988; Plint and Nummedal, 2000; Posamentier and Morris, 2000; Tesson et al.,

2000; Zecchin and Tosi, 2014; Catuneanu, 2019a; Sweet et al., 2019).

With the reality seemingly going out of fashion, and to paraphrase Krynine (1941), uncalibrated modeling has become the new "triumph of interpretation over facts and common sense." The pitfalls of uncalibrated modeling can be exacerbated by basic misconceptions incorporated in the numerical models (e.g., confusions between "sediment supply" and "sedimentation," or between "accommodation" and "sedimentation"), as well as by the incorrect application of the methodology. For example, the 3D variability of accommodation rates across an underfilled basin makes it necessary to consider the fact that only the relative sea-level changes *at the shoreline* are relevant to the timing of stratigraphic sequences and component systems tracts. Failure to do so (e.g., by using reconstructions of relative sea-level changes in the deep-water setting as a reference for the development of sequences) leads to overblown and ultimately false conclusions. Such "developments" leave the practitioner confused, and therefore at a loss.

Numerical modeling is particularly beneficial when used in conjunction with field data (e.g., Euzen et al., 2004; Rabineau et al., 2005, 2006; Csato et al., 2013, 2015; Leroux et al., 2014). A critical parameter in the numerical model is the selection of the mode of sediment transport, deposition, and erosion. In different models (e.g., geometric vs. diffusive vs. process-based), this parameter is implemented in different ways, leading to very different results. Therefore, the selection of this parameter (e.g., the diffusive coefficient "K" in the Dionisos modeling software) affects significantly the model results, to the extent that it can even outweigh the effects of the tested controls (i.e., eustasy, tectonics, and sediment supply). This prompts a note of caution, since this coefficient remains entirely subjective, and possibly unrealistic, unless calibrated with real data. The abusive use of numerical modeling overshadows its potential value and erodes its credibility, which is counterproductive. The confusion between modeling and methodology is also counterproductive, as it obscures the observation-based workflow of the methodology.

Calibration with field data provides an essential reality check on numerical models, and ensures geologically realistic model parameters. Since the model results depend on the selection of input parameters, the use of uncalibrated simulations to "demonstrate" how nature works is circular reasoning at best. While simulations help understand how sedimentary systems *may* respond to various controls, numerical models are not a substitute for field data. In cases where insufficient data enable alternative interpretations (e.g., forced vs. normal regression), numerical simulations can "prove" either option, but only the acquisition of more data can demonstrate what option is true. Overreliance on numerical models, particularly in the face of field data which point to different conclusions, implies that modeling results are superior to real data, leading to the argument that the sequence stratigraphic methodology needs to incorporate numerical modeling. In actuality, numerical modeling plays no role in the sequence stratigraphic methodology. Methodology and modeling are two independent lines of stratigraphic research, which follow different workflows and serve different purposes (Figs. 1.2 and 3.3).

Case studies document the common trends in the rock record (e.g., the development of subaerial unconformities during forced regression, or the tendency of fluvial systems to aggrade during transgression; Shanley et al., 1992; Shanley and McCabe, 1993, 1994; Wright and Marriott, 1993; Kerr et al., 1999; Catuneanu et al., 2006; Fanti and Catuneanu, 2010), as well as the exceptions from the common trends (e.g., fluvial aggradation during forced regression, or fluvial incision during transgression; Zecchin et al., 2009a, 2010b, 2011; Leckie, 1994). The sequence stratigraphic methodology does not rely on statistical norms, but on the observation of stratal stacking patterns in the sedimentary record (Posamentier and Morris, 2000; Løseth and Helland-Hansen, 2001; Zecchin, 2007; Helland-Hansen and Hampson, 2009; Catuneanu et al., 2011; Zecchin and Catuneanu, 2013). Not all depositional trends have the same degree of stratigraphic significance; in this example, fluvial processes of aggradation or erosion are not diagnostic to the definition of systems tracts (Figs. 4.23 and 5.9). The use of non-diagnostic variability as "evidence" that sequence stratigraphy fails to work reflects a poor understanding of the methodology.

Source-to-sink studies expand the scope of numerical modeling to include the production of extrabasinal and intrabasinal sediment, and the delivery systems that link the source areas to the depocenters. This involves an appraisal of extrabasinal sediment sources, weathering efficiency in relation to paleoclimates, distances and means of sediment transport, and the location of sediment entry points into the marine or lacustrine basins. Intrabasinal sediment sources are equally important, as they explain sedimentation patterns that are unrelated to extrabasinal sediment supply. In this context, the sequence stratigraphic framework provides an anchor for the calibration of model results with field data (Fig. 1.2). The development of source-to-sink models does not change nor replace the need for sequence stratigraphic work. Sequence stratigraphy continues to provide the methodology to analyze the stratigraphic relationships within a sedimentary basin, in a data-driven manner that is independent of model assumptions. The results of sequence stratigraphic analysis provide a reality check for numerical simulations, and need to be used

to constrain realistic input parameters for the source-to-sink models.

For any practical purposes, sequence stratigraphy remains a data-based methodology founded on observation rather than experiment. The separation of methodology from modeling is as fundamental as the distinction between reality and virtual reality. The methodology leads to the construction of sequence stratigraphic frameworks based on the observation of stratal stacking patterns, at scales afforded by the resolution of the data available (e.g., seismic vs. outcrop). The sequence stratigraphic framework depicts the architecture of sequences and systems tracts in an objective manner that is independent of the interpretation of the underlying controls (Figs. 1.1 and 3.3). The reliability of the constructed framework depends on (1) the amount and quality of the data available; and (2) the ability of the practitioner to restore stratal geometries at syn-depositional time, and to recognize the diagnostic stacking patterns. These are issues that may impact the outcome of sequence stratigraphic work, but are not a pitfall of the methodology. Rather, these practical limitations are the reason why sequence stratigraphic frameworks may require revisions as new data or skills are acquired.

Confusions between modeling and methodology triggered unwarranted questions about the "future of sequence stratigraphy." In reality, the future is already here in terms of a standard methodology. The methodology is not intended to elucidate the controls on sequence development, whose modeling and testing may continue indefinitely after the construction of a sequence stratigraphic framework. In many cases it is difficult if not impossible to identify the underlying controls (e.g., eustasy vs. tectonics; allogenic vs. autogenic), or to quantify their relative contributions to the stratigraphic architecture. That is beyond the scope of the model-independent approach to sequence stratigraphy, whereby the construction of local stratigraphic frameworks based on the data available replaces the assumption of global correlations or any other assumptions regarding the controls on sequence development. Instead, the methodology provides guidelines that enable the identification of all physical elements (systems tracts and bounding surfaces) that build the stratigraphic framework (Figs. 1.1 and 3.3).

9.3.3 Standard methodology and nomenclature

The standard methodology and nomenclature is based on a robust platform of simple and objective guidelines that enable a consistent application of sequence stratigraphy irrespective of geological setting, stratigraphic scale, and the types of data available. The existence of several competing approaches (Figs. 1.9 and 1.10) hindered, for decades, the definition of a standard methodology and the inclusion of sequence stratigraphy in international stratigraphic guides. These competing approaches differ in terms of the nomenclature of systems tracts and bounding surfaces (Figs. 1.9 and 1.10); the selection of the "sequence boundary" (Fig. 1.10); the classification and nomenclature of stratigraphic cycles; and the assertions of the dominant controls on sequence development (e.g., global eustasy: Vail et al., 1977b, 1991; Haq et al., 1987; tectonism: Embry, 1995; accommodation: Neal and Abreu, 2009; or the interplay of accommodation and sediment supply: Schlager, 1993). The standard methodology transcends these differences and relies on core principles that underlie all competing approaches.

The identification of systems tracts and bounding surfaces fulfills the practical purpose of sequence stratigraphy. Within this framework, the practitioner can explain and predict the patterns of sediment distribution across a basin, at the resolution afforded by the data available. Beyond this, the delineation of sequences (i.e., the selection of the "sequence boundary"; Fig. 1.10) becomes a matter of model-dependent organization of systems tracts into stratigraphic cycles (Fig. 7.40). On practical grounds, the delineation of sequences is guided by the development and mappability of the different kinds of sequence stratigraphic surfaces within the study area, which depend on geological setting and the types of data available (e.g., maximum flooding surfaces may or may not be present within the studied section, and where they are, it is typically easier to pinpoint them on well logs than on seismic lines). As a result, no specific approach to the delineation of sequences provides a universal solution to all case studies (Catuneanu et al., 2011). Notably, the definition of all types of sequences and systems tracts is based on the observation of stratal stacking patterns and not on the interpreted origin of cycles (Catuneanu and Zecchin, 2013). The emphasis on stacking patterns promotes objectivity in the construction of sequence stratigraphic frameworks, which may consist of variable combinations of systems tracts (Csato and Catuneanu, 2012, 2014, Figs. 1.23 and 4.33). The flexibility afforded by this model-independent approach frees the practitioner from the expectation to fulfill model predictions, and provides the foundation for a standard methodology.

The model-independent guidelines are simpler than the requirements of any specific model. Significant progress has been made in outlining the common ground in sequence stratigraphy (Catuneanu et al., 2009, 2010), which led to the publication of formal recommendations by the International Commission of Stratigraphy (Catuneanu et al., 2011): "The definition of the common ground in sequence stratigraphy

should promote flexibility with respect to the choice of approach that is best suited to a specific set of conditions as defined by tectonic setting, depositional setting, data available, and scale of observation" (Catuneanu et al., 2011, p. 176); "A standard methodology can be defined based on the common ground between the different approaches, with emphasis on the observation of stratal stacking patterns in the rock record" (Catuneanu et al., 2011, p. 233). It has become clear that none of the competing models provides the "best practice" under all circumstances, as defined by different geological settings and types of data available.

A key aspect of the methodology and nomenclature is the scale at which sequences can be defined. In the context of seismic stratigraphy, the definition of a sequence as a "relatively conformable succession" (Mitchum, 1977, Fig. 1.22) inadvertently linked the scale of a sequence to the resolution of the seismic data. The subsequent definition of other types of stratigraphic cycles at smaller and larger scales (e.g., "parasequences" at smaller scales, and "composite sequences" and "megasequences" at larger scales; Van Wagoner et al., 1988, 1990; Mitchum and Van Wagoner, 1991; Sprague et al., 2003; Neal and Abreu, 2009; Abreu et al., 2010) led to nomenclatural inconsistency, since the scale of the reference unit (i.e., the "relatively conformable" sequence) varies with the resolution of the data available (Fig. 7.6; details in Chapter 7). Sequences do not occupy any specific niche within a framework of nested stratigraphic cycles, and can be observed at all stratigraphic scales, depending on the geological setting (i.e., local conditions of accommodation and sedimentation), the resolution of the data available (e.g., seismic vs. well data), and the scope of the study (e.g., petroleum exploration vs. production development) (details in Chapters 5 and 7).

The sequence stratigraphic framework consists of the same types of building blocks (i.e., sequences, systems tracts, and depositional systems) at all stratigraphic scales, which enables a scale-independent approach to the sequence stratigraphic methodology and nomenclature (Figs. 4.3, 4.70, 5.1, 5.2, 5.3, 5.5, and 7.1–7.5). Depending on the scale of observation and the style of delineation of sedimentary cycles, systems tracts may be subdivided into sequence stratigraphic cycles (higher frequency sequences), allostratigraphic cycles (parasequences), or sedimentological cycles (beds and bedsets). Systems tracts which consist only of sedimentological cycles are the smallest sequence stratigraphic units at any location. At each scale of observation (i.e., hierarchical level), systems tracts are defined by the stacking pattern of their component sedimentary cycles, and changes in stacking pattern mark the position of sequence stratigraphic surfaces (i.e., systems tract boundaries; Figs. 1.1, 1.2, and 7.40). The relative ranking

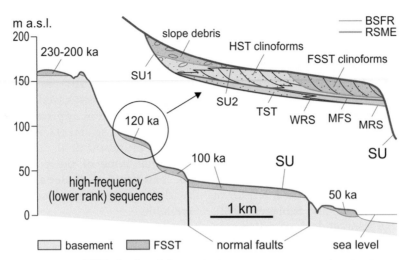

FIGURE 9.6 Falling-stage systems tract (FSST) developed during the long-term regional uplift in the Crotone Basin, Italy (modified after Zecchin et al., 2010b). The FSST consists of a series of downstepping and detached lower rank sequences formed in response to higher frequency glacio-eustatic cycles, which include their own transgressive, highstand and falling-stage systems tracts. The FSST is bounded at the top by a subaerial unconformity (SU) formed during the long-term forced regression. Subaerial unconformities of lower hierarchical rank also formed in relation to the higher frequency transgressions (SU1) and forced regressions (SU2) (see Zecchin et al., 2022, for details). Abbreviations: FSST—falling-stage systems tract; TST—transgressive systems tract; HST—highstand systems tract; SU—subaerial unconformity; MRS—maximum regressive surface; WRS—wave-ravinement surface; MFS—maximum flooding surface; BSFR—basal surface of forced regression; RSME—regressive surface of forced regression.

of sequences is defined by their nested architecture, whereby lower rank sequences build the systems tracts of higher rank sequences (Figs. 4.3, 4.6, 4.70, 5.1, 5.5, 7.16, 8.55, and 9.6).

In the absence of temporal or physical standards for the definition of hierarchical ranks, the classification of sequences observed at different scales is only meaningful in relative terms, within the context of each sedimentary basin (details in Chapter 7; Figs. 4.70, 7.1, 7.16, and 8.55). As such, sequences of the same hierarchical ranks in different basins may differ in terms of their temporal and physical scales. However, systems tracts of all hierarchical ranks are defined by the same types of stratal stacking patterns (e.g., a transgressive systems tract of any hierarchical rank is always defined by a retrogradational stacking pattern, which can be observed at different scales; Figs. 5.1, 5.3, 5.5, 5.105, 6.57, and 7.3—7.5). The degree of internal complexity of sequences and systems tracts changes with the hierarchical rank, and can be used to describe and predict the sedimentological makeup of stratigraphic units that develop at different scales within a sedimentary basin (e.g., Ainsworth et al., 2017). However, such patterns are tied to and vary with the geological setting, and therefore cannot be used to define a universal system of sequence classification. Moreover, the different types of data that are available in different basins affect the degree of stratigraphic and sedimentological detail that can be observed in each study.

Despite the nested architecture of stratigraphic cycles and the scale invariance of stratal stacking patterns, the stratigraphic architecture is not truly "fractal," as sequences of different scales may have different controls and internal composition of systems tracts (Fig. 7.16). Sequences of different scales and origins may also consist of similar combinations of systems tracts (Figs. 7.1—7.5), and sequences of similar scales may vary in terms of origin and component systems tracts (Fig. 4.33). The scales, the origins, and the systems-tract composition of sequences that develop at any hierarchical level are basin specific, reflecting the local conditions of accommodation and sedimentation that lead to the formation of stratigraphic units at each scale of observation. Therefore, the observation of stratal stacking patterns, at scales selected by the practitioner or imposed by the resolution of the data available, takes precedence over any model assumptions in the process of constructing a sequence stratigraphic framework.

The standard methodology is enabled by the fact that, despite the broad natural variability of the stratigraphic architecture in terms of the possible combinations of depositional trends, there is only a limited number of stacking patterns that are diagnostic to the definition of systems tracts (Figs. 4.1, 4.2). The distinction between diagnostic features and the non-diagnostic variability is a fundamental aspect of the sequence stratigraphic methodology (Fig. 1.2). The identification of systems tracts based on the observation of the diagnostic stacking patterns, at scales selected by the practitioner or imposed by the resolution of the data available, defines the standard approach to sequence stratigraphy. The construction of sequence stratigraphic frameworks is independent of the interpretation of underlying controls, which is the purpose of stratigraphic modeling. It is

409

therefore important to separate methodology from modeling in sequence stratigraphy (Fig. 1.2). A standard methodology does not prevent future developments in the field of stratigraphic modeling, as the interpretation and testing of the possible controls on sequence development may continue indefinitely after the construction of a sequence stratigraphic framework.

The development of basin-specific stratigraphic frameworks, each with potentially unique timing, scales, origins, and composition of sequences, demonstrates that the methodology and nomenclature cannot be tied to scale, the resolution of the data available, or any other local variables, as that would result in inconsistencies from one sedimentary basin to another. Only a scale-independent approach can provide the methodological and nomenclatural consistency required by a standard application of sequence stratigraphy across the entire range of geological settings, stratigraphic scales, and types of data available. The scale-independent methodology and nomenclature ensures maximum flexibility and objectivity, and the simplest approach to sequence stratigraphy. The construction of basin-specific stratigraphic frameworks based on local data is a major departure from the early model-driven approaches centered on global standards. While global correlations can still be tested following the construction of local stratigraphic frameworks, global cycle charts are no longer part of the sequence stratigraphic workflow and methodology.

A standard methodology does not imply a freeze on developments. While the standard approach to sequence stratigraphy is already defined, future developments are still needed to refine and diversify the field criteria that afford the identification of systems tracts and bounding surfaces in the sedimentary record (e.g., the emerging trends of using geochemical and paleontological proxies in sequence stratigraphy, in conjunction with information derived from other independent methods and datasets; details in Chapter 2). Parallel progress in stratigraphic modeling will also continue in order to improve the accuracy of testing of sedimentary responses to the variety of possible controls on sequence development. However, methodology and modeling remain two distinct lines of stratigraphic research, with different goals and underlying data-driven vs. model-driven principles (Fig. 1.2).

9.4 Workflow of sequence stratigraphy

The basin-specific nature of the sequence stratigraphic framework makes the acquisition of local data (i.e., collected from the basin under analysis) the essential prerequisite in the sequence stratigraphic workflow. The resolution and reliability of the constructed sequence stratigraphic framework depend on the amounts, types, and quality of the data available. It is advisable to acquire and integrate as many different types of data as possible (e.g., seismic, petrophysical, biostratigraphic, sedimentological, geochemical, etc.), as each type of data contributes with specific insights toward the construction of the sequence stratigraphic framework (Figs. 1.1 and 2.2). The minimum amount of data that affords a sequence stratigraphic analysis is typically encountered in offshore "frontier" basins (i.e., prior to drilling), where only seismic data are available. In such cases, stratal stacking patterns are observed at scales above the seismic resolution, on the basis of seismic reflection terminations and architecture (i.e., seismic stratigraphy; Figs. 4.7 and 4.8).

Data integration is critical in sequence stratigraphy, as the mutual calibration of different datasets (e.g., outcrop, core, well-log and seismic) leads to the most comprehensive and reliable results (Fig. 1.1). For example, information from scattered outcrops benefits greatly from integration with the continuous subsurface imaging provided by seismic data, wherever possible. At the same time, the use of seismic data without calibration with well data can lead to erroneous interpretations, especially where the architecture of seismic reflections is not diagnostic of any particular depositional system (e.g., the interpretation of depositional systems in Fig. 2.67 is difficult without calibration with well data). Similarly, the lack of calibration of well logs with rock data (cuttings, core, or nearby outcrops), and their correlation without the support provided by seismic imaging, can also lead to erroneous interpretations, as different depositional systems can share the same log motifs (Figs. 2.55—2.58 and 2.60). The integration of various datasets is therefore key to the most effective and reliable application of sequence stratigraphy.

The construction of a sequence stratigraphic framework is best approached from the "big picture" to the detail, as the basinal setting provides the context to rationalize the smaller scale elements of the sedimentary record. For example, knowledge of the strike and dip directions within the basin at syn-depositional time helps devising the best strategy to observe shallow-water systems (i.e., as clinoforms are best visualized along dip directions), fluvial incised valley (i.e., best visualized along strike directions), or leveed channels of turbidity flows in a deep-water setting (i.e., best visualized along strike directions). Also, knowledge of the shelf-edge location during the evolution of a shelf-slope system helps to avoid confusion between fluvial and turbidite channels, which can look similar on seismic horizon slices. The workflow in the petroleum exploration of a sedimentary basin, from a "frontier" stage when only seismic data are available to a "mature" stage when well data become available as well, exemplifies the practice of starting with the coarser, larger scale stratigraphic framework, and gradually increasing the degree of detail as higher resolution data become available. Sequence stratigraphic frameworks are typically "work in progress" as they are constantly improved and refined with the acquisition of new resolution data.

The sequence stratigraphic workflow is independent of scale, with the same aim of identifying stratal stacking patterns from the scales of seismic stratigraphy to the scales of high-resolution sequence stratigraphy. At any scale of observation, stratal stacking patterns and changes thereof are identified on the basis of physical criteria, both at specific locations (e.g., in individual outcrops or boreholes) and at regional scales (e.g., on 2D seismic lines or well-log cross sections) (Fig. 1.1). Time control enhances the reliability of correlations, but the lack thereof (e.g., in most Precambrian and many Phanerozoic case studies) does not prevent the application of sequence stratigraphy. The lack of age data may be compensated by the identification of reliable stratigraphic markers (e.g., volcanic ash beds, regional coal seams, or regional maximum flooding surfaces) and/or a good knowledge of the facies relationships within the study area (Eriksson et al., 1998, 2001, 2004, 2005a,b, 2013; Catuneanu and Biddulph, 2001; Catuneanu et al., 2005, 2012; Sarkar et al., 2005; Magalhaes et al., 2015; Kunzmann et al., 2020).

The best practice requires a three-dimensional control of the stratigraphic architecture, which integrates the observation of vertical profiles afforded by outcrop and well data with section views (e.g., seismic reflection terminations and architecture on 2D lines) and plan views (e.g., geomorphological features on seismic horizon slices; Figs. 2.71, 2.81, and 9.7). This approach is possible where sufficient data are available, including 3D seismic surveys. Following the initial low-resolution screening of the available data, smaller areas of interest can be defined for higher resolution studies. Improvements in the quality of subsurface data acquisition and processing afford not only an imaging of the geometry of depositional elements but also insights into depositional processes (Figs. 2.82 and 2.83). Sequence stratigraphy relies on process sedimentology, which is a prerequisite for a process-based approach to stratigraphy, as well as on integration with other disciplines including all types of classical stratigraphy, geophysics, geomorphology, and basin analysis. For example, the identification of sequence stratigraphic surfaces requires knowledge of the underlying and overlying depositional systems, in the case of conformable contacts, or of the overlying system in the case of unconformities (Figs. 6.5 and 6.6).

The workflow of sequence stratigraphy follows three logical steps which are devised to minimize the risk of

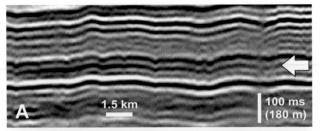

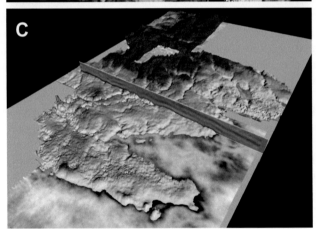

FIGURE 9.7 Devonian fluvial system in the Western Canada Sedimentary Basin (images courtesy of Henry Posamentier). The nature of the depositional system is difficult to infer from the 2D seismic transect (A) without the 3D seismic horizon slice (B). The perspective view (C) further enhances the visualization of the fluvial system by combining 2D and 3D seismic data. The channels are c. 300 m wide.

errors in data analysis: Step 1—clarification of the tectonic setting, to provide the context for the sequence stratigraphic study (i.e., type of sedimentary basin, basin physiography, subsidence mechanisms, dip and strike directions); Step 2—clarification of the depositional setting, to constrain the meaning of depositional trends (e.g., coarsening- vs. fining-upward trends, whose sedimentological or stratigraphic significance varies with the depositional system); and Step 3—construction of the sequence stratigraphic framework (i.e., identification of systems tracts and bounding surfaces, based on the observation of stratal stacking patterns at scales defined by the purpose of study or the resolution of the data available; Figs. 1.1, 1.2, and 7.40). Preliminary studies of the tectonic setting (including the location of the study area relative to the main physiographic elements of the host basin, such as the basin margin or the shelf edge) and depositional setting (depositional environments and paleogeography) ensure that the data are understood in the correct geological context, which affords the most reliable outcome of a sequence stratigraphic study.

9.4.1 Step 1: Tectonic setting

The type of sedimentary basin that hosts the stratigraphic section under analysis is a fundamental variable that provides the context for the sequence stratigraphic study. Each tectonic setting is potentially unique in terms of key parameters, including subsidence mechanisms, subsidence rates, and the pattern of change in subsidence rates across the basin, basin physiography, and sediment supply. These factors control the overall stratigraphic architecture, the geometry of sequences, and the relative development of the different types of systems tracts within sequences. Knowledge of the dip and strike directions at syn-depositional is also important, as it provides clues about the orientation of sediment fairways that operate in different environments (e.g., fluvial systems, longshore currents, gravity flows, etc.), and generally about the patterns of sediment dispersal within the basin. For all these reasons, the study of the tectonic setting is an important step before the construction of a sequence stratigraphic framework.

The types and distribution of depositional systems within a sedimentary basin reflect the interplay between accommodation and sedimentation during the evolution of the basin (i.e., underfilled vs. overfilled accommodation, and the location of shorelines and depocenters; Figs. 3.11, 3.12, 8.2, and 8.5–8.7). For example, subsidence rates across divergent continental margins increase in a distal direction, in response to the thermal cooling and sinking of oceanic crust (Fig. 8.2). In this context, fluvial to shallow-water systems occur on the continental shelf, and deep-water systems (slope and basin floor) develop beyond the shelf edge. In contrast, foreland systems which form by the flexural downwarping of the lithosphere under the weight of orogens show opposite trends, with subsidence rates increasing in a proximal direction (Figs. 8.5–8.7). As a result, the deep-water systems of an underfilled foredeep occupy a proximal position relative to the fluvial to shallow-water systems in the forebulge and backbulge areas (Catuneanu et al., 2002, Figs. 8.5–8.7). Retroarc forelands also evolve predictably from underfilled to overfilled states, in response to the long-term shift in the balance between accommodation and sedimentation (Catuneanu, 2019c). More

restricted basins, such as grabens and rifts, are difficult to generalize as they may host a variety of depositional environments, from fully continental to deep-water (Leeder and Gawthorpe, 1987).

The amounts of extrabasinal and intrabasinal sediment supply to sedimentary basins can also be linked to the tectonic setting, in addition to the influence of climate and geological age, as discussed above in Chapters 8 and 9. For example, foreland basins are prone to siliciclastic sedimentation, due to their association with orogens which provide high amounts of extrabasinal sediment. Extensional basins such as grabens and rifts, as well as pull-apart basins related to strike-slip faults, are also prone to siliciclastic sedimentation due to the fault-controlled steep gradients along the basin margins. In contrast, intracratonic basins are often conducive to chemical precipitation, particularly in the absence of surrounding highlands, as their formation does not require the uplift of the basin margins, and the seafloor and adjacent landscape gradients are typically low, limiting the influx of extrabasinal sediment. Other types of basins, such as divergent continental margins, can offer variable conditions in terms of the production of extrabasinal vs. intrabasinal sediment, depending on the syn-depositional climate and the physiography of the basin and surrounding areas. Therefore, even though knowledge of the tectonic setting narrows down the range of possible paleodepositional environments within the basin under analysis, the study of the depositional setting is still required as another step in the sequence stratigraphic workflow.

The reconstruction of the tectonic setting is typically based on regional data, among which 2D seismic surveys stand out as they can provide continuous imaging of a basin fill. The preliminary study of regional 2D seismic transects yields basic information on the strike and dip directions within the basin, the location and types of faults, the general structural style, and the overall stratal architecture of the basin fill. Fig. 2.68 provides an example of a 2D seismic line which shows the overall progradation of a divergent continental margin. In this case, the position of the shelf edge can be mapped for different timesteps, which affords the distinction between fluvial to shallow-water systems (i.e., first-order topset, updip from the shelf edge) and deep-water systems (i.e., first-order foreset and bottomset, downdip from the shelf edge; Fig. 4.60). Following the initial 2D seismic survey, 3D seismic volumes can be acquired in areas of particular interest, to constrain the location and nature of specific physiographic and morphological elements (e.g., the location of paleoshorelines, and the identification of depositional elements and erosional features; Fig. 9.8). The preliminary regional work provides the context for the next steps in the sequence stratigraphic workflow.

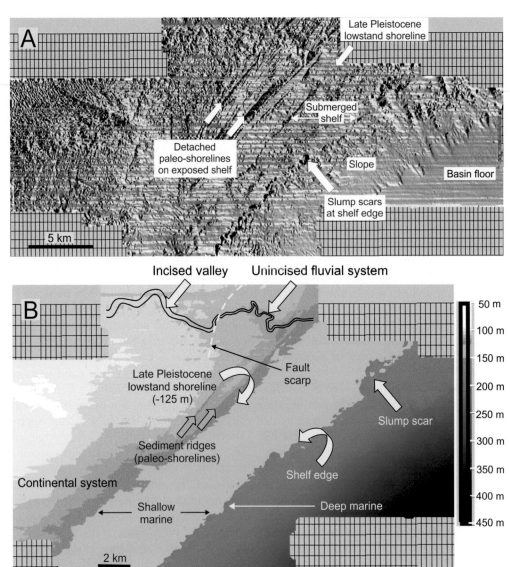

FIGURE 9.8   Seismic azimuth map (A) and depth-structure map (B) showing depositional systems during the Late Pleistocene relative sea-level lowstand (images courtesy of Henry Posamentier; offshore east Java, Indonesia). The detached shorelines on the shelf reflect stages of forced regression during relative sea-level fall. The slump scars indicate instability at the shelf edge. The lowstand shoreline remained inboard of the shelf edge, which explains the presence of unincised fluvial systems on the continental shelf. In this example, the change from incised to unincised fluvial systems is controlled by a fault scarp, as rivers incised into the more elevated footwall of the seaward-plunging normal fault.

9.4.2 Step 2: Depositional setting

Once the tectonic setting is clarified, the next step in the workflow of sequence stratigraphy is to constrain the nature of the depositional systems that are present within the study area. This can be accomplished in various ways, depending on the types of data available (details in Chapter 2). The most common protocols for the reconstruction of paleodepositional environments include the collection of litho- and biofacies data from outcrops or wells; the plan-view visualization of depositional elements on 3D seismic horizon slices, coupled with knowledge of the scales that are reasonable for

specific depositional and erosional processes; and the identification of subaerial and subaqueous clinoform rollovers on 2D seismic transects, to constrain the location of key physiographic elements such as the shoreline and the shelf edge (Fig. 4.19). The shoreline may also coincide with the shelf edge, when high-magnitude falls in relative sea level expose the entire shelf (Figs. 4.6, 4.20, and 8.53). In such cases, the offlapping shelf edge indicates both the location of the shoreline and the elevation of the relative sea level during forced regression (Figs. 4.6 and 4.60).

Paleoenvironmental reconstructions are important for several reasons, both within and outside the scope of sequence stratigraphy. From a sequence stratigraphic perspective, knowledge of the depositional setting is critical to understand the meaning of depositional trends (e.g., coarsening- and fining-upward, which may indicate depositional elements in fluvial settings, shoreline shifts in shallow-water settings, or the lateral migration of depositional elements in deep-water settings; Figs. 2.55–2.61), and to identify sequence stratigraphic surfaces (e.g., a "subaerial unconformity" is an unconformity that preserves continental deposits on top; Figs. 6.5 and 6.6). The physical and temporal relationships of depositional systems, including their shift directions through time within a sedimentary basin, are essential criteria that afford the identification of systems tracts and bounding surfaces. In this genetic approach, the origin and distribution of petroleum and water reservoirs, coal seams or placed deposits can be assessed and rationalized in light of the sedimentary processes that operate within each depositional environment. The identification of specific depositional elements of depositional systems (e.g., channels, splays, beaches, etc.) is also important at this stage, both for paleogeographic reconstructions and the economic evaluation of the stratigraphic units of interest.

The reliability of paleoenvironmental reconstructions depends on the data available. Integration and mutual calibration of independent datasets (e.g., seismic, well-log, sedimentological, biostratigraphic, geochemical) ensures the best results, as each type of data has advantages and limitations. For example, geophysical data (seismic, well-log) provide continuous but indirect geological information, whereas rock data from cuttings, core, and outcrops afford direct geological information but from discrete stratigraphic levels and locations (details in Chapter 2). At this stage in the workflow, the 3D seismic data add significant value to the information obtained from the 2D seismic transects. The latter are ideal to reveal structural styles and the large-scale stratigraphic architecture (Fig. 2.68), but fall short in terms of the detailed analysis of depositional systems and component depositional elements. Beyond the insights afforded by the 2D seismic transects, the 3D seismic horizon slices highlight geomorphological details that help constrain the paleodepositional environment (Figs. 9.7 and 9.8). However, the features observed on the 3D seismic data still need to be placed in the proper tectonic setting, to avoid confusion between depositional elements that share a similar morphology (e.g., fluvial channels on a shelf vs. turbidite channels on a basin floor). Paleoecological information from fossils is also critical to constrain the depositional setting and to place all other types of data in a proper context.

The results of paleoenvironmental reconstructions may be presented on paleogeographic maps which show the main physiographic elements at particular times in the evolution of the basin (Fig. 9.8). In underfilled basins, the shoreline is one of the most important features on such paleogeographic maps, because (1) its trajectory is key to the definition of systems tracts and bounding surfaces, and (2) its position marks the location of the riverborne sediment entry points into the marine basin relative to other important physiographic elements such as the basin margin and the shelf edge, which defines the extent and distribution of the different environments within the basin. In the context of shelf-slope systems, the position of the shoreline relative to the shelf edge has a significant influence on the sediment supply to the slope and basin-floor settings, and hence on the development of deep-water submarine fans (Figs. 4.20 and 7.9; details in Chapter 8). The shoreline also controls the downdip extent of coal seams and placer deposits, and the distribution of petroleum or water reservoirs of different origins (e.g., fluvial, coastal, or marine). Paleogeographic maps can be constructed with various degrees of accuracy for different time windows, depending on the resolution of the stratigraphic study (Fig. 4.61).

Paleoshorelines are typically best preserved within normal regressive (lowstand and highstand) systems tracts (Fig. 6.56), have a good preservation potential in the open-shoreline (interfluve) areas of falling-stage systems tracts (Figs. 5.29, 5.30, and 9.8), and are at least in part eroded by wave ravinement processes during transgressions (Fig. 4.57). Therefore, the subaerial clinoform rollovers which mark the position of paleoshorelines are most commonly observed in regressive settings, although the usual pattern of stepped retrogradation also allows inferring the position of transgressive shorelines (Figs. 4.29 and 5.5). Paleoshoreline locations in transgressive settings may also be inferred from coastal onlap, which develops immediately downdip of the area of shoreface erosion (i.e., the onlap defined by the landward-shifting shallow-water healing-phase wedges; Figs. 4.55 and 4.57). The correct reconstruction of paleoshoreline locations may be affected by confusions between subaerial and subaqueous clinoform rollovers (Fig. 4.19), or between coastal onlap and other types of onlap (Fig. 4.8). Such errors can be eliminated by the clarification of the tectonic and depositional settings (i.e., steps 1 and 2 of the workflow), which needs to precede the construction of the sequence stratigraphic framework.

9.4.3 Step 3: Sequence stratigraphic framework

The sequence stratigraphic framework describes changes in the patterns of sediment distribution between different depositional environments during the evolution of a sedimentary basin, over various temporal and physical scales (Figs. 4.3, 4.61, 5.4, 5.27, 5.28, 5.48,

5.63, 5.64, and 5.80). This framework affords the most efficient strategies of exploration for natural resources, and subsequent production development, as facies predictability is an intrinsic element of the process-based approach to stratigraphic analysis. Beyond economic applications, the sequence stratigraphic framework also affords the reconstruction of the record of change in accommodation and sedimentation conditions through time across a sedimentary basin, which can be used in more general basin analysis studies.

The construction of the sequence stratigraphic framework is guided by the observation of stratal stacking patterns at scales defined by the purpose of study or the resolution of the data available. An important part of this process is the separation of stacking patterns that are diagnostic to the definition of systems tracts from the non-diagnostic variability that can accompany the formation of any systems tract (Fig. 1.2; details in Chapters 4 and 5). The same types of diagnostic stacking patterns form and afford the identification of systems tracts and bounding surfaces at all stratigraphic scales, which brings consistency to the sequence stratigraphic methodology and nomenclature (Figs. 4.3, 4.70, 5.1, 5.5, 7.1–7.5, and 8.55). The stratigraphic framework can be further constrained with age data, wherever available, to quantify the magnitude of hiatuses and the timing of stratal units and bounding surfaces (Fig. 1.1). However, sequence stratigraphic frameworks can also be constructed in the absence of age data (e.g., in frontier basins), based on physical criteria (stacking patterns) and the relative chronology of stratal units and bounding surfaces (Fig. 8.49).

9.4.3.1 Stratal stacking patterns

Stratal stacking patterns describe the architecture of the stratigraphic record, as defined by the geometrical relationships of strata or seismic reflections in cross-sectional views or by vertical changes in grain size and depositional elements (Figs. 1.1, 4.1, 4.2, 4.7, and 4.8; see Chapter 4 for details and examples). Therefore, depending on the types of data available, stacking patterns may be expressed as 1D vertical profiles (e.g., fining- vs. coarsening-upward trends in outcrops or wells) or 2D stratal geometries (e.g., seismic-reflection terminations and architecture), which can be integrated within a 3D stratigraphic framework (details in Chapter 7). The stratigraphic meaning of stacking patterns is best constrained in a regional context relative to the location of key physiographic elements such as the basin margin, the shoreline, and the shelf edge, following the completion of steps 1 and 2 of the sequence stratigraphic workflow. For example, non-diagnostic fining-upward trends and onlapping reflection terminations may develop in both continental and marine environments (fluvial vs. coastal or marine onlap, respectively; Fig. 4.8). In this case, their stratigraphic significance (e.g., the topset of a

regressive unit vs. the "healing-phase" wedge of a transgressive unit) can be determined by placing them in the correct depositional setting.

Stratal stacking patterns are central to the sequence stratigraphic methodology, as they provide the criteria to define sequences, systems tracts, and the bounding sequence stratigraphic surfaces, in both downstream- and upstream-controlled settings (Fig. 3.11). In downstream-controlled settings, the diagnostic stacking patterns are defined in relation to shoreline trajectories (normal regressions, forced regressions, transgressions; Fig. 4.1). In upstream-controlled settings, stratal stacking patterns are defined by the dominant depositional element of the continental system (e.g., channels vs. floodplains in fluvial systems; Fig. 4.2). All types of stacking patterns can be observed at different scales: forced regressions (Fig. 8.55); normal regressions (Figs. 2.68, 5.1, and 5.5); transgressions (Figs. 5.1, 5.2, 5.3, 5.5, and 5.105); and shoreline-independent stacking patterns in upstream-controlled settings (Fig. 4.70). The scale of observation may be selected by the practitioner according to the scope of the study (e.g., petroleum exploration vs. production development), or it may be imposed by the resolution of the data available.

The definition of all types of stacking patterns is based solely on physical criteria, and it does not rely on age data (Figs. 4.1 and 4.2). In the absence of time control, stacking patterns can be used not only to describe the physical architecture of the stratigraphic record but also to establish the relative chronology of systems tracts within the study area (Fig. 8.49). Key physical features in downstream-controlled settings include the trajectory of subaerial clinoform rollovers and stratal terminations (Figs. 4.6, 4.7, 4.8, and 4.21). The former describe geometrical rather than depositional trends (Figs. 4.23 and 4.24), whereas the latter can describe both geometrical relationships and depositional processes (Figs. 4.7 and 4.8). Stratal terminations may provide clues regarding the conformable vs. unconformable nature of stratigraphic surfaces (e.g., truncation is diagnostic of erosion), whereas others are more equivocal artifacts generated by topsets or bottomsets that fall below the seismic resolution (Fig. 4.13). Some stratal terminations are diagnostic for particular shoreline trajectories, such as coastal onlap for transgression or offlap for forced regression, while others may develop independently of shoreline trajectories (e.g., fluvial onlap may form during both regressions and transgressions; Fig. 4.8).

The distinction between the different types of stacking patterns is important on both practical and theoretical grounds. On practical grounds, changes in stratal stacking patterns across systems tract boundaries modify the distribution of sediment between the depositional environments of a sedimentary basin, due to changes in the balance between accommodation and sedimentation at syn-depositional time (details in Chapters 4 and 5). For example, falling-stage systems tracts typically include the best conventional petroleum reservoirs in deep-water settings because the transfer efficiency of river-borne sediment to the shelf edge is highest during forced regressions (e.g., the submarine fans observed at different scales within the Mississippi Fan Complex consist primarily of riverborne sediment delivered to the shelf edge during forced regressions of corresponding magnitudes; Fig. 8.55). On theoretical grounds, stratal stacking patterns afford the definition and identification of all units and surfaces that build the sequence stratigraphic framework.

9.4.3.2 Sequences and systems tracts

Sequences and component systems tracts are the stratal units of sequence stratigraphy, which can be observed at all stratigraphic scales, depending on the purpose of study or the resolution of the data available (Fig. 4.3; details in Chapter 7). A sequence corresponds to a stratigraphic cycle of change in stratal stacking pattern, defined by the recurrence of the same type of sequence stratigraphic surface in the sedimentary record (Fig. 1.22). Systems tracts are subdivisions of sequences, defined by specific stratal stacking patterns and bounding surfaces (details in Chapter 5). Depositional systems are units of sedimentology, but play an important role in the stratigraphic framework as they define the internal composition of stratigraphic units and the link between the scopes of sedimentology and stratigraphy at all stratigraphic scales (i.e., the end result of a sedimentological study, and the starting point of a stratigraphic study). The overlap between the scales of sedimentology and stratigraphy reflects the different degrees of stratigraphic resolution (details in Chapters 5 and 7).

At each stratigraphic scale, depositional systems are the sedimentological building blocks of systems tracts, and systems tracts are the stratigraphic building blocks of sequences. The observation of depositional systems at different scales (i.e., depositional systems *sensu stricto* and *sensu lato*; Catuneanu, 2019b) explains the scale-independent nature of sequences and systems tracts. Depositional systems *sensu stricto* consist strictly of process-related facies accumulated in specific environments, and develop commonly at scales below the resolution of seismic stratigraphy, particularly within the transit area of a shoreline where changes in depositional environment are most frequent (Figs. 5.2 and 5.5). At larger scales (higher hierarchical ranks), depositional systems *sensu lato* reflect dominant depositional trends, but may record higher frequency changes in depositional environment (Fig. 5.1); these are typically the depositional systems identified at the scales of seismic stratigraphy. There are no standards for the physical or temporal scales of sequences and component

systems tracts and depositional systems. The scale of the smallest stratigraphic sequences at any location depends on the geological setting (i.e., local conditions of accommodation and sedimentation). At the other end of the spectrum, the scale of the largest sequences and component systems tracts and depositional systems is defined by the scale of the basin fill (Figs. 5.3 and 7.1).

The "grey area" between scales of sedimentology *sensu stricto* and seismic stratigraphy is now clarified by the high-resolution sequence stratigraphy, whereby high-frequency sequences develop at sub-seismic scales of 10^0-10^1 m and 10^2-10^5 yrs. At the smallest stratigraphic scales, systems tracts and component depositional systems consist solely of sedimentological cycles (i.e., beds and bedsets; Figs. 3.24, 3.25, 4.3, and 4.62). The scale of the lowest rank systems tracts at any location defines the highest resolution that can be achieved with a stratigraphic study. At any larger stratigraphic scales, systems tracts and component depositional systems consist of higher frequency (lower rank) stratigraphic cycles (i.e., sequences; Figs. 5.1, 5.5, and 9.6). The smallest stratigraphic scale that can be identified at any location depends on the resolution of the data available. For example, the smallest "sequence" that can be identified in the context of seismic stratigraphy is typically not the smallest stratigraphic cycle within the study area, but the smallest stratigraphic cycle that is above the resolution of the seismic data. The inability to identify the smallest sequence in every study indicates that the classification of stratigraphic cycles is best approached starting with the "first-order" basin fill as the anchor for the definition of hierarchical ranks (details in Chapter 7).

9.4.3.3 Sequence stratigraphic surfaces

Sequence stratigraphic surfaces are stratigraphic contacts which serve, at least in part, as systems tract boundaries (i.e., they mark a change in stratal stacking pattern). This is the attribute that sets a sequence stratigraphic surface apart from any other type of stratigraphic contact (e.g., lithostratigraphic, allostratigraphic, chronostratigraphic, etc.). Sequence stratigraphic surfaces may or may not coincide with other types of stratigraphic contacts (Figs. 1.14 and 1.15), and in some cases their origins may be related. For example, both allostratigraphic and sequence stratigraphic surfaces may form in relation to the same transgressions (i.e., flooding surfaces during transgressions, and maximum flooding surfaces at the end of transgressions, respectively; Fig. 5.100). However, not all transgressions result in the formation of flooding surfaces, but every transgression ends with a maximum flooding surface. Moreover, maximum flooding surfaces can be mapped in all depositional environments, whereas flooding surfaces are restricted to coastal and shallow-water systems (details in Chapter 5 and 6). For these reasons, sequence stratigraphic surfaces (maximum flooding surfaces in this example), and the systems tracts they separate, are more reliable for correlation than the litho- or allostratigraphic units and surfaces defined by lithological criteria.

The origin of sequence stratigraphic surfaces is linked to specific conditions of accommodation and sedimentation at syn-depositional time (e.g., a correlative conformity forms at the point of balance between negative and positive accommodation; a maximum flooding surface forms at the point of balance between accommodation and sedimentation at the end of transgression; details in Chapter 6). As both accommodation and sedimentation conditions are variable across a sedimentary basin, sequence stratigraphic surfaces are inherently diachronous (Fig. 6.6; details in Chapter 7). For this reason, sequence stratigraphy is not chronostratigraphy, and sequence stratigraphic surfaces are not time lines. The amounts of diachroneity depend on the degree of variability in the rates of accommodation and sedimentation at syn-depositional time, and may be below or above the resolution of the data available. Examples of highly diachronous shoreline shifts have been documented with ammonite biostratigraphy in the Upper Cretaceous Western Interior Seaway of North America (Gill and Cobban, 1973), and this reality should be regarded as the norm rather than the exception.

The identification of sequence stratigraphic surfaces relies on physical criteria, including (1) the conformable or unconformable nature of the surface; (2) the depositional systems below and above the surface; and (3) the stratal stacking patterns below and above the surface (Fig. 6.6). All sequence stratigraphic surfaces can be observed at different scales (hierarchical ranks), depending on the resolution of the stratigraphic study (Figs. 4.70, 5.1, 5.5, 5.105, 7.1—7.5, and 7.16). At each scale of observation, sequence stratigraphic surfaces provide the most reliable means of correlation within the genetic context of paleodepositional environments. For this reason, correlation work needs to start with the identification and mapping of sequence stratigraphic surfaces. Once a sequence stratigraphic framework is in place, other types of contacts (e.g., lithostratigraphic, allostratigraphic) can be mapped and rationalized within that genetic context.

In the case of laterally continuous stratigraphic sections that preserve their syn-depositional architecture (e.g., seismic transects or large-scale outcrops in tectonically "passive" settings), the mapping of sequence stratigraphic surfaces can be straight forward. In the case of structurally complex basins, independent time control (e.g., provided by biostratigraphy, radiochronology, magnetostratigraphy, chronostratigraphic markers) can help constrain the correlations across structural elements. Time control is also useful where the correlation work is based on data collected from discrete locations within the basin, with significant data gaps in between (e.g., well data or isolated outcrops). Irrespective of situation, the integration of all data available always ensures the highest degree of reliability in the process of constructing a sequence stratigraphic framework (Fig. 1.1).

CHAPTER

10

Conclusions

Sequence stratigraphy is a type of stratigraphy that relies on the observation of stacking patterns for the definition, nomenclature, classification, and correlation of strata! units and bounding surfaces. The same types of stacking patterns can develop at all stratigraphic scales, as a result of the interplay between multiple local and global controls on accommodation and sedimentation (Figs. 3.1, 3.3, and 3.11; details in Chapter 3). Changes in strata! stacking patterns at different scales define a nested architecture of stratigraphic sequences (Figs. 4.3, 4.70, 5.1, 5.5, and 7.16). Interpretations that are intrinsic to the sequence stratigraphic methodology pertain to the sedimentary processes that generate the observed stacking patterns, which can be rationalized by placing the data in the correct tectonic and depositional settings (steps 1 and 2 of the sequence stratigraphic workflow; details in Chapter 9). In turn, this process-based approach to stratigraphic analysis enables facies predictions within the sequence stratigraphic framework, at scales afforded by the resolution of the data available (Fig. 10.1). The observations that lead to the construction of the sequence stratigraphic framework are independent of the interpretation of the possible controls on sequence development; the latter defines the scope of stratigraphic modeling. Methodology and modeling are two distinct lines of stratigraphic research, with different goals and underlying data-driven vs. model-driven principles (Fig. 1.2).

Confusions between modeling and methodology raised unwarranted questions about the "future of sequence stratigraphy." In fact, a standard methodology emerged based on model-independent principles and unbiased data analyses. Despite the variability of the stratigraphic architecture, only a limited number of strata! stacking patterns is diagnostic to the definition of systems tracts and bounding surfaces, which can be recognized at all stratigraphic scales (Fig. 1.2). The identification of the diagnostic stacking patterns, at scales selected by the practitioner or imposed by the resolution of the data available, defines the standard approach to sequence stratigraphy. A standard methodology does not prevent future developments in stratigraphic

modeling, as the testing of the possible controls on sequence development may continue indefinitely following the construction of a sequence stratigraphic framework. While a standard methodology is now in place, the field criteria that afford the identification of systems tracts and bounding surfaces will also continue to be refined and diversified, which is at the forefront of current and future work in sequence stratigraphy (details in Chapter 9).

10.1 Stratigraphic framework

The stratigraphic framework is defined by a nested architecture of stratigraphic cycles (Fig. 4.3). Attempts have been made to recognize different types of units at different scales (e.g., bedsets < parasequences < sequences < composite sequences < megasequences), by designating "relatively conformable" sequences as the anchor for stratigraphic classification (Van Wagoner et al., 1988, 1990; Mitchum and Van Wagoner, 1991; Sprague et al., 2003; Neal and Abreu, 2009; Abreu et al., 2010). The pitfall of this approach is that the scale of a "relatively conformable" sequence depends on the resolution of the stratigraphic study, and therefore it does not provide a reproducible anchor for classification (e.g., seismic-scale "sequences" become "composite sequences" or "megasequences" in higher resolution studies, which strips this terminology of stratigraphic meaning; Fig. 7.6). The solution is a scale-invariant approach to stratigraphic nomenclature, which is independent of data resolution and any other local variables (details in Chapter 7). This enables a standard application of sequence stratigraphy despite the scale variability introduced by the geological setting and the types of data available.

Stratigraphic sequences display a 3D temporal and physical variability driven by changes in accommodation and sedimentation conditions across a sedimentary basin, leading to the diachronous development of systems tracts and bounding surfaces along dip and strike directions (Figs. 6.6 and 7.37; details in Chapter 7).

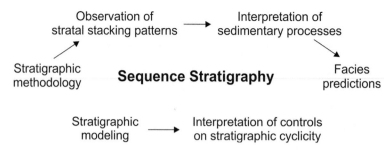

FIGURE 10.1 Observation vs. interpretation in sequence stratigraphy. The methodology relies on the observation of stratal stacking patterns at scales afforded by the resolution of the data available. Different aspects of interpretation tackle (1) the underlying controls on sequence development (e.g., the relative contributions of tectonism, eustasy, climate, and autogenic processes to stratigraphic cyclicity) and (2) the sedimentary processes that generate the observed stacking patterns (e.g., the relative contributions of traction currents, gravity-driven processes, and suspension fallout to sedimentation). The latter are intrinsically linked to the sequence stratigraphic methodology and give meaning to stratigraphic observations, whereas the former define the scope of stratigraphic modeling and can be tested independently of the data-driven construction of the sequence stratigraphic framework.

This 3D variability of the stratigraphic framework led to the paradigm shift from a model-driven approach, underlain by assumptions of global correlations and isochrony of sequence stratigraphic surfaces, to a data-driven approach that honors the reality of each sedimentary basin. Local data take precedence over any model-driven predictions, and lead to the construction of basin-specific and potentially unique stratigraphic frameworks. The sequence stratigraphic methodology is now decoupled from global standards and any preconceived assumptions regarding the architecture of the stratigraphic framework and its underlying controls.

The degree of preservation of the stratigraphic record typically increases from continental to marine settings (Fig. 5.15). Nevertheless, the stacking patterns of the preserved units in any depositional setting can be used to define the sequence stratigraphic framework, at the scales afforded by the preserved record (Figs. 4.70, 5.1, 8.55, and 9.3). For example, stratigraphic cyclicity in fluvial settings is commonly observed at coarser scales than in marine settings, and stratigraphic resolution is generally poorer in siliciclastic settings compared to carbonate settings (details in Chapters 7 and 8). The methodology remains consistent despite this variability, since the same types of stacking patterns develop at all stratigraphic scales. The scale-independent approach to methodology and nomenclature is key to the standard application of sequence stratigraphy across the entire range of geological settings, stratigraphic scales, and the types and resolution of the data available.

The stratigraphic record consists of different types of stratal units, including sedimentological (beds, bedsets, and depositional systems), allostratigraphic (e.g., parasequences), and sequence stratigraphic (sequences and systems tracts). Due to differences in their defining attributes, these units are not always organized according to orderly patterns defined by mutually exclusive scales (e.g., parasequences < sequences), as inferred by

the scale-variant approaches to stratigraphic classification. The best practice in sequence stratigraphy requires the consistent use of sequence stratigraphic criteria at all stratigraphic scales. This is enabled by the increase in stratigraphic resolution, whereby high-frequency sequences replace parasequences in the sequence stratigraphic workflow. Key concepts such as "relative sea level," "base level," "sedimentary cycle," "stacking pattern," and "shoreline trajectory" apply to both stratigraphic and sub-stratigraphic (i.e., sedimentological *sensu stricto*) scales (Figs. 3.23, 3.24, and 4.62). Unless otherwise specified, the concepts herein are used in a stratigraphic sense.

10.1.1 Sequences, systems tracts, and depositional systems

Sequences, systems tracts, and depositional systems are the building blocks of the sequence stratigraphic framework at all stratigraphic scales. The scale of observation may be selected by the practitioner according to the purpose of study (e.g., petroleum exploration vs. production development), or it may be constrained by local parameters such as the geological setting and the resolution of the data available. There are no temporal or physical standards for the scale of any type of unit or surface of sequence stratigraphy; the scales of sequences, systems tracts, and bounding surfaces can vary significantly between and within sedimentary basins, reflecting the natural variability in local accommodation and sedimentation conditions (Fig. 3.1). Within the nested architecture of stratigraphic cycles, sequences and unconformities can be observed at all stratigraphic scales, both below and above the seismic resolution (Figs. 7.1–7.6), depending on the geological setting (i.e., local conditions of accommodation and sedimentation), the resolution of the data available (e.g., seismic vs. well data), and the scope of the study (e.g., petroleum exploration vs. production development) (details in Chapter 7).

Sequence stratigraphic frameworks are basin-specific in terms of the architecture, timing, scales, origin, and composition of sequences, reflecting the local conditions of accommodation and sedimentation. This prevents the definition of a general template that could describe stratigraphic frameworks worldwide. The resolution of the data available also fails to provide a reproducible anchor for the classification of stratigraphic cycles within the nested architecture of a basin fill (Fig. 7.6). In the absence of universal standards for the definition of hierarchical ranks, the classification of sequences observed at different scales is only meaningful in relative terms, within the context of each sedimentary basin (Figs. 4.70, 5.104, 5.106, 7.16, and 8.55). Where a study is placed in a basinal context, the "first-order" basin fill can be used as an anchor for classification and the definition of hierarchical orders (Figs. 5.3, 6.57, and 7.1). Sequences of the same hierarchical ranks in different basins may differ in terms of timing, temporal and physical scales, sedimentological and systems-tract composition, and underlying controls (details in Chapter 7).

Depositional systems are the building blocks of systems tracts. Within the transit area of a shoreline, where changes in depositional environment are most frequent, only the lowest rank depositional systems consist strictly of process-related facies accumulated in specific environments; these depositional systems *sensu stricto* develop commonly at scales below the resolution of seismic stratigraphy (Figs. 5.2 and 5.5). At larger scales, depositional systems *sensu lato* reflect dominant depositional trends, but may record higher frequency changes in depositional environment (Figs. 5.1 and 5.5). The observation of depositional systems at different scales affords the usage of the "systems tract" concept at all hierarchical levels (e.g., Figs. 5.1, 5.3, and 5.5; Catuneanu, 2019b). Therefore, sequences and systems tracts are not restricted to any specific scale within the nested architecture of stratigraphic cycles; instead, the same types of building blocks (sequences, systems tracts, and depositional systems) can be observed at all stratigraphic scales (Figs. 4.3, 4.70, 5.1, 5.3, 5.5, and 7.1–7.5; details in Chapter 7).

Stratigraphic resolution has increased to the extent that refined criteria are now required to define the limit between stratigraphy and sedimentology *sensu stricto*, and to discriminate between stratigraphic and sedimentological cycles (Figs. 3.24 and 4.62). Depositional systems provide the link between sedimentology and stratigraphy at all stratigraphic scales, as the largest units of sedimentology but only the building blocks of systems tracts in stratigraphy at each hierarchical level (Fig. 5.1). The observation of depositional systems at

different hierarchical levels defines the overlap between the scales of sedimentology and stratigraphy (Fig. 4.62; details in Chapters 5 and 7). At the smallest stratigraphic scales, systems tracts and component depositional systems consist solely of sedimentological cycles (i.e., beds and bedsets; Figs. 3.24, 3.25, and 4.3). At any larger scales (higher hierarchical ranks), systems tracts and component depositional systems consist of lower rank stratigraphic cycles (i.e., sequences) (Figs. 4.3, 5.1, 5.5, and 9.6; details in Chapter 7).

The scale and the degree of sedimentological complexity of the lowest rank systems tracts and component depositional systems at any location depend on the geological setting (i.e., local accommodation and sedimentation conditions) and the time involved in the development of the defining stacking pattern (e.g., 10^2–10^3 vs. 10^4–10^5 yrs). Therefore, the sedimentological makeup of stratigraphic units cannot be defined by any standard model, although various classifications have been proposed for specific depositional settings (e.g., Miall, 1996; for fluvial settings; Gardner et al., 2003; for deep-water settings; Ainsworth et al., 2017; for coastal to shallow-water settings). However, these classifications reflect the particular conditions of the areas from which the supporting data were derived, and cannot be extrapolated to all sedimentary basins. Similarly, no orderly pattern can be defined to describe a standard architecture of stratigraphic sequences worldwide. As a safe norm, every study needs to start from the premise that stratigraphic frameworks, and their sedimentological makeup, are basin or even sub-basin specific, reflecting the particular conditions of each geological setting.

10.1.2 Parasequences

Parasequences were introduced as the building blocks of seismic-scale systems tracts in the context of low-resolution seismic stratigraphy. The concept of parasequence is commonly applied at scales of 10^0–10^1 m and 10^2–10^5 yrs, which coincide with the scales of high-resolution sequence stratigraphy (Figs. 4.62 and 5.4). In contrast, the sequences of seismic stratigraphy are typically recognized at scales of 10^1–10^2 m and 10^5–10^6 yrs; (Vail et al., 1991, Duval et al., 1998; Schlager, 2010; Catuneanu, 2019a,b). The observation of sequences at seismic scales led to the proposal of a scale-variant hierarchy system which postulates orderly patterns in the sedimentary record (i.e., bedsets < parasequences < sequences; Van Wagoner et al., 1990; Sprague et al., 2003; Neal and Abreu, 2009; Abreu et al., 2010). However, this scheme does not provide a reproducible standard, as parasequences and depositional sequences of equal hierarchical ranks can coincide (e.g., in the case of orbital cycles; Strasser et al., 1999; Fielding et al., 2008; Tucker et al., 2009) or coexist within tectonically active basins (Catuneanu et al., 2002; Gawthorpe et al., 2003; Zecchin, 2010) (Fig. 5.106).

The critical aspect of the definition of a parasequence is the flooding surface, which is a facies discontinuity generated by abrupt water deepening (Fig. 5.94). Flooding surfaces are allostratigraphic contacts which may be represented by wave-ravinement surfaces in areas transgressed by a shoreline, or by maximum regressive surfaces, within-trend facies contacts or maximum flooding surfaces in fully marine successions (Fig. 5.100). Where flooding is preceded by forced regression, parasequence boundaries can also incorporate subaerial unconformities and stages of relative sea-level fall (Figs. 5.99 and 5.100C). Flooding surfaces may also be missing from shallow-water systems where transgressions and water deepening are gradual. Beyond the invariant bathymetric meaning of flooding surfaces, parasequences are variable in terms of scales (most commonly in a range of meters to tens of meters), composition (any ratio of transgressive to regressive deposits), and significance (from bedsets to stratigraphic cycles; Fig. 6.41).

Whether expressed as bedsets or stratigraphic cycles, parasequences are always bounded by flooding surfaces. In the case of bedsets, parasequences and flooding surfaces form during uninterrupted transgression with variable rates (higher during the formation of flooding surfaces; Figs. 5.5 and 6.41). In the case of stratigraphic cycles, the formation of parasequences involves changes in the direction of shoreline shift and coastal environments, which is the case with the "classic" parasequences of Van Wagoner et al. (1990) (Figs. 5.5 and 6.41). Where parasequences are reduced to bedsets, they only have a sedimentological significance. Only parasequences of stratigraphic significance can be substituted with high-frequency sequences. The types of sequences that can be used to replace parasequences ("depositional," "transgressive—regressive," or "genetic stratigraphic") depend on the sequence stratigraphic surfaces that can be mapped in the study area at the parasequence scales (i.e., subaerial unconformities, maximum regressive surfaces, or maximum flooding surfaces, respectively; Figs. 5.5, 5.99, 5.100, 5.106, 5.107, 5.108, and 9.6).

The allostratigraphic nature of flooding surfaces is at odds with the criteria that define systems tract boundaries, which rely on changes in stratal stacking patterns (e.g., from retrogradation to progradation in the case of maximum flooding surfaces) rather than lithological discontinuities. Both flooding surfaces and systems tract boundaries can form at the same stratigraphic scales, in relation to the same transgressions (Figs. 5.100, 5.103, 5.107—5.109). Irrespective of the scale of the transgression that leads to the formation of a flooding surface,

the backstepping of the shoreline starts from a time of maximum regression and ends with a maximum flooding. Therefore, the difference between flooding surfaces and maximum flooding surfaces is a matter of definition rather than scale. Both flooding surfaces and maximum flooding surfaces can form at multiple scales (Figs. 5.104—5.106), and maximum flooding surfaces can also be identified at lower hierarchical ranks than flooding surfaces where higher frequency genetic stratigraphic sequences are nested within parasequences (Fig. 5.104).

The mismatch between the criteria that define flooding surfaces and systems tract boundaries results in different degrees of mappability of parasequences and systems tracts. Sequences and systems tracts observed at parasequence scales extend beyond the areas of development of parasequences, in both landward and basinward directions (Figs. 5.99 and 5.107; details in Chapter 5). These limitations prevent the dependable use of the parasequence concept in the methodological workflow of sequence stratigraphy. Sequences and systems tracts are the only types of stratigraphic building blocks that can be mapped in all depositional settings. Systems tract boundaries, whether or not in breach of Walther's Law, extend over larger areas than any lithological discontinuities observed at the same scales, and therefore, are more reliable for regional correlation (e.g., everywhere flooding surfaces form, maximum flooding surfaces of equal hierarchical rank form as well, and with a larger extent; Fig. 6.61). For this reason, the sequence stratigraphic workflow starts with the identification of systems tract boundaries, which define a broader framework within which all other types of stratigraphic contacts can be rationalized. Where flooding surfaces coincide with systems tract boundaries, the nomenclature of the latter constrains the precise sequence stratigraphic meaning of the contact.

Despite its limitations, the parasequence concept is still routinely forced upon stratigraphic successions, whether or not flooding surfaces can be identified, in order to fulfill the predictions of the scale-variant model of classification of stratigraphic cycles which requires parasequences as the building blocks of sequences and systems tracts. The pitfalls of the parasequence concept lead to the conclusion that other types of units (bedsets in the case of sedimentological cycles, and high-frequency sequences in the case of stratigraphic cycles) provide better alternatives to define sedimentary cyclicity at parasequence scales. At stratigraphic scales, high-frequency sequences are more dependable for the construction of the sequence stratigraphic framework, both within and outside of coastal to shallow-water settings, rendering parasequences obsolete. Flooding surfaces remain important to define facies relationships associated with water deepening, and their stratigraphic meaning needs to be rationalized on a case-by-case basis in the context of the sequence stratigraphic framework. The use of sequences and systems tracts in high-resolution studies ensures methodological and nomenclatural consistency irrespective of geological setting, stratigraphic scales, and the types and resolution of the data available.

10.2 Standard approach to sequence stratigraphy

The existence of competing models (Figs. 1.9 and 1.10) impeded for decades, from the 1970s to the 2000s (Fig. 1.17), the communication of ideas and results between practitioners embracing alternative approaches to stratigraphic analysis. Despite the variety of opinions, common ground does exist since all practitioners, regardless of their model affiliation and preferences, describe the same stratigraphic successions, only using different nomenclature and definitions of sequences and systems tracts (Fig. 1.10). The model-independent principles that define the common ground enable a standard approach to sequence stratigraphy, which was ratified by the International Commission on Stratigraphy in the early 2010s (Catuneanu et al., 2011, Figs. 1.9 and 1.17).

The standard approach to sequence stratigraphy involves the observation of stratal stacking patterns in a manner that is independent of scale and the interpretation of the underlying controls Fig. 1.2). At each stratigraphic scale, cycles of changes in stratal stacking patterns define sequences, which consist of systems tracts and depositional systems. The classification of sequences of different scales reflects the relative magnitude of stratigraphic cycles within the nested architecture of a basin fill, related to base-level cycles in upstream-controlled settings and shoreline transit cycles in downstream-controlled settings (Figs. 4.70, 5.1, 5.5, 5.87, 7.1–7.5, 7.14, and 7.16). This flexible approach provides a robust platform of simple and objective guidelines that enable a consistent application of sequence stratigraphy irrespective of the local variability introduced by the geological setting and the resolution of the data available. The identification of sequences and systems tracts is the main objective of sequence stratigraphy, at the scales defined by the scope of the stratigraphic study.

Systems tracts are the stratigraphic building blocks of sequences at all stratigraphic scales (hierarchical levels; Fig. 4.3). Depending on the scale of observation, systems tracts may consist of sedimentological cycles (beds and bedsets) or stratigraphic cycles (higher frequency sequences) (Fig. 9.6). Systems tracts that consist only of sedimentological cycles (i.e., the systems tracts of the lowest rank sequences) are the smallest stratigraphic

units at any location (Figs. 4.3 and 4.62). On practical grounds, the identification of systems tracts is more important than the delineation of sequences. Systems tracts afford insights into the sediment dispersal patterns across a basin at different stages of sequence development, which helps to rationalize and predict the distribution of facies and natural resources in the sedimentary record. The construction of a framework of systems tracts and bounding surfaces fulfills the practical purpose of sequence stratigraphy. Beyond this, the delineation of sequences becomes a matter of model-dependent organization of systems tracts into stratigraphic cycles (Fig. 1.10).

10.2.1 Guidelines for a standard nomenclature

Systems tracts in all geological settings, whether downstream- or upstream-controlled, are defined by stratal stacking patterns and stratigraphic relationships. All conventional sequence stratigraphic models account for the presence of unfilled accommodation (i.e., an interior seaway, or a lake) within the basin under analysis, and highlight the importance of shoreline trajectories within the downstream-controlled portion of the underfilled basin (i.e., the area where the stratigraphic architecture is influenced by changes in relative sea level; Figs. 1.10, 3.11, 5.5, and 5.9). Outside of the area of influence of relative sea-level changes, sequences and systems tracts in upstream-controlled settings are defined on the basis of field criteria that are independent of shoreline trajectories (Figs. 3.11 and 4.70).

In downstream-controlled settings, the formation of systems tracts and bounding surfaces is controlled by shoreline trajectories. This justifies a systems-tract nomenclature that makes reference to transgressions and regressions, with the latter being classified into "forced" or "normal," depending on the trajectory of the regressive shoreline (i.e., downstepping vs. upstepping subaerial clinoform rollovers, respectively; Fig. 4.1). Forced regression is always driven by relative sea-level fall, whereas normal regression is driven by sediment supply in excess of accommodation during relative sea-level rise (details in Chapter 4). In general terms, shoreline trajectories are controlled by the interplay of relative sea-level changes (i.e., accommodation) and base-level changes (i.e., sedimentation) *at the shoreline* (Figs. 4.32 and 4.33). Both elements of this "dual control" are critically important, and none is a constant during the development of stratigraphic cycles.

With emphasis on shoreline trajectories, transgression defines the transgressive systems tract, and forced regression defines the falling-stage systems tract. Any intervening stages of normal regression are classified into "lowstand" (i.e., a normal regression that follows a forced regression of equal hierarchical rank) and "highstand" (i.e., a normal regression that follows a transgression of equal hierarchical rank), which define the lowstand and highstand systems tracts, respectively (Fig. 5.9). It can be noted that all systems tracts are defined by local stacking patterns and stratigraphic relationships, independently of any global standards and interpretations. The reference to the falling stage, or to the lowstands and highstands in relative sea level, does not detract from the objective definition of the systems tracts. This nomenclature acknowledges the contribution of the relative sea level to the stratigraphic architecture, which can be demonstrated and quantified with stratigraphic and sedimentological data (e.g., the amounts of upstepping and downstepping of the subaerial clinoform rollovers and associated subtidal facies; Figs. 4.23, 4.40, and 4.41; Plint, 1988; Plint and Nummedal, 2000; Posamentier and Morris, 2000; Tesson et al., 2000; Zecchin and Tosi, 2014; Catuneanu, 2019a; Sweet et al., 2019).

Proposals have been made to decouple the nomenclature of systems tracts from any reference to the relative sea level, and emphasize instead on the depositional trends that accompany the shifts of the shoreline: i.e., progradation, retrogradation, aggradation, and degradation (Neal and Abreu, 2009; Abreu et al., 2010). However, the underlying assumption is that aggradation and degradation are still controlled by relative sea-level rise and fall, respectively. The flaw of this assumption is that the depositional trends in coastal systems may not follow the geometrical trends of the shoreline (i.e., shoreline trajectories), as the two types of trends are controlled by different processes (i.e., base-level changes vs. relative sea-level changes, respectively; Fig. 4.23). The original nomenclature that makes reference to shoreline trajectories still remains more accurate in terms of describing the defining attributes of systems tracts (Fig. 5.9; see Chapter 5 for details on systems-tract nomenclature). Muddling the distinction between relative sea-level rise and fall is a setback in the sequence stratigraphic analysis, as it leads to a critical loss of insight into key aspects of the stratigraphic architecture.

The relative sea level has tangible effects on the stratigraphic architecture (e.g., the development of stratigraphic foreshortening during relative sea-level fall; Fig. 4.36), shelf processes (e.g., the deposition of compressed forced regressive vs. expanded normal regressive shorefaces with sharp vs. gradational bases, respectively; Fig. 4.30), the transfer of sediment from the shelf to the deep-water setting (e.g., increased efficiency of riverborne sediment transfer to the shelf edge during relative sea-level fall; Fig. 5.34), and the distribution of organic facies within stratigraphic sequences (Fig. 2.51) (Posamentier and Morris, 2000; Zecchin and Tosi, 2014; Catuneanu, 2019a, 2020a; Dominguez et al., 2020). The distinction between accommodation and sedimentation at the shoreline affords the most in-depth insights into the rationale of sediment distribution across a basin, which is fundamental to all practical applications. The separation of lowstand and highstand normal regressions, as well as of normal regressions and forced regressions, are also critical to the proper understanding of the evolution of sedimentary basins. The identification of lowstands and highstands in relative sea level is based on local stratigraphic relationships, and not on correlations with global sea-level changes. Global cycle charts are no longer part of the sequence stratigraphic methodology.

The distinction between lowstands and highstands in relative sea level explains paleogeographic trends (e.g., the relative locations of lowstand and highstand shorelines, due to the intervening stages of forced regression and transgression; Fig. 5.87), changes in the shelf capacity to retain extrabasinal sediment (i.e., lower during relative fall; Fig. 4.6), the production of intrabasinal sediment (e.g., reduced carbonate factories at times of lowstand; Figs. 6.26 and 8.19), and the consequent sediment supply to the deep-water setting (e.g., "lowstand shedding" in siliciclastic settings, and "highstand shedding" in carbonate settings; Schlager et al., 1994; Sweet et al., 2019, Figs. 5.34 and 5.83). Non-diagnostic variability in the development and composition of systems tracts (e.g., depositional trends in fluvial and coastal systems; variations in the production of extrabasinal and intrabasinal sediment; the increase in riverborne sediment supply to the shelf edge in the case of narrow shelves or shelves dissected by unfilled incised valleys; the influence of contour currents on the sedimentological makeup of systems tracts in deep-water settings) reflects the unique tectonic and depositional settings of each sedimentary basin, and needs to be rationalized on a case-by-case basis.

In upstream-controlled settings, the definition and nomenclature of systems tracts is based on the dominant depositional elements (e.g., channels vs. floodplain deposits in fluvial settings; Figs. 4.2 and 4.70). In such cases, the relative sea level no longer plays any role in the formation of systems tracts, and the temporary base level becomes the sole control on the development of unconformity-bounded depositional sequences (details in Chapters 4 and 5). As accommodation is only one of the several controls that interplay to generate stacking patterns in fully continental settings (Fig. 3.11), any direct link between accommodation and the nomenclature of systems tracts is to be avoided. This led to a change in nomenclature from terms that make reference to inferred accommodation conditions (i.e., low- vs. high-accommodation) to terms that describe the observed stacking patterns (i.e., high- vs. low-amalgamation; Catuneanu, 2017, 2019a) (details in Chapters 4 and 5).

10.2.2 *Guidelines for a standard methodology*

"First let us find the rule, then we will try to explain the exceptions" (Eco, 1980). Despite the seemingly unlimited variability of the stratigraphic architecture, there are "rules" (i.e., objective guidelines that afford the construction of the sequence stratigraphic framework based on the identification of stacking patterns that are diagnostic to the definition of systems tracts and bounding surfaces; Figs. 4.1, 4.2, 5.9, and 7.40) and there are "exceptions" (i.e., non-diagnostic features that are common among different systems tracts, and therefore with no bearing on the sequence stratigraphic methodology; Figs. 1.2 and 5.9). The diagnostic elements must be identified first, and once a framework of systems tracts is in place, the non-diagnostic variability can be rationalized within that context.

In downstream-controlled settings, sequences are defined by shoreline transit cycles which occur at different scales. At each scale of observation (hierarchical level), changes in the type of shoreline trajectory result in the formation of systems-tract boundaries (Fig. 5.5). Non-diagnostic variables include the depositional trends of aggradation or degradation in fluvial to coastal systems; the production of extrabasinal vs. intrabasinal sediment; the overall increase in riverborne sediment supply to the shelf edge in the case of narrow shelves or shelves dissected by unfilled incised valleys; and the influence exerted by contour currents on the sedimentological makeup of systems tracts in deep-water settings. In upstream-controlled settings, sequences of different scales (hierarchical ranks) generated by base-level cycles of different magnitudes consist of systems tracts defined by the dominant depositional elements (Fig. 4.70). In this case, a non-diagnostic variability may develop in response to changes in the sedimentation conditions (e.g., topographic gradients and fluvial styles) across the basin, which modifies the timing, the thickness, and the sedimentological makeup of systems tracts.

Both elements of the "dual control" on the stratigraphic architecture record variable rates through time and across a basin. The distinction between accommodation and sedimentation is only meaningful in downstream-controlled settings, and specifically *at the shoreline* where their interplay controls the timing of systems tracts and bounding surfaces. The emphasis on shoreline trajectories and shoreline transit cycles defines the common thread for all stratigraphic studies in downstream-controlled settings, irrespective of the non-diagnostic variability. The separation of relative sea-level changes (accommodation) and base-level changes (sedimentation) at the shoreline affords the critical insights into the rationale behind both diagnostic and non-diagnostic elements of the sequence

stratigraphic framework (Figs. 4.23 and 5.9). The indiscriminate mix of diagnostic and non-diagnostic features leads to the perception of randomness in the stratigraphic record, as often inferred by uncalibrated modeling, whereby no "rules" can be defined for a consistent methodological approach. The methodology lends order to chaos, by providing the criteria to separate the diagnostic elements of the sequence stratigraphic framework from the non-diagnostic variability.

While accommodation and sedimentation are equally important controls on the "conventional" stratigraphic architecture, the reconstruction of relative sea-level changes at the shoreline remains significant on several grounds, as discussed above. The identification of lowstands and highstands in relative sea level is decoupled from global standards, and is based on local stratal stacking patterns and stratigraphic relationships (i.e., lowstand and highstand systems tracts are defined by normal regressive stacking patterns that follow forced regressions and transgressions of equal hierarchical ranks, respectively; Fig. 5.87). Sedimentological and stratigraphic criteria are well established in coarser grained successions (e.g., Figs. 4.40, 4.41, and 4.54; Plint, 1988; Plint and Nummedal, 2000; Posamentier and Morris, 2000; Tesson et al., 2000; Zecchin and Tosi, 2014; Catuneanu, 2019a; Sweet et al., 2019), and additional biostratigraphic and geochemical proxies can be used in finer grained successions where evidence of shoreline shifts is more subtle (e.g., Gutierrez Paredes et al., 2017; Dong et al., 2018; Harris et al., 2018; Playter et al., 2018; LaGrange et al., 2020; Zecchin et al., 2021).

The following principles underlie the standard approach to sequence stratigraphy:

1. The acquisition of local data is the first essential step in the sequence stratigraphic workflow. Stratigraphic frameworks are basin specific, reflecting the local conditions of accommodation and sedimentation. Global standards play no role in the sequence stratigraphic methodology.
2. Preliminary studies of the tectonic and depositional settings are necessary to ensure that data are understood in the correct geological context. The integration and mutual calibration of independent datasets afford the most effective and reliable application of sequence stratigraphy.
3. The sequence stratigraphic methodology relies on the observation of stratal stacking patterns at scales afforded by the resolution of the data available, independently of the interpretation of the underlying controls (Figs. 1.2 and 3.3). The same types of stacking patterns can form at all stratigraphic scales.
4. At each stratigraphic scale, systems tracts are defined by specific types of stacking patterns; changes in the stacking pattern mark the position of

sequence stratigraphic surfaces, and the recurrence of the same types of sequence stratigraphic surfaces delineates sequences.

5. Within a framework of nested sequences, hierarchical ranks are only meaningful in a relative sense, in the context of each sedimentary basin. Sequences of equal ranks from different basins may differ in terms of timing, scales, composition, and underlying controls.

6. Despite the nested architecture of stratigraphic cycles, the sequence stratigraphic framework is not truly "fractal," as sequences of different scales, as well as sequences of similar scales, may differ in terms of underlying controls and internal composition of systems tracts.

7. The range of stratigraphic scales extends from the lowest rank systems tracts, which consist solely of sedimentological cycles, to entire basin fills which consist of lower rank stratigraphic cycles. Changes in the tectonic setting mark the position of first-order sequence boundaries.

8. The sequence stratigraphic workflow requires methodological and nomenclatural consistency at all stratigraphic scales, for an objective approach that is independent of any local variables, such as those defined by the geological setting and the resolution of the data available.

9. At each scale of observation, the construction of a framework of systems tracts and bounding surfaces fulfills the practical purpose of sequence stratigraphy. Within this framework, one can explain and predict the patterns of sediment distribution within a sedimentary basin.

10. Beyond the identification of systems tracts, the model-dependent selection of the "sequence boundary" is of secondary importance, and it may be made as a function of the development and mappability of the different types of sequence stratigraphic surfaces within the studied section.

While the standard approach to sequence stratigraphy is already defined, future developments are still needed to refine and diversify the field criteria that afford the identification of systems tracts and bounding surfaces in the sedimentary record (e.g., the emerging trends of using geochemical and paleontological proxies in sequence stratigraphy, in conjunction with information derived from other independent datasets; details in Chapter 2). Parallel progress in stratigraphic modeling will also continue in order to improve the accuracy of testing of sedimentary responses to the variety of possible controls on sequence development. However, methodology and modeling remain two distinct lines of stratigraphic research, with different goals and underlying data-driven vs. model-driven principles (Figs. 1.2 and 3.3; details in Chapter 9).

10.3 Summary of key points

Sequence stratigraphy examines stratigraphic cyclicity and the related changes in sedimentation regimes at scales afforded by the resolution of the data available. Stratigraphic cycles and the sedimentological makeup of their corresponding sequences are in a two-way process—response relationship. Stratigraphic cyclicity may drive changes in the sedimentation regimes (e.g., relative sea-level changes modify the shelf capacity to retain extrabasinal sediment or to produce intrabasinal sediment, in which case sedimentation is controlled by stratigraphic cycles), but changes in sedimentation regimes can also generate stratigraphic cycles (e.g., changes in the rates of carbonate production or siliciclastic influx due to any allocyclic or autocyclic processes can influence the direction of shoreline shift, in which case stratigraphic cycles are driven by changes in sedimentation regimes). Therefore, the stratigraphic and sedimentological attributes of sequences are interdependent. For this reason, the analysis of sedimentary processes that generate the observed stacking patterns is inherently linked to the sequence stratigraphic methodology (Figs. 1.4 and 10.1; details in Chapters 4—6).

Scale is among the most subtle and critical aspects of sequence stratigraphy, with implications for methodology and nomenclature. Scale-variant systems of classification of sedimentary cycles (e.g., bedsets < parasequences < sequences < composite sequences < megasequences) assume orderly patterns whereby the scales of the different types of cycles are mutually exclusive at any location. Pitfalls of this approach include (1) The scale of a "relatively conformable" sequence depends on the resolution of the stratigraphic study; e.g., seismic-scale "sequences" become "composite sequences" or even "megasequences" in higher resolution studies, which introduces nomenclatural inconsistency (Fig. 7.6); (2) the scales of sequences and parasequences are not mutually exclusive: sequences and parasequences of equal hierarchical ranks can coincide (Fig. 5.106) or coexist within sedimentary basins that are subject to coeval subsidence and uplift (Catuneanu et al., 2002; Gawthorpe et al., 2003; Zecchin, 2010), and lower rank sequences can be nested within larger scale parasequences (Fig. 5.105); and (3) parasequences may develop at both sequence and bedset scales (Fig. 6.41; details in Chapter 5).

The pitfalls of the scale-variant systems of stratigraphic nomenclature stem from the model-driven assumptions with respect to the origin and the relative scales of the different types of cycles, as well as from mixing allostratigraphic and sequence stratigraphic criteria of cycle delineation (details in Chapter 5). Stratigraphic units are defined by their bounding surfaces rather than physical or temporal scales, which allows for the development of different types of cycles at overlapping scales (Fig. 5.103); this invalidates the orderly patterns predicted by the scale-variant systems of classification. The model-independent approach to sequence stratigraphy is inherently simpler than the requirements of any specific model, and provides a flexible platform for the standard application of the method across the entire range of geological settings, stratigraphic scales, and types of data available. A standard methodology and nomenclature must remain independent of parameters that vary with the case study, such as the types and resolution of the data available, and the scales imposed by the local conditions of accommodation and sedimentation. Only the simplest, scale-invariant approach to methodology and nomenclature can provide a common denominator to all case studies, for a consistent application of sequence stratigraphy in a manner that is independent of all local variables.

Key milestones in the development of sequence stratigraphy include a de-emphasis on global cycle charts in the methodological workflow, and the construction of basin-specific sequence stratigraphic frameworks based on the collection of local data. This highlights the shift in paradigm from a model-driven approach underlain by assumptions regarding the dominant role of eustasy on sequence development, with consequent assertions of global correlations (Vail et al., 1977b; Haq et al., 1987), to a data-driven approach that promotes the use of local data and unbiased geological reasoning (Miall, 1986, 1991, 1992; Catuneanu et al., 2011). The following points outline the principles of the model-independent sequence stratigraphy.

10.3.1 Stratigraphic architecture

• Sequence stratigraphic framework:

The sequence stratigraphic framework records a nested architecture of stratigraphic cycles that can be observed at different scales, depending on the purpose of the study and the resolution of the data available (Figs. 4.70, 5.1, 7.1–7.5, and 8.55). At each scale of observation, stratigraphic cycles delineated by the recurrence of the same types of sequence stratigraphic surfaces define sequences, which consist of systems tracts and component depositional systems. These strata units are the building blocks of the sequence stratigraphic framework at all stratigraphic scales. High-frequency sequences (and component systems tracts and depositional systems) are commonly observed at scales of 10^0–10^1 m and 10^2–10^5 yrs, which defines the scope of high-resolution sequence stratigraphy (Figs. 4.62 and 5.4). The stacking pattern of higher frequency sequences defines depositional

systems and systems tracts of higher hierarchical ranks in lower resolution studies (Figs. 4.3, 4.70, 5.1, and 9.6). Despite the nested architecture of stratigraphic cycles, the sequence stratigraphic framework is not truly fractal as sequences of different scales, as well as sequences of similar scales, may differ in terms of underlying controls and internal composition of systems tracts (details in Chapter 7).

• Sequence stratigraphic hierarchy:

Sequence stratigraphic frameworks are basin specific in terms of timing and scales of stratigraphic cycles, reflecting the interplay of local and global controls on accommodation and sedimentation (Fig. 3.1). There are no standards for the temporal or physical scales of sequences and systems tracts, nor for the degree of stratigraphic complexity of a basin fill (i.e., number of hierarchical ranks that can be expected with a basin, which depends on the tectonic and depositional settings). Stratigraphic cycles of different scales define sequences of different hierarchical orders, whose numerical designation starts from the basin fill as a reference (i.e., a "first-order" sequence that relates to a specific tectonic setting; Figs. 5.3, 6.57, and 7.1). Where data are insufficient to observe the entire basin fill, the ranking of sequences that develop at different scales can be indicated in relative terms (e.g., higher vs. lower hierarchical ranks; Figs. 4.70, 5.106, and 7.16). Hierarchical orders have no time or thickness connotations, but only a relative stratigraphic significance within each basin. Sequences of equal hierarchical ranks in different basins may differ in terms of timing, scales, origin, and internal stratigraphic and sedimentological makeup (details in Chapter 7).

• Controls on stratigraphic cyclicity:

Sequence stratigraphic cycles (i.e., stratigraphic sequences) should not be assumed a priori as allocycles, as multiple allogenic and autogenic processes may interplay to generate stratigraphic cyclicity (Figs. 3.1 and 3.11; details in Chapter 3). The observation of stratal stacking patterns, which the sequence stratigraphic methodology is based upon, needs to be decoupled from any assumptions regarding the underlying controls (Figs. 1.2 and 3.3). The discrimination between the various allogenic and autogenic processes is a matter of interpretation and is tentative at best in many instances. The rates and the relative importance of these processes may vary with stratigraphic age and geological setting, and may be difficult to reconstruct in every case study; however, their interplay defines two fundamental variables, namely accommodation and sedimentation, which explain the formation of stratal stacking patterns in the stratigraphic record irrespective of the dominant controls at syn-depositional time (Fig. 3.3). This keeps the methodology focused on observational (field) criteria, in a manner that is independent of the interpretation of the underlying controls (Fig. 1.2).

- Accommodation vs. sedimentation:

  Accommodation is space made available for sediments to fill, by relative sea-level changes (i.e., subsidence and changes in sea level) in downstream-controlled settings, and by subsidence in upstream-controlled settings (Fig. 3.11). Accommodation may be underfilled or overfilled (Fig. 3.12). Sedimentation depends on all factors that modify the balance between sediment supply and energy flux at any location, including accommodation, climate, source-area tectonism, and autocyclicity (Fig. 3.11). The rates of sedimentation or erosion reflect the rates of change in the temporary base level, from sedimentological to stratigraphic scales (Figs. 1.19, 3.22, and 3.23). Stratigraphic cyclicity forms in response to base-level changes in upstream-controlled settings, and the interplay of base-level and relative sea-level changes in downstream-controlled settings. The distinction between relative sea-level changes and base-level changes affords critical insights into all aspects of the stratigraphic architecture and the rationale behind the patterns of sediment distribution across a basin during the different stages of sequence development. "Relative sea level" can be substituted with "relative lake level" in lacustrine settings (details in Chapter 3).

- Relative sea level:

  The relative sea level plays an important role as one of the "dual controls" on the stratigraphic architecture (Fig. 3.3). Both accommodation and sedimentation contribute in discernable ways to the architecture and makeup of the sequence stratigraphic framework. There are no true "stillstands" in relative sea level at any stratigraphic scale; the relative sea level is constantly shifting, and changes in the direction of shift are "events" at stratigraphic scales that result in the formation of sequence stratigraphic surfaces. The identification of lowstands and highstands in relative sea level is decoupled from global standards, and is based on local stratal stacking patterns and stratigraphic relationships (i.e., lowstand and highstand systems tracts are defined by normal regressive stacking patterns that follow forced regressions and transgressions of equal hierarchical ranks, respectively; Fig. 5.87). Independent datasets, such as biostratigraphic and geochemical, can also be used to further constrain relative sea-level changes in fine-grained successions, where evidence of shoreline shifts is more subtle (e.g., Gutierrez Paredes et al., 2017; Dong et al., 2018; Harris et al., 2018; Playter et al., 2018; LaGrange et al., 2020; details in Chapters 2—4).

- Stratal stacking patterns:

  Stratal stacking patterns provide the basis for the definition of all units and surfaces of sequence stratigraphy, in a manner that is independent of temporal and physical scales, and of the inferred underlying controls (details in Chapter 4). The same types of stacking patterns may form at all stratigraphic scales, in relation to stratigraphic cycles of different magnitudes (Figs. 4.70, 5.1, 5.3, 5.5, 6.57, 7.1—7.5, 8.55, and 9.6). At each scale of observation, specific stacking patterns define systems tracts, and changes in the type of stacking pattern across systems tract boundaries mark the position of sequence stratigraphic surfaces. Full cycles of change in stratal stacking patterns, as defined by the recurrence of the same types of sequence stratigraphic surfaces in the sedimentary record, delineate sequences. The identification of systems tracts and bounding surfaces at scales afforded by the resolution of the data available aids rationalizing the patterns of sediment distribution within a basin, which fulfills the practical purpose of sequence stratigraphy. Beyond this, the model-dependent selection of the "sequence boundary" is of secondary importance, and it may be guided by development and mappability of the different types of sequence stratigraphic surfaces within the study areas (Fig. 1.10).

- Diagnostic vs. non-diagnostic stacking patterns:

  The stratigraphic record includes a complex array of stacking patterns generated by various combinations of geometrical and depositional trends, only some of which are diagnostic to the definition of systems tracts (Figs. 1.2 and 4.23). In downstream-controlled settings, the diagnostic stacking patterns relate to the trajectory of subaerial clinoform rollovers (i.e., geometrical trends that define shoreline trajectories; Fig. 4.1), in which both accommodation and sedimentation play a critical role. In addition to these diagnostic geometrical trends, non-diagnostic depositional trends of coastal aggradation or erosion may accompany different types of shoreline trajectory, generating a stratigraphic variability that is irrelevant to the identification of systems tracts (Fig. 5.9). The distinction between diagnostic and non-diagnostic trends is key to the sequence stratigraphic methodology. Once a sequence stratigraphic framework is constructed, based on the observation of the diagnostic stacking patterns, the non-diagnostic variability can be rationalized within the context of systems tracts, to fully constrain the patterns of sediment distribution across the basin (details in Chapter 4).

- 3D stratigraphic variability:

The timing of systems tracts and bounding surfaces in downstream-controlled settings is controlled by the interplay of accommodation and sedimentation *at the shoreline* (Fig. 4.32). Variations in the rates of accommodation and sedimentation along the shoreline explain the coeval deposition of different systems tracts and the development of diachronous sequence stratigraphic surfaces (Figs. 7.37, 7.38, and 7.39; details in Chapter 7). Additional variability is introduced by changes in the rates of accommodation and sedimentation away from the shoreline, which modify the architecture systems tracts. For example, forced regressions may be accompanied by stratigraphic foreshortening in shelf settings (Posamentier and Morris, 2000; Zecchin and Tosi, 2014; Zecchin et al., 2017c, Fig. 4.36), or by stratigraphic expansion in ramp settings (e.g., Lickorish and Butler, 1996; Gawthorpe et al., 1997, 2000; Castelltort et al., 2003; Zecchin et al., 2003; Zecchin, 2007) or in areas affected by growth faults (e.g., Edwards, 1995; Zecchin et al., 2006) (Fig. 4.38). Variability in the timing and architecture of systems tracts is also recorded in upstream-controlled settings, in response to changes in sedimentation conditions along both strike and dip directions (Fig. 4.64).

- Stratigraphic sequences:

The definition of a "sequence" evolved in response to conceptual advances and improvements in stratigraphic resolution (Fig. 1.22). In an all-encompassing view, a sequence corresponds to a stratigraphic cycle of change in stratal stacking patterns, defined by the recurrence of the same type of sequence stratigraphic surface in the rock record. This definition is independent of model, scale, and data resolution. Sequences develop at different hierarchical levels, and their position within the nested architecture of stratigraphic cycles is indicated by hierarchical orders (Figs. 6.57 and 7.1−7.5) or simply in relative terms (Figs. 4.70, 5.106, and 7.16). At each hierarchical level, sequences may include unconformities of equal and/or lower hierarchical ranks, which break their "relatively conformable" character within the area of development of the unconformities (e.g., a genetic stratigraphic sequence may include a subaerial unconformity of equal hierarchical rank); therefore, sequences may or may not be "relatively conformable successions," but they always consist of "genetically related strata" that belong to the same stratigraphic cycle of change in stratal stacking

patterns (Galloway, 1989; Catuneanu et al., 2011; details in Chapters 5 and 7).

- Parasequences:

Parasequences were introduced as the building blocks of seismic-scale systems tracts in the context of low-resolution seismic stratigraphy (details in Chapter 5). Pitfalls of this concept relate to the definition of parasequence boundaries as lithological discontinuities that mark episodes of abrupt water deepening (Fig. 5.94). With this general meaning, "flooding surfaces" may be facies contacts within transgressive deposits, or may coincide with different types of sequence stratigraphic surfaces (maximum regressive, transgressive ravinement, or maximum flooding; Fig. 5.100) (details in Chapter 5 and 6). In all cases, flooding surfaces are allostratigraphic contacts restricted to coastal and shallow-water settings, where evidence of abrupt water deepening can be demonstrated. Flooding surfaces may also be absent from shallow-water successions, where the water deepening during transgression is gradual. It follows that (1) parasequences have smaller extent than systems tracts, and (2) systems tracts do not always consist of stacked parasequences. These limitations prevent the dependable use of the parasequence concept in sequence stratigraphy.

Advances in high-resolution sequence stratigraphy show that sequences and parasequences may form at the same scales, providing alternative approaches to the delineation of stratigraphic cycles (Fig. 5.103). Every transgression that results in the formation of a flooding surface starts from a maximum regressive surface and ends with a maximum flooding surface observed at the scale of that transgression (Fig. 5.104). These sequence stratigraphic surfaces are invariably more extensive than the flooding surface that may form during the transgression, in both updip and downdip directions (Fig. 6.61). Therefore, sequences that develop at parasequence scales provide a better and more reliable alternative for regional correlation, which renders parasequences obsolete. Flooding surfaces remain relevant to the description of facies relationships that may develop during transgressions. The use of sequences and systems tracts in high-resolution studies provides consistency in methodology and nomenclature at all stratigraphic scales, and eliminates the confusion created by mixing allostratigraphic and sequence stratigraphic methods in sequence stratigraphy (details in Chapters 5, 6, and 9).

- Subaerial vs. subaqueous clinoform rollovers:

Clinoforms and clinoform rollovers play a key role in sequence stratigraphy (e.g., Figs. 4.7, 4.8, 4.39, and

6.41). The distinction between subaerial and subaqueous clinoform rollovers (i.e., shoreline vs. a subaqueous "delta" or a submerged shelf edge; Fig. 4.19) is critical to the proper identification of sequence stratigraphic units and bounding surfaces in downstream-controlled settings (e.g., it is the backstepping of the shoreline, and not of the shelf edge, that defines a transgressive systems tract; Fig. 4.20). The trajectory of subaerial clinoform rollovers observed at stratigraphic scales (Figs. 3.25 and 4.58) provides the basis for the definition of "conventional" systems tracts (Figs. 4.1 and 5.9). Changes in the type of shoreline trajectory result in the formation of systems-tract boundaries, and full shoreline transit cycles generate sequences. Shoreline shifts can also be observed at sub-stratigraphic scales, without changes in coastal environment (e.g., in relation to tidal or fairweather-storm cycles), in which case they only result in the formation of sedimentological cycles (i.e., beds and bedsets within the lowest rank systems tracts and component depositional systems; Figs. 3.24, 3.25, and 4.62) (details in Chapters 3–7).

The shelf edge may also become a subaerial clinoform rollover during times of high-magnitude relative sea-level fall, when shelves are exposed and the shoreline coincides with the shelf edge (Figs. 4.6 and 8.53). The trajectory of the shelf edge is the first-order shoreline trajectory observed at the basin scale (Fig. 4.60). At any scale of observation, the limit between the topset and the foreset of a prograding system represents the trajectory of maximum regressive shorelines of immediately lower hierarchical rank (Figs. 4.58 and 4.60). This is true from the scale of deltas (in which case the limit between topset and foreset corresponds to the low tide level: i.e., the tidal lowstand; Fig. 3.25) to the scale of shelf-slope systems (in which case the limit between topset and foreset corresponds to the shelf edge: i.e., the first-order shoreline trajectory, which connects the lowstand shorelines of the second-order sequences; Fig. 4.60). Therefore, the transit area of a shoreline is located updip from the shoreline trajectory of higher hierarchical rank (Figs. 4.58 and 4.60). The shoreline shifts within a transit area reflect higher frequency (lower rank) transgressions and regressions within a higher rank topset (Figs. 4.6, 4.58, 4.60, and 5.87; more details on the timescales of shoreline shifts and the sedimentological composition of topsets in Chapters 4 and 9).

• Geometrical vs. depositional trends:

The architecture of the downstream-controlled stratigraphic framework may be described in terms of geometrical trends of the shoreline (i.e., shoreline trajectories: forestepping, backstepping, upstepping,

downstepping) or in terms of depositional trends at the shoreline (i.e., coastal processes that accompany the shoreline shifts: progradation, retrogradation, aggradation, degradation) (Figs. 4.23 and 4.24). The lateral components of the two types of trends are controlled by the same interplay of relative sea-level changes (accommodation) and base-level changes (sedimentation), and therefore are equivalent to each other and can be used interchangeably. The vertical trends, however, are controlled by different processes and can be out of sync: the geometrical trends of upstepping and downstepping are controlled by relative sea-level changes, whereas the depositional trends of aggradation or degradation are controlled by base-level changes. Diagnostic to the definition of systems tracts are the geometrical trends that describe shoreline trajectories, irrespective of the non-diagnostic depositional trends of aggradation or degradation that accompany the shoreline shifts (Figs. 1.2 and 5.9; details in Chapter 4).

• Forced vs. normal regressions:

Forced and normal regressions differ in terms of driving mechanism (relative sea-level fall vs. sediment supply during relative sea-level rise), rates of progradation (higher vs. lower), and the direction of shift of the wave base during progradation (downstepping vs. upstepping), leading to differences in stratigraphic architecture (downstepping vs. upstepping clinoforms and subaerial clinoform rollovers; Figs. 4.6, 4.30, 4.40, and 4.41), facies relationships (sharp- vs. gradationally based shorefaces; Fig. 4.30), facies development (compressed vs. expanded shoreface successions; Fig. 5.31), clinoform composition (coarser vs. finer, and poorer vs. better sediment sorting), and sediment supply to the shelf edge (Fig. 4.20). The distinction between forced and normal regressions is therefore important on several grounds, from the patterns of sediment distribution across a basin to the reconstruction of relative sea-level changes within the basin under analysis (Figs. 4.24 and 5.9). The defining features of forced and normal regressions in energy- and supply-dominated settings are summarized in Fig. 4.54. Normal regressions are further classified into "lowstand" and "highstand," based on local stratigraphic relationships (i.e., normal regressions that follow forced regressions vs. transgressions, respectively; Fig. 5.87; details in Chapter 4).

• Stratigraphic architecture in deep-water settings:

Stratigraphic cyclicity in the deep-water setting reflects the interplay of accommodation and sedimentation on the shelf, which controls shoreline trajectories, sediment supply to the shelf edge, and the timing of all elements of the sequence stratigraphic framework. Stratigraphic trends defined by changes in the types, volume, and composition of gravity flows during the shoreline transit cycles on the shelf provide the diagnostic criteria for the identification of deep-water systems tracts and bounding surfaces (Figs. 8.41, 8.49, and 8.54). Contour currents may also contribute to the identification of systems tracts and bounding surfaces, in cases where variations in their sediment composition are linked to relative sea-level changes and shoreline shelf-transit cycles (Fig. 8.44). Non-diagnostic variability in the sedimentological makeup of systems tracts (e.g., extrabasinal vs. intrabasinal sediment, or the increase in riverborne sediment supply to deep water in the case of narrow shelves or shelves dissected by unfilled incised valleys) reflects the unique tectonic and depositional settings of each basin, and needs to be rationalized on a case-by-case basis (details in Chapter 8).

- Stratigraphic architecture in upstream-controlled settings:

  The stratigraphic architecture in upstream-controlled settings is defined by nested depositional sequences that form in response to cycles of aggradation and degradation (i.e., base-level cycles) of different scales (Figs. 4.2 and 4.70). Sequences of all hierarchical ranks can be subdivided into systems tracts defined by the dominant depositional elements (e.g., channels vs. floodplains in fluvial systems; Fig. 4.70). The origin of these stacking patterns may be linked to any combination of controls on sedimentation (i.e., accommodation, climate, source-area tectonism, and autocyclicity; Figs. 3.11 and 4.63). Therefore, the nomenclature of stacking patterns and associated systems tracts needs to make reference to observational features (e.g., high vs. low degree of channel amalgamation in fluvial systems), rather than inferred underlying controls (e.g., low vs. high accommodation conditions). This is particularly evident in overfilled basins, where the rates of sedimentation may exceed the rates of subsidence at all stratigraphic scales, leading to the development of systems tracts in excess of the amounts of accommodation generated at syn-depositional time (details in Chapters 4 and 5).

10.3.2 Stratigraphic scales

- Sedimentological vs. stratigraphic scales:

  The observation of depositional systems at different hierarchical levels (Figs. 5.1, 5.3, and 5.5) defines the overlap between the scales of sedimentology and stratigraphy (Fig. 4.62). At each hierarchical level, depositional systems are the largest units of sedimentology, and the building blocks of systems tracts in stratigraphy. However, sedimentological cycles (i.e., beds and bedsets; Figs. 3.24 and 3.25) are always nested within the lowest rank systems tracts and component depositional systems (Fig. 4.3). The scale and the sedimentological makeup of the lowest rank systems tracts and component depositional systems vary with the geological setting (i.e., local conditions of accommodation and sedimentation), and the time involved in the development of their defining stacking patterns. Systems tracts and component depositional systems of higher hierarchical ranks consist of higher frequency (lower rank) stratigraphic cycles (i.e., sequences; Figs. 4.3, 5.1, and 5.5). The scale of the lowest rank systems tracts at any location defines the highest resolution that can be achieved with a stratigraphic study, and the limit between stratigraphy and sedimentology *sensu stricto* (Figs. 4.62 and 5.2; details in Chapters 5 and 7).

- Scales in sequence stratigraphy:

  Sequences, systems tracts, and depositional systems may be observed from 10^0 m outcrop scales (Fig. 5.2) to 10^3 m basin scales (Fig. 5.3). The timescales involved in the formation of a sequence stratigraphic framework range from $10^2 - 10^3$ yrs (i.e., the minimum amount of time required to form depositional systems and systems tracts) to $\leq 10^8$ yrs (i.e., the lifespans of entire sedimentary basins) (Figs. 4.62, 5.3, 5.5, and 9.1). Stratigraphic cyclicity in a range of $10^0 - 10^1$ yrs may also exist (Schlager, 2010), but these timescales become too short to afford the formation of geologically recognizable soil horizons as evidence of subaerial unconformities, and the accumulation of depositional elements that define depositional systems. Therefore, realistic timescales for a sequence stratigraphic analysis start with $10^2 - 10^3$ yrs (Fig. 4.62; Catuneanu, 2017, 2019a,b). The physical scales of the lowest rank sequences, systems tracts, and component depositional systems depend on the geological setting, and are commonly below the resolution of seismic stratigraphy (Figs. 5.2 and 5.5). Basins isolated from the global ocean are more susceptible to short-term sea/lake-level changes, and therefore prone to the development of higher frequency stratigraphic cycles, starting with centennial scales (Fig. 9.1; details in Chapters 5 and 7).

- Scale of paleogeographic changes:

  Paleogeographic reconstructions capture the relationship between different paleodepositional environments at discrete timesteps during the evolution of a sedimentary basin, with the degrees of detail and accuracy afforded by the resolution of the data available. Paleogeographic changes occur at all stratigraphic scales, with magnitudes that may or may not correlate with the hierarchical rank of the observed stratigraphic cycle (Fig. 4.61). In some cases, the degree of paleogeographic change is proportional to the hierarchical rank (e.g., Fig. 5.1: the facies shifts associated with the fourth-order transgressions are greater than the facies shifts associated with the fifth-order transgressions; Fig. 4.61A). In other cases, the opposite is true (e.g., Fig. 4.34: the facies shifts associated with the higher frequency glacial–interglacial cycles are greater than the facies shifts described by the longer term shoreline trajectories, as the shoreline was more stable in the long term; Fig. 4.61B). The relative degrees of paleogeographic changes associated with stratigraphic cycles of different scales depend on the syn-depositional conditions of accommodation and sedimentation, and need to be rationalized on a case-by-case basis (details in Chapters 4 and 9).

- Scale of depositional systems:

Depositional systems are the sedimentological building blocks of systems tracts, and may be observed at multiple hierarchical levels, from 10^0 to 10^3 m scales (Figs. 5.1–5.3). The formation of depositional systems requires typically timescales of minimum 10^2 yrs (Fig. 4.62), and it may be sustained for as long as the defining environments are maintained as dominant (but not necessarily exclusive) sediment fairways (Fig. 5.1). The largest (highest rank) depositional systems are defined by depositional trends observed at first-order basin scales of $\leq 10^8$ yrs and $\leq 10^3$ m (Fig. 5.3). The lowest rank depositional systems (i.e., depositional systems sensu stricto) consist solely of sedimentological cycles (i.e., beds and bedsets), and develop commonly at scales below the resolution of seismic stratigraphy (Fig. 5.2); at any larger scales (higher hierarchical ranks), depositional systems sensu lato reflect the dominant depositional trends but include lower rank stratigraphic cycles (i.e., higher frequency sequences; Figs. 5.1, 5.5). The observation of depositional systems at different scales affords the usage of the "systems tract" concept at different hierarchical levels (Fig. 4.3; details in Chapters 5 and 7).

- Scale of sequences and systems tracts:

Sequences and component systems tracts are the stratigraphic building blocks of the sequence stratigraphic framework, at all hierarchical levels and stratigraphic scales; i.e., the concepts of "sequence"

and "systems tract" are not restricted to any specific position within the nested architecture of stratigraphic cycles (Figs. 4.3, 5.1–5.3; details in Chapters 7 and 9). The definition of sequences and systems tracts is based on the observation of stratal stacking patterns, irrespective of scales and the interpretation of the underlying controls (Fig. 1.2). The same types of stacking patterns may form at all stratigraphic scales, in relation to stratigraphic cycles of different magnitudes. At each scale of observation (i.e., hierarchical rank), systems tracts are defined by specific stacking patterns, and sequences correspond to cycles of change in stratal stacking patterns (Fig. 1.22). Sequences consist of systems tracts of equal hierarchical rank which accumulated during the same stratigraphic cycle of change in stratal stacking patterns. There is no correlation between scale and the systems-tract composition of sequences, which can be highly variable depending on the local conditions of accommodation and sedimentation (Figs. 1.23 and 4.33).

- Unconformities vs. diastems:

Hiatal surfaces form at all temporal and physical scales, ranging in relevance from the scope of sedimentology sensu stricto (i.e., bedding planes separating beds and bedsets) to the scope of stratigraphy (i.e., unconformable sequence stratigraphic surfaces). Across a continuum of scales, the distinction between "major" (unconformities) and "minor" (diastems) breaks in sedimentation becomes arbitrary, and dependent on the resolution of the data available (Fig. 7.8). There are no temporal or physical standards for the scales of unconformities and diastems. Moreover, the amount of missing time associated with a hiatal surface is often unknown, and the physical expression of hiatal surfaces is not a reliable proxy for the magnitude of the time gap. The ambiguity in the distinction between diastems and unconformities requires either restricting the usage of diastems to sedimentology sensu stricto (i.e., bed and bedset boundaries; Figs. 3.24 and 4.62) or eliminating the concept of diastem from sedimentary geology altogether. Ultimately, "every bedding plane is, in effect, an unconformity" (Ager, 1993) (details in Chapters 1 and 7).

- Intra-sequence unconformities:

Sequences of any scale may include unconformities of equal and/or lower hierarchical ranks. Therefore, sequences may or may not be "relatively conformable" successions, even though they consist of genetically related strata that belong to the same stratigraphic cycle of change in stratal stacking patterns. Internal unconformities of lower hierarchical ranks are negligible relative to the scale of the host sequence, thus maintaining the relatively conformable character of the succession (Figs. 7.6 and 7.12). Internal unconformities of equal hierarchical rank may break the continuity in the paleogeographic evolution observed at the scale of the host sequence, in which case the succession is no longer "relatively conformable" within the area of development of the unconformities (e.g., abrupt shifts in depositional systems across subaerial unconformities within genetic stratigraphic sequences, or across transgressive surfaces of erosion within depositional sequences). However, beyond the termination of unconformities, correlative conformable surfaces preserve the genetic relationship between the strata below and above, thus maintaining the integrity of the sequence as one stratigraphic cycle (Fig. 5.92; details in Chapter 7).

- Relatively conformable successions:

  Relatively conformable successions can be observed at different scales, depending on the resolution of the stratigraphic study (Fig. 7.6). Likewise, the detection of unconformities in a stratigraphic succession depends on the resolution of the data available. Unconformities that are not resolvable with a low-resolution dataset break the relatively conformable character of the succession in higher resolution studies (Fig. 7.6). Stratigraphic resolution has increased to the extent that most if not all "relatively conformable" successions in the sense of Mitchum (1977) are now restricted to the sub-seismic scales of high-resolution sequence stratigraphy (Figs. 5.2, 5.4, and 5.5), which would render seismic stratigraphy obsolete. However, relatively conformable successions *sensu lato* can be defined relative to the scale of observation rather than the resolution of the data available (i.e., units with negligible rather than unresolvable internal unconformities), thus expanding the application of Mitchum's (1977) definition of a depositional sequence to all stratigraphic scales (Figs. 7.6 and 7.12). True relatively conformable successions *sensu stricto* may only be found at the scale of the lowest rank systems tracts and component depositional systems, which consist solely of beds or bedsets (Fig. 5.2; details in Chapter 7).

10.3.3 Methodology and nomenclature

- Uses and abuses in sequence stratigraphy:

  The greatest hurdle to overcome in sequence stratigraphy is dogma. Bias toward a model may predispose practitioners to force fit the data into theoretical templates (e.g., by seeking the ideal succession of systems tracts in every sequence). Such practice may stem from a lack of adequate knowledge and self-confidence, or from the desire to fulfill the predictions of a particular model. An open mind, along with the realization that each basin is potentially unique in terms of stratigraphic architecture, is key to an objective approach that supersedes any preconceived ideas. This frees the practitioner from model-driven expectations, and enables an objective construction of local stratigraphic frameworks. Such frameworks may or may not correlate with the frameworks of other basins or even sub-basins of the same sedimentary basin, due to the influence of the local controls on accommodation and sedimentation. The "magic" of building local frameworks on the basis of global standards and information extrapolated from other basins is now replaced by the more tedious but also more realistic practice of acquisition and analysis of local data. The data define the model, not the other way around (details in Chapters 1 and 8–10).

- Observation vs. interpretation in sequence stratigraphy:

  The observation of stratal stacking patterns that afford the identification of systems tracts and bounding surfaces is fundamental to the sequence stratigraphic methodology (Figs. 1.2, 3.3, and 7.40). Interpretations in sequence stratigraphy relate to (1) the underlying controls on sequence development (Fig. 3.1); and (2) the sedimentary processes that generate the observed stacking patterns (Figs. 4.23 and 4.63), which enable the prediction of facies in areas away from data control points such as outcrops or wells (Fig. 10.1). The interpretation of the underlying controls on sequence development (i.e., the relative contributions of the various allogenic and autogenic controls on stratigraphic cyclicity; Figs. 3.1 and 3.11; details in Chapter 3) is beyond the scope of the sequence stratigraphic methodology, and can continue indefinitely after the construction of a sequence stratigraphic framework (Fig. 1.2; details in Chapter 9).

  Facies predictions are rationalized within the context of systems tracts, which establish genetic relationships between same-age depositional systems (e.g., the link between shelf processes and the sediment supply to the deep-water setting; details in Chapter 8). The modes of sediment transport also play an important role in the prediction of facies, as they control the sediment dispersal patterns and the location and orientation of depositional elements (e.g., gravity-flow deposits vs. contourites in deep-water settings; Fig. 8.45). The sedimentary processes that generate stacking patterns, which may involve suspension fallout, chemical and biochemical precipitation, and various types of traction currents and gravity-driven processes, are best understood by placing the data in the correct depositional setting (see the workflow of sequence stratigraphy in Chapter 9). The clarification of these processes, which is intrinsic to the construction of sequence stratigraphic frameworks, brings meaning to stratigraphic observations and enables a genetic approach that sets sequence stratigraphy apart from any other type of stratigraphy (Figs. 1.1 and 1.4; details in Chapter 4–6).

- Methodology vs. modeling in sequence stratigraphy:

Methodology and modeling are two distinct lines of stratigraphic research, with different goals and underlying data-driven vs. model-driven principles (Figs. 1.2 and 3.3). Mixing the two aspects of stratigraphic research was a pitfall since the inception of sequence stratigraphy, which took decades of work to correct (e.g., decoupling the methodology from global standards and the use of global cycle charts). The methodology guides the construction of basin-specific stratigraphic frameworks based on the observation of stratal stacking patterns, in a manner that is independent of the interpretation and testing of the underlying controls (Fig. 1.2). Proposals that numerical models should become part of the sequence stratigraphic workflow are not only impractical, but a setback as they bring confusion between modeling and methodology. This undermines the progress made in the development of sequence stratigraphy as a data-driven methodology, with an objective workflow that relies on observations rather than model-driven assumptions. The modeling and testing of the possible controls on sequence development may continue indefinitely after the construction of a sequence stratigraphic framework (details in Chapter 9).

- Standard methodology and nomenclature:

The sequence stratigraphic workflow requires methodological and nomenclatural consistency at all stratigraphic scales, for an objective approach that is independent of the variability introduced by the geological setting and the resolution of the data available. The identification of stacking patterns that afford the definition of systems tracts and bounding surfaces, at scales selected by the practitioner or imposed by the resolution of the data available, defines the standard approach to sequence stratigraphy. The standard workflow is simpler than the requirements of any specific model, as it eliminates the need to find a particular type of "sequence boundary" or to define the scale of a "relatively conformable" succession as a reference for the classification of stratal units. Only a scale-invariant approach to methodology and nomenclature can provide the flexibility needed for a standard application of sequence stratigraphy in a manner that is independent of all local variables. The standard approach to sequence stratigraphy enables a consistent application of the method across the entire range of geological settings, stratigraphic scales, and types of data available (details in Chapters 9 and 10).

References

Abbott, S.T., 1998. Transgressive systems tracts and onlap shellbeds from mid-Pleistocene sequences, Wanganui Basin, New Zealand. J. Sediment. Res. 68, 253–268.

Abbott, S.T., Carter, R.M., 1994. The sequence architecture of mid-Pleistocene (c.1.1.-0.4 Ma) cyclothems from New Zealand: facies development during a period of orbital control on sea-level cyclicity. In: De Boer, P.L., Smith, D.G. (Eds.), Orbital Forcing and Cyclic Sequences, vol. 19. International Association of Sedimentologists, pp. 367–394. Special Publication.

Abrantes, F.R., Basilici, G., Theodoro Soares, M.V., 2020. Mesoproterozoic erg and sand sheet system: architecture and controlling factors (Galho do Miguel Formation, SE Brazil). Precambrian Res. 338, 105592.

Abreu, V., Neal, J.E., Bohacs, K.M., Kalbas, J.L., 2010. Sequence stratigraphy of siliciclastic systems – the ExxonMobil methodology. SEPM Conc. Sediment. & Paleontol. 226. #9.

Ager, D.V., 1981. The Nature of the Stratigraphic Record, second ed. John Wiley, New York, p. 151.

Ager, D.V., 1993. The New Catastrophism. Cambridge University Press, Cambridge, p. 231.

Agterberg, F.P., Gradstein, F.M., 1999. The RASC method for ranking scaling of biostratigraphic events. Earth-Sci. Rev. 46, 1–25.

Aigner, T., 1984. Dynamic stratigraphy of epicontinental carbonates, Upper Muschelkalk (M. Triassic) South German basin. Neues Jahrbuch für Geol. und Palaeontol. Abhandlungen 169, 127–159.

Ainsworth, R.B., 1992. Sedimentology and sequence stratigraphy of the Upper Cretaceous. In: Bearpaw-Horseshoe Canyon transition, Drumheller, Alberta. American Association of Petroleum Geologists Annual Convention, Calgary, Field Trip Guidebook, vol. 7, p. 118.

Ainsworth, R.B., 1994. Marginal marine sedimentology and high resolution sequence analysis; Bearpaw - Horseshoe Canyon transition, Drumheller, Alberta. Bull. Can. Petrol. Geol. 42 (1), 26–54.

Ainsworth, R.B., 2005. Sequence stratigraphic-based analysis of reservoir connectivity: influence of depositional architecture – a case study from a marginal marine depositional setting. Pet. Geosci. 11, 257–276.

Ainsworth, R.B., Pattison, S.A.J., 1994. Where have all the lowstands gone? Evidence for attached lowstand systems tracts in the Western Interior of North America. Geology 22, 415–418.

Ainsworth, R.B., Walker, R.G., 1994. Control of estuarine valley-fill deposition by fluctuations of relative sea-level, Cretaceous Bearpaw-Horseshoe Canyon transition, Drumheller, Alberta, Canada. In: Dalrymple, R.G., Boyd, R., Zaitlin, B.A. (Eds.), Incised-valley Systems: Origin and Sedimentary Sequences. SEPM (Society for Sedimentary Geology), pp. 159–174. Special Publication No. 51.

Ainsworth, R.B., Vakarelov, B.K., MacEachern, J.A., Rarity, F., Lane, T.I., Nanson, R.A., 2017. Anatomy of a shoreline regression: implications for the high-resolution stratigraphic architecture of deltas. J. Sediment. Res. 87, 1–35.

Ainsworth, R.B., McArthur, J.B., Lang, S.C., Vonk, A.J., 2018. Quantitative sequence stratigraphy. Am. Assoc. Pet. Geol. Bull. 102, 1913–1939.

Aitken, J.D., 1978. Revised models for depositional grand cycles, Cambrian of the Southern Rocky Mountains, Canada. Bull. Can. Pet. Geol. 26, 515–542.

Aitken, J.F., Flint, S.S., 1994. High-frequency sequences and the nature of incised-valley fills in fluvial systems of the Breathitt Group (Pennsylvanian), Appalachian foreland basin, Eastern Kentucky. In: Dalrymple, R.W., Boyd, R., Zaitlin, B.A. (Eds.), Incised Valley Systems: Origin and Sedimentary Sequences, vol. 51. SEPM, pp. 353–368. Special Publication.

Aitken, J.F., Flint, S.S., 1996. Variable expressions of interfluvial sequence boundaries in the Breathitt Group (Pennsylvanian), Eastern Kentucky, USA. In: Howell, J.A., Aitken, J.F. (Eds.), High Resolution Sequence Stratigraphy: Innovations and Applications, vol. 104. Geological Society of London, pp. 193–206. Special Publication.

Algeo, T.J., Ingall, E., 2007. Sedimentary C_{org}:P ratios, paleocean ventilation, and phanerozoic atmospheric pO_2. Palaeogeogr. Palaeoclimatol. Palaeoecol. 256, 130–155.

Algeo, T.J., Wilkinson, B.H., 1988. Periodicity of mesoscale phanerozoic sedimentary cycles and the role of Milankovitch orbital modulation. J. Geol. 96, 313–322.

Allen, D.R., 1975. Identification of sediments – their depositional environment and degree of compaction – from well logs. In: Chilingarian, G.V., Wolf, K.H. (Eds.), Compaction of Coarse-Grained Sediments I. Elsevier, New York, pp. 349–402.

Allen, G.P., 1991. Sedimentary processes and facies in the Gironde estuary: a recent model for macrotidal estuarine systems. In: Smith, D.G., Reinson, G.E., Zaitlin, B.A., Rahmani, R.A. (Eds.), Clastic Tidal Sedimentology, vol. 16. Canadian Society of Petroleum Geologists, Memoir, pp. 29–40.

Allen, G.P., Posamentier, H.W., 1993. Sequence stratigraphy and facies model of an incised valley fill: the Gironde estuary, France. J. Sediment. Pet. 63 (3), 378–391.

Allen, G.P., Posamentier, H.W., 1994. Transgressive facies and sequence architecture in mixed tide- and wave-dominated incised valleys: example from the Gironde Estuary, France. In: Dalrymple, R.W., Boyd, R., Zaitlin, B.A. (Eds.), Incised Valley Systems: Origin and Sedimentary Sequences, vol. 51. SEPM, pp. 225–240. Special Publication.

Allen, J.R.L., 1978. Studies in fluviatile sedimentation: an exploratory quantitative model for the architecture of avulsion-controlled alluvial suites. Sediment. Geol. 21, 129–147.

Allen, J.R.L., 1979. Studies in fluviatile sedimentation: an elementary geometrical model for the connectedness of avulsion-related channel sand bodies. Sediment. Geol. 24, 253–267.

Allen, P.A., Allen, J.R., 2013. Basin Analysis: Principles and Application to Petroleum Play Assessment, third ed. Wiley-Blackwell, p. 632.

Allen, P.A., Eriksson, P.G., Alkmim, F.F., Betts, P.G., Catuneanu, O., Mazumder, R., Meng, Q., Young, G.M., 2015. Classification of sedimentary basins, with special reference to Proterozoic examples. In: Mazumder, R., Eriksson, P.G. (Eds.), Precambrian Basins of India: Stratigraphic and Tectonic Context, vol. 43. Geological Society, London, Memoirs, pp. 5–28.

Alonso-Zarza, A.M., 2003. Paleoenvironmental significance of palustrine carbonates and calcretes in the geological record. Earth-Sci. Rev. 60, 261–298.

Amorosi, A., Centineo, M.C., Colalongo, M.L., Fiorini, F., 2005. Millennial-scale depositional cycles from the Holocene of the Po Plain, Italy. Mar. Geol. 222–223, 7–18.

Amorosi, A., Ricci Lucchi, M., Rossi, V., Sarti, G., 2009. Climate change signature of small-scale parasequences from Lateglacial-Holocene transgressive deposits of the Arno valley fill. Palaeogeogr. Palaeoclimat. Palaeoecol. 273, 142–152.

Amorosi, A., Bruno, L., Campo, B., Morelli, A., Rossi, V., Scarponi, D., Hong, W., Bohacs, K.M., Drexler, T.M., 2017. Global sea-level control on local parasequence architecture from the Holocene record of the Po Plain, Italy. Mar. & Pet. Geol. 87, 99–111.

Anderson, J.B., Wolfteich, C., Wright, R., Cole, M.L., 1982. Determination of depositional environments of sand using vertical grain size progressions. Gulf Coast Assoc. Geol. Soc. Transac. 32, 565–577.

Andresen, N., Reijmer, J.J.G., Droxler, A.W., 2003. Timing and distribution of calciturbidites around a deeply submerged carbonate platform in a seismically active setting (Pedro Bank, Northern Nicaragua Rise, Caribbean Sea). Int. J. Earth Sci. 92 (4), 573–592.

Armentrout, J.M., 1987. Integration of biostratigraphy and seismic stratigraphy: Pliocene–Pleistocene, Gulf of Mexico. In: Innovative Biostratigraphic Approaches to Sequence Analysis: New Exploration Opportunities. 8th Annual Research Conference. Gulf Coast Section, SEPM, pp. 6–14.

Armentrout, J.M., 1991. Paleontologic constraints on depositional modeling: examples of integration of biostratigraphy and seismic stratigraphy, Pliocene–Pleistocene, Gulf of Mexico. In: Weimer, P., Link, M.H. (Eds.), Seismic Facies and Sedimentary Processes of Submarine Fans and Turbidite Systems. Springer-Verlag, New York, pp. 137–170.

Armentrout, J.M., 1996. High resolution sequence biostratigraphy: examples from the Gulf of Mexico Plio-Pleistocene. In: Howell, J.A., Aitken, J.F. (Eds.), High Resolution Sequence Stratigraphy: Innovations and Applications. Geological Society, pp. 65–86. Special Publication No. 104.

Armentrout, J.M., Echols, R.J., Lee, T.D., 1991. Patterns of foraminiferal abundance and diversity: implications for sequence stratigraphic analysis. In: Armentrout, J.M., Perkins, B.F. (Eds.), Sequence Stratigraphy as an Exploration Tool: Concepts and Practices, Eleventh Annual Conference, June 2–5, Texas Gulf Coast Section, SEPM, pp. 53–58.

Arndorff, L., 1993. Lateral relations of deltaic palaeosols from the lower Jurassic Ronne formation on the island of Bornholm, Denmark. Palaeogeogr. Palaeoclimat. Palaeoecol. 100, 235–250.

Arnott, R.W.C., 1995. The parasequence definition — are transgressive deposits inadequately addressed? J. Sediment. Res. B65, 1–6.

Arnott, R.W.C., Zaitlin, B.A., Potocki, D.J., 2002. Stratigraphic response to sedimentation in a net-accommodation-limited setting, Lower Cretaceous Basal Quartz, South-central Alberta. Bull. Can. Pet. Geol. 50 (1), 92–104.

Arnott, R.W.C., Dumas, S., Southard, J.B., 2004. Hummocky cross-stratification and other shallow-marine structures: 25 years and still the debate continues. In: Geological Association of Canada – Mineralogical Association of Canada Joint Annual Meeting. St. Catharines, Abstract on CD-ROM.

Arthur, M.A., Sageman, B.B., 1994. Marine black shales: depositional mechanisms and environments of ancient deposits. Ann. Rev. Earth & Planet. Sci. 22, 499–551.

Arthur, M.A., Sageman, B.B., 2005. Sea-level control on source-rock development: perspectives from the Holocene Black Sea, the mid-Cretaceous Western Interior Basin of North America, and the Late Devonian Appalachian Basin. In: Harris, N.B. (Ed.), The Deposition of Organic-Carbon-Rich Sediments: Models, Mechanisms, and Consequences, vol. 82. Society for Sedimentary Geology (SEPM), pp. 35–59. Special Publication.

Avarjani, S., Mahboubi, A., Moussavi-Harami, R., Amiri-Bakhtiar, H., Brenner, R.L., 2015. Facies, depositional sequences, and biostratigraphy of the Oligo-Miocene Asmari formation in Marun oilfield, North Dezful Embayment, Zagros Basin, SW Iran. Palaeoworld 24, 336–358.

Bailey, R.J., Smith, D.G., 2010. Scaling in stratigraphic data series: implications for practical stratigraphy. First Break 28, 57–66.

Banerjee, I., Kidwell, S.M., 1991. Significance of molluscan shell beds in sequence stratigraphy: an example from the Lower Cretaceous Mannville Group of Canada. Sedimentology 38, 913–934.

Banerjee, I., Kalkreuth, W., Davies, E.H., 1996. Coal seam splits and transgressive-regressive coal couplets: a key to stratigraphy of high-frequency sequences. Geology 24 (11), 1001–1004.

Bard, B., Hamelin, R.G., Fairbanks, R., 1990. U-Th obtained by mass spectrometry in corals from Barbados: sea level during the past 130.000 years. Nature 346, 456–458.

Barka, A., Serdar Akyuz, H., Cohen, H.A., Watchorn, F., 2000. Tectonic evolution of the Niksar and Tasova–Erbaa pull-apart basins, North Anatolian fault zone: their significance for the motion of the Anatolian block. Tectonophysics 322, 243–264.

Barrell, J., 1917. Rhythms and the measurements of geological time. Geol. Soc. Am. Bull. 28, 745–904.

Barron, J.A., Gladenkov, A.Y., 1995. Early Miocene to Pleistocene diatom stratigraphy of Leg 145. In: Rea, D.K., Basov, I.A., Scholl, D.W., Allan, J.F. (Eds.), Proc. ODP Sci. Results, vol. 145. Ocean Drilling Program), College Station, TX, pp. 3–19.

Bart, P.J., De Santis, L., 2012. Glacial intensification during the Neogene: a review of seismic stratigraphic evidence from the Ross Sea, Antarctica, continental shelf. Oceanography 25, 166–183.

Bartek, L.R., Vail, P.R., Anderson, J.B., Emmet, P.A., Wu, S., 1991. Effect of Cenozoic ice sheet fluctuations in Antarctica on the stratigraphic signature of the Neogene. J. Geophys. Res. 96, 6753–6778.

Bartek, L.R., Andersen, J.L.R., Oneacre, T.A., 1997. Substrate control on distribution of subglacial and glaciomarine seismic facies based on stochastic models of glacial seismic facies deposition on the Ross Sea continental margin, Antarctica. Mar. Geol. 143, 223–262.

Bassetti, M.A., Berne, S., Jouet, G., Taviani, M., Dennielou, B., Flores, J.A., Gaillot, A., Gelfort, R., Lafuerza, S., Sultan, N., 2008. The 100-ka and rapid sea level changes recorded by prograding shelf sand bodies in the Gulf of Lions (Western Mediterranean Sea). Geochem. Geophys. Geosyst. 9, 1–27.

Bastia, R., Radhakrishna, M., Das, S., Kale, A.S., Catuneanu, O., 2010. Delineation of the 85°E ridge and its structure in the Mahanadi Offshore Basin, Eastern Continental Margin of India (ECMI), from seismic reflection imaging. Mar. & Pet. Geol. 27, 1841–1848.

Bates, R.L., Jackson, J.A. (Eds.), 1987. Glossary of Geology, third ed. American Geological Institute, Alexandria, Virginia, p. 788.

Beaubouef, R.T., Rossen, C., Zelt, F.B., Sullivan, M.D., Mohrig, D.C., Jennette, D.C., 1999. Deep-water Sandstones, Brushy Canyon Formation, West Texas. AAPG Continuing Education Course Note Series 40. American Association of Petroleum Geologists, Tulsa, Oklahoma, p. 48.

Beckvar, N., Kidwell, S.M., 1988. Hiatal shell concentrations, sequence analysis, and sealevel history of a Pleistocene coastal alluvial fan, Punta Chueca, Sonora. Lethaia 21, 257–270.

Beerbower, J.R., 1964. Cyclothems and cyclic depositional mechanisms in alluvial plain sedimentation. In: Merriam, D.F. (Ed.), Symposium on Cyclic Sedimentation. Kansas Geological Survey Bulletin, vol. 169, pp. 31–42.

Bekker, A., Slack, J.F., Planavsky, N., Krapez, B., Hofmann, A., Konhauser, K.O., Rouxel, O.J., 2010. Iron formation: the sedimentary product of a complex interplay among mantle, tectonic, oceanic, and biospheric processes. Econ. Geol. 105, 467–508.

Belknap, D.F., Kraft, J.C., 1981. Preservation potential of transgressive coastal lithosomes on the U.S. Atlantic shelf. Mar. Geol. 42, 429–442.

Belopolsky, A.V., Droxler, A.W., 2003. Imaging Tertiary carbonate system — the Maldives, Indian Ocean: insights into carbonate sequence interpretation. Lead. Edge (Soc. Explor. Geophys.) 22 (7), 646–652.

Berg, O.R., 1982. Seismic detection and evaluation of delta and turbidite sequences: their application to exploration for the subtle trap. Am. Assoc. Pet. Geol. Bull. 66, 1271–1288.

Berger, A.L., Loutre, M.F., 1994. Astronomical forcing through geological time. In: de Boer, P.L., Smith, D.G. (Eds.), Orbital Forcing and Cyclic Sequences, vol. 19. International Association of Sedimentologists, pp. 15–24. Special Publication.

Berggren, W.A., Kent, D.V., Aubry, M.P., Hardenbol, J., 1995. Geochronology, time scales and global stratigraphic correlation. Soc. Sediment. Geol. (SEPM) 54. https://doi.org/10.2110/pec.95.04.

Beukes, N.J., Cairncross, B., 1991. A lithostratigraphic-sedimentological reference profile for the late Archaean Mozaan Group, Pongola sequence: application to sequence stratigraphy and correlation with the Witwatersrand Supergroup. South Afr. J. Geol. 94, 44–69.

Bhandari, A., 2003. Ostracode bioevents in Tertiary beds of West Coast of India. Gondwana Geol. Mag. 6, 139–166.

Bhattacharya, J., 1988. Autocyclic and allocyclic sequences in river- and wave-dominated deltaic sediments of the Upper Cretaceous Dunvegan Formation, Alberta: core examples. In: James, D.P., Leckie, D.A. (Eds.), Sequences, Stratigraphy, Sedimentology: Surface and Subsurface, vol. 15. Canadian Society of Petroleum Geologists Memoir, pp. 25–32.

Bhattacharya, J., 1991. Regional to sub-regional facies architecture of river-dominated deltas, Upper Cretaceous Dunvegan Formation, Alberta subsurface. In: Miall, A.D., Tyler, N. (Eds.), The Three-Dimensional Facies Architecture of Terrigenous Clastic Sediments and its Implications for Hydrocarbon Discovery and Recovery, Society of Economic Paleontologists and Mineralogists, Concepts in Sedimentology and Paleontology, vol. 3, pp. 189–206.

Bhattacharya, J.P., 1993. The expression and interpretation of marine flooding surfaces and erosional surfaces in core; examples from the Upper Cretaceous Dunvegan Formation in the Alberta foreland basin. In: Summerhayes, C.P., Posamentier, H.W. (Eds.), Sequence Stratigraphy and Facies Associations, vol. 18. International Association of Sedimentologists, pp. 125–160. Special Publication.

Bhattacharya, J., Walker, R.G., 1991. Allostratigraphic subdivision of the Upper Cretaceous Dunvegan, Shaftesbury, and Kaskapau formations in the Northwestern Alberta subsurface. Bull. Can. Pet. Geol. 39, 145–164.

Bhattacharya, J.P., Walker, R.G., 1992. Deltas. In: Walker, R.G., James, N.P. (Eds.), Facies Models: Response to Sea Level Change, vol. 1. Geological Association of Canada, GeoText, pp. 157–178.

Birgenheier, L.P., Horton, B., McCauley, A.D., Johnson, C.L., Kennedy, A., 2017. A depositional model for offshore deposits of the lower Blue Gate Member, Mancos Shale, Uinta Basin, Utah, USA. Sedimentology 64 (5), 1402–1438.

Bjarnadóttir, L.R., Rüther, D.C., Winsborrow, M.C.M., Andreassen, K., 2013. Grounding-line dynamics during the last deglaciation of Kveithola, W Barents Sea, as revealed by seabed geomorphology and shallow seismic stratigraphy. Boreas 42, 84–107.

Bjarnadóttir, L.R., Winsborrow, M.C.M., Andreassen, K., 2014. Deglaciation of the central Barents Sea. Quat. Sci. Rev. 92, 208–226.

Blackwelder, E., 1909. The valuation of unconformities. J. Geol. 17, 289–299.

Blum, M.D., 1991. Systematic controls of genesis and architecture of alluvial sequences: a late Quaternary example. In: Leckie, D.A., Posamentier, H.W., Lovell, R.W. (Eds.), NUNA Conference on High-Resolution Sequence Stratigraphy, Program with Abstracts, pp. 7–8.

Blum, M.D., 1994. Genesis and architecture of incised valley fill sequences: a late Quaternary example from the Colorado River, Gulf Coastal Plain of Texas. In: Weimer, P., Posamentier, H.W. (Eds.), Siliciclastic Sequence Stratigraphy: Recent Developments and Applications, vol. 58. American Association of Petroleum Geologists Memoir, pp. 259–283.

Blum, M.D., Price, D.M., 1998. Quaternary alluvial plain construction in response to glacio-eustatic and climatic controls, Texas Gulf Coastal Plain. In: Shanley, K.W., McCabe, P.J. (Eds.), Relative Role of Eustasy, Climate, and Tectonism in Continental Rocks, vol. 59. SEPM, pp. 31–48. Special Publication.

Blum, M.D., Törnqvist, T.E., 2000. Fluvial responses to climate and sea-level change: a review and look forward. Sedimentology 47, 2–48.

Bohacs, K., Suter, J., 1997. Sequence stratigraphic distribution of coaly rocks: fundamental controls and paralic examples. Am. Assoc. Pet. Geol. Bull. 81 (10), 1612–1639.

Bohacs, K.M., Grabowski, G.J., Carroll, A.R., Mankiewicz, P.J., Gerhardt, K.J., Schwalbach, J.R., Wegner, M.B., Simo, J.A., 2005. Production, destruction, and dilution – the many paths to source rock development. In: Harris, N.B. (Ed.), The Deposition of Organic-Carbon-Rich Sediments: Models, Mechanisms, and Consequences, vol. 82. Society for Sedimentary Geology (SEPM), pp. 61–101. Special Publication.

Bohacs, K.M., Lazar, O.R., Demko, T.M., 2014. Parasequence types in shelfal mudstone strata – quantitative observations of lithofacies and stacking patterns, and conceptual link to modern depositional regimes. Geology 42 (2), 131–134.

Bolli, H.M., Saunders, J.B., 1985. Oligocene to Recent low latitude planktic foraminifera. In: Bolli, H.M., Saunders, J.B., Perch-Nielsen, K. (Eds.), Plankton Stratigraphy. Cambridge University Press, New York, pp. 155–262.

Bookman, R., Enzel, Y., Agnon, A., Stein, M., 2004. Late Holocene lake levels of the Dead Sea. Bull. Geol. Soc. Am. 116, 557–571.

Bookman, R., Bartov, Y., Enzel, Y., Stein, M., 2006. Quaternary lake levels in the Dead Sea basin: two centuries of research. Geol. Soc. Am. Special Pap. 401, 155.

Borcovsky, D., Egenhoff, S., Fishman, N., Maletz, J., Boehlke, A., Lowers, H., 2017. Sedimentology, facies architecture, and sequence stratigraphy of a Mississippian black mudstone succession—the upper member of the Bakken Formation, North Dakota, United States. Am. Assoc. Pet. Geol. Bull. 101 (10), 1625–1673.

Bordy, E.M., Catuneanu, O., 2001. Sedimentology of the upper Karoo fluvial strata in the Tuli Basin, South Africa. J. Afr. Earth Sci. 33/3–4, 605–629.

Bordy, E.M., Catuneanu, O., 2002a. Sedimentology and palaeontology of upper Karoo aeolian strata (Early Jurassic) in the Tuli Basin, South Africa. J. Afr. Earth Sci. 35/2, 301–314.

Bordy, E.M., Catuneanu, O., 2002b. Sedimentology of the lower Karoo fluvial strata in the Tuli Basin, South Africa. J. Afr. Earth Sci. 35/4, 503–521.

Bordy, E.M., Bumby, A.J., Catuneanu, O., Eriksson, P.G., 2004. Advanced early Jurassic termite (Insecta: Isoptera) nests: evidence from the Clarens formation in the Tuli Basin, Southern Africa. Palaios 19, 68–78.

Bordy, E.M., Bumby, A.J., Catuneanu, O., Eriksson, P.G., 2009. Possible trace fossils of putative termite origin in the Lower Jurassic (Karoo Supergroup) of South Africa and Lesotho. South Afr. J. Sci. 105, 356–362.

Boulton, G.S., 1990. Sedimentary and sea level changes during glacial cycles and their control on glacimarine facies architecture. In: Dowdeswell, J.A., Scourse, J.D. (Eds.), Glacimarine Environments: Processes and Sediments, vol. 53. Geological Society, London, pp. 15–52. Special Publication.

Bover-Arnal, T., Salas, R., Moreno-Bedmar, J.A., Bitzer, K., 2009. Sequence stratigraphy and architecture of a late Early-Middle Aptian carbonate platform succession from the Western Maestrat Basin (Iberian Chain, Spain). Sediment. Geol. 219, 280–301.

Bowen, D.W., Weimer, P.W., Scott, A.J., 1993. The relative success of sequence stratigraphic concepts in exploration: examples from incised valley fill and turbidite systems reservoirs. In: Weimer, P., Posamentier, H.W. (Eds.), Siliciclastic Sequence Stratigraphy – Recent Developments and Applications, vol. 58. American Association of Petroleum Geologists, Memoir, pp. 15–42.

Bown, T.M., Kraus, M.J., 1981. Lower Eocene alluvial paleosols (Willwood Formation, Northwest Wyoming, USA) and their significance for paleoecology, paleoclimatology, and basin analysis. Palaeogeogr. Palaeoclimatol. Palaeoecol. 34, 1–30.

Boyd, R., Williamson, P., Haq, B.U., 1993. Seismic stratigraphy and passive-margin evolution of the southern Exmouth Plateau. In: Posamentier, H.W., Summerhayes, C.P., Haq, B.U., Allen, G.P. (Eds.), Sequence Stratigraphy and Facies Associations, vol. 18. International Association of Sedimentologists, pp. 581–603. Special Publication.

Boyd, R., Diessel, C.F.K., Wadsworth, J., Chalmers, G., Little, M., Leckie, D., Zaitlin, B., 1999. Development of a nonmarine sequence stratigraphic model. In: American Association of Petroleum Geologists Annual Meeting. Official Program, San Antonio, Texas, USA, p. A15.

Boyd, R., Diessel, C.F.K., Wadsworth, J., Leckie, D., Zaitlin, B.A., 2000. Organization of non marine stratigraphy. In: Boyd, R., Diessel, C.F.K., Francis, S. (Eds.), Advances in the Study of the Sydney Basin. Proceedings of the 34th Newcastle Symposium. University of Newcastle, Callaghan, New South Wales, Australia, pp. 1–14.

Brackenridge, R., Stow, D.A.V., Hernández-Molina, F.J., 2011. Contourites within a deepwater sequence stratigraphic framework. Geo Mar. Lett. 31, 343–360.

Braga, J.C., Martin, J.M., Alcala, B., 1990. Coral reefs in coarse-terrigenous sedimentary environments, Upper Tortonian, Granada Basin, Southern Spain. Sediment. Geol. 66, 135–150.

Brandano, M., Civitelli, G., 2007. Non-seagrass meadow sedimentary facies of the Pontinian Islands, Tyrrhenian Sea: a modern example of mixed carbonate-siliciclastic sedimentation. Sediment. Geol. 201, 286–301.

Brandao, A., Vidigal-Souza, P., Holz, M., 2020. Evaporite occurrence and salt tectonics in the Cretaceous Camamu-Almada Basin, Northeastern Brazil. J. South Am. Earth Sci. 97 (102421), 1–12.

Breckenridge, J., Maravelis, A.G., Catuneanu, O., Ruming, K., Holmes, E., Collins, W.J., 2019. Outcrop analysis and facies model of an Upper Permian tidally-influenced fluvio-deltaic system: Northern Sydney Basin, Southeast Australia. Geol. Mag. 156, 1715–1741.

Brett, C.E., 1995. Sequence stratigraphy, biostratigraphy, and taphonomy in shallow marine environments. Palaios 10, 597–616.

Brett, C.E., 1998. Sequence stratigraphy, paleoecology, and evolution: biotic clues and responses to sea-level fluctuations. Palaios 13, 241–262.

Brett, C.E., Baird, G.C., 1986. Comparative taphonomy: a key to paleoenvironmental interpretation based on fossil preservation. Palaios 1, 207–227.

Breyer, J.A., 1995. Sedimentary facies in an incised valley in the Pennsylvanian of Beaver County, Oklahoma. J. Sediment. Res. B65, 338–347.

Bridge, J.S., Leeder, M.R., 1979. A simulation model of alluvial stratigraphy. Sedimentology 26, 617–644.

Bridge, J.S., Mackey, S., 1993a. A theoretical study of fluvial sandstone body dimensions. In: Flint, S., Bryant, I.D. (Eds.), Quantitative Description and Modelling of Clastic Hydrocarbon Reservoirs and Outcrop Analogues, vol. 15. International Association of Sedimentologists, pp. 213–236. Special Publication.

Bridge, J.S., Mackey, S., 1993b. A revised alluvial stratigraphy model. In: Marzo, M., Puide-Fabregas, C. (Eds.), Alluvial Sedimentology, vol. 17. International Association of Sedimentologists, pp. 101–114. Special Publication.

Brister, B.S., Stephens, W.C., Norman, G.A., 2002. Structure, stratigraphy, and hydrocarbon system of a Pennsylvanian pull-apart basin in North-central Texas. AAPG Bull. 86 (1), 1–20.

Bristow, C.S., Best, J.L., 1993. Braided rivers: perspectives and problems. In: Best, J.L., Bristow, C.S. (Eds.), Braided Rivers. Geological Society, pp. 1–11. Special Publication No. 75.

Bromley, R.G., 1975. Trace fossils at omission surfaces. In: Frey, R.W. (Ed.), The Study of Trace Fossils. Springer-Verlag, New York, pp. 399–428.

Bromley, R.G., Ekdale, A.A., 1984. Composite ichnofabrics and tiering of burrows. Geol. Mag. 123, 59–65.

Bromley, R.G., Pemberton, S.G., Rahmani, R.A., 1984. A Cretaceous woodground: the Teredolites ichnofacies. J. Paleontol. 58, 488–498.

Brookfield, M.E., 1977. The origin of the bounding surfaces in ancient eolian sandstones. Sedimentology 24, 303–332.

Brookfield, M.E., 1992. Eolian systems. In: Walker, R.G., James, N.P. (Eds.), Facies Models — Response to Sea Level Change. Geological Association of Canada, pp. 143–156.

Brown, A.R., 1991. Interpretation of 3-Dimensional Seismic Data, vol. 42. American Association of Petroleum Geologists Memoir, p. 341.

Brown Jr., L.F., Fisher, W.L., 1977. Seismic stratigraphic interpretation of depositional systems: examples from Brazilian rift and pull apart basins. In: Payton, C.E. (Ed.), Seismic Stratigraphy - Applications to Hydrocarbon Exploration, vol. 26. American Association of Petroleum Geologists Memoir, pp. 213–248.

Brown Jr., L.F., Benson, J.M., Brink, G.J., Doherty, S., Jollands, A., Jungslager, E.H.A., Keenan, J.H.G., Muntingh, A., van Wyk, N.J.S., 1995. Sequence Stratigraphy in Offshore South African Divergent Basins: an Atlas on Exploration for Cretaceous Lowstand Traps by Soekor (Pty) Ltd. American Association of Petroleum Geologists, Studies in Geology, Tulsa, Oklahoma, p. 184. #41.

Bruno, L., Bohacs, K.M., Campo, B., Drexler, T.M., Rossi, V., Sammartino, I., Scarponi, D., Hong, W., Amorosi, A., 2017. Early Holocene transgressive palaeogeography in the Po coastal plain (Northern Italy). Sedimentology 64, 1792–1816.

Bruun, P., 1962. Sea-level rise as a cause of shore erosion. Ame. Soc. Civil Eng. Proc. J. Waterways & Harb. Div. 88, 117–130.

Buatois, L.A., Echevarria, C., 2019. Ichnofabrics from a Cretaceous eolian system of Western Argentina: expanding the application of core ichnology to desert environments. Palaios 34, 190–211.

Buatois, L.A., Mangano, M.G., 1995. The paleoenvironmental and paleoecological significance of the lacustrine Mermia ichnofacies: an archetypical subaqueous nonmarine trace fossil assemblage. Ichnos 4, 151–161.

Buatois, L.A., Mangano, M.G., 1998. Trace fossil analysis of lacustrine facies and basins. Palaeogeogr. Palaeoclimatol. Palaeoecol. 140, 367–382.

Buatois, L.A., Mangano, M.G., 2002. Trace fossils from Carboniferous floodplain deposits in Western Argentina: implications for ichnofacies models of continental environments. Palaeogeogr. Palaeoclimatol. Palaeoecol. 183, 71–86.

Buatois, L.A., Mangano, M.G., 2004. Animal-substrate Interactions in Freshwater Environments: Applications of Ichnology in Facies and Sequence Stratigraphic Analysis of Fluvio-Lacustrine Successions, vol. 228. Geological Society, London, pp. 311–333. Special Publication.

Buatois, L.A., Mangano, M.G., 2009. Applications of ichnology in lacustrine sequence stratigraphy: potential and limitations. Palaeogeogr. Palaeoclimatol. Palaeoecol. 272, 127–142.

Buatois, L.A., Mangano, M.G., 2011. Ichnology: Organism-Substrate Interactions in Space and Time. Cambridge University Press, New York, p. 358.

Buatois, L.A., Mangano, M.G., Wu, X., Zhang, G., 1996. Trace fossils from Jurassic lacustrine turbidites of the Anyao formation (Central China) and their environmental and evolutionary significance. Ichnos 4, 287–303.

Buatois, L.A., Mangano, M.G., Alissa, A., Carr, T.R., 2002. Sequence stratigraphic and sedimentologic significance of biogenic structures from a late Paleozoic marginal to open-marine reservoir, Morrow Sandstone, subsurface of Southwest Kansas, USA. Sediment. Geol. 152, 99–132.

Buatois, L.A., Netto, R.G., Mangano, M.G., Balistieri, P.R.M.N., 2006. Extreme freshwater release during the late Paleozoic Gondwana deglaciation and its impact on coastal ecosystems. Geology 34, 1021–1024.

Buchs, D.M., Cukur, D., Masago, H., Garbe-Schonberg, D., 2015. Sediment flow routing during formation of forearc basins: constraints from integrated analysis of detrital pyroxenes and stratigraphy in the Kumano Basin, Japan. Earth & Planet. Sci. Lett. 414, 164–175.

Bumby, A.J., Eriksson, P.G., Catuneanu, O., Nelson, D.R., Rigby, M.J., 2012. Meso-Archaean and Palaeo-Proterozoic sedimentary sequence stratigraphy of the Kaapvaal Craton. Mar. & Pet. Geol. 33, 92–116.

Burchette, T.P., Wright, V.P., 1992. Carbonate ramp depositional systems. Sediment. Geol. 79, 3–57.

Burgess, P.M., Hovius, N., 1998. Rates of delta progradation during highstands; consequences for timing of deposition in deep-marine systems. J. Geol. Soc. 155, 217–222. London.

Busby, C.J., Ingersoll, R.V. (Eds.), 1995. Tectonics of Sedimentary Basins. Blackwell Science, p. 579.

Butcher, S.W., 1990. The nickpoint concept and its implications regarding onlap to the stratigraphic record. In: Cross, T.A. (Ed.), Quantitative Dynamic Stratigraphy. Prentice-Hall, Englewood Cliffs, pp. 375–385.

Campbell, C.V., 1967. Lamina, laminaset, bed and bedset. Sedimentology 8 (1), 7–26.

Cande, S.C., Kent, D.V., 1992. A new geomagnetic polarity time-scale for the Late Cretaceous and Cenozoic. J. Geophys. Res. 97, 13917–13951.

Cant, D., 1992. Subsurface facies analysis. In: Walker, R.G., James, N.P. (Eds.), Facies Models: Response to Sea Level Change, vol. 1. Geological Association of Canada, GeoText, pp. 27–45.

Cant, D., 2004. The Real Significance of Sequence Stratigraphy to Subsurface Geological Work, vol. 31. Reservoir, Canadian Society of Petroleum Geologists, p. 14.

Cantalamessa, G., Di Celma, C., 2004. Sequence response to syndepositional regional uplift: insights from high-resolution sequence stratigraphy of late Early Pleistocene strata, Periadriatic Basin, Central Italy. Sediment. Geol. 164, 283–309.

Cantalamessa, G., Di Celma, C., Ragaini, L., 2005. Sequence stratigraphy of the middle unit of the Jama Formation (Early Pleistocene, Ecuador): insights from integrated sedimentologic, taphonomic and paleoecologic analysis of molluscan shell concentrations. Palaeogeogr. Palaeoclimatol. & Palaeoecol. 216, 1–25.

Cantalamessa, G., Di Celma, C., Ragaini, L., Valleri, G., Landini, W., 2007. Sedimentology and high-resolution sequence stratigraphy of the late middle to late Miocene Angostura formation (Western Borbon Basin, Northwestern Ecuador). J. Geol. Soc. 164, 653–665. London.

Carter, R.M., Abbott, S.T., Fulthorpe, C.S., Haywick, D.W., Henderson, R.A., 1991. Application of global sea-level and sequence-stratigraphic models in southern hemisphere Neogene strata. In: Macdonald, D.I.M. (Ed.), Sedimentation, Tectonics and Eustasy: Sea-Level Changes at Active Margins, vol. 12. International Association of Sedimentologists, pp. 41–65. Special Publication.

Carter, R.M., Fulthorpe, C.S., Naish, T.R., 1998. Sequence concepts at seismic and outcrop scale: the distinction between physical and conceptual stratigraphic surfaces. Sediment. Geol. 122, 165–179.

Carvajal, C.R., Steel, R.J., 2006. Thick turbidite successions from supply-dominated shelves during sea-level highstand. Geology 34, 665–668.

Castelltort, S., Guillocheau, F., Robin, C., Rouby, D., Nalpas, T., Lafont, F., Eschard, R., 2003. Fold control on the stratigraphic record: a quantified sequence stratigraphic study of the Pico del Aguila anticline in the South-Western Pyrenees (Spain). Basin Res. 15, 527–551.

Cathro, D.L., Austin, J.A., Moss, G.D., 2003. Progradation along a deeply submerged Oligocene-Miocene heterozoan carbonate shelf: how sensitive are clinoforms to sea level variations? Am. Assoc. Pet. Geol. Bull. 87, 1547–1574.

Cattaneo, A., Steel, R.J., 2003. Transgressive deposits: a review of their variability. Earth-Sci. Rev. 62, 187–228.

Catuneanu, O., 2001. Flexural partitioning of the late Archaean Witwatersrand foreland system, South Africa. Sediment. Geol. 141–142, 95–112.

Catuneanu, O., 2002. Sequence stratigraphy of clastic systems: concepts, merits, and pitfalls. J. African Earth Sci. 35, 1–43.

Catuneanu, O., 2003. Sequence stratigraphy of clastic systems. Geol. Assoc. Can. Short Cour. Notes 16, 248.

Catuneanu, O., 2004a. Basement control on flexural profiles and the distribution of foreland facies: the Dwyka Group of the Karoo Basin, South Africa. Geology 32, 517–520.

Catuneanu, O., 2004b. Retroarc foreland systems – evolution through time. J. Afr. Earth Sci. 38, 225–242.

Catuneanu, O., 2006. Principles of Sequence Stratigraphy. Elsevier, Amsterdam, p. 375.

Catuneanu, O., 2007. Sequence stratigraphic context of microbial mat features. In: Schieber, J., Bose, P.K., Eriksson, P.G., Banerjee, S., Sarkar, S., Altermann, W., Catuneanu, O. (Eds.), Atlas of Microbial Mat Features Preserved within the Siliciclastic Rock Record, Atlases in Geosciences, vol. 2. Elsevier, Amsterdam, pp. 276–283.

Catuneanu, O., 2017. Sequence stratigraphy: guidelines for a standard methodology. In: Montenari, M. (Ed.), Stratigraphy and Timescales, vol. 2. Academic Press, UK, pp. 1–57.

Catuneanu, O., 2019a. Model-independent sequence stratigraphy. Earth-Sci. Rev. 188, 312–388.

Catuneanu, O., 2019b. Scale in sequence stratigraphy. Mar. & Pet. Geol. 106, 128–159.

Catuneanu, O., 2019c. First-order foreland cycles: interplay of flexural tectonics, dynamic loading, and sedimentation. J. Geodyn. 129, 290–298.

Catuneanu, O., 2020a. Sequence stratigraphy of deep-water systems. Sequence stratigraphy of deep-water systems. Mar. & Pet. Geol. 114, 1–13, 104238.

Catuneanu, O., 2020b. Sequence stratigraphy in the context of the 'modeling revolution'. Mar. & Pet. Geol. 116, 1–19, 104309.

Catuneanu, O., 2020c. Sequence stratigraphy. In: Scarselli, N., Adam, J., Chiarella, D., Roberts, D.G., Bally, A.W. (Eds.), Regional Geology and Tectonics, Principles of Geologic Analysis, second ed., vol. 1. Elsevier, pp. 605–686.

Catuneanu, O., Biddulph, M.N., 2001. Sequence stratigraphy of the Vaal Reef facies associations in the Witwatersrand foredeep, South Africa. Sediment. Geol. 141–142, 113–130.

Catuneanu, O., Bowker, D., 2001. Sequence stratigraphy of the Koonap and Middleton fluvial formations in the Karoo foredeep, South Africa. J. Afr. Earth Sci. 33, 579–595.

Catuneanu, O., Dave, A., 2017. Cenozoic sequence stratigraphy of the Kachchh Basin, India. Mar. & Pet. Geol. 86, 1106–1132.

Catuneanu, O., Elango, H.N., 2001. Tectonic control on fluvial styles: the Balfour formation of the Karoo Basin, South Africa. Sediment. Geol. 140, 291–313.

Catuneanu, O., Eriksson, P.G., 1999. The sequence stratigraphic concept and the Precambrian rock record: an example from the 2.7-2.1 Ga Transvaal Supergroup, Kaapvaal Craton. Precambrian Res. 97, 215–251.

Catuneanu, O., Eriksson, P.G., 2002. Sequence stratigraphy of the Precambrian Rooihoogte - Timeball Hill rift succession, Transvaal Basin, South Africa. Sediment. Geol. 147, 71–88.

Catuneanu, O., Eriksson, P.G., 2007. Sequence stratigraphy of the Precambrian. Gondwana Res. 12, 560–565.

Catuneanu, O., Sweet, A.R., 1999. Maastrichtian-Paleocene foreland basin stratigraphies, Western Canada: a reciprocal sequence architecture. Can. J. Earth Sci. 36, 685–703.

Catuneanu, O., Sweet, A.R., 2005. Fluvial sequence stratigraphy and sedimentology of the uppermost Cretaceous to Paleocene, Alberta Foredeep. In: American Association of Petroleum Geologists Annual Convention, Calgary, 23–25 June 2005, Field Trip Guidebook, p. 68.

Catuneanu, O., Zecchin, M., 2013. High-resolution sequence stratigraphy of clastic shelves II: controls on sequence development. Mar. & Pet. Geol. 39, 26–38.

Catuneanu, O., Zecchin, M., 2016. Unique vs. non-unique stratal geometries: relevance to sequence stratigraphy. Mar. & Pet. Geol. 78, 184–195.

Catuneanu, O., Zecchin, M., 2020. Parasequences: allostratigraphic misfits in sequence stratigraphy. Earth-Sci. Rev. 208, 1–17, 103289.

Catuneanu, O., Sweet, A.R., Miall, A.D., 1997a. Reciprocal architecture of Bearpaw T-R sequences, uppermost Cretaceous, Western Canada sedimentary basin. Bull. Can. Pet. Geol. 45 (1), 75–94.

Catuneanu, O., Beaumont, C., Waschbusch, P., 1997b. Interplay of static loads and subduction dynamics in foreland basins: reciprocal stratigraphies and the "missing" peripheral bulge. Geology 25 (12), 1087–1090.

Catuneanu, O., Willis, A.J., Miall, A.D., 1998a. Temporal significance of sequence boundaries. Sediment. Geol. 121, 157–178.

Catuneanu, O., Hancox, P.J., Rubidge, B.S., 1998b. Reciprocal flexural behaviour and contrasting stratigraphies: a new basin development model for the Karoo retroarc foreland system, South Africa. Basin Res. 10, 417–439.

Catuneanu, O., Sweet, A.R., Miall, A.D., 1999. Concept and styles of reciprocal stratigraphies: Western Canada foreland basin. Terra Nova 11, 1–8.

Catuneanu, O., Sweet, A.R., Miall, A.D., 2000. Reciprocal stratigraphy of the Campanian-Paleocene Western Interior of North America. Sediment. Geol. 134, 235–255.

Catuneanu, O., Hancox, P.J., Cairncross, B., Rubidge, B.S., 2002. Foredeep submarine fans and forebulge deltas: orogenic off-loading in the underfilled Karoo Basin. J. Afr. Earth Sci. 35 (4), 489–502.

Catuneanu, O., Heredia, E., Ortega, V., Robles, J., Toledo, C., Boll, L.P., Lopez, F., 2003. Seismic stratigraphy of the Neogene sequences in the Gulf of Mexico. In: American Association of Petroleum Geologists Annual Convention, vol. 12. Official Program, Salt Lake City, pp. A27–A28.

Catuneanu, O., Embry, A.F., Eriksson, P.G., 2004. Concepts of sequence stratigraphy. In: Eriksson, P.G., Altermann, W., Nelson, D., Mueller, W., Catuneanu, O. (Eds.), The Precambrian Earth: Tempos and Events, Developments in Precambrian Geology, vol. 12. Elsevier Science Ltd., Amsterdam, pp. 685–705.

Catuneanu, O., Martins-Neto, M., Eriksson, P.G., 2005. Precambrian sequence stratigraphy. Sediment. Geol. 176, 67–95.

Catuneanu, O., Khalifa, M.A., Wanas, H.A., 2006. Sequence stratigraphy of the lower Cenomanian Bahariya formation, Bahariya Oasis, Western Desert, Egypt. Sediment. Geol. 190, 121–137.

Catuneanu, O., Abreu, V., Bhattacharya, J.P., Blum, M.D., Dalrymple, R.W., Eriksson, P.G., Fielding, C.R., Fisher, W.L., Galloway, W.E., Gibling, M.R., Giles, K.A., Holbrook, J.M., Jordan, R., Kendall, C.G.S.C., Macurda, B., Martinsen, O.J., Miall, A.D., Neal, J.E., Nummedal, D., Pomar, L., Posamentier, H.W., Pratt, B.R., Sarg, J.F., Shanley, K.W., Steel, R.J., Strasser, A., Tucker, M.E., Winker, C., 2009. Towards the standardization of sequence stratigraphy. Earth-Sci. Rev. 92, 1–33.

Catuneanu, O., Bhattacharya, J.P., Blum, M.D., Dalrymple, R.W., Eriksson, P.G., Fielding, C.R., Fisher, W.L., Galloway, W.E., Gianolla, P., Gibling, M.R., Giles, K.A., Holbrook, J.M., Jordan, R., Kendall, C.G.S.C., Macurda, B., Martinsen, O.J., Miall, A.D., Nummedal, D., Posamentier, H.W., Pratt, B.R., Shanley, K.W.,

Steel, R.J., Strasser, A., Tucker, M.E., 2010. Sequence stratigraphy: common ground after three decades of development. First Break 28, 21–34.

Catuneanu, O., Galloway, W.E., Kendall, C.G.S.C., Miall, A.D., Posamentier, H.W., Strasser, A., Tucker, M.E., 2011. Sequence stratigraphy: methodology and nomenclature. Newslett. Stratigr. 44/3, 173–245.

Catuneanu, O., Martins-Neto, M.A., Eriksson, P.G., 2012. Sequence stratigraphic framework and application to the Precambrian. Mar. & Pet. Geol. 33, 26–33.

Caudill, M.R., Driese, S.G., Mora, C.I., 1997. Physical compaction of vertic palaeosols: implications for burial diagenesis and palaeoprecipitation estimates. Sedimentology 44, 673–685.

Cecil, C.B., 1990. Paleoclimate controls on stratigraphic repetition of chemical and siliciclastic rocks. Geology 18, 533–536.

Chan, M.A., Kocurek, G., 1988. Complexities in eolian and marine interactions: processes and eustatic controls on erg development. Sediment. Geol. 56, 283–300.

Chang, K.I.I., 1975. Unconformity-bounded stratigraphic units. Geol. Soc. Am. Bull. 86, 1544–1552.

Chaplin, J.R., 1996. Ichnology of transgressive-regressive surfaces in mixed carbonate-siliciclastic sequences, Early Permian Chase Group, Oklahoma. In: Witzke, B.J., Ludvigson, G.A., Day, J. (Eds.), Paleozoic Sequence Stratigraphy; Views from the North American Craton. Geological Society of America, Special Paper 306, pp. 399–418.

Christie-Blick, N., 1991. Onlap, offlap, and the origin of unconformity-bounded depositional sequences. Mar. Geol. 97, 35–56.

Christie-Blick, N., Driscoll, N.W., 1995. Sequence stratigraphy. Ann. Rev. Earth & Planet. Sci. 23, 451–478.

Christie-Blick, N., Grotzinger, J.P., Von der Borch, C.C., 1988. Sequence stratigraphy in Proterozoic successions. Geology 16, 100–104.

Christie-Blick, N., Mountain, G.S., Miller, K.G., 1990. Seismic stratigraphic record of sea-level change. In: Revelle, R. (Ed.), Sea-level Change, National Research Council, Studies in Geophysics. National Academy Press, Washington, pp. 116–140.

Clauzon, G., Suc, J.-P., Do Couto, D., Jouannic, G., Melinte-Dobrinescu, M.C., Jolivet, L., Quillevere, F., Lebret, N., Mocochain, L., Popescu, S.-P., Martinell, J., Domenech, R., Rubino, J.-L., Gumiaux, C., Warny, S., Bellas, S.M., Gorini, C., Bache, F., Rabineau, M., Estrada, F., 2015. New insights on the Sorbas Basin (SE Spain): the onshore reference of the Messinian Salinity Crisis. Mar. & Petrol. Geol. 66, 71–100.

Clemmensen, L.B., Hegner, J., 1991. Eolian sequences and erg dynamics: the Permian Corrie Sandstone, Scotland. J. Sediment. Petrol. 61, 768–774.

Clemmensen, L.B., Oxnevad, I.E.I., de Boer, P.L., 1994. Climatic controls on ancient desert sedimentation: some late Paleozoic and Mesozoic examples from NW Europe and the Western Interior of the USA. In: de Boer, P.L., Smith, D. (Eds.), Orbital Forcing and Cyclic Sequences, vol. 19. International Association of Sedimentologists, pp. 439–457. Special Publication.

Clift, P.D., 2006. Controls on the erosion of Cenozoic Asia and the flux of clastic sediment to the ocean. Earth & Planet. Sci. Lett. 241 (3–4), 571–580.

Clifton, H.E., 2006. A reexamination of facies models for clastic shorelines. In: Posamentier, H.W., Walker, R.G. (Eds.), Facies Models Revisited, vol. 84. SEPM, pp. 293–337. Special Publication.

Cloetingh, S., 1988. Intraplate stress: a new element in basin analysis. In: Kleinspehn, K., Paola, C. (Eds.), New Perspectives in basin Analysis. Springer-Verlag, New York, pp. 205–230.

Cloetingh, S., McQueen, H., Lambeck, K., 1985. On a tectonic mechanism for regional sea level variations. Earth & Planet. Sci. Lett. 75, 157–166.

Cloetingh, S., Kooi, H., Groenewoud, W., 1989. Intraplate stresses and sedimentary basin evolution. In: Price, R.A. (Ed.), Origin and

Evolution of Sedimentary Basins and Their Energy and Mineral Resources, vol. 48. American Geophysical Union, Geophysical Monographs, pp. 1—16.

Coffey, B.P., Read, J.F., 2007. Subtropical to temperate facies from a transition zone, mixed carbonate-siliciclastic system, Palaeogene, North Carolina, USA. Sedimentology 54, 339—365.

Cohen, C.R., 1982. Model for a passive to active continental margin transition: implications for hydrocarbon exploration. Amer. Assoc. Pet. Geol. Bull. 66, 708—718.

Collinson, J.D., 1986. Deserts. In: Reading, H.G. (Ed.), Sedimentary Environments and Facies, second ed. Blackwell Scientific Publications, Oxford, pp. 95—112.

Colombera, L., Mountney, N.P., 2020. On the geological significance of clastic parasequences. Earth-Sci. Rev. 201, 103062.

Coniglio, M., Dix, G.R., 1992. Carbonate slopes. In: Walker, R.G., James, N.P. (Eds.), Facies Models: Response to Sea Level Change, vol. 1. Geological Association of Canada, GeoText, pp. 349—373.

Coniglio, M., Myrow, P., White, T., 2000. Stable carbon and oxygen isotope evidence of Cretaceous sea-level fluctuations recorded in septarian concretions from Pueblo, Colorado, U.S.A. J. Sediment. Res. 70, 700—714.

Contreras, J., Zühlke, R., Bowman, S., Bechstädt, T., 2010. Seismic stratigraphy and subsidence analysis of the Southern Brazilian margin (Campos, Santos and Pelotas basins). Mar. & Pet. Geol. 27 (9), 1952—1980.

Corcoran, P.L., Mueller, W.U., 2002. The Effects of Weathering, Sorting and Source Composition in Archean High-Relief Basins: Examples from the Slave Province, Northwest Territories, Canada, vol. 33. International Association of Sedimentologists, pp. 183—211. Special Publication.

Corcoran, P.L., Mueller, W.U., 2004. Aggressive Archean weathering. In: Eriksson, P.G., Altermann, W., Nelson, D.R., Mueller, W.U., Catuneanu, O. (Eds.), The Precambrian Earth: Tempos and Events, Developments in Precambrian Geology, vol. 12. Elsevier, Amsterdam, pp. 494—511.

Cotter, E., Driese, S.G., 1998. Incised-valley fills and other evidence of sea-level fluctuations affecting deposition of the Catskill Formation (Upper Devonian), Appalachian foreland basin, Pennsylvania. J. Sediment. Res. 68, 347—361.

Covault, J.A., Graham, S.A., 2010. Submarine fans at all sea-level stands: tectonomorphologic and climatic controls on terrigenous sediment delivery to the deep sea. Geology 38, 939—942.

Covault, J.A., Normark, W.R., Romans, B.W., Graham, S.A., 2007. Highstand fans in the California borderland: the overlooked deep-water depositional systems. Geology 35, 783—786.

Crabaugh, M., Kocurek, G., 1993. Entrada Sandstone — example of a wet eolian system. In: Pye, K. (Ed.), The Dynamics and Environmental Context of Aeolian Sedimentary Systems, vol. 72. Geological Society, pp. 103—126. Special Publication.

Crampton, S.L., Allen, P.A., 1995. Recognition of forebulge unconformities associated with early stage foreland basin development: example from North Apline foreland basin. Am. Assoc. Pet. Geol. Bull. 79, 1495—1514.

Creaney, S., Passey, Q.R., 1993. Recurring patterns of total organic carbon and source rock quality within a sequence stratigraphic framework. AAPG Bull. 77, 386—401.

Cross, T.A., 1991. High-resolution stratigraphic correlation from the perspectives of base-level cycles and sediment accommodation. In: Dolson, J. (Ed.), Unconformity Related Hydrocarbon Exploration and Accumulation in Clastic and Carbonate Settings. Rocky Mountain Association of Geologists, Short Course Notes, pp. 28—41.

Csato, I., Catuneanu, O., 2012. Systems tract successions under variable climatic and tectonic regimes: a quantitative approach. Stratigraphy 9 (2), 109—130.

Csato, I., Catuneanu, O., 2014. Quantitative conditions for the development of systems tracts. Stratigraphy 11 (1), 39—59.

Csato, I., Granjeon, D., Catuneanu, O., Baum, G.R., 2013. A three-dimensional stratigraphic model for the Messinian crisis in the Pannonian Basin, eastern Hungary. Basin Res. 25, 121—148.

Csato, I., Catuneanu, O., Granjeon, D., 2014. Millennial-scale sequence stratigraphy: numerical simulation with Dionisos. J. Sediment. Res. 84, 394—406.

Csato, I., Toth, S., Catuneanu, O., Granjeon, D., 2015. A sequence stratigraphic model for the Upper Miocene-Pliocene basin fill of the Pannonian Basin, eastern Hungary. Mar. & Pet. Geol. 66, 117—134.

Csato, I., Homonnai, O., Zadravecz, C., Catuneanu, O., 2021. Lower Visean sea-level changes in the Northern Precaspian Basin. Mar. & Pet. Geol. 132, 105186.

Dalrymple, G.B., 1994. The Age of the Earth. Stanford University Press, Stanford, California, ISBN 9780804723312.

Dalrymple, R.W., 1999. Tide-dominated Deltas: Do They Exist or Are They All Estuaries? American Association of Petroleum Geologists Annual Convention, San Antonio, Official Program, p. A29.

Dalrymple, R.W., Zaitlin, B.A., Boyd, R., 1992. Estuarine facies models: conceptual basis and stratigraphic implications. J. Sediment. Petrol. 62, 1130—1146.

Dalrymple, R.W., Zaitlin, B.A., Boyd, R. (Eds.), 1994. Incised Valley Systems: Origin and Sedimentary Sequences, vol. 51. SEPM, p. 391. Special Publication.

D'Argenio, B., 2001. From megabanks to travertines - the independence of carbonate rock growth-forms from scale and organismal templates through time. Proc. Int. School Earth Planet. Sci. Siena 109—130.

Dashtgard, S.E., MacEachern, J.A., 2016. Unburrowed mudstones may record only slightly lowered oxygen conditions in warm, shallow basins. Geology 44 (5), 371—374.

Davies, P.J., Symonds, P.A., Feary, D.A., Pigram, C.J., 1989. The evolution of the carbonate platforms of Northeast Australia. In: Crevello, P.G., Wilson, J.J., Sarg, J.F., Read, J.F. (Eds.), Carbonate Platform and Basin Development, Soc. Econ. Petrol. Min., vol. 44. (Society for Sedimentary Geology) Spec. Publ., pp. 233—257

Davies, R.J., Posamentier, H.W., Wood, L.J., Cartwright, J.A. (Eds.), 2007. Seismic Geomorphology: Applications to Hydrocarbon Exploration and Production, vol. 277. Geological Society, London, p. 274. Special Publication.

Davies, S.J., Gibling, M.R., 2003. Architecture of coastal and alluvial deposits in an extensional basin: the Carboniferous Joggins Formation of Eastern Canada. Sedimentology 50, 415—439.

Davydov, V., Crowley, J.L., Schmitz, M.D., Poletaev, V., 2010. High-precision U-Pb zircon age calibration of the global Carboniferous time-scale and Milankovitch band cyclicity in the Donets Basin, Eastern Ukraine. Geochem. Geophys. Geosyst. 11 (online at A.G.U).

De Gasperi, A., Catuneanu, O., 2014. Sequence stratigraphy of the Eocene turbidite reservoirs in Albacora field, Campos Basin, offshore Brazil. Am. Assoc. Pet. Geol. (AAPG) Bull. 98 (2), 279—313.

De la Beche, H.T., 1830. Sections and Views Illustrative of Geological Phenomena. Treuttel & Wurtz, London, p. 177.

Demaison, G.J., Moore, G.T., 1980. Anoxic environments and oil source bed genesis. Am. Assoc. Pet. Geol. Bull. 64, 1179—1209.

Demarest, J.M., Kraft, J.C., 1987. Stratigraphic record of Quaternary sea levels: implications for more ancient strata. In: Nummedal, D., Pilkey, O.H., Howard, J.D. (Eds.), Sea Level Fluctuation and Coastal Evolution, vol. 41. SEPM, pp. 223—239. Special Publication.

Denton, G.H., 2000. Does an asymmetric thermohaline-ice-sheet oscillator drive 100 000-yr glacial cycles? J. Quat. Sci. 15, 301—318.

Devlin, W.J., Rudolph, K.W., Shaw, C.A., Ehman, K.D., 1993. The effect of tectonic and eustatic cycles on accommodation and sequence-stratigraphic framework in the Upper Cretaceous foreland basin of Southwestern Wyoming. In: Posamentier, H.W., Summerhayes, C.P., Haq, B.U., Allen, G.P. (Eds.), Sequence Stratigraphy and Facies Associations, vol. 18. International Association of Sedimentologists, pp. 501–520. Special Publication.

Di Celma, C., Cantalamessa, G., 2007. Sedimentology and high-frequency sequence stratigraphy of a forearc extensional basin: the Miocene Caleta Herradura Formation, Mejillones Peninsula, Northern Chile. Sediment. Geol. 198, 29–52.

Di Celma, C., Ragaini, L., Cantalamessa, G., Landini, W., 2005. Basin Physiography and Tectonic Influence on the Sequence Architecture and Stacking Pattern: Pleistocene Succession of the Canoa Basin (central Ecuador), vol. 117. Geological Society of America Bulletin, pp. 1226–1241.

Dickinson, W.R., 1995. Forearc basins. In: Busby, C.J., Ingersoll, R.V. (Eds.), Tectonics of Sedimentary Basins. Blackwell, Cambridge, Massachusetts, pp. 221–261.

Diessel, C., Boyd, R., Wadsworth, J., 2000. On balanced and unbalanced accommodation/peat accumulation ratios in the Cretaceous coals from the Gates Formation, Western Canada, and their sequence-stratigraphic significance. Int. J. Coal Geol. 43, 143–186.

Dominguez, J.M.L., Wanless, H.R., 1991. Facies architecture of a falling sea-level strandplain, Doce River coast, Brazil. In: Swift, D.J.P., Oertel, G.F., Tillman, R.W., Thorne, J.A. (Eds.), Shelf Sand and Sandstone Bodies: Geometry, Facies and Sequence Stratigraphy, vol. 14. International Association of Sedimentologists, pp. 259–281. Special Publication.

Dominguez, R.F., Catuneanu, O., 2017. Regional stratigraphic framework of the Vaca Muerta − Quintuco system in the Neuquén Embayment, Argentina. In: The 20th Geological Congress of Argentina, Tucuman, Argentina, 7–11 August 2017, pp. 1–10.

Dominguez, R.F., Continanzia, M.J., Mykietiuk, K., Ponce, C., Pérez, G., Guerello, R., Lanusse Noguera, I., Caneva, M., Di Benedetto, M., Catuneanu, O., Cristallini, E., 2016. Organic-rich stratigraphic units in the Vaca Muerta Formation, and their distribution and characterization in the Neuquén Basin (Argentina). In: Unconventional Resources Technology Conference. URTeC, pp. 1–12. https://doi.org/10.15530/urtec-2016-2456851.

Dominguez, R.F., Catuneanu, O., Reijenstein, H., Notta, R., Posamentier, H.W., 2020. Sequence stratigraphy and the three-dimensional distribution of organic-rich units. In: Minisini, D., Fantin, M., Lanusse Noguera, I., Leanza, H.A. (Eds.), Integrated Geology of Unconventionals: The Case of the Vaca Muerta Play, Argentina. AAPG Memoir 121, pp. 163–200.

Donaldson, W.S., Plint, A.G., Longstaffe, F.J., 1998. Basement tectonic control on distribution of the shallow marine Bad Heart Formation: Peace River Arch area, Northwest Alberta. Bull. Can. Pet. Geol. 46 (4), 576–598.

Donaldson, W.S., Plint, A.G., Longstaffe, F.J., 1999. Tectonic and eustatic control on deposition and preservation of Upper Cretaceous ooidal ironstone and associated facies; Peace River Arch area, NW Alberta, Canada. Sedimentology 46 (6), 1159–1182.

Dong, T., Harris, N.B., Ayranci, K., 2018. Relative sea-level cycles and organic matter accumulation in shales of the Middle and Upper Devonian Horn River Group, northeastern British Columbia, Canada: Insights into sediment flux, redox conditions, and bioproductivity. Geol. Soc. Am. Bull. 130 (5/6), 859–880.

Donovan, A.D., 1993. The use of sequence stratigraphy to gain new insights into stratigraphic relationships in the Upper Cretaceous of the US Gulf Coast. In: Posamentier, H.W., Summerhayes, C.P., Haq, B.U., Allen, G.P. (Eds.), Sequence Stratigraphy and Facies Associations, vol. 18. International Association of Sedimentologists, pp. 563–577. Special Publication.

Donovan, A.D., Gardner, R.D., Pramudito, A., Staerker, T.S., Wehner, M., Corbett, M.J., Lundquist, J.J., Romero, A.M., Henry, L.C., Rotzien, J.R., Boling, K.S., 2015. Chronostratigraphic relationships of the Woodbine and Eagle Ford groups across Texas. Gulf Coast Assoc. Geol. Soc. J. 4, 67–87.

D'Orbigny, A., 1842. Paleontologie Francaise. Description zoologique et geologique des tous les animaux mollusques et rayonnes fossiles de France, 2 (Gastropoda). Victor Masson, Paris, p. 456.

Dorsey, R.J., Umhoefer, P.J., Falk, P.D., 1997. Earthquake clustering inferred from Pliocene Gilbert-type fan deltas in the Loreto basin. Baja Calif. Sur Mexico Geol. 25, 679–682.

Droxler, A.W., Schlager, W., 1985. Glacial versus interglacial sedimentation rates and turbidite frequency in the Bahamas. Geology 13, 799–802.

Drummond, C.N., Wilkinson, B.H., 1996. Stratal thickness frequencies and the prevalence of orderedness in stratigraphic sequences. J. Geol. 104, 1–18.

Dunbar, G.B., Naish, T.R., Barrett, P.J., Fielding, C.R., Powell, R.D., 2008. Constraining the amplitude of late Oligocene bathymetric changes in Western Ross Sea during orbitally-induced oscillations in the East Antarctic Ice Sheet: (1) Implications for glacimarine sequence stratigraphic models. Palaeogeogr. Palaeoclimatol. Palaeoecol. 260, 50–65.

Duval, B., Cramez, C., Vail, P.R., 1998. Stratigraphic cycles and major marine source rocks. In: De Graciansky, P.C., Hardenbol, J., Jacquin, T., Vail, P.R. (Eds.), Mesozoic and Cenozoic Sequence Stratigraphy of European Basins, vol. 60. Society for Sedimentary Geology, pp. 43–51. Special Publication.

Dyke, A.S., Peltier, W.R., 2000. Forms, response times and variability of relative sea-level curves, glaciated North America. Geomorphology 32, 315–333.

Eberli, G.P., Kendall, C.G.S.C., Moore, P., Whittle, G.L., Cannon, R., 1994. Testing a seismic interpretation of Great Bahama Bank with a computer simulation. Am. Assoc. Pet. Geol. Bull. 78 (6), 981–1004.

Eberth, D.A., O'Connell, S., 1995. Note on changing paleoenvironments across the Cretaceous- Tertiary boundary (Scollard Fomation) in the Red Deer River valley of Southern Alberta. Bull. Can. Pet. Geol. 43 (1), 44–53.

Eco, U., 1980. The Name of the Rose. Harcourt, San Diego, California, p. 512.

Edwards, M.B., 1995. Differential subsidence and preservation potential of shallow-water Tertiary sequences, Northern Gulf Coast Basin, U.S.A. In: Plint, A.G. (Ed.), Sedimentary Facies Analysis, vol. 22. International Association of Sedimentologists, pp. 265–281. Special Publication.

Einsele, G., 2000. Sedimentary Basins: Evolution, Facies, and Sediment Budget, second ed. Springer-Verlag Berlin Heidelberg, p. 792.

Einsele, G., Ricken, W., Seilacher, A. (Eds.), 1991. Cycles and Events in Stratigraphy. Springer-Verlag, p. 955.

Ekdale, A.A., Bromley, R.G., Pemberton, S.G., 1984. Ichnology: the use of trace fossils in sedimentology and stratigraphy. Soc. Econ. Paleontol. & Mineral. 317. Short Course Notes Number 15.

Elliott, T., 1975. The sedimentary history of a delta lobe from a Yoredale (Carboniferous) cyclothem. Yorkshire Geol. Soc. Proc. 40, 505–536.

Embry, A.F., 1993. Transgressive-regressive (T-R) sequence analysis of the Jurassic succession of the Sverdrup Basin, Canadian Arctic Archipelago. Can. J. Earth Sci. 30, 301–320.

Embry, A.F., 1995. Sequence boundaries and sequence hierarchies: problems and proposals. In: Steel, R.J., Felt, V.L., Johannessen, E.P., Mathieu, C. (Eds.), Sequence Stratigraphy on the Northwest European Margin. Norwegian Petroleum Society (NPF), pp. 1–11. Special Publication 5.

Embry, A.F., 2001. The six surfaces of sequence stratigraphy. In: A.A.P.G. Hedberg Research Conference on "Sequence Stratigraphic and Allostratigraphic Principles and Concepts", Dallas, August 26–29, Program and Abstracts Volume, pp. 26–27.

Embry, A.F., Catuneanu, O., 2001. Practical Sequence Stratigraphy: Concepts and Applications. Canadian Society of Petroleum Geologists, short course notes, p. 167.

Embry, A.F., Catuneanu, O., 2002. Practical Sequence Stratigraphy: Concepts and Applications. Canadian Society of Petroleum Geologists, short course notes, p. 147.

Embry, A.F., Johannessen, E.P., 1992. T–R sequence stratigraphy, facies analysis and reservoir distribution in the uppermost Triassic-Lower Jurassic succession, Western Sverdrup Basin, Arctic Canada. In: Vorren, T.O., Bergsager, E., Dahl-Stamnes, O.A., Holter, E., Johansen, B., Lie, E., Lund, T.B. (Eds.), Arctic Geology and Petroleum Potential, vol. 2. Norwegian Petroleum Society (NPF), pp. 121–146. Special Publication.

Emery, D., Myers, K.J. (Eds.), 1996. Sequence Stratigraphy. Blackwell Science, p. 297.

Enge, H.D., Howell, J.A., Buckley, S.J., 2010. Quantifying clinothem geometry in a forced-regressive river-dominated delta, Panther Tongue Member, Utah, USA. Sedimentology 57, 1750–1770.

Enos, P., 1991. Sedimentary parameters for computer modeling. Kansas Geol. Surv. Bull. 233, 63–99.

Eriksson, K.A., Krapez, B., Fralick, P.W., 1994. Sedimentology of greenstone belts: signature of tectonic evolution. Earth Sci. Rev. 37, 1–88.

Eriksson, P.G., Catuneanu, O., 2004a. Third-order sequence stratigraphy in the Palaeoproterozoic Daspoort Formation (Pretoria Group, Transvaal Supergroup), Kaapvaal Craton. In: Eriksson, P.G., Altermann, W., Nelson, D., Mueller, W., Catuneanu, O. (Eds.), The Precambrian Earth: Tempos and Events, Developments in Precambrian Geology, vol. 12. Elsevier, Amsterdam, pp. 724–735.

Eriksson, P.G., Catuneanu, O., 2004b. A commentary on Precambrian plate tectonics. In: Eriksson, P.G., Altermann, W., Nelson, D., Mueller, W., Catuneanu, O. (Eds.), The Precambrian Earth: Tempos and Events, Developments in Precambrian Geology, vol. 12. Elsevier, Amsterdam, pp. 201–213.

Eriksson, P.G., Condie, K.C., Tirsgaard, H., Mueller, W.U., Altermann, W., Miall, A.D., Aspler, L.B., Thamm, A., Catuneanu, O., Chiarenzelli, J.R., 1998. Precambrian clastic sedimentation systems. Sediment. Geol. 120, 5–53.

Eriksson, P.G., Altermann, W., Catuneanu, O., van der Merwe, R., Bumby, A.J., 2001. Major influences on the evolution of the 2.67-2.1 Ga Transvaal basin, Kaapvaal Craton. Sediment. Geol. 141–142, 205–231.

The precambrian earth: tempos and events. In: Eriksson, P.G., Altermann, W., Nelson, D.R., Mueller, W.U., Catuneanu, O. (Eds.), 2004. Developments in Precambrian Geology, vol. 12. Elsevier, Amsterdam), p. 941.

Eriksson, P.G., Catuneanu, O., Nelson, D., Popa, M., 2005a. Controls on Precambrian sea-level change and sedimentary cyclicity. Sediment. Geol. 176 (Issues 1–2), 43–65.

Eriksson, P.G., Catuneanu, O., Els, G., Bumby, A.J., van Rooy, J.L., Popa, M., 2005b. Kaapvaal Craton: changing first- and second-order controls on sea level from c. 3 Ga to 2 Ga. Sediment. Geol. 176 (Issues 1–2), 121–148.

Eriksson, P.G., Mazumder, R., Catuneanu, O., Bumby, A.J., Ilondo, B.O., 2006. Precambrian continental freeboard and geological evolution: a time perspective. Earth-Sci. Rev. 79, 165–204.

Eriksson, P.R., Porada, H., Banerjee, S., Bouougri, E., Sarkar, S., Bumby, A.J., 2007. Mat-destruction features. In: Schieber, J., Bose, P.K., Eriksson, P.G., Banerjee, S., Sarkar, S., Altermann, W., Catuneanu, O. (Eds.), Atlas of Microbial Mat Features Preserved within the Siliciclastic Rock Record, Atlases in Geosciences, vol. 2. Elsevier, Amsterdam, pp. 76–105.

Eriksson, P.G., Long, D.G.F., Bumby, A.J., Eriksson, K.A., Simpson, E.L., Catuneanu, O., Claassen, M., Mtimkulu, M.N., Mudziri, K.T., Brumer, J.J., van der Neut, M., 2008. Palaeohydrological data from the c. 2.0 to 1.8 Ga Waterberg Group, South Africa: discussion of a possibly unique Palaeoproterozoic fluvial style. South Afr. J. Geol. 111, 281–304.

Eriksson, P.G., Sarkar, S., Samanta, P., Banerjee, S., Porada, H., Catuneanu, O., 2010. Paleoenvironmental context of microbial mat-related structures in siliciclastic rocks. In: Seckbach, J., Oren, A. (Eds.), Microbial Mats: Modern and Ancient Microorganisms in Stratified Systems. Springer, Berlin, pp. 73–108.

Eriksson, P.G., Bartman, R., Catuneanu, O., Mazumder, R., Lenhardt, N., 2012. A case study of microbial mat-related features in coastal epeiric sandstones from the Paleoproterozoic Pretoria Group (Transvaal Supergroup, Kaapvaal Craton, South Africa); The effect of preservation (reflecting sequence stratigraphic models) on the relationship between mat features and inferred paleoenvironment. Sediment. Geol. 263–264, 67–75.

Eriksson, P.G., Banerjee, S., Catuneanu, O., Corcoran, P.L., Eriksson, K.A., Hiatt, E.E., Laflamme, M., Lenhardt, N., Long, D.G.F., Miall, A.D., Mints, M.V., Pufahl, P.K., Sarkar, S., Simpson, E.L., Williams, G.E., 2013. Secular changes in sedimentation systems and sequence stratigraphy. Gondwana Res. 24, 468–489.

Eschard, R., Desaubliaux, G., Lecomte, J.C., van Buchem, F.S.P., Tveiten, B., 1993. High resolution sequence stratigraphy and reservoir prediction of the Brent Group (Tampen Spur area) using an outcrop analogue (Mesaverde Group, Colorado). In: Eschard, R., Doligez, B. (Eds.), Subsurface Reservoir Characterization from Outcrop Observations, Seventh Institute Francais du Petrole, Exploration and Production Research Conference, Scarborough, England, April 12–17, 1992, Proceedings, pp. 35–52.

Ethridge, F.G., Germanoski, D., Schumm, S.A., Wood, L.J., 2001. The morphologic and stratigraphic effects of base-level change: a review of experimental studies. In: Seventh International Conference on Fluvial Sedimentology, Lincoln, August 6–10, Program and Abstracts, p. 95.

Euzen, T., Joseph, P., du Fornel, E., Guillocheau, F., 2004. Three-dimensional stratigraphic modelling of the Gres d'Annot system, Eocene-Oligocene, SE France. Geol. Soc. London 221 (1), 161–180. Special Publication.

Fanti, F., Catuneanu, O., 2009. Stratigraphy of the upper Cretaceous Wapiti formation, West-central Alberta, Canada. Can. J. Earth Sci. 46, 263–286.

Fanti, F., Catuneanu, O., 2010. Fluvial sequence stratigraphy: the Wapiti formation, West-central Alberta, Canada. J. Sediment. Res. 80, 320–338.

Fastovsky, D.E., McSweeney, K., 1987. Paleosols spanning the Cretaceous-Paleogene transition, eastern Montana and Western North Dakota. Geol. Soc. of Am. Bull. 99, 66–77.

Feldman, H.R., Franseen, E.K., Joeckel, R.M., Heckel, P.H., 2005. Impact of longer-term modest climate shifts on architecture of high-frequency sequences (cyclothems), Pennsylvanian of Midcontinent U.S.A. J. Sediment. Res. 75, 350–368.

Fielding, C.R., Naish, T.R., Woolfe, K.J., Lavelle, M.A., 2000. Facies analysis and sequence stratigraphy of CRP-2/2A, Victoria Land Basin, Antarctica. Terra Antart. 7, 323–338.

Fielding, C.R., Naish, T.R., Woolfe, K.J., 2001. Facies architecture of the CRP-3 drillhole, Victoria Land Basin, Antarctica. Terra Antart. 8, 217–224.

Fielding, C.R., Bann, K.L., MacEachern, J.A., Tye, S.C., Jones, B.G., 2006. Cyclicity in the nearshore marine to coastal, Lower Permian, Pebbley Beach Formation, Southern Sydney Basin, Australia: a record of relative sea-level fluctuations at the close of the Late Palaeozoic Gondwanan ice age. Sedimentology 53, 435–463.

Fielding, C.R., Frank, T.D., Birgenheier, L.P., Rygel, M.C., Jones, A.T., Roberts, J., 2008. Stratigraphic record and facies associations of the late Paleozoic ice age in Eastern Australia (New South Wales and Queensland). In: Fielding, C.R., Frank, T.D., Isbell, J.L. (Eds.), Resolving the Late Paleozoic Ice Age in Time and Space, vol. 441. Geological Society of America Special Paper, pp. 41–57.

Fielding, C.R., Browne, G.H., Field, B., Florindo, F., Harwood, D.M., Krissek, L.A., Levy, R.H., Panter, K.S., Passchier, S., Pekar, S.F., 2011. Sequence stratigraphy of the ANDRILL AND-2A drillcore, Antarctica: a long-term, ice-proximal record of Early to Mid-Miocene climate, sea-level and glacial dynamism. Palaeogeogr. Palaeoclimatol. Palaeoecol. 305, 337—351.

Fildani, A., Hessler, A.M., Graham, S.A., 2008. Trench-forearc interactions reflected in the sedimentary fill of Talara basin, Northwest Peru. Basin Res. 20, 305—331.

Fillon, R.H., 2007. Biostratigraphy and condensed sections in deepwater settings. In: Weimer, P., Slatt, R. (Eds.), Introduction to the Petroleum Geology of Deepwater Settings, AAPG Studies in Geology 57 AAPG/Datapages Discovery Series 8, pp. 499—577.

Fischer, A.G., Bottjer, D.J., 1991. Orbital forcing and sedimentary sequences (introduction to special issue). J. Sediment. Petrol. 61, 1063—1069.

Fisher, W.L., McGowen, J.H., 1967. Depositional Systems in the Wilcox Group of Texas and Their Relationship to Occurrence of Oil and Gas, vol. 17. Gulf Coast Association Geological Society, Transactions, pp. 105—125.

Fisher, W.L., Galloway, W.E., Steel, R.J., Olariu, C., Kerans, C., Mohrig, D., 2021. Deep-water depositional systems supplied by shelf-incising submarine canyons: recognition and significance in the geological record. Earth-Sci. Rev. 214 (103531), 1—62.

Fleck, S., Michels, R., Ferry, S., Malartre, F., Elion, P., Landais, P., 2002. Organic geochemistry in a sequence stratigraphic framework: The siliciclastic shelf environment of Cretaceous series, SE France. Org. Geochem. 33, 1533—1557.

Folkestad, A., Satur, N., 2008. Regressive and transgressive cycles in a rift-basin: depositional model and sedimentary partitioning of the Middle Jurassic Hugin Formation, Southern Viking Graben, North Sea. Sediment. Geol. 207, 1—21.

Fonnesu, M., Palermo, D., Galbiati, M., Marchesini, M., Bonamini, E., Bendias, D., 2020. A new world-class deep-water play type, deposited by the syndepositional interaction of turbidity flows and bottom currents: the giant Eocene Coral Field in Northern Mozambique. Mar. & Pet. Geol. 111, 179—201.

Fouke, B.W., Farmer, J.D., Des Marais, D.J., Pratt, L., Sturchio, N.C., Burns, P.C., Discipulo, M.K., 2000. Depositional facies and aqueous-solid geochemistry of travertine-depositing hot springs (Angel Terrace, Mammoth Hot Springs, Yellowstone National Park, USA). J. Sediment. Res. 70, 265—285.

Frazier, D.E., 1974. Depositional Episodes: Their Relationship to the Quaternary Stratigraphic Framework in the Northwestern Portion of the Gulf Basin, vols. 71—1. University of Texas at Austin Bureau of Economic Geology Geological Circular, p. 28.

Frey, R.W., Pemberton, S.G., 1987. The Psilonichnus ichnocoenose and its relationship to adjacent marine and nonmarine ichnocoenoses along the Georgia coast. Bull. Can. Pet. Geol. 35, 333—357.

Frey, R.W., Pemberton, S.G., Fagerstrom, J.A., 1984. Morphological, ethological, and environmental significance of ichnogenera Scoyenia and Ancorichnus. J. Paleontol. 58, 511—528.

Frey, R.W., Howard, J.D., Hong, J.S., 1987. Prevalent lebensspuren on a modern macrotidal flat, Inchon, Korea; ethological and environmental significance. Palaios 2, 571—593.

Frey, R.W., Pemberton, S.G., Saunders, T.D.A., 1990. Ichnofacies and bathymetry: a passive relationship. J. Paleontol. 64, 155—158.

Frostick, L.E., Steel, R.J., 1993a. Tectonic signatures in sedimentary basin fills: an overview. In: Frostick, L.E., Steel, R.J. (Eds.), Tectonic Controls and Signatures in Sedimentary Successions, vol. 20. Special Publication of the International Association of Sedimentologists, pp. 1—9.

Frostick, L.E., Steel, R.J., 1993b. Sedimentation in divergent-plate basins. In: Frostick, L.E., Steel, R.J. (Eds.), Tectonic Controls and Signatures in Sedimentary Successions, vol. 20. Special Publication of the International Association of Sedimentologists, pp. 111—128.

Fursich, F.R., Mayr, H., 1981. Non-marine Rhizocorallium (trace fossils) from the Upper Freshwater Molasse (Upper Miocene) of Southern Germany. Neues Jahrbuch Fur Geologie und Palaontologie, Monatshefte 6, 321—333.

Galloway, W.E., 1989. Genetic stratigraphic sequences in basin analysis, I. Architecture and genesis of flooding-surface bounded depositional units. Am. Assoc. Pet. Geol. Bull. 73, 125—142.

Galloway, W.E., 2001. The many faces of submarine erosion: theory meets reality in selection of sequence boundaries. In: A.A.P.G. Hedberg Research Conference on "Sequence Stratigraphic and Allostratigraphic Principles and Concepts", Dallas, August 26—29, Program and Abstracts Volume, pp. 28—29.

Galloway, W.E., 2004. Accommodation and the sequence stratigraphic paradigm. Reserv. Can. Soc. Pet. Geol. 31 (5), 9—10.

Galloway, W.E., Hobday, D.K., 1996. Terrigenous Clastic Depositional Systems, second ed. Springer-Verlag, p. 489.

Galloway, W.E., Sylvia, D.A., 2002. The many faces of erosion: theory meets data in sequence stratigraphic analysis. In: Armentrout, J.M., Rosen, N.C. (Eds.), Sequence Stratigraphic Models for Exploration and Production: Evolving Methodology, Emerging Models and Application Histories, 22nd Annual Gulf Coast Section SEPM Foundation, Bob F. Perkins Research Conference, Conference Proceedings, Program and Abstracts, p. 8.

Gardner, M.H., Borer, J.M., Melick, J.J., Mavilla, N., Dechesne, M., Wagerle, R.N., 2003. Stratigraphic process-response model for submarine channels and related features from studies of Permian Brushy Canyon outcrops, West Texas. Mar. & Pet. Geol. 20, 757—787.

Gardner, M., Borer, J.M., Romans, B., Baptista, N., Kling, E.K., Hanggoro, D., Melick, J., Wagerle, R., Dechesne, M., Carr, M., Amerman, R., Atan, S., 2009. Stratigraphic models for deep-water sedimentary systems. In: Schofield, K., Rosen, N.C., Pfeiffer, D., Johnson, S. (Eds.), Answering the Challenges of Production from Deep-Water Reservoirs: Analogues and Case Histories to Aid a New Generation, 28th Annual Gulf Coast Section SEPM Foundation, Bob F. Perkins Research Conference, 2008, pp. 77—175.

Gastaldo, R.A., Demko, T.M., 2011. The relationship between continental landscape evolution and the plant-fossil record: long term hydrologic controls on preservation. In: Allison, P.A., Bottjer, D.J. (Eds.), Taphonomy: Process and Bias through Time. Springer, Dordrecht, pp. 249—285.

Gastaldo, R.A., Demko, T.M., Liu, Y., 1993. Application of sequence and genetic stratigraphic concepts to carboniferous coal-bearing strata: an example from the Black Warrior Basin, U.S.A. Geol. Rundschau 82, 212—226.

Gawthorpe, R.L., Fraser, A.J., Collier, R.E.L., 1994. Sequence stratigraphy in active extensional basins: implications for the interpretation of ancient basin-fills. Mar. & Pet. Geol. 11, 642—658.

Gawthorpe, R.L., Sharp, I., Underhill, J.R., Gupta, S., 1997. Linked sequence stratigraphic and structural evolution of propagating normal faults. Geology 25, 795—798.

Gawthorpe, R.L., Hall, M., Sharp, I., Dreyer, T., 2000. Tectonically enhanced forced regressions: examples from growth folds in extensional and compressional settings, the Miocene of the Suez rift and the Eocene of the Pyrenees. In: Hunt, D., Gawthorpe, R.L. (Eds.), Sedimentary Responses to Forced Regressions, vol. 172. Geological Society, pp. 177—191. Special Publication.

Gawthorpe, R.L., Hardy, S., Ritchie, B., 2003. Numerical modeling of depositional sequences in half-graben rift basins. Sedimentology 50, 169—185.

Genise, J.F., 2017. Ichnoentomology — Insect Traces in Soils and Paleosols. Springer, p. 695.

Genise, J.F., Bown, T.M., 1994. New trace fossils of termites (Insecta: Isoptera) from the late Eocene-early Miocene of Egypt, and the reconstruction of ancient isopteran social behavior. Ichnos 3, 155—183.

Genise, J.F., Mangano, M.G., Buatois, L.A., Laza, J.H., Verde, M., 2000. Insect trace fossil associations in paleosols: the *Coprinisphaera* ichnofacies. Palaios 15, 49–64.

Genise, J.F., Melchor, R.N., Bellosi, E.S., Verde, M., 2010. Chapter 7: invertebrate and vertebrate trace fossils from continental carbonates. In: Alonso-Zarza, A.M., Tanner, L.H. (Eds.), Carbonates in Continental Settings: Facies, Environments, and Processes, Developments in Sedimentology, vol. 61. Elsevier, pp. 319–369.

Genise, J.F., Bedatou, E., Bellosi, E.S., Sarzetti, L.C., Sánchez, M.V., Krause, J.M., 2016. The phanerozoic four revolutions and evolution of paleosol ichnofacies. In: Mángano, M.G., Buatois, L.A. (Eds.), The Trace-Fossil Record of Major Evolutionary Events, Topics in Geobiology, vol. 40. Springer, Dordrecht, pp. 301–370.

Ghibaudo, G., Grandesso, P., Massari, F., Uchman, A., 1996. Use of trace fossils in delineating sequence stratigraphic surfaces (Tertiary Venetian Basin, Northeastern Italy). Palaeogeogr. Palaeoclimatol. Palaeoecol. 120, 261–279.

Gibling, M.R., Bird, D.J., 1994. Late Carboniferous cyclothems and alluvial paleovalleys in the Sydney Basin, Nova Scotia. Geol. Soc. Am. Bull. 106, 105–117.

Gibling, M.R., Wightman, W.G., 1994. Palaeovalleys and protozoan assemblages in a Late Carboniferous cyclothem, Sydney Basin, Nova Scotia. Sedimentology 41, 699–719.

Gibling, M.R., Tandon, S.K., Sinha, R., Jain, M., 2005. Discontinuity-bounded alluvial sequences of the Southern Gangetic Plains, India: aggradation and degradation in response to monsoonal strength. J. Sediment. Res. 75 (3), 369–385.

Gilbert, G.K., 1895. Sedimentary measurement of geologic time. J. Geol. 3, 121–127.

Giles, K.A., Bocko, M., Lawton, T.F., 1999. Stacked Late Devonian lowstand shorelines and their relation to tectonic subsidence at the Cordilleran hingeline, Western Utah. J. Sediment. Res. 69, 1181–1190.

Gill, J.R., Cobban, W.A., 1973. Stratigraphy and Geologic History of the Montana Group and Equivalent Rocks, Montana, Wyoming, and North and South Dakota: United States Geological Survey. Professional Paper 776, p. 73.

Gingras, M.K., MacEachern, J.A., Pickerill, R.K., 2004. Modern perspectives on the *Teredolites* ichnofacies: observations from Willapa Bay, Washington. Palaios 19, 79–88.

Gladenkov, A.Y., 2007. Late Cenozoic Detailed Stratigraphy and marine Ecosystems of the North Pacific Region (Based on Diatoms). GEOS, Moscow.

Gladenkov, Y.B., 2010. Zonal biostratigraphy in the solution of the fundamental and applied problems of geology. Stratigr. & Geol. Correl. 18 (6), 660–673.

Gleick, J., 1992. Genius: The Life and Science if Richard Feynman. Pantheon Books, New York, p. 489.

Glumac, B., Walker, K.R., 2000. Carbonate deposition and sequence stratigraphy of the terminal Cambrian grand cycle in the Southern Appalachians. J. Sed. Res. 70, 952–963.

Grabau, A.W., 1905. Physical characters and history of some New York formations. Science 22, 528–535.

Grabau, A.W., 1906. Types of sedimentary overlap. Geol. Soc. Am. Bull. 17, 567–636.

Gradstein, F., Ogg, J., Schmitz, M., Ogg, G. (Eds.), 2012. The Geologic Time Scale 2012. Elsevier B.V., ISBN 978-0-444-59425-9

Gressly, A., 1838. Observations geologiques sur le Jura Soleurois. Nouv. Mem. Soc. Helv. Sci. Nat. 2, 1–112.

Grotzinger, J.P., Knoll, A.H., 1999. Stromatolites: evolutionary mileposts or environmental dipsticks? Ann. Rev. Earth & Planet. Sci. 27, 313–358.

Gurbuz, A., 2010. Geometric characteristics of pull-apart basins. Lithosphere 2, 199–206.

Gutierrez Paredes, H.C., Catuneanu, O., Romano, U.H., 2017. Sequence stratigraphy of the Miocene section, Southern Gulf of Mexico. Mar. & Pet. Geol. 86, 711–732.

Gutierrez Paredes, H.C., Peterson Rodriguez, R.H., Catuneanu, O., Romano, U.H., 2018. Tectonic influence on the morphology, facies and distribution of Miocene reservoirs, Southern Gulf of Mexico. J. South Am. Earth Sci. 88, 399–414.

Gutzmer, J., Beukes, N.J., 1998. Earliest laterites and possible evidence for terrestrial vegetation in the Early Proterozoic. Geology 26, 263–266.

Hajek, E.A., Heller, P.L., Sheets, B.A., 2010. Significance of channel-belt clustering in alluvial basins. Geology 38, 535–538.

Hamblin, A.P., 1997. Regional distribution and dispersal of the Dinosaur Park Formation, Belly River Group, surface and subsurface of Southern Alberta. Bull. Can. Pet. Geol. 45 (3), 377–399.

Hamilton, D.S., Tadros, N.Z., 1994. Utility of coal seams as genetic stratigraphic sequence boundaries in non-marine basins: an example from the Gunnedah basin, Australia. Am. Assoc. Pet. Geol. Bull. 78, 267–286.

Hampson, G.J., 2000. Discontinuity surfaces, clinoforms, and facies architecture in a wave-dominated, shoreface-shelf parasequence. J. Sediment. Res. 70, 325–340.

Hampson, G.J., Premwichein, K., 2017. Sedimentological character of ancient muddy subaqueous-deltaic clinoforms: down cliff clay member, Bridgeport sand formation, Wessex Basin, UK. J. Sediment. Res. 87, 951–966.

Hampson, G.J., Storms, J.E.A., 2003. Geomorphological and sequence stratigraphic variability in wave-dominated, shoreface-shelf parasequences. Sedimentology 50, 667–701.

Hampson, G.J., Rodriguez, A.B., Storms, J.E.A., Johnson, H.D., Meyer, C.T., 2008. Geomorphology and high-resolution stratigraphy of progradational wave-dominated shoreline deposits: impact on reservoir-scale facies architecture. In: Hampson, G.J., Steel, R.J., Burgess, P.M., Dalrymple, R.W. (Eds.), Recent Advances in Models of Siliciclastic Shallow-Marine Stratigraphy, vol. 90. SEPM, pp. 117–142. Special Publication.

Hampson, G.J., Gani, M.R., Sharman, K.E., Irfan, N., Bracken, B., 2011. Along-strike and down-dip variations in shallow-marine sequence stratigraphic architecture: Upper Cretaceous Star Point Sandstone, Wasatch Plateau, Central Utah, U.S.A. J. Sediment. Res. 81, 159–184.

Haq, B.U., Hardenbol, J., Vail, P.R., 1987. Chronology of fluctuating sea levels since the Triassic (250 million years ago to present). Science 235, 1156–1166.

Haq, B.U., Hardenbol, J., Vail, P.R., 1988. Mesozoic and Cenozoic chronostratigraphy and cycles of sea-level change. In: Wilgus, C.K., Hastings, B.S., C, G., Kendall, S.C., Posamentier, H.W., Ross, C.A., Van Wagoner, J.C. (Eds.), Sea Level Changes — an Integrated Approach, vol. 42. SEPM, pp. 71–108. Special Publication.

Harms, J.C., Fahnestock, R.K., 1965. Stratification, bed forms, and flow phenomena (with an example from the Rio Grande) In: Middleton, G.V. (Ed.), Primary Sedimentary Structures and Their Hydrodynamic Interpretation, vol. 12. Society of Economic Paleontologists and Mineralogists, pp. 84–115. Special Publication.

Harris, N.B., Freeman, K.H., Pancost, R.D., White, T.S., Mitchell, G.D., 2004. The character and origin of lacustrine source rocks in the Lower Cretaceous synrift section, Congo Basin, West Africa. Am. Assoc. Pet. Geol. Bull. 88, 1163–1184.

Harris, N.B., Mnich, C.A., Selby, D., Korn, D., 2013. Minor and trace element and Re-Os chemistry of the Upper Devonian Woodford Shale, Permian Basin, West Texas: insights into metal abundance and basin processes. Chem. Geol. 356, 76–93.

Harris, N.B., McMillan, J.M., Knapp, L.V., Mastalerz, M., 2018. Organic matter accumulation in the Upper Devonian Duvernay Formation, Western Canada Sedimentary Basin, from sequence stratigraphic analysis and geochemical proxies. Sediment. Geol. 376, 185–203.

Hart, B.S., 2000. 3-D seismic interpretation: a primer for geologists. SEPM (Soc. Sediment. Geol.) 123. Short Course No. 48.

Hart, B.S., 2015. The Greenhorn cyclothem of the Cretaceous Western Interior Seaway: lithology trends, stacking patterns, log signatures, and application to the Eagle Ford of West Texas. Gulf Coast Assoc. Geol. Soc. Transac. 65, 155–174.

Hart, B.S., Plint, A.G., 1993. Origin of an erosion surface in shoreface sandstones of the Kakwa Member (Upper Cretaceous Cardium Formation, Canada): importance for reconstruction of stratal geometry and depositional history. In: Posamentier, H.W., Summerhayes, C.P., Haq, B.U., Allen, G.P. (Eds.), Sequence Stratigraphy and Facies Associations, vol. 18. International Association of Sedimentologists, pp. 451–467. Special Publication.

Hasiotis, S.T., Chan, M.A., Totman Parrish, J., 2021. Defining bounding surfaces within and between eolian and non-eolian deposits, Lower Jurassic Navajo Sandstone, Moab area, Utah: implications for subdividing erg system strata. J. Sediment. Res. 91, 1275–1304.

Havholm, K.G., Kocurek, G., 1994. Factors controlling eolian sequence stratigraphy: clues from super bounding surfaces features in the Middle Jurassic Page Sandstone. Sedimentology 41, 913–934.

Hay, W.W., 1995. Paleoceanography of marine organic-carbon-rich sediments. In: Huc, A.Y. (Ed.), Paleogeography, Paleoclimate, and Source Rocks, Petroleum Geologists, Studies in Geology, vol. 40. American Association or, pp. 21–59.

Hayward, B.W., 1976. Lower Miocene bathyal and submarine canyon ichnocoenoses from Northland, New Zealand. Lethaia 9, 149–162.

Heckel, P.H., 1986. Sea-level curve for Pennsylvanian eustatic marine transgressive-regressive depositional cycles along midcontinent outcrop belt, North America. Geology 14, 330–334.

Heckel, P.H., Gibling, M.R., King, N.R., 1998. Stratigraphic model for glacial-eustatic Pennsylvanian cyclothems in highstand nearshore detrital regimes. J. Geol. 106, 373–383.

Hedberg, H.D., 1951. Nature of Time-Stratigraphic Units and Geologic Time Units, 76–1c. American Association of Petroleum Geologists Bulletin, pp. 191–193. Report of Activities.

Hedberg, H.D., 1976. International stratigraphic guide: A guide to stratigraphic classification, terminology, and procedure. In: International Union of Geological Sciences, Commission on Stratigraphy, International Subcommission on Stratigraphic Classification. Wiley, New York, p. 200.

Heezen, B.C., Tharp, M., Ewing, M., 1959. The Floors of the Oceans, Vol. 1: The North Atlantic. The Geological Society of America Special Paper, No. 65, 122 pp.

Helenes, J., De Guerra, C., Vasquez, J., 1998. Palynology and chronostratigraphy of the Upper Cretaceous in the subsurface of the Barinas area, Western Venezuela. A.A.P.G. Bull. 82, 757–772.

Helland-Hansen, W., Gjelberg, J.G., 1994. Conceptual basis and variability in sequence stratigraphy: a different perspective. Sediment. Geol. 92, 31–52.

Helland-Hansen, W., Hampson, G.J., 2009. Trajectory analysis: concepts and applications. Basin Res. 21, 454–483.

Helland-Hansen, W., Martinsen, O.J., 1996. Shoreline trajectories and sequences: description of variable depositional-dip scenarios. J. Sediment. Res. 66 (4), 670–688.

Helland-Hansen, W., Lomo, L., Steel, R., Ashton, M., 1992. Advance and retreat of the Brent delta: recent contributions of the depositional model. In: Morton, A.C., Haszeldine, R.S., Giles, M.R., Brown, S. (Eds.), Geology of the Brent Group. Geological Society of London, pp. 109–127. Special Publication No 61.

Hempton, M.R., Dunne, L.A., 1984. Sedimentation in pull-apart basins: active examples in eastern Turkey. J. Geol. 92, 513–530.

Hernandez, R.M., Gomez Omil, R., Boll, A., July 2008. Estratigrafia, tectonica y potencial petrolero del rift Cretacico en la provincia de Jujuy. Relatorio del XVII Congreso Geologico Argentino, pp. 207–232.

Hernandez-Molina, F.J., 1993. Dinamica sedimentaria y evolucion durante el Pleistoceno terminal – Holoceno del Margen Noroccidental del Mar de alboran: Modelo de estratigrafia secuencial de muy alta resolution en plataformas continentales. Ph.D. Thesis. University of Granada, Spain, p. 618.

Hessler, A.M., Sharman, G.R., 2018. Subduction zones and their hydrocarbon systems. Geosphere 14, 2044–2067.

Hine, A.C., Wilber, R.J., Bane, J.M., Neumann, A.C., Lorenson, K.R., 1981. Offbank transport of carbonate sands along leeward bank margins, Northern Bahamas. Mar. Geol. 42, 327–348.

Hine, A.C., Locker, S.D., Tedesco, L.P., Mullins, H.T., Hallock, P., Belknap, D.F., Gonzales, J.L., Neumann, A.C., Snyder, S.W., 1992. Megabreccia shedding from modern low-relief carbonate platforms, Nicaraguan Rise. Geol. Soc. Am. Bull. 104, 928–943.

Hines, B.R., Gazley, M.F., Collins, K.S., Bland, K.J., Crampton, J.S., Ventura, G.T., 2019. Chemostratigraphic resolution of widespread reducing conditions in the Southwest Pacific Ocean during the Late Paleocene. Chem. Geol. 504, 236–252.

Hofmann, M.H., Wroblewski, A., Boyd, R., 2011. Mechanisms controlling the clustering of fluvial channels and the compensational stacking of cluster belts. J. Sediment. Res. 81, 670–685.

Holbrook, J.M., 1996. Complex fluvial response to low gradients at maximum regression: a genetic link between smooth sequence-boundary morphology and architecture of overlying sheet sandstone. J. Sediment. Res. 66 (4), 713–722.

Holbrook, J.M., Bhattacharya, J.P., 2012. Reappraisal of the sequence boundary in time and space: case and considerations for a SU (subaerial unconformity) that is not a sediment bypass surface, a time barrier, or an unconformity. Earth-Sci. Rev. 113, 271–302.

Holbrook, J.M., Scott, R.W., Oboh-Ikuenobe, F.E., 2006. Base-level buffers and buttresses: a model for upstream versus downstream control on fluvial geometry and architecture within sequences. J. Sediment. Res. 76, 162–174.

Holland, S.M., 1995. The stratigraphic distribution of fossils. Paleobiology 21, 92–109.

Holland, S.M., 2000. The quality of the fossil record—a sequence stratigraphic perspective. In: Erwin, D.H., Wing, S.L. (Eds.), Deep Time: Paleobiology's Perspective. The Paleontological Society, Lawrence, Kansas, pp. 148–168.

Holland, S.M., 2020. The stratigraphy of mass extinctions and recoveries. Ann. Rev. Earth & Planet. Sci. 48, 75–97.

Holland, S.M., Loughney, K.M., 2020. The stratigraphic paleobiology of nonmarine systems. In: Sumrall, C.D. (Ed.), Elements of Paleontology. Cambridge University Press, pp. 1–79.

Holland, S.M., Patzkowsky, M.E., 1999. Models for simulating the fossil record. Geology 27, 491–494.

Holland, S.M., Patzkowsky, M.E., 2002. Stratigraphic variation in the timing of first and last occurrences. Palaios 17, 134–146.

Holland, S.M., Patzkowsky, M.E., 2015. The stratigraphy of mass extinction. Palaeontology 58, 903–924.

Holland, S.M., Miller, A.I., Meyer, D.L., Dattilo, B.F., 2001. The detection and importance of subtle biofacies within a single lithofacies: the Upper Ordovician Kope Formation of the Cincinnati, Ohio region. Palaios 16, 205–217.

Holz, M., Kalkreuth, W., Banerjee, I., 2002. Sequence stratigraphy of paralic coal-bearing strata: an overview. Int. J. Coal Geol. 48, 147–179.

Genetic stratigraphy on the exploration and production scales: case studies from the Upper Devonian of Alberta and the Pennsylvanian of the Paradox Basin. In: Homewood, P.W., Eberli, G.P. (Eds.), Elf Explor. Prod. Ed. Memoirs 24, 290.

Homewood, P.W., Guillocheau, F., Eschard, R., Cross, T.A., 1992. Correlations haute resolution et stratigraphie genetique: une demarche integree. Bulletin des Centres de Recherches Explor. Prod. Elf-Aquitaine 16, 357–381.

Howell, J.A., Flint, S.S., 1996. A model for high resolution sequence stratigraphy within extensional basins. In: Howell, J.A., Aitken, J.F. (Eds.), High Resolution Sequence Stratigraphy: Innovations and Applications, vol. 104. Geological Society, pp. 129–137. Special Publication.

Hubbard, S.M., MacEachern, J.A., Bann, K.L., 2012. Slopes. In: Knaust, D., Bromley, R.G. (Eds.), Trace Fossils as Indicators of Sedimentary Environments, vol. 64. Elsevier, Developments in Sedimentology, pp. 607–642.

Sedimentary responses to forced regressions. In: Hunt, D., Gawthorpe, R.L. (Eds.), Geol. Soc. London Spec. Publ. 172, 383.

Hunt, D., Tucker, M.E., 1992. Stranded parasequences and the forced regressive wedge systems tract: deposition during base-level fall. Sediment. Geol. 81, 1–9.

Hunt, D., Tucker, M.E., 1995. Stranded parasequences and the forced regressive wedge systems tract: deposition during base-level fall – reply. Sediment. Geol. 95, 147–160.

Husinec, A., 2016. Sequence stratigraphy of the Red River Formation, Williston Basin, USA: Stratigraphic signature of the Ordovician Katian greenhouse to icehouse transition. Mar. & Petrol. Geol. 77, 487–506.

Hutton, J., 1788. Theory of the Earth: Or an Investigation of the Laws Observable in the Composition, Dissolution, and Restoration of Land upon the globe, vol. 1. Trans. Royal Society, Edinburgh, pp. 209–304.

Hutton, J., 1795. Theory of the Earth with Proofs and Illustrations, vols. I and II (Edinburgh).

Ichaso, A.A., Dalrymple, R.W., Martinius, A.W., 2016. Basin analysis and sequence stratigraphy of the Synrift Tilje Formation (Lower Jurassic), Halten terrace giant oil and gas fields, offshore mid-Norway. AAPG Bull. 100 (8), 1329–1375.

Imbrie, J., 1985. A theoretical framework for the Pleistocene ice ages. J. Geol. Soc. 142, 417–432.

Imbrie, J., Imbrie, K.P., 1979. Ice Ages: Solving the Mystery. Enslow, Hillside, New Jersey, p. 224.

Ineson, J.R., Surlyk, F., 2000. Carbonate mega-breccias in a sequence stratigraphic context: evidence from the Cambrian of North Greenland. In: Hunt, D., Gawthorpe, R.L. (Eds.), Sedimentary Responses to Forced Regressions, vol. 172. Geol. Soc. of London Spec. Publ., pp. 47–68

Isbell, J.L., Cole, D.I., Catuneanu, O., 2008. Carboniferous-Permian glaciation in the main Karoo Basin, South Africa: stratigraphy, depositional controls, and glacial dynamics. In: Fielding, C.R., Frank, T.D., Isbell, J.L. (Eds.), Resolving the Late Paleozoic Ice Age in Time and Space, vol. 441. Geological Society of America Special Paper, pp. 71–82.

Issler, D.R., Willett, S.D., Beaumont, C., Donelick, R.A., Grist, A.M., 1999. Paleotemperature history of two transects across the Western Canada Sedimentary Basin: constraints from apatite fission track analysis. Bull. Can. Pet. Geol. 47 (4), 475–486.

Ito, M., O'Hara, S., 1994. Diachronous evolution of systems tracts in a depositional sequence from the middle Pleistocene palaeo-Tokyo Bay, Japan. Sedimentology 41, 677–697.

Ito, M., Nishikawa, T., Sugimoto, H., 1999. Tectonic control of high-frequency depositional sequences with duration shorter than Milankovitch cyclicity: an example from the Pleistocene paleo-Tokyo Bay, Japan. Geology 27, 763–766.

Jackson, J.M., Simpson, E.L., Eriksson, K.A., 1990. Facies and sequence stratigraphic analysis in an intracratonic, thermal-relaxation basin: the middle Proterozoic, lower Quilalar Formation, Mount Isa Orogen, Australia. Sedimentology 37, 1053–1078.

James, N.P., Jones, B., 2015. Origin of Carbonate Sedimentary Rocks. Willey, p. 464.

James, N.P., Kendall, A.C., 1992. Introduction to carbonate and evaporite facies models. In: Walker, R.G., James, N.P. (Eds.), Facies Models: Response to Sea Level Change, vol. 1. Geological Association of Canada, GeoText, pp. 265–275.

James, N.P., Kendall, A.C., Pufahl, P.K., 2010. Introduction to Bioelemental Sediments. In: James, N.P., Dalrymple, R.W. (Eds.), Facies Models, fourth ed. Geological Association of Canada, pp. 323–340.

Jameson, R., 1805. A Mineralogical Description of the County of Dumfries (Edinburgh).

Jervey, M.T., 1988. Quantitative geological modeling of siliciclastic rock sequences and their seismic expression. In: Wilgus, C.K., Hastings, B.S., St, C.G., Kendall, C., Posamentier, H.W., Ross, C.A., Van Wagoner, J.C. (Eds.), Sea Level Changes – an Integrated Approach, vol. 42. Society of Economic Paleontologists and Mineralogists (SEPM), pp. 47–69. Special Publication.

Johnson, J.G., Murphy, M.A., 1984. Time-rock model for Siluro-Devonian continental shelf, Western United States. Geol. Soc. Am. Bull. 95, 1349–1359.

Johnson, M.R., 1991. Discussion on "Chronostratigraphic subdivision of the Witwatersrand Basin based on a Western Transvaal composite column". South Afr. J. Geol. 94, 401–403.

Jones, B., Desrochers, A., 1992. Shallow platform carbonates. In: Walker, R.G., James, N.P. (Eds.), Facies Models: Response to Sea Level Change, vol. 1. Geological Association of Canada, GeoText, pp. 277–301.

Jordan, O.D., Mountney, N.P., 2010. Styles of interaction between aeolian, fluvial and shallow marine environments in the Pennsylvanian to Permian lower Cutler beds, South-east Utah, USA. Sedimentology 57, 1357–1385.

Jordan, O.D., Mountney, N.P., 2012. Sequence stratigraphic evolution and cyclicity of an ancient coastal desert system: the Pennsylvanian-Permian Lower Cutler Beds, Paradox Basin, Utah, U.S.A. J. Sediment. Res. 82, 755–780.

Kamola, D.L., Van Wagoner, J.C., 1995. Stratigraphy and facies architecture of parasequences with examples from the Spring Canyon Member, Blackhawk Formation, Utah. In: Van Wagoner, J.C., Bertram, G.T. (Eds.), Sequence Stratigraphy of Foreland Basin Deposits, vol. 64. American Association of Petroleum Geologists Memoir, pp. 27–54.

Karner, G.D., 1986. Effects of lithospheric in-plane stress on sedimentary basin stratigraphy. Tectonics 5, 573–588.

Karner, G.D., Driscoll, N.W., Weissel, J.K., 1993. Response of the lithosphere to in-plane force variations. Earth & Planet. Sci. Lett. 114, 397–416.

Katz, B., 2005. Controlling factors on source rock development – A review of productivity, preservation, and sedimentation rate. In: Harris, N.B. (Ed.), The Deposition of Organic-Carbon-Rich Sediments: Models, Mechanisms, and Consequences, vol. 82. Society for Sedimentary Geology (SEPM), pp. 7–16. Special Publication.

Kerans, C., Loucks, R.G., 2002. Stratigraphic setting and controls on occurrence of high-energy carbonate beach deposits: Lower Cretaceous of the Gulf of Mexico. Transac. Gulf Coast Assoc. Geol. Soc. 52, 517–526.

Kerans, C., Fitchen, W., Zahm, L., Kempter, K., 1995. High-frequency Sequence Framework of Cretaceous (Albian) Carbonate Ramp Reservoir Analog Outcrops, Pecos River Canyon, Northwestern Gulf of Mexico Basin. The University of Texas at Austin, Bureau of Economic Geology, Field Trip Guidebook, p. 105.

Kerr, D., Ye, L., Bahar, A., Kelkar, B.G., Montgomery, S., 1999. Glenn Pool field, Oklahoma: a case of improved prediction from a mature reservoir. Am. Assoc. Pet. Geol. Bull. 83 (1), 1–18.

Ketzer, J.M., Morad, S., Amorosi, A., 2003a. Predictive diagenetic clay-mineral distribution in siliciclastic rocks within a sequence stratigraphic framework. In: Worden, R.H., Morad, S. (Eds.), Clay Mineral Cements in Sandstones. International Association of Sedimentologists, pp. 43–61. Special Publication No. 4.

Ketzer, J.M., Holz, M., Morad, S., Al-Aasm, I.S., 2003b. Sequence stratigraphic distribution of diagenetic alterations in coal-bearing, paralic sandstones: evidence from the Rio Bonito Formation (early Permian), southern Brazil. Sedimentology 50, 855–877.

Khalifa, M.A., 1983. Origin and occurrence of glauconite in the green sandstone associated with unconformity, Bahariya Oases, Western Desert, Egypt. J. Afr. Earth Sci. 31 (3/4), 321–325.

Khidir, A., Catuneanu, O., 2003. Sedimentology and diagenesis of the Scollard sandstones in the Red Deer Valley area, Central Alberta. Bull. Can. Petrol. Geol. 51 (1), 45−69.

Khidir, A., Catuneanu, O., 2005. Predictive diagenetic clay-mineral distribution in siliciclastic rocks within a nonmarine sequence stratigraphic framework: The Coalspur Formation, West-central Alberta. In: American Association of Petroleum Geologists Annual Convention, Calgary, vol. 14, p. A73. Abstracts.

Kidwell, S.M., 1984. Outcrop features and origin of basin margin unconformities in the Lower Chesapeake Group (Miocene), Atlantic Coastal Plain. In: Schlee, J.S. (Ed.), Interregional Unconformities and Hydrocarbon Accumulation, vol. 36. American Association of Petroleum Geologists Memoir, pp. 37−58.

Kidwell, S.M., 1986. Models for fossil concentrations: paleobiologic implications. Paleobiology 12, 6−24.

Kidwell, S.M., 1989. Stratigraphic condensation of marine transgressive records: origin of major shell deposits in the Miocene of Maryland. J. Geol. 97, 1−24.

Kidwell, S.M., 1991a. The stratigraphy of shell concentrations. In: Allison, P.A., Briggs, D.E.G. (Eds.), Taphonomy: Releasing the Data Locked in the Fossil Record. Plenum Press, New York, pp. 211−290.

Kidwell, S.M., 1991b. Condensed deposits in siliciclastic sequences: expected and observed features. In: Einsele, G., Ricken, W., Seilacher, A. (Eds.), Cycles and Events in Stratigraphy. Springer-Verlag, Berlin, pp. 682−695.

Kidwell, S.M., 1993. Taphonomic expressions of sedimentary hiatuses: field observations on bioclastic concentrations and sequence anatomy in low, moderate and high subsidence settings. Geologische Rundschau 82, 189−202.

Kidwell, S.M., 1997. Anatomy of extremely thin marine sequences landward of a passive margin hinge-zone: Neogene Calvert Cliffs succession, Maryland, USA. J. Sediment. Res. 67, 322−340.

Kidwell, S.M., 2013. Time-averaging and fidelity of modern death assemblages: building a taphonomic foundation for conservation palaeobiology. Palaeontology 56, 487−522.

Kidwell, S.M., Bosence, D.W.J., 1991. Taphonomy and time-averaging of marine shelly faunas. In: Allison, P.A., Briggs, D.E.G. (Eds.), Taphonomy: Releasing the Data Locked in the Fossil Record. Plenum Press, New York, pp. 115−209.

Kidwell, S.M., Brenchley, P.J., 1994. Patterns in bioclastic accumulations through the Phanerozoic: changes in input or in destruction? Geology 22, 1139−1143.

Kidwell, S.M., Flessa, K.W., 1996. The quality of the fossil record: populations, species, and communities. Ann. Rev. Earth & Planet. Sci. 24, 433−464.

Kidwell, S.M., Holland, S.M., 1991. Field description of coarse bioclastic fabrics. Palaios 6, 426−434.

Kidwell, S.M., Holland, S.M., 2002. The quality of the fossil record: implications for evolutionary analyses. Ann. Rev. Ecol. & Syst. 33, 561−588.

Kidwell, S.M., Jablonski, D., 1983. Taphonomic feedback: ecological consequences of shell accumulation. In: Tevesz, M.J.S., McCall, P.L. (Eds.), Biotic Interactions in Recent and Fossil Benthic Communities. Plenum Press, New York, pp. 195−248.

Kidwell, S.M., Fürsich, F.T., Aigner, T., 1986. Conceptual framework for the analysis and classifications of fossil concentrations. Palaios 1, 228−238.

Kitazawa, T., Murakoshi, N., 2016. Tidal ravinement surfaces in the Pleistocene macrotidal tide-dominated Dong Nai estuary, Southern Vietnam. In: Tessier, B., Reynaud, J.-Y. (Eds.), Contributions to Modern and Ancient Tidal Sedimentology: Proceedings of the Tidalites 2012 Conference. International Association of Sedimentologists. John Wiley & Sons, pp. 233−241.

Knapp, L.J., Harris, N.B., McMillan, J.M., 2019. A sequence stratigraphic model for the organic-rich Upper Devonian Duvernay Formation, Alberta, Canada. Sediment. Geol. 387, 152−181.

Knaust, D., Bromley, R.G., 2012. Trace fossils as indicators of sedimentary environments. Develop. Sediment. 64, 924. Elsevier.

Knaust, D., Curran, H.A., Dronov, A.V., 2012. Shallow-marine carbonates. In: Knaust, D., Bromley, R.G. (Eds.), Trace Fossils as Indicators of Sedimentary Environments, Developments in Sedimentology, vol. 64. Elsevier, pp. 705−750.

Kocurek, G., 1981. Significance of interdune deposits and bounding surfaces in aeolian dune sands. Sedimentology 28, 753−780.

Kocurek, G., 1988. First-order and super bounding surfaces in eolian sequences. Sediment. Geol. 56, 193−206.

Kocurek, G., 1991. Interpretation of ancient eolian sand dunes. Ann. Rev. Earth & Planet. Sci. 19, 43−75.

Kocurek, G., 1996. Desert aeolian systems. In: Reading, H.G. (Ed.), Sedimentary Environments: Processes, Facies and Stratigraphy, third ed. Blackwell Science, pp. 125−153.

Kocurek, G., 1998. Aeolian system response to external forcing factors − a sequence stratigraphic view of the Saharan Region. In: Alsharan, A.S., Glennie, K.W., Whittle, G.L., Kendall, C.G.S.C. (Eds.), Quaternary Deserta and Climatic Change. Balkema, Rotterdam Brookfield, pp. 327−338.

Kocurek, G., Havholm, K.G., 1993. Eolian sequence stratigraphy − a conceptual framework. In: Weimer, P., Posamentier, H.W. (Eds.), Siliciclastic Sequence Stratigraphy, vol. 58. AAPG Memoir, pp. 393−409.

Kocurek, G., Nielson, J., 1986. Conditions favourable for the formation of warm-climate aeolian sand sheets. Sedimentology 33, 795−816.

Kolla, V., Bandyopadhyay, A., Gupta, P., Mukherjee, B., Ramana, D.V., 2012. Morphology and internal structure of a recent upper Bengal fan-valley complex. In: Prather, B.E., Deptuck, M.E., Mohrig, D., Van Hoorn, B., Wynn, R.B. (Eds.), Application of the Principles of Seismic Geomorphology to Continental-Slope and Base-Of-Slope Systems − Case Studies from Seafloor and Near-Seafloor Analogues. SEPM, pp. 347−369. Special Publication No. 99.

Komar, P.D., 1976. Beach Processes and Sedimentation. Prentice-Hall, New Jersey, p. 429.

Kominz, M.A., Pekar, S.F., 2001. Oligocene eustasy from two-dimensional sequence stratigraphic backstripping. Geol. Soc. Am. Bull. 113, 291−304.

Kominz, M.A., Miller, K.G., Browning, J.V., 1998. Long-term and short-term global Cenozoic sea-level estimates. Geology 26, 311−314.

Kondo, Y., Abbott, S.T., Kitamura, A., Kamp, P.J.J., Naish, T.R., Kamataki, T., Saul, G.S., 1998. The relationship between shellbed type and sequence architecture: examples from Japan and New Zealand. Sediment. Geol. 122, 109−127.

Koss, J.E., Ethridge, F.G., Schumm, S.A., 1994. An experimental study of the effects of base-level change on fluvial, coastal plain, and shelf systems. J. Sediment. Res. B64, 90−98.

Kraft, J.C., Chrzastowski, M.J., Belknap, D.F., Toscano, M.A., Fletcher III, C.H., 1987. The transgressive barrier-lagoon coast of Delaware: morphostratigraphy, sedimentary sequences, and responses to relative rise in sea level. In: Nummedal, D., Pilkey, O.H., Howard, J.D. (Eds.), Sea Level Fluctuation and Coastal Evolution, vol. 41. SEPM, pp. 129−143. Special Publication.

Krapez, B., 1993. Sequence stratigraphy of the Archaean Supracrustal-belts of the Pilbara Block, Western Australia. Precambrian Res. 60, 1−45.

Krapez, B., 1996. Sequence-stratigraphic concepts applied to the identification of basin-filling rhythms in Precambrian successions. Austr. J. Earth Sci. 43, 355−380.

Krapez, B., 1997. Sequence-stratigraphic concepts applied to the identification of depositional basins and global tectonic cycles. Austr. J. Earth Sci. 44, 1−36.

Krapovickas, V., Mangano, M.G., Buatois, L.A., Marsicano, C.A., 2016. Integrated ichnofacies models for deserts: Recurrent patterns and megatrends. Earth-Sci. Rev. 157, 61−85.

Kraus, M.J., 1999. Paleosols in clastic sedimentary rocks: their geological applications. Earth-Sci. Rev. 47, 41–70.

Krawinkel, H., Seyfried, H., 1996. Sedimentologic, palaeoecologic, taphonomic and Ichnologic criteria for high-resolution sequence analysis; a practical guide for the identification and interpretation of discontinuities in shelf deposits. Sediment. Geol. 102 (1–2), 79–110.

Kress, P.R., Catuneanu, O., Gerster, R., Bolatti, N., 2021. Tectonic and stratigraphic evolution of the Cretaceous Western South Atlantic. Mar. & Pet. Geol. 134, 105197.

Krumbein, W.C., Sloss, L.L., 1951. Stratigraphy and Sedimentation. Freeman, San Francisco, p. 497.

Krynine, P.D., 1941. Differentiation of sediments during the life history of a landmass. Abstracts for the Boston meeting of the Geological Society of America. Geol. Soc. Am. Bull. 52, 1915–1916.

Kunzmann, M., Crombez, V., Catuneanu, O., Blaikie, T.N., Barth, G., Collins, A.S., 2020. Sequence stratigraphy of the ca. 1730 Ma Wollogorang Formation, McArthur Basin, Australia. Mar. & Petr. Geol. 116, 104297.

LaBrecque, J.L., Kent, D.V., Cande, S.C., 1977. Revised magnetic polarity time scale for Late Cretaceous Cenozoic time. Geology 5, 330–335.

LaGrange, M.T., Konhauser, K.O., Catuneanu, O., Harris, B.S., Playter, T.L., Gingras, M.K., 2020. Earth-Sci. Rev. 203, 103137.

Lambeck, K., 1980. The Earth's Variable Rotation. Cambridge University Press, Cambridge, p. 449.

Lambeck, K., Rouby, H., Purcell, A., Sun, Y., Sambridge, M., 2014. Sea level and global ice volumes from the Last Glacial Maximum to the Holocene. Proc. Nat. Acad. Sci. USA 111 (43), 15296–15303.

Lander, R.H., Bloch, S., Mehta, S., Atkinson, C.D., 1991. Burial diagenesis of paleosols in the giant Yacheng gas field, People's Republic of China: bearing on illite reactivation pathways. J. Sediment. Petrol. 61, 256–268.

Langereis, C.G., Dekkers, M.J., De Lange, G.J., Paterne, M., Van Santvoort, P., 1997. Magnetostratigraphy and astronomical calibration of the last 1.1 Myr from an eastern Mediterranean piston core and dating of short events in the Brunhes. Geophys. J. Int. 129, 75–94.

Langereis, C.G., Krijgsman, W., Muttoni, G., Menning, M., 2010. Magnetostratigraphy – concepts, definitions, and applications. Newslett. Stratigr. 43/3, 207–233.

Larcombe, P., Costen, A., Woolfe, K.J., 2001. The hydrodynamic and sedimentary setting of nearshore coral reefs, central Great Barrier Reef shelf, Australia, Paluma Shoals, a case study. Sedimentology 48, 811–835.

Leckie, D.A., 1994. Canterbury Plains, New Zealand – Implications for sequence stratigraphic models. AAPG Bull. 78, 1240–1256.

Leckie, D.A., Boyd, R., 2003. Towards a nonmarine sequence stratigraphic model. In: American Association of Petroleum Geologists Annual Convention, Salt Lake City, 11–14 May 2003, vol. 12. Official Program, p. A101.

Leckie, D.A., Fox, C., Tarnocai, C., 1989. Multiple paleosols of the late Albian Boulder Creek Formation, British Columbia, Canada. Sedimentology 36, 307–323.

Leckie, D.A., Wallace-Dudley, K.E., Vanbeselaere, N.A., James, D.P., 2004. Sedimentation in a low-accommodation setting: nonmarine (Cretaceous) Mannville and marine (Jurassic) Ellis Groups, Manyberries Field, southeastern Alberta. Am. Assoc. Pet. Geol. Bull. 88 (10), 1391–1418.

Leeder, M.R., 1978. A quantitative stratigraphical model for alluvium, with specific reference to channel deposit density and interconnectedness. In: Miall, A.D. (Ed.), Fluvial Sedimentology, vol. 5. Memoirs of the Canadian Society of Petroleum Geologists, pp. 587–596.

Leeder, M.R., Gawthorpe, R.L., 1987. Sedimentary models for extensional tilt-block/half-graben basins. In: Coward, M.P., Dewey, J.F., Hancock, P.L. (Eds.), Continental Extension Tectonics, vol. 28. Geological Society, London, pp. 139–152. Special Publication.

Leeder, M.R., Harris, T., Kirkby, M.J., 1998. Sediment supply and climate change: implications for basin stratigraphy. Basin Res. 10, 7–18.

Legaretta, L., Uliana, M.A., Larotonda, C.A., Meconi, G.R., 1993. Approaches to nonmarine sequence stratigraphy – theoretical models and examples from Argentine basins. In: Eschard, R., Doliez, B. (Eds.), Subsurface Reservoir Characterization from Outcrop Observations, vol. 51. Editions Technip, Collection Colloques et Seminaires, Paris, pp. 125–145.

Lehrmann, D.J., Goldhammer, R.K., 1999. Secular variation in parasequence and facies stacking patterns of platform carbonates: a guide to application of stacking-patterns analysis in strata of diverse ages and settings. In: Harris, P.M., Saller, A.H., Simo, J.A. (Eds.), Advances in Carbonate Sequence Stratigraphy: Application to Reservoirs, Outcrops and Models, vol. 63. SEPM (Society for Sedimentary Geology), pp. 187–225. Special Publication.

Lenoble, J.L., Canerot, J., 1993. Sequence stra-tigraphy of the Clansayesian (Uppermost Aptian) formations in the Western Pyrenees (France). In: Posamentier, H.W., Summerhayes, C.P., Haq, B.U., Allen, G.P. (Eds.), Sequence Stratigraphy and Facies Associations. Special Publication International Association of Sedimentologists, vol. 18. Blackwell Scientific Publications, Oxford, pp. 283–294.

Leopold, L.B., Bull, W.B., 1979. Base level, aggradation, and grade. Am. Philos. Soc. Proc. 123, 168–202.

Lerbekmo, J.F., Sweet, A.R., 2000. Magnetostratigraphy and biostratigraphy of the continental Paleocene in the Calgary area, Southwestern Alberta. Bull. Can. Petrol. Geol. 48, 285–306.

Lerbekmo, J.F., Sweet, A.R., 2008. Magnetobiostratigraphy of the continental Paleocene upper Coalspur and Paskapoo formations near Hinton, Alberta. Bull. Can. Petrol. Geol. 56, 118–146.

Lerbekmo, J.F., Sweet, A.R., Braman, D.R., 1995. Magnetobiostratigraphy of late Maastrichtian to early Paleocene sediments of the Hand Hills, South-Central Alberta, Canada. Bull. Can. Petrol. Geol. 43, 35–43.

Leroux, E., Rabineau, M., Aslanian, D., Granjeon, D., Droz, L., Gorini, C., 2014. Stratigraphic simulations of the shelf of the Gulf of Lions: testing subsidence rates and sea-level curves during the Pliocene and Quaternary. Terra Nova 26, 230–238.

Li, L., Keller, G., Adatte, T., Stinnesbeck, W., 2000. Late Cretaceous sea-level changes in Tunisia: a multidisciplinary approach. J. Geol. Soc. London 157, 447–458.

Li, X., Liang, X., Sun, M., Guan, H., Malpas, J.G., 2001. Precise $^{206}Pb/^{238}U$ age determination on zircons by laser ablation microprobe-inductively coupled plasma-mass spectrometry using continuous linear ablation. Chem. Geol. 175 (3–4), 209–219.

Li, Y., Shao, L., Eriksson, K.A., Tong, X., Gao, C., Chen, Z., 2014. Linked sequence stratigraphy and tectonics in the Sichuan continental foreland basin, Upper Triassic Xujiahe Formation, Southwest China. J. Asian Earth Sci. 88, 116–136.

Lickorish, W.H., Butler, R.W.H., 1996. Fold amplification and parasequence stacking patterns in syntectonic shoreface carbonates. Geol. Soc. Am. Bull. 108, 966–977.

Lindsay, J.F., Kennard, J.M., Southgate, P.N., 1993. Application of sequence stratigraphy in an intracratonic setting, Amadeus Basin, Central Australia. In: Posamentier, H.W., Summerhayes, C.P., Haq, B.U., Allen, G.P. (Eds.), Sequence Stratigraphy and Facies Associations, vol. 18. International Association of Sedimentologists, pp. 605–631. Special Publication.

Llave, E., Hernandez-Molina, F.J., Somoza, L., Diaz-del-Rio, V., Stow, D.A.V., Maestro, A., Alveirinho Dias, J.M., 2001. Seismic stacking pattern of the Faro-Albufeira contourite system (Gulf of Cadiz): a Quaternary record of paleoceanographic and tectonic influences. Mar. Geophys. Res. 22, 487–508.

Llave, E., Schonfeld, J., Hernández-Molina, F.J., Mulder, T., Somoza, L., del Rio, V.D., Sanchez-Almazo, I., 2006. High-resolution stratigraphy of the Mediterranean outflow contourite system in the Gulf of Cadiz during the late Pleistocene: the impact of Heinrich event. Mar. Geol. 227, 241−262.

Lobo, F.J., Tesson, M., Gensous, B., 2004. Stratal architectures of late Quaternary regressive-transgressive cycles in the Roussillon Shelf (SW Gulf of Lions, France). Mar. & Petrol. Geol. 21, 1181−1203.

Locker, S.D., Hine, A.C., Tedesco, L.P., Shinn, E.A., 1996. Magnitude and timing of episodic sea-level rise during the last deglaciation. Geology 24, 827−830.

Long, D.G.F., 1993. Limits on late Ordovician eustatic sea-level change from carbonate shelf sequences: and example from Anticosti Island, Quebec. In: Posamentier, H.W., Summerhayes, C.P., Haq, B.U., Allen, G.P. (Eds.), Sequence Stratigraphy and Facies Associations, vol. 18. International Association of Sedimentologists, pp. 487−499. Special Publication.

Long, D.G.F., 2011. Architecture and depositional style of fluvial systems before land plants: a comparison of Precambrian, early Paleozoic and modern river deposits. In: Davidson, S.K., Leleu, S., North, C.P. (Eds.), From River to Rock Record: The Preservation of Fluvial Sediments and Their Subsequent Interpretation, vol. 97. SEPM, pp. 37−61. Special Publication.

Long, D.G.F., 2019. Archean fluvial deposits: a review. Earth-Sci. Rev. 188, 148−175.

Long, D.G.F., 2021. Trickling down the paleoslope: an empirical approach to paleohydrology. Earth-Sci. Rev. 220, 1−25, 103740.

Longwell, C.R., 1949. Sedimentary facies in geologic history. Geol. Soc. Am. Memoir 39, 171.

Loope, D.B., 1985. Episodic deposition and preservation of eolian sands: a Late Palaeozoic example from Southeastern Utah. Geology 13, 73−76.

Løseth, T.M., Helland-Hansen, W., 2001. Predicting the pinchout distance of shoreline tongues. Terra Nova 13, 241−248.

Loughney, K.M., Fastovsky, D.E., Parker, W.G., 2011. Vertebrate fossil preservation in blue paleosols from the Petrified Forest National Park, Arizona, with implications for vertebrate biostratigraphy in the Chinle Formation. Palaios 26, 700−719.

Lourens, L.J., Hilgen, F.J., Laskar, J., Shackleton, N.J., Wilson, D., 2004. The Neogene period. In: Gradstein, F.M., Ogg, J.G., Smith, A.G. (Eds.), A Geologic Time Scale 2004. Cambridge University Press, p. 500.

Loutit, T.S., Hardenbol, J., Vail, P.R., Baum, G.R., 1988. Condensed sections: the key to age-dating and correlation of continental margin sequences. In: Wilgus, C.K., Hastings, B.S., St, C.G., Kendall, C., Posamentier, H.W., Ross, C.A., Van Wagoner, J.C. (Eds.), Sea Level Changes − an Integrated Approach, vol. 42. SEPM, pp. 183−213. Special Publication.

Lucchi, F., 2009. Late-Quaternary terraced marine deposits as tools for wide-scale correlation of unconformity-bounded units in the volcanic Aeolian archipelago (Southern Italy). Sediment. Geol. 216, 158−178.

Ludvigsen, R., Westrop, S.R., Pratt, B.R., Tuffnell, P.A., Young, G.A., 1986. Dual biostratigraphy: zones and biofacies. Geosci. Canada 13, 139−154.

Lyell, C., 1830. Principles of geology, being an attempt to explain the former changes of the Earth's surface. In: Reference to Causes Now in Operation, vol. 1. John Murray, London, p. 511.

Lyell, C., 1833. Principles of geology, being an attempt to explain the former changes of the Earth's surface. In: Reference to Causes Now in Operation, vol. 3. John Murray, London, p. 109.

MacEachern, J.A., Bann, K.L., 2020. The *Phycosiphon* ichnofacies and the *Rosselia* ichnofacies: two new ichnofacies for marine deltaic environments. J. Sediment. Res. 90 (8), 855−886.

MacEachern, J.A., Raychaudhuri, I., Pemberton, S.G., 1992. Stratigraphic applications of the Glossifungites ichnofacies: delineating discontinuities in the rock record. In: Pemberton, S.G. (Ed.), Applications of Ichnology to Petroleum Exploration, Society of Economic

Paleontologists and Mineralogists, Core Workshop Notes, vol. 17, pp. 169−198.

MacEachern, J.A., Pemberton, S.G., Zaitlin, B.A., 1999. A late lowstand tongue, Viking Formation of the Joffre Field, Alberta. In: Bergman, K. (Ed.), Isolated Marine Sand Bodies: Sequence Stratigraphic Analysis and Sedimentological Interpretation, vol. 64. Society of Economic Paleontologists and Mineralogists, pp. 273−296. Special Publication.

MacEachern, J.A., Bann, K.L., Pemberton, S.G., Gingras, M.K., 2009a. The Ichnofacies Paradigm: High-Resolution Paleoenvironmental Interpretation of the Rock Record. SEPM (Society for Sedimentary Geology) Applied Ichnology SC52, pp. 27−64.

MacEachern, J.A., Pemberton, S.G., Bann, K.L., Gingras, M.K., 2009b. Departures from the Archetypal Ichnofacies: Effective Recognition of Physico-Chemical Stresses in the Rock Record. Society for Sedimentary Geology Applied Ichnology SC52, pp. 65−93.

MacEachern, J.A., Pemberton, S.G., Gingras, M.K., Bann, K.L., 2010. Ichnology and facies models. In: James, N.P., Dalrymple, R.W. (Eds.), Facies Models 4, vol. 6. Geological Association of Canada, St. John's, Geotext, pp. 19−58.

MacEachern, J.A., Dashtgard, S.E., Knaust, D., Catuneanu, O., Bann, K.L., Pemberton, S.G., 2012. Sequence stratigraphy. In: Knaust, D., Bromley, R.G. (Eds.), Trace Fossils as Indicators of Sedimentary Environments, Developments in Sedimentology, vol. 64. Elsevier, pp. 157−194.

Machel, H.G., 2004. Concepts and models of dolomitization: a critical reappraisal. In: Braithwaite, C.J.R., Giancarlo, R., Darke, G. (Eds.), The Geometry and Petrogenesis of Dolomite Hydrocarbon Reservoirs, vol. 235. Geological Society London, pp. 7−63. Special Publication.

Mack, G.H., James, W.C., Monger, H.C., 1993. Classification of paleosols. Geol. Soc. Am. Bull. 105, 129−136.

Mackey, S.D., Bridge, J.S., 1995. Three-dimensional model of alluvial stratigraphy: theory and applications. J. Sediment. Res. Section B: Stratig. & Glob. Stud. 65B (No. 1), 7−31.

MacNeil, A.J., Jones, B., 2006. Sequence stratigraphy of a Late Devonian ramp-situated reef system in the Western Canada Sedimentary Basin: dynamic responses to sea-level change and regressive reef development. Sedimentology 53, 321−359.

Madof, A.S., Harris, A.D., Connell, S.D., 2016. Nearshore along-strike variability: is the concept of the systems tract unhinged? Geology 44, 315−318.

Madof, A.S., Harris, A.D., Baumgardner, S.E., Sadler, P.M., Laugier, F.J., Christie-Blick, N., 2019. Stratigraphic aliasing and the transient nature of deep-water depositional sequences: Revisiting the Mississippi Fan. Geology 47 (6), 545−549.

Magalhaes, A.J.C., Raja Gabaglia, G.P., Scherer, C., Ballico, M., Guadagnin, F., Bento Freire, E., Silva Born, L., Catuneanu, O., 2015. Sequence hierarchy in the Mesoproterozoic Tombador Formation, Chapada Diamantina, Brazil. Basin Res. 1−40.

Magalhaes, A.J.C., Raja Gabaglia, G.P., Fragoso, D.G.C., Bento Freire, E., Lykawka, R., Arregui, C.D., Silveira, M.M.L., Carpio, K.M.T., De Gasperi, A., Pedrinha, S., Artagao, V.M., Terra, G.J.S., Bunevich, R.B., Roemers-Oliveira, E., Gomes, J.P., Hernandez, J.I., Hernandez, R.M., Bruhn, C.H.L., 2020. High-resolution sequence stratigraphy applied to reservoir zonation and characterisation, and its impact on production performance − shallow marine, fluvial downstream, and lacustrine carbonate settings. Earth-Sci. Rev. 210, 103325.

Maliva, R., Knoll, A.H., Siever, R., 1989. Secular change in chert distribution: a reflection of evolving biological participation in the silica cycle. Palaios 4, 519−532.

Maliva, R., Knoll, A.H., Simonson, B.M., 2005. Secular change in the Precambrian silica cycle: insights from chert petrology. Geol. Soc. Am. Bull. 117, 835−845.

Manyeruke, T.D., Blenkinsop, T.G., Buchholz, P., Love, D., Oberthür, T., Vetter, U.K., Davis, D.W., 2004. The age and petrology of the

Chimbadzi Hill Intrusion, NW Zimbabwe: first evidence for early Paleoproterozoic magmatism in Zimbabwe. J. Afr. Earth Sci. 40 (5), 281–292.

Maravelis, A.G., Boutelier, D., Catuneanu, O., Seymour, K.S., Zelilidis, A., 2016. A review of tectonics and sedimentation in a forearc setting: Hellenic Thrace Basin, North Aegean Sea and Northern Greece. Tectonophysics 674, 1–19.

Maravelis, A.G., Catuneanu, O., Nordsvan, A., Landerberger, B., Zelilidis, A., 2017. Interplay of tectonism and eustasy during the Early Permian icehouse: Southern Sydney Basin, Southeast Australia. Geol. J. 1–32. https://doi.org/10.1002/gj.2962.

Marriott, S.B., Wright, V.P., 1993. Paleosols as indicators of geomorphic stability in two Old Red Sandstones alluvial suites, South Wales. J. Geol. Soc. London 150, 1109–1120.

Martini, E., 1971. Standard Tertiary and Quaternary calcareous nanno-plankton zonation. In: Farinacci, A. (Ed.), Proceedings 2nd Planktonic Conference, Roma, 1970, vol. 2, pp. 739–785.

Martins-Neto, M.A., Catuneanu, O., 2010. Rift sequence stratigraphy. Mar. & Petrol. Geol. 27, 247–253.

Martinsen, O.J., Helland-Hansen, W., 1995. Strike variability of clastic depositional systems: does it matter for sequence stratigraphic analysis? Geology 23, 439–442.

Martinsen, O.J., Ryseth, A., Helland-Hansen, W., Flesche, H., Torkildsen, G., Idil, S., 1999. Stratigraphic base level and fluvial architecture: Ericson Sandstone (Campanian), Rock Springs Uplift, SW Wyoming, USA. Sedimentology 46, 235–259.

Martinsen, R.S., Christensen, G., 1992. A Stratigraphic and Environmental Study of the Almond Formation, Mesaverde Group, Greater Green River Basin, Wyoming. Wyoming Geological Association Guidebook, Forty-third Field Conference, pp. 171–190.

Massari, F., Chiocci, F., 2006. Biocalcarenite and mixed cool-water prograding bodies of the Mediterranean Pliocene and Pleistocene: architecture, depositional setting and forcing factors. In: Pedley, H.M., Carannante, G. (Eds.), Cool-Water Carbonates: Depositional Systems and Palaeoenvironmental Controls, vol. 255. Geological Society, pp. 95–120. Special Publication.

Massari, F., D'Alessandro, A., 2012. Facies partitioning and sequence stratigraphy of a mixed siliciclastic-carbonate ramp stack in the Gelasian of Sicily (S Italy): a potential model for Icehouse, distally-steepened Heterozoan ramps. Rivista Italiana di Paleontologia e Stratigrafia 118, 503–534.

Massari, F., Capraro, L., Rio, D., 2007. Climatic modulation of timing of systems-tract development with respect to sea-level changes (middle Pleistocene of Crotone, Calabria, Southern Italy). J. Sediment. Res. 77, 461–468.

Masterson, W.D., Eggert, J.T., 1992. Kuparuk River field, North Slope, Alaska. In: Beaumont, E.A., Foster, N.H. (Eds.), Atlas of Oil and Gas Fields. American Association of Petroleum Geologists Treatise of Petroleum Geology, pp. 257–284.

Mawson, M., Tucker, M.E., 2009. High-frequency cyclicity (millennial-scale and Milankovitch) in slope carbonates (Zechstein) Permian, NE England. Sedimentology 56, 1905–1936.

McKay, R., Browne, G., Carter, L., Cowan, E., Dunbar, G., Krissek, L., Naish, T., Powell, R., Reed, J., Talarico, F., Wilch, T., 2009. The stratigraphic signature of the late Cenozoic Antarctic Ice Sheets in the Ross Embayment. GSA Bull. 121, 1537–1561.

McMullen, S.K., Holland, S.M., O'Keefe, F.R., 2014. The occurrence of vertebrate and invertebrate fossils in a sequence stratigraphic context: the Jurassic Sundance Formation, Bighorn Basin, Wyoming, USA. Palaios 29, 277–294.

McMurray, L.S., Gawthorpe, R.L., 2000. Along-strike variability of forced regressive deposits: late Quaternary, Northern Peloponnesos, Greece. In: Hunt, D., Gawthorpe, R.L. (Eds.), Sedimentary Responses to Forced Regressions, vol. 172. Geol. Soc. of London Spec. Publ., pp. 363–377.

McNeill, D.F., Cunningham, K.J., Guertin, L.A., Anselmetti, F.S., 2004. Depositional themes of mixed carbonate–siliciclastics in the South Florida Neogene: application to ancient deposits. In: Grammer, G.M., Harris, P.M., Eberli, G.P. (Eds.), Integration of Outcrop and Modern Analogs in Reservoir Modeling, vol. 80. American Association of Petroleum Geologists, Memoir, pp. 23–43.

McNeill, D.F., Klaus, J.S., Budd, A.F., Lutz, B.P., Ishman, S.E., 2012. Late Neogene chronology and sequence stratigraphy of mixed carbonate-siliciclastic deposits of the Cibao Basin, Dominican Republic. GSA Bull. 124, 35–58.

Menegazzo, M., Catuneanu, O., Chang, H.K., 2016. The South American retroarc foreland system: The development of the Bauru Basin in the back-bulge province. Mar. & Pet. Geol. 73, 131–156.

Meng, X., Ge, M., Tucker, M.E., 1997. Sequence stratigraphy, sea-level changes and depositional systems in the Cambro–Ordovician of the North China carbonate platform. Sed. Geol. 114, 189–223.

Merrill, R.T., McFadden, P.L., 1994. Geomagnetic field stability: reversal events and excursions. Earth & Planet. Sci. Lett. 121, 57–69.

Mestdagh, T., Lobo, F.J., Llave, E., Hernandez-Molina, F.J., Van Rooij, D., 2019. Review of the late Quaternary stratigraphy of the Northern Gulf of Cadiz continental margin: New insights into controlling factors and global implications. Earth-Sci. Rev. 198, 102944.

Miall, A.D., 1986. Eustatic sea-level change interpreted from seismic stratigraphy: a critique of the methodology with particular reference to the North Sea Jurassic record. Am. Assoc. Pet. Geol. Bull. 70, 131–137.

Miall, A.D., 1990. Principles of Sedimentary Basin Analysis, second ed. Springer, p. 668.

Miall, A.D., 1991. Stratigraphic sequences and their chronostratigraphic correlation. J. Sediment. Petrol. 61, 497–505.

Miall, A.D., 1992. The Exxon global cycle chart: an event for every occasion? Geology 20, 787–790.

Miall, A.D., 1994. Sequence stratigraphy and chronostratigraphy: problems of definition and precision in correlation, and their implications for global eustasy. Geosci. Can. 21, 1–26.

Miall, A.D., 1995. Whither stratigraphy? Sediment. Geol. 100, 5–20.

Miall, A.D., 1996. The Geology of Fluvial Deposits: Sedimentary Facies, Basin Analysis and Petroleum Geology. Springer-Verlag Inc., Heidelberg, p. 582.

Miall, A.D., 1997. The Geology of Stratigraphic Sequences. Springer-Verlag, p. 433.

Miall, A.D., 1999. Cryptic Sequence Boundaries in Braided Fluvial Successions. American Association of Petroleum Geologists Annual Meeting, San Antonio, p. A93. Official Program.

Miall, A.D., 2000. Principles of Sedimentary Basin Analysis, third ed. Springer, p. 616.

Miall, A.D., 2002. Architecture and sequence stratigraphy of Pleistocene fluvial systems in the Malay Basin, based on seismic time-slice analysis. Am. Assoc. Pet. Geol. Bull. 86 (7), 1201–1216.

Miall, A.D., 2010. The Geology of Stratigraphic Sequences, second ed. Springer-Verlag, Berlin, p. 522.

Miall, A.D., 2015. Updating uniformitarianism: stratigraphy as just a set of "frozen accidents". In: Smith, D.G., Bailey, R.J., Burgess, P., Fraser, A. (Eds.), Strata and Time, vol. 404. Geological Society, London, pp. 11–36. Special Publication.

Miall, A.D., 2016. The valuation of unconformities. Earth-Sci. Rev. 163, 22–71.

Miall, A.D., Miall, C.E., 2001. Sequence stratigraphy as a scientific enterprise: the evolution and persistence of conflicting paradigms. Earth-Sci. Rev. 54, 321–348.

Miall, A.D., Catuneanu, O., Vakarelov, B.K., Post, R., 2008. The western interior basin. In: Miall, A.D. (Ed.), The Sedimentary Basins of the United States and Canada. Sedimentary Basins of the World (Series Editor: K.J. Hsu). Elsevier, Amsterdam, pp. 329–362.

Middleton, G.V., 1973. Johannes Walther's law of the correlation of facies. Geol. Soc. Am. Bull. 84, 979–988.

Milankovitch, M., 1930. Mathematische klimalehre und astronomische theorie der klimaschwankungen. In: Koppen, W., Geiger, R. (Eds.), Handbuch der Klimatologie, I (A). Gebruder Borntraeger, Berlin.

Milankovitch, M., 1941. Kanon der Erdbestrahlung und seine Anwendung auf das Eiszeitenproblem. Akad Royale Serbe 133, 633.

Millar, R.A., 2021. Sequence Stratigraphy and Underlying Tectonism of the Northern Richardson Mountains and Adjacent Mackenzie Delta Related to the Formation of the Arctic Ocean. M.Sc. Thesis. University of Alberta, p. 142.

Miller, D.J., Eriksson, K.A., 1999. Linked sequence development and global climate change: the upper Mississippian record in the Appalachian basin. Geology 27, 35–38.

Miller, D.J., Eriksson, K.A., 2000. Sequence stratigraphy of Upper Mississippian strata in the central Appalachians: a record of glacioeustasy and tectonoeustasy in a foreland basin setting. Amer. Assoc. Pet. Geol. Bull. 84 (2), 210–233.

Miller, K.G., Wright, J.D., Fairbanks, R.G., 1991. Unlocking the ice house: Oligocene-Miocene oxygen isotopes, eustasy, and margin erosion. J. Geophys. Res. 96, 6829–6848.

Miller, K.G., Mountain, G.S., 1996. Drilling and dating New Jersey Oligocene-Miocene sequences: ice volume, global sea level, and Exxon records. Science 271, 1092–1094. The Leg 150 Shipboard Party, and Members of the New Jersey Coastal Plain Drilling Project.

Miller, K.G., Mountain, G.S., Browning, J.V., Kominz, M.A., Sugarman, P.J., Christie-Blick, N., Katz, M.E., Wright, J.D., 1998. Cenozoic global sea-level, sequences, and the New Jersey transect: results from coastal plain and slope drilling. Rev. Geophys. 36, 569–601.

Miller, K.G., Sugarman, P.J., Browning, J.V., Kominz, M.A., Hernandez, J.C., Olsson, R.K., Wright, J.D., Feigenson, M.D., Van Sickel, W., 2003. Late Cretaceous chronology of large, rapid sea-level changes: glacioeustacy during the greenhouse world. Geology 31, 585–588.

Miller, K.G., Sugarman, P.J., Browning, J.V., Kominz, M.A., Olsson, R.K., Feigenson, M.D., Hernandez, J.C., 2004. Upper Cretaceous sequences and sea-level history, New Jersey Coastal Plain. Geol. Soc. Am. Bull. 116 (3/4), 368–393.

Mitchum Jr., R.M., 1977. Seismic stratigraphy and global changes of sea level, part 11: glossary of terms used in seismic stratigraphy. In: Payton, C.E. (Ed.), Seismic Stratigraphy — Applications to Hydrocarbon Exploration, vol. 26. American Association of Petroleum Geologists Memoir, pp. 205–212.

Mitchum Jr., R.M., Vail, P.R., 1977. Seismic stratigraphy and global changes of sea-level, part 7: stratigraphic interpretation of seismic reflection patterns in depositional sequences. In: Payton, C.E. (Ed.), Seismic Stratigraphy — Applications to Hydrocarbon Exploration, vol. 26. American Association of Petroleum Geologists Memoir, pp. 135–144.

Mitchum Jr., R.M., Van Wagoner, J.C., 1991. High-frequency sequences and their stacking patterns: sequence stratigraphic evidence of high-frequency eustatic cycles. Sediment. Geol. 70, 131–160.

Mitchum Jr., R.M., Vail, P.R., Thompson III, S., 1977. Seismic stratigraphy and global changes of sea level, Part 2: The depositional sequence as a basic unit for stratigraphic analysis. In: Payton, C.E. (Ed.), Seismic Stratigraphy - Applications to Hydrocarbon Exploration, vol. 26. American Association of Petroleum Geologists, Memoir, pp. 53–62.

Mora, J.A., Oncken, O., Le Breton, E., Mora, A., Veloza, G., Velez, V., de Freitas, M., 2018. Controls on forearc basin formation and evolution: insights from Oligocene to recent tectono-stratigraphy of the Lower Magdalena Valley basin of Northwest Colombia. Mar. & Petrol. Geol. 97, 288–310.

Moran, M.G., 2020. Density Flows and Microsequences in the Nahal Darga Delta, West Bank, Dead Sea: Anatomy of a Lowstand Wedge Turned Falling Stage. M.Sc. thesis. Texas Christian University, Forth Worth, Texas, USA.

Morrison, R.B., 1978. Quaternary soil stratigraphy — concepts, methods, and problems. In: Mahaney, W.C. (Ed.), Quaternary Soils. Geo Abstracts, Norwich, pp. 77–108.

Mort, H., Jacquat, O., Adatte, T., Steinmann, P., Follmi, K., Matera, V., Berner, Z., Stuben, D., 2007. The Cenomanian/Turonian anoxic event at the Bonarelli level in Italy and Spain: enhanced productivity and/or better preservation? Cretac. Res. 28, 597–612.

Mount, J.F., 1984. Mixing of siliciclastic and carbonate sediments in shallow shelf environments. Geology 12, 432–435.

Mountney, N.P., 2006. Periodic accumulation and destruction of aeolian erg sequences: the Cedar Mesa Sandstone, White Canyon, Southern Utah. Sedimentology 53, 789–823.

Mountney, N.P., 2012. A stratigraphic model to account for complexity in aeolian dune and interdune successions. Sedimentology 59, 964–989.

Mountney, N.P., Jagger, A., 2004. Stratigraphic evolution of an Aeolian erg margin system: the Permian Cedar Mesa Sandstone, SE Utah, USA. Sedimentology 51, 713–743.

Muhs, D.R., Bettis III, E.A., 2003. Quaternary loess-paleosol sequences as examples of climate-driven sedimentary extremes. In: Chan, M.A., Archer, A.W. (Eds.), Extreme Depositional Environments: Mega End Members in Geologic Time, vol. 370. Geological Society of America Special Paper, pp. 53–74.

Mumpy, A.J., Catuneanu, O., 2019. Controls on accommodation during the early-middle Campanian in Southern Alberta, western Canada foreland system. J. Geodyn. 129, 178–201.

Muto, T., Steel, R.J., 1997. Principles of regression and transgression: the nature of the interplay between accommodation and sediment supply. J. Sediment. Res. 67, 994–1000.

Muto, T., Steel, R.J., 2002. Role of autoretreat and A/S changes in the understanding of deltaic shoreline trajectory: a semi-quantitative approach. Basin Res. 14, 303–318.

Muto, T., Steel, R.J., Swenson, J.B., 2007. Autostratigraphy: A framework norm for genetic stratigraphy. J. Sediment. Res. 77, 2–12.

NACSN (North American Commission on Stratigraphic Nomenclature), 1983. North American stratigraphic code. Am. Assoc. Pet. Geol. Bull. 67, 841–875.

Naish, T., Kamp, P.J.J., 1997. Foraminiferal depth palaeoecology of Late Pliocene shelf sequences and systems tracts, Wanganui Basin, New Zealand. Sediment. Geol. 110, 237–255.

Naish, T.R., Woolfe, K.J., Barrett, P.J., Wilson, G.S., Atkins, C., Bohaty, S.M., Bucker, C.J., Claps, M., Davey, F.J., Dunbar, G.B., Dunn, A.G., Fielding, C.R., Florindo, F., Hannah, M.J., Harwood, D.M., Henrys, S.A., Krissek, L.A., Lavelle, M.A., van der Meer, J., McIntosh, W.C., Niessen, F., Passchier, S., Powell, R.D., Roberts, A.P., Sagnotti, L., Scherer, R.P., Strong, C.P., Talarico, F., Verosub, K.L., Villa, G., Watkins, D.K., Webb, P.-N., Wonik, T., 2001. Orbitally induced oscillations in the East Antarctic ice sheet at the Oligocene/Miocene boundary. Nature 413, 719–723.

Naish, T.R., Wilson, G.S., Dunbar, G.B., Barrett, P.J., 2008. Constraining the amplitude of late Oligocene bathymetric changes in western Ross Sea during orbitally-induced oscillations in the East Antarctic Ice Sheet: (2) Implications for global sea-level changes. Palaeogeogr. Palaeoclimatol. Palaeoecol. 260, 66–76.

Naish, T., Powell, R., Levy, R., Wilson, G., Scherer, R., Talarico, F., Krissek, L., Niessen, F., Pompilio, M., Wilson, T., Carter, L., DeConto, R., Huybers, P., McKay, R., Pollard, D., Ross, J., Winter, D., Barrett, P., Browne, G., Cody, R., Cowan, E., Crampton, J., Dunbar, G., Dunbar, N., Florindo, F., Gebhardt, C., Graham, I., Hannah, M., Hansaraj, D., Harwood, D., Helling, D., Henrys, S., Hinnov, L., Kuhn, G., Kyle, P., Läufer, A., Maffioli, P., Magens, D., Mandernack, K., McIntosh, W., Millan, C., Morin, R., Ohneiser, C., Paulsen, T., Persico, D., Raine, I., Reed, J., Riesselman, C., Sagnotti, L., Schmitt, D., Sjunneskog, C., Strong, P., Taviani, M., Vogel, S., Wilch, T., Williams, T., 2009. Obliquity-paced Pliocene West Antarctic Ice Sheet oscillations. Nature 458, 322–328.

Nalin, R., Massari, F., 2009. Facies and stratigraphic anatomy of a temperate carbonate sequence (Capo Colonna Terrace, late Pleistocene, Southern Italy). J. Sediment. Res. 79, 210–225.

Nalin, R., Massari, F., Zecchin, M., 2007. Superimposed cycles of composite marine terraces: the example of Cutro terrace (Calabria, Southern Italy). J. Sediment. Res. 77, 340–354.

Nanson, R.A., Vakarelov, B.K., Ainsworth, R.B., Williams, F.M., Price, D.M., 2013. Evolution of a Holocene, mixed-process, forced regressive shoreline: The Mitchell River delta, Queensland, Australia. Mar. Geol. 339, 22–43.

Nawrot, R., Scarponi, D., Azzarone, M., Dexter, T.A., Kusnerik, K.M., Wittmer, J.M., Amorosi, A., Kowalewski, M., 2018. Stratigraphic signatures of mass extinctions: ecological and sedimentary determinants. Proc. R. Soc. London B Biol. Sci. 285. Paper 20181191.

Neal, J., Abreu, V., 2009. Sequence stratigraphy hierarchy and the accommodation succession model. Geology 37 (9), 779–782.

Neal, J.E., Stein, J.A., Gamber, J.A., 1999. Nested stratigraphic cycles and depositional systems of the Paleogene Central North Sea. In: De Graciansky, P.C., Hardenbol, J., Jacquin, T., Vail, P.R. (Eds.), Mesozoic and Cenozoic Sequence Stratigraphy of European Basins, vol. 60. SEPM, pp. 261–288. Special Publication.

Nesbitt, H.W., Young, G.M., 1982. Early Proterozoic climates and plate motions inferred from major element chemistry of lutites. Nature 299, 715–717.

Netto, R.G., Benner, J.S., Buatois, L.A., Uchman, A., Mangano, M.G., Ridge, J.C., Kazakauskas, V., Gaigalas, A., 2012. Glacial environments. In: Knaust, D., Bromley, R.G. (Eds.), Trace Fossils as Indicators of Sedimentary Environments, Developments in Sedimentology, vol. 64. Elsevier, pp. 299–327.

Neuendorf, K.K.E., Mehl Jr., J.P., Jackson, J.A. (Eds.), 2005. Glossary of Geology, fifth ed. American Geological Institute, Alexandria, Virginia, p. 779.

Nichols, D.J., Sweet, A.R., 1993. Biostratigraphy of Upper Cretaceous non marine palynomorphs in north-south transect of Western Interior Basin. In: Caldwell, W.G.E., Kauffman, E.G. (Eds.), The Evolution of the Western Interior Basin, vol. 39. Geological Association of Canada Special Paper, pp. 539–584.

Nixon, F.C., England, J.H., Lajeunesse, P., Hanson, M.A., 2014. Deciphering patterns of postglacial sea level at the junction of the Laurentide and Innuitian Ice Sheets, Western Canadian High Arctic. Quat. Sci. Rev. 91, 165–183.

Nottvedt, A., Gabrielsen, R.H., Steel, R.J., 1995. Tectonostratigraphy and sedimentary architecture of rift basins, with reference to the Northern North Sea. Mar. & Petrol. Geol. 12, 881–901.

Nummedal, D., 1992. The falling sea-level systems tract in ramp settings. In: SEPM Theme Meeting, Fort Collins, Colorado, p. 50 (Abstracts).

Nummedal, D., Swift, D.J.P., 1987. Transgressive stratigraphy at sequence-bounding unconformities: some principles derived from Holocene and Cretaceous examples. In: Nummedal, D., Pilkey, O.H., Howard, J.D. (Eds.), Sea-level Fluctuation and Coastal Evolution, vol. 41. Society of Economic Paleontologists and Mineralogists (SEPM), pp. 241–260. Special Publication.

Nummedal, D., Riley, G.W., Templet, P.L., 1993. High-resolution sequence architecture: a chronostratigraphic model based on equilibrium profile studies. In: Posamentier, H.W., Summerhayes, C.P., Haq, B.U., Allen, G.P. (Eds.), Sequence Stratigraphy and Facies Associations, vol. 18. International Association of Sedimentologists, pp. 55–68. Special Publication.

Nutz, A., Ghienne, J.-F., Schuster, M., Dietrich, P., Roquin, C., Hay, M.B., Bouchette, F., Cousineau, P.A., 2015. Forced regressive deposits of a deglaciation sequence: example from the late Quaternary succession in the Lake Saint-Jean basin (Québec, Canada). Sedimentology 62, 1573–1610.

Oberthür, T., Davis, D.W., Blenkinsop, T.G., Hoehndorf, A., 2002. Precise U–Pb mineral ages, Rb–Sr and Sm–Nd systematics for the Great Dyke, Zimbabwe—constraints on late Archean events in the Zimbabwe craton and Limpopo belt. Precambrian Res. 113 (3–4), 293–306.

Obradovich, J.D., 1993. A Cretaceous time scale. In: Caldwell, W.G.E., Kauffman, E.G. (Eds.), Evolution of the Western Interior Basin, Geological Association of Canada, Special Paper 39, pp. 379–396.

Ogg, J.G., 2012. Geomagnetic polarity time scale. In: Gradstein, F.M., Ogg, J.G., Schmitz, M.D., Ogg, G.M. (Eds.), The Geologic Time Scale 2012. Elsevier, pp. 85–113.

Olsen, T., Steel, R., Høgseth, K., Skar, T., Røe, S.L., 1995. Sequential architecture in a fluvial succession: sequence stratigraphy in the Upper Cretaceous Mesaverde Group, Price Canyon, Utah. J. Sediment. Res. B65, 265–280.

Olsson, R.K., Miller, K.G., Browning, J.V., Wright, J.D., Cramer, B.S., 2002. Sequence stratigraphy and sea level change across the Cretaceous-Tertiary boundary on the New Jersey passive margin. In: Koeberl, C., MacLeod, K.G. (Eds.), Catastrophic Events and Mass Extinctions: Impacts and Beyond, vol. 356. Geological Society of America, pp. 97–108. Special Paper.

Opdyke, N.D., Channell, J.E.T., 1996. Magnetic Stratigraphy. Academic Press, San Diego, USA, p. 346.

Oppel, A., 1856. Die Juraformation Englands, Frankreichs und des sudwestlichen Deutschlands. Wurttemb. Naturwiss. Verein Jahresh vol. xii, 438. Stuttgart.

Orme, D.A., Surpless, K.D., 2019. The birth of a forearc: The basal Great Valley Group, California, USA. Geology 47 (8), 757–761.

Paola, C., 2000. Quantitative models of sedimentary basin filling. Sedimentology 47 (Suppl. 1), 121–178.

Paola, C., Heller, P.L., Angevine, C.L., 1992. The large-scale dynamics of grain-size variation in alluvial basins, 1: theory. Basin Res. 4, 73–90.

Paola, C., Mullin, J., Ellis, C., Mohrig, D.C., Swenson, J.B., Parker, G., Hickson, T., Heller, P.L., Pratson, L., Syvitski, J., Sheets, B., Strong, N., 2001. Experimental stratigraphy. GSA Today 11 (7), 4–9.

Parizot, M., Eriksson, P.G., Aifa, T., Sarkar, S., Banerjee, S., Catuneanu, O., Altermann, W., Bumby, A.J., Bordy, E.M., van Rooy, J.L., Boshoff, A.J., 2005. Suspected microbial mat-related crack-like sedimentary structures in the Palaeoproterozoic Magaliesberg Formation sandstones, South Africa. Precambrian Res. 138, 274–296.

Park, R.K., 2011. The impact of sea-level change on ramp margin deposition: lessons from the Holocene sabkhas of Abu Dhabi, United Arab Emirates. In: Kendall, C.G.S.C., Alsharhan, A.S., Jarvis, I., Stevens, T. (Eds.), Quaternary Carbonate and Evaporite Sedimentary Facies and Their Ancient Analogues, vol. 43. International Association of Sedimentologists, pp. 89–112. Special Publication.

Passchier, S., Browne, G., Field, B., Fielding, C.R., Krissek, L.A., Panter, K., Pekar, S.F., ANDRILL-SMS Science Team, 2011. Early and middle Miocene Antarctic glacial history from the sedimentary facies distribution in the AND-2A drill hole, Ross Sea, Antarctica. GSA Bull. 123, 2352–2365.

Pattison, S.A.J., 1995. Sequence stratigraphic significance of sharp-based lowstand shoreface deposits, Kenilworth Member, Book Cliffs, Utah. AAPG Bull. 79, 444–462.

Pattison, S.A.J., 2019. Re-evaluating the sedimentology and sequence stratigraphy of classic Book Cliffs outcrops at Tusher and Thompson canyons, Eastern Utah, USA: Applications to correlation, modelling, and prediction in similar nearshore terrestrial to shallow marine subsurface settings worldwide. Mar. & Petrol. Geol. 102, 202–230.

Patzkowsky, M.E., Holland, S.M., 2012. Stratigraphic Paleobiology: Understanding the Distribution of Fossil Taxa in Time and Space. The University of Chicago Press, Chicago, p. 259.

Payton, C.E. (Ed.), 1977. Seismic Stratigraphy — Applications to Hydrocarbon Exploration, vol. 26. American Association of Petroleum Geologists Memoir, p. 516.

Pazos, P.J., 2002. The late Carboniferous glacial to postglacial transition: facies and sequence stratigraphy, Western Paganzo Basin, Argentina. Gondwana Res. 5, 467—487.

Pearce, T.J., Besly, B.M., Wray, D.S., Wright, D.K., 1999. Chemostratigraphy: a method to improve inter-well correlation in barren sequences — a case study using onshore Duckmantian/Stephanian sequences (West Midlands, U.K.). Sed. Geol. 124, 197—220.

Pedersen, T.F., Calvert, S.E., 1990. Anoxia vs. productivity: What controls the formation of organic-carbon-rich sediments and sedimentary rocks? Am. Assoc. Petrol. Geol. Bull. 74, 454—466.

Pekar, S.F., Kominz, M.A., 2001. Two-dimensional paleoslope modeling: a new method for estimating water depths for benthic foraminiferal biofacies and paleo shelf margins. J. Sediment. Res. 71, 608—620.

Pekar, S.F., Christie-Blick, N., Kominz, M.A., Miller, K.G., 2001. Evaluating the stratigraphic response to eustasy from Oligocene strata in New Jersey. Geology 29, 55—58.

Pellegrini, C., Maselli, V., Gamberi, F., Asioli, A., Bohacs, K.M., Drexler, T.M., Trincardi, F., 2017. How to make a 350-m-thick lowstand systems tract in 17,000 years: The Late Pleistocene Po River (Italy) lowstand wedge. Geology 45 (4), 327—330.

Pellegrini, C., Asioli, A., Bohacs, K.M., Drexler, T.M., Feldman, H.R., Sweet, M.L., Maselli, V., Rovere, M., Gamberi, F., Dalla Valle, G., Trincardi, F., 2018. The Late Pleistocene Po River lowstand wedge in the Adriatic Sea: controls on architecture variability and sediment partitioning. Mar. & Petrol. Geol. 96, 16—50.

Pemberton, S.G., Frey, R.W., 1985. The Glossifungites ichnofacies: modern examples from the Georgia coast, U.S.A. In: Curran, H.A. (Ed.), Biogenic Structures: Their Use in Interpreting Depositional Environments. Society of Economic Paleontologists and Mineralogists, pp. 237—259. Special Publication No. 35.

Pemberton, S.G., MacEachern, J.A., 1995. The sequence stratigraphic significance of trace fossils: examples from the Cretaceous foreland basin of Alberta, Canada. In: Van Wagoner, J.C., Bertram, G. (Eds.), Sequence Stratigraphy of Foreland basin Deposits — Outcrop and Subsurface Examples from the Cretaceous of North America, vol. 64. American Association of Petroleum Geologists Memoir, pp. 429—475.

Pemberton, S.G., Spila, M., Pulham, A.J., Saunders, T., MacEachern, J.A., Robbins, D., Sinclair, I.K., 2001. Ichnology and sedimentology of shallow to marginal marine systems: Ben Nevis and Avalon reservoirs, Jeanne d'Arc Basin. Geol. Assoc. Can. 15, 343. Short Course Notes.

Pemberton, S.G., Gingras, M.K., Dashtgard, S.E., Bann, K.L., MacEachern, J.A., 2009. The Teichichnus Ichnofacies: A Temporally and Spatially Recurring Ethological Grouping Characteristic of Brackish-Water Conditions. American Association of Petroleum Geologists, Annual Convention (Denver).

Pemberton, S.G., MacEachern, J.A., Dashtgard, S.E., Bann, K.L., Gingras, M.K., Zonneveld, J.-P., 2012. Shorefaces. In: Knaust, D., Bromley, R.G. (Eds.), Trace Fossils as Indicators of Sedimentary Environments, Developments in Sedimentology, vol. 64. Elsevier, pp. 563—603.

Peper, T., 1994. Tectonic and eustatic control on Albian shallowing (Viking and Paddy Formations) in the Western Canada foreland basin. Geol. Soc. Am. Bull. 106, 253—264.

Peper, T., Cloetingh, S., 1992. Lithosphere dynamics and tectono-stratigraphic evolution of the Mesozoic Betic rifted margin (Southeastern Spain). Tectonophysics 203, 345—361.

Peper, T., Cloetingh, S., 1995. Autocyclic perturbations of orbitally forced signals in the sedimentary record. Geology 23, 937—940.

Peper, T., Beekman, F., Cloetingh, S., 1992. Consequences of thrusting and intraplate stress fluctuations for vertical motions in foreland basins and peripheral areas. Geophys. J. Int. 111, 104—126.

Peper, T., Van Balen, R., Cloetingh, S., 1995. Implications of orogenic growth, intraplate stress variations, and eustatic sea-level change for foreland basin stratigraphy — inferences from numerical modeling. In: Dorobek, S.L., Ross, G.M. (Eds.), Stratigraphic Evolution of Foreland Basins. SEPM (Society for Sedimentary Geology), pp. 25—35. Special Publication No. 52.

Peters, S.E., Husson, J.M., 2017. Sediment cycling on continental and oceanic crust. Geology 45 (4), 323—326.

Peters, S.E., Antar, M.S.M., Zalmout, I.S., Gingerich, P.D., 2009. Sequence stratigraphic control on preservation of Late Eocene whales and other vertebrates at Wadi Al-Hitan, Egypt. Palaios 25, 290—302.

Pettijohn, F.J., 1975. Sedimentary Rocks, third ed. Harper & Row, New York, p. 628.

Pitman, W.C., Golovchenko, X., 1988. Sea-level changes and their effect on the stratigraphy of Atlantic-type margins. In: Sheridan, R.E., Grow, J.A. (Eds.), The Geology of North America, The Atlantic Continental Margin, United States, vols. I-2. Geological Society of America, Boulder, Colorado, pp. 545—565.

Platt, N.H., Keller, B., 1992. Distal alluvial deposits in a foreland basin setting — the lower freshwater Molasse (lower Miocene), Switzerland: sedimentology, architecture and palaeosols. Sedimentology 39, 545—565.

Playford, P.E., Hocking, R.M., Cockbain, A.E., 2009. Devonian reef complexes of the canning basin, Western Australia. Geol. Surv. West. Austr. Bull. 145, 444.

Playter, T., Corlett, H., Konhauser, K., Robbins, L., Rohais, S., Crombez, V., MacCormack, K., Rokosh, D., Prenoslo, D., Furlong, C.M., Pawlowicz, J., Gingras, M., Lalonde, S., Lyster, S., Zonneveld, J.-P., 2018. Clinoform identification and correlation in fine-grained sediments: A case study using the Triassic Montney Formation. Sedimentology 65, 263—302.

Playton, T.E., Kerans, C., 2002. Slope and toe-of-slope deposits shed from a late Wolfcampian tectonically active carbonate ramp margin. Gulf Coast Assoc. Geol. Soc. Transac. 52, 811—820.

Plint, A.G., 1988. Sharp-based shoreface sequences and "offshore bars" in the Cardium Formation of Alberta; their relationship to relative changes in sea level. In: Wilgus, C.K., Hastings, B.S., Kendall, C.G.S.C., Posamentier, H.W., Ross, C.A., Van Wagoner, J.C. (Eds.), Sea Level Changes — an Integrated Approach, vol. 42. SEPM, pp. 357—370. Special Publication.

Plint, A.G., 1991. High-frequency relative sea-level oscillations in Upper Cretaceous shelf clastics of the Alberta foreland basin: possible evidence for a glacio-eustatic control? In: Macdonald, D.I.M. (Ed.), Sedimentation, Tectonics and Eustasy: Sea-Level Changes at Active Margins, vol. 12. International Association of Sedimentologists, Blackwell, pp. 409—428. Special Publication.

Plint, A.G., 2000. Sequence stratigraphy and paleogeography of a Cenomanian deltaic complex: the Dunvegan and lower Kaskapau formations in subsurface and outcrop, Alberta and British Columbia, Canada. Bull. Can. Pet. Geol. 48 (1), 43—79.

Plint, A.G., Nummedal, D., 2000. The falling stage systems tract: recognition and importance in sequence stratigraphic analysis. In: Hunt, D., Gawthorpe, R.L. (Eds.), Sedimentary Response to Forced Regressions, vol. 172. Geological Society of London, pp. 1—17. Special Publication.

Plint, A.G., Eyles, N., Eyles, C.H., Walker, R.G., 1992. Control of sea level change. In: Walker, R.G., James, N.P. (Eds.), Facies Models: Response to Sea Level Change, vol. 1. Geological Association of Canada, GeoText, pp. 15—25.

Plint, A.G., Hart, B.S., Donaldson, W.S., 1993. Lithospheric flexure as a control on stratal geometry and facies distribution in upper Cretaceous rocks of the Alberta foreland basin. Basin Res. 5, 69—77.

Pomar, L., Kendall, C.G.S.C., 2008. Carbonate platform architecture; a response to hydrodynamics and evolving ecology. In: Lukasik, J., Simo, T. (Eds.), Controls on Carbonate Platform and Reef Development, vol. 89. SEPM, pp. 187—216. Special Publication.

Pomar, L., Tropeano, M., 2001. The Calcarenite di Gravina Formation in Matera (Southern Italy): new insights for coarse-grained, large-scale, cross-bedded bodies encased in offshore deposits. AAPG Bull. 85, 661–689.

Porebski, S.J., Steel, R.J., 2006. Deltas and sea-level change. J. Sediment. Res. 76, 390–403.

Posamentier, H.W., 2000. Seismic stratigraphy into the next millennium; a focus on 3D seismic data. In: American Association of Petroleum Geologists Annual Convention, New Orleans, vol. 9, p. A118. Abstracts Volume.

Posamentier, H.W., 2001. Lowstand alluvial bypass systems: incised vs. unincised. Am. Assoc. Pet. Geol. Bull. 85 (10), 1771–1793.

Posamentier, H.W., 2002. Ancient shelf ridges — a potentially significant component of the transgressive systems tract: case study from offshore Northwest Java. Am. Assoc. Pet. Geol. Bull. 86/1, 75–106.

Posamentier, H.W., 2003. Depositional elements associated with a basin floor channel-levee system: case study from the Gulf of Mexico. Mar. & Pet. Geol. 20, 677–690.

Posamentier, H.W., 2004. Seismic geomorphology: imaging elements of depositional systems from shelf to deep basin using 3D seismic data: implications for exploration and development. In: Davies, R.J., Cartwright, J.A., Stewart, S.A., Lappin, M., Underhill, J.R. (Eds.), 3D Seismic Technology: Application to the Exploration of Sedimentary Basins, vol. 29. Geological Society, London, Memoir, pp. 11–24.

Posamentier, H.W., Allen, G.P., 1993. Variability of the sequence stratigraphic model: effects of local basin factors. Sediment. Geol. 86, 91–109.

Posamentier, H.W., Allen, G.P., 1999. Siliciclastic sequence stratigraphy: concepts and applications. SEPM Conc. Sediment. & Paleontol. 7, 210.

Posamentier, H.W., Chamberlain, C.J., 1993. Sequence stratigraphic analysis of Viking Formation lowstand beach deposits at Joarcam Field, Alberta, Canada. In: Posamentier, H.W., Summerhayes, C.P., Haq, B.U., Allen, G.P. (Eds.), Sequence Stratigraphy and Facies Associations, vol. 18. International Association of Sedimentologists, pp. 469–485. Special Publication.

Posamentier, H.W., James, D.P., 1993. Sequence stratigraphy — uses and abuses. In: Posamentier, H.W., Summerhayes, C.P., Haq, B.U., Allen, G.P. (Eds.), Sequence Stratigraphy and Facies Associations, vol. 18. International Association of Sedimentologists, pp. 3–18. Special Publication.

Posamentier, H.W., Kolla, V., 2003. Seismic geomorphology and stratigraphy of depositional elements in deep-water settings. J. Sediment. Res. 73 (3), 367–388.

Posamentier, H.W., Morris, W.R., 2000. Aspects of the stratal architecture of forced regressive deposits. In: Hunt, D., Gawthorpe, R.L. (Eds.), Sedimentary Responses to Forced Regressions, vol. 172. Geological Society, pp. 19–46. Special Publication.

Posamentier, H.W., Vail, P.R., 1985. Eustatic controls on depositional stratal patterns. In: Soc. Econ. Paleont. Mineral. (SEPM) Research Conference No. 6, Sea Level Changes — an Integrated Approach. Houston, October 20–23, 1985 (Abstract and Poster).

Posamentier, H.W., Vail, P.R., 1988. Eustatic controls on clastic deposition. II. Sequence and systems tract models. In: Wilgus, C.K., Hastings, B.S., Kendall, C.G.S.C., Posamentier, H.W., Ross, C.A., Van Wagoner, J.C. (Eds.), Sea Level Changes––An Integrated Approach, vol. 42. SEPM, pp. 125–154. Special Publication.

Posamentier, H.W., Jervey, M.T., Vail, P.R., 1988. Eustatic controls on clastic deposition. I. Conceptual framework. In: Wilgus, C.K., Hastings, B.S., Kendall, C.G.S.C., Posamentier, H.W., Ross, C.A., Van Wagoner, J.C. (Eds.), Sea Level Changes––An Integrated Approach, vol. 42. SEPM, pp. 110–124. Special Publication.

Posamentier, H.W., Allen, G.P., James, D.P., 1992a. High resolution sequence stratigraphy — the East Coulee Delta, Alberta. J. Sediment. Petrol. 62 (2), 310–317.

Posamentier, H.W., Allen, G.P., James, D.P., Tesson, M., 1992b. Forced regressions in a sequence stratigraphic framework: concepts, examples, and exploration significance. Am. Assoc. Pet. Geol. Bull. 76, 1687–1709.

Poulton, S.W., Fralick, P.W., Canfield, D.E., 2010. Spatial variability in oceanic redox structure 1.8 billion years ago. Nat. Geosci. 3 (7), 486–490.

Powell, J.W., 1875. Exploration of the Colorado River of the West and its Tributaries. U.S. Government Printing Office, Washington, D.C., p. 291

Powell, R.D., Cooper, J.M., 2002. A glacial sequence stratigraphic model for temperate glaciated continental shelves. In: Dowdeswell, J.A., Ó Cofaigh, C. (Eds.), Glacier-Influenced Sedimentation on High-Latitude Continental Margins, vol. 203. Geological Society, London, pp. 215–244. Special Publication.

Pratt, B.R., Haidl, F.M., 2008. Microbial patch reefs in Upper Ordovician Red River strata, Williston Basin, Saskatchewan: Signal of heating in a deteriorating epeiric sea. In: Pratt, B.R., Holmden, C. (Eds.), Dynamics of Epeiric Seas, vol. 48. Geological Association of Canada, pp. 303–340. Special Publication.

Pratt, B.R., James, N.P., Cowan, C.A., 1992. Peritidal carbonates. In: Walker, R.G., James, N.P. (Eds.), Facies Models: Response to Sea Level Change, vol. 1. Geological Association of Canada, GeoText, pp. 303–322.

Press, F., Siever, R., 1986. Earth, fourth ed. W.H. Freeman and Company, p. 656.

Press, F., Siever, R., Grotzinger, J., Jordan, T.H., 2004. Understanding Earth, fourth ed. W.H. Freeman and Company, New York, p. 567.

Price, E.D., 1999. The evidence and implication of polar ice during the Mesozoic. Earth Sci. Rev. 48, 183–210.

Price, R.A., 1994. Cordilleran tectonics and the evolution of the Western Canada Sedimentary Basin. In: Mossop, G.D., Shetsen, I. (Eds.), (Compilers) Geological Atlas of the Western Canada Sedimentary Basin. Canadian Society of Petroleum Geologists and Alberta Research Council, pp. 13–24.

Prosser, S., 1993. Rift-related Linked Depositional Systems and Their Seismic Expression, vol. 71. Geological Society, pp. 35–66. Special Publication.

Pulham, A.J., 1989. Controls on internal structure and architecture of sandstone bodies within Upper Carboniferous fluvial-dominated deltas, County Clare, Western Ireland. In: Whateley, M.K.G., Pickering, K.T. (Eds.), Deltas: Sites and Traps for Fossil Flues, vol. 41. Geological Society, pp. 179–203. Special Publication.

Purdy, E.G., Gischler, E., 2003. The Belize margin revisited. 1. Holocene marine facies. Int. J. Earth Sci. 92, 532–551.

Qayyum, F., de Groot, P., Hemstra, N., Catuneanu, O., 2014. 4D Wheeler diagrams: concept and applications. In: Smith, D.G., Bailey, R.J., Burgess, P.M., Fraser, A.J. (Eds.), Strata and Time: Probing the Gaps in Our Understanding, vol. 404. The Geological Society of London, p. 10. Special Publication.

Qayyum, F., Catuneanu, O., de Groot, P., 2015. Historical developments in Wheeler diagrams and future directions. Basin Res. 27, 336–350.

Qayyum, F., Betzler, C., Catuneanu, O., 2017. The Wheeler diagram, flattening theory, and time. Mar. & Pet. Geol. 86, 1417–1430.

Rabineau, M., Berné, S., Aslanian, D., Olivet, J.-L., Joseph, P., Guillocheau, F., Bourillet, J.-F., Ledrezen, E., Granjeon, D., 2005. Sedimentary sequences in the Gulf of Lion: a record of 100,000 years climatic cycles. Mar. & Pet. Geol. 22, 775–804.

Rabineau, M., Berné, S., Olivet, J.-L., Aslanian, D., Guillocheau, F., Joseph, P., 2006. Paleo sea levels reconsidered from direct observation of paleoshoreline position during Glacial Maxima (for the last 500,000 yr). Earth & Planet. Sci. Lett. 252, 119–137.

Rahmani, R.A., 1988. Estuarine tidal channel and nearshore sedimentation of a Late Cretaceous epicontinental sea, Drumheller, Alberta, Canada. In: de Boer, P.L., van Gelder, A., Nio, S.D. (Eds.), Tide-influenced Sedimentary Environments and Facies. D. Reidel, Dordrecht, The Netherlands, pp. 433–471.

Ramaekers, P., Catuneanu, O., 2004. Development and sequences of the Athabasca Basin, Early Proterozoic, Saskatchewan and Alberta, Canada. In: Eriksson, P.G., Altermann, W., Nelson, D., Mueller, W., Catuneanu, O. (Eds.), The Precambrian Earth: Tempos and Events, Developments in Precambrian Geology, vol. 12. Elsevier Science Ltd., Amsterdam, pp. 705–723.

Ramsbottom, W.H.C., 1979. Rates of transgression and regression in the Carboniferous of NW Europe. J. Geol. Soc. 136, 147–153. London.

Rankey, E.C., Bachtel, S.L., Kaufman, J., 1999. Controls on stratigraphic architecture of icehouse mixed carbonate—siliciclastic systems: a case study from the Holder Formation (Pennsylvanian, Virgillian), Sacramento Mountains, New Mexico. In: Harris, P.M., Saller, A.H., Simo, J.A. (Eds.), Advances in Sequence Stratigraphy: Application to Reservoirs Outcrops and Models, Soc. Econ. Petrol. Min., vol. 63. (Society for Sedimentary Geology) Spec. Publ., pp. 127–150

Ratcliffe, K.T., Wright, A.M., Schmidt, K., 2012. Application of inorganic whole—rock geochemistry to shale resource plays: an example from the Eagle Ford Shale Formation, Texas. Sediment. Rec. 10, 4–9.

Raymo, M.E., 1997. The timing of major climate terminations. Paleoceanography 12, 577–585.

Reading, H.G. (Ed.), 1978. Sedimentary Environments and Facies, second ed. Blackwell Scientific Publications, p. 615.

Reading, H.G. (Ed.), 1996. Sedimentary Environments: Processes, Facies and Stratigraphy, third ed. Blackwell Science, p. 688.

Reading, H.G., Collinson, J.D., 1996. Clastic coasts. In: Reading, H.G. (Ed.), Sedimentary Environments: Processes, Facies and Stratigraphy, third ed. Blackwell Science, pp. 154–231.

Reid, S.K., Dorobek, S.L., 1993. Sequence stratigraphy and evolution of a progradational foreland carbonate ramp, Lower Mississippian Mission Canyon Formation and stratigraphic equivalents, Montana and Idaho. In: Loucks, R.G., Sarg, J.F. (Eds.), Carbonate Sequence Stratigraphy, Recent Developments and Applications, vol. 57. American Association of Petroleum Geologists, Memoir, pp. 327–352.

Reijenstein, H.M., Posamentier, H.W., Bande, A., Lozano, F.A., Dominguez, R.F., Wilson, R., Catuneanu, O., Galeazzi, S., 2020. Seismic geomorphology, depositional elements, and clinoform sedimentary processes: Impact on unconventional reservoir prediction. In: Minisini, D., Fantin, M., Lanusse Noguera, I., Leanza, H.A. (Eds.), Integrated Geology of Unconventionals: The Case of the Vaca Muerta Play, Argentina. AAPG Memoir 121, pp. 237–266.

Reimer, P.J., Baillie, M.G.L., Bard, E., Bayliss, A., Beck, J.W., Bertrand, C.J.H., Blackwell, P.G., Buck, C.E., Burr, G.S., Cutler, K.B., et al., 2004. IntCal04 Terrestrial radiocarbon age calibration, 0–26 cal kyr BP. Radiocarbon 46 (3), 1029–1058.

Reinson, G.E., 1992. Transgressive barrier island and estuarine systems. In: Walker, R.G., James, N.P. (Eds.), Facies Models – Response to Sea Level Change, vol. 1. Geological Association of Canada, Geo-Text, pp. 179–194.

Rich, J.L., 1951. Three critical environments of deposition and criteria for recognition of rocks deposited in each of them. Geol. Soc. Am. Bull. 62, 1–20.

Rider, M.H., 1990. Gamma Ray log shape used as a facies indicator: critical analysis of an oversimplified methodology. In: Hurst, A., Lovell, M.A., Morton, A.C. (Eds.), Geological Applications of Wireline Logs. Geological Society of London, pp. 27–37. Special Publication Classics.

Rimmer, S.M., Thompson, J.A., Goodnight, S.A., Robl, T.L., 2004. Multiple controls on the preservation of organic matter in Devonian—Mississippian marine black shales: geochemical and petrographic evidence. Palaeogeogr. Palaeoclimatol. Palaeoecol. 215, 125–154.

Roberts, A.P., Winkelhofer, M., 2004. Why are geomagnetic excursions not always recorded in sediments? Constraints from post-depositional remanent magnetisation lock-in modelling. Earth & Planet. Sci. Lett. 227, 345–359.

Rodriguez-Lopez, J.P., Melendez, N., de Boer, P.L., Soria, A.R., 2012. Controls on marine-erg margin cycle variability: aeolian-marine interaction in the mid-Cretaceous Iberian Desert System, Spain. Sedimentology 59, 466–501.

Rodriguez-Lopez, J.P., Clemmensen, L.B., Lancaster, N., Mountney, N.P., Veiga, G.D., 2014. Archean to Recent aeolian sand systems and their sedimentary record: current understanding and future prospects. Sedimentology 61, 1487–1534.

Rogers, R.R., Kidwell, S.M., 2000. Associations of vertebrate skeletal concentrations and discontinuity surfaces in terrestrial and shallow marine records: a test in the Cretaceous of Montana. J. Geol. 108, 131–154.

Rosas, J.C., Korenaga, J., 2021. Archaean seafloors shallowed with age due to radiogenic heating in the mantle. Nat. Geosci. 14, 51–56.

Rossetti, D.F., 1998. Facies architecture and sequential evolution of an incised-valley estuarine fill: the Cujupe Formation (Upper Cretaceous to Lower Tertiary), Sao Luis Basin, Northern Brazil. J. Sediment. Res. 68, 299–310.

Roveri, M., Taviani, M., 2003. Calcarenite and sapropel deposition in the Mediterranean Pliocene: shallow- and deep-water record of astronomically driven climatic events. Terra Nova 15, 279–286.

Roveri, M., Lugli, S., Manzi, V., Reghizzi, M., Rossi, F.P., 2020. Stratigraphic relationships between shallow-water carbonates and primary gypsum: insights from the Messinian succession of the Sorbas Basin (Betic Cordillera, Southern Spain). Sediment. Geol. 404 (105678), 1–18.

Rubidge, B.S. (Ed.), 1995. Biostratigraphy of the Beaufort Group (Karoo Supergroup). Council for Geoscience, Geological Survey of South Africa, p. 46.

Ruffell, A.H., Price, G.D., Mutterlose, J., Kessels, K., Baraboshkin, E., Grocke, D.R., 2002a. Palaeoclimate indicators (clay minerals, calcareous nannofossils, stable isotopes) compared from two successions in the late Jurassic of the Volga Basin (SE Russia). Geol. J. 37, 17–33.

Ruffell, A.H., McKinley, J.M., Worden, R.H., 2002b. Comparison of clay mineral stratigraphy to other proxy palaeoclimate indicators in the Mesozoic of NW Europe. Phil. Trans. Roy. Soc 360, 675–693. London.

Rüther, D.C., Andreassen, K., Spagnolo, M., 2013. Aligned glaciotectonic rafts on the central Barents Sea seafloor revealing extensive glacitectonic erosion during the last deglaciation. Geophys. Res. Lett. 40, 6351–6355.

Ryang, W.H., Chough, S.K., 1997. Sequential development of alluvial/lacustrine system: Southeastern Eumsung Basin (Cretaceous), Korea. J. Sediment. Res. 67 (2), 274–285.

Sadler, P.M., Kemple, W.G., Kooser, M.A., 2003. CONOP9 programs for solving the stratigraphic correlation and seriation problems as constrained optimization. In: Harries, P. (Ed.), High Resolution Approaches in Stratigraphic Paleontology. Kluwer Academic, Dordrecht, pp. 461–465.

Safronova, P.A., Henriksen, S., Andreassen, K., Laberg, J.S., Vorren, T.O., 2014. Evolution of shelf-margin clinoforms and deep-water fans during the middle Eocene in the SØrvestsnaget Basin, Southwest Barents Sea. AAPG Bull. 98 (3), 515–544.

Sageman, B.B., Murphy, A.E., Werne, J.P., Ver Straeten, C.A., Hollander, D.J., Lyons, T.W., 2003. A tale of shales: the relative roles of production, decomposition, and dilution in the accumulation of organic-rich strata, Middle-Upper Devonian, Appalachian Basin. Chem. Geol. 195, 229–273.

Saha, S., Burley, S.D., Banerjee, S., Ghosh, A., Saraswati, P.K., 2016. The morphology and evolution of tidal sand bodies in the macrotidal Gulf of Khambhat, Western India. Mar. & Petrol. Geol. 77, 1–17.

Salazar, M., Moscardelli, L., Wood, L., 2016. Utilising clinoform architecture to understand the drivers of basin margin evolution: a case study in the Taranaki Basin, New Zealand. Basin Res. 28, 840–865.

Saller, A.H., Barton, J.W., Barton, R.E., 1989. Slope sedimentation associated with a vertically building shelf, Bone Spring Formation, Mescalero Escarpe Field, southeastern New Mexico. In: Crevello, P.G., Wilson, J.J., Sarg, J.F., Read, J.F. (Eds.), Carbonate Platform and Basin Development, vol. 44. SEPM, pp. 275–288. Special Publication.

Sanders, D., Höfling, R., 2000. Carbonate deposition in mixed siliciclastic-carbonate environments on top of an orogenic wedge (Late Cretaceous, Northern Calcareous Alps, Austria). Sediment. Geol. 137, 127–146.

Sano, J.L., Ratcliffe, K.T., Spain, D.R., 2013. Chemostratigraphy of the Haynesville Shale. In: Hammes, U., Gales, J. (Eds.), Geology of the Haynesville Gas Shale in East Texas and West Louisiana, vol. 105. U.S.A. American Association of Petroleum Geologists Memoir, pp. 137–154.

Santisteban, C., Taberner, C., 1988. Sedimentary models of siliciclastic deposits and coral reef interactions. In: Doyle, L.J., Roberts, H.H. (Eds.), Carbonate–Clastic Transitions, Developments in Sedimentology, vol. 42. Elsevier, Amsterdam, pp. 35–76.

Sarg, J.F., 1988. Carbonate sequence stratigraphy. In: Wilgus, C.K., Hastings, B.S., Kendall, C.G.S.C., Posamentier, H.W., Ross, C.A., Van Wagoner, J.C. (Eds.), Sea Level Changes — An Integrated Approach, vol. 42. SEPM, pp. 155–182. Special Publication.

Sarg, J.F., 2001. The sequence stratigraphy, sedimentology, and economic importance of evaporite-carbonate transitions: a review. Sediment. Geol. 140, 9–42.

Sarkar, S., Banerjee, S., Eriksson, P.G., Catuneanu, O., 2005. Microbial mat control on siliciclastic Precambrian sequence stratigraphic architecture. Sediment. Geol. 176, 195–209.

Sarmiento Rojas, L.F., 2001. Mesozoic Rifting and Cenozoic basin Inversion History of the Eastern Cordillera, Colombian Andes. Ph.D. Thesis. Vrije University, Amsterdam, The Netherlands, p. 319.

Satkoski, A.M., Fralick, P.W., Beard, B.L., Johnson, C.M., 2017. Initiation of modern-style plate tectonics recorded in Mesoarchean marine chemical sediments. Geochem. Acta 209, 216–232.

Sattler, U., Immenhauser, A., Hillgartner, H., Esteban, M., 2005. Characterization, lateral variability and lateral extent of discontinuity surfaces on a carbonate platform (Barremian to Lower Aptian, Oman). Sedimentology 52, 339–361.

Saucier, R.T., 1974. Quaternary geology of the lower Mississippi valley. Arkansas Archeol. Surv. 26. Research Series No. 6.

Saul, G., Naish, T.R., Abbott, S.T., Carter, R.M., 1999. Sedimentary cyclicity in the marine Pliocene-Pleistocene of the Wanganui Basin (New Zealand): sequence stratigraphic motifs characteristic of the past 2.5 m.y. Geol. Soc. Am. Bull. 111, 524–537.

Savrda, C.E., 1991. Teredolites, wood substrates, and sea-level dynamics. Geology 19, 905–908.

Savrda, C.E., Bottjer, D.J., 1989. Trace-fossil model for reconstructing oxygenation histories of ancient marine bottom waters: application to Upper Cretaceous Niobrara Formation, Colorado. Palaeogeogr. Palaeoclimatol. Palaeoecol. 74, 49–74.

Savrda, C.E., Blanton-Hooks, A.D., Collier, J.W., Drake, R.A., Graves, R.L., Hall, A.G., Nelson, A.I., Slone, J.C., Williams, D.D., Wood, H.A., 2000. Taenidium and associated ichnofossils in fluvial deposits, Cretaceous Tuscaloosa Formation, Eastern Alabama, Southeastern USA. Ichnos 7, 227–242.

Scarponi, D., Kowalewski, M., 2004. Stratigraphic paleoecology: bathymetric signatures and sequence overprint of mollusk associations from upper Quaternary sequences of the Po Plain, Italy. Geology 32, 989–992.

Scherer, C.M.S., Lavina, E.L.C., Dias Filho, D.C., Oliveira, F.M., Bongiolo, D.E., Aguiar, E.S., 2007. Stratigraphy and facies architecture of the fluvial–aeolian–lacustrine Sergi Formation (Upper Jurassic), Reconcavo Basin, Brazil. Sediment. Geol. 194, 169–193.

Atlas of microbial mat features preserved within the siliciclastic rock record. In: Schieber, J., Bose, P.K., Eriksson, P.G., Banerjee, S., Sarkar, S., Altermann, W., Catuneanu, O. (Eds.), 2007. Atlases in Geoscience, vol. 2. Elsevier, Amsterdam), p. 311.

Schlager, W., 1981. The paradox of drowned reefs and carbonate platforms. GSA Bullet. 92, 197–211.

Schlager, W., 1989. Drowning unconformities on carbonate platforms. In: Crevello, P.D., Wilson, J.L., Sarg, J.F., Read, J.F. (Eds.), Controls on Carbonate Platform and Basin Development, vol. 44. SEPM (Society of Economic Paleontologists and Mineralogists), pp. 15–25. Special Publication.

Schlager, W., 1992. Sedimentology and Sequence Stratigraphy of Reefs and Carbonate Platforms. Continuing Education Course Note Series #34. American Association of Petroleum Geologists, p. 71.

Schlager, W., 1993. Accommodation and supply — a dual control on stratigraphic sequences. In: Cloetingh, S., Sassi, W., Horvath, F., Puigdefabregas, C. (Eds.), Basin Analysis and Dynamics of Sedimentary Basin Evolution, vol. 86. Sedimentary Geology, pp. 111–136.

Schlager, W., 2004. Fractal nature of stratigraphic sequences. Geology 32, 185–188.

Schlager, W., 2005. Carbonate sedimentology and sequence stratigraphy. SEPM Conc. Sedimentol. & Paleontol. #8, 200.

Schlager, W., 2010. Ordered hierarchy versus scale invariance in sequence stratigraphy. Int. J. Earth Sci. 99, S139–S151.

Schlager, W., Reijmer, J.J.G., Droxler, A., 1994. Highstand shedding of carbonate platforms. Jour. of Sed. Res. B64, 270–281.

Schreiber, B.C., Hsu, K.J., 1980. Evaporites. In: Hobson, G.D. (Ed.), Developments in Petroleum Geology — 2. Applied Science Publishers Ltd., London, pp. 87–138.

Schultz, S.K., MacEachern, J.A., Catuneanu, O., Dashtgard, S.E., 2020. Coeval deposition of transgressive and normal regressive stratal packages in a structurally controlled area of the Viking Formation, Central Alberta, Canada. Sedimentology 67, 2974–3002.

Schumm, S.A., 1993. River response to baselevel change: implications for sequence stratigraphy. J. Geol. 101, 279–294.

Schwarz, E., Veiga, G.D., Alvarez Trentini, G., Spalletti, L., 2016. Climatically versus eustatically controlled, sediment-supply-driven cycles: carbonate-siliciclastic, high-frequency sequences in the Valanginian of the Neuquén Basin (Argentina). J. Sediment. Res. 86, 312–335.

Schwarzacher, W., 1993. Cyclostratigraphy and the Milankovitch Theory. Developments in Sedimentology 52. Elsevier, Amsterdam, p. 225.

Scott, J.J., Renaut, R.W., Buatois, L.A., Owen, R.B., 2009. Biogenic structures in exhumed surfaces around saline lakes: and example from Lake Bogoria, Kenya Rift Valley. Palaeogeogr. Palaeoclimatol. Palaeoecol. 272, 176–198.

Scott, J.J., Buatois, L.A., Mangano, M.G., 2012. Lacustrine environments. In: Knaust, D., Bromley, R.G. (Eds.), Trace Fossils as Indicators of Sedimentary Environments, Developments in Sedimentology, vol. 64. Elsevier, pp. 379–417.

Seilacher, A., 1964. Biogenic sedimentary structures. In: Imbrie, J., Newell, N. (Eds.), Approaches to Paleoecology. John Wiley, New York, pp. 296–316.

Seilacher, A., 1967. Bathymetry of trace fossils. Mar. Geol. 5, 413–428.

Seilacher, A., 1978. Use of trace fossil assemblages for recognizing depositional environments. In: Basan, P.B. (Ed.), Trace Fossil Concepts. Society of Economic Paleontologists and Mineralogists, Short Course 5, pp. 185–201.

Selley, R.C., 1978a. Ancient Sedimentary Environments, second ed. Cornell University Press, Ithaca, New York, p. 287.

Selley, R.C., 1978b. Concepts and Methods of Subsurface Facies Analysis, vol. 9. American Association of Petroleum Geologists, Continuing Education Course Notes Series, p. 82.

Semtner, A.K., Klitzsch, E., 1994. Early Paleozoic paleogeography of the northern Gondwana margin: new evidence for Ordovician-Silurian glaciation. Geol. Rundsch 83, 743–751.

Serra, O., Abbott, H.T., 1982. The contribution of logging data to sedimentology and stratigraphy. Soc. Pet. Eng. J. 22 (1), 117–131.

Sethi, P.S., Leithold, E.L., 1994. Climatic cyclicity and terrigenous sediment influx to the early Turonian Greenhorn sea, Southern Utah. J. Sediment. Res. B64, 26–39.

Shanley, K.W., McCabe, P.J., 1991. Predicting facies architecture through sequence stratigraphy — an example from the Kaiparowits Plateau, Utah. Geology 19, 742–745.

Shanley, K.W., McCabe, P.J., 1993. Alluvial architecture in a sequence stratigraphic framework: a case history from the Upper Cretaceous of Southern Utah, U.S.A. In: Flint, S., Bryant, I. (Eds.), Quantitative Modeling of Clastic Hydrocarbon Reservoirs and Outcrop Analogues, vol. 15. International Association of Sedimentologists, pp. 21–55. Special Publication.

Shanley, K.W., McCabe, P.J., 1994. Perspectives on the sequence stratigraphy of continental strata. Am. Assoc. Pet. Geol. Bull. 78, 544–568.

Shanley, K.W., McCabe, P.J. (Eds.), 1998. Relative Role of Eustasy, Climate, and Tectonism in Continental Rocks. SEPM, p. 234. Special Publication No. 59.

Shanley, K.W., McCabe, P.J., Hettinger, R.D., 1992. Significance of tidal influence in fluvial deposits for interpreting sequence stratigraphy. Sedimentology 39, 905–930.

Shanmugam, G., Poffenberger, M., Alava, J.T., 2000. Tide-dominated estuarine facies in the Hollin and Napo ("T" and "U") formations (Cretaceous), Sacha Field, Oriente Basin, Ecuador. Am. Assoc. Pet. Geol. Bull. 84 (5), 652–682.

Shaw, A.B., 1964. Time in Stratigraphy. McGraw-Hill, New York, p. 365.

Shields, G.A., 2007. A normalized seawater strontium isotope curve: possible implications for Neoproterozoic-Cambrian weathering rates and the further oxygenation of the earth. eEarth 2, 35–42.

Siggerud, E.I.H., Steel, R.J., 1999. Architecture and trace-fossil characteristics of a 10,000-20,000 year, fluvial-to-marine sequence, SE Ebro Basin, Spain. J. Sediment. Res. 69B, 365–383.

Simo, A., 1989. Upper Cretaceous platform to basin depositional sequence development, Tremp Basin, South Central Pyrenees, Spain. In: Crevello, P.G., Wilson, J.J., Sarg, J.F., Read, J.F. (Eds.), Carbonate Platform and Basin Development, vol. 44. Soc. Econ. Petrol. Min. Spec. Publ., pp. 233–257.

Simons, D.B., Richardson, E.V., Nordin, C.F., 1965. Sedimentary structures generated by flow in alluvial channels. In: Middleton, G.V. (Ed.), Primary Sedimentary Structures and Their Hydrodynamic Interpretation, vol. 12. Society of Economic Paleontologists and Mineralogists, pp. 34–52. Special Publication.

Simpson, E., Eriksson, K.A., 1990. Early Cambrian progradational and transgressive sedimentation patterns in Virginia: an example of the early history of a passive margin. J. Sediment. Pet. 60, 84–100.

Sinclair, H.D., Tomasso, M., 2002. Depositional evolution of confined turbidite basins. J. Sediment. Res. 72, 451–456.

Slingerland, R., Kump, L.R., Arthur, M.A., Fawcett, P.J., Sageman, B.B., Barron, E.J., 1996. Estuarine circulation in the Turonian western interior seaway of North America. Geol. Soc. Am. Bull. 108, 941–952.

Sloss, L.L., 1962. Stratigraphic models in exploration. Am. Assoc. Petrol. Geol. Bull. 46, 1050–1057.

Sloss, L.L., 1963. Sequences in the cratonic interior of North America. Geol. Soc. Am. Bull. 74, 93–114.

Sloss, L.L., Krumbein, W.C., Dapples, E.C., 1949. Integrated facies analysis. In: Longwell, C.R. (Ed.), Sedimentary Facies in Geologic History, vol. 39. Geological Society of America Memoir, pp. 91–124.

Slowakiewicz, M., Mikolajewski, Z., 2009. Sequence stratigraphy of the Upper Permian Zechstein Main Dolomite carbonates in Western Poland: a new approach. J. Pet. Geol. 32, 215–234.

Smith, A.B., Gale, A.S., Monks, N.E.A., 2001. Sea-level change and rock-record bias in the Cretaceous: a problem for extinction and biodiversity studies. Paleobiology 27, 241–253.

Smith, L.B., Read, J.F., 2001. Discrimination of local and global effects on Upper Mississippian stratigraphy, Illinois Basin, U.S.A. J. Sed. Res. 71, 985–1002.

Smith, W., 1819. Strata Identified by Organized Fossils, pp. 1816–1819. London.

Snedden, J.W., 1984. Validity of the use of the spontaneous potential curve shape in the interpretation of sandstone depositional environments. Gulf Coast Assoc. Geol. Soc. Transac. 34, 255–263.

Snedden, J.W., 1991. Origin and sequence stratigraphic significance of large dwelling traces in the Escondido Formation (Cretaceous, Texas, USA). Palaios 6 (6), 541–552.

Snedden, J.W., Tillman, R.W., Kreisa, R.D., Schweller, W.J., Culver, S.J., Winn Jr., R.D., 1994. Stratigraphy and genesis of a modern shoreface-attached sand ridge, Peahala Ridge, New Jersey. J. Sediment. Res. B64, 560–581.

Soil Survey Staff, 1975. Soil Taxonomy, vol. 436. U.S. Department of Agriculture Handbook, p. 754.

Soil Survey Staff, 1998. Keys to Soil Taxonomy, eighth ed. U.S. Department of Agriculture, Natural Resources Conservation Services, p. 327.

Somme, T.O., Howell, J.A., Hampson, G.J., Storms, J.E.A., 2008. Genesis, architecture, and numerical modeling of intra-parasequence discontinuity surfaces in wave-dominated deltaic deposits: Upper Cretaceous Sunnyside Member, Blackhawk Formation, Book Cliffs, Utah, USA. In: Hampson, G.J., Steel, R.J., Burgess, P.M., Dalrymple, R.W. (Eds.), Recent Advances in Models of Siliciclastic Shallow-Marine Stratigraphy, vol. 90. SEPM, pp. 421–441. Special Publication.

Soreghan, G.S., 1997. Walther's law, climate change and upper Paleozoic cyclostratigraphy in the ancestral Rocky Mountains. J. Sed. Res. 67, 1001–1004.

Soreghan, G.S., Elmore, R.D., Katz, B., Cogoini, M., Banerjee, S., 1997. Pedogenically enhanced magnetic susceptibility variations preserved in Paleozoic loessite. Geology 25, 1003–1006.

Southgate, P.N., Kennard, J.M., Jackson, M.J., O'Brien, P.E., Sexton, M.J., 1993. Reciprocal lowstand clastic and highstand carbonate sedimentation, subsurface Devonian Reef Complex, Canning Basin, Western Australia. In: Loucks, R.G., Sarg, J.F. (Eds.), Carbonate Sequence Stratigraphy, vol. 57. AAPG Memoir, pp. 157–179.

Spence, G.H., Tucker, M.E., 1997. Genesis of limestone megabreccias and their significance in carbonate sequence stratigraphic models: a review. Sediment. Geol. 112, 163–193.

Spence, G.H., Tucker, M.E., 2007. A proposed integrated multi-signature model for peritidal cycles in carbonates. J. Sediment. Res. 77, 797–808.

Spencer-Cervato, C., Thierstein, H.R., Lazarus, D.B., Beckmann, J.P., 1994. How synchronous are Neogene marine plankton events? Paleoceanography 9, 739–763.

Sprague, A.R., Patterson, P.E., Sullivan, M.D., Campion, K.M., Jones, C.R., Garfield, T.R., Sickafoose, D.K., Jennette, D.C., Jensen, G.N., Beaubouef, R.T., Goulding, F.J., Van Wagoner, J.C., Wellner, R.W., Larue, D.K., Rossen, C., Hill, R.E., Geslin, J.K., Feldman, H.R., Demko, T.M., Abreu, V., Zelt, F.B., Ardill, J., Porter, M.L., 2003. Physical stratigraphy of clastic strata: a hierarchical approach to the analysis of genetically related stratigraphic elements for improved reservoir prediction. South Texas Geol. Soc. Bull. 44, 7.

Stefani, M., Vincenzi, S., 2005. The interplay of eustasy, climate and human activity in the late Quaternary depositional evolution and sedimentary architecture of the Po Delta system. Mar. Geol. 222 (1), 19–48.

Steno, N., 1669. De Solido Intra Solidum Naturaliter Contento Dissertationis Prodromus (Firenze).

Stoll, H.M., Schrag, D.P., 1996. Evidence for glacial control of rapid sea level changes in the Early Cretaceous. Science 272, 1771–1774.

Stouthamer, E., Berendsen, H.J.A., 2007. Avulsion: the relative roles of autogenic and allogenic processes. Sediment. Geol. 198 (Issues 3–4), 309–325.

Strasser, A., 2016. Hiatuses and condensation: an estimation of time lost on a shallow carbonate platform. Deposit. Rec. 1 (2), 91–117.

Strasser, A., 2018. Cyclostratigraphy of shallow-marine carbonates — limitations and opportunities. In: Montenari, M. (Ed.), Stratigraphy & Timescales, vol. 3. Academic Press, UK, pp. 151–187.

Strasser, A., Pittet, B., Hillgärtner, H., Pasquier, J.-B., 1999. Depositional sequences in shallow carbonate-dominated sedimentary systems: concepts for a high-resolution analysis. Sediment. Geol. 128, 201–221.

Strauss, H., 1993. The sulfur isotopic record of Precambrian sulfates: new data and a critical evaluation of the existing record. Precambrian Res. 63 (3–4), 225–246.

Summerfield, M.A., 1985. Plate tectonics and landscape development on the African continent. In: Morisawa, M., Hack, J. (Eds.), Tectonic Geomorphology. Allen and Unwin, Boston, pp. 27–51.

Summerfield, M.A., 1991. Global Geomorphology: An Introduction to the Study of Landforms. Wiley, New York, p. 537.

Sumner, D.Y., Grotzinger, J.P., 2004. Implications for Neoarchaean ocean chemistry from primary carbonate mineralogy of the Campbellrand-Malmani Platform, South Africa. Sedimentology 51, 1273–1299.

Suppe, J., Chou, G.T., Hook, S.C., 1992. Rates of folding and faulting determined from growth strata. In: McClay, K.R. (Ed.), Thrust Tectonics. Chapman and Hall, London, pp. 105–121.

Suter, J.R., Berryhill Jr., H.L., Penland, S., 1987. Late Quaternary sea level fluctuations and depositional sequences, Southwest Louisiana continental shelf. In: Nummedal, D., Pilkey, O.H., Howard, J.D. (Eds.), Sea-level Changes and Coastal Evolution. SEPM (Society for Sedimentary Geology), pp. 199–222. Special Publication No. 41.

Sweet, A.R., Long, D.G.F., Catuneanu, O., 2003. Sequence boundaries in fine-grained terrestrial facies: biostratigraphic time control is key to their recognition. In: Geological Association of Canada—Mineralogical Association of Canada Joint Annual Meeting, Vancouver, May 25–28, vol. 28, p. 165. Abstracts.

Sweet, A.R., Catuneanu, O., Lerbekmo, J.F., 2005. Uncoupling the position of sequence-bounding unconformities from lithological criteria in fluvial systems. In: American Association of Petroleum Geologists Annual Convention, 19–22 June 2005, Calgary, Alberta, vol. 14, p. A136. Abstracts Volume.

Sweet, K., Knoll, A.H., 1989. Marine pisolites from Upper Proterozoic carbonates of East Greenland and Spitsbergen. Sedimentology 36, 75–93.

Sweet, M.L., 2020. Controls on the grain size of sediment routed to deep water: insights from Quaternary systems. Res. Insights 2 (4), 1–13.

Sweet, M.L., Blum, M.D., 2016. Connections between fluvial to shallow marine environments and submarine canyons: Implications for sediment transfer to deep water. J. Sediment. Res. 86, 1147–1162.

Sweet, M.L., Gaillot, G.T., Jouet, G., Rittenour, T.M., Toucanne, S., Marsset, T., Blum, M.D., 2019. Sediment routing from shelf to basin floor in the Quaternary Golo System of Eastern Corsica, France, Western Mediterranean Sea. Geol. Soc. Am. Bull. 132, 1217–1234.

Swift, D.J.P., 1968. Coastal erosion and transgressive stratigraphy. J. Geol. 76, 444–456.

Swift, D.J.P., 1975. Barrier-island genesis: evidence from the Central Atlantic shelf, Eastern U.S.A. Sediment. Geol. 14, 1–43.

Swift, D.J.P., 1976. Coastal sedimentation. In: Stanley, D.J., Swift, D.J.P. (Eds.), Marine Sediment Transport and Environmental Management. John Wiley & Sons, New York, pp. 255–311.

Swift, D.J.P., Field, M.F., 1981. Evolution of a classic ridge field, Maryland sector, North American inner shelf. Sedimentology 28, 461–482.

Swift, D.J.P., Thorne, J.A., 1991. Sedimentation on continental margins I: a general model for shelf sedimentation. In: Swift, D.J.P., Oertel, G.F., Tillman, R.W., Thorne, J.A. (Eds.), Shelf Sand and Sandstone Bodies — Geometry, Facies, and Sequence Stratigraphy, vol. 14. International Association of Sedimentologists, Blackwell, Oxford, pp. 3–31. Special Publication.

Swift, D.J.P., Kofoed, J.W., Saulsbury, F.B., Sears, P.C., 1972. Holocene evolution of the shelf surface, Central and Southern Atlantic shelf of North America. In: Swift, D.J.P., Duane, D.B., Pilkey, O.H. (Eds.), Shelf Sediment Transport: Process and Pattern, Stroudsburg, Pennsylvania. Dowden, Hutchinson & Ross, pp. 499–574.

Swift, D.J.P., Parsons, B.S., Foyle, A., Oertel, G.F., 2003. Between beds and sequences: stratigraphic organization at intermediate scales in the Quaternary of the Virginia coast, USA. Sedimentology 50, 81–111.

Sylvia, D., Galloway, B., 2001. Response of the Brazos River (Texas USA) to latest Pleistocene climatic and eustatic change. In: Seventh International Conference on Fluvial Sedimentology, Lincoln, August 6–10, Program and Abstracts, p. 264.

Tabor, N.J., Myers, T.S., Michel, L.A., 2017. Sedimentologist's guide for recognition, description, and classification of paleosols. In: Zeigler, K.E., Parker, W.G. (Eds.), Terrestrial Depositional Systems: Deciphering Complexities through Multiple Stratigraphic Methods. Elsevier, pp. 165–208.

Takashima, R., Kawabe, F., Nishi, H., Moriya, K., Wani, R., Ando, H., 2004. Geology and stratigraphy of forearc basin sediments in Hokkaido, Japan: Cretaceous environmental events on the North-West Pacific margin. Cretac. Res. 25, 365–390.

Tandon, S.K., Gibling, M.R., 1994. Calcrete and coal in late Carboniferous cyclothems of Nova Scotia, Canada: Climate and sea-level changes linked. Geology 22 (8), 755–758.

Tandon, S.K., Gibling, M.R., 1997. Calcretes at sequence boundaries in upper Carboniferous cyclothems of the Sydney Basin, Atlantic Canada. Sediment. Geol. 112, 43–67.

Taylor, A.M., Gawthorpe, R.L., 1993. Application of sequence stratigraphy and trace fossil analysis to reservoir description: examples from the Jurassic of the North Sea. In: Parker, J.R. (Ed.), Petroleum Geology of Northwest Europe. Geological Society of London, pp. 317–336.

Tedeschi, L.R., Jenkyns, H.C., Robinson, S.A., Sanjinés, A.E., Viviers, M.C., Quintaes, C.M., Vazquez, J.C., 2017. New age constraints on Aptian evaporites and carbonates from the South Atlantic: implications for Oceanic Anoxic Event 1a. Geology 45 (6), 543–546.

Tesson, M., Gensous, B., Allen, G.P., Ravenne, C., 1990. Late Quaternary deltaic lowstand wedges on the rhone Continental Shelf, France. Mar. Geol. 91, 325–332.

Tesson, M., Posamentier, H.W., Gensous, B., 2000. Stratigraphic organization of Late Pleistocene deposits of the western part of the Rhone shelf (Languedoc shelf) from high resolution seismic and core data. Am. Assoc. Petrol. Geol. Bull. 84, 119–150.

Thorne, J., 1995. On the scale independent shape of prograding stratigraphic units. In: Barton, C.B., La Pointe, P.R. (Eds.), Fractals in Petroleum Geology and Earth Processes. Plenum Press, New York, pp. 97–112.

Tibert, N.E., Gibling, M.R., 1999. Peat accumulation on a drowned coastal braidplain: the Mullins Coal (Upper Carboniferous), Sydney Basin, Nova Scotia. Sediment. Geol. 128, 23–38.

Tipper, J.C., 1997. Modeling carbonate platform sedimentation—lag comes naturally. Geology 25 (6), 495–498.

Tomkeieff, S.I., 1962. Unconformity - an historical study. Proc. Geol. Assoc. 73, 383–417.

Trendall, A.F., 2002. The significance of iron-formation in the Precambrian stratigraphic record. In: Altermann, W., Corcoran, P.L. (Eds.), Precambrian Sedimentary Environments: A Modern Approach to Ancient Depositional Systems, vol. 33. International Association of Sedimentologists, Blackwell, Oxford, pp. 33—66. Special Publication.

Tucker, M.E., 1991. Sequence stratigraphy of carbonate-evaporite basins: models and applications to the Upper Permian (Zechstein) of Northeast England and adjoining North Sea. J. Geol. Soc. London 148, 1019—1036.

Tucker, M.E., 2003. Mixed clastic carbonate successions: Quaternary of Egypt, Mid Carboniferous, NE England. Geol. Croat. 56, 19—37.

Tucker, M.E., Garland, J., 2010. High-frequency cycles and their sequence stratigraphic context: orbital forcing and tectonic controls on Devonian cyclicity, Belgium. Geol. Belg. 13/3, 213—240.

Tucker, M.E., Wright, V.P., 1990. Carbonate Sedimentology. Blackwell Science Ltd., p. 482

Tucker, M.E., Calvet, F., Hunt, D., 1993. Sequence stratigraphy of carbonate ramps: systems tracts, models and application to the Muschelkalk carbonate platforms of Eastern Spain. In: Posamentier, H.W., Summerhayes, C.P., Haq, B.U., Allen, G.P. (Eds.), Sequence Stratigraphy and Facies Associations, vol. 18. International Association of Sedimentologists, pp. 397—415. Special Publication.

Tucker, M.E., Gallagher, J., Leng, M., 2009. Are beds millennial-scale cycles? An example from the Carboniferous of NE England. Sediment. Geol. 214, 19—34.

Turner, B.W., Slatt, R.M., 2016. Assessing bottom water anoxia within the Late Devonian Woodford Shale in the Arkoma Basin, Southern Oklahoma. Mar. & Pet. Geol. 78, 536—546.

Turner, B.W., Molinares-Blanco, C.E., Slatt, R.M., 2015. Chemostratigraphic, palynostratigraphic, and sequence stratigraphic analysis of the Woodford Shale, Wyche Farm Quarry, Pontotoc County, Oklahoma. Interpretation 3 (1), SH1—SH9.

Turner, B.W., Tréanton, J.A., Slatt, R.M., 2016. The use of chemostratigraphy to refine ambiguous sequence stratigraphic correlations in marine mudrocks. An example from the Woodford Shale, Oklahoma, USA. J. Geol. Soc. 173, 854—868.

Tyson, R.V., 2001. Sedimentation rate, dilution, preservation, and total organic carbon: Some results of a modelling study. Org. Geochem. 32, 333—339.

Tyson, R.V., 2005. The "productivity versus preservation" controversy: cause, flaws, and resolution. In: Harris, N.B. (Ed.), The Deposition of Organic-Carbon-Rich Sediments: Models, Mechanisms, and Consequences, vol. 82. Society for Sedimentary Geology (SEPM), pp. 17—33. Special Publication.

Tyson, R.V., Pearson, T.H., 1991. Modern and ancient continental shelf anoxia: An overview. In: Tyson, R.V., Pearson, T.H. (Eds.), Modern and Ancient Continental Shelf Anoxia, vol. 58. Geological Society, London, pp. 1—24. Special Publication.

Uchman, A., Wetzel, A., 2012. Deep-sea fans. In: Knaust, D., Bromley, R.G. (Eds.), Trace Fossils as Indicators of Sedimentary Environments, Developments in Sedimentology, vol. 64. Elsevier, pp. 643—671.

Udden, J.A., 1912. Geology and mineral resources of the Peoria Quadrangle. U.S. Geol. Surv. Bull. 506, 103.

Underhill, J.R., 1991. Controls on Late Jurassic seismic sequences, Inner Moray Firth, UK North Sea: a critical test of a key segment of Exxon's original global cycle chart. Basin Res. 3, 79—98.

Vail, P.R., 1975. Eustatic cycles from seismic data for global stratigraphic analysis (abstract). Am. Assoc. Pet. Geol. Bull. 59, 2198—2199.

Vail, P.R., 1987. Seismic stratigraphic interpretation using sequence stratigraphy. Part 1: seismic stratigraphy interpretation procedure. In: Bally, A.W. (Ed.), Atlas of Seismic Stratigraphy, vol. 27. American Association of Petroleum Geologists, Studies in Geology, pp. 1—10. Series 1.

Vail, P.R., Mitchum Jr., R.M., Thompson III, S., 1977a. Seismic stratigraphy and global changes of sea level, Part 3: relative changes of sea level from coastal onlap. In: Payton, C.E. (Ed.), Seismic Stratigraphy — Applications to Hydrocarbon Exploration, vol. 26. American Association of Petroleum Geologists, Memoir, pp. 63—81.

Vail, P.R., Mitchum Jr., R.M., Thompson III, S., 1977b. Seismic stratigraphy and global changes of sea level, Part 4: global cycles of relative changes of sea level. In: Payton, C.E. (Ed.), Seismic Stratigraphy — Applications to Hydrocarbon Exploration, vol. 26. American Association of Petroleum Geologists, Memoir, pp. 83—97.

Vail, P.R., Hardenbol, J., Todd, R.G., 1984. Jurassic unconformities, chronostratigraphy and sea-level changes from seismic stratigraphy and biostratigraphy. In: Schlee, J.S. (Ed.), Interregional Unconformities and Hydrocarbon Accumulation, vol. 36. American Association of Petroleum Geologists Memoir, pp. 129—144.

Vail, P.R., Audemard, F., Bowman, S.A., Eisner, P.N., Perez-Cruz, C., 1991. The stratigraphic signatures of tectonics, eustasy and sedimentology — an overview. In: Einsele, G., Ricken, W., Seilacher, A. (Eds.), Cycles and Events in Stratigraphy. Springer-Verlag, pp. 617—659.

van der Merwe, W.C., Flint, S.S., Hodgson, D.M., 2010. Sequence stratigraphy of an argillaceous, deepwater basin-plain succession: Vischkuil Formation (Permian), Karoo Basin, South Africa. Marine & Petrol. Geol. 27, 321—333.

van Loon, A.J., 2000. The stolen sequence. Earth-Sci. Rev. 52, 237—244.

Van Siclen, D.C., 1958. Depositional topography — examples (Louisiana and Texas) and theory. Am. Assoc. Pet. Geol. Bull. 42, 1896—1913.

Van Wagoner, J.C., 1985. Reservoir Facies Distribution as Controlled by Sea-Level Change. SEPM Mid-Year Meeting, Golden, Colorado, pp. 91—92.

Van Wagoner, J.C., 1995. Sequence stratigraphy and marine to nonmarine facies architecture of Foreland Basin Strata, Book Cliffs, Utah, U.S.A. In: Van Wagoner, J.C., Bertram, G.T. (Eds.), Sequence Stratigraphy of Foreland Basin Deposits, vol. 64. American Association Petroleum Geologists Memoir, pp. 137—223.

Van Wagoner, J.C., Posamentier, H.W., Mitchum, R.M., Vail, P.R., Sarg, J.F., Loutit, T.S., Hardenbol, J., 1988. An overview of sequence stratigraphy and key definitions. In: Wilgus, C.K., Hastings, B.S., Kendall, C.G.S.C., Posamentier, H.W., Ross, C.A., Van Wagoner, J.C. (Eds.), Sea Level Changes——An Integrated Approach, vol. 42. SEPM, pp. 39—45. Special Publication.

Van Wagoner, J.C., Mitchum Jr., R.M., Campion, K.M., Rahmanian, V.D., 1990. Siliciclastic sequence stratigraphy in well logs, core, and outcrops: concepts for high-resolution correlation of time and facies. Am. Assoc. Petr. Geol. Methods Explor. Series 7, 55.

Van Wagoner, J.C., Hoyal, D.C.J.D., Adair, N.L., Sun, T., Beaubouef, R.T., Deffenbaugh, M., Dunn, P.A., Huh, C., Li, D., 2003. Energy Dissipation and the Fundamental Shape of Siliciclastic Sedimentary Bodies. American Association of Petroleum Geologists, Search Discovery. Article #40080.

Varban, B.L., Plint, A.G., 2008. Sequence stacking patterns in the Western Canada foredeep: influence of tectonics, sediment loading and eustasy on deposition of the Upper Cretaceous Kaskapau and Cardium formations. Sedimentology 55, 395—421.

Vecsei, A., Duringer, P., 2003. Sequence stratigraphy of Middle Triassic carbonates and terrigenous deposits (Muschelkalk and Lower Keuper) in the SW Germanic Basin: maximum flooding versus maximum depth in intracratonic basins. Sediment. Geol. 160, 81—105.

Veizer, J., Hoefs, J., Lowe, D.R., Thurston, P.C., 1989. Geochemistry of Precambrian carbonates: II. Archean greenstone belts and Archean sea water. Geochim. et Cosmochim. Acta 53 (4), 859—871.

Verde, M., Ubilla, M., Jimenez, J.J., Genise, J.F., 2007. A new earthworm trace fossil from paleosols: aestivation chambers from the Late

Pleistocene Sopas Formation of Uruguay. Palaeogeogr. Palaeoclimatol. Palaeoecol. 243, 339—347.

Verdier, A.C., Oki, T., Suardy, A., 1980. Geology of the Handil field (East Kalimantan, Indonesia). In: Halbouty, M.T. (Ed.), Giant Oil and Gas Fields of the Decade: 1968—1978, vol. 30. American Association of Petroleum Geologists Memoir, pp. 399—421.

Vinn, O., Wilson, M.A., 2010. Microconchid-dominated hardground association from the late Pridoli (Silurian) of Saaremaa, Estonia. Palaeontol. Electr. 2010 (2), 13.2.9A.

Visser, J.N.J., 1997. Deglaciation sequences in the Permo-Carboniferous Karoo and Kalahari basins of Southern Africa: a tool in the analysis of cyclic glaciomarine basin fills. Sedimentology 44, 507—521.

Wadsworth, J., Boyd, R., Diessel, C., Leckie, D., Zaitlin, B., 2002. Stratigraphic style of coal and non-marine strata in a tectonically influenced intermediate accommodation setting: the Mannville Group of the Western Canadian Sedimentary Basin, South-central Alberta. Bull. Can. Pet. Geol. 50 (4), 507—541.

Wadsworth, J., Boyd, R., Diessel, C., Leckie, D., 2003. Stratigraphic style of coal and non-marine strata in a high accommodation setting: Falher Member and Gates Formation (Lower Cretaceous), Western Canada. Bull. Can. Pet. Geol. 51 (3), 275—303.

Waldron, J.W.F., Rygel, M.C., 2005. Role of evaporite withdrawal in the preservation of a unique coal-bearing succession: Pennsylvanian Joggins Formation, Nova Scotia. Geology 33 (5), 337—340.

Walker, J.C.G., Zahnle, K., 1986. The lunar nodal tide and the distance to the moon during the Precambrian Era. Nature 320, 600—602.

Walker, R.G., 1992. Facies, facies models and modern stratigraphic concepts. In: Walker, R.G., James, N.P. (Eds.), Facies Models: Response to Sea Level Change, vol. 1. Geological Association of Canada, GeoText, pp. 1—14.

Walker, R.G., Plint, A.G., 1992. Wave- and storm-dominated shallow marine systems. In: Walker, R.G., James, N.P. (Eds.), Facies Models: Response to Sea Level Change, vol. 1. Geological Association of Canada, GeoText, pp. 219—238.

Walther, J., 1894. Einleitung in die Geologie als historische Wissenschaft; Beobachtungen uber die Bildung der Gesteine und ihrer organischen Einschlusse. Jena G. Fischer, p. 1055.

Wanas, H.A., 2003. An authigenesis of glauconite in association with unconformity: a case study from the Albian/Cenomanian boundary at Gabal Shabrawet, Egypt. Geochemistry 5, 570—576.

Wanless, H.R., Shepard, E.P., 1936. Sea level and climatic changes related to Late Paleozoic cycles. Geol. Soc. Am. Bull. 47, 1177—1206.

Wanless, H.R., Weller, J.M., 1932. Correlation and extent of Pennsylvanian cyclothems. Geol. Soc. Am. Bull. 43, 1003—1016.

Warren, J., 2000. Dolomite: occurrence, evolution and economically important associations. Earth-Sci. Rev. 52, 1—81.

Warren, J.K., 2006. Evaporites: Sediments, Resources and Hydrocarbons. Springer-Verlag Berlin, p. 1035.

Webb, G.E., 1994. Paleokarst, paleosol, and rocky-shore deposits at the Mississippian-Pennsylvanian unconformity, Northwestern Arkansas. Geol. Soc. Am. Bull. 106, 634—648.

Webber, A.J., Hunda, B.R., 2007. Quantitatively comparing morphological trends to environment in the fossil record (Cincinnatian Series; Upper Ordovician). Evolution 61, 1455—1465.

Wehr, F.L., 1993. Effects of variations in subsidence and sediment supply on parasequence stacking patterns. In: Weimer, P., Posamentier, H.W. (Eds.), Siliciclastic Sequence Stratigraphy — Recent Developments and Applications, vol. 58. American Association of Petroleum Geologists Memoir, pp. 369—378.

Wei, H.Y., Chen, D.Z., Wang, J.G., Yu, H., Tucker, M.E., 2012. Organic accumulation in the lower Chihsia Formation (Middle Permian) of South China: Constraints from pyrite morphology and multiple geochemical proxies. Palaeogeogr. Palaeoclimatol. Palaeoecol. 353, 73—86.

Weimer, P., 1990. Sequence stratigraphy, facies geometries, and depositional history of the Mississippi Fan, Gulf of Mexico. Bull. Am Assoc. Pet. Geol. 74, 425—453.

Weimer, P., Dixon, B.T., 1994. Regional sequence stratigraphic setting of the Mississippi Fan complex, Northern deep Gulf of Mexico: Implications for evolution of the Northern Gulf basin margin. In: Weimer, P., et al. (Eds.), Submarine Fans and Production Characteristics. Society for Sedimentary Geology (SEPM), Gulf Coast Section, 15th Annual Research Conference, Houston, Texas, pp. 373—381.

Weimer, R.J., 1966. Time-stratigraphic analysis and petroleum accumulations, Patrick Draw field, Sweetwater County, Wyoming. Bull. Am. Assoc. Pet. Geol. 50 (10), 2150—2175.

Weissert, H., Joachimski, M., Sarnthein, M., 2008. Chemostratigraphy. Newslett. Stratigr. 42 (3), 145—179.

Weller, J.M., 1930. Cyclical sedimentation of the Pennsylvanian period and its significance. J. Geol. 38, 97—135.

Westaway, R., 1993. Quaternary uplift of Southern Italy. J. Geophys. Res. 98, 21741—21772.

Wetzel, A., Uchman, A., 2012. Hemipelagic and Pelagic Basin Plains. In: Knaust, D., Bromley, R.G. (Eds.), Trace Fossils as Indicators of Sedimentary Environments, Developments in Sedimentology, vol. 64. Elsevier, pp. 673—701.

Wheeler, H.E., 1958. Time stratigraphy. Am. Assoc. Pet. Geol. Bull. 42, 1047—1063.

Wheeler, H.E., 1959. Unconformity bounded units in stratigraphy. Am. Assoc. Petrol. Geol. Bull. 43, 1975—1977.

Wheeler, H.E., 1964. Baselevel, lithosphere surface, and time-stratigraphy. Geol. Soc. Am. Bull. 75, 599—610.

Wilgus, C.K., Hastings, B.S., Kendall, C.G.S.C., Posamentier, H.W., Ross, C.A., Van Wagoner, J.C. (Eds.), 1988. Sea Level Changes——An Integrated Approach, vol. 42. SEPM, p. 407. Special Publication.

Wilkinson, B.H., Merrill, G.K., Kivett, S.J., 2003. Stratal Order in Pennsylvanian Cyclothems, vol. 115. Bulletin Geological Society of America, pp. 1068—1087.

Williams, D.F., 1988. Evidence for and against sea-level changes from the stable isotopic record of the Cenozoic. In: Wilgus, C.K., Hastings, B.S., Kendall, C.G.S.C., Posamentier, H.W., Ross, C.A., Van Wagoner, J.C. (Eds.), Sea Level Changes——An Integrated Approach, vol. 42. SEPM, pp. 31—36. Special Publication.

Willis, A., Wittenberg, J., 2000. Exploration significance of healing-phase deposits in the Triassic Doig Formation, Hythe, Alberta. Bull. Can. Pet. Geol. 48 (3), 179—192.

Willis, B., 1910. Principles of palaeogeography. Science 33, 248—251.

Wilson, J.L., 1967. Cyclic and reciprocal sedimentation in Virgilian strata of Southern New Mexico. Geol. Soc. Am. Bull. 78, 805—818.

Wilson, M.A., Palmer, T.J., 1992. In: Hardgrounds and Hardground Faunas, vol. 9. Institute of Earth Studies Publications, University of Wales, Aberystwyth, pp. 1—131.

Wilson, M.E.J., Lockier, S.W., 2002. Siliciclastic and volcaniclastic influences on equatorial carbonates: insights from the Neogene of Indonesia. Sedimentology 49, 583—601.

Wing, S.L., Alroy, J., Hickey, L.J., 1995. Plant and mammal diversity in the Paleocene to Early Eocene of the Bighorn Basin. Palaeogeogr. Palaeoclimatol. Palaeoecol. 115, 117—155.

Winter, H.R., 1984. Tectonostratigraphy, as applied to the analysis of South African Phanerozoic basins. Transac. Geol. Soc. South Afr. 87, 169—179.

Winter, H.R., Brink, M.R., 1991. Chronostratigraphic subdivision of the Witwatersrand Basin based on a Western Transvaal composite column. South Afr. J. Geol. 94, 191—203.

Wood, L.J., Ethridge, F.G., Schumm, S.A., 1993. The effects of rate of base-level fluctuations on coastal plain, shelf and slope depositional systems: an experimental approach. In: Posamentier, H.W., Summerhayes, C.P., Haq, B.U., Allen, G.P. (Eds.), Sequence Stratigraphy and Facies Associations, vol. 18. International Association of Sedimentologists, pp. 43—53. Special Publication.

Wright, V.P., 1994. Paleosols in shallow marine carbonate sequences. Earth-Sci. Rev. 35, 367—395.

Wright, V.P., Marriott, S.B., 1993. The sequence stratigraphy of fluvial depositional systems: the role of floodplain sediment storage. Sediment. Geol. 86, 203–210.

Wright, V.P., Platt, N.H., 1995. Seasonal wetland carbonate sequences and dynamic catenas; a re-appraisal of palustrine limestones. Sediment. Geol. 99, 65–71.

Wu, J.E., McClay, K., Whitehouse, P., Dooley, T., 2009. 4D analogue modelling of transtensional pull-apart basins. Mar. & Pet. Geol. 26 (8), 1608–1623.

Xue, C., 1993. Historical changes in the Yellow River delta, China. Mar. Geol. 113, 321–330.

Yang, W., Escalona, A., 2011. Tectonostratigraphic evolution of the Guyana Basin. AAPG Bull. 95 (8), 1339–1368.

Ye, L., 1995. Paleosols in the Upper Guantae Formation (Miocene) of the Gudong oil field and their application to the correlation of fluvial deposits. Am. Assoc. Petr. Geol. Bull. 79, 981–988.

Ye, L., Kerr, D., 2000. Sequence stratigraphy of the Middle Pennsylvanian Bartlesville Sandstone, Northeastern Oklahoma: a case of an underfilled incised valley. Am. Assoc. Pet. Geol. Bull. 84 (8), 1185–1204.

Yoshida, S., Willis, A., Miall, A.D., 1996. Tectonic control of nested sequence architecture in the Castlegate Sandstone (Upper Cretaceous), Book Cliffs, Utah. J. Sediment. Res. 66, 737–748.

Yoshida, S., Miall, A.D., Willis, A., 1998. Sequence stratigraphy and marine to nonmarine facies architecture of foreland basin strata, Book Cliffs, Utah, U.S.A.: discussion. Am. Assoc. Pet Geol. Bull. 82, 1596–1606.

Young, G.M., 2013. Precambrian supercontinents, glaciations, atmospheric oxygenation, metazoan evolution and an impact that may have changed the second half of Earth history. Geosci. Front. 4 (3), 247–261.

Zaitlin, B.A., Dalrymple, R.W., Boyd, R., 1994. The stratigraphic organization of incised-valley systems associated with relative sea-level change. In: Dalrymple, R.W., Zaitlin, B.A. (Eds.), Incised-valley Systems: Origin and Sedimentary Sequences, vol. 51. SEPM, pp. 45–60. Special Publication.

Zaitlin, B.A., Potocki, D., Warren, M.J., Rosenthal, L., Boyd, R., 2000. Sequence stratigraphy in low accommodation Foreland Basins: an example from the lower Cretaceous Basal Quartz formation of Southern Alberta. In: Abstracts, GeoCanada 2000 Conference. Canadian Society of Petroleum Geologists, CD-ROM.

Zaitlin, B.A., Warren, M.J., Potocki, D., Rosenthal, L., Boyd, R., 2002. Depositional styles in a low accommodation foreland setting: an example from the Basal Quartz (Lower Cretaceous), Southern Alberta. Bull. Can. Pet. Geol. 50 (1), 31–72.

Zecchin, M., 2005. Relationships between fault-controlled subsidence and preservation of shallow-marine small-scale cycles: example from the lower Pliocene of the Crotone Basin (Southern Italy). J. Sed. Res. 75, 300–312.

Zecchin, M., 2007. The architectural variability of small-scale cycles in shelf and ramp clastic systems: the controlling factors. Earth-Sci. Rev. 84, 21–55.

Zecchin, M., 2010. Towards the standardization of sequence stratigraphy: Is the parasequence concept to be redefined or abandoned? Earth-Sci. Rev. 102, 117–119.

Zecchin, M., Caffau, M., 2011. Key features of mixed carbonate-siliciclastic shallow-marine systems: the case of the Capo Colonna terrace (Southern Italy). Ital. J. Geosci. 130, 370–379.

Zecchin, M., Catuneanu, O., 2013. High-resolution sequence stratigraphy of clastic shelves I: units and bounding surfaces. Mar. & Pet. Geol. 39, 1–25.

Zecchin, M., Catuneanu, O., 2015. High-resolution sequence stratigraphy of clastic shelves III: applications to reservoir geology. Mar. & Petrol. Geol. 62, 161–175.

Zecchin, M., Catuneanu, O., 2017. High-resolution sequence stratigraphy of clastic shelves VI: mixed siliciclastic-carbonate systems. Mar. & Petrol. Geol. 88, 712–723.

Zecchin, M., Catuneanu, O., 2020. High-resolution sequence stratigraphy of clastic shelves VII: 3D variability of stacking patterns. Mar. & Petrol. Geol. 121, 104582.

Zecchin, M., Tosi, L., 2014. Multi-sourced depositional sequences in the Neogene to Quaternary succession of the Venice area (Northern Italy). Mar. & Petrol. Geol. 56, 1–15.

Zecchin, M., Massari, F., Mellere, D., Prosser, G., 2003. Architectural styles of prograding wedges in a tectonically active setting, Crotone Basin, Southern Italy. J. Geol. Soc. 160, 863–880. London.

Zecchin, M., Mellere, D., Roda, C., 2006. Sequence stratigraphy and architectural variability in growth fault-bounded basin fills: a review of Plio-Pleistocene stratal units of the Crotone Basin (Southern Italy). J. Geol. Soc. 163, 471–486. London.

Zecchin, M., Brancolini, G., Tosi, L., Rizzetto, F., Caffau, M., Baradello, L., 2009a. Anatomy of the Holocene succession of the southern Venice Lagoon revealed by very high resolution seismic data. Conti. Shelf Res. 29, 1343–1359.

Zecchin, M., Civile, D., Caffau, M., Roda, C., 2009b. Facies and cycle architecture of a Pleistocene marine terrace (Crotone, Southern Italy): a sedimentary response to late Quaternary, high-frequency glacio-eustatic changes. Sediment. Geol. 216, 138–157.

Zecchin, M., Caffau, M., Civile, D., Roda, C., 2010a. Anatomy of a late Pleistocene clinoformal sedimentary body (Le Castella, Calabria, Southern Italy): a case of prograding spit system? Sediment. Geol. 223, 291–309.

Zecchin, M., Caffau, M., Tosi, L., Civile, D., Brancolini, G., Rizzetto, F., Roda, C., 2010b. The impact of late Quaternary glacio-eustasy and tectonics on sequence development: evidence from both uplifting and subsiding settings in Italy. Terra Nova 22, 324–329.

Zecchin, M., Civile, D., Caffau, M., Sturiale, G., Roda, C., 2011. Sequence stratigraphy in the context of rapid regional uplift and high-amplitude glacio-eustatic changes: the Pleistocene Cutro Terrace (Calabria, Southern Italy). Sedimentology 58, 442–477.

Zecchin, M., Catuneanu, O., Rebesco, M., 2015. High-resolution sequence stratigraphy of clastic shelves IV: high-latitude settings. Mar. & Pet. Geol. 68, 427–437.

Zecchin, M., Caffau, M., Ceramicola, S., 2016. Interplay between regional uplift and glacio-eustasy in the Crotone Basin (Calabria, Southern Italy) since 0.45 Ma: a review. Glob. & Planet. Chang. 143, 196–213.

Zecchin, M., Catuneanu, O., Caffau, M., 2017a. High-resolution sequence stratigraphy of clastic shelves V: criteria to discriminate between stratigraphic sequences and sedimentological cycles. Mar. & Pet. Geol. 85, 259–271.

Zecchin, M., Caffau, M., Catuneanu, O., Lenaz, D., 2017b. Discrimination between wave-ravinement surfaces and bedset boundaries in Pliocene shallow-marine deposits, Crotone Basin, Southern Italy: An integrated sedimentological, micropaleontological and mineralogical approach. Sedimentology 64, 1755–1791.

Zecchin, M., Donda, F., Forlin, E., 2017c. Genesis of the Northern Adriatic Sea (Northern Italy) since early Pliocene. Mar. & Petr. Geol. 79, 108–130.

Zecchin, M., Catuneanu, O., Caffau, M., 2019. Wave-ravinement surfaces: Classification and key characteristics. Earth-Sci. Rev. 188, 210–239.

Zecchin, M., Caffau, M., Catuneanu, O., 2021. Recognizing maximum flooding surfaces in shallow-water deposits: An integrated sedimentological and micropaleontological approach (Crotone Basin, Southern Italy). Mar. & Pet. Geol. 133, 105225.

Zecchin, M., Catuneanu, O., Caffau, M., 2022. High-resolution sequence stratigraphy of clastic shelves VIII: full-cycle subaerial unconformities. Mar. & Petrol. Geol. 135, 105425.

Zentmyer, R.A., Pufahl, P.K., James, N.L., Hiatt, E.E., 2011. Dolomitization on an evaporitic Paleoproterozoic ramp: widespread synsedimentary dolomite in the Denault Formation, Labrador Trough, Canada. Sediment. Geol. 238, 116–132.

Zhang, Y., Swift, D.J.P., Niedoroda, A.W., Reid, C.W., Thorne, J.A., 1997. Simulation of sedimentary facies on the Northern California shelf: implications for an analytical theory of facies differentiation. Geology 27, 635–638.

Zonneveld, J.P., Gingras, M.K., Pemberton, S.G., 2001. Trace fossil assemblages in a Middle Triassic mixed siliciclastic-carbonate marginal marine depositional system, British Columbia. Palaeogeogr. Palaeoclimatol. Palaeoecol. 166, 249–276.

Zubalich, R., Capozzi, R., Fanti, F., Catuneanu, O., 2021. Evolution of the Bearpaw seaway in West-central Alberta (late Campanian, Canada): implications for hydrocarbon exploration. Mar. & Pet. Geol. 124, 104779.

Glossary of terms

A

Accommodation: Space made available for potential sediment accumulation, by subsidence, sea-level rise, or a combination of these processes (Jervey, 1988; Fig. 3.11). Accommodation may be underfilled or overfilled, depending on the balance between accommodation (creation of space) and sedimentation (consumption of space) (Figs. 3.12 and 3.22; details in Chapter 3).

Aggradation: A depositional trend of vertical sediment accumulation; i.e., the building-up of the Earth's surface by means of deposition (Neuendorf et al., 2005). Aggradation may occur in any depositional environment, during either relative sea-level rise or fall (details in Chapter 4).

Allogenic: Allogenic processes are those external to a depositional system, such as eustasy, tectonism, and climate, which modify the total energy and sediment budget of the depositional environments hosted within a sedimentary basin (Einsele et al., 1991; details in Chapter 3).

Allostratigraphy: A type of stratigraphy which relies on lithological discontinuities as the basis for stratigraphic correlation (Fig. 1.6).

Angular unconformity: An unconformity where the strata below and above dip at a different angle (Fig. 1.18).

Autogenic: Autogenic processes are those internal to a depositional system, which lead to the redistribution of sediment within a depositional environment without changes in the total energy and sediment budget of that environment (Einsele et al., 1991; details in Chapter 3).

B

Backstepping: A geometrical trend that describes the landward shift of a depositional system (e.g., an estuary, or a submarine fan complex) or of a morphological element (e.g., a shoreline, or a shelf edge). The terms "backstepping" and "retrogradation" are interchangeable (Fig. 4.23). The backstepping of a shoreline occurs during relative sea-level rise (Fig. 4.1), but other elements may backstep during either relative sea-level rise or fall (details in Chapter 4).

Basal surface of forced regression: A sequence stratigraphic surface that marks a change in stratal stacking pattern from normal regressive to forced regressive at the base of the falling-stage systems tract; i.e., the paleoseafloor at the onset of relative-level fall (Figs. 4.6 and 6.7; details in Chapter 6).

Base level: A surface of equilibrium which sedimentary processes strive to attain, at which neither erosion nor deposition occurs (Barrell, 1917; Figs. 1.19, 3.22, and 3.23). The base level is a descriptor of sedimentation; base-level changes account for all factors controlling sedimentation, including but not restricted to accommodation (Fig. 3.11; details in Chapter 3).

Base-level cycles: Cycles of base-level rise and fall, leading to depositional trends of aggradation and degradation, respectively. Base-level cycles can be observed at multiple scales; sediment preservation is possible when the periods of deposition occur during a longer term base-level rise (Fig. 1.19).

Baselap: A stratigraphic relationship that marks the termination of strata or seismic reflections against an underlying surface. Types of baselap terminations include downlap, onlap, and offlap (Figs. 4.7 and 4.8).

Bathymetric trends: Trends of water-depth changes (i.e., shallowing or deepening; details in Chapter 7).

Bathymetry: The measurement of water depth, from the sea/lake floor to the sea/lake level; a proxy for how much accommodation is still unfilled in a marine or lacustrine basin (Fig. 3.22).

Bayhead delta: A delta at the head of a bay or estuary into which a river discharges (Neuendorf et al., 2005; Fig. 5.71). In contrast to actual deltas, bayhead deltas display a retrogradational stacking pattern as being part of backstepping coastal systems, typically in wave-dominated settings (Figs. 5.68 and 6.39).

Bed: A relatively conformable succession of genetically related laminae or lamina sets bounded by bedding planes (Campbell, 1967). Beds and bedsets are sedimentological units that serve as the building blocks of the lowest rank systems tracts (Figs. 3.24 and 4.62; details in Chapters 3–7).

Bedset: A relatively conformable succession of genetically related beds bounded by surfaces which mark a change in the stacking pattern of beds (Fig. 6.67; Campbell, 1967). Beds and bedsets are sedimentological units that serve as the building blocks of the lowest rank systems tracts (Figs. 3.24 and 4.62; details in Chapters 3–7).

Biostratigraphy: A type of stratigraphy which relies on the fossil content of strata as the basis for stratigraphic correlation (Fig. 1.6).

Bottomset: A horizontal or gently dipping succession of strata that accumulates in front of a prograding foreset; e.g., the prodelta sediments of a delta system. Bottomsets may form during either relative sea-level rise or fall, in relation to normal regressions or forced regressions, respectively (details in Chapter 4).

Bypass: The process of sediment transport across an area, without accumulation; often used to refer to sediment transport along a graded profile, when neither erosion nor deposition occurs.

C

Carbonate factory: An area of significant carbonate production, prone to a dominantly carbonate sedimentation regime (details in Chapter 8).

Chemostratigraphy: A type of stratigraphy which relies on the chemical properties of strata as the basis for stratigraphic correlation (Fig. 1.6).

Chronostratigraphic chart: *see* **Wheeler diagram**.

Chronostratigraphy: A type of stratigraphy which relies on the absolute ages of strata as the basis for stratigraphic correlation (Fig. 1.6).

Clinoform: An inclined stratal unit that is part of the foreset of a prograding system. Clinoforms may be observed from the 10^0–10^1 m scales of deltas (Fig. 4.5) to the 10^2–10^3 m scales of shelf-slope systems (Figs. 8.3 and 8.4).

Clinoform rollover: The starting point of a clinoform surface, where the seafloor records a sharp increase in slope gradient; examples include the shoreline (i.e., a subaerial clinoform rollover) and the shelf edge (i.e., a subaqueous clinoform rollover) (Fig. 4.19).

Clinoform surface: A sloping paleoseafloor that separates adjacent clinoforms within a foreset (Fig. 4.58).

Coastal onlap: Coastal to shallow-water sediment onlapping the transgressive surface of erosion during transgression (Fig. 4.8);

most commonly, shallow-water "healing-phase" deposits onlapping the wave-ravinement surface (Fig. 4.57).

Coastal plain: A relatively flat area bordering a coastline and extending inland to the nearest elevated land (Bates and Jackson, 1987); the depositional surface at the top of coastal prisms, which may extend tens to hundreds of kilometers updip from the shoreline (Fig. 4.12).

Coastal prism: The continental to shallow-water topset and foreset of a prograding system, most commonly formed during lowstand and highstand normal regressions (Fig. 4.12).

Coastline: *see* **Shoreline.**

Cohesive debris flows: *see* **Mudflows.**

Compaction: Reduction in bulk volume or thickness of a body of sediment in response to the increase in lithostatic pressure during burial or to compressional stress driven by tectonism. Compaction typically leads to a reduction of the original porosity of the sediment (Neuendorf et al., 2005).

Composite sequence: A succession of genetically related sequences in which the individual sequences stack into lowstand, transgressive, and highstand sequence sets (Van Wagoner, 1995). The lack of reproducible standards for the definition of a "sequence" as an anchor for stratigraphic classification undermines this scale-variant nomenclature, as the terms are stripped of a consistent stratigraphic meaning (details in Chapter 7).

Composite sequence set: A set of composite sequences organized into lowstand, transgressive, or highstand stacking patterns; a subdivision of "megasequences" (*cf.* Van Wagoner, 1995). The lack of reproducible standards for the definition of a "sequence" as an anchor for stratigraphic classification undermines this scale-variant nomenclature, as the terms are stripped of a consistent stratigraphic meaning (details in Chapter 7).

Compressional ridges: Compressional structures that form in cohesive debris-flow deposits, as a result of the internal shear strength that causes the flow to freeze on deceleration (Fig. 5.40; details in Chapters 5 and 8).

Condensed section: A conformable succession that accumulates at low rates, "condensing" time within the stratigraphic section; typically the product of pelagic or hemipelagic sedimentation during transgressions, often hosting the greatest abundance of microfossils, authigenic minerals, and organic matter (Loutit et al., 1988; Fig. 4.20).

Conformable: A stratigraphic relationship between sedimentary strata that accumulate in continuity of sedimentation (i.e., no hiatus; Fig. 1.18).

Conformable transgressive surface: *see* **Maximum regressive surface.**

Conformity: A conformable stratigraphic surface.

Constructional systems tract: A systems tract in eolian settings defined by an increase with time in the thickness of dune–interdune cycles and the angle of climb of interdune migration surfaces. This stacking pattern corresponds to a stage of growth in the development of eolian ergs (Fig. 6.54; details in Chapters 6 and 8).

Contractional systems tract: A systems tract in eolian settings defined by a decrease with time in the thickness of dune–interdune cycles and the angle of climb of interdune migration surfaces. This stacking pattern corresponds to a stage of decline in the development of eolian ergs (Fig. 6.54; details in Chapters 6 and 8).

Correlative conformity: A sequence stratigraphic surface that marks a change in stratal stacking pattern from forced regressive to lowstand normal regressive at the top of the falling-stage systems tract; i.e., the paleoseafloor at the end of relative sea-level fall (Figs. 4.6 and 6.7; details in Chapter 6).

Cyclostratigraphy: A type of stratigraphy which relies on orbital forcing and the resulting cyclicity as the basis for stratigraphic correlation (Fig. 1.6).

Cyclothem: A small-scale unconformity-bounded unit that describes the type of sedimentary cycle that prevailed during the

Carboniferous in the mid-continental US (Weller, 1930; Wanless and Weller, 1932); precursor of the concept of "stratigraphic sequence."

D

Debris flow: A type of gravity flow in which the grain-support mechanism is provided by grain collision (noncohesive debris flows, or grainflows) or grain cohesion (cohesive debris flows, or mudflows) (Figs. 5.42 and 8.42; details in Chapters 5 and 8).

Deflation surface: A subaerial unconformity that forms in an eolian environment by means of wind erosion (Figs. 3.18 and 3.19).

Degradation: The wearing down of the Earth's surface by processes of erosion (Neuendorf et al., 2005). Degradation may occur in any depositional environment, from continental to deep water (e.g., eolian deflation, current scouring, or mass wasting; details in Chapter 4).

Denudation: The wearing away that leads to the progressive lowering of the Earth's surface by means of weathering and erosion.

Depositional sequence: A type of stratigraphic sequence bounded by subaerial unconformities and their correlative conformities. Four types of depositional sequences have been defined, depending on their systems-tract composition and the timing of the correlative conformity (Figs. 1.9 and 1.10; details in Chapter 5).

Depositional system: A three-dimensional assemblage of lithofacies, genetically linked by active or inferred processes and environments (Fisher and McGowen, 1967; Fig. 1.8). Depositional systems may be observed at different scales, depending on the resolution of the stratigraphic study and of the data available (Figs. 5.1–5.3 and 5.5; details in Chapters 5 and 7).

Depositional trend: The trend recorded by a depositional surface in response to processes of aggradation, degradation, progradation, or retrogradation. Note the difference between depositional and geometrical trends (Figs. 4.23 and 4.24; details in Chapter 4).

Diachronous: A stratal unit or a stratigraphic surface that is of different ages in different areas (i.e., time-transgressive; Fig. 7.17; details in Chapter 7).

Diagenesis: All the chemical, physical, and biological changes undergone by sediments after deposition, during and after lithification, exclusive of surficial alteration (weathering) and metamorphism (Neuendorf et al., 2005).

Diastem: A relatively short interruption in sedimentation, involving only a brief interval of time, with little or no erosion before deposition resumes; a depositional break of lesser magnitude than a paraconformity, or a paraconformity of very small time value (Neuendorf et al., 2005). Due to the ambiguity in the distinction between diastems and unconformities, which grade into each other (Barrell, 1917), it is recommended to either restrict the usage of diastems to sedimentology (i.e., bed or bedset boundaries) or eliminate this concept from sedimentary geology altogether. Ultimately, "every bedding plane is, in effect, an unconformity" (Ager, 1993) (details in Chapter 7).

Differential subsidence: A subsidence pattern which involves variable rates across a sedimentary basin.

Disconformity: An unconformity in which the bedding planes of the strata above and below the contact are parallel to each other and the surface is marked by evidence of erosion (Fig. 1.18).

Discontinuity: Any interruption in sedimentation, whatever its cause and magnitude, usually a manifestation of nondeposition and accompanying erosion; an unconformity (Neuendorf et al., 2005).

Distal: A depositional area that is located relatively far from the source of sediment.

Downcutting: Process of erosion in which the cutting is directed downwards (as opposed to lateral erosion), quantified by the amount of vertical incision into the underlying substrate.

Downdip: A basinward direction, parallel to the dip of the depositional surface.

Downlap: Termination of inclined strata against an underlying lower angle surface (Fig. 4.8).

Downlap surface: A surface downlapped by the overlying strata; most commonly, a maximum flooding surface at seismic scales (Fig. 6.4), or a within-trend facies contact at sedimentological scales at the limit between prograding inner-shelf deposits and the underlying outer-shelf fines (Figs. 6.2 and 6.64; details in Chapter 6).

Downstepping: A geometrical trend that describes a decrease in the elevation of a depositional system (e.g., a delta; Fig. 4.16) or of a morphological element (e.g., a shoreline or a shelf edge; Fig. 4.6). Where used to describe a shoreline trajectory, downstepping is the product of relative sea-level fall (Fig. 4.1), and it can be accompanied by either aggradation or degradation (Fig. 4.23). Note the difference between geometrical and depositional trends (details in Chapter 4).

Downstream-controlled setting: An area within a sedimentary basin in which stratal stacking patterns relate to shoreline trajectories, and are at least in part controlled by relative sea-level changes (Fig. 3.11; details in Chapters 3 and 4).

Drowning surface: A stratigraphic surface that marks abrupt water deepening, commonly referred to as a "flooding surface" in siliciclastic settings, or a "drowning unconformity" in carbonate settings. Drowning surfaces are allostratigraphic contacts which may or may not coincide with sequence stratigraphic surfaces (Figs. 6.61 and 6.62; details in Chapter 6).

Drowning unconformity: A type of drowning surface which forms as a result of abrupt water deepening in carbonate settings (Fig. 8.21; details in Chapters 6 and 8).

E

Energy flux: *see* **Environmental energy flux.**

Environmental energy flux: The energy of the sediment-transport agents (e.g., wind, rivers, subaqueous currents), which enables the transfer of particles from one area to another. The balance between sediment load and energy flux at any location determines the direction of shift of the temporary base level (i.e., processes of sedimentation or erosion; details in Chapter 3).

Erosion: The removal of material from a substrate by the action of the sediment-transport agents that operate in active environments (e.g., wind, rainfall, rivers, waves, subaqueous traction currents, gravity-driven processes, or moving ice). Erosion can be vertical (i.e., downcutting) or lateral (e.g., cut-bank erosion in rivers or subaqueous currents).

Eustasy: The global sea level, with reference to the center of Earth or a satellite in fixed orbit around the Earth (Fig. 3.13; details in Chapter 3). Eustatic fluctuations may be caused by changes in the global hydrologic balance (water stored in glacier ice, groundwater, or soil moisture), volumetric changes due to ocean temperature variations, or changes in the capacity of ocean basins due to plate-tectonic activity (variations in volcanic activity along spreading centers, or crustal deformation at subduction zones) (Neuendorf et al., 2005).

Eustatic: Pertaining to eustasy.

F

Facies: The distinctive characteristics of a rock unit, usually reflecting its origin, which can be used to differentiate the unit from adjacent or associated units (Neuendorf et al., 2005).

Facies model: An association of facies that is representative of a particular depositional process, depositional element, or depositional system. Facies models can be defined at different scales, depending on the data available and the purpose of study.

Fairweather wave base: The maximum water depth up to which the wave energy can still move sediments on the seafloor during fairweather; i.e., the limit between shoreface and inner shelf (Figs. 4.11 and 8.14). The depth of the fairweather wave base is variable, depending on the wind regime at each region, from meters to tens of meters, with an average of 10–15 m.

Falling-stage systems tract: A systems tract defined by a forced regressive stacking pattern, which accumulates during relative sea-level fall (details and examples in Chapters 4 and 5).

Final transgressive surface: *see* **Maximum flooding surface.**

Firmground: A semiconsolidated substrate which is firm but unlithified. Firmgrounds may form in a variety of depositional settings, from continental to fully marine, but the classification of ichnofacies that colonize them is based on the environment in which the tracemakers lived; e.g., the *Glossifungites* Ichnofacies reflects marine colonization of firmgrounds (Figs. 2.47 and 2.48; details in Chapter 2).

Flooding surface: A stratigraphic surface that forms as a result of abrupt water deepening, defining the parasequence boundary (Fig. 5.94). The term is most commonly used in the context of siliciclastic settings (Figs. 5.96, 5.100, and 6.63). Flooding surfaces are allostratigraphic contacts which may or may not coincide with sequence stratigraphic surfaces (Figs. 6.61 and 6.62; details in Chapter 6). *Cf.* **Drowning surface.**

Fluidized flow: A type of gravity flow in which the grain-support mechanism is provided by the water-escape process. Fluidized flows are typically part of the "traction carpet" of turbidity flows (Figs. 5.42 and 8.42; details in Chapters 5 and 8).

Fluvial entrenchment: *see* **Subaerial unconformity.**

Fluvial knickpoint: A point of abrupt change in gradient along the longitudinal profile of a stream, marking the updip limit of an erosional or depositional area; knickpoints typically migrate in an upstream direction (Figs. 4.42, 4.43, and 5.16; details in Chapters 4 and 5).

Fluvial onlap: Fluvial systems onlapping the subaerial unconformity as the area of fluvial sedimentation expands in a landward direction, most commonly during normal regressions and transgressions (Figs. 4.8 and 5.13).

Forced regression: A type of shoreline trajectory defined by progradation and downstepping (Figs. 4.1, 4.16, and 4.34). Forced regressions are driven by relative sea-level fall (Figs. 4.21 and 4.32; details in Chapter 4).

Forced regressive wedge systems tract: *see* **Falling-stage systems tract.**

Forebulge unconformity: An unconformity that forms as a result of the flexural uplift of a forebulge in a foreland setting (Figs. 8.5 and 8.6; details in Chapter 8).

Foreset: A succession of inclined strata which form the steeper part of a prograding system; e.g., the delta front of a delta, or the clinoforms of any other subaqueous prograding system. Foresets may form during either relative sea-level rise or fall, in relation to normal regressions or forced regressions, respectively (details in Chapter 4).

Forestepping: A geometrical trend that describes the basinward shift of a depositional system (e.g., a delta, or a submarine fan complex) or of a morphological element (e.g., a shoreline, or a shelf edge). The terms "forestepping" and "progradation" are interchangeable (Fig. 4.23). Forestepping may occur during either relative sea-level rise or fall (Fig. 4.1; details in Chapter 4).

G

Genetic stratigraphic sequence: A type of stratigraphic sequence bounded by maximum flooding surfaces (Fig. 1.10; details in Chapter 5).

Geometrical trend: The direction of shift of a depositional system or morphological element such as a clinoform rollover; i.e., upstepping, downstepping, forestepping, or backstepping (Fig. 4.1). Note the difference between geometrical and depositional trends (Figs. 4.23 and 4.24; details in Chapter 4).

Glacial: Pertaining to the presence and activity of ice or glaciers in glaciated settings (details in Chapter 8).

Glaciated setting: A geological setting in which the stratigraphic architecture is shaped by the advance and retreat of permanent ice sheets (details in Chapter 8).

Glaciation: A period of time during which glaciers were more extensive than at present; a climatic regime during which extensive glaciers developed, attained a maximum extent, and receded (Neuendorf et al., 2005).

Glacio-eustasy: The global changes in sea level produced by the successive withdrawal and return of water in the ocean due to the formation and melting of ice sheets, respectively (Neuendorf et al., 2005).

Glacio-isostasy: Crustal adjustment to loading and unloading that is caused by the addition and removal of glacier ice, respectively (Neuendorf et al., 2005).

Global cycle chart: A chart depicting global changes in sea level (Fig. 5.88). Global cycle charts are no longer used in the workflow of sequence stratigraphy, as stratigraphic sequences are basin specific, with a timing that reflects the interplay of global and local controls on accommodation and sedimentation (details in Chapters 7 and 9).

Gradationally based shoreface: A shoreface succession that is typical of normal regressions, in which the shift from shelf to shoreface facies is gradual, due to the upstepping of the shoreline during progradation (Fig. 4.30; details in Chapter 4). Cf. **Sharp-based shoreface**.

Graded profile: A surface of equilibrium between sedimentation and erosion, along which sediment load and the energy of the sediment-transport agents are in perfect balance; e.g., a graded fluvial profile, or the hydrodynamic equilibrium profile of a seafloor (Figs. 3.22 and 3.23; details in Chapter 3).

Grading: see **Sediment grading**.

Grainflow: A type of gravity flow in which the grain-support mechanism is provided by the collision between grains; a noncohesive debris flow, dominated by sand-size clasts. Grainflows may operate as independent flows, or may be part of the "traction carpet" of turbidity flows (Figs. 5.42 and 8.42; details in Chapters 5 and 8).

Gravity flow: A type of gravity-driven process that involves the transport of loose sediment (Fig. 8.42). Different kinds of gravity flows can be defined depending on the rheology of the flow and the grain-support mechanism (Figs. 5.42 and 8.42; details in Chapters 5 and 8).

Greenhouse: A climatic regime characterized by higher temperatures, the absence of permanent ice sheets, and higher global sea levels; e.g., during the lower and middle Paleozoic, and the Mesozoic (details in Chapters 2 and 8).

H

Hardground: A fully lithified substrate which may form in a variety of depositional settings, from continental to fully marine. The classification of ichnofacies that colonize hardgrounds is based on the environment in which the tracemakers lived; e.g., the *Trypanites* Ichnofacies reflects marine colonization of a lithified substrate (Fig. 2.49; details in Chapter 2).

Healing-phase wedge: A wedge of sediment that accumulates during transgression in shallow- or deep-water environments, "healing" the bathymetric profile of the seafloor. Healing-phase wedges include mainly sediment accumulated from suspension, which onlaps the wave-ravinement surface in shallow-water settings (i.e.,

coastal onlap; Figs. 4.8 and 4.15) or the maximum regressive surface in slope settings (i.e., marine onlap; Figs. 4.6 and 4.8); details in Chapter 4.

Hemipelagic: A mix of fine-grained terrigenous and pelagic sediment that forms condensed sections in deep-water environments. A dominantly pelagic or hemipelagic sedimentation is most common at or around the time of formation of maximum flooding surfaces (Figs. 4.20 and 8.46; details in Chapters 4, 5, 6, and 8).

Hiatus: A break in the continuity of the sedimentary record, due to nondeposition or erosion; a gap in sedimentation, defining an unconformity (Fig. 1.8). Cf. **Unconformity, Diastem**.

Hierarchy: see **Stratigraphic hierarchy**.

High-accommodation systems tract: see **Low-amalgamation systems tract**.

High-amalgamation systems tract: A systems tract in upstream-controlled fluvial settings, defined by a high degree of channel amalgamation (Fig. 4.70). The high-amalgamation systems tract replaces the "low-accommodation" systems tract, as factors other than accommodation may also control the degree of channel amalgamation (Fig. 4.63; details in Chapters 4 and 5).

Highstand prism: The continental to shallow-water topset and foreset of a highstand systems tract (Fig. 5.22). Cf. **Coastal prism**.

Highstand shedding: In carbonate settings, most sediment is typically shed into the adjacent basin at times of highstand in relative sea level, when the carbonate factory is most productive (Fig. 5.83).

Highstand systems tract: A systems tract defined by a normal regressive stacking pattern which follows a transgression of equal hierarchical rank (details and examples in Chapters 4 and 5).

Hyperpycnal flow: A gravity flow produced at a river mouth when the density of the incoming sediment-laden river flow is greater than the density of the water in the receiving basin (a lake or a sea); most commonly expressed as turbidites in delta-front settings (Figs. 7.30 and 7.31).

Hypopycnal plume: A mass of sediment transported by flowing water that is less dense than the body of water it enters; e.g., a river entering a sea. Hypopycnal plumes float at the surface, until processes such as flocculation enable the settling of the fine-grained sediments from suspension.

I

Icehouse: A climatic regime characterized by lower temperatures, the presence of permanent ice sheets, and lower global sea levels; e.g., during the Neoproterozoic, Permo-Carboniferous, and Plio-Pleistocene (details in Chapters 2 and 8).

Ichnofabric: A record of animal-sediment interactions at the bed scale, such as their abundance and disposition, which characterize the texture and internal structure of the deposit (Bromley and Ekdale, 1984; details in Chapter 2).

Ichnofacies: A summary of a paleocommunity's response to the depositional environment, regardless of specific trace-fossil genera, outlining a facies model at depositional-system scales (details in Chapter 2).

Incised valley: A valley cut by fluvial erosion during forced regressions or transgressions in downstream-controlled settings (Figs. 1.12 and 4.39), or during stages of base-level fall in upstream-controlled settings (Fig. 3.21); details in Chapters 3 and 4.

Initial transgressive surface: see **Maximum regressive surface**.

Interfluve: The overbank area between rivers flowing in the same general direction. Interfluves may be intermittently flooded and subject to aggradation in the case of unincised rivers, or dry and subject to pedogenesis or eolian deflation in the case of incised valleys.

Interglacial: Pertaining to the period of time between glaciations, conducive to nonglaciated settings (details in Chapter 8).

Isostasy: The condition of equilibrium, comparable to floating, of lithospheric plates above the asthenosphere. Crustal loading, such as by ice, water, sediments, or volcanic flows, leads to isostatic depression (downwarping); removal of load leads to isostatic rebound (upwarping) (Neuendorf et al., 2005).

Isostatic rebound: The process of lithospheric uplift as a result of removal of supracrustal loads such as during deglaciation or denudation.

K

Karst: A type of topography that is formed on limestone, gypsum, and other soluble rocks, primarily by dissolution as a result of exposure to rivers and meteoric water. Karsts usually include sinkholes, caves, and underground drainage (Fig. 6.17; Neuendorf et al., 2005).

Knickpoint: A point of abrupt change in slope gradient; typically used in the context of fluvial systems: *see* **Fluvial knickpoint**.

L

Lag deposit: A concentration of coarser sediment on a surface after the dispersal of the finer grained fractions by the sediment-transport agents; commonly, on top of unconformities (Figs. 5.101 and 6.47). Average thicknesses are in a range of cm–dm, but meter-scale lags may also form in areas of exceptionally high energy (Fig. 6.48). Lags of economic value are "placer" deposits.

Lapout: A general term that indicates the lateral termination of strata or seismic reflections at the depositional limit of sedimentary units. Lapout terminations can be against an underlying surface (i.e., baselap: downlap, onlap, or offlap), or against an overlying surface (i.e., toplap). Stratal terminations produced by truncation do not qualify as lapout since they mark erosional rather than depositional limits (Figs. 4.7 and 4.8; details in Chapter 4).

Liquefied flow: A type of gravity flow in which the grain-support mechanism is provided by the pore-water pressure. Liquefied flows are typically part of the "traction carpet" of turbidity flows (Figs. 5.42 and 8.42; details in Chapters 5 and 8).

Lithostratigraphy: A type of stratigraphy which relies on the lithological character of strata as the basis for stratigraphic correlation (Fig. 1.6).

Local flooding surface: A type of flooding surface produced by sediment starvation and erosion in a shelf environment during transgression (Figs. 5.109 and 6.3; details in Chapter 6). The local flooding surface corresponds to the "offshore marine erosion diastem" of Nummedal and Swift (1987), and may record variable degrees of diachroneity (Carter et al., 1998).

Looseground: A sandy type of softground substrate in continental settings, colonized by the *Scoyenia* Ichnofacies (details in Chapter 2).

Low-accommodation systems tract: *see* **High-amalgamation systems tract**.

Low-amalgamation systems tract: A systems tract in upstream-controlled fluvial settings, defined by a low degree of channel amalgamation (Fig. 4.70). The low-amalgamation systems tract replaces the "high-accommodation" systems tract, as factors other than accommodation may also control the degree of channel amalgamation (Fig. 4.63; details in Chapters 4 and 5).

Lowstand fan: A part of a submarine fan complex accumulated during a lowstand in relative sea level (Fig. 5.34; details in Chapters 5 and 8). *Cf.* **Lowstand shedding**.

Lowstand shedding: In siliciclastic settings, most riverborne sediment is typically shed into the adjacent basin at times of lowstand in relative sea level, when accommodation on the shelf is minimum (Fig. 5.34).

Lowstand systems tract: A systems tract defined by a normal regressive stacking pattern which follows a forced regression of equal hierarchical rank (details and examples in Chapters 4 and 5).

Lowstand unconformity: *see* **Subaerial unconformity**.

Lowstand wedge: see **Lowstand systems tract**.

M

Magnetic polarity chron: *see* **Polarity chron**.

Magnetostratigraphy: A type of stratigraphy which relies on the magnetic polarity of strata as the basis for stratigraphic correlation (Fig. 1.6).

Marine onlap: Deep-water sediment onlapping a ramp or the slope of a shelf-slope system (Figs. 4.6 and 4.8).

Mass-transport deposit (MTD): A general term that includes all deposits which accumulate in deep-water settings as a result of gravity-driven submarine mass failures. Some authors exclude turbidites from MTDs, although flow transformations enable the development of turbidity flows from debris flows, and also, sandy debris flows can be part of the traction carpet of turbidity flows (Figs. 5.42 and 8.42). Therefore, the term "MTD" can be used safely for all undifferentiated mass-failure deposits. Where specific processes can be identified, the more precise terms (e.g., slides, slumps, debris-flow deposits, or turbidites) are preferred.

Mass-transport processes: All gravity-driven processes which result in the accumulation of mass-transport deposits; examples include the transport of lithified (rock falls, slides), semi-lithified (slumps), and loose (gravity flows) sediments (Fig. 8.42). *Cf.* **Mass-transport deposit**.

Maximum depth interval: Interval of deepest water at the time of sedimentation, which can be identified with paleontological and geochemical data. The maximum water depth is often recorded above the maximum flooding surface, within the highstand systems tract (Figs. 2.28, 2.29, 7.28, and 7.32; details in Chapters 2 and 7).

Maximum expansion level: A sequence stratigraphic surface in eolian settings which marks a change from an increase to a decrease in the thickness of dune–interdune cycles and the angle of climb of interdune migration surfaces, at the limit between stages of growth and decline of eolian ergs; i.e., the boundary between constructional and contractional systems tracts (Fig. 6.54; details in Chapters 6 and 8).

Maximum flooding surface: A sequence stratigraphic surface that marks a change in stratal stacking pattern from transgression to regression at the top of the transgressive systems tract; i.e., the paleoseafloor, and its correlative surface within the continental realm, at the end of transgression (Figs. 6.7 and 6.61; details in Chapter 6).

Maximum progradation surface: *see* **Maximum regressive surface**.

Maximum regressive surface: A sequence stratigraphic surface that marks a change in stratal stacking pattern from regression to transgression at the base of the transgressive systems tract; i.e., the paleoseafloor, and it correlative surface within the continental realm, at the end of regression (Fig. 6.7; details in Chapter 6).

Maximum transgressive surface: *see* **Maximum flooding surface**.

Megasequence: A succession of genetically related composite sequences in which the individual composite sequences stack into lowstand, transgressive, and highstand composite sequence sets (*cf.* Van Wagoner, 1995). The lack of reproducible standards for the definition of a "sequence" as an anchor for stratigraphic classification undermines this scale-variant nomenclature, as the terms are stripped of a consistent stratigraphic meaning (details in Chapter 7).

Milankovitch band/cycles: *see* **Orbital forcing**.

Mudflow: A type of gravity flow in which the grain-support mechanism is provided by the density of the matrix; a cohesive debris flow, dominated by fine-grained particles (Figs. 5.42 and 8.42; details in Chapters 5 and 8).

N

Non-cohesive debris flow: *see* **Grainflow**.

Nonconformity: An unconformity between igneous or metamorphic basement rocks and the overlying sedimentary rocks (Fig. 1.18).

Non-glaciated setting: An ice-free geological setting in which the stratigraphic architecture is controlled by the interplay of accommodation and sedimentation, with no interference from ice (details in Chapter 8).

Normal regression: A type of shoreline trajectory defined by progradation and upstepping (Figs. 4.1, 4.17, and 4.50). Normal regressions are driven by sediment supply outpacing the rates of relative sea-level rise at the shoreline (Figs. 4.21 and 4.32; details in Chapter 4).

O

Offlap: Termination of low-angle strata against an underlying steeper surface, in which each successively younger unit leaves exposed a portion of the older unit on which it lies; a lapout with downstepping in the younging direction (Fig. 4.8).

Offshore marine erosion diastem: *see* **Local flooding surface**.

Omission surface: A discontinuity surface of the most minor nature, which marks a temporary halt in deposition but little or no erosion (Bromley, 1975); i.e., a diastem. *Cf.* **Diastem**.

Onlap: Termination of low-angle strata against an underlying steeper surface, in which each successively younger unit extends beyond the limit of the older unit on which it lies; a lapout with upstepping in the younging direction (Fig. 4.8).

Orbital forcing: The control exerted by the Earth's orbital parameters on sedimentation and stratigraphic cyclicity. Orbital cycles at scales of 10^4–10^5 yrs relate to eccentricity, obliquity, and precession (Fig. 2.15).

P

Paleoflow: The direction of a flow at syndepositional time, which can be reconstructed from sedimentary structures (e.g., cross-stratification; Fig. 2.10) or other physical features such as the orientation of basal grooves in deep-water settings (Fig. 5.37).

Paleosol: An ancient soil horizon formed by means of pedogenic processes in the geological past (details in Chapter 2).

Palynological marine index: The ratio between marine and continental palynomorphs, which can be used to identify the position of sequence stratigraphic surfaces such as the maximum regressive and the maximum flooding surfaces in continental settings that still record a marine influence (details in Chapter 6).

Paraconformity: An unconformity in which the bedding planes of the strata above and below the contact are parallel to each other and the surface lacks evidence of erosion (Fig. 1.18). In the absence of physical evidence of erosion, other methods (e.g., biostratigraphy, radiochronology) provide the evidence for the missing time. "Minor" paraconformities are referred to as diastems.

Parasequence: A succession of genetically related beds and bedsets bounded by flooding surfaces (Fig. 5.94). Parasequences were introduced as the building blocks of seismic-scale systems tracts, but proved problematic due to their restricted applicability to coastal and shallow-water settings, and the allostratigraphic rather than sequence stratigraphic nature of flooding surfaces. Parasequences have become obsolete with the advent of high-resolution sequence stratigraphy, as high-frequency sequences that develop at parasequence scales provide a better and more reliable alternative for stratigraphic correlation (details in Chapter 5).

Peat: An unconsolidated deposit of semicarbonized plant remains accumulated in a water-saturated environment; an early stage in the development of coal (Neuendorf et al., 2005).

Pedogenesis: The process of soil formation (Fig. 2.14; details in Chapter 2).

Pedology: The study of soil morphology, formation, and classification (details in Chapter 2).

Pelagic: A type of fine-grained marine sediment that consists mainly of skeletal remains of microorganisms such as plankton. Pelagic deposits are typical of deep-water environments deprived of terrigenous sediment influx. *Cf.* **Hemipelagic**.

Peneplanation: The flattening of a landscape as a result of long-term subaerial erosion.

Placer deposit: A lag deposit of economic value; a sedimentary ore deposit containing gold nuggets, diamonds, or other valuable minerals. *Cf.* **Lag deposit**.

Polarity chron: The time interval between polarity reversals of Earth's magnetic field.

Progradation: A depositional trend of basinward shift of a depositional system (e.g., a delta, or a submarine fan complex) or of a morphological element (e.g., a shoreline, or a shelf edge). Progradation may occur during either relative sea-level rise or fall (details in Chapter 4).

Proximal: A depositional area that is located relatively close to the source of sediment.

R

Radiochronology: A method that measures the ratios between radioactive isotopes to determine the absolute ages of rocks.

Ramp: An area of water deepening with no significant breaks in seafloor gradients between the shoreline and the deepest parts of the sedimentary basin (Fig. 8.1). *Cf.* **Shelf-slope system**.

Ravinement surface: *see* **Transgressive ravinement surface** and **Regressive ravinement surface**.

Regression: A seaward shift of the shoreline (Fig. 4.31).

Regressive ravinement surface: *see* **Regressive surface of marine erosion**.

Regressive surface of fluvial erosion: A subaerial unconformity formed during regression.

Regressive surface of marine erosion: A sequence stratigraphic surface that forms by means of wave or tidal scouring of the seafloor during relative sea-level fall (Figs. 4.44A and 4.45). The regressive surface of marine erosion invariably reworks part of the basal surface of forced regression, thus becoming a systems-tract boundary (Fig. 6.23; details in Chapter 6).

Regressive systems tract: A systems tract that includes all sediments deposited during shoreline regression (Fig. 1.10). The mixing of undifferentiated forced and normal regressive deposits within the same systems tract reduces the resolution of the stratigraphic study, and hampers the full understanding of the patterns of sediment distribution across a basin (details in Chapter 5).

Regressive wave ravinement: *see* **Regressive surface of marine erosion**.

Relative sea level: Sea level relative to a subsurface reference horizon, which measures changes in accommodation independently of sedimentation (Figs. 3.13 and 3.22). Relative sea-level changes relevant to the formation of systems tracts are proxied by relative changes in the elevation of the shoreline (i.e., upstepping or downstepping; Figs. 4.1 and 4.23). The usage of the "relative sea level" concept eliminates the need to interpret the underlying controls on accommodation (e.g., eustasy vs. tectonics; Fig. 3.3; details in Chapter 3).

Relatively conformable: A relatively conformable succession is a stratal unit with no resolvable or negligible internal unconformities. The scale of a relatively conformable succession depends on the resolution of the stratigraphic study (Fig. 7.6; details in Chapter 7).

Retrogradation: A depositional trend of landward shift of a depositional system (e.g., an estuary, or a submarine fan complex) or of a morphological element (e.g., a shoreline, or a shelf edge). The retrogradation of a shoreline occurs during relative sea-level rise (Figs. 4.1 and 4.32), but other elements may retrograde during either relative sea-level rise or fall (details in Chapter 4).

S

Scale-invariant nomenclature: A system of stratigraphic nomenclature and classification in which the same types of units are recognized at all stratigraphic scales (i.e., sequences and systems tracts of different hierarchical ranks; Fig. 7.16). This approach provides consistency in the application of sequence stratigraphy, irrespective of the resolution of the stratigraphic study and of the data available (details in Chapters 7 and 9).

Scale-variant nomenclature: A system of stratigraphic nomenclature and classification in which different types of units are defined at different scales (e.g., parasequences < sequences < composite sequences < megasequences). The lack of reproducible standards for the definition of a "sequence" as an anchor for stratigraphic classification undermines the scale-variant approach, as the terms are stripped of a consistent stratigraphic meaning (details in Chapters 7 and 9).

Sea level: The height of the sea surface relative to the center of Earth or a satellite in fixed orbit around the Earth (Fig. 3.13). Sea levels can be global or local, in basins connected to or isolated from the ocean, respectively (details in Chapter 3). Local sea-level changes may be out-of-phase with the global eustatic changes due to the influence of local controls (e.g., evaporation can lead to local sea-level fall, while to global sea level is rising; Fig. 9.1).

Sediment bypass: *see* **Bypass**.

Sediment grading: A gradual change of grain size in an upward direction; i.e., coarsening-upward (normal grading) vs. fining-upward (inverse grading).

Sediment load: The solid material that is transported by natural agents such as wind, rivers, or subaqueous flows. Sediment load includes bedload, saltation load, and suspended load, and is quantified by the dry mass of all sediment that passes a given point in a given period of time (e.g., measured in metric tons per year); i.e., sediment supply.

Sediment supply: Mass of sediment per time unit delivered to any specific area of bypass or deposition. Sediment supply and sedimentation are two distinct parameters, whose relationship is mediated by the energy of the transport agents (details in Chapter 3). *Cf.* **Sedimentation**.

Sedimentation: Process of sediment accumulation due to sediment supply in excess of the energy of the sediment-transport agent (details in Chapter 3).

Sedimentological cycles: Sedimentary cycles at substratigraphic scales, bounded by bedding planes (i.e., beds and bedsets). Sedimentological cycles are the building blocks of the lowest rank systems tracts and component depositional systems (Figs. 3.24, 4.62, and 6.67; details in Chapter 6).

Sedimentology: The study of sediments and sedimentary rocks, and of the processes by which they form (Fig. 1.4).

Seismic geomorphology: The plan-view imaging of depositional elements based on 3D seismic data (Posamentier, 2004; details in Chapter 2).

Seismic stratigraphy: A type of stratigraphy which relies on the seismic-reflection architecture as the basis for stratigraphic correlation (Fig. 1.6).

Sequence: *see* **Stratigraphic sequence**.

Sequence boundary: Sequence stratigraphic surface(s) selected to delineate stratigraphic sequences. Different kinds of sequence stratigraphic surfaces may serve as sequence boundaries, depending on the type of stratigraphic sequence (Figs. 1.9 and 1.10; details in Chapter 5).

Sequence set: A set of sequences organized into lowstand, transgressive, or highstand stacking patterns; a subdivision of "composite sequences" (Van Wagoner, 1995). The lack of reproducible standards for the definition of a "sequence" as an anchor for stratigraphic classification undermines this scale-variant nomenclature, as the terms are stripped of a consistent stratigraphic meaning (details in Chapter 7).

Sequence stratigraphic surface: A type of stratigraphic surface that marks a change in stratal stacking pattern; i.e., a systems tract boundary (details in Chapter 6).

Sequence stratigraphy: A type of stratigraphy which relies on stacking patterns for the definition, nomenclature, classification, and correlation of stratal units and bounding surfaces (Fig. 1.6).

Sharp-based shoreface: A shoreface succession that is typical of forced regressions, in which the shift from shelf to shoreface facies is abrupt, due to the downstepping of the shoreline during progradation (Fig. 4.30; details in Chapter 4). *Cf.* **Gradationally based shoreface**.

Shelf: The low gradient portion of a shelf-slope system (Fig. 8.1), which is prone to frequent changes between continental and shallow-water environments in response to shoreline shelf-transit cycles (Fig. 4.60). *Cf.* **Shoreline shelf-transit cycles**.

Shelf break: *see* **Shelf edge**.

Shelf edge: The limit between the shelf and the slope of a shelf-slope system (Fig. 8.1). Shelf edges typically prograde and upstep in a long term, but may also backstep in response to tectonic or hydraulic instability that lead to the collapse of the upper slope (Figs. 2.68, 4.60, and 9.8).

Shelf-edge delta: A delta located at the shelf edge, commonly at times of lowstand in relative sea level (Fig. 8.53). Under special conditions (e.g., narrow shelves and high sediment supply), highstand deltas may also prograde to the shelf edge. The distinction between lowstand and highstand deltas is based on the observation of the preceding shoreline trajectory (i.e., forced regressive vs. transgressive, respectively) (Fig. 5.87; details in Chapter 4).

Shelf-margin systems tract: The equivalent of a lowstand systems tract in a "type-2" sequence. The "shelf-margin" systems tract is no longer in use following the abandonment of "type-1" vs. "type-2" sequences (details in Chapter 5). *Cf.* **Type-1 sequence**, **Type-2 sequence**.

Shelf ridge: A sand-prone macroform that forms commonly in relation to tidal reworking on a shelf during transgression (Figs. 5.64, 5.74–5.79; details in Chapter 5).

Shelf-slope system: A physiographic element with distinct shelf and slope, built by multiple depositional systems (continental to shallow-water in the shelf topset, and deep-water in the slope foreset; Fig. 4.60). Shelf-slope systems develop in response to long-term progradation in relatively stable basins, with clinoform heights in a range of 10^2–10^3 m (details in Chapter 8). Seafloor gradients are typically less than 1:1000 ($0.05°$) on the shelf, and more than 1:40 ($1.4°$) on the slope (Heezen et al., 1959).

Shell bed: A concentration of shell fragments, commonly in areas of low sediment supply or high environmental energy. Types of shell beds include onlap shell beds, backlap shell beds, downlap shell beds, and toplap shell beds (Kidwell, 1991; Figs. 6.3 and 6.46).

Shore: A coastal environment bordering a sea or a lake, between the low tide level and the storm flooding; i.e., the beach, subdivided into a foreshore and a backshore (Fig. 4.11).

Shoreface: The shallow-water area in front of an open shore, between the low tide level and the fairweather wave base (Fig. 4.11).

Shoreline: The intersection of a plane of water (sea level or lake level) with the land surface (Fig. 3.22). The shoreline migrates from the sedimentological scales of tidal cycles to the stratigraphic scales of sequences and systems tracts (Fig. 4.62). For the purpose of stratigraphic studies, the terms "shoreline" and "coastline" can be used interchangeably (details in Chapter 4).

Shoreline shelf-transit cycles: Cycles of change in the direction of shoreline shift across a shelf, commonly observed on stratigraphic scales of 10^2-10^5 yrs (Figs. 5.5, 5.23, and 9.1). *Cf.* **Shelf**.

Shoreline trajectory: The direction of shoreline shift, defined by combinations of upstepping, backstepping, forestepping, and downstepping (Fig. 4.1). Note the difference between the geometrical trends that define shoreline trajectories and the concurrent depositional trends in fluvial and coastal systems (Figs. 4.23 and 4.24). Only the geometrical trends are diagnostic to the definition of systems tracts (details in Chapters 4 and 5).

Slide: A type of mass-transport deposit which involves the movement of lithified sediment. Slides preserve the original stratification of the sediment, undeformed (Fig. 8.42).

Slump: A type of mass-transport deposit which involves the movement of semi-lithified sediment. The original stratification of the sediment is preserved but deformed (Fig. 8.42).

Softground: A type of substrate in a conformable succession which records active sedimentation (low to high rates) on a moist to fully subaqueous depositional surface. Soft substrates support the development of ichnofacies at syn-depositional time (details in Chapter 2).

Stacking pattern: *see* **Stratal stacking pattern**.

Standard sequence stratigraphy: A standard approach to sequence stratigraphic methodology and nomenclature which is consistent irrespective of geological setting, stratigraphic scales, and the types and resolution of the data available (details in Chapters 9 and 10).

Storm wave base: The maximum water depth up to which the wave energy can still move sediments on the seafloor during storms; i.e., the limit between inner and outer shelf (Figs. 4.11 and 8.14). The depth of the storm wave base is variable, depending on the intensity of the storms, commonly in a range of tens of meters (e.g., 55 m in the case of Hurricane Katrina in 2005). The storm wave base marks the distal depositional limit of tempestites.

Stratal stacking pattern: The architecture of sedimentary strata, defined by geometrical trends, depositional trends, or the ratio between the elements of a depositional system (Figs. 4.1, 4.2, and 4.23). Not all stacking patterns are diagnostic to the definition of systems tracts and bounding surfaces (Fig. 1.2; details in Chapter 4).

Stratigraphic correlation: The process by which stratigraphic units in different areas are shown to be laterally equivalent in terms of geological age, fossil content, lithological character, or any other attributes of strata (Fig. 1.6). In sequence stratigraphy, correlation is based on the stacking patterns that define systems tracts and bounding surfaces (details in Chapter 4).

Stratigraphic cycles: Sedimentary cycles defined by the recurrence of the same types of sequence stratigraphic surfaces in the sedimentary record; i.e., stratigraphic sequences (Fig. 1.22). Within the nested architecture of stratigraphic cycles, stratigraphic sequences can be observed at different scales (Fig. 4.3; details in Chapter 7).

Stratigraphic cyclicity: Pertaining to the development of stratigraphic cycles.

Stratigraphic hierarchy: A system of classification of stratigraphic cycles on the basis of their absolute or relative scales (details in Chapter 7).

Stratigraphic sequence: A cycle of change in stratal stacking patterns, defined by the recurrence of the same type of sequence stratigraphic surface in the sedimentary record (Fig. 1.22). Different types of sequences have been defined, depending on the selection of the sequence boundary (Fig. 1.10; details in Chapter 5).

Stratigraphy: The science of rock strata; all attributes of strata which afford insights into their origin, classification, correlation, and physical and temporal relationships.

Subaerial unconformity: A sequence stratigraphic surface that forms by means of erosion or sediment bypass in a continental environment (details in Chapter 6). Subaerial unconformities form during stages of base-level fall in upstream-controlled settings, and during stages of forced regression or transgression in downstream-controlled settings (Fig. 4.39; details in Chapter 4).

Submarine canyon: A steep-sided, commonly V-shaped erosional feature generated by the passage of mass-transported deposits in a deep-water environment (Fig. 8.50); typical of slope settings, often connected with incised valleys on the shelf (Fig. 5.82).

Submarine fan: An accumulation of mass-transported sediment in a deep-water setting (Figs. 8.41 and 8.49). Submarine fans typically include the products of several mass-transport processes representing different stages of fan development (Fig. 8.47), and can be observed at different scales in relation to stratigraphic cycles of different magnitudes (Fig. 8.55; details in Chapter 8).

Subsidence: Downwarping of the Earth's crust in response to processes such as tectonism, ice or sediment loading, compaction, subsurface dissolution or diapirism, volcanism (loading by lava flows or emptying of magma chambers), and human activity (subsurface mining or extraction of hydrocarbons or groundwater). Types of subsidence accounted for in the classification of sedimentary basins include extensional, thermal, flexural, and dynamic (details in Chapter 8).

Supersequence: Informal term that describes a large-scale sequence in the sense of Sloss et al. (1949), or a group of seismic-scale sequences (Mitchum et al., 1977) (Fig. 5.4). The term is no longer in use in the formal classification of stratigraphic cycles (details in Chapter 7).

Surf diastem: A facies contact generated by the seaward migration of longshore troughs and rip channels during coastal progradation, which separates lower shoreface facies below from trough cross-bedded upper shoreface deposits above (Fig. 6.65). Surf diastems form during both forced and normal regressions (details in Chapter 6).

Systems tract: A linkage of contemporaneous depositional systems, forming the subdivision of a sequence (Brown and Fisher, 1977). Systems tracts are stratal units defined by specific stacking patterns and bounding surfaces, which can be observed at all stratigraphic scales; systems tracts of the lowest hierarchical rank consist of sedimentological cycles (beds and bedsets); systems tracts of higher hierarchical ranks consist of stratigraphic cycles (sequences) (Fig. 4.3; details in Chapters 5 and 7).

T

Tectonism: A term that describes all movements of the crust produced by tectonic processes, which result in the formation of sedimentary basins and extrabasinal sediment sources.

Tempestite: A storm deposit; sediment typically produced by the erosion of coastal systems during storms, and redeposited from storm currents as their energy declines towards the storm wave base. Tempestites display normal grading and hummocky cross-stratification, and relate to traction (rather than gravity-driven) currents. *Cf.* **Turbidites**.

Terrigenous: Sediment derived from the land or continent; material eroded from the land surface (Neuendorf et al., 2005).

Tidal-ravinement surface: A type of transgressive surface of erosion that forms by means of tidal scouring; typically preserved at the base of estuary mouth-complex deposits (details and examples in Chapter 6).

Top-amalgamation surface: A sequence stratigraphic surface that forms in upstream-controlled fluvial settings at the limit between high- and low-amalgamation systems tracts (Fig. 4.70; details in Chapter 6).

Toplap: Termination of inclined strata against an overlying lower angle surface that shows no evidence of erosion (Figs. 4.7 and 4.8).

Topset: A gently dipping succession of strata that accumulates behind the clinoform rollovers of a prograding system; e.g., the delta plain sediments of a delta system, or the shelf sediments of a shelf-slope system (Figs. 4.49 and 4.60). Topsets form commonly during normal regressions, but possibly during forced regressions as well when the fall in relative sea-level is accompanied by fluvial and coastal aggradation (Figs. 4.24 and 4.40; details in Chapter 4).

Transgression: A landward shift of the shoreline (Fig. 4.31); a type of shoreline trajectory defined by retrogradation and upstepping (Figs. 4.1, 4.18, and 4.29). Transgressions are driven by relative sea-level rise outpacing the rates of sedimentation at the shoreline (Figs. 4.21 and 4.32; details in Chapter 4).

Transgressive lag: A type of lag deposit that overlies a transgressive surface of erosion, most commonly a wave-ravinement surface (Figs. 6.45 and 6.47; details in Chapter 6).

Transgressive ravinement surface: *see* **Transgressive surface of erosion**.

Transgressive–regressive (T–R) sequence: A type of stratigraphic sequence bounded by maximum regressive surfaces (Fig. 1.10; details in Chapter 5).

Transgressive surface: A term synonymous with "maximum regressive surface," which emphasizes the onset of transgression rather than the end of regression (details in Chapter 6).

Transgressive surface of erosion: A sequence stratigraphic surface that forms by means of wave or tidal scouring during transgression (Fig. 4.55). The transgressive surface of erosion invariably reworks part of the maximum regressive surface, thus becoming a systems-tract boundary (details and examples in Chapter 6).

Transgressive systems tract: A systems tract defined by a transgressive stacking pattern (Fig. 4.1; details and examples in Chapters 4 and 5).

Truncation: Termination of strata against an overlying erosional surface (Figs. 4.7 and 4.8).

Turbidite: The product of sedimentation from a turbidity flow. Turbidites ("Bouma sequences") display normal grading, as the sediment accumulates during the decline in the energy of the flow.

Turbidity current: *see* **Turbidity flow**.

Turbidity flow: A type of gravity flow in which the grain-support mechanism is provided by the turbulence of the fluid between grains (Figs. 5.42 and 8.42; details in Chapters 5 and 8).

Type-1 sequence: A depositional sequence bounded by "major" sub-aerial unconformities which extend across the entire shelf. This nomenclature is no longer in use, since the type-1 and type-2 sequences are end members of a stratigraphic continuum (details in Chapter 5). *Cf.* **Type-2 sequence**.

Type-2 sequence: A depositional sequence bounded by "minor" sub-aerial unconformities restricted to the inner portion of a shelf. This nomenclature is no longer in use, since the type-1 and type-2 sequences are end members of a stratigraphic continuum (details in Chapter 5). *Cf.* **Type-1 sequence**.

U

Unconformable: A stratigraphic relationship which involves a break in sedimentation between stratal units that are juxtaposed across a physical contact. Three types of sequence stratigraphic surfaces are invariably unconformable (Fig. 6.5).

Unconformity: A break (hiatus) in the stratigraphic record, whatever its cause and magnitude, with or without accompanying erosion (Bates and Jackson, 1987). See Fig. 1.8 for the various types of unconformities.

Updip: A landward direction, parallel to the dip of the depositional surface.

Upstepping: A geometrical trend that describes an increase in the elevation of a depositional system (e.g., a coastal system; Fig. 5.1) or of a morphological element (e.g., a shoreline or a shelf edge; Fig. 4.6). Where used to describe a shoreline trajectory, upstepping is the product of relative sea-level rise (Fig. 4.1), and it can be accompanied by either aggradation or degradation (Figs. 4.23 and 4.39). Note the difference between geometrical trends and depositional trends (details in Chapter 4).

Upstream-controlled setting: An area within a sedimentary basin in which stratal stacking patterns form under the influence of upstream controls, independently of shoreline trajectories and relative sea-level changes (Fig. 3.11; details in Chapters 3 and 4).

W

Walther's Law: A law of facies relationships, which defines the correspondence between vertical and lateral facies shifts within conformable successions (Fig. 1.5).

Wave-ravinement surface: A type of transgressive surface of erosion that forms by means of wave scouring in a shoreface environment; typically preserved at the base of shallow-water transgressive "healing-phase" deposits (details and examples in Chapter 6).

Weathering: All physical and chemical processes that affect the Earth's surface as a result of exposure to air, water, or biological activity; i.e., mechanical breakdown, dissolution, or decomposition of pre-existing rocks. Weathering occurs *in situ*, and excludes the removal of the resulting material. *Cf.* **Erosion**, **Denudation**.

Wheeler diagram: A chronostratigraphic chart describing stratigraphic relationships in a time domain (Wheeler, 1958, 1964; Figs. 5.89 and 7.34).

Within-trend facies contact: A lithological discontinuity within a systems tract; a surface of lithostratigraphy or allostratigraphy (Fig. 6.1; details in Chapter 6). Within-trend facies contacts have been defined in all downstream-controlled systems tracts, in relation to normal regressions (within-trend normal regressive surfaces), forced regressions (within-trend forced regressive surfaces), and transgressions (within-trend flooding surfaces).

Within-trend flooding surface: A facies contact which marks an abrupt increase in water depth. Flooding surfaces form during shoreline transgression in coastal to shallow-water settings where abrupt water deepening can be demonstrated (Figs. 5.94 and 6.60; details in Chapter 6).

Within-trend forced regressive surface: A conformable facies contact that develops during forced regressions at the limit between the foreset (delta front) and the bottomset (prodelta) of river-dominated deltas (Figs. 6.6 and 6.58; details in Chapter 6).

Within-trend normal regressive surface: A conformable or scoured facies contact that develops during normal regressions at the limit between the topsets and the foresets of lowstand and highstand systems tracts (Figs. 4.49, 4.50, 6.6, and 6.56; details in Chapter 6).

Woodground: An *in situ* and laterally extensive carbonaceous substrate, such as peat or coal, which forms in continental to marginal-marine settings. The classification of ichnofacies that colonize woodgrounds is based on the environment in which the tracemakers lived; e.g., the *Teredolites* Ichnofacies reflects colonization in marine or marginal-marine environments (Fig. 2.50; details in Chapter 2).

Workflow of sequence stratigraphy: A logical workflow of data analysis that leads to the construction of a sequence stratigraphic framework. Steps in the workflow include the study of the tectonic and depositional settings, and the identification of systems tracts and bounding surfaces based on the observation of stratal stacking patterns at scales afforded by the resolution of the data available (details in Chapter 9).

Author index

Elliott, T., 101
Ellis, C., 8—9
Elmore, R.D., 30
Els, G., 3, 96, 312, 390—391, 397—398
Embry, A.F., 8, 164, 196, 201, 206f, 231, 236,
        244f, 252f, 264f, 268f, 274, 280f—281f,
        310, 313, 313f, 341, 406—407
Emmet, P.A., 3, 98—99, 389
Enge, H.D., 296
England, J.H., 15—16, 157—158, 169,
        175—177, 300, 390—391, 395—396
Enos, P., 364
Enzel, Y., 175, 393f
Eriksson, K.A., 97, 260f, 341, 390—391,
        396—399
Eriksson, P.G., 3, 8, 19f, 20, 44—45, 54f,
        68—69, 94—96, 150f, 221, 229—230,
        235, 251, 312, 318, 338, 345, 390—391,
        396—400, 410
Eriksson, P.R., 399
Escalona, A., 341
Eschard, R., 14, 97, 139, 208, 427
Esteban, M., 243—244, 316
Estrada, F., 375f
Ethridge, F.G., 8, 182, 216—217
Euzen, T., 14, 405
Eyles, C.H., 33f
Eyles, N., 33f
Fagerstrom, J.A., 50
Fahnestock, R.K., 3
Fairbanks, R.G., 96—97, 187f, 389
Falk, P.D., 97
Fanti, F., 3, 77, 161, 241f, 338, 406
Farmer, J.D., 314
Fastovsky, D.E., 30, 43
Fawcett, P.J., 66, 385—386
Feary, D.A., 366
Feigenson, M.D., 96—97
Feldman, H.R., 15—16, 24, 98, 155—158,
        169, 174—177, 231, 237, 243—244,
        247, 300, 302, 306, 308—310, 313,
        315—316, 395—396, 401, 407, 417,
        419—420
Ferry, S., 66—67
Field, B., 390
Field, M.F., 208
Fielding, C.R., 3, 19f, 20, 97—100, 150f,
        229—231, 235, 243, 249, 251, 350, 387,
        389—390, 392f, 419—420
Fildani, A., 344
Fillon, R.H., 385
Fiorini, F., 3, 15—16, 24, 101, 103f, 169, 173f,
        177, 231, 239, 300, 395—396
Fischer, A.G., 97
Fisher, W.L., 9f, 14, 19f, 20, 150f, 169,
        176—177, 229—230, 235, 251,
        383—384, 387
Fishman, N., 401—402
Fitchen, W., 359
Fleck, S., 66—67
Flesche, H., 43
Flessa, K.W., 41
Fletcher III, C.H., 208
Flint, S.S., 30, 97, 175f, 217—219, 309, 376,
        379—381, 389

Flores, J.A., 15—16, 169, 300, 395—396
Florindo, F., 98—99, 389—390
Folkestad, A., 345
Follmi, K., 66
Fonnesu, M., 377f, 380f
Forlin, E., 139, 140f, 427
Fouke, B.W., 314
Fox, C., 30, 34
Foyle, A., 295—296
Fragoso, D.G.C., 231, 301
Fralick, P.W., 97, 399
Frank, T.D., 3, 97, 230—231, 243, 249, 350,
        387, 419—420
Franseen, E.K., 98
Fraser, A.J., 97, 341
Frazier, D.E., 231, 272, 306
Freeman, K.H., 66—67
Frey, R.W., 44—45, 47, 47f, 50, 56, 61—62,
        231, 272, 306
Frostick, L.E., 95, 345, 401
Fulthorpe, C.S., 173f, 274—275, 295, 314,
        396
Furlong, C.M., 66—68, 423, 426
Fursich, F.R., 61—62
Fürsich, F.T., 35—36, 36f

G
Gabrielsen, R.H., 345
Gaigalas, A., 44—45
Gaillot, A., 15—16, 169, 300, 395—396
Gaillot, G.T., 96, 189—193, 220—221, 224,
        359, 381, 387, 405, 421—423
Galbiati, M., 377f, 380f
Gale, A.S., 40, 366—367
Galeazzi, S., 67, 67f, 342—343
Gallagher, J., 366—367
Galloway, B., 335, 336f
Galloway, W.E., 6—8, 14, 19f, 20, 73, 75,
        150f, 158, 177, 229—230, 235, 251,
        267, 306, 335, 352, 355, 383—384, 387,
        396, 406, 427
Gamberi, F., 15—16, 24, 155—158, 169,
        174—177, 231, 244, 300, 308—309,
        395—396
Gani, M.R., 96
Gao, C., 341
Garbe-Schonberg, D., 341
Gardner, M.H., 309, 387, 419
Gardner, R.D., 385
Garfield, T.R., 237, 243, 247, 302, 306, 310,
        313, 315—316, 395—396, 401, 407,
        417, 419—420
Garland, J., 241
Gastaldo, R.A., 44, 274
Gawthorpe, R.L., 96—97, 99—100, 139, 237,
        279, 338, 341, 345, 411, 419—420
Gazley, M.F., 227
Ge, M., 366—367
Gebhardt, C., 390
Gelfort, R., 15—16, 169, 300, 395—396
Genise, J.F., 52—55, 54f
Gensous, B., 15—16, 143f, 224, 231, 300, 387,
        405, 421—423
Gerhardt, K.J., 66
Germanoski, D., 182

Gerster, R., 341, 394f
Geslin, J.K., 237, 243, 247, 302, 306, 310,
        313, 315—316, 395—396, 401, 407,
        417, 419—420
Ghibaudo, G., 279
Ghienne, J.-F., 390—391
Ghosh, A., 208, 219f, 355
Gianolla, P., 20
Gibling, M.R., 19f, 20, 30—31, 31f, 33,
        97—98, 150f, 229—230, 235, 251, 341
Gilbert, G.K., 14
Giles, K.A., 19f, 20, 150f, 229—230, 235, 251,
        341
Gill, J.R., 336, 348, 416
Gingerich, P.D., 40
Gingras, M.K., 44—45, 58—59, 61—62,
        66—68, 227, 241f, 279, 368—370, 376,
        385—386, 423, 426
Gischler, E., 368—370
Gladenkov, A.Y., 69—70
Gladenkov, Y.B., 68—70
Gleick, J., 405
Glumac, B., 366—367
Goldhammer, R.K., 173f
Golovchenko, X., 142, 144f
Gomes, J.P., 231, 301
Gomez Omil, R., 300f
Gonzales, J.L., 358
Goodnight, S.A., 66—67
Gorini, C., 14, 375f, 405
Goulding, F.J., 237, 243, 247, 302, 306, 310,
        313, 315—316, 395—396, 401, 407,
        417, 419—420
Grabau, A.W., 16, 119—121, 303
Grabowski, G.J., 66
Gradstein, F.M., 40—41, 71
Graham, I., 390
Graham, S.A., 344, 380—381
Grandesso, P., 279
Granjeon, D., 14—16, 21, 94f, 95, 109,
        134—135, 174—178, 193, 249, 267,
        300, 314, 342—343, 370, 405, 407
Graves, R.L., 50
Grist, A.M., 320
Grocke, D.R., 68
Groenewoud, W., 29, 93—94
Grotzinger, J.P., 68—69, 399
Guadagnin, F., 3, 14—16, 169, 171f,
        300—301, 395—396, 410
Guan, H., 71—72
Guerello, R., 67f, 319f
Guertin, L.A., 368—370
Guillocheau, F., 14, 97, 139, 405, 427
Gumiaux, C., 375f
Gupta, P., 195f
Gupta, S., 97, 139
Gurbuz, A., 345
Gutierrez Paredes, H.C., 68, 96, 241f,
        273—274, 376, 381, 423
Gutzmer, J., 34

H
Haidl, F.M., 371
Hajek, E.A., 94—95, 168, 393
Hall, A.G., 50

Subject index

Note: 'Page numbers followed by "f" indicate figures and "t" indicate tables.'

A

Accommodation
 changes in, 106
 definition, 375—376
 proxies, 140, 227
 subaerial, 103, 109—110
 subaqueous, 103—106
Acoustic impedance. *See* Seismic data
Active basins, 3, 29, 95, 343—345
Aggradation. *See* Depositional trends
Allogenic/allogenic processes, 93, 95—101,
 115
Allostratigraphic, 12—14, 231, 240, 251,
 415—416, 420
Allostratigraphy, 6, 12—14, 288
Alluvial plain, 52, 155
Angular unconformity, 16
Apparent downlap, 142f—143f
Apparent onlap, 81f, 123
Astronomical forcing. *See* Orbital forcing
Athabasca Basin, 29—30, 400
Autogenic/autogenic processes, 34, 93,
 100—101, 368

B

Backbulge. *See* Foreland system
Backshore, 49, 55, 73
Backstepping. *See* Geometrical trends
Bad Heart Formation, 294f
Balfour Formation, 28f, 160f, 225f, 260f
Basal surface of forced regression
 definition. *See* Sequence stratigraphic
 surfaces
 examples, 261—262
 outcrop/core expression, 261—262
 seismic expression, 261—262
 well-log expression, 264f
Base level, 14, 16, 32—34
Base-level cycles, 33, 113, 115
Basement, 16, 111f, 371f
Basin center, 325—326, 370, 373—374
Basin floor, 61, 83, 90—92,
 110—111
Basin margin, 109, 151, 274, 334, 411, 414
Basinward/downdip direction, 77—78, 97,
 139, 149, 285—286
Bathymetric profile, 154—155, 273, 279
Bathymetric trends, 135—136, 288—289,
 325, 333
Bathymetry, 44, 47, 172f—173f, 193, 387
Bayhead delta, 205—206, 212f, 214f—215f
Bayline, 124f, 356f

Beach cycle, 356f
Beach subenvironments, open shoreline
 settings, 55, 126f, 280—281, 291,
 345—346
Bearpaw Formation, 69f, 79f, 270f, 280f,
 282f
Beds. *See* Sedimentological cycles
Bedsets. *See* Sedimentological cycles
Belly River Group, 152f
Berm crest, 356f
Biofacies, 34—35, 41—42
Biostratigraphy, 68—70
Biozones, 69
Blackhawk Formation, 146f, 192f, 296f,
 402f
Blocky sand, 74f, 203
Body fossils, 34—44
Book Cliffs, 24—25, 60f—61f, 241f
Bottomset, 28f, 36—37, 67f, 120f, 145—146,
 275, 290f, 291, 414
Box sand. *See* Blocky sand
Bypass, sediment, 293—294, 351

C

Camborygma Ichnofacies, 50, 53—55, 55f
Canterbury Plains, New Zealand,
 154—155, 276—277
Carbonate/carbonates
 basin/basins, 82f
 environments, 52—53
 factory/factories, 193
 platform/platforms, 44, 193, 327—329
 setting/settings, 141f, 193, 224
 slope/slopes, 361f
Cardium Formation, 61f, 134f, 285f, 293f
Castlegate Formation, 34f, 205f, 207f
Celliforma Ichnofacies, 50, 52—53
Channel fills, amalgamated, 73—75, 92f,
 100—101, 151, 161
Chronostratigraphic chart. *See* Wheeler
 diagram
Chronostratigraphic framework, 9—11
Chronostratigraphy, 6, 320, 416
Clean limestone/limestones, 237—238,
 256f, 262f, 362f
Clean sand/sands, 59, 203
Climate, 33, 53—55, 97—100, 351, 389
Climate cycles, 26—28, 168, 257, 311, 350
Climatic fluctuations, 97
Climatic regimes, 96, 341, 349—350
 sequences in glaciated settings, 390—391
 sequences in non-glaciated settings, 390

Clinoform, 26f, 77—78, 83, 120f, 126—127,
 146, 343f, 393—394
Clinoform rollover, 112f, 117, 126—127,
 127f, 307, 412, 421, 427—428
Clinoform surface, 77—78, 120f, 181f,
 262—263
Coal, 77
Coal beds, 3, 34, 77, 226f, 276—277
Coarsening-upward, 17—19, 47—48, 73—75,
 146
Coastal erosion, 140, 155
Coastal onlap, 36—37, 123, 202—203, 229f
Coastal plain, 30, 32f, 96, 124f, 144f, 196,
 346—348
Coastal prism, 124f, 346—348
 highstand coastal prism, 124f
 lowstand coastal prism, 124f
Coastal settings, 30, 127—128, 133, 172—174
Coastline/coastlines, 40, 50, 121—122, 135,
 155
Cohesive debris flows. *See* Gravity-driven
 processes
Colorado River, 163f
Compaction, 92f, 94—95, 94f, 106, 208f
Composite sequence, 173f, 302, 306, 315,
 395—396, 407
Composite sequence set, 302, 315—316
Condensed section/sections, 44, 55,
 77—78, 251, 273—274
Conformable, 4, 223, 231—232, 251, 267
Conformable transgressive surface. *See*
 Maximum regressive surface
Conformity, 19—20, 135
Continental settings, 348—354, 393—394
Continental shelf, 83, 184—186, 388f
Continental slope, 60—61, 123, 379f, 394f
Convergent margin/margins, 342f, 344
Coprinisphaera Ichnofacies, 48—49, 52—53
Correlation. *See* Stratigraphic correlation
Correlative conformity
 definition. *See* Sequence stratigraphic
 surfaces
 outcrop/core expression, 1f, 4f, 24—25,
 76
 seismic expression, 125f, 253f
 types, 179f, 232f, 229
 well-log expression, 134f, 148f, 256f, 264f
Crevasse splays, 53—55, 73—75, 203
Cross-plots, 78—79
Cross-sections, 77—80
Cruziana Ichnofacies, 46—47, 59—61, 65—66
Cycle/cycles, 310, 317, 319—320, 351

485

Tides, 56, 75, 128—129, 145—146, 208, 276, 282, 292, 356
Tilt, 29—30, 96—97, 123, 333, 351, 400
Time-barrier, 335
Time/temporal attributes, 260—261, 321, 335—336
Top-amalgamation surface. *See* Sequence stratigraphic surfaces
Toplap. *See* Stratal terminations
Topographic breaks, 182
Topset, 121—122, 125f, 140, 142, 146—147, 149—151, 157, 163, 176, 194—195, 203, 219, 223, 234—236, 395
Transgression. *See* Shoreline trajectories
Transgressive lag/lags, 208, 239, 243f, 277—280, 367, 403
Transgressive ravinement surface. *See* Transgressive surface of erosion
Transgressive shales, 77—78
Transgressive slope aprons, 202—203
Transgressive surface. *See* Maximum regressive surface
Transgressive surface of erosion, 135, 181—182, 193—194, 196, 201, 211—212, 215—216, 236, 255, 263—264, 267, 276—282, 333—334, 391, 430—431
  definition. *See* Sequence stratigraphic surfaces
  tidal-ravinement surface, 280—282
  wave-ravinement surface, 279—280
Transgressive systems tract. *See* Systems tracts
Transgressive-regressive (T-R) sequence. *See* Stratigraphic sequences
Truncation. *See* Stratal terminations
Truncation surfaces, 414
Trypanites Ichnofacies, 41—42, 48—50, 62—63, 65f, 279
Turbidite, 60—61, 189—193, 197—200, 409—410
Turbidity flows/currents. *See* Gravity-driven processes
Turbidity-flow elements, 193, 197—200
  channels, 193, 197—200, 409—410
  levees, 193, 197—200
  sediment waves, 197—200

splays, 381—383
Type-1 sequence, 180
Type-2 sequence, 180

U

Unconformable, 4, 11—12, 14, 179, 246, 253—255, 253f, 279—280, 289, 294—295, 414, 416
Unconformities vs. diastems, 303—306
Unconformity, 4, 16, 19—20, 28—29, 28f
Unconformity-bounded units, 16—17, 19
Updip. *See* Landward
Upstepping. *See* Geometrical trends
Upstream controls, 117—118, 140—142, 151—154, 166—168, 187, 221, 395
Upstream-controlled settings, 106—110, 115, 117—119, 151, 158, 161, 164, 166, 168, 172—174, 176—177, 221, 223—224, 227—228, 232, 235, 255, 257—260, 286—287, 307, 310, 317, 334, 337, 348, 393, 395, 414, 422—423
Upward, coarsening, 17—19, 26—28, 73—75, 78—79, 78f, 134—135, 139, 146, 189—193, 196—197, 200, 219, 238—239, 239f, 268, 271—272, 279—280, 288—289, 329—333
Upward, deepening, 268, 271
Upward, fining, 73—75, 78—79, 100—101, 119, 134—135, 151, 154—155, 164—165, 168, 195—196, 202, 207—208, 210, 216—217, 221, 274—275, 279—280, 309, 324—325, 327—329, 384—385, 389, 412—414
Upward, shallowing, 50—52, 237—238, 320, 400—401

V

Variability of stratigraphic sequences, 341—342
  with the climatic regime, 389—392
  with the depositional setting, 345—389
  with the tectonic setting, 341—345
Vertical profiles, grading trends, 26—28

W

Walther's law, 65—66, 239—241, 293, 348, 401—404, 420

Water deepening, 66, 108—109, 135—136, 238—239, 244—246, 249, 292—294, 325, 327—331, 333f, 359, 374, 385, 401—402, 420—421
Water shallowing, 135, 188, 325—326
Water-depth changes, 327—329, 329f
Wave base. *See* Fairweather wave base and Storm wave base
Wave-dominated estuary, 64f
Wave-dominated settings, 267—268, 284
Wave-ravinement surface. *See* Transgressive ravinement surface
Weathering, 26, 98, 221, 351, 370, 374—375, 379—380, 396, 399
Well-log data, 72—80, 151, 273—274
  geological uncertainties, 73—75
  well-log interpretations, 76—80
  well-log motifs, 76—80
Western Canada foreland system, 320
Western Canada Sedimentary Basin, 320
Wheeler diagram, 182f
Within-trend facies contacts, 239, 251, 288—297, 337, 403, 420
  bedset boundaries, 288, 296—297
  downlap surface, 251—252, 288, 295
  surf diastem, 288, 295—296, 296f
  within-trend flooding surface, 292—295, 294f, 337
  within-trend forced regressive surface, 290f, 291—292, 337
  within-trend normal regressive surface, 288—291, 288f—289f, 337
Woodground, 48—49, 63—66, 279
Workflow of sequence stratigraphy, 249, 251, 288, 316—317, 339f, 401, 403—404, 409—416, 420, 431—432
  Step 1: Tectonic setting, 2—3, 119, 357, 398—399, 407
  Step 2: Depositional setting, 2—3, 23, 98, 119, 135, 169, 230—231, 341, 357, 395, 410—413, 417
  Step 3: Sequence stratigraphic framework, 1—3, 14, 21—23, 115, 169, 310, 372—373, 381—387, 410—411, 413—417, 419, 425

Z

Zoophycos Ichnofacies, 47, 60—61, 62f

Printed in the United States
by Baker & Taylor Publisher Services